FISH PHYSIOLOGY

Volume VIII

Bioenergetics and Growth

CONTRIBUTORS

FRED W. ALLENDORF

J. R. BRETT

C. B. COWEY

EDWARD M. DONALDSON

ULF H. M. FAGERLUND

RAGNAR FÄNGE

J. R. GOLD

DAVID GROVE

T. D. D. GROVES

DAVID A. HIGGS

KIM D. HYATT

J. R. McBRIDE

RICHARD E. PETER

W. E. RICKER

J. R. SARGENT

CHARLES TERNER

FRED M. UTTER

FISH PHYSIOLOGY

Edited by

W. S. HOAR
DEPARTMENT OF ZOOLOGY
UNIVERSITY OF BRITISH COLUMBIA
VANCOUVER, CANADA

D. J. RANDALL
DEPARTMENT OF ZOOLOGY
UNIVERSITY OF BRITISH COLUMBIA
VANCOUVER, CANADA

and

J. R. BRETT
PACIFIC BIOLOGICAL STATION
FISHERIES AND MARINE SERVICE
DEPARTMENT OF FISHERIES OF CANADA
NANAIMO, BRITISH COLUMBIA, CANADA

Volume VIII

Bioenergetics and Growth

ACADEMIC PRESS, INC.
(Harcourt Brace Jovanovich, Publishers)

Orlando   San Diego   San Francisco   New York   London
Toronto   Montreal   Sydney   Tokyo   São Paulo

ACADEMIC PRESS, INC.
Orlando, Florida 32887

United Kingdom Edition published by
ACADEMIC PRESS, INC. (LONDON) LTD.
24/28 Oval Road, London NW1 7DX

Library of Congress Cataloging in Publication Data

Hoar, William Stewart, Date
      Fish physiology.

      Includes bibliographies and index.
      CONTENTS: v. 1. Excretion, ionic regulation,
and metabolism.--v. 2. The endocrine system.--[etc.]
--v. 8. Bioenergetics and growth.
      1. Fishes--Physiology. I. Randall, D. J.,
joint author. II. Conte, Frank P., Date
III. Title.
QL639.1.H6   597'.01   76-84233
ISBN 0-12-350408-2 (v. 8)

PRINTED IN THE UNITED STATES OF AMERICA

84 85 86 87   9 8 7 6 5 4 3 2

CONTENTS

8. Population Genetics
Fred W. Allendorf and Fred M. Utter

9. Hormonal Enhancement of Growth
Edward M. Donaldson, Ulf H. M. Fagerlund, David A. Higgs, and J. R. McBride

10. Environmental Factors and Growth
J. R. Brett

11. Growth Rates and Models
W. E. Ricker

LIST OF CONTRIBUTORS

Numbers in parentheses indicate the pages on which the authors' contributions begin.

FRED W. ALLENDORF* (407), *Department of Zoology, University of Montana, Missoula, Montana 59801*

J. R. BRETT (279, 599), *Pacific Biological Station, Fisheries and Marine Service, Department of Fisheries of Canada, Nanaimo, British Columbia V9R 5K6, Canada*

C. B. COWEY (1), *The Natural Environment Research Council Institute of Marine Biochemistry, Aberdeen AB1 3RA, United Kingdom*

EDWARD M. DONALDSON (455), *Nutrition and Applied Endocrinology Program, Resource Services Branch, Fisheries and Marine Service, Department of Fisheries and the Environment, West Vancouver, British Columbia V7V 1N6, Canada*

ULF H. M. FAGERLUND (455), *Nutrition and Applied Endocrinology Program, Resource Services Branch, Fisheries and Marine Service, Department of Fisheries and the Environment, West Vancouver, British Columbia V7V 1N6, Canada*

RAGNAR FÄNGE (161), *Department of Zoophysiology, University of Göteborg, Fack, S-40033 Göteborg, Sweden*

J. R. GOLD (353), *Genetics Section, Texas A & M University, College Station, Texas 77843*

DAVID GROVE (161), *Department of Marine Biology, Marine Science Laboratories, Menai Bridge, Anglesey, LL 59 5EH Wales, United Kingdom*

T. D. D. GROVES (279), *Apex BioResources, Ltd., Duncan, British Columbia, Canada*

DAVID A. HIGGS (455), *Nutrition and Applied Endocrinology Program, Resource Services Branch, Fisheries and Marine Service, Department of Fisheries and Environment, West Vancouver, British Columbia V7V 1N6, Canada*

* Present address: Genetics Research Unit, University Hospital, Clifton Boulevard, Nottingham NG7 2UH, England.

KIM D. HYATT (71), *Institute of Animal Resource Ecology, Vancouver, British Columbia V6T 1W5, Canada*

J. R. MCBRIDE (455), *Nutrition and Applied Endocrinology Program, Resource Services Branch, Fisheries and Marine Service, Department of Fisheries and the Environment, West Vancouver, British Columbia V7V 1N6, Canada*

RICHARD E. PETER (121), *Department of Zoology, The University of Alberta, Edmonton, Alberta T6G 2E9, Canada*

W. E. RICKER (677), *Pacific Biological Station, Fisheries and Marine Service, Department of Fisheries of Canada, Nanaimo, British Columbia V9R 5K6, Canada*

J. R. SARGENT (1), *The Natural Environment Research Council Institute of Marine Biochemistry, Aberdeen AB1 3RA, United Kingdom*

CHARLES TERNER (261), *Department of Biology, Boston University, Boston, Massachusetts 02215*

FRED M. UTTER (407), *Northwest and Alaska Fisheries Center, National Marine Fisheries Service, Seattle, Washington 98112*

PREFACE

As the title indicates, this volume was inspired by the classic text on "Bioenergetics and Growth" of domestic animals written by Samuel Brody in 1945. Fisheries science has lacked a comparable, comprehensive text dealing with physiological and biochemical processes concerning the acquisition and utilization of food for energy and growth. Despite the fact that an immense volume of literature on size frequency, growth rate, and stomach contents of fishes has formed an integral part of population dynamics, studies on functional relations have been slow to provide an understanding of the energetics of growth. Undoubtedly this can be attributed in part to the fact that hunting rather than farming of fish has provided man with the major portion of his food from the sea. Fish farming, particularly as freshwater pond culture of carp, has been practiced for over 2000 years—but as an art rather than a science. In this century the expanding development of fish hatcheries and the recent advances in commercial aquaculture have brought a great increase in the need for synthesized knowledge of the physiology and nutrition of fishes.

The chapters have been organized to proceed from a review of the most recent advances in nutrition, with emphasis on energy and growth requirements from the diet, through the stages of acquisition of food, to its metabolic utilization for maintenance and activity. Energy conversion during developmental stages, environmental effects, and hormonal influence on growth are complemented by considerations of aspects of cytological and population genetics. A final chapter on mathematical approaches to growth modeling completes this volume.

Physiological studies on the bioenergetics and growth of fish can be expected to expand greatly in the future. The diversity of form and function of fishes, their prodigious reproductive potential, their adaptive radiation from polar seas to coral reefs, and their important contribution as a food resource to man are bound to bring a greater transition from hunting to farming. Just as agriculture has been aided by the studies in animal husbandry, so aquaculture will be advanced by similar measures in fish husbandry. Understanding the effective and efficient transformation of biological energy has become an ever more pressing issue in world affairs.

W. S. HOAR
D. J. RANDALL
J. R. BRETT

CONTENTS OF OTHER VOLUMES

1

NUTRITION

C. B. COWEY and J. R. SARGENT

I. INTRODUCTION

In 1972 we published a review in *Advances in Marine Biology* which summarized the current situation in fish nutrition as we saw it at that time (Cowey and Sargent, 1972). Our approach, both from inclination and conviction, was biochemical. We intend now to build on the foundations of that review, referring to it where necessary rather than repeating facts or arguments which have already been stated, and describing advances which have occurred in the interim in this ex-

1

FISH PHYSIOLOGY, VOL. VIII
Copyright © 1979 by Academic Press, Inc.
All rights of reproduction in any form reserved.
ISBN 0-12-350408-2

panding field. A comprehensive treatment of the subject appeared at about the time of our earlier review (Halver, 1972a) and statements of nutrient requirements of different species have appeared or are in the process of preparation (e.g., National Research Council, 1973). Finally, the first volume of this series contains a paper on nutrition, digestion, and energy utilization in fish (Phillips, 1969).

The preponderance of fish species in the natural environment are carnivorous and we shall be concerned with the metabolic and biochemical characteristics of this habit, together with any effects of changes in dietary intake on metabolism. To an extent we are considering the adaptation of the fish to particular dietary regimes during culture; cultivation, however, involves not only some adaptation to diet but also to environmental parameters such as temperature, light intensity, and the chemistry of the medium. These factors will therefore be touched on where necessary.

The value of a biochemical approach to animal and particularly fish nutrition is frequently questioned. Is there not already enough information available on nutrient requirements to permit a satisfactory level of fish cultivation? It is not our business, at this point in time, to rebut this view in detail but we would point out that there is no sharp dividing line between good nutrition and inadequate nutrition. It seems abundantly clear to us that progress toward attaining optimal nutrition depends, at the end of the day, on an understanding of the metabolism of both the whole organism and of the cell. The structure of this review reflects that belief, the contents sustain it.

II. PROTEINS

A. Essential Amino Acids

The most significant recent advance has been the quantitative definition of the essential amino acids required by the Japanese eel (*Anguilla japonica*) and by the carp (*Cyprinus carpio*) by Drs. Arai and Nose. As their values determined for these species show important differences from those of Halver and his colleagues (Mertz, 1969) relating to chinook salmon (*Oncorhynchus tshawytscha*) they are shown in Table I. The values given were measured at the optimal dietary protein intake for each species as the requirement for at least some amino acids increased with dietary protein level up to the optimal intake of the latter (Bressani and Mertz, 1958). For the species

Table I
Amino Acid Requirements of Certain Species of Fish and of the Rat[a]

Amino acid	Chinook salmon	Japanese eel	Carp	Rat
Arginine	2.4	1.7	1.6	0.2
Histidine	0.7	0.8	0.8	0.4
Isoleucine	0.9	1.5	0.9	0.5
Leucine	1.6	2.0	1.3	0.9
Lysine	2.0	2.0	2.2	1.0
Methionine	1.6[b]	1.9[b]	1.2[c]	0.6[c]
Phenylalanine	2.1[d]	2.2[d]	2.5[d]	0.9[d]
Threonine	0.9	1.5	1.5	0.5
Tryptophan	0.2	0.4	0.3	0.2
Valine	1.3	1.5	1.4	0.4

[a] Data for chinook salmon and rat are from Mertz (1969); those for Japanese eel and carp are from Nose and Arai (unpublished data). Values are grams per 100 g of dry diet.
[b] Methionine + cystine.
[c] In the absence of cystine.
[d] In the absence of tyrosine.

shown these protein levels were in g/100 g dry diet: chinook salmon, 40; Japanese eel, 37.7; carp, 38.5; and rat, 13.19.

The requirements of the eel for isoleucine, tryptophan, threonine and sulfur-containing amino acids are markedly higher than those of chinook salmon; the requirement for arginine is much lower than that of salmon. These differences make for something of a dilemma in allocating allowances of amino acids when formulating the protein component of rations for other species whose amino acid requirements are as yet undetermined (all marine species). Perhaps the safest course is to use, for each amino acid, the highest level required by any of those species for which data are available. The situation with regard to sulfur-containing amino acids requires further clarification because the sparing action of cystine on methionine requirement is very marked. In fact cystine appears to be superior to methionine as a supplemental sulfur-containing amino acid in diets for eel (Arai et al., 1971). This is borne out by the findings of Halver et al. (1959) that at low cystine levels (0.05% of the diet) chinook salmon failed to show maximal growth with methionine levels at 1.6% of the diet; when 1.0% cystine was present in the diet the methionine requirement was 0.6%. The relative amounts of methionine and cystine in the eel diet (Table I) were 0.9 and 1.0%, respectively (Arai and Nose, personal communication).

The estimation of methionine in proteins presents some problems because methionine sulfur is readily oxidized during acid hydrolysis. The hydrolysate may thus contain a mixture of methionine, methionine sulfone, and methionine sulfoxide. While methionine may be converted quantitatively to the sulfone by oxidation with performic acid prior to hydrolysis, this method is not readily applicable to heterogeneous materials such as feed proteins. Thus the recent application of the cyanogen bromide method to the quantitative determination of methionine in proteins (Finlayson and MacKenzie, 1976; Ellinger and Duncan, 1976) should prove extremely useful in assessing the methionine content of fish diets reliably.

The requirements of fish for many amino acids are more than twice those of the omnivorous terrestrial mammal. As net nitrogen retention of protein by fish is not greater than that of mammals this suggests important differences in amino acid catabolism between carnivorous fish and omnivorous land mammals. These differences are considered in detail later.

Few amino acid deficiencies are known to lead to characteristic pathological symptoms. Rather, nearly all give rise to a loss of appetite, resulting in low food intake with a consequent fall in growth rate and activity. Recently, however, more specific symptoms have been described in rainbow trout given diets lacking tryptophan (Kloppel and Post, 1975). Trout deficient in tryptophan showed transient scoliosis, and the notochord of all scoliotic fish was affected by a protrusion of fibrous material, on the concave side of the fish, between the cartilaginous processes. Deficient fish were hyperemic (judged subjectively) and there was an abnormal deposition of calcium both in the bony plates surrounding the notochord and the kidney. Some of these symptoms may be explicable in biochemical terms but no primary lesion has yet been identified.

B. Biological Value of Proteins

1. METHODS

No generally accepted method for measuring the biological value of proteins for fish has been established. Most workers state PER (protein efficiency ratio, grams live weight gain/grams protein fed) or supply sufficient data to permit this value to be calculated; the usefulness and limitations of PER were dealt with previously (Cowey and Sargent, 1972).

An innovation has been the method proposed by Ogino *et al.* (1973) which permits separate collection of fecal and metabolic end products. Water from the fish tank is continuously siphoned through a trap where feces are retained and precipitated, and thence to a large column (5.5 × 60 cm) containing a strongly basic cation exchange resin (Amberlite IR-120H) where cationic substances are retained. These latter substances are eluted from the column with 5% (w/v) HCl at the end of the experiment. The method has the advantage that nitrogen balance can be measured in fish that are not (obviously) stressed, the only limit to flow rate being imposed by the size and capacity of the resin column. The method obviously cannot be applied to marine fish, and equally obviously any neutral or anionic substances (urea, taurine) are not retained quantitatively by the resin. Ogino and his colleagues claim that urea is not excreted by carp and spurious nitrogen balance measurements do not result from application of the method to this fish. As rotary evaporators of capacity up to 50 liters with continuous feed-in are available, it appears that collection, concentration, and analysis of post-trap water in Ogino's apparatus would lead to values for excreted N which could not be questioned. Moreover, the method could then be extended to fish which excrete significant amounts of urea.

Several groups (Zeitoun *et al.*, 1973; Rumsey and Ketola, 1975) have found the apparent efficiency of deposition of dietary proteins as tissue protein: (carcass protein at end − carcass protein initially)/protein fed: to be a simple, useful, and practical expression of protein efficiency. Although this scheme does not attempt to allow for maintenance requirements, it may prove sufficiently reproducible and give the same relative ratings to various proteins as are obtained with more complex procedures. If some such method of assessment of protein quality were generally accepted then it might be possible to eliminate some of the other variables that exist between one group of investigators and another. These variables include dietary protein level, dietary energy level, duration of the experiment, food consumption, and environmental factors. While certain factors such as species and water temperature are bound to vary, much might be gained if more of the other factors were constant. Thus, optimal dietary protein level is not that different for a number of species (Cowey, 1975) and a value of 40% crude protein would cover most species currently being studied. In the same way dietary energy level and the duration of the experiment could readily be standardized. In the absence of *ad libitum* feeding, ration size might be specified. As protein quality is at best a nonspecific measurement, the elimination of as many experimental variables as possible is essential if meaningful comparisons between different proteins and different species are to be made.

2. RESULTS

When omnivorous land mammals are fed high levels of dietary proteins the biological values obtained are very similar. Thus, Miller and Payne (1961) showed that net protein utilization (NPU) values of beef powder, casein, and wheat gluten for the rat are very similar when given at a dietary level of 40% total calories although they differ markedly when fed at a level (10% total calories) where amino acid composition is the limiting feature. Presumably, at high protein intakes, the essential amino acid requirements of the rat are met even by proteins of low nutritional quality provided the protein intake is sufficiently high (obviously excluding those proteins such as gelatin and zein totally lacking an essential amino acid).

It was observed recently (Cowey, 1975) that, in contrast to the omnivorous mammal, differences in the nutritional values of food proteins for fish are evident even at high protein intake. This was already evident from the data of Ogino and Chen (1973) on carp and of Cowey *et al.* (1974a) on plaice, *Pleuronectes platessa.* Similar differences (Table II) in nutritional quality are evident when casein and an isolated fish protein were fed to rainbow trout, *Salmo gairdneri* (40% protein in diet), by Rumsey and Ketola (1975). These differences in nutritional value between different proteins given at high dietary levels must be due to the high essential amino acid requirements of the fish. Most of the methods currently applied to the evaluation of nutritional quality of proteins for fish measure nitrogen retention to a greater or lesser extent (some procedures allow for digestibility and maintenance requirement, others do not) but these methods should rate proteins in the same relative order for a given species. It is noteworthy that markedly different relative nutritive values have been obtained for one or two proteins with different species. Rumsey and Ketola (1975) obtained a relatively high value for an isolated fish protein given to Atlantic salmon, *Salmo salar,* but Cowey *et al.* (1974a) obtained a relatively low value for a similar protein given to plaice. Equally casein has given high nutritive values with rainbow trout and carp but low values with Atlantic salmon (Table II). It is as yet too early to attempt to resolve these differences because quantitative data on the essential amino acid requirements of these species are not yet available; it has already been noted that these requirements may vary between species.

Table II

Nutritional Value of Proteins for Different Species of Fish

Species	Protein	Dietary protein level (g/100 g)	Method used	Value (%)	Reference
Salmo gairdneri	Casein	Up to 53.5	Biological value	50.8	Nose (1971)
	Whitefish meal	Up to 37.2	Biological value	44.5	Nose (1971)
	Soybean meal	Up to 35.3	Biological value	25.0	Nose (1971)
Cyprinus carpio	Casein	43	Biological value	63	Ogino and Chen (1973)
	Whitefish meal	43	Biological value	64	Ogino and Chen (1973)
	Corn gluten meal	43	Biological value	39	Ogino and Chen (1973)
Salmo salar	Casein	40	Apparent efficiency, protein retention	13	Rumsey and Ketola (1975)
	Casein + amino acids	40	Apparent efficiency, protein retention	24	Rumsey and Ketola (1975)
	Isolated fish protein	40	Apparent efficiency, protein retention	24	Rumsey and Ketola (1975)

In general, however, measurements of nutritional quality of likely feed proteins show a fair measure of agreement in that fishmeal or proteins derived therefrom are appreciably superior to other sources such as cereal proteins and certain single cell proteins (Nose, 1971; Cowey *et al.*, 1971; Orme and Lemm, 1973; Ogino and Chen, 1973; Windell *et al.*, 1974). As implied above, the absolute values obtained for different proteins by the methods applied vary considerably, but fishmeal has given consistently higher values than other feedstuff proteins by virtually all the methods applied.

3. VALUE OF FISH PROTEINS

The high biological value of fishmeal for many species has led to a heavy dependence on this commodity in diets used for high density fish production. Many attempts have been made to substitute other proteins either wholely or in part for fishmeal but until quite recently such substitutions have resulted in a marked fall in growth rate.

Thus, when menhaden meal was replaced isonitrogenously by soybean meal in diets (35% protein) for channel catfish (*Ictalurus punctatus*) weight gain decreased from 188 to 130% (Andrews and Page, 1974); replacement of half the menhaden meal with soybean meal led to less dramatic fall in weight gain (168%). Andrews and Page identified sulfur-containing amino acids and lysine as the first limiting amino acids for catfish in the soybean meal, presumably by reference to the quantitative requirements of chinook salmon. However, supplementation of diets (in which menhaden meal had been substituted by soybean meal) with lysine (to a level of 1.9% in the diet) and with sulfur-containing amino acids (to give a total sulfur amino acid level in the dietary protein of between 3.08 and 4.52%) did not enhance growth rate.

As supplementation of soya-substituted diets with limiting amino acids did not restore growth rate to levels attained when the fish were given diets containing menhaden meal, Andrews and Page sought unidentified growth factors in the menhaden meal. However supplementation of the soybean diets with fat-soluble, water-soluble, or mineral fractions of the menhaden meal were without effect on growth rate. Only when the soybean diets were supplemented with the residue remaining after ether or chloroform/methanol extraction of the menhaden meal was growth of the fish restored to that in the control diet.

Among possible explanations suggested for the failure of lysine and sulfur-containing amino acids to enhance soybean meal-substituted diets are (i) growth factors present in the lipid-free, ash-free protein fraction although no such factors have been mooted in analyses of fishmeal (e.g., Regier *et al.*, 1974); (ii) low biological availability of other amino acids in the soybean meal; (iii) the presence of a delicate balance of amino acids in the fishmeal; (iv) free methionine and lysine are not efficiently utilized by catfish.

The latter possibility tends to have been ruled out on the grounds that a growth response to crystalline amino acids has been obtained both on catfish (Dupree and Halver, 1970) and other species. In these experiments, however, all or many of the amino acids were supplied in crystalline form and would all be rapidly absorbed to arrive at sites of protein synthesis more or less simultaneously. When only one or two amino acids are used to supplement a protein, then these amino acids will be assimilated much more rapidly by the fish than those linked by peptide bonds in the food protein. Tissue concentrations of these supplementary acids may thus be transiently elevated and they may be catabolized rather than used for protein synthesis, a process which requires the availability of many amino acids at about the same time. Thus, it would be interesting to examine, in the type of experiment carried out by Andrews and Page, the effect of slowing down the release of supplementary amino acids in the gut. This might be achieved by some form of microencapsulation. An alternative might be to predigest diets supplemented with free amino acids with mixtures of proteases.

By contrast with the experiments of Andrews and Page (1974), Rumsey and Ketola (1975) achieved marked improvements in the growth rate of rainbow trout when a diet containing soybean meal (40% crude protein) was supplemented with amino acids. The object was to raise the levels of amino acids in the protein (soybean meal) component of the ration to those occurring in composite protein from trout carcasses, or in the whole egg protein of rainbow trout, or in an isolated fish protein. To this end nine diets were prepared by supplementing the basic diet with essential amino acids either singly (four diets), in pairs (two diets), or in greater numbers (one diet each of three, five and eight supplementary amino acids).

Only those groups of trout given diets containing five (methionine, leucine, lysine, valine, and threonine) or eight (methionine, leucine, lysine, valine, threonine, histidine, tryptophan, and tyrosine) supplementary amino acids grew significantly better than trout given the basic soybean diet. The growth rates of trout given the two diets containing several supplementary amino acids were not themselves significantly different. Rumsey and Ketola interpret their results as indicating that the amino acid requirements of fingerling trout are approximated by the amino acid patterns in isolated fish protein or by the protein in trout eggs. Alternatively they consider that soybean meal contains disproportionate levels of amino acids which reduce growth rate and can be corrected by the addition of specific amino acids.

Rumsey and Ketola were able to repeat their findings in a similar experiment in which it was shown that casein was inferior to an isolated fish protein in supporting weight gain and protein utilization of salmon fry. Supplementation of the casein with arginine, lysine, cystine, methionine, threonine, and tryptophan (so that it resembled isolated fish protein in overall amino acid composition) led to a significant increase in weight gain of the fry while protein utilization increased to a level equivalent to that of isolated fish protein.

The results of Rumsey and Ketola are important because they demonstrate that the biological value of two proteins can be markedly enhanced by multiple supplementation with amino acids. Presumably the value of other proteins can be similarly improved in the same way, thereby diminishing the present reliance on fishmeal. Their explanation of this achievement is less convincing because it begs the question of why supplementation of the protein component of a diet with the first, or first and second, limiting amino acids only, fails to elicit any enhancement of growth rate or protein utilization. The basic concept of biological value is that it is a function of the first limiting amino acid in the protein and rectifying this deficiency will increase the biological value of the protein. Mutual supplementation between two or more food proteins is dependent on this concept. Conversely, when a moderate luxus of dietary amino acids is supplied to an animal the excess, that is, those not rapidly used in protein synthesis, is catabolized.

It is difficult to accept that soybean meal (and presumably by the same token casein) contain disproportionate amounts of amino acids especially as both give reasonable growth rates with several species of fish. The deleterious effects of a moderate imbalance of dietary amino acids usually occur only when omnivorous birds and mammals are fed low levels of imbalanced proteins (Harper et al., 1970). Rumsey and Ketola cite experiments on poultry nutrition as evidence that supplementation of casein with amino acids improves its utilization, but these experiments (e.g., Dean and Scott, 1965) in no way conflict with the fact that the biological value of casein for chicks may be improved by supplementation with only one (the first limiting) amino acid. These experiments were not primarily designed to improve the utilization of casein but were intended to evaluate the nutritional requirement of chicks for amino acids (casein provided only 10% of the protein in a diet containing a total of nearly 18% crude protein). At the same time there is no ready physiological explanation for the lack of effect of single supplementary amino acids and the marked effect of a batch of amino acids in enhancing the biological value of proteins. Perhaps when more is known about the metabolism of amino acids in fish and especially about any control of this process a satisfactory explanation may materialize.

Finally Cho *et al.* (1974) showed that the herring meal content of an open formula diet for rainbow trout could be reduced (from 35 to 18% with soybean meal being appropriately increased) without concomitant loss in protein utilization. When however fermentation products, corn gluten meal and alfalfa meal were omitted from the original diet (herring meal remaining at 35% but soybean meal and wheat middlings being increased) growth rate of the fish improved significantly. Thus, the main contribution of Cho *et al.* (1974) was a simplification of the original diet, for when the herring meal in the simplified, improved diet was reduced by half (soybean meal replacing it) the value of the diet to fish fell significantly.

C. Food Energy and Protein Utilization

By contrast with birds and mammals, which expend large amounts of energy in maintaining body temperature, the energy requirements of fish, which do not are likely to be small. It would therefore appear at first sight that energy conversion by the fish should be more efficient than that of warm-blooded animals. A recent comparison of energy conversion in the two types of animal, chick and carp (Nijkamp *et al.*, 1974) indicates that this does not hold good for the growing animal. This comparison is shown in Table III and it is evident that while the maintenance requirement of the chick is about five times that of carp (growing at 23°C) growth in terms of energy retention per unit metabolic size is approximately fivefold higher for the bird than for the fish. The gross efficiency of energy conversion in the two animals is thus of a similar order. In addition the ratio N retention : N intake was maximally about 30% in carp but usually greater than 50% for the chick (Nijkamp *et al.*, 1974). Blaxter (1975) observes that cultivation (as opposed to hunting) of fish is advantageous in terms of economy of human effort; a complementary view might be that lack of conservation of natural fish stocks necessitates increased production by artificial means.

Table III
Food Energy Utilization by Chicken and by Carp[a]

Utilization of energy for	Chicken	Carp
Maintenance of weight (kcal/kg$^{0.75}$)	85	13–14
Daily growth (kcal/kg$^{0.75}$)	90–124	20–24
Gross efficiency (%)	30	27

[a] Data are from Nijkamp et al. (1974).

The relationship between energy content and protein content of diets has been examined for several species of fish (Ringrose, 1971; Lee and Putnam, 1973; Page and Andrews, 1973; Cowey et al., 1975a; Takeda et al., 1975). Certain trends or findings were common to most of these studies: an increase in the ratio digestible energy : protein led to an increased deposition of lipid in the fish; an increase in energy level at constant dietary protein level resulted in improved feed efficiency; protein efficiency ratio was negatively correlated with the ratio dietary protein : energy; starch or dextrin up to a level of 25% of the diet was effective as an energy source; fish eat to meet their energy requirement.

Close comparison of results is complicated, not only because different species are involved, but also because there is not always common ground between different investigators over the metabolizable energy content of food components. Thus Ringrose (1971) used the metabolizable energy values of feedstuffs as determined with poultry; Lee and Putnam (1973) used conventional mammalian values (4 kcal/g of carbohydrate, 4 kcal/g of protein, 9 kcal/g of fat) but assumed that starch contained 90% carbohydrate and a casein–gelatin mixture contained 90% protein; Page and Andrews (1973) combined digestibility studies (using the inert indicator Cr_2O_3) with indirect calorimetry to estimate the metabolizable energy of components; Smith (1971) used a metabolism chamber together with indirect calorimetry.

By comparing energy values among various diets Page and Andrews (1973) obtained a digestible energy value for cornstarch fed to channel catfish of 2.7 kcal/g (25% starch in diet). This compares with values of 2.14 kcal/g for cooked starch and 0.72 for raw starch (50% starch in diet) obtained by Smith (1971) using rainbow trout; Smith's values for glucose and dextrin were 3.12 and 3.03 kcal/g, respectively.

The value of 6.78 kcal/g for the digestible energy of fat (Page and Andrews, 1973) is a good deal lower than that usually accorded this material.

It is not possible (by comparing energy values among various diets) to derive a metabolizable energy value for protein from the data of Page and Andrews (1973) mainly because the protein components in their diets changed disproportionately between treatments. In fact their claim that a constant amino acid balance was maintained throughout their diets appears to be incorrect.

Smith (1971) obtained a value of 4.5 kcal/g for metabolizable energy of protein consumed by rainbow trout; this value also is high in comparison with that for mammals. Smith ascribed this to the fact that the main end product of nitrogen metabolism in fish is ammonia (cf. Cowey and Sargent, 1972). The most recent data on this point (Brett and Zala, 1975) showed that when food was withheld from *Oncorhynchus nerka* urea was excreted at a mean rate of 2.2 mg of N/kg/hr and ammonia at 8.2 mg of N/kg/hr; following a meal urea excretion rate did not change but ammonia excretion rose to a peak of 35 mg of N/kg/hr about 4 hr after the start of feeding falling thereafter. Another factor affecting metabolizable energy of protein is specific dynamic action. This is considered in a separate chapter in this volume.

Ultimately the object of studies on dietary protein/energy relationships is to obtain estimates of the concentration of protein, relative to that of energy, which could be commended for diets designed to allow rapid growth. In this context it can be shown from the data of Ringrose (1971) that *Salvelinus fontinalis* require 7.5 kcal of metabolizable energy for each gram of protein in the feed; from the data of Takeda *et al.* (1975), who used *Seriola quinqueradiata*, it can be shown that maximum protein retention is obtained with diets which provide 9.0 kcal of metabolizable energy for each gram of protein; maximum energy retention occurred with diets containing 7.0 kcal for each gram of protein.

When considering this question there is a case for stating the protein content of the diet in terms of the proportion of energy it contributes. Protein acts both as a nutrient and as an energy source. Consequently addition of energy to a diet not only increases energy intake but also lowers the protein energy : total energy ratio. Both factors are likely to influence nitrogen balance and their effects, although related, may be separate. Lack of attention to this point has led to loose statements "increasing the protein level with constant energy . . . always resulted in improved feed efficiency" (Page and Andrews, 1973). In fact the energy levels (kcal/kg) of the diets in question were not constant but stated as 1878 versus 2941 and 2330 versus 3343, the second diet in each pair having the increased protein level. It should thus be emphasized that, *above maintenance level* and at constant dietary energy density, the law of diminishing returns applies— increase in dietary protein content results in lower protein utilization.

Few investigators have been able to include sufficient different total energy and protein energy levels in their experiments to permit a useful prediction of the ratio necessary in diets designed for rapid growth. Perhaps the results of Lee and Putnam (1973) are most complete. We have recalculated the energy contents of their diets using a calorific value of 4.5 kcal/g of protein but retaining the factors applied by Lee and Putnam (1973) to the carbohydrate content of starch and the protein content of casein–gelatin (Table IV). Values for protein retention were not given in Lee and Putnam's paper and have been calculated from other data they supply. It is noteworthy that these values are much greater than those of Nijkamp *et al.* (1974) mentioned earlier.

Within the confines of their experiment (35–53% protein; 300–450 kcal/100 g) Lee and Putnam found no significant differences in weight gain unless both protein and energy in the diet were low. The relationship between protein retention and protein energy : total energy is given by

$$N = 51.76 - 0.27P \qquad (r = -0.849, P < 0.01)$$

where N is protein retention and P is % protein calories. For their high energy diets (450 kcal/100 g) alone the corresponding equation is

$$N = 68.98 - 0.69P \qquad (r = -0.999, P < 0.01)$$

For this species these equations permit the prediction of protein utilization within reasonably wide limits of protein concentration and (in the first case) energy density in the diets. The equations appear to us to provide an acceptable approach to deciding a desirable protein energy : total energy ration in practical diets.

D. Protein Synthesis

Relatively little attention has been focused on protein biosynthesis in fish although recent studies by Haschemeyer (see below) suggest that significant findings can be derived from research on mechanisms of protein biosynthesis in aquatic species.

Table IV
Effect of Variation in Dietary, Protein Energy: Total Energy Ratio on
Protein Utilization by Rainbow Trout[a]

Total energy in diet (kcal/100 g)	Protein level in diet (g/100 g)	Protein energy as percentage of total energy	Protein retention[b]	PER[c]
312	35	46	37.0	2.28
316	44	57	36.9	2.27
320	53	67	34.6	2.10
384	35	38	39.7	2.28
388	44	46	39.8	2.47
392	53	55	36.9	2.29
456	35	32	45.5	2.97
460	44	39	42.0	2.70
464	53	46	36.9	2.43

[a] From Lee and Putnam (1973), recalculated.
[b] 100 × protein retained/protein intake.
[c] Protein efficiency ratio (grams live weight gain/grams protein eaten).

The incorporation of [14C]leucine, injected intraperitoneally, into the muscle proteins of *Fundulus heteroclitus* has been investigated by Jackim and LaRoche (1973). Amino acid incorporation into proteins increased with temperature up to a critical point at 26°–29°C, beyond which it decreased sharply; in these experiments fish were acclimatized to different temperatures for 17–21 days before being studied. Amino acid incorporation was decreased by fasting, by reducing dissolved oxygen concentration below 2.5 mg/liter and by stress (confinement in small volumes of water). Insulin treatment increased amino acid incorporation but exercise, darkness, fish size, and sex had no significant effect. The [14C]leucine in this work was injected together with large amounts of carrier leucine to offset possible changes in amino acid pool size during various treatments. Pool sizes, however, were not measured directly and alterations in the pool sizes of essential amino acids other than leucine could alter the observed rate of incorporation of leucine, especially if such amino acids were rate limiting for protein biosynthesis.

The main object of Haschemeyer was to decide "whether a control over protein synthesis might provide a common basis for the increased levels of enzymes of respiratory metabolism and other pathways responsible for physiological adaptation to low temperatures" (Haschemeyer, 1969b). The studies involved incorporation of radioactive amino acids (leucine, phenylalanine, or an amino acid mixture injected into the portal vein) into the proteins of the liver of toadfish, *Opsanus tau*. During the course of the work, methods permitting simultaneous measurement of L-amino acid uptake, activation, and incorporation into growing polypeptide chains by liver *in vivo* were innovated. As a result Haschemeyer and Persell (1973) were able to show that the overall rate of protein synthesis in toadfish liver, measured at 24°C after the fish had been acclimatized at sensibly the same temperature (22°C) was 1.6 mg/g liver/hr, about one-fifth of the mammalian rate; the authors infer that "when temperature and ribosome concentration differences are taken into account, it would appear that the protein synthetic system of toadfish liver is comparable to the mammalian liver with respect to levels of substances involved in control of elongation rates."

Concerning temperature compensation, it is evident that when measurements of protein synthesis were made at a given temperature (either 24° or 11°C) fish acclimatized at 11°C had approximately 50% greater rates of protein synthesis in their livers than fish acclimatized at 22° irrespective of the temperature selected for measurement (Haschemeyer and Persell, 1973). Earlier Haschemeyer (1969a) had demonstrated that the rate of protein biosynthesis is controlled at the stages of polypeptide chain elongation and release, and that the rate increase in cold-acclimatized fish was paralleled by elevated levels of elongation factor I, the enzyme that promotes binding of amino acyl tRNA units at codon recognition sites (Haschemeyer, 1969b). Finally it has been shown that neither the rate of accumulation of amino acids into the liver nor the rate of formation of amino acyl tRNA units limited the rate of protein biosynthesis (Haschemeyer and Persell, 1973).

It is known that the levels of plasma proteins in fish do not change seasonally with temperature changes. In line with this, Haschemeyer (1973) has shown that the levels of circulating plasma proteins in cold-, and warm-acclimatized fish do not differ. Likewise the rate of turnover of plasma proteins is not significantly altered by cold acclimation ($t_{1/2} = 9.2 \pm 2.1$ days at acclimation temperature 20°C; $t_{1/2} = 8.0 \pm 1.0$ days at acclimation temperature 10°C). Haschemeyer (1973) concluded that, under conditions of an increased rate of protein biosynthesis in liver, control over the levels of circulating plasma proteins is exercised at a stage between completion of biosynthesis and secretion of the protein. The rate of degradation of circulating proteins in serum is unaltered. It would be highly interesting to extend these studies to the effects of cold acclimation on those constitutive proteins of liver, for example, aminotransferases, that are not secreted from the liver cells.

E. Amino Acid Catabolism

Protein metabolism contrasts with carbohydrate and fat metabolism in that when a surfeit of protein is ingested there is no form or organ in which protein may be stored in major quantities. Carbohydrate can be stored as glycogen in the liver or muscle, and fat can be stored as triacylglycerols in the various adipose tissues. The labile protein of soft tissues (liver, intestine, kidney) varies in amount with the state of nitrogen balance but, when soft tissues are replete, any amino acids assimilated over and above those needed immediately for protein synthesis are deaminated and the carbon residue either oxidized or stored as fat (lipogenesis) or as carbohydrate (gluconeogenesis).

Two factors control the rate of degradation of amino acids in omnivorous mammals (Krebs, 1972) and these have been considered as coarse and fine controls over the process. Coarse control involves a marked increase in activities (or concentrations) of amino acid-degrading enzymes when animals are adapted to a high protein diet. Fine control is vested in the K_m values of the enzymes concerned; enzymatic catalysis will proceed slowly if the tissue concentration of the substrate is less than the K_m value of the enzyme concerned, but will increase very rapidly as the substrate concentration rises (e.g., postprandially) over and above the K_m of the enzyme.

In omnivorous mammals the tissue levels of many amino acid-degrading enzymes increase several fold when the animals are transferred from a low to a high dietary protein regime. This is especially true of enzymes which degrade essential amino acids and very low concentrations of these enzymes are found when conservation of essential amino acids is necessary to meet dietary protein restriction. In addition, the total liver content of all urea cycle enzymes is directly proportional to daily protein consumption in rats (Schimke, 1962). Alanine aminotransferase and aspartate aminotransferase showed a similar response to the urea cycle enzymes but, interestingly, glutamate dehydrogenase which is considered necessary to supply about half of the nitrogen found in urea was not affected by a change in protein intake.

Few comparable data are available for fish. However, a response similar to that in mammals was observed in carp by Sakaguchi and

Kawai (1970) with regard to two enzymes concerned in the degradation of histidine. The activities of histidine deaminase and urocanase in carp given a diet containing 80% casein were 19.2 and 21.8 μmol/hr/g of tissue, but the activities in carp given diets containing 5% casein were reduced to 2.0 and 4.8 μmol/hr/g. Addition of histidine (2.7%) to the low-casein diet resulted in intermediate enzyme activities of 4.6 and 12.9 μmol/hr/g. The latter observation is slightly surprising because enzymes are not normally induced in animals given a restricted protein intake and 5% casein would be a very restricted protein intake for a fish. In a later paper (Sakaguchi and Kawai, 1974) it was shown that liver histidine deaminase and urocanase could be induced by intraperitoneal injection (three doses at 12-hr intervals) of histidine; similar injections of urocanate were without effect on urocanase activity. It appears probable that the induction of both histidine deaminase and urocanase in carp liver is dependent mainly (or solely) on histidine intake. This contrasts with the induction of threonine dehydrase in rats which requires the presence in the diet of four essential amino acids, threonine, valine, methionine, and tryptophan (Mauron et al., 1973).

Cowey et al. (1974b) did not find any significant changes in total hepatic glutamate dehydrogenase, aspartic aminotransferase, or alanine aminotransferase activities in plaice fed high or low protein diets for several weeks. Nagai and Ikeda (1973) were similarly unable to show any effect of dietary protein level on the activities of these aminotransferases in carp liver. It should be noted that all these enzymes deaminate nonessential amino acids and that the aminotransferases have important functions (e.g., transport of four carbon units between mitochondria and cytosol) other than amino acid degradation.

The lack of effect of varying dietary protein intake on glutamate dehydrogenase activity is surprising because fish excrete nitrogen mainly as ammonia, whose main pathway of formation is considered to be transdeamination (Forster and Goldstein, 1969; Watts and Watts, 1974) where glutamate dehydrogenase plays a central role. Glutamate dehydrogenase is an allosteric enzyme and it may be that enhanced activity in both mammals and fish given high protein diets is due to allosteric effects (high ADP concentration or altered pyridine nucleotide concentrations related to changes in mitochondrial activity); their effects were not considered by Schimke (1962) or Cowey et al. (1974b). One other possibility is that glutamate dehydrogenase is not primarily involved in ammonia production since McGivan and Chappell (1975) have recently suggested that the main function of the enzyme concerns nitrogen storage. The latter possibility implies that most of the ammonia excreted by fish is formed by some other pathway such as the purine nucleotide cycle (Lowenstein, 1972) but there is little evidence that this pathway is active in the liver which must act as a major site of amino acid degradation. The tissue distribution of ammonia-forming reactions in rainbow trout is shown in Table V. The inferences from these data are that amino acid deamination in the liver (the repository for amino acids postabsorption) occurs through the mediation of glutamate dehydrogenase whereas catabolism in muscle occurs via the purine nucleotide cycle.

Fine control of amino acid catabolism (via K_m) has been convincingly illustrated in omnivorous mammals (Krebs, 1972). The tissue concentrations of most amino acids are normally less than 1 mM and K_m values of enzymes which degrade amino acids are millimolar or more. These relative levels militate against amino acid degradation but any increase in amino acid concentrations (e.g., postabsorptive or by increasing protein intake) will lead to a rapid increase in amino acid breakdown. Tissue levels of amino acids in fish may vary with species but recent analyses on channel catfish (Wilson and Poe, 1974) show that in liver, gills, and kidney they are greater than 1 mM. There are no data available for fish on the K_m values of enzymes which deaminate amino acids but another factor affecting the issue in cold blooded animals is temperature. For many enzymes from fish the most common effect of a decrease in environmental temperature is a reduction in apparent K_m (Hochachka and Somero, 1973) although, to date, most of the results bear on glycolytic enzymes. The resultant increase in enzyme–substrate affinity tends to compensate at nonsaturating substrate concentrations for the fall in reaction rate with temperature. The net effect of a change in environmental temperature on the metabolic fate of amino acids depends on whether changes in K_m of amino acid deaminating and amino acid activating enzymes are in a

Table V

Activity of Ammonia-Forming Enzymes in Tissues of Rainbow Trout[a]

	Liver	Kidney	Gill	Muscle
Glutamate dehydrogenase[b]	0.95 ± 0.17	0.78 ± 0.13	0.31 ± 0.06	Not detected
Glutaminase	3.37 ± 0.99	1.93 ± 0.15	2.07 ± 0.14	Not detected
AMP deaminase	Not detected	Not detected	9 ± 4	226 ± 45

[a] Data are from Walton and Cowey (1977). Activity is expressed as micromoles per gram wet weight per minute at 15°C.

[b] Glutamate deamination in the absence of adenosine diphosphate.

similar direction and of a similar magnitude. The optimal temperature for cultivation of a given species will clearly be influenced by these effects.

The ability of the rat to adapt metabolically to variations in protein intake, presumably by mechanisms such as those discussed above, has been demonstrated by experiments in which the oxidation of isotopically labeled essential and nonessential amino acids has been followed. Thus, McFarlane and von Holt (1969) gave ^{14}C-labeled amino acids intraperitoneally to rats which had previously received either high or low protein diets; the production of $^{14}CO_2$ in the expired air was monitored for 3 hr afterward. The nonessential amino acids, glutamate and alanine, were oxidized rapidly irrespective of dietary protein intake; oxidation of the essential amino acids, leucine and phenylalanine, was very markedly reduced, however, in those animals which received low protein diets. The rat thus possesses ability to conserve essential amino acids under conditions where their supply in the diet is restricted.

Similar experiments have been carried out on plaice and turbot. Results for turbot appear in Tables VI and VII. Glutamate and alanine were oxidized rapidly—over 50% of the radioactivity being expired in 24 hr. This is to be anticipated since carbon residues from both acids rapidly enter the tricarboxylic acid cycle; the results are in agreement with similar experiments carried out on carp (Nagai and Ikeda, 1972, 1973). Thus, the oxidation of these nonessential amino acids is gener-

Table VI

Incorporation of Radioactivity from L-[1-^{14}C]Leucine, L-[1-^{14}C]Phenylalanine, L-[1-^{14}C]-Alanine, and L-[1-^{14}C]Glutamic Acid *in Vivo* into Liver Protein, Carcass Protein, and Carbon Dioxide of Turbot Given Either a High- or Low-Protein Diet[a]

Amino acid	Protein level in diet (g/100 g)	Oxidation as $^{14}CO_2$ (% dose given)	Incorporation into liver protein (% dose given)	Incorporation into carcass protein (% dose given)
Leucine	6	23.5	1.1	13.6
	50	28.9	0.9	13.7
Phenylalanine	6	24.1	0.9	11.4
	50	19.9	1.1	13.7
Glutamic acid	6	56.3	0.08	1.1
	50	59.1	0.08	1.0
Alanine	6	56.6	0.11	1.2
	50	48.5	0.11	1.8

[a] From Knox and Cowey (unpublished data).

Table VII

Effect of Dietary Protein Level on Incorporation of Radioactivity into Carbon Dioxide, Liver Protein, and Carcass Protein from Intraperitoneally Injected L-[1-^{14}C]Leucine, L-[1-^{14}C]Phenylalanine, L-[1-^{14}C]Glutamic Acid, and L-[1-^{14}C]Alanine in Turbot[a]

Amino acid	Oxidation as $^{14}CO_2$	Incorporation into liver protein	Incorporation into carcass protein
Leucine	0.82	1.23	1.01
Phenylalanine	1.21	0.82	0.83
Glutamic acid	0.95	1.00	1.06
Alanine	1.17	1.00	0.66

[a] From Knox and Cowey (unpublished data). Results are given as the ratio of radioactivity incorporated by turbot fed a low-protein diet (6 g crude protein/100 g) to that of turbot fed a high-protein diet (50 g crude protein/100 g).

ally similar in the carnivorous, poikilothermous fish and in the omnivorous, warm-blooded mammal.

Appreciable quantities of the essential amino acids, leucine and phenylalanine, were also oxidized irrespective of the dietary protein level. This contrasts with the reduction in phenylalanine and leucine oxidation in the rat when dietary protein intake was reduced.

While it appears that incorporation of essential amino acids into the tissue protein of the turbot (especially the carcass protein) is greater than that of nonessential amino acids, direct comparison of the metabolism of different amino acids in this type of experiment is hazardous because results can be greatly influenced by factors such as pool size and compartmentation.

However, as the same amino acid pool probably serves both catabolic and anabolic processes, effects of variation in dietary protein level on the metabolism of any amino acid may be assessed by examining the ratio of amino acid oxidation on low and high protein diets and comparing it with the ratio of amino acid incorporation into protein at the different dietary protein levels. These ratios appear in Table VII and it is evident that there is no significant reduction in the oxidation of either essential or nonessential amino acids in response to dietary protein restriction. Neither is there an increase in the rate of incorporation of amino acids into tissue proteins in response to this change in diet composition. Thus turbot do not appear to adapt to restriction of protein intake in the manner in which rats were found to adapt by McFarlane and von Holt (1969); there is no conservation of at least two essential amino acids in the turbot when their supply in the diet is limited.

III. CARBOHYDRATES

A. Carbohydrate Utilization

It has already been seen that carbohydrate up to levels of about 25% in the diet is as effective as fat as an energy source for several species of fish (channel catfish, rainbow trout, plaice). This value fits well with the apparent digestibility of starch for rainbow trout (Singh and Nose, 1967) in that higher levels of starch result in a sharp fall in digestibility for this species. Recent high values for the digestibility of starch (Chiou and Ogino, 1975) by carp therefore contrast with this picture; on average 85% of ingested α-starch was digested by this species at dietary levels between 19 and 48%. These values are surprising because other evidence (see below) does not suggest that carp metabolize carbohydrate appreciably more rapidly than do other species (although grass carp may do so). Chiou and Ogino (1975) found the digestibility of β-starch by carp to be lower than that of α-starch.

Cellulase activity has been demonstrated (Stickney and Shumway, 1974) in the digestive tracts of several species of estuarine fish from the southeastern coast of the United States and also in the freshwater channel catfish reared in an intensive indoor culture system. In the latter case it was shown that the artificial pelleted diet contained no cellulase activity; starved catfish exposed to streptomycin (200 mg/liter) lost their cellulase activity while untreated, starved controls did not. Thus the cellulase activity was associated with a microflora in the alimentary tract. Of the 148 wild elasmobranch and teleost fish examined, 16 (all estuarine) possessed cellulase activity. This demonstration that a cellulolytic microflora can be established in the alimentary tract of captive fish given pelleted food has very significant consequences for aquaculture.

The two main energy yielding sequences from glucose (Embden–Meyerhof pathway and pentose phosphate pathway) have been discussed previously in a nutritional context (Cowey and Sargent, 1972).

Little information is available on the relative importance of these cycles in fish. On the basis of the distribution of certain pentose phosphate pathway enzymes in the tissues of the barracuda, yellowtail, and carp, Shimeno and Takeda (1972, 1973) suggested that the pathway was active in liver and kidney. Earlier estimates (Hochachka, 1969) indicate a minor role for the pentose phosphate pathway; in liver the role of this pathway is largely the production of NADPH for the reductive reactions in fatty acid synthesis. Catabolism of glucose by either sequence requires prior phosphorylation by hexokinase. If omnivorous mammals are subjected to a high carbohydrate intake, a hepatic glucokinase (which has a high K_m for glucose of 15–20 mM and is not inhibited by glucose 6-phosphate) is induced. The hepatic hexokinase isoenzymes of six species of fish were studied by Nagayama and Ohshima (1974) but no isoenzyme which resembled mammalian glucokinase was detected. At the blood glucose levels characteristic of fish (generally less than 4 mM) such an enzyme would not function efficiently.

A glucose dehydrogenase which oxidizes glucose directly has been isolated from the liver of rainbow trout and other salmonids (Shatton *et al.*, 1971). The enzyme has a requirement for Mg^{2+} ions and is active with either NAD^+ or $NADP^+$; it is present in both particulate and cytosolic fractions of liver cells. Physiologically the enzyme is something of an enigma. It is optimally active at pH 10, has a K_m for glucose of 40–80 mM at physiological pH (depending on the coenzyme), and moreover, is, active toward several other sugars and sugar phosphates; the K_m for glucose 6-phosphate at physiological pH is 0.01–0.02 mM.

Glucose dehydrogenase activity in livers of carnivorous fish is said to be four to seven times that in mammalian livers, whereas the activity in herbivorous fish (grass carp, silver carp) is similar to that of mammals (Nagayama *et al.*, 1973). This claim rather ignores differences between herbivorous and carnivorous mammals as well as the fact that measurements for mammalian and piscean enzymes were made at pH 7.5 (Ballard, 1965) and pH 10, respectively. Later studies (Nagayama *et al.*, 1975a,b) on several species of fish confirm the kinetic values obtained for glucose dehydrogenase by Shatton *et al.* (1971); in grass carp, rainbow trout, and yellowtail glucose-6-phosphate dehydrogenase and glucose dehydrogenase were isolated from the liver as separate proteins. Glucose-6-phosphate dehydrogenase only, having glucose dehydrogenase activity, was isolated from livers of carp and eel. Altogether we are some way from being able to evaluate the significance of glucose dehydrogenase.

Hexokinase activity in different organs from various species of fish has been measured by several workers. Data of Nagayama *et al.* (1972) are shown in Table VIII where comparable data for some rat tissues (Newsholme and Start, 1973) are included; fish and rat enzymes were measured at 20° and 25°C, respectively. Hexokinase (+ glucokinase) activity in rat liver is about tenfold higher than that in the livers of fish, that in rat kidney about threefold greater than fish kidney. This appears one of the prime reasons for the inability of fish to metabolize glucose rapidly. White muscle of omnivorous mammals has very low

Table VIII

Hexokinase Activity in Tissues of Fish and Rat[a]

	Liver	Heart	Kidney	Muscle
Grass carp	0.28	3.58	0.99	0.19
Carp	0.29	1.99	0.88	0.29
Rainbow trout	0.52	1.45	0.99	0.19
Eel	0.25	3.14	0.94	0.17
Rat	2.5	6.10	2.8	2.0

[a] Data on fish are from Nagayama *et al.* (1972); those on rat are from Newsholme and Start (1973). Activity is expressed as micromoles of substrate transformed per minute per gram of tissue.

hexokinase activity (Burleigh and Schimke, 1968). Red muscle of both mammals and fish is richer than white in hexokinase activity. The higher proportion red : white muscle in mammals compared to fish is another factor in their greater glucose utilization.

Nagayama *et al.* (1972) have also attempted to assess the relative ability ("latent activity") of four species of fish to metabolize glucose on the basis of the total hepatic activity of certain enzymes in fish of unit weight. On this criterion species rated in the order grass carp, eel, carp, and rainbow trout. It would be useful to have this order checked by measurements of glucose oxidation under controlled conditions because with the possible exception of hexokinase none of the enzymes measured are known to be regulatory in the sense of Newsholme and Start (1973).

When omnivorous mammals are deprived of food the carbohydrates in the body are used up in a short period of time; glycogen constitutes a reserve from which glucose can be formed rapidly when required. Thus when food is withheld from rats, liver glycogen falls to very low levels within 24 hr (Freedland, 1967; Newsholme and Start, 1973). By contrast fish do not mobilize liver glycogen rapidly when they are starved. Nagai and Ikeda (1971a) found that blood glucose and liver glycogen levels of carp which had been starved for 22 days were not significantly different from those of carp which had been given diets varying markedly in gross composition (starved fish, 10.65% liver glycogen; fed fish, 7.5–10.9% liver glycogen depending on diet). Even after 100 days without food appreciable amounts of glycogen (1.5%) remained in the carp liver. This phenomenon is common to other species such as European and Japanese eel (Larsson and Lewander, 1973; Hayashi and Ooshiro, 1975a).

The conversion of glycogen to glucose 1-phosphate is catalyzed by phosphorylases. The inability of fish to mobilize liver glycogen

rapidly under circumstances such as starvation suggests either a paucity of phosphorylase or that metabolic and hormonal factors restrict the
activity of the enzyme. The total phosphorylase activity in red, white,
and cardiac muscle of rainbow trout is much greater than that in liver
(Yamamoto, 1968) although all the hepatic phosphorylase was apparently in the active *a* form. Liver phosphorylase activity of rainbow
trout (Yamamoto, 1968) appears to be of a similar order to that in the
omnivorous rat (Freedland, 1967) although the fish, as well as the rat,
measurements were made at 37°C. Very little work has been carried
out on the phosphorylases of liver. The properties of the enzymes
concerned (phosphorylase *a* and *b*, and the interconverting enzymes)
are probably exceedingly complex and it is not yet possible to attempt
to offer an explanation of the factors which control glycogenolysis in
the liver either in the feeding or fasting fish.

The fact that even in starving fish the oxidation of substrates other
than glucose takes precedence over the mobilization and hydrolysis of
glycogen and oxidation of the resulting glucose suggests that the capacity of fish to oxidize glucose aerobically is somewhat limited. It
also indicates that the demands of any tissues (presumably brain and
nervous tissue) which catabolize glucose as a primary fuel are met by
gluconeogenesis rather than by glycogenolysis, blood glucose levels
being maintained by the former process. Why this should be so is by
no means clear but the fact that in more extreme cases of starvation
(e.g., the spawning migration of *Oncorhynchus nerka*) the glycogen
content of the liver is actually elevated, after a transient initial decrease, from 18 to 32 mg/g in males and from 2 to 4 mg/g in females
(Chang and Idler, 1960) suggests that glucose production by
gluconeogenesis may occasionally or marginally exceed glucose demand, the excess being stored as glycogen.

Attempts to measure the rate of glucose oxidation by fish and to
assess the effect of diet composition on glucose oxidation have involved the use of radioactive glucose, usually administered intraperitoneally. Inferences from such experiments should be viewed
critically because of factors such as pool size and compartmentation
which could be crucially changed by giving diets of very different
composition over a long period. From such experiments Nagai and
Ikeda (1971b) have concluded that glucose oxidation by carp given
diets containing 50% protein was significantly lower than in carp
given low (10%) protein, high starch diets. Later these authors, Nagai
and Ikeda (1972) compared the oxidation of glutamate and of glucose
by carp given one of two diets, a high carbohydrate, low protein diet or
a diet containing approximately 50% starch and 50% protein. Carp
subjected to either regime oxidized glutamate rapidly and glucose

slowly. The authors deduced that amino acids are a superior energy source to glucose for carp and that energy utilization in the fish resembles that in a diabetic mammal. In fact glutamate is rapidly oxidized by most animals (as it may enter the tricarboxylic acid cycle on deamination) and MacFarlane and von Holt (1969) observed high rates of glutamate oxidation in the rat irrespective of dietary protein intake. Thus fish and omnivorous mammals do not differ in their facility to oxidize nonessential amino acids; where they may differ is in the facility with which they oxidize glucose.

Cowey *et al.* (1975a) found that between 12 and 23% (depending on dietary history) of a trace dose of radioactive glucose was oxidized by plaice in 18 hr at 15°C. Up to 10% of the radioactive glucose (again depending on diet) was converted to glycogen in this time while 57% of the dose remained as glucose, or small acid-soluble molecules derived from glucose, in the acid-soluble fraction. By contrast about 80% of a trace dose of glucose was oxidized by mice in 8 hr (Vrba, 1966); in addition virtually none of the injected radioactivity remained in the acid-soluble fraction at this time, all the radioactive glucose having been metabolized. Cowey *et al.* (1975a) claim that these differences in the rate of glucose catabolism between mice and plaice are greater than can be accounted for by Q_{10} effects and that glucose oxidation by the omnivorous mammal is substantially more rapid than in the carnivorous plaice.

B. Gluconeogenesis

Indirect evidence that gluconeogenesis plays a significant role in maintaining blood sugar levels in fasting fish has already been mentioned. It is likely that in cultured or wild fish consuming high protein diets gluconeogenesis, probably from amino acids and triacylglycerol glycerol, occurs to a greater or lesser extent. As glycerol may derive from glucose itself; amino acids are the only source of new glucose in the strictest sense. The characteristics of the pathway in fish, the main substrates, and the effects of diet composition on the process are only now beginning to command attention.

In a recent series of papers Inui and Yokote have examined some aspects of the process in eels. Their approach probably suffers from being indirect, relying on correlations between changes in liver glycogen, plasma amino acids, liver enzymes, and other parameters rather than actual measurement of glucose formation per se in liver slices, hepatocytes, or a perfused liver. In the fasting eel (Inui and

Yokote, 1974) prolonged monitoring (80 days) of some of these parameters showed that blood glucose concentration remained relatively constant, liver glycogen decreased at first but then gradually increased, hepatic alanine and aspartate aminotransferase activity increased while plasma α-amino nitrogen levels increased transiently and then remained constant. The authors inferred that the main substrates of gluconeogenesis were plasma amino acids derived from body protein.

Subsequent papers (Inui and Yokote, 1975a,b) indicated some control over gluconeogenesis in eel by insulin. Insulin deficiency, induced by treatment with alloxan, accelerated gluconeogenesis by increasing the rate of immobilization of amino acids from body protein. Insulin treatment (intramuscular injection) suppressed gluconeogenesis by decreasing plasma α-amino nitrogen; it did not however affect the activity of the gluconeogenic enzyme fructose diphosphatase in the liver; aminotransferase activities in the liver were similarly unaffected by insulin treatment.

The effects of hydrocortisone injection on gluconeogenesis in the eel (Inui and Yokote, 1975c,d) are not clear. It is apparent that the differences in liver glycogen levels between intact (noninjected) and saline (injected daily with 0.75% saline) controls are as great as are the differences between saline controls and hydrocortisone-treated fish. As the liver glycogen levels are clearly affected by daily saline injections this would not seem to be an ideal parameter on which to base comparisons of gluconeogenic rates. Hydrocortisone had little effect on amino acid mobilization but was said to accelerate gluconeogenesis; a surprising feature of the liver enzyme measurements was that fructose diphosphatase activity was only slightly increased in the hydrocortisone-treated eel, but phosphofructokinase activity (a glycolytic enzyme) was markedly increased.

Finally Inui et al. (1975) grouped amino acids into three classes according to their response (level in plasma) to alloxan treatment, hepatectomy, and insulin treatment but it is not clear whether one or other of these groups was more gluconeogenic than others.

Direct measurements of the rate of gluconeogenesis in the isolated perfused liver of the eel have been achieved by Hayashi and Ooshiro (1975a). Glucose production from lactate was linear for periods of up to 2 hr in this preparation and the rate of glucose production 0.9 μmol/g wet liver/min at 25°C was similar to that of rat liver measured at 35°C (Hems et al., 1966). In the perfused eel liver lactate removal was initially 4.5 μmol/g wet liver/min but decreased to 2 μmol/g wet liver after 30 min; oxygen consumption was 5.7 fold lower than that of perfused rat liver.

It has already been observed that glycogen is only slowly removed from fish livers when food is withheld. Hayashi and Ooshiro found the liver glycogen levels in their eels, which had been starved 3–6 days, were 23 ± 8 mg/g liver; this could complicate measurement of gluconeogenesis (cf. Krebs et al., 1963). We ourselves were unable to reduce the glycogen content of rainbow trout livers to very low levels by phlorhizin treatment as Krebs et al. (1966) had done on mouse livers. Hayashi and Ooshiro certainly had difficulty in assessing the effects of epinephrine and cyclic AMP on gluconeogenesis because of the complicating presence of glycogen in the liver.

In a subsequent paper Hayashi and Ooshiro (1975b) showed that the mean time for conversion of [^{14}C]lactate to glucose in isolated perfused eel liver was about 60 sec, comparing with a reported time of 75 sec for the perfused rat liver (Exton and Park, 1967).

The distribution of key gluconeogenic enzymes in the tissues of hatchery-reared trout (Table IX) indicate that the main sites of gluconeogenesis in the fish are the liver and kidney. The data do not support the view of Driedzic and Kiceniuk (1976) that a large fraction of the lactate produced by trout during strenuous exercise is reconverted to glucose in the gills. The presence of fructose disphosphatase in skeletal muscle is probably connected not with gluconeogenesis, but with the operation of a cycle between fructose 6-phosphate and fructose diphosphate which increases the sensitivity of the initial stages of glycolysis to adenosine monophosphate concentration (Newsholme and Crabtree, 1970).

The effect of diet composition on gluconeogenesis in rainbow trout has been examined (De La Higuera and Cowey, unpublished data) by measuring the conversion of radioactive alanine to glucose in the whole fish. Glucose space was first measured by the kinetic method described by Holmes and Donaldson (1969) following injection of a

Table IX
Activities of Gluconeogenic Enzymes in Tissues of Hatchery-Reared Rainbow Trout[a]

Tissue	Fructose-1,6-diphosphatase	Pyruvate carboxylase	Phosphoenolpyruvate carboxykinase
Liver	1.60	0.79	1.32
Kidney	0.44	0.11	0.33
Gill	0.07	0.04	0.01
White muscle	0.22	0.01	Not detected

[a] Activity is expressed as micromoles per gram fresh weight per minute at 15°C. Mean values for six fish. Data are from Cowey et al. (1977).

trace dose of high specific activity [U-^{14}C]glucose. It amounted to 13.7% of the body weight, a value similar to that reported for other species (Mazeaud, 1973). After the glucose space was determined, gluconeogenesis from alanine was followed by isolating glucose from the blood, measuring its radioactivity and calculating alanine conversion to glucose from the relationship:

% Dose incorporated = body wt (g) × glucose diffusion space
                     × (dpm per ml blood/dpm alanine injected)

The dose used contained 1 mmol of carrier alanine so obviating spurious effects from a large and variable dilution of the [U-^{14}C]alanine by endogenous alanine in the fish subjected to different dietary treatments.

The results in Table X show that gross changes in the composition of the diet have significant effects on the rate of gluconeogenesis in trout. Under the conditions little or no synthesis of glycogen from radioactive alanine occurred, so that incorporation of radioactivity into blood glucose served as a reliable measure of gluconeogenesis uncomplicated by glycogen. The highest rates of gluconeogenesis were observed in trout given a high protein diet (from which carbohydrate was absent) and in trout subjected to starvation. The feeding of a high carbohydrate, low protein diet depressed gluconeogenesis markedly. By comparison in normal fed rats (diet unspecified) gluconeogenesis

Table X

Effect of Dietary Treatment on Gluconeogenesis in Rainbow Trout. Incorporation of Radioactivity from [U-^{14}C]Alanine into Blood Glucose and Liver Glycogen[a]

	Blood glucose		Liver glycogen	
Diet treatment[b]	Concentration (mg/100 ml)	Percentage of administered radioactivity	Concentration (mg/g wet wt)	Percentage of administered radioactivity
Fasted	54	2.9	2.6	0.005
60% Protein, 8% lipid, zero carbohydrate	71	3.1	4.6	0.04
10% Protein, 13% lipid, 55% carbohydrate	96	0.5	83.5	0.024

[a] Gluconeogenesis from alanine is expressed as the percentage of the administered radioactivity detected in blood glucose or liver glycogen 6 hr after administration of [U-^{14}C]alanine.

[b] Diet treatment lasted 4 weeks. Data are from Cowey et al. (1977).

measured in this way proceeded relatively slowly, 2.2% of an administered dose being incorporated into glucose; high rates of gluconeogenesis were however recorded in 48 hr-fasted rats, 9.8% of an administered dose being converted to blood glucose (Friedmann *et al.*,
1965).

The suppression of gluconeogenesis in trout in response to a high
carbohydrate, low protein diet suggests that some metabolic control
over glucose metabolism is exercised. Whether this adaptation is
primarily a result of a luxus of glucose or a paucity of amino acids is not
clear. Little attempt has been made to distinguish between the two
effects in mammals when the reverse dietary change (transfer to high
protein, low carbohydrate diet with enhanced gluconeogenesis) has
been practiced (Schoolworth *et al.*, 1974). Protein deficiency (necessitating amino acid conservation) may not be the sole or main cause
because starving fish exhibit quite high rates of gluconeogenesis,
though this is a different physiological situation.

IV. LIPIDS

In common with other animals, fish require lipids as a source of
metabolic energy (ATP) and to maintain the structure and integrity of
cellular membranes. Historically, lipids have always featured promi-
nently in the chemistry of aquatic and especially marine animals. This
situation arose because marine oils are characteristically different from
the fats of terresterial animals in that they are rich in the fat-soluble
vitamins and highly unsaturated fatty acids, properties that in the past
made marine oils attractive for commercial exploitation. Moreover,
many marine animals contain large amounts of oil that are easily ac-
cessible and extractable, for example, cod liver oil. Both the unsaturation of fish oils and their presence in large amounts in fish have important consequences for fish nutrition. As we shall see, the high degree of
unsaturation imposes stringent dietary requirements on marine fish for
certain long chain, polyunsaturated fatty acids. The presence of large
amounts of oil in fish means that lipid rather than carbohydrate is the
favored energy reserve of most aquatic animals in their natural environment. These facts apply to all fish but especially to marine fish.

Not only are marine fish exposed to large amounts of highly unsaturated lipid in their natural diets but the class composition of the
lipid often differs strikingly from that of terrestrial animals, especially
mammals. Thus, triacylglycerols are the major if not the sole neutral
dietary lipid of land mammals whereas wax esters are the normal

dietary lipid of very many species of commercially important fish including herring, sprats, sardines, pilchards, menhaden, anchovies, mackerel, and young salmonids. The natural diet of these fishes is predominantly zooplankton among which the calanoid copepods are particularly prominent. The latter organisms contain wax esters as their major lipid reserve; these generally account for one-third and can reach up to two-thirds of their dry body weight (Lee *et al.*, 1971a; Lee and Hirota, 1973; Benson *et al.*, 1972; Sargent *et al.*, 1976). Wax esters, however, are not present to any appreciable extent in the freshwater environment and are not generally consumed by freshwater fishes.

The above considerations indicate that the lipid nutrition of fish is likely to differ significantly from that of the land animals, especially the mammals. It is important, therefore, to understand the basic lipid metabolism of fish so that aspects of lipid nutrition unique to fish are fully appreciated. In this section, therefore, we present initially an outline of special aspects of lipid metabolism in fish in the expectation that this will sharpen our understanding of the overall lipid nutrition of fish which follows as our main concern.

A. Lipid Metabolism

1. DIGESTION, ABSORPTION, AND FURTHER METABOLISM

Morphologically the intestine of many teleost fish is characterized by numerous pyloric ceca as well as by a pancreas that is frequently diffuse (Barrington, 1957). These features at once imply the existence of major differences between fish and land mammals in the processes of lipid digestion and metabolism.

Difficulty has frequently been experienced in the past in detecting triacylglycerol lipase in the diffuse pancreas of marine fish (Brockerhoff, 1966a; Overnell, 1973; Patton *et al.*, 1975) although the enzyme has been detected in and partially purified from the diffuse pancreas of the trout (Leger and Bauchart, 1972; Leger, 1972). Triacylglycerol lipase activity is readily detected in the discrete pancreas of the skate (Brockerhoff and Hoyle, 1965) and the leopard shark (Patton, 1975). The low activity of triacylglycerol lipase apparent in marine fish pancreas is but one side of the problem since the enzyme from mammalian and elasmobranch pancreas alike shows relatively little activity in cleaving the long chain, polyunsaturated fatty acyl groups concentrated on position-2 of the glycerol backbone of the triacylglycerol molecule and present in large amounts in marine oils. The positional specificity of triacylglycerol lipase served as a corner stone in the "Brockerhoff hypothesis" that marine triacylglycerols were absorbed from the intestine of fish and deposited in adipose tissues such as the liver by processes that did not involve cleavage of the fatty acyl bond at position-2 of the triacylglycerols (Brockerhoff, 1966b). Subsequent experiments, however, showed that, at least in the cod *Gadus morhua*, all the ester bonds in a fed triacylglycerol were broken and re-formed at sites between the intestinal lumen and the liver (Brockerhoff and Hoyle, 1967). This conclusion has been amply confirmed for rainbow trout (Leger *et al.*, 1977).

Recent work has clarified many of the above problems. An active lipase can be readily detected in fresh or lyophilized intestinal fluids from a number of marine fish including northern anchovy, *Engraulis mordax*, pink salmon, *Oncorhynchus gorbuscha*, striped bass, *Morone saxatilis*, jack mackerel, *Trachurus symmetricus,* and Pacific mackerel, *Scomber japonicus* (Patton *et al.*, 1975). Unlike triacylglycerol lipase the lipase present in intestinal fluids is nonspecific in that it hydrolyzes fatty acyl groups with equal facility from all three positions of the glycerol backbone of triacylglycerols, that is, the lipase hydrolyzes primary as well as secondary alcoholic esters. Moreover, the intestinal fluid lipase readily hydrolyzes polyunsaturated as well as saturated and monounsaturated fatty acyl groups from triacylglycerols. The enzyme is probably the same as the carboxylic ester hydrolase of mammalian pancreas that hydrolyzes a variety of esters including cholesterol esters (Borgstrom, 1974). Carboxylic acid esterase is greatly stimulated by added bile and the intestinal fluids used by Patton *et al.* (1975) contain bile as well as lipid hydrolase activity.

It is noteworthy that Patton (1975) described a pancreatic lipase in the leopard shark with a positional specificity similar to that of mammalian pancreatic lipase; addition of shark bile to shark pancreatic extracts induced the appearance of the nonspecific lipase activity referred to above. Thus, a bile-activated carboxylic ester hydrolase exists in shark pancreas. At the same time nonspecific lipase activity is present in the bile itself of some fish including the anchovy and jack mackerel (Patton *et al.*, 1975). This is not a general property of the bile of marine teleosts since the biles of bass, salmon, and jack mackerel lack nonspecific lipase activity (Patton *et al.*, 1975). The nonspecific lipase activity in the intestinal fluids of the latter species presumably originates in pancreas. A particularly noteworthy finding in the experiments of Patton *et al.* (1975) was that some 50% of the 2-monoacylglycerols formed during hydrolyses of triacylglycerols by nonspecific lipase were themselves completely degraded to fatty acids. This contrasts markedly with the situation in mammalian intestine where 2-monoacylglycerols are the major end product of lipid digestion. That is, fish intestines have a marked capacity to reduce dietary triacylglycerols completely to free fatty acids.

The northern anchovy, *Engraulis mordax*, was featured in the work of Patton *et al.* (1975) because the natural diet of this fish is mainly calanoid copepods that can contain up to 70% of their dry body weight as wax esters.

The concern with intestinal hydrolysis of wax esters stems from the knowledge that these lipids are considerably more hydrophobic than triacylglycerols and consequently more difficult to emulsify; for example, triacylglycerols form monolayers at water–air interfaces but wax esters do not (Carey and Small, 1970). This difficulty may be exacerbated by the extremely hydrophobic product of wax ester hydrolysis, free fatty alcohol, being a potent inhibitor of pancreatic triacylglycerol lipase (Mattson *et al.*, 1970).

Intestinal fluids from the northern anchovy hydrolyze wax esters although the rate of hydrolysis is much less than that of triacyglycerols. Patton and Benson (1975) demonstrated that seven species of marine fish representing five families all had at least some ability to hydrolyze dietary wax esters and to incorporate the products of intestinal hydrolysis into tissue lipids. Digestion of wax esters, however, varied markedly in the different species. Patton and Benson (1975) argue that digestion of wax esters is facilitated in fish not so much by increasing the activities or amounts of hydrolases in the intestine but rather by morphological characteristics of the intestinal tract such as the presence of numerous pyloric ceca. Such ceca are considered to ensure a long retention time for food in the intestine so as to ensure the eventual hydrolysis of "awkward" dietary constituents such as the wax esters. Such a strategy which relies on time rather than catalytic efficiency to effect a biochemical reaction can only be adopted by animals that experience quite long gaps between meals.

We have seen above that fish differ from land mammals in their pronounced ability to reduce large quantities of dietary triacylglycerols completely to free fatty acids and glycerol. This finding implies that, at a practical level, lipid can be readily presented in the diet of fish as triacylglycerols, simple methyl esters of fatty acids, or even as free fatty acids. It also implies, however, that substantial differences could exist between fish and land mammals with respect to the mechanisms of absorption of lipid from the intestine.

Experiments by Robinson and Mead (1973) present good evidence that such differences exist. These authors fed radioactive palmitic acid to rainbow trout and examined the time course of appearance of radioactivity in the lipids in trout blood. For the first 2 hr after feeding nearly all the radioactivity in blood was present as free fatty acids. It was only after 2 hr that radioactivity in free fatty acids declined, to be replaced by radioactivity in triacylglycerols. Similarly, Kayama and Iijima (1976) have shown that radioactivity from either [^{14}C]palmitic acid or [^{14}C]tripalmitin fed by stomach tube to carp first appears in the plasma as free fatty acids associated with albumin. The triacylglycerols of the pre-β- and β-lipoproteins in carp plasma become labeled only at a later period in the time course. The simplest conclusion from these studies is that free fatty acids are absorbed directly from the intestinal lumen into the blood. Robinson and Mead (1973), however, propose a more subtle conclusion, namely, that free fatty acids originating in the intestinal lumen are absorbed into the intestinal mucosal cells and are converted there to triacylglycerols; the triacylglycerols are then exported into the extracellular serosal space to reach the underlying lamina propria, which in fish is known to be rich in lipid (Barrington, 1957). Radioactive fatty acids appearing rapidly in the plasma in the experiments of Robinson and Mead (1973) and Kayama and Iijima (1976) are presumed to be derived from radioactive triacylglycerols deposited in the lamina propria and cleaved either there or at the surfaces of the blood capillaries that drain the lamina propria. It is noteworthy that triacylglycerols are known not to pass cellular membranes including capillary membranes in terrestrial mammals without being cleaved to glycerol and free fatty acids by lipoprotein lipase (Robinson and Wing, 1971). The radioactivity appearing later in fish plasma as triacylglycerols could derive from free fatty acids absorbed from the lamina propria into the portal blood, which are delivered to the liver and synthesized to triacylglycerols. However, direct absorption of triacylglycerols from the intestine by the lymphatic route is not excluded.

Underlying the above considerations is the knowledge that fish do not have the well-developed lymphatic (lacteal) system or the thoracic duct that in mammals delivers triacylglycerols directly from the intestine to the peripheral blood as chylomicron particles. This situation is in line with the failure of Lee and Puppione (1972) to detect chylomicrons in sardine serum even after the fish had consumed very fatty meals. Nonetheless, some lipoprotein particles similar in size to chylomicrons or very low density lipoproteins have been detected by electron microscope studies on trout intestine (Bergot and Fléchon, 1970). These particles appear within the intestinal cells, in the spaces between the cells, in the lamina propria, and in the blood capillaries that drain the intestine. Bergot and Flechon (1970) consider that the particles leave the intestine by two routes: the portal route, which is well developed in some animals including the chicken, and the lymphatic route, which is well developed in others including terrestrial mammals. One synthesis from all the above findings, both metabolic and structural, is that fish differ from terrestrial mammals in that they depend heavily on the portal blood drainage system for lipid assimilation, much more so than on a lymphatic drainage system.

In a terrestrial mammal triacylglycerols absorbed from the intestine into the blood are freely distributed in the peripheral circulation to the various adipose depots in the animal. In fish it would appear that the intestinal lamina propria may serve as a buffer zone between dietary lipid and the peripheral adipose depots such as the liver and musculature. Whether the latter situation imposes restrictions on the rate of deposition of dietary fat into the peripheral adipose tissues of fish remains to be investigated. It should be noted that adipose depots in fish are frequently present in tissues that have important physiological roles other than fat stores, for example, in liver and muscle.

Little is known about the mechanisms controlling lipid mobilization from the adipose stores of fish although Bilinski *et al.* (1971) detected a lysosomal triacylglycerol lipase in the lateral line tissue of rainbow trout. Such a lipase is undoubtedly involved in intracellular degradation of triacylglycerols but it is possible that the enzyme could be exported from the cells to act on the extracellular lipid known to be present in fish muscle. It is noteworthy in this context that the lipid depots in red muscle of fish are both intra- and extracellular while the depots in white muscle are predominantly extracellular (Greene, 1919, 1926; Nag, 1972). In the experiments of Robinson and Mead (1973) the lipid reserves of the white muscle were depleted much more rapidly than those of the red muscle in starving trout. The translocation of lipids from white to red muscle under these conditions is a distinct possibility. Finally it should be noted that the liver of trout is not normally rich in lipid. Thus the appearance of radioactivity in blood triacylglycerols after the decline of radioactivity in free fatty acids is consistent with the transport of absorbed free fatty acids in the portal blood to the liver where synthesis and export of triacylglycerols follow (Robinson and Mead, 1973).

It is pertinent at this point to consider the mechanisms whereby fish that ingest large amounts of wax esters metabolize the free fatty alcohol produced in their intestinal lumen. Rahn *et al.* (1973) and Sand *et al.* (1973) have studied the metabolism of wax esters and free fatty alcohols fed to the freshwater gourami, *Trichosgaster cosby*, a species that produces eggs rich in wax esters. The female eats unfertilized eggs so that wax esters figure significantly in the normal diet of the fish although the major neutral lipid in the body is triacylglycerol. It was shown that free fatty alcohol produced in the intestine of the gourami is converted into triacylglycerols in the intestinal mucosa (Rahn et al., 1973; Sand et al., 1973), a finding later confirmed by Patton and Benson (1975) for a variety of marine fish. These findings show that fatty alcohol is oxidized in the intestines of fish. When [1-^{3}H]fatty alcohol is fed to gouramis much of the tritium label is incorporated into all three positions of the glycerol backbone of triacylglycerols (Rahn et al., 1973). This finding is consistent with the production of glycerol from both amino acids (glyceroneogenesis) and glucose (glycerogenesis). The implication is that the oxidation of fatty alcohols in the intestine may be coupled to the reduction of either glucose or amino acids to glycerophosphate. The latter can then combine with fatty acids to produce triacylglycerols in the usual way.

This inference has been confirmed by recent work in our laboratory on the metabolism of various substrates in isolated preparations of pyloric ceca from different species of fish. The fish used were herring (Clupea harengus) fed a synthetic diet containing 25% copepod oil (rich in wax esters), sea trout (Salmo trutta) fed a natural diet of squid muscle which contains little fat and virtually no carbohydrate, and brown trout (Salmo trutta) fed a commercial pelleted diet containing substantial amounts of both carbohydrate and triacylglycerols. Transverse rings (1 mm length) of pyloric ceca from each of these species were incubated separately (4 hr at 15°C) with radioactive glucose, aspartic acid, and hexadecanol (0.1 mM; 1.0 μCi/ml) following which triacylglycerols were isolated and assayed for radioactivity. With glucose and aspartic acid as substrates more than 90% of the radioactivity in triacylglycerols was present in the glycerol moiety; with hexadecanol more than 90% was present in the fatty acid moiety. The results (Table XI) show that all three species are capable of oxidizing fatty alcohol to fatty acid and incorporating the latter into triacylglycerols, and also possess both glycerogenic and glyceroneogenic activities. Herring had the greatest capacity for oxidizing fatty alcohol to fatty acid and incorporating the latter into triacylglycerols. The ratio glyceroneogenesis : glycerogenesis was highest in herring and lowest in brown trout. In general, the ability to oxidize fatty alcohol is accompanied by a relatively high glyceroneogenic activity and a relatively low glycerogenic activity.

Such experiments may have implications for the nutrition of wild fish whose diet is largely the wax ester-rich calanoid copepods. In these fish the problem of fatty alcohol metabolism in the intestine is

Table XI

Conversion of [U-¹⁴C]Glucose, [U-¹⁴C]Aspartic Acid, and [U-¹⁴C]Hexadecanol into
[¹⁴C]Triacylglycerols in Isolated Preparations of Pyloric Ceca of Herring, Sea
Trout, and Brown Trout Fed Diets Containing Different Energy Supplements

Species	Energy supplement	Substrate	Radioactivity in triacylglycerols[a] (cpm)	B/A	C/B
Herring	Protein, wax ester	A. Glucose B. Aspartic acid C. Hexadecanol	6,100 ± 78 S.E. 4,100 ± 64 S.E. 12,500 ± 112 S.E.	0.67	3.00
Sea trout	Protein	A. Glucose B. Aspartic acid C. Hexadecanol	7,840 ± 180 S.E. 4,700 ± 130 S.E. 2,820 ± 510 S.E.	0.60	0.60
Brown trout	Protein, tri-acylglycerol, carbohydrate	A. Glucose B. Aspartic acid C. Hexadecanol	27,500 ± 2,200 S.E. 12,700 ± 800 S.E. 5,860 ± 570 S.E.	0.46	0.46

[a] Results are given as cpm incorporated into triacylglycerols/100 mg wet weight tissue/4 hr in 1.0 ml assays containing each substrate at a concentration of 0.1 mM and 1.0 μCi/ml. Values for sea trout and brown trout are the means of many experiments; those for herring are derived for material pooled from seven individual fish (Sargent, Bauermeister, Gatten, and Blaxter, 1976 unpublished data.)

particularly acute since, not only are the calanoids very rich in wax esters, but they contain negligible amounts of carbohydrate (Raymont, 1963) other than chitin. That is, there is little if any preformed glycerol in the diet, and carbohydrate, a potential source of glycerol in the intestinal mucosa, is equally scarce. The likely source of triacylglycerol–glycerol in the intestine of such fish is, therefore, dietary amino acids. That is, dietary wax esters can only be converted to triacylglycerols in the intestines of the fish at the expense of catabolizing some dietary protein. The amounts of protein involved are not great as the glycerol backbone of triacylglycerols accounts for only some 10% of the total molecule. What is not known, however, is the rate at which these reactions occur and the extent, if any, to which they may limit the clearance of free fatty alcohol from the intestine and so the rate of wax ester digestion.

There is good evidence for mammals (Hansen and Mead, 1965) and for gouramis (Sand et al., 1973) that high levels of dietary wax esters will overload the metabolic capacity of the intestine. Under these circumstances the liver intervenes to metabolize absorbed wax esters. No trace of wax esters, however, can be found in the tissues or serum of fish such as the sardine when it is gorging copepods (Lee and Puppione, 1972). We presume that the intestines of these fish have a high capacity to metabolize dietary wax esters. At the same time it is well known that young salmonids actively feeding on copepods produce numerous floating fecal pellets rich in wax esters (Benson *et al.*, 1972). Herring fed a diet containing 25% dry weight of copepod oil produce fecal pellets containing 40% lipid of which some 10% is wax ester. At the present time the efficiency of assimilation of wax esters by fish is not known, although it would be surprising if it were not very high. Such information is clearly essential in determining the quantities of food consumed by clupeoid fish and young salmonids in their natural environments and bears not only on the nutrition of these fish in the wild but also on the efficiency of very critical steps in the marine food chain.

2. POLYUNSATURATED FATTY ACIDS

Most fish are different from terrestrial animals in that their tissues contain predominantly polyunsaturated fatty acids of the $\omega3$ series rather than the $\omega6$ series. In fact, $20:5\ \omega3$ and $22:6\ \omega3$ are the major polyunsaturated fatty acids of fish whereas linoleic acid, $18:2\ \omega6$, and arachidonic acid, $20:4\ \omega6$, are the major polyunsaturates of land animals. We shall see later that fish can either obtain preformed C_{20} and C_{22} polyunsaturates from their diet or they can chain elongate and further desaturate C_{18} polyunsaturates also originating in the diet. Our concern here is with the biochemical mechanisms underlying the marked preference of fish for $\omega3$ rather than $\omega6$ polyunsaturated fatty acids. Nutritional evidence to be described later clearly shows that most fish species specifically require $\omega3$ polyunsaturated fatty acids. Biochemical evidence now strongly suggests that this specific requirement reflects a major characteristic of the aquatic and especially the marine environment, namely that it is an environment where relatively low and constant temperatures prevail. Earlier work had shown that increasing the environmental temperature of fish resulted in tissue fatty acids becoming less unsaturated (Knipprath and Mead, 1968; Smith, 1967; Smith and Kemp, 1969; Kemp and Smith, 1970). In more recent work natural populations of freshwater eels, *Anguilla anguilla*, subjected to an environmental temperature of 15–20°C had gill lipids with an $\omega3/\omega6$ fatty acid ratio of 0.6 (Thomson *et al.*, 1975). These eels could be fully adapted to seawater at 15°C with no change in the $\omega3/\omega6$ ratio indicating that seawater per se did not require increased amounts of $\omega3$ fatty acids. In contrast, populations of eels from a seawater environment at 5–10°C had gill fatty acids with a ratio $\omega3/\omega6$ of 2.3. Maintaining these eels in seawater at 15°C resulted in a replacement of $\omega3$ polyunsaturates with saturated and monounsaturated fatty acids (Thomson *et al.*, 1975). In short, temperature and not salinity changes the degree of unsaturation of gill lipids in the euryhaline eel.

The above findings can readily be explained on the basis of membrane fluidity at different temperatures. Current ideas view biomembranes as "proteins floating in a sea of lipid" (Singer and Nicholson, 1972) where movements of enzymatic protein molecules are essential to the function of the membrane and require a degree of fluidity of the lipid phase. Such fluidity is determined to a large extent by the degree of unsaturation of the polyunsaturated fatty acids preferentially esterified at position-2 of the glycerol backbone of the membrane phospholipids. Thus, it has been shown clearly that the degree of fluidity of the phospholipid phase of a typical biomembrane enzyme, (Na^+, K^+, Mg^{2+})-dependent ATPase, is directly related to the degree of unsaturation of the fatty acyl moieties in the phospholipids (Kimelberg and Papahadjopoulos, 1974). The temperature at which phase transitions occur in the lipid phase of such enzymes can be determined from the position of discontinuity in Arrhenius plots (activity versus reciprocal of temperature). The discontinuity in the gill (Na^+, K^+, Mg^{2+})-dependent ATPase of freshwater eels living at temperatures of 15–20°C was approximately 18°C (Thomson *et al.*, 1975), that is, the same as for mammalian and amphibian enzymes (Kimelberg and Papahadjopoulos, 1974; Tanaka and Teruya, 1973). In seawater eels living at 5–10°C, the position of discontinuity was approximately 12°C; adapting these fish to 15°C increased the position of discontinuity to 18°C. Thus, the fluidity of the biomembranes in the gill epithelium is altered by changes in the degree of unsaturation of the gill lipids at different temperatures. Low temperatures are characterized by $(\omega3)$ polyunsaturates, high temperatures by $(\omega6)$ polyunsaturates.

Recent work has given some insight into how these changes may be brought about. Ninno *et al.* (1974) demonstrated that the freshwater fish, *Pimelodus maculatus*, like the rat, has fatty acid desaturases in its liver microsomes capable of desaturating oleic acid, linoleic acid, and linolenic acid ($\Delta9$, $\Delta6$, and $\Delta5$ desaturases, respectively). De Torrengo and Brenner (1976) have shown that *Pimelodus* maintained at 14–15°C had desaturases (as well as chain elongating enzymes) with higher activities than *Pimelodus* kept at 29–30°C. The ratios of $\Delta9$, $\Delta6$, and $\Delta5$ desaturases remained the same at the two temperatures. The enhanced activities at lower temperatures were brought about by increasing the maximal velocity (V_{max}) of the reaction rather than by altering the affinity of the enzyme for its substrate (K_m). These experiments are the first to show that temperature per se affects the activities of fatty acid desaturases which are themselves biomembrane enzymes. In a recent study Leslie and Buckley (1976) isolated liver microsomes from goldfish (*Carrasius auratus*) maintained at 18°C on a diet containing soya oil. Fatty acids incorporated into phospholipids by the isolated microsomes were considerably more unsaturated at an incubation temperature of 10°C than at 30°C. Thus, temperature itself immediately determines not only which fatty acid is biosynthesized but also which fatty acid is inserted into biomembrane phospholipids. Farkas and Csengeri (1976) demonstrated that these effects apply *in vivo*; the extent to which injected [^{14}C]acetate was incorporated into saturated, monounsaturated, and polyunsaturated fatty acids of both total lipids and phospholipids of the carp (*Cyprinus carpio*) depended on the immediate experimental temperature and not the previous temperature history of the fish. These immediate temperature effects could involve either the protein or lipid moieties of the enzyme systems involved in lipid biosynthesis. The fluidity of fatty acids will clearly affect the activity of biomembrane enzymes such as fatty acid desaturases and ester synthetases. Equally the fluidity of fatty acids is likely to affect the activity of soluble enzymes such as fatty acid synthetase where fatty acids may be substrates or products of the reactions. It may also be noted that the existence of immediate temperature effects in lipid biosynthesis does not rule out the existence of more long-term mechanisms in temperature adaptation. It remains eminently possible that hormonal effects leading to changes in either the amounts or activities of enzymes involved in lipid biosynthesis also play a role in long-term adaptation to temperature. Indications of such a possibility were noted by Farkas and Csengeri (1976).

Fatty acid desaturases of fish are affected in the same way by the inhibitor sterculic acid as are fatty acid desaturases of mammals. Sterculic acid is a cyclopropenoid fatty acid occurring naturally in triacylglycerols from plants of the order Malvales (e.g., cottonseed oil). Roehm *et al.* (1969) fed rainbow trout diets containing salmon oil supplemented with up to 200 ppm of sterculic acid. The inhibitor caused increased ratios of 16 : 0/16 : 1 and 18 : 0/18 : 1 as well as a decreased level of 22 : 6ω3 in the trout. These changes were accompanied by liver pathology and a decreased growth rate during the initial stages of feeding; slight recovery was observed during the later stages of growth.

An important implication from the above consideration of the effects of temperature on fatty acid unsaturation is that fish reared at warm temperatures are likely to have different polyunsaturated fatty acid requirements from fish raised at lower temperatures. The following section will establish conclusively that many fish raised at relatively low temperatures (10°–20°C) have specific requirements for $\omega 3$ polyunsaturates. It is known, however, that channel catfish can be successfully raised at 30°C, their optimal temperature, on diets containing either beef tallow, olive oil, or menhaden oil as sole lipid supplements (Stickney and Andrews, 1972). The former two lipids are sources of $\omega 6$ fatty acids, the latter is rich in $\omega 3$ fatty acids. The implication from these findings is that channel catfish will grow optimally at 30°C given a dietary supply of either $\omega 6$ or $\omega 3$ polyunsaturated fatty acids. Whether this is a general feature of fish grown at warmer temperatures or a unique finding for the catfish remains to be investigated.

As a final cautionary note we may add that polyunsaturated fatty acids are not the only lipid factors determining biomembrane fluidity *in vivo*. It is well known that cholesterol decreases the fluidity of artificial lipid biomolecular leaflets (liposomes) by decreasing the surface area of the membrane (Van Deenen *et al.*, 1972; Brockerhoff, 1975). It has also been demonstrated that fat-free diets or diets supplemented with cholesterol can alter the kinetic properties (Hill numbers) of enzymes in the erythrocyte membranes of mammals (Farias *et al.*, 1975). The situation is complicated by the fact that while cholesterol decreases the solute permeability of synthetic membranes when the latter are fluid, cholesterol increases the solute permeability of gelled (frozen) membranes (De Kruyff, 1975). Clearly complex interactions between cholesterol and polyunsaturated fatty acids are possible and these may well be important in nutritional disorders related to dietary lipids. At the same time little if anything is known about the interactions between dietary polyunsaturates and cholesterol in poikilotherms including fish. This represents a rich field for future studies.

B. Quantitative Requirements

1. Energy

Dietary lipids serve as sources of both metabolic energy and of polyunsaturated fatty acids essential for membrane structure and function. When fish mobilize lipids for energy, however, they do not discriminate between the polyunsaturates that originate solely in the diet and the saturates and monounsaturates that can either originate in the diet or from the animal's internal biosynthetic activity. This has been clearly shown in fish which have a pronounced ability to oxidize polyunsaturated fatty acids in their tissue mitochondria (Brown and Tappel, 1959). The essential fatty acid requirements of fish will be discussed later in this section and we shall assume for the present that the lipid used by fish for metabolic energy is "high quality" lipid, that is, lipid with a sufficient complement of polyunsaturated fatty acids to meet the animal's "essential fatty acid" requirements.

Carnivorous fish preying on zooplankton have a natural diet that contains about one-third of its dry weight as wax esters. The clupeoid fish generally have neutral lipid (oil) consisting exclusively of triacylglycerols and accounting for approximately 10–20% of their dry body weight. The latter figure may be taken as the normal dietary intake of lipid by fish such as the gadoids that prey on the small clupeoid fishes. The lipid contents of the natural diets of marine flatfish and freshwater salmonids are not known with certainty but there is no reason to believe that they are notably different from these values.

Experiments with cultured fish have shown that the optimal lipid intake is essentially similar to that for wild fish. As discussed earlier a major reason for supplementing the diets of cultured fish with lipid is to spare the oxidation of dietary protein as an energy source. At the same time a luxus of dietary lipid will result in excessive deposition of lipid within the fish and this may be 'unacceptable to the consumer. Lee and Putnam (1973) raised rainbow trout at 12°C on diets containing up to 24% dry weight as herring oil with excellent feed and energy conversions as well as growth rates. More recently Adron et al. (1976) fed diets to turbot containing up to 9% dry weight lipid added largely as capelin oil. The temperature of the experiment was 15°C and both the weight gain and protein utilization of the fish increased up to the maximum level of lipid in the diet. The lipid levels in the fish carcasses did not exceed those of wild turbot. Yone et al. (1971) showed that crude diets containing 10% lipid gave superior growth and feed efficiency than similar diets containing 20% lipid. Channel catfish have been successfully grown on diets containing up to 10% fat added either as beef tallow, olive oil, or menhaden oil (Stickney and Andrews, 1972). Later experiments used diets containing up to 12% dry weight as lipid added as a mixture of 90% beef tallow and 10% menhaden oil (Stickney and Andrews, 1972). As observed earlier, increasing the energy level (fat or carbohydrate) of the diet at constant dietary protein always resulted in improved protein utilization. All these experiments with catfish were conducted at 27°–30°C, the temperature of optimum growth (Stickney and Andrews, 1971). The high temperature, of course, greatly aids the digestion of a hard fat such as beef tallow.

The general consensus from all these experiments seems to be that not less than about 10% and not more than approximately 20% of lipid can be added to fish diets with excellent results. Higher levels of lipid can probably be incorporated with few untoward metabolic effects but at the risk of gross alteration of carcass composition by deposition of excess lipid.

2. POLYUNSATURATED FATTY ACIDS

Earlier work on "essential fatty acids" in fish was reviewed by Cowey and Sargent (1972). The situation at that time could be summarized thus: Fatty acids of the $\omega6$ series, especially linoleic acid ($18:2$ $\omega6$) and arachidonic acid ($20:4$ $\omega6$), are the predominant polyunsaturated fatty acids in the tissues of terrestrial animals. The latter two acids have full essential fatty acid activity in small omnivorous mammals. Dietary linoleic acid is readily chain elongated and further desaturated to arachidonic acid by the pathways outlined in Fig. 1. Fatty acids of the $\omega3$ series have partial essential fatty acid activity in small terrestrial mammals. Fish differ strikingly from land animals in that the major polyunsaturated fatty acids in their tissues are of the $\omega3$ series, especially $20:5$ $\omega3$ and $22:6$ $\omega3$ fatty acids. This applies to freshwater and marine fish alike. Both groups of fish have much smaller quantities of $\omega6$ polyunsaturated fatty acids although freshwater fish have higher concentrations of this series than marine fish. The latter situation reflects the fact that part of the diet of freshwa-

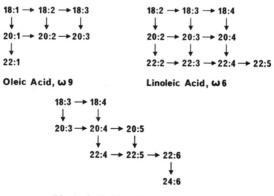

Fig. 1. Pathways of polyunsaturated biosynthesis in fish. Vertical arrows show chain elongation reactions; horizontal arrows show desaturation reactions. (Based on Mead and Kayama, 1967; Sinnhuber, 1969.)

ter fish especially is derived from terrestrial sources. Early work had shown that guppies could readily convert $20:5\ \omega3$ fatty acids in their brine shrimp diet to $22:6\ \omega3$ fatty acid (Kayama *et al.*, 1963a). Likewise the marine fish *Parablax clathratus* was capable of converting linolenic acid to $20:5\ \omega3$ and $22:6\ \omega3$ fatty acids (Kayama *et al.*, 1963b). The pathways involved in these chain elongation and desaturation reactions are outlined in Fig. 1.

Feeding experiments had clearly shown that the $\omega3$ series of fatty acids, specifically linolenic acid $18:3\ \omega3$, had essential fatty acid activity in salmonids (Nicolaides and Woodall, 1962; Higashi *et al.*, 1966; Sinnhuber, 1969). The marine fishes *Mugil cephalus* and *Fundulus grandis* could convert dietary linolenic acid to higher polyunsaturated fatty acids including $22:6\ \omega3$ (Reiser *et al.*, 1963). In many of these experiments linoleic acid was very much less active in preventing symptoms of essential fatty acid deficiency than was linolenic acid.

More recent work has amply confirmed and extended these conclusions. A particularly fine study by Castell *et al.* (1972a,b,c) clearly showed that linolenic acid was superior to linoleic acid in stimulating the growth and improving feed conversion in rainbow trout fed a diet lacking polyunsaturated fatty acids. The requirement of rainbow trout for linolenic acid was 1% by weight of the diet or approximately 2.7% of the total dietary calories. Symptoms of essential fatty acid deficiency revealed in these experiments included cessation of growth, caudal fin erosion, a shock syndrome, occasional heart myopathy, and fatty livers. These symptoms had been frequently observed in earlier work in several species of fish. The mitochondrial fragility observed in essential fatty acid deficiency in small land mammals was also observed in rainbow trout and manifested as a greatly increased swelling rate of isolated mitochondria (Castell *et al.*, 1972b). Yu and Sinnhuber (1972) showed that $22:6\ \omega3$ fatty acid was equally as effective as linolenic acid in reversing essential fatty acid deficiency in rainbow trout. This contrasts somewhat with the situation in mammals where arachidonic acid has greater essential fatty acid activity than does linoleic acid. In the experiments of Castell *et al.* (1972a,b,c) much of the dietary linolenic acid was converted to $22:6\ \omega3$ acid in the fish. These findings have since been amply confirmed by Watanabe *et al.* (1974a,b,c) with the slight difference that the latter authors found 0.88–1.66% to be the optimal requirement of rainbow trout for linolenic acid. All the aforementioned work on trout was performed at 11°–12°C (Castell *et al.*, 1972a,b,c) or 17°–20°C (Watanabe *et al.*, 1974a,b,c).

A characteristic of small mammals given diets deficient in polyunsaturated fatty acids is that they chain elongate and further desaturate oleic acid 18 : 1 ω9. The product of this biochemical activity, the fatty acid 20 : 3 ω9, is therefore characteristic of essential fatty acid deficiency. A similar phenomenon has also been recorded for the trout (Castell et al., 1972a,b,c; Watanabe et al., 1974a,b,c). Mohrhauer and Holman (1963a,b) noted that the ratio 20 : 3 ω3/20 : 4 ω6 in the tissue lipids of a terrestrial animal provides an index of ω6 polyunsaturated fatty acid deficiency. Alfin-Slater and Aftergood (1968) later proposed that the ratio 20 : 3 ω9/20 : 5 ω3 be taken as an index of deficiency of ω3 polyunsaturated fatty acids in land animals. Since the major polyunsaturated fatty acid in trout is 22 : 6 ω3, Castell et al. (1972a,b,c) proposed the ratio 20 : 3 ω9/22 : 6 ω3 as an index of deficiency of ω3 polyunsaturated fatty acids in rainbow trout. These authors concluded from their feeding experiments that when this ratio is 0.4 or greater then essential fatty acid deficiency is indicated. Slightly different results were obtained by Watanabe et al. (1974a,b,c) in that, on diets containing 1% linolenic acid, rainbow trout showed a ratio 20 : 3 ω9/22 : 6 ω3 of 0.5; on diets containing 2% linolenic acid the ratio fell to 0.1. These findings are in line with the earlier statement that the optimal requirements of rainbow trout in the experiments of Watanabe et al. (1974a,b,c) are somewhat greater than 1% linolenic acid.

Concerning the use of ratios of tissue fatty acids as an index of the dietary supply of essential fatty acids it should be emphasized that in the original experiments of Mohrhauer and Holman (1963a,b) the break-points of the curves for increase in tissue arachidonic acid and decrease in tissue eicosatrienoic acid occurred at the same level of dietary linoleic acid. The measured ratio at that point for eicosatrienoate/arachidonate was 0.4. A close correspondence between increasing tissue concentrations of docosahexaenoic acid and decreasing tissue concentrations of eicosatrienoic acid in response to increasing dietary concentrations of linolenic acid has yet to be demonstrated in fish. Nor has it been demonstrated that tissue docosahexaenoate and eicosatrienoate reach steady concentrations at the same level of dietary linolenate and, if they do, that the ratio eicosatrienoate/docosahexaenoate at that point is 0.4. Thus, the value of 0.4 for fish appears to have been selected largely on the basis of the data presented by Mohrhauer and Holman (1963a,b) for rats.

The above experiments were carried out on rainbow trout, a freshwater carnivore. Interesting experiments on a freshwater herbivore, the carp Cyprinus carpio, have recently been carried out by Watanabe et al. (1975a,b). Diets free of fat or deficient in polyunsaturated fatty acids resulted in retarded growth of carp fry at

20–25°C. Either methyl linoleate or methyl linolenate improved growth. The authors concluded that both $\omega6$ and $\omega3$ fatty acids are essential dietary factors and suggest that both of the ratios 20 : 3 $\omega9/$ 20 : 4 $\omega6$ and 20 : 3 $\omega9/22$: 6 $\omega3$ be considered in determining optimal essential fatty acid requirements for the carp. When both ratios are applied at values of 0.4 or less, satisfactory essential fatty acid requirements of carp are met by diets containing either 3% soybean oil + 2% cod liver oil, or 4% corn oil + 1% methyl linolenate, or 3% corn oil + 2% methyl linolenate (Watanabe *et al.*, 1975a,b).

In considering freshwater species it should be noted that Stickney and Andrews (1972) obtained highest weight gains with channel catfish grown at 30°C on diets containing either beef tallow, olive oil, or menhaden oil. The implications from these experiments are that either $\omega3$ or $\omega6$ fatty acids can meet the essential fatty acid requirements of the channel catfish. This important implication clearly deserves to be established by directly determining the essential fatty acid requirements of the catfish.

When we consider a marine carnivore the situation changes again. Owen *et al.* (1975) studied the chain elongation and further desaturation of various radioactive fatty acids fed to turbot, *Scophthalmus maximus,* that had been maintained on a fat-free diet for 16 weeks at 15°C. Neither 18 : 1 $\omega9$, 18 : 2 $\omega6$, nor 18 : 3 $\omega3$ was significantly converted to higher polyunsaturated fatty acids. In particular, 18 : 3 $\omega3$ was not converted to 22 : 6 $\omega3$. In these experiments rainbow trout subjected to essentially the same dietary treatment as turbot converted some 70% of dietary radioactive 18 : 3 $\omega3$ to 22 : 6 $\omega3$. The results are intriguing since the natural diet of wild turbot consists primarily of smaller fish that have a luxus of preformed 22 : 6 $\omega3$ fatty acid in their tissues. In short, turbot is very unlikely to experience a deficiency of 22 : 6 $\omega3$ fatty acid in its natural environment and certainly will not experience a diet rich in 18 : 3 $\omega3$. Owen *et al.* (1975) point out that the absence or inactivity of desaturating and further elongating pathways in the turbot would not place this carnivorous fish at a disadvantage in the sea. It could of course be argued that the turbot does have the relevant pathways but they operate at much reduced rates as compared to, say, rainbow trout. While this is possible the argument is academic since the net result is still an apparent inability of the turbot to elongate and desaturate short chain dietary polyunsaturates. This is not a unique situation since it has recently been demonstrated that the cat, *Felis catus,* is also unable to desaturate dietary essential fatty acids so that 20 : 4 $\omega6$ as well as 18 : 2 $\omega6$ must be supplied in the diet of this species if freedom from pathology and normal growth are to be achieved (Rivers *et al.*, 1975).

The work of Owen *et al.* (1975) is also noteworthy in that, of the administered radioactive oleic acid, some 90% was recovered as oleic acid and radioactive 20:3 ω9 acid was not detected. These results were obtained on a fat-free diet. Therefore the ratio 20:3 ω9/22:6 ω3 is not applicable as an index of essential fatty acid deficiency in the turbot.

Additional evidence for the lack of detectable elongating and desaturating enzyme activity in turbot comes from feeding diets containing different lipids (Cowey *et al.*, 1976). Diets containing the highly saturated hydrogenated coconut oil did not give rise to 20:3 ω9 in body lipids; likewise diets containing corn oil rich in 18:2 ω6 did not lead to deposition of arachidonic acid in tissue lipids. The latter finding agrees with that for the plaice, *Pleuronectes platessa*, (Owen *et al.*, 1972) and contrasts with that of Castell *et al.* (1972a,b,c) who noted significant conversion of linoleic acid to arachidonic acid in trout fed corn oil diets.

The case that marine carnivores differ in their essential fatty acid requirements from freshwater carnivores receives support from the work of Yone and Fujii (1975a,b) on red sea bream, *Chrysophrys major*. These authors demonstrated that corn oil supplemented with 1% linolenic acid did not improve the growth rate of the fish at 25°C. Corn oil supplemented with 2% polyunsaturated fatty acids (a mixture of 20:5 ω3 and 22:6 ω3), however, markedly improved both the growth rate and the food conversion efficiency. Even higher growth rates were obtained when bream were fed a diet containing pollack residual oil as sole lipid supplement. Analyses of liver phospholipids in fish fed corn oil supplemented with linolenic acid indicated some conversion of dietary 18:3 ω3 to both 20:5 ω3 and 22:6 ω3. The liver phospholipids of fish fed diets containing corn oil supplemented with 20:5 ω3 and 22:6 ω3 fatty acids were characterized by (i) increased levels of monounsaturated fatty acids of chain length greater than C_{18} and (ii) decreased amounts of ω3 and ω6 polyunsaturates of chain length less than C_{20}. These changes in fatty acid composition could be correlated with improved feed conversion efficiency on diets containing 20:5 ω3 and 22:6 ω3 fatty acids. The work shows overall that red sea bream can to a degree convert dietary 18:3 ω3 fatty acid to 22:6 ω3 fatty acid but preformed dietary 22:6 ω3 is superior to 18:3 ω3 in promoting growth.

In attempting to deduce some general guide lines concerning essential fatty acid requirements of fish the following points should be considered. At the outset we consider that the preference of fish for ω3 compared to ω6 polyunsaturated fatty acids reflects fundamentally the low temperatures (generally less than 20°C) of the majority of aquatic environments. Likewise we consider shorter chain polyunsaturates to be characteristic more of a terrestrial than an aquatic environment. However, freshwater fish consume natural food that is partly terrestrial in origin, a situation that applies to herbivores, omnivores, and to a lesser extent carnivores. Thus, freshwater fish have higher levels of $\omega6$ fatty acids in their tissues than marine fish (Ackman, 1967). Furthermore, freshwater fish have higher levels of short chain $\omega3$ polyunsaturates (e.g., $18:3\ \omega3$) than marine fish (Gruger et al., 1964). For these reasons freshwater fish might be expected a priori to chain elongate shorter chain polyunsaturates of the $\omega6$ and especially the $\omega3$ series rather efficiently. This is observed experimentally.

While the marine biosphere is characterized by large amounts of $22:6\ \omega3$, not all marine fish consume this as their sole or even their major dietary polyunsaturated fatty acid. It is true that the phytoplankton contain large amounts of both C_{20} and C_{22} polyunsaturates but these organisms also abound in C_{18} and C_{16} polyunsaturates (Ackman et al., 1968; Chuecas and Riley, 1969). It is quite possible, therefore, that herbivorous marine fish such as the Peruvian anchovy are active in chain elongating and further desaturating short chain dietary polyunsaturated fatty acids. The major consumers of phytoplankton are, of course, the zooplankton among which the calanoid copepods are particularly abundant. The major polyunsaturated fatty acid in a variety of copepods is $20:5\ \omega3$ closely followed by $22:6\ \omega3$ (Lee et al., 1971b; Sargent et al., 1976). Fish such as the herring, capelin, menhaden, and young salmon have oils rich in $20:5\ \omega3$ as well as $22:6\ \omega3$ but the tissue phospholipids of these fish contain largely $22:6\ \omega3$ fatty acid (Cowey and Sargent, 1972). It is probable these fish preferentially incorporate dietary $22:6\ \omega3$ directly into their phospholipids and the fate of dietary $20:5\ \omega3$ (i.e., whether oxidized or converted to $22:6\ \omega3$) depends on the intake of $22:6\ \omega3$.

It is only the larger extreme carnivores further up the marine food chain such as the gadoids (and the turbot) that have a natural diet where $22:6\ \omega3$ is the major polyunsaturated fatty acid. In short, the further we proceed up the marine food chain the less obvious is the need to chain elongate and further desaturate shorter chain dietary polyunsaturated fatty acids. The substrates for the necessary enzymatic machinery are present in low amounts while the products are present in large amounts. We believe that the presently available data on essential fatty acid requirements of fish can be fitted into the foregoing simple framework. At the same time we are very conscious that, as yet, too few species of defined natural food habits have been studied to make our general conclusion solid. Clearly this is a rich field for future studies.

V. VITAMINS

The recommended dietary levels of vitamins for certain cold water fish are included in a recent publication (National Research Council, 1973); those for warm water fish will shortly form part of a future publication from the same source. As they are also dealt with in accompanying chapters of this volume they will not be recapitulated. It should however be noted that quoted values are not *sensuo stricto* the requirements of the fish but are the amounts which should be included in a ration for maximal growth and freedom from pathology. Thus it has been observed (Murai, 1966, quoted by Murai and Andrews, 1975) that the pantothenate content of a trout diet (500 mg/kg dry diet, pellets 1 mm diameter) was reduced to half within 10 sec and completely disappeared after 1 min in the water. This is an extremely small pellet with a very high surface area : weight ratio, no doubt the effect will be considerably reduced in larger pellets. Nevertheless the point stands. The stated requirement for a water-soluble vitamin thus appears to be partly a function of the water stability of the pellet and of the time for which the pellet is exposed to the water before being eaten: to some extent a function of the husbandry. This dictum no doubt applies to all water-soluble compounds used in experimental diets but, because they are included at very low levels, water-soluble vitamins are especially vulnerable to solution in the water prior to consumption by the fish. It is obviously hazardous to draw comparisons between fish and terrestrial animals with respect to water-soluble vitamins.

By and large few of the pathologies arising from vitamin deficiency are diagnostic. There is thus a case for utilizing biochemical parameters wherever possible to assess the nutritional status of the fish with respect to vitamins (Cowey and Sargent, 1972). This section is therefore restricted to recent research on the nutritional biochemistry of vitamins.

A. Pyridoxine

The functional form of pyridoxine is pyridoxal phosphate which acts as a coenzyme for many enzymes including aminotransferases. The effects of pyridoxine deficiency on tissue alanine aminotransferase activity in rainbow trout of mean weight 150 g have been examined by Smith *et al.* (1974). Enzyme activity in erythrocytes from trout given the pyridoxine-deficient diet for 9 weeks was significantly lower than in erythrocytes from control fish (20.1 units/ml against 43.7

units/ml). Enzyme levels in the erythrocytes of deficient fish were restored to normal within 3 weeks by feeding a complete diet.

It was also of interest that preincubation of erythrocytes from deficient fish with pyridoxal phosphate increased alanine aminotransferase activity by 40%. Surprisingly the apoenzyme in fish given the complete diet was not fully saturated with coenzyme, the same preincubation procedure increasing alanine aminotransferase activity by 22%.

Whether erythrocyte alanine aminotransferase activity can be adapted to assess pyridoxine status in fish under cultivation remains to be seen. It would certainly be interesting to know the effects of graded dietary levels of pyridoxine on enzyme activity.

Alanine aminotransferase activity in the livers of pyridoxine-deficient trout was not significantly changed from that in control fish. The absence of effect was related by Smith *et al.* (1974) to the physiology of starvation in the fish. Trout fed the deficient diet grew much less than did those given a normal diet and at necropsy their stomachs were empty. If the deficient fish were taking progressively less feed and eventually effectively fasting then it is possible that blood sugar levels would be maintained by gluconeogenesis and alanine aminotransferase being a gluconeogenic enzyme would be essential to the maintenance of this pathway.

B. Thiamine

Cowey *et al.* (1975b) measured the effect of graded dietary levels of thiamine on growth and erythrocyte transketolase activity in turbot; their results suggest that the latter measurement can be used with advantage as an indicator of dietary thiamine sufficiency.

Optimal growth occurred at dietary thiamine levels of 0.6 mg/kg or more, erythrocyte transketolase activity was, however, suboptimal when dietary thiamine levels were less than 2.6 mg/kg. At or above this dietary thiamine level, the erythrocyte transketolase apoenzyme was saturated with coenzyme, preincubation of the hemolysate with thiamine pyrophosphate (the active form of thiamine) did not enhance erythrocyte transketolase activity. Preincubation with thiamine pyrophosphate of hemolysates from turbot given diets containing less than 2.6 mg thiamine/kg restored their erythrocyte transketolase activity to levels pertaining in fish given diets containing high levels of thiamine.

There is a causal relationship between dietary thiamine level and the degree of saturation of erythrocyte transketolase with thiamine pyrophosphate. Moreover there is satisfactory agreement between the optimal dietary level of thiamine for growth and for saturation of erythrocyte transketolase. Relationships of this sort could prove useful in monitoring vitamin status in fish under cultivation.

C. Ascorbic Acid

The essentiality of ascorbic acid for channel catfish has recently been conclusively demonstrated by growth studies (Lovell, 1973; Andrews and Murai, 1975), and by the inability of microsomes from the liver and kidney of catfish to synthesize ascorbic acid from D-glucurono-γ-lactone (Wilson, 1973). An absolute requirement for ascorbic acid had previously been shown for several other species including yellowtail, carp, coho salmon, and rainbow trout (Halver *et al.*, 1975).

A practical problem associated with the supply of ascorbic acid (although not unique to it) is its lability. Andrews and Murai (1975) showed a very large fall in ascorbic acid activity when a practical diet was stored at 20°C for 16 weeks. Moist pellets are likely to be even more vulnerable to such losses even if stored at temperatures below zero. A possible solution to this problem may reside in the use of ascorbate 2-sulfate as a dietary source of ascorbic acid for fish. Ascorbate 2-sulfate is stable for weeks, even in aqueous solution, and detailed studies by Halver *et al.* (1975) have demonstrated that this compound is an adequate source of ascorbic acid for rainbow trout; on a molar basis ascorbate 2-sulfate was as effective as L-ascorbic acid in preventing ascorbic acid deficiency symptoms. Trout given 160 mg of dipotassium ascorbic 2-sulfate dihydrate/kg dry diet showed normal growth, low mortality, and no deficiency symptoms in an experiment lasting 20 weeks.

An enzyme that hydrolyzes ascorbate 2-sulfate to ascorbic acid, namely, ascorbate 2-sulfohydrolase, has been detected in the liver and several other tissues of rainbow trout (Halver *et al.*, 1975). It is possible that the enzyme hydrolyzes metabolically inactive ascorbate 2-sulfate to generate metabolically active ascorbic acid; but the presence of this enzyme is not the sole or main factor involved in the utilization of ascorbate 2-sulfate by animals. Thus, guinea pig tissues contain ascorbate 2-sulfohydrolase activity but ascorbate 2-sulfate will not cure scurvy in the guinea pig. Clearly interesting questions in the comparative nutrition of ascorbic acid await solution.

The metabolic function of ascorbic acid is obscure. Tyrosine hydroxylase activity in the adrenal gland of scorbutic guinea pigs is markedly reduced and recovers on administration of ascorbic acid; however ascorbic acid did not activate the enzyme directly (Nakashima *et al.*, 1972). Other experiments (Degkwitz and Staudinger, 1974) indicate an influence of ascorbic acid on the biosynthesis of heme but no specific coenzyme function for ascorbic acid has been identified. There is also evidence that ascorbic acid intake affects the electron transport components of guinea pig liver microsomes; NADPH–cytochrome *P*-450 reductase activity is markedly reduced in liver microsomes from weanling guinea pigs given diets deficient in ascorbic acid (Zannoni and Sato, 1975).

Most recent experiments on fish have been concerned with the role of ascorbic acid in mediating collagen and bone formation, perhaps because of the striking abnormalities in the vertebral column observed in scoliosis. Halver (1972b) showed that ascorbic acid is rapidly fixed in rainbow trout tissues wherever collagen is being formed. Wilson and Poe (1973) observed that the collagen content of vertebrae from scorbutic catfish was markedly lower than that of vertebrae from normal catfish. In addition the vertebrae from scorbutic fish were comparatively brittle and easily broken and the collagen isolated from them contained significantly fewer hydroxyproline residues than that prepared from vertebrae of normal fish. Wilson and Poe (1973) also found a marked decrease in serum alkaline phosphatase activity in scorbutic catfish. By analogy with mammalian work this was interpreted in terms of a reduction in the activity of osteoblasts resulting in a decrease in bone formation.

VI. MINERALS

The mineral requirements of fish are complicated by the intimate relationship between the animal and its environment. Freshwater fish experience a small but steady loss of NaCl in the copious hypotonic urine they secrete as a result of an ingress of water across their gill epithelium from their markedly hypotonic external medium. Urinary loss of NaCl is compensated by an active uptake of NaCl across the gills into the plasma. These processes have been reviewed comprehensively in recent years (Conte, 1969; Motais and Garcia-Romeu, 1972; Maetz, 1971, 1974) and will not be considered in detail here. It is noteworthy with respect to the nutritional requirements of freshwater fish for NaCl that the NaCl pump in gill is remarkably efficient, being capable of pumping NaCl into plasma from freshwater concentrations in the millimolar range and lower. Similar active transport mechanisms relating to other ions which the fish is known to absorb from the water (Ichikawa and Oguri, 1961; Templeton and Brown, 1963) may exist.

A wealth of evidence supports the idea that, as in amphibian skin, the active uptake of NaCl is coupled to an efflux of either HCO_3^- together with H^+ or HCO_3^- together with NH_4^+. The latter ions are derived from the CO_2 produced in oxidative metabolism and the NH_3 produced during amino acid catabolism according to the relationship

$$CO_2 + H_2O \rightarrow H_2CO_3 \rightarrow HCO_3^- + H^+ \underset{NH_3}{\rightarrow} NH_4^+$$

This mechanism does not in any way imply that all of the NH_3 produced in amino acid catabolism and all of the CO_2 produced in oxidative metabolism are necessarily excreted as $H^+HCO_3^-$ or $NH_4^+HCO_3^-$ in exchange for Na^+Cl^- in the external medium. That is, clearance of CO_2 and NH_3 in a fish rapidly metabolizing amino acids will not be readily limited by available environmental NaCl. This situation arises because the un-ionized CO_2 and NH_3 can readily diffuse across cellular membranes and be easily excreted by the gill by passive diffusion down their concentration gradients. The actual proof of such a mechanism, however, is complicated by the fact that all of the ammonia excreted by the fish, whether as NH_3 or NH_4^+, appears in the external medium as NH_4^+ due to the fact that freshwater invariably has a pH value of 7 or less; likewise much of the excreted carbon dioxide, whether excreted as CO_2 or HCO_3^- will appear as environmental HCO_3^-. It is obviously a common experience that freshwater fish thrive in natural freshwaters without supplementation with NaCl.

Marine fish experience an excess of NaCl in their external medium so that the problem is one of avoiding desiccation through water loss across the gills. These fish constantly drink small amounts of water (which may incidentally meet many of their nutritional needs for cations) and pump the incoming NaCl across the gill epithelium into the external seawater. The seawater fish produces a sparse and concentrated urine to offset water loss although fish shocked or stressed (e.g., by handling) can produce a copious urine with resultant loss of NaCl. The latter loss can, of course, be readily made up by drinking seawater. It is certain that ammonia and carbon dioxide produced in catabolism are cleared across the gills of the seawater fish, probably by the same mechanism as occurs in the freshwater fish. In seawater, however, any incoming NaCl will be entrained into the NaCl pump that is mainly concerned with pumping excess NaCl, derived from intestinal seawater via plasma, out of the gill and into the external medium.

A. Total Mineral Intake

As there is a continuous net outward flux of ions from freshwater fish the animal is very dependent on an adequate mineral supply in its food. Attempts to estimate optimal levels for different fish have involved the feeding of graded levels of balanced salt mixtures and examination of the effect on the growth response curve. Ogino and Kamizono (1975) used McCollum's salt mixture (formulated for rats) supplemented with trace elements (Halver and Coates, 1957). This was fed to young carp at stepwise levels of 0–8% in the diet and to young rainbow trout at levels of 0–4%. For both species weight gain was best when the salt mixture constituted 4% of the diet. Mineral deficiency resulted in marked depression of growth and feed efficiency in rainbow trout but carp were relatively insensitive to the mineral level in the diet (weight gain 131% in 7 weeks with no mineral supplement; 156% with 4% mineral supplement in diet).

This lack of sensitivity of carp to dietary mineral level was also evident in the absence of overt deficiency symptoms. By contrast rainbow trout suffered from anemia (due to the deficiency of iron) as well as lordosis, scoliosis, convulsions, and a malformation of the head when fed diets lacking a mineral supplement.

Arai *et al.* (1974) fed eel on diets containing graded amounts (0–8%) of the salt mixture of Halver (1957). The optimum dietary level of this mixture was ajudged from growth rate and feed efficiency to be 2%. The protein component of the diets used by Arai and his colleagues was white fishmeal (70% in all diets). As this material provides considerable amounts of ash the results obtained with the eel are probably in line with those of Ogino and Kamizono on carp and rainbow trout.

B. Calcium, Phosphorus, and Magnesium

Andrews *et al.* (1973) drew attention to the fact that the inclusion of high levels of fishmeal in rations for cultivated fish obviated the need for supplements of calcium and phosphorus. They pointed out that the substitution of other proteinaceous materials necessitated more precise information on calcium and phosphorus requirements, a point amply demonstrated later by Ketola (1975a). Andrews *et al.* (1973) also noted that as fish possess the ability to absorb calcium through the gill, the requirement for this metal may be affected by its concentration in the water.

In experiments with channel catfish Andrews *et al.* (1973) varied both dietary levels (calcium 0.5–2.0%, phosphorus 0.5–1.2%) and dietary ratios of these minerals. Maximal growth of this species occurred when fish were given diets containing 1.5% calcium, higher levels leading to reduced growth. The optimal level of available phosphorus was 0.8–1.0%, levels of 0.5% giving rise to deficiency symptoms and diets containing 1.2% phosphorus reducing growth rate. The recommended calcium and phosphorus levels are similar to those used by Deyoe and Tiemeier (1968) in feed trials with the channel catfish. Andrews and his colleagues obtained no growth response to calcium phytate and inferred that the phosphorus of phytin is not available.

The calcium requirement of marine fish is apparently very different from that of freshwater fish (Sakamoto and Yone, 1973). Marine teleosts drink seawater which contains considerable amounts of calcium but little phosphorus, and excrete the excess of divalent ions absorbed (Hickman and Trump, 1969). Marine fish, therefore, require a dietary calcium : phosphorus ratio of less than unity. Sakamoto and Yone (1973) found that red sea bream grew best when given diets containing 0.34% calcium with a calcium : phosphorus ratio of 1 : 2.

Calcium and phosphorus requirements are generally considered together because the metabolism of the two elements is intimately connected. The interplay between minerals is not however restricted to these elements and an imbalance in the dietary level of a mineral may frequently have repercussions on the metabolism of others. It is well known that a magnesium deficiency in rats leads to soft tissue calcinosis and may also be accompanied by a loss of cellular potassium. Similarly in rainbow trout magnesium deficiency may give rise to renal calcinosis (Cowey, 1976). When magnesium-deficient diets were given to rainbow trout renal calcinosis occurred at dietary calcium levels of 2.6% (calcium : phosphorus, 1 : 1) or more. At a lower dietary calcium level (1.4%) kidney calcium concentrations were not grossly elevated even when dietary magnesium levels were limiting. Equally, no renal calcinosis occurred at dietary calcium levels of 2.6% when supplemental magnesium (0.1%) was provided in the diet. Thus dietary calcium levels of 2.6% were not obviously deleterious to trout when a sufficiency of magnesium was present; metabolic disturbances of one element may be affected by the intake of another. In these experiments the growth rate was controlled by magnesium intake (200% increase in initial weight with 0.1% magnesium, 100% increase with 0.006% magnesium, over 16 weeks).

Finally it should be noted that derangements of calcium

metabolism or excretion occur in response to other dietary deficiencies or imbalances. Thus abnormal calcium deposition was observed in tryptophan-deficient trout (Kloppel and Post, 1975) and calcium oxalate crystals have been found in the kidneys of trout given diets lacking pyridoxine (Smith *et al.*, 1974). Oxaluria characterizes pyridoxine deficiency in a number of mammals, and in the experiments of Smith *et al.* (1974) the condition may have been exacerbated by the use of gelatin (resulting in a high glycine intake and thus elevated oxalate production) in the diet.

VII. ADDENDUM

Since this chapter was completed, a number of papers relevant to the subject matter have appeared. This section briefly outlines the main results.

Zeitoun *et al.* (1976) compared various methods for quantifying the protein requirement of rainbow trout fingerlings from dose–response data; these included analysis of variance and multiple comparisons of the means of response, a "broken line" analysis, and a polynomial regression analysis. These authors favored polynomial regression analysis (it is continuous and should be more accurate than other methods when the stepwise increments in dose level are large) and from the use of it they introduced the concept of economic nutrient (in this instance protein) requirement.

Several other data have been obtained on essential amino acid requirements of fish. Wilson *et al.* (1977) showed that the lysine requirement of fingerling channel catfish is 1.23% of the dry diet; they also showed that the pH of amino acid test diets must be adjusted to 7.0 (probably mainly a matter of neutralizing amino acid hydrochlorides, although dicarboxylic and basic amino acids would also affect the pH) if maximum growth rate and maximum feed conversion are to be obtained. Harding *et al.* (1977) found that channel catfish require 0.56 g methionine/100 g dry diet, in the absence of cystine, for maximal growth; on a molar sulfur basis cystine could replace 60% of the methionine. Harding *et al.* (1977) used serum methionine concentration as an adjunct to weight gain measurements. Similarly, Kaushik (1977) used both weight gain measurements and free amino acid concentrations of plasma and muscle to evaluate the arginine requirement of rainbow trout, which was shown to amount to 1.2% of the dry diet.

Recent findings on amino acid supplementation of proteins with free amino acids appear to confirm that this procedure is effective with rainbow trout but ineffective with channel catfish. Andrews *et al.* (1977) increased the arginine levels of partially purified channel catfish diets from 1.1 to 1.7% by isonitrogenous substitution of a portion of the casein with gelatin and obtained significant enhancement of growth. The addition of free arginine to the casein diet, however, had little effect. Poston *et al.* (1977), on the other hand, markedly enhanced the nutritional value for trout of a diet based on promine R (a soya protein isolate) by supplementing it with the first limiting amino acid, methionine. Dabrowska and Wojno (1977) improved rainbow trout diets containing fishmeal, meat and bone meal, and yeast and soybean meal (as the protein component) by supplementation with 1.0% cystine and 0.5% tryptophan. Finally, Tiews *et al.* (1976) demonstrated that fishmeal could be entirely replaced in diets for rainbow trout by a mixture of poultry by-product meal and feather meal supplemented with 1.7% lysine HCl, 0.48% DL-methionine, and 0.14% DL-tryptophan.

An interesting method for measuring the methionine in proteins, described by Njaa (1977), makes use of an iodopalmitate reagent before and after reducing the methionine with titanium trichloride, thus giving values for both methionine and the sulfoxide in the original protein.

A significant contribution to the interaction between food energy and protein utilization in channel catfish was made by Garling and Wilson (1976), who were able to include sufficient different treatments in their experiments to permit predictions concerning the relationship between protein utilization and protein energy : total energy ratio. A diet containing 275 kcal/100 g produced optimal growth with 24% protein present; a diet containing 341 kcal/100 g produced optimal growth at a dietary protein level of 28–32%. Murray *et al.* (1977) claim that improved protein conversion occurred when the protein content of channel catfish diets was increased from 25 to 35%; in fact, this was probably due to the much higher energy density of the 35% protein diet in conjunction with a limited rate of feeding (3% biomass per day).

The lability of ascorbic acid in practical diets has been examined by Hilton *et al.* (1977a). They confirm that losses of ascorbic acid due to water addition to a mash, cold pelleting, and drying are severe. In a later paper (Hilton *et al.*, 1977b) they suggest that liver ascorbic acid concentrations may serve as a useful index of ascorbic acid in trout, levels of 20 μg/g or lower indicating that supplementation is necessary.

Recent experiments on mineral nutrition in fish have shown a magnesium requirement of 0.04–0.05% in the dry diet of carp (Ogino and Chiou, 1976). When this species was given diets containing different concentrations of calcium and phosphorus (Ogino and Takeda, 1976), growth rate correlated positively with dietary phosphorus concentration but not with dietary calcium concentration. Water concentrations of phosphorus and calcium were 0.002 and 20 ppm, respectively, and the authors suggest that the carp can compensate for a low calcium intake by absorption of this ion from the water. Under the conditions of the experiment, maximal growth of carp occurred when diets contained 0.6–0.7% phosphorus. A similar level (0.6% of the diet) of supplemental inorganic phosphorus improved the weight gain, feed utilization, and bone ash content of Atlantic salmon given diets containing 0.7% of the phosphorus from plant protein sources (Ketola, 1975b). Ketola suggests that, compared to inorganic phosphorus or that present in materials derived from animals, plant sources of phosphorus are largely unavailable to the fish.

ACKNOWLEDGMENTS

We are greatly indebted to Drs. Takeshi Nose and Shigeru Arai of the Freshwater Fisheries Research Laboratory, Hino-shi, Tokyo, who kindly allowed us to use their data on essential amino acid requirements of carp and eels prior to publication. We are similarly indebted to Miss Anne Bauermeister and Drs. R. R. Gatten, M. J. Walton, M. de la Higuera, and Mr. D. Knox of the Institute of Marine Biochemistry, all of whom allowed us to use unpublished data. We thank Mrs. Linda Shepherd and Mrs. Glenda Ward for patiently and carefully typing the manuscript. Finally, we acknowledge the many helpful discussions with Dr. P. T. Grant during the course of the work described in this chapter.

REFERENCES

Ackman, R. G. (1967). Characteristics of the fatty acid composition and biochemistry of some freshwater fish oils and lipids in comparison with marine oils and lipids. *Comp. Biochem. Physiol.* **22**, 907–992.

Ackman, R. G., Tocker, C. S., and McCachlan, J. (1968). Marine phytoplankton fatty acids. *J. Fish. Res. Board Can.* **25**, 1603–1620.

Adron, J. W., Blair, A., Cowey, C. B., and Shanks, A. M. (1976). Effects of dietary energy level and dietary energy source on growth, feed conversion and body composition of turbot (*Scophthalmus maximus*, L.). *Aquaculture* **7**, 125–132.

Alfin-Slater, R., and Aftergood, L. (1968). Essential fatty acids reinvestigated. *Physiol. Rev.* **48**, 758–784.

Andrews, J. W., and Murai, T. (1975). Studies on the vitamin C requirements of channel catfish (*Ictalurus punctatus*). *J. Nutr.* **105**, 557–561.

Andrews, J. W., and Page, J. W. (1974). Growth factors in the fishmeal component of catfish diets. *J. Nutr.* **104**, 1091–1096.

Andrews, J. W., Murai, T., and Campbell, C. (1973). Effects of dietary calcium and phosphorous on growth, food conversion, bone ash and haematocrit levels of catfish. *J. Nutr.* **103**, 766–771.

Andrews, J. W., Page, J. W., and Murray, M. W. (1977). Supplementation of a semipurified casein diet for catfish with free amino acids and gelatin. *J. Nutr.* **107**, 1153–1156.

Arai, S., Nose, T., and Hashimoto, Y. (1971). A purified test diet for the eel, *Anguilla japonica*. *Tansuiku Suisan Kenkyusho Kenkyu Hokoku* **21**, 161–178.

Arai, S., Nose, T., and Kawatsu, H. (1974). Effects of minerals supplemented to the fishmeal diet on growth of eel, *Anguilla japonica*. *Tansuiku Suisan Kenkyusho Kenkyu Hokoku* **24**, 95–100.

Ballard, F. J. (1965). Glucose utilization in mammalian liver. *Comp. Biochem. Physiol.* **14**, 437–443.

Barrington, E. J. W. (1957). Alimentary canal and digestion. In "The Physiology of Fishes" (M. E. Brown, ed.), Vol. 1, pp 109–161. Academic Press, New York.

Benson, A. A., Lee, R. F., and Nevenzel, J. C. (1972). Wax esters: Major marine metabolic energy reserves. *Biochem. Soc. Symp.* No. 35, pp. 175–187.

Bergot, P., and Fléchon, J.-E. (1970). Forme et voie d'absorption intestinale des acides gras à chaine longue chez la truite arc-en-ciel (*Salmo gairdnerii* Rich). *Ann. Biol. Anim. Biochem. Biophys.* **10**, 459–472.

Bilinski, E., Jonas, R. E. E., and Lau, Y. C. (1971). Lysosomal triglyceride lipase from the lateral line tissue of rainbow trout (*Salmo gairdneri*). *J. Fish. Res. Board Can.* **28**, 1015–1018.

Blaxter, K. L. (1975). Conventional and unconventional farmed animals. *Proc. Nutr. Soc.* **34**, 51–56.

Borgstrom, B. (1974). Fat digestion and absorption. In "Biomembranes" (L. A. Manson, ed.), Vol. 4B, pp. 555–620. Plenum, New York.

Bressani, R., and Mertz, E. T. (1958). Relationship of protein level to the minimum lysine requirement of the rat. *J. Nutr.* **65**, 481–491.

Brett, J. R., and Zala, C. A. (1975). Daily pattern of nitrogen excretion and oxygen consumption of sockeye salmon (*Oncorhynchus nerka*) under controlled conditions. *J. Fish. Res. Board Can.* **32**, 2479–2486.

Brockerhoff, H. (1966a). Digestion of fat by cod. *J. Fish. Res. Board Can.* **23**, 1835–1839.

Brockerhoff, H. (1966b). Fatty acid distribution patterns of animal depot fat. *Comp. Biochem. Physiol.* **19**, 1–12.

Brockerhoff, H. (1975). Model of interactions of polar lipids, cholesterol, and proteins in biological membranes. *Lipids* **9**, 645–650.

Brockerhoff, H., and Hoyle, R. J. (1965). Hydrolysis of triglycerides by the pancreatic lipase of a skate. *Biochim. Biophys. Acta* **98**, 435–436.

Brockerhoff, H., and Hoyle, R. J. (1967). Conversion of dietary triglycerides into depot fat in fish and lobster. *Can. J. Biochem.* **45**, 1365–1370.

Brown, W. D., and Tappel, A. L. (1959). Fatty acid oxidation by carp liver mitochondria. *Arch. Biochem. Biophys.* **85**, 149–158.

Burleigh, I. G., and Schimke, R. T. (1968). On the activities of some enzymes concerned with glycolysis and glycogenolysis in extracts of rabbit skeletal muscles. *Biochem. Biophys. Res. Commun.* **31**, 831–836.

Carey, M. C., and Small, D. M. (1970). Characteristics of mixed micellar solutions with particular reference to bile. *Am. J. Med.* **49**, 590–608.

Castell, J. D., Sinnhuber, R. O., Wales, J. H., and Lee, D. J. (1972a). Essential fatty acids in the diet of rainbow trout (*Salmo gairdneri*): Growth, feed conversion, and some gross deficiency symptoms. *J. Nutr.* **102**, 77–86.

Castell, J. D., Sinnhuber, R. O., Lee, D. J., and Wales, J. H. (1972b). Essential fatty acids in the diet of rainbow trout (*Salmo gairdneri*): Physiological symptoms of essential fatty acid deficiency. *J. Nutr.* **102**, 87–92.

Castell, J. D., Lee, D. J., and Sinnhuber, R. O. (1972c). Essential fatty acids in the diet of rainbow trout (*Salmo gairdneri*): Lipid metabolism and fatty acid composition. *J. Nutr.* **102**, 93–100.

Chang, V. M., and Idler, D. R. (1960). Biochemical studies on sockeye salmon during spawning migration. XII. Liver glycogen. *Can. J. Biochem. Physiol.* **38**, 553–558.

Chiou, J. Y., and Ogino, C. (1975). Digestibility of starch in carp. *Nippon Suisan Gakkaishi* **41**, 465–466.

Cho, C. Y., Bayley, H. S., and Slinger, S. J. (1974). Partial replacement of herringmeal with soybean meal and other changes in a diet for rainbow trout (*Salmo gairdneri*). *J. Fish. Res. Board Can.* **31**, 1523–1528.

Chuecas, L., and Riley, J. P. (1969). Component fatty acids of the total lipids of some marine phytoplankton. *J. Mar. Biol. Assoc. U.K.* **49**, 97–116.

Conte, F. P. (1969). Salt secretion. In "Fish Physiology" (W. S. Hoar and D. J. Randall, eds.), Vol. 1, pp. 241–292. Academic Press, New York.

Cowey, C. B. (1975). Aspects of protein utilization by fish. *Proc. Nutr. Soc.* **34**, 57–63.

Cowey, C. B. (1976). The use of synthetic diets and biochemical criteria in the assessment of nutrient requirements of fish. *J. Fish. Res. Board Can.* **33**, 1040–1045.

Cowey, C. B., and Sargent, J. R. (1972). Fish nutrition. *Adv. Mar. Biol.* **10**, 383–492.

Cowey, C. B., Pope, J. A., Adron, J. W., and Blair, A. (1971). Studies on the nutrition of marine flatfish. Growth of plaice, *Pleuronectes platessa*, on diets containing proteins derived from plants and other sources. *Mar. Biol.* **10**, 145–153.

Cowey, C. B., Adron, J., Blair, A., and Shanks, A. M. (1974a). Studies on the nutrition of marine flatfish. Utilization of various dietary proteins by plaice (*Pleuronectes platessa*). *Br. J. Nutr.* **31**, 297–306.

Cowey, C. B., Brown, D. A., Adron, J. W., and Shanks, A. M. (1974b). Studies on the nutrition of marine flatfish. The effect of dietary protein content on certain cell components and enzymes in the liver of *Pleuronectes platessa*. *Mar. Biol.* **28**, 207–213.

Cowey, C. B., Adron, J. W., Brown, D. A., and Shanks, A. M. (1975a). Studies on the nutrition of marine flatfish. The metabolism of glucose by plaice (*Pleuronectes platessa*) and the effect of dietary energy source on protein utilization in plaice. *Br. J. Nutr.* **33**, 219–231.

Cowey, C. B., Adron, J. W., Knox, D., and Ball, G. T. (1975b). Studies on the nutrition of marine flatfish. The thiamine requirement of turbot (*Scopthalmus maximus*). *Br. J. Nutr.* **34**, 383–390.

Cowey, C. B., Adron, J. W., Owen, J. M., and Roberts, R. J. (1976). The effect of different dietary oils on tissue fatty acids and tissue pathology in turbot *Scopthalmus maximus*. *Comp. Biochem. Physiol. B* **53**, 399–403.

Cowey, C. B., de la Higuera, M., and Adron, J. W. (1977). The effect of dietary composition and of insulin and gluconeogenesis in rainbow trout. *Br. J. Nutr.* **38**, 385–395.

Dabrowska, H., and Wojno, T. (1977). Studies on the utilization by rainbow trout (*Salmo gairdneri*) of feed mixtures containing soya bean meal and an addition of amino acids. *Aquaculture* **10**, 297–310.

Dean, W. F., and Scott, H. M. (1965). The development of an amino acid reference diet for the early growth of chicks. *Poult. Sci.* **44**, 803–808.

Degkwitz, E., and Staudinger, H. (1974). Role of vitamin C on microsomal cytochromes. In "Vitamin C. Recent Aspects of Its Physiological and Technological Importance" (G. G. Birch and K. Parker, eds.), pp. 161–178. Appl. Sci. Publ., London.

De Kruyff, B. (1975). Lipid–sterol interactions in liposomes and membranes. Biochem. Soc. Trans. 3, 618–621.

De Torrengo, M. P., and Brenner, R. R. (1976). Influence of environmental temperature on the fatty acid desaturation and elongation activity of fish (Pimelodus maculatus) liver microsomes. Biochim. Biophys. Acta 424, 36–44.

Deyoe, C. W., and Tiemeier, O. W. (1968). Nutritional requirements for channel catfish fingerlings. Feedstuffs 40, 48–49.

Driedzic, W. R., and Kiceniuk, J. W. (1976). Blood lactate levels in free-swimming rainbow trout (Salmo gairdneri) before and after strenuous exercise resulting in fatigue. J. Fish. Res. Board Can. 33, 173–176.

Dupree, H. K., and Halver, J. E. (1970). Amino acids essential for the growth of channel catfish, Ictalurus punctatus. Trans. Am. Fish. Soc. 99, 90–92.

Ellinger, G. M., and Duncan, A. (1976). The determination of methionine in proteins by gas–liquid chromatography. Biochem. J. 155, 615–621.

Exton, J. H., and Park, C. R. (1967). Control of gluconeogenesis in liver. I. General features of gluconeogenesis in the perfused livers of rats. J. Biol. Chem. 242, 2622–2636.

Farias, R. N., Bloj, B., Morero, R. D., Sineriz, F., and Trucco, R. E. (1975). Regulation of allosteric membrane-bound enzymes through changes in membrane lipid composition. Biochim. Biophys. Acta 415, 231–252.

Farkas, T., and Csengeri, I. (1976). Biosynthesis of fatty acids by the carp, Cyprinus carpio L., in relation to environmental temperature. Lipids 11, 401–407.

Finlayson, A. J., and MacKenzie, S. L. (1976). A rapid method for methionine determination in plant materials. Anal. Biochem. 70, 397–402.

Forster, R. P., and Goldstein, L. (1969). Formation of excretory products. In "Fish Physiology" (W. S. Hoar and D. J. Randall, eds.), Vol. 1, pp. 313–350. Academic Press, New York.

Freedland, R. A. (1967). Effect of progressive starvation on rat liver enzyme activities. J. Nutr. 91, 489–495.

Friedmann, B., Goodman, E. H., and Weinhouse, S. (1965). Dietary and hormonal effects on gluconeogenesis in the rat. J. Biol. Chem. 240, 3729–3735.

Garling, D. L., and Wilson, R. P. (1976). Optimum dietary protein to energy ratio for channel catfish fingerlings, Ictalurus punctatus. J. Nutr. 106, 1368–1375.

Greene, C. W. (1919). Biochemical changes in the muscle tissue of king salmon during the fast of spawning migration. J. Biol. Chem. 39, 435–456.

Greene, C. W. (1926). The physiology of the spawning salmon. Physiol. Rev. 6, 201–241.

Gruger, G. H., Nelson, R. W., and Stansby, M. (1964). Fatty acid composition of oils from twenty-one species of marine fish, freshwater fish, and shellfish. J. Am. Oil Chem. Soc. 41, 662–667.

Halver, J. E. (1957). Nutrition of salmonoid fishes. III. Water-soluble vitamin requirements of chinook salmon. J. Nutr. 62, 225–243.

Halver, J. E. (1972a). "Fish Nutrition." Academic Press, New York.

Halver, J. E. (1972b). The role of ascorbic acid in fish disease and tissue repair. Nippon Suisan Gakkaishi 31, 79–92.

Halver, J. E., and Coates, J. A. (1957). A vitamin test diet for long-term feeding studies. Prog. Fish Cult. 19, 112–118.

Halver, J. E., DeLong, D. C., and Mertz, E. T. (1959). Methionine and cystine requirements of chinook salmon. Fed. Proc., Fed. Am. Soc. Exp. Biol. 18, 2076.

Halver, J. E., Smith, R. R., Tolbert, B. M., and Baker, E. M. (1975). Utilization of ascorbic acid in fish. *Ann. N.Y. Acad. Sci.* **258**, 81–102.

Hansen, I. A., and Mead, J. F. (1965). The fate of dietary wax esters in the rat. *Proc. Soc. Exp. Biol. Med.* **120**, 527–532.

Harding, D. E., Allen, O. W., and Wilson, R. P. (1977). Sulfur amino acid requirement of channel catfish: L-Methionine and L-cystine. *J. Nutr.* **107**, 2031–2035.

Harper, A. E., Benevenga, J., and Wohlhueter, R. M. (1970). Effects of ingestion of disproportionate amounts of amino acids. *Physiol. Rev.* **50**, 428–558.

Haschemeyer, A. E. V. (1969a). Rates of polypeptide chain assembly in liver *in vivo*: Relation to the mechanism of temperature acclimation in *Opsanus tan. Proc. Natl. Acad. Sci. U.S.A.* **62**, 128–135.

Haschemeyer, A. E. V. (1969b). Studies on the control of protein synthesis in low temperature acclimation. *Comp. Biochem. Physiol.* **28**, 535–552.

Haschemeyer, A. E. V. (1973). Kinetic analysis of synthesis and secretion of plasma proteins in a marine teleost. *J. Biol. Chem.* **248**, 1643–1649.

Haschemeyer, A. E. V., and Persell, R. (1973). Kinetic studies on amino acid uptake and protein synthesis in liver of temperature acclimated toadfish. *Biol. Bull. (Woods Hole, Mass.)* **145**, 472–481.

Hayashi, S., and Ooshiro, Z. (1975a). Gluconeogenesis and glycolysis in isolated perfused liver of the eel. *Nippon Suisan Gakkaishi* **41**, 201–208.

Hayashi, S., and Ooshiro, Z. (1975b). Incorporation of [^{14}C]lactate into glucose by perfused eel liver. *Nippon Suisan Gakkaishi* **41**, 791–796.

Hems, R., Ross, B. D., Berry, M. N., and Krebs, H. A. (1966). Gluconeogenesis in the perfused rat liver. *Biochem. J.* **101**, 284–292.

Hickman, C. P., and Trump, B. J. (1969). The kidney. *In* "Fish Physiology" (W. S. Hoar and D. J. Randall, eds.), Vol. 1, pp. 91–239. Academic Press, New York.

Higashi, H., Kaneko, T., Ishui, S., Ushiyama, M., and Sugihasi, T. (1966). Effect of ethyl linoleate, ethyl linolenate, and ethylesters of highly unsaturated fatty acids on essential fatty acid deficiency in rainbow trout. *J. Vitaminol.* **12**, 74–79.

Hilton, J. W., Cho, C. Y., and Slinger, S. J. (1977a). Factors affecting the stability of supplemental ascorbic acid in practical trout diets. *J. Fish. Res. Board Can.* **34**, 683–687.

Hilton, J. W., Cho, C. Y., and Slinger, S. J. (1977b). Evaluation of the ascorbic acid status of rainbow trout (*Salmo gairdneri*). *J. Fish. Res. Board Can.* **34**, 2207–2210.

Hochachka, P. W. (1969). Intermediary metabolism in fishes. *In* "Fish Physiology" (W. S. Hoar and D. J. Randall, eds.), Vol. 1, pp. 351–389. Academic Press, New York.

Hochachka, P. W., and Somero, G. N. (1973). "Strategies of Biochemical Adaptation," 358 pp. Saunders, Philadelphia, Pennsylvania.

Holmes, W. N., and Donaldson, E. M. (1969). The body compartments and the distribution of electrolytes. *In* "Fish Physiology" (W. S. Hoar and D. J. Randall, eds.), Vol. 1, pp. 1–89. Academic Press, New York.

Ichikawa, R., and Oguri, M. (1961). Metabolism of radionuclides in fish. I. Strontium–calcium discrimination in gill absorption. *Nippon Suisan Gakkaishi* **27**, 351–356.

Inui, Y., and Yokote, M. (1974). Gluconeogenesis in eel. I. Gluconeogenesis in the fasted eel. *Tansuiku Suisan Kenkyusho Kenkyu Hokoku* **24**, 33–46.

Inui, Y., and Yokote, M. (1975a). Gluconeogenesis in the eel. II. Gluconeogenesis in the alloxanised eel. *Nippon Suisan Gakkaishi* **41**, 291–300.

Inui, Y., and Yokote, M. (1975b). Gluconeogenesis in the eel. III. Effects of mammalian insulin on the carbohydrate metabolism of the eel. *Nippon Suisan Gakkaishi* **41**, 965–972.

Inui, Y., and Yokote, M. (1975c). Gluconeogenesis in the eel. IV. Gluconeogenesis in the hydrocortisone-administered eel. *Nippon Suisan Gakkaishi* **41**, 973–981.

Inui, Y., and Yokote, M. (1975d). Gluconeogenesis in the eel. V. Effects of alloxan and hydrocortisone administration on amino acid mobilization in the hepatectomized eel. *Nippon Suisan Gakkaishi* **41**, 1101–1104.

Inui, Y., Arai, S., and Yokote, M. (1975). Gluconeogenesis in the eel. VI. Effects of hepatectomy, alloxan, and mammalian insulin on the behaviour of plasma amino acids. *Nippon Suisan Gakkaishi* **41**, 1105–1111.

Jackim, E., and LaRoche, G. (1973). Protein synthesis in *Fundulus heteroclitus* muscle. *Comp. Biochem. Physiol. A* **44**, 851–856.

Kaushik, S. J. (1977). Influence de la salinité sur le metabolisme azoté et le besoin en arginine chez la truite arc-en-ciel. Ph.D. Thesis, L'Universite Bretagne Occidentale, Brest 230 pp.

Kayama, M., and Iijima, N. (1976). Studies on lipid transport mechanism in fish. *Bull. Jpn. Soc. Sci. Fish.* **42**, 987–996.

Kayama, M., Tsuchiya, Y., and Mead, J. F. (1963a). A model experiment of aquatic food chain with special significance in fatty acid conversion. *Nippon Suisan Gakkaishi* **29**, 452–458.

Kayama, M., Tsuchiya, Y., Nevenzel, J. C., Fulso, A., and Mead, J. F. (1963b). Incorporation of linolenic-[1-^{14}C]acid into eicosapentaenoic and docosahexaenoic acids in fish. *J. Am. Oil Chem. Soc.* **40**, 499–502.

Kemp, P., and Smith, M. W. (1970). Effect of temperature acclimatization on the fatty acid composition of goldfish intestinal lipids. *Biochem. J.* **117**, 9–15.

Ketola, H. G. (1975a). Mineral supplementation of diets containing soybean meal as a source of protein for rainbow trout. *Prog. Fish Cult.* **37**, 73–75.

Ketola, H. G. (1975b). Requirement of Atlantic salmon for dietary phosphorus. *Trans. Am. Fish. Soc.* **104**, 548–551.

Kimelberg, H. K., and Papahadjopoulos, D. (1974). Effects of phospholipid acyl chain fluidity, phase transitions, and cholesterol on $(Na^+ + K^+)$-stimulated adenosine triphosphatase. *J. Biol. Chem.* **249**, 1071–1080.

Kloppel, T. M., and Post, G. (1975). Histological alterations in tryptophan-deficient rainbow trout. *J. Nutr.* **105**, 861–866.

Knipprath, W. G., and Mead, J. F. (1968). The effect of the environmental temperature on the fatty acid composition and on the *in vivo* incorporation of [1-^{14}C]acetate in goldfish (*Carrasius auratus* L.). *Lipids* **3**, 121–128.

Krebs, H. A. (1972). Some aspects of the regulation of fuel supply in omnivorous animals. *Adv. Enzyme Regul.* **10**, 397–420.

Krebs, H. A., Bennett, D. A. H., DeGasquet, P., Gascoyne, T., and Yoshida, T. (1963). Renal gluconeogenesis. The effect of diet on the gluconeogenic capacity of rat kidney cortex slices. *Biochem. J.* **86**, 22–27.

Krebs, H. A., Notton, B. M., and Hems, R. (1966). Gluconeogenesis in mouse liver slices. *Biochem. J.* **101**, 607–617.

Larsson, A., and Lewander, K. (1973). Metabolic effect of starvation in the eel, *Anguilla anguilla* L. *Comp. Biochem. Physiol. A* **44**, 367–374.

Lee, D. J., and Putnam, G. B. (1973). The response of rainbow trout to varying protein/ energy ratios in a test diet. *J. Nutr.* **103**, 916–922.

Lee, R. F., and Hirota, J. (1973). Wax esters in tropical zooplankton and nekton and the geographical distribution of wax esters in marine copepods. *Limnol. Oceanogr.* **18**, 227–239.

Lee, R. F., and Puppione, D. L. (1972). Serum lipoproteins of the Pacific sardine (*Sardinops caerulea* Girard). *Biochim. Biophys. Acta* **270**, 272–278.

Lee, R. F., Hirota, J., and Barnet, A. M. (1971a). Distribution and importance of wax esters in marine copepods and other zooplankton. *Deep Sea Res.* **18**, 1147–1165.

Lee, R. F., Nevenzel, J. C., and Paffenhofer, G. A. (1971b). Importance of wax esters and other lipids in the marine food chain: Phytoplankton and copepods. *Mar. Biol.* **9**, 99–108.

Leger, C. (1972). Essai de purification de la lipase du tissu intercaecal de la truite (*Salmo gairdneri* Rich). *Ann. Biol. Anim., Biochim., Biophys.* **12**, 341–345.

Leger, C., and Bauchart, D. (1972). Hydrolyse de triglycérides par le système lipasique du pancréas de truite (*Salmo gairdneri* Rich). Mise en evidence d'un nouveau type de specificite d'action. *C. R. Acad. Sci., Ser. D* **275**, 2419–2422.

Leger, C., Bergot, P., Luquet, P., Flanzy, J., and Meurot, J. (1977). Specific distribution of fatty acids in the triglycerides of rainbow trout adipose tissue. Influence of temperature. *Lipids* **12**, 538–543.

Leslie, J. M., and Buckley, J. T. (1976). Phospholipid composition of goldfish (*Carassius auratus* L.) liver and brain and temperature dependence of phosphatidyl choline synthesis. *Comp. Biochem. Physiol. B* **53**, 335–337.

Lovell, R. T. (1973). Essentiality of vitamin C in feeds for intensively fed caged channel catfish. *J. Nutr.* **103**, 134–138.

Lowenstein, J. M. (1972). Ammonia production in muscle and other tissues: The purine nucleotide cycle. *Physiol. Rev.* **52**, 382–413.

McFarlane, I. G., and von Holt, C. (1969). Metabolism of amino acids in protein–calorie deficient rats. *Biochem. J.* **111**, 557–563.

McGivan, J. D., and Chappell, J. B. (1975). On the metabolic function of glutamate dehydrogenase in rat liver. *FEBS Lett.* **52**, 1–5.

Maetz, J. (1971). Fish gills: Mechanisms of salt transfer in freshwater and sea water. *Philos. Trans. R. Soc. London, Ser. B* **262**, 209–251.

Maetz, J. (1974). Aspects of adaptation to hypo-osmotic and hyperosmotic environments. *In* "Biochemical and Biophysical Perspectives in Marine Biology" (D. C. Malins and J. R. Sargent, eds.), Vol. 1, pp. 1–167. Academic Press, New York.

Mattson, F. H., Volpenheim, R. A., and Benjamin, L. (1970). Inhibition of lipolysis by normal alcohols. *J. Biol. Chem.* **245**, 5335–5340.

Mauron, J., Mottu, F., and Spohr, G. (1973). Reciprocal induction and repression of serine dehydratase and phosphoglycerate dehydrogenase by proteins and dietary essential amino acids in rat liver. *Eur. J. Biochem.* **32**, 331–342.

Mazeaud, F. (1973). Recherches sur la régulation des acides gras libres plasmatiques et de la glycémie chez les poissons. Ph.D. Thesis, Univ. de Paris, Paris, 107 pp.

Mead, J. F., and Kayama, M. (1967). Lipid metabolism in fish. *In* "Fish Oils" (M. E. Stansby, ed.), pp. 289–299. Avi Publ. Co., Westport, Connecticut.

Mertz, E. T. (1969). Amino acid and protein requirements of fish. *In* "Fish in Research" (O. W. Neuhaus and J. E. Halver, eds.), pp. 233–244. Academic Press, New York.

Miller, D. S., and Payne, P. R. (1961). Problems in the prediction of protein values in diets. The influence of protein concentration. *Br. J. Nutr.* **15**, 11–19.

Mohrhauer, H., and Holman, R. T. (1963a). The effect of dose level of essential fatty acids upon fatty acid composition of the rat liver. *J. Lipid Res.* **4**, 151–159.

Mohrhauer, H., and Holman, R. T. (1963b). The effect of dietary essential fatty acids upon composition of polyunsaturated fatty acids in depot fat and erythrocytes of the rat. *J. Lipid Res.* **4**, 346–350.

Motais, R., and Garcia-Romeu, F. (1972). Transport mechanisms in the teleostean gill and amphibian skin. *Annu. Rev. Physiol.* **34**, 141–176.

Murai, T., and Andrews, J. W. (1975). Pantothenic acid supplementation of diets for catfish fry. *Trans. Am. Fish. Soc.* **104**, 313–316.

Murray, M. W., Andrews, J. W., and DeLoach, H. L. (1977). Effects of dietary lipids, dietary protein and environmental temperatures on growth, feed conversion and body composition of channel catfish. *J. Nutr.* **107**, 272–280.

Nag, A. C. (1972). Ultrastructure and adenosine triphosphatase activity of red and white muscle fibers of a fish, *Salmo gairdneri*. *J. Cell Biol.* **55**, 42–57.

Nagai, M., and Ikeda, S. (1971a). Carbohydrate metabolism in fish. I. Effects of starvation and dietary composition on the blood glucose level and the hepatopancreatic glycogen and lipid contents in carp. *Nippon Suisan Gakkaishi* **37**, 404–409.

Nagai, M., and Ikeda, S. (1971b). Carbohydrate metabolism in fish. II. Effect of dietary composition on metabolism of glucose-6[^{14}C] in carp. *Nippon Suisan Gakkaishi* **37**, 410–414.

Nagai, M., and Ikeda, S. (1972). Carbohydrate metabolism in fish. III. Effect of dietary composition on metabolism of glucose[U-^{14}C] and glutamate[U-^{14}C]. *Nippon Suisan Gakkaishi* **38**, 137–143.

Nagai, M., and Ikeda, S. (1973). Carbohydrate metabolism in fish. IV. Effect of dietary composition on metabolism of acetate[U-^{14}C] and L-alanine[U-^{14}C] in carp. *Nippon Suisan Gakkaishi* **39**, 633–643.

Nagayama, F., and Ohshima, H. (1974). Studies on the enzyme system of carbohydrate metabolisn in fish. I. Properties of liver hexokinase. *Nippon Suisan Gakkaishi* **40**, 285–290.

Nagayama, F., Ohshima, H., and Umezawa, K. (1972). Distribution of glucose 6-phosphate metabolizing enzymes in fish. *Nippon Suisan Gakkaishi* **38**, 589–593.

Nagayama, F., Ohshima, H., and Umezawa, K. (1973). Activities of hexokinase and glucose dehydrogenase in fish liver. *Nippon Suisan Gakkaishi* **39**, 1349.

Nagayama, F., Ohshima, H., and Takeuchi, T. (1975a). Studies on the enzyme system of carbohydrate metabolism in fish. II. Purification of glucose-6-phosphate dehydrogenase and glucose dehydrogenase. *Nippon Suisan Gakkaishi* **41**, 1063–1067.

Nagayama, F., Ohshima, H., and Takeuchi, T. (1975b). Studies on the enzyme system of carbohydrate metabolism in fish. III. Properties of glucose-6-phosphate dehydrogenase and glucose dehydrogenase. *Nippon Suisan Gakkaishi* **41**, 1069–1074.

Nakashima, Y., Suzue, R., Sanada, H., and Kawada, S. (1972). Effect of ascorbic acid on tyrosine hydroxylase activity *in vivo*. *Arch. Biochem. Biophys.* **152**, 515–520.

National Research Council (1973). "Nutrient Requirements of Domestic Animals," No. 11, "Nutrient Requirements of Trout, Salmon and Catfish." Natl. Res. Counc., Washington, D.C.

Newsholme, E. A., and Crabtree, B. (1970). The role of fructose-1,6-diphosphatase in the regulation of glycolysis in skeletal muscle. *FEBS Lett.* **7**, 195–198.

Newsholme, E. A., and Start, C. (1973). "Regulation in Metabolism," 349 pp. Wiley, New York.

Nicolaides, N., and Woodall, A. N. (1962). Impaired pigmentation in chinook salmon fed diets deficient in essential fatty acids. *J. Nutr.* **78**, 431–437.

Nijkamp, H. J., van Es, A. J. H., and Huisman, A. E. (1974). Retention of nitrogen, fat, ash, carbon, and energy in growing chickens and carp. *Eur. Assoc. Anim. Prod., Publ.* **14**, 277–280.

Ninno, R. E., De Torrengo, M. A. P., Castuma, J. C., and Brenner, R. (1974). Specificity of 5- and 6- fatty acid desaturases in rat and fish. *Biochim. Biophys. Acta* **360**, 124–133.

Njaa, L. R. (1977). A method for determining the methionine sulfoxide content in protein concentrates. *Proc. 11th Meet. Fed. Eur. Biochem. Soc.* **A3–8**, 933.

Nose, T. (1971). Determination of nutritive value of food protein in fish. III. Nutritive value of casein, white fishmeal and soybean meal in rainbow trout fingerlings. *Tansuiku Suisan Kenkyusho Kenkyu Hokoku* 21, 85–98.

Ogino, C., and Chen, M. S. (1973). Protein nutrition in fish. IV. Biological value of dietary proteins in carp. *Nippon Suisan Gakkaishi* 39, 797–800.

Ogino, C., and Chiou, J. Y. (1976). Mineral requirements in fish. II. Magnesium requirement of carp. *Nippon Suisan Gakkaishi* 41, 71–75.

Ogino, C., and Kamizono, M. (1975). Mineral requirements in fish. I. Effects of dietary salt mixture levels on growth, mortality, and body composition in rainbow trout and carp. *Nippon Suisan Gakkaishi* 41, 429–434.

Ogino, C., and Takeda, H. (1976). Mineral requirements in fish. III. Calcium and phosphorus requirements in carp. *Nippon Suisan Gakkaishi* 42, 793–799.

Ogino, C., Kakino, J., and Chen, M. S. (1973). Protein nutrition in fish. II. Determination of metabolic faecal nitrogen and endogenous nitrogen excretion of carp. *Nippon Suisan Gakkaishi* 39, 519–523.

Orme, L. E., and Lemm, C. (1973). Use of dried sludge from paper processing wastes in trout diets. *Feedstuffs* 45, 28–30.

Overnell, J. (1973). Digestive enzymes of the pyloric caeca and of their associated mesentery in the cod (*Gadus morhua*). *Comp. Biochem. Physiol. B* 46, 519–531.

Owen, J. M., Adron, J. W., Sargent, J. R., and Cowey, C. B. (1972). Studies on the nutrition of marine flatfish. The effect of dietary fatty acids on the tissue fatty acids of the plaice *Pleuronectes platessa*. *Mar. Biol.* 13, 160–166.

Owen, J. M., Adron, J. W., Middleton, C., and Cowey, C. B. (1975). Elongation and desaturation of dietary fatty acids in turbot *Scophthalmus maximus* L., and rainbow trout, *Salmo gairdneri Rich*. *Lipids* 10, 528–531.

Page, J. W., and Andrews, J. W. (1973). Interactions of dietary levels of protein and energy on channel catfish (*Ictalurus punctatus*). *J. Nutr.* 103, 1339–1346.

Patton, J. S. (1975). High levels of pancreatic nonspecific lipase in rattlesnake and leopard shark. *Lipids* 10, 562–564.

Patton, J. S., and Benson, A. A. (1975). A comparative study of wax ester digestion in fish. *Comp. Biochem. Physiol. B* 52, 111–116.

Patton, J. S., Nevenzel, J. C., and Benson, A. A. (1975). Specificity of digestive lipases in hydrolysis of wax esters and triglycerides studied in anchovy and other selected fish. *Lipids* 10, 575–583.

Phillips, A. M. (1969). Nutrition, digestion and energy utilization. *In* "Fish Physiology" (W. S. Hoar and D. J. Randall, eds.), Vol. 1, pp. 391–432. Academic Press, New York.

Poston, H. A., Riis, R. C., Rumsey, G. L., and Ketola, H. G. (1977). The effect of supplemental dietary amino acids, minerals and vitamins on salmonids fed catarogenic diets. *Cornell Vet.* 67, 472–509.

Rahn, C. H., Sand, D. M., and Schlenk, H. (1973). Wax esters in fish. Metabolism of dietary palmityl palmitate in the gourami (*Trichogaster cosby*). *J. Nutr.* 103, 1441–1447.

Raymont, J. E. G. (1963). "Plankton and Productivity in the Oceans," pp. 533–535. Pergamon Press, Oxford.

Regier, L. W., Jangaard, P. M., Power, H. E., March, B. E., and Biely, J. (1974). Composition and nutritive characteristics of Atlantic Canadian white fishmeals. *J. Fish. Res. Board Can.* 31, 201–204.

Reiser, R., Stevenson, B., Kayama, M., and Choudhury, R. B. R. (1963). The influence of dietary fatty acids and environmental temperature on the fatty acid composition of teleost fish. *J. Am. Oil Chem. Soc.* 40, 507–513.

Ringrose, R. C. (1971). Calorie-to-protein ratio for brook trout (*Salvelinus fontinalis*). *J. Fish. Res. Board Can.* 28, 1113–1117.

Rivers, J. P. W., Sinclair, A. J., and Crawford, M. A. (1975). Inability of the cat to desaturate essential fatty acids. *Nature (London)* **258**, 171–173.

Robinson, D. S., and Wing, D. R. (1971). Studies on tissue clearing factor lipase related to its role in the removal of lipoprotein triglyceride from the plasma. *Biochem. Soc. Symp.* No. 33, pp. 123–135.

Robinson, J. S., and Mead, J. F. (1973). Lipid absorption and deposition in rainbow trout (*Salmo gairdneri*). *Can. J. Biochem.* **51**, 1050–1058.

Roehm, J. N., Lee, D. J., Wales, J. H., Politylca, S. D., and Sinnhuber, R. O. (1969). The effect of dietary sterculic acid on the hepatic lipids of rainbow trout. *Lipids* **5**, 80–84.

Rumsey, G. L., and Ketola, H. G. (1975). Amino acid supplementation of casein diets of Atlantic salmon (*Salmo salar*) fry and of soybean meal for rainbow trout (*Salmo gairdneri*) fingerlings. *J. Fish. Res. Board Can.* **32**, 422–426.

Sakaguchi, M., and Kawai, A. (1970). Histidine metabolism in fish. V. The effect of protein deficiency and fasting on the activities of histidine deaminase and urocanase in carp liver. *Nippon Suisan Gakkaiki* **36**, 783–787.

Sakaguchi, M., and Kawai, A. (1974). Some responses of histidine deaminase and urocanase in carp liver. *Kyoto Daigaku Shokuryo Kagaku Kenkyusho Hokoku* **37**, 28–31.

Sakamoto, S., and Yone, Y. (1973). Effect of dietary calcium/phosphorous ratio upon growth, feed efficiency, and blood serum calcium and phosphorous level in red sea bream. *Nippon Suisan Gakkaishi* **39**, 343–348.

Sand, D. M., Rahn, C. H., and Schlenk, H. (1973). Wax esters in fish: Absorption and metabolism of oleyl alcohol in the gourami (*Trichogaster cosby*). *J. Nutr.* **103**, 600–607.

Sargent, J. R., Lee, R. F., and Nevenzel, J. C. (1976). Marine waxes. *In* "Chemistry and Biochemistry of Natural Waxes" (P. E. Kolattuckudy, ed.), pp. 50–91. Elsevier/ North-Holland Publ., Amsterdam.

Schimke, R. T. (1962). Adaptive characteristics of urea cycle enzymes in the rat. *J. Biol. Chem.* **237**, 459–468.

Schoolworth, A. C., Blondin, J., and Klahr, S. (1974). Renal gluconeogenesis. Influence of diet and hydrogen ions. *Biochim. Biophys. Acta* **372**, 274–284.

Shatton, J. B., Halver, J. E., and Weinhouse, S. (1971). Glucose (hexose-6-phosphate) dehydrogenase in liver of rainbow trout. *J. Biol. Chem.* **246**, 4878–4885.

Shimeno, S., and Takeda, M. (1972). Studies on hexose monophosphate shunt of fishes. I. Properties of hepatic glucose-6-phosphate dehydrogenase of barracuda. *Nippon Suisan Gakkaishi* **38**, 645–650.

Shimeno, S., and Takeda, M. (1973). Studies on hexose monophosphate shunt of fishes. II. Distribution of glucose-6-phosphate dehydrogenase. *Nippon Suisan Gakkaishi* **39**, 461–466.

Singer, S. J., and Nicolson, G. L. (1972). Fluid mosaic model of the structure of cell membranes. *Science* **175**, 720–731.

Singh, R. P., and Nose, T. (1967). Digestibility of carbohydrate in young rainbow trout. *Tansuiku Suisan Kenkyusho Kenkyu Hokoku* **17**, 21–25.

Sinnhuber, R. O. (1969). The role of fats. *In* "Fish in Research" (O. Neuhaus and J. E. Halver, eds.), pp. 245–259. Academic Press, New York.

Smith, C. E., Brin, M., and Halver, J. E. (1974). Biochemical, physiological, and pathological changes in pyridoxine-deficient rainbow trout (*Salmo gairdneri*). *J. Fish. Res. Board Can.* **31**, 1893–1898.

Smith, M. W. (1967). Influence of temperature acclimatization on the temperature dependence and ouabain sensitivity of goldfish intestinal adenosine triphosphatase. *Biochem. J.* **105**, 65–71.

Smith, M. W., and Kemp, P. (1969). Phospholipase C-induced changes in intestinal adenosine triphosphatase prepared from goldfish acclimatized to different temperatures. *Biochem. J.* **114**, 659–661.

Smith, R. R. (1971). A method for measuring digestibility and metabolizable energy of fish feeds. *Prog. Fish Cult.* **33**, 132–134.

Stickney, R. R., and Andrews, J. W. (1971). Combined effects of dietary lipids and environmental temperature on growth, metabolism, and body composition of channel catfish (*Ictalurus punctatus*). *J. Nutr.* **101**, 1703–1710.

Stickney, R. R., and Andrews, J. W. (1972). Effects of dietary lipids on growth, food conversion, lipid and fatty acid composition of channel catfish. *J. Nutr.* **102**, 249–258.

Stickney, R. R., and Shumway, S. E. (1974). Occurrence of cellulase activity in the stomachs of fishes. *J. Fish Biol.* **6**, 779–790.

Takeda, M., Shimeno, S., Hosokawa, H., Kajiyama, H., and Kaisyo, T. (1975). The effect of dietary calorie-to-protein ratio on the growth, feed conversion, and body composition of young yellowtail. *Nippon Suisan Gakkaishi* **41**, 443–447.

Tanaka, R., and Teruya, A. (1973). Lipid dependence of activity–temperature relationship of (Na$^+$, K$^+$)-activated ATPase. *Biochim. Biophys. Acta* **323**, 584–591.

Templeton, W. L., and Brown, V. M. (1963). Accumulation of calcium and strontium by brown trout from waters in the United Kingdom. *Nature (London)* **198**, 198–200.

Thomson, A. J., Sargent, J. R., and Owen, J. M. (1975). Effect of environmental changes on the lipid composition and (Na$^+$, K$^+$)-dependent adenosine triphosphatase in the gills of the eel, *Anguilla anguilla*. *Biochem. Soc. Trans.* **3**, 668.

Tiews, K., Gropp, J., and Koops, H. (1976). On the development of optimal rainbow trout pellet feeds. *Arch. Fischereiwiss, Beih.* **27**, 1–29.

Van Deenen, L. L. M., De Gier, J., and Demel, R. A. (1972). Relations between lipid composition and permeability of membranes. *Biochem. Soc. Symp.* No. 35, pp. 377–382.

Vrba, R. (1966). Effects of insulin-induced hypoglycaemia on the fate of glucose carbon atoms in the mouse. *Biochem. J.* **99**, 367–380.

Walton, M. J., and Cowey, C. B. (1977). Aspects of ammoniogenesis in rainbow trout, *Salmo gairdneri*. *Comp. Biochem. Physiol. B* **57**, 143–149.

Watanabe, T., Takashima, F., and Ogino, C. (1974a). Effect of dietary methyl linolenate on growth of rainbow trout. *Nippon Suisan Gakkaishi* **40**, 181–188.

Watanabe, T., Kobayashi, I., Utsue, O., and Ogino, C. (1974b). Effect of dietary methyl linolenate on fatty acid composition of lipids in rainbow trout. *Nippon Suisan Gakkaishi* **40**, 387–392.

Watanabe, T., Ogino, C., Koshiishi, Y., and Matsunaga, T. (1974c). Requirements of rainbow trout for essential fatty acids. *Nippon Suisan Gakkaishi* **40**, 493–499.

Watanabe, T., Utsue, O., Kobayashi, I., and Ogino, C. (1975a). Effect of dietary methyl linoleate and linolenate on growth of carp. I. *Nippon Suisan Gakkaishi* **41**, 257–262.

Watanabe, T., Takeuchi, T., and Ogino, C. (1975b). Effect of dietary methyl linoleate and linolenate on growth of carp. II. *Nippon Suisan Gakkaishi* **41**, 263–269.

Watts, R. L., and Watts, D. C. (1974). Nitrogen metabolism in fishes. *In* "Chemical Zoology" (M. Florkin and B. T. Scheer, eds.), Vol. 8, pp. 369–446. Academic Press, New York.

Wilson, R. P. (1973). Absence of ascorbic acid synthesis in channel catfish, *Ictalurus punctatus*, and blue catfish, *Ictalurus frucatus*. *Comp. Biochem. Physiol. B* **46**, 635–638.

Wilson, R. P., and Poe, W. E. (1973). Impaired collagen formation in the scorbutic channel catfish. *J. Nutr.* **103**, 1359–1364.

Wilson, R. P., and Poe, W. E. (1974). Nitrogen metabolism in channel catfish, *Ictalurus punctatus*. III. Relative pool sizes of free amino acids and related compounds in various tissues of the catfish. *Comp. Biochem. Physiol. B* **48**, 545–556.

Wilson, R. P., Harding, D. E., and Garling, D. L. (1977). Effect of pH on amino acid utilization and the lysine requirement of fingerling channel catfish. *J. Nutr.* **107**, 166–170.

Windell, J. T., Armstrong, R., and Clinebell, J. R. (1974). Substitution of brewer's single cell protein into pelleted fish feed. *Feedstuffs* **46**, 22–23.

Yamamoto, M. (1968). Fish muscle glycogen phophorylase. *Can. J. Biochem.* **46**, 423–432.

Yone, Y., and Fujii, M. (1975a). Studies on nutrition of red sea bream. XI. Effect of ω3 fatty acid supplement in a corn oil diet on growth rate and feed efficiency. *Nippon Suisan Gakkaishi* **41**, 73–77.

Yone, Y., and Fujii, M. (1975b). Studies on the nutrition of red sea bream. XII. Effect of ω3 fatty acid supplement in a corn oil diet on fatty acid composition of fish. *Nippon Suisan Gakkaishi* **41**, 79–86.

Yone, Y., Furuichi, M., and Sakamoto, S. (1971). Studies on nutrition of red sea bream. III. Nutritive value and optimum content of lipids in diet. *Rep. Fish. Res. Lab., Kyushu Univ.* No. 1, pp. 49–60.

Yu, T. C., and Sinnhuber, R. O. (1972). Effect of dietary linolenic acid and docosahexaenoic acid on growth and fatty acid composition of rainbow trout (*Salmo gairdneri*). *Lipids* **7**, 450–454.

Zannoni, V. G., and Sato, P. H. (1975). Effects of ascorbic acid on microsomal drug metabolism. *Ann. N.Y. Acad. Sci.* **258**, 119–131.

Zeitoun, I. H., Halver, J. E., Ullrey, D. E., and Tack, P. I. (1973). Influence of salinity on protein requirements of rainbow trout (*Salmo gairdneri*) fingerlings. *J. Fish. Res. Board Can.* **30**, 1867–1873.

Zeitoun, I. H., Ullrey, D. E., Magee, W. T., Gill, J. L., and Bergen, W. G. (1976). Quantifying nutrient requirements of fish. *J. Fish. Res. Board Can.* **33**, 167–172.

2

FEEDING STRATEGY

KIM D. HYATT

I. INTRODUCTION

Several contributions to this volume have focused on the internal processes of digestion, energy metabolism, and growth in a few species of fishes under highly controlled conditions. However, fishes as a group have evolved under highly variable conditions in which efficiency of detection, capture, and ingestion of natural foods often limits the level of performance of these internal processes. Although careful study of the factors that determine natural feeding habits is usually regarded as the concern of ecologists dealing with field popu-

FISH PHYSIOLOGY, VOL. VIII
Copyright © 1979 by Academic Press, Inc.
All rights of reproduction in any form reserved.
ISBN 0-12-350408-2

lations, there is evidence that many of these factors substantially influence the conversion of food to fish flesh under the extremely simplified conditions of laboratory culture. Paloheimo and Dickie (1966), in an extensive review of fish feeding and growth experiments, observed that growth efficiency varies dramatically with different food types (hatchery mash, minnows, crustaceans). They concluded that different growth efficiencies were due to differences in foraging efficiency on different food types, rather than to major differences in the physiological conversion of foods. These authors suggested that relative food size was the major factor determining efficient utilization of rations but of course many other characteristics of food influence foraging efficiency, as will become apparent later in this chapter.

The chapter is divided into three main sections. The first surveys the range of dietary specialization of fishes in a number of temperate and tropical ecosystems; the second examines physiological, behavioral, and morphological characteristics that simultaneously affect overall foraging efficiencies and predispose fishes to acquire particular foods; the third looks briefly at some recent attempts to quantify and explain the "selective" exploitation of foods by fishes.

II. TROPHIC DIVERSITY

A. Major Forage Groups

Fishes differ greatly in the character of the food they consume, and a number of very general accounts on the subject are available (Norman and Greenwood, 1963; Marshall, 1966; Nikolsky, 1963). Both the size and systematic position of the food organisms show great variability and the range of foodstuffs consumed by fishes is greater than that for other groups of vertebrates (Nikolsky, 1963). A number of excellent monographs supply detailed information on the dietary habits of fishes (Suyehiro, 1942; Bigelow and Schroeder, 1953; Hiatt and Strasburg, 1960; Knöppel, 1970; Fryer and Iles, 1972; Scott and Crossman, 1973; Lowe-McConnell, 1975). A preliminary step in bringing order to this enormous diversity is to group fishes into a limited number of basic foraging types in order to examine the representation of the various types in major aquatic ecosystems distributed over broad geographic areas.

The scheme presented (Table I) is based on a number of major food types to which fish commonly have access. The majority of fishes, from

Table I

Classification of Fishes from Temperate and Tropical Ecosystems according to Major Food Groups Exploited

	Ecosystem											
	North temperate				Tropical freshwater						Tropical marine	
Food category[a]	Freshwater, Canada		Marine Gulf of Maine		Subtemperate Estuarine, Lake Pontchartrain		South America, Lake Redondo, Amazon		Africa, Lake Victoria		Marshall Island coral reefs	
	%	N	%	N	%	N	%	N	%	N	%	N
Herbivores												
Phytoplankton	0.9	2	0.7	1	5.4	2	2.4	1	2.0	2	0	
Benthic diatoms	3.2	7	0.0				4.8	2			1.5	3
Filamentous algae	4.5	10	0.0		5.4	2	7.1	3	12.0	12	16.0	33
Vascular plants and thallose algae	0.9	2	0.0				7.1	3			8.7	18
Detritivores	1.4	3	0.7	1	18.9	7	9.5	4	—[b]		3.9	8
Carnivores												
Zooplankton	18.5	41	16.9	25	5.4	2	4.8	2	1.0	1	6.3	13
Benthic invertebrates	43.2	96	41.2	61	35.1	13	19.0	8	34.0	34	54.9	113
Terrestrial insects	2.7	6	0		0		4.8	2	0		0	
Fish	18.0	40	39.2	58	21.6	8	11.9	5	41.0	41	—[c]	
Omnivores	6.8	15	2.0	3	8.1	3	28.6	12	10.0	10	8.9	18
Source	Scott and Crossman (1973)		Bigelow and Schroeder (1953)		Darnell (1961)		Marlier (1968, cited in Lowe-McConnell, 1975)		Fryer and Iles (1972); Corbet (1961)		Hiatt and Strasburg (1960)	

[a] In some cases single species of fish were included in more than a single food category.

[b] No fishes in these reports were considered to be strictly detritus feeders but many of those using sedimented algae as a food source could be classed as such.

[c] Piscivorous fishes were included with those that feed on benthic invertebrates in this study.

all ecosystems examined, are carnivorous and the importance of complex bottom faunas and other fish in supplying the major energy sources is clearly apparent. At the other extreme, phytoplankton are only rarely used by fish as a major food source although most of the assemblages contain one or two fish species that are exceptions.

A few of the observed differences of food utilization are related to the differences in food present in lake, river, estuarine, and marine ecosystems. Vascular plant material and terrestrial insects are unique supplies in freshwater, coral polyps and thallose algae in marine systems. In large lakes (such as Lake Victoria, Africa) terrestrial insects are not expected to supply a major food source for the majority of fish; in many of the Southeast Asian forest rivers, however, as many fishes may feed on terrestrial insects as feed on aquatic insects and benthos (Lowe-McConnell, 1975). An abundance of detritus feeders is often present in river systems and their estuaries, although this may not be true of the north temperate zone. Darnell (1961) and Odum (1970) have stressed the role of detritus as a food source of fishes in tropical and subtropical estuaries.

Consideration of a limited number of food categories may produce the impression that the major difference in the patterns of food acquisition by fishes in tropical and temperate systems is quantitative, that is, fish species from tropical ecosystems are more equitably distributed with respect to exploitation of food resources (Table I). However, a number of important qualitative differences appear when a more detailed examination of feeding niches is conducted.

B. Trophic Diversity in Tropical and Temperate Fishes

Comparisons of tropical with temperate ecosystems usually stress that tropical ecosystems have had longer evolutionary histories accompanied by relatively stable and/or predictable environmental changes. These conditions have apparently permitted community interactions to favor the development of diverse tropical faunas composed of species with narrow niches. The many examples of extreme trophic specialization, unique to fishes in tropical communities, have led a number of authors (Hartley, 1948; Larkin, 1956; Keast and Webb, 1966) to emphasize the more generalized foraging habits of fishes in temperate zone communities. However, comparisons at the family and species levels suggest that fishes in temperate communities not only exhibit a less diverse range of trophic specialization, but also have failed to achieve the degree of flexibility and adaptability commonly attained by many groups of tropical fishes. At the family level, tropical groups such as the Chaetodontidae (butterfly fish) and Pomacentridae (damselfish) of marine ecosystems or the Cichlidae and Characidae of freshwater ecosystems include species that are strictly carnivorous, strictly herbivorous, and truly omnivorous (Hiatt and Strasburg, 1960; Fryer and Iles, 1972; Lowe-McConnell, 1975). In the temperate zone, even the families that exhibit a striking degree of adaptive radiation (e.g., Cottidae) fail to contain species that span a similar range from extreme specialists to extreme generalists.

1. SPECIALISTS

At the species level, the development of trophic specializations by herbivores, benthic carnivores, planktivores, and piscivores has proceeded much further in tropical ecosystems than in temperate ones.

In north temperate ecosystems, the chiselmouth chub (*Arocheilus alutaceus*) is regarded as a trophic specialist because it is among the few species that collect a diet of diatoms and filamentous algae by scraping rock surfaces (Moodie and Lindsey, 1972). In tropical ecosystems herbivorous fish that use epilithic algae as a source of food are common. Fryer's study (1959) of a single fish community in Lake Malawi, Africa, revealed the presence of seventeen species of cichlids specialized to obtain their food by scraping, rasping, combing, biting, and nibbling diatoms and/or filamentous algae from rock substrates.

The greatest proportion of fishes from temperate ecosystems feed on various classes of benthic invertebrates, but even here they do not rival fish from tropical communities in degree of specialization. For example, both tropical and temperate water ecosystems contain fish species that have specialized dentition which enables them to feed on mollusks by crushing them, but only fish species from tropical ecosystems appear to have mastered the task of extracting mollusks without crushing the shells (e.g., *Haplochromis sauvagei*). Fryer and Iles observed that in Lakes Tanganyika and Malawi insectivorous species of fish are often so specialized for the collection of food from one particular situation that they are incapable of dealing with equally suitable prey species living under slightly different environmental conditions (e.g., in association with different substrates). By contrast, Keast and Webb (1966) observed that few fishes in a north temperate lake (Lake Opinicon, Canada) exhibited morphological or behavioral specializations so extreme as to prevent their owners from exploiting at least three broad classes of invertebrate foods, under a variety of environmental conditions. The majority of high latitude fishes conform to this level of generalized food habits.

The utilization of fishes as a source of food by other fishes is equally common in tropical and temperate ecosystems but in tropical species there is a trend toward specialized processing of prey in advance of ingestion. Piranhas (*Serrasalmus* spp.) of the Amazon and tiger fish (*Hydrocynus* spp.) of Africa have shearing teeth to bite pieces of flesh from their prey (Foxx, 1972; Lewis, 1974). At least ten species of Amazon fishes (Roberts, 1970), nine African cichlids (Fryer and Iles, 1972), and a tropical marine blenny (Losey, 1972) make their living solely by removing scales and fins of other fishes. These tropical species clearly qualify as unique specialists on the basis of diet morphology or behavior.

2. GENERALISTS

Numerous species of fish in temperate and tropical ecosystems include a wide range of food types in their diets and possess few unique morphological characteristics that are obviously associated with food gathering. Consequently, many species have inherited a poorly defined designation as "food generalists" or "opportunistic foragers." Because a variety of criteria have been used in assigning species to these categories, basic differences between species designated as generalists in temperate versus tropical fish communities have often been overlooked. Scott and Crossman (1973) characterized the majority of Canadian freshwater fishes as generalists because they exploit a wide variety of invertebrate taxa. Similarly, Keast and Webb (1966) labeled the bluegills (*Lepomis macrochirus*) in a temperate lake as highly generalized foragers because up to nine invertebrate taxa were commonly present as 5% or more of the diet by volume and because there was a great deal of overlap in food habits between different age classes of fish. However, these fish appear relatively specialized when compared with the extremely diverse food habits of some tropical species. The diet of reef-dwelling triggerfish (*Balistopus undulatus*) contains substantial amounts of branched coral, coral polyps, fish, crabs, shrimp, gastropods, thallose algae, filamentous algae, coralline algae, echinoderms, tunicates, and pelecypods (Hiatt and Strasburg, 1960). The wide range of foodstuffs present on tropical reefs compared to temperate zone habitats will explain in part the more catholic dietary habits of some tropical species. However, the potential for adaptation to omnivorous food habits has not been exploited by as large a proportion of the fishes in temperate communities as those in tropical communnities (Table I).

III. CHARACTERISTICS THAT INFLUENCE PATTERNS OF FOOD ACQUISITION

Because of the overwhelming trophic diversity displayed by fishes, it is unlikely that simple schemes of diet classification will provide much insight into general principles that determine natural patterns of food acquisition. The approach taken in the following section is to examine the characteristics of fishes and their foods which interact to produce the tremendous diversity of dietary patterns in this group. Because the composition of fish diets may be the outcome of interactions that occur during food search, approach, handling, or ingestion, I will discuss characteristics of fishes and their foods that operate during each of these stages of exploitation. The emphasis throughout these discussions is to review the differences between fishes that account for their exploitation of different foods and to comment on the differences between food items that alter their relative probabilities of ingestion by any single fish species.

A. Sensory Modes

To process food, predators must first locate it. For effective food location they must possess specialized receptors and respond in a specific manner to stimuli generated by "prey." Sensory capacities of even similar fishes diverge so widely that the characteristics of food items to which one forager responds may be quite different from those by which another animal recognizes food. The location of food by fish is commonly mediated by sensory systems that process visual, electrical, mechanical (turbulence and sound), or chemical stimuli. Since a complex set of optical, acoustic, tactile, chemical, and electrical stimuli may be associated with each potential food item, well-defined responses of fishes are usually a consequence of the summation of a complex of signals received through a combination of sensory pathways. However, for most species particular sensory channels are more important than others in food search.

1. VISUAL DETECTION

Extensive research into feeding periodicities based on diel changes in stomach contents (Woodhead, 1966) confirms that in many species food is present only during the daylight hours, thus implying that many fishes rely predominantly on vision in searching for food. Laboratory experiments, involving controlled manipulation of food supply and light conditions (Ali, 1959; Girsa, 1961; Blaxter, 1970), surgical alteration of sensory systems (Tesch, 1975), or careful manipulation of models with different stimulus characteristics (De Groot, 1969) have established the essential role of visual cues in successful food location. Of thirty-seven freshwater and marine species treated experimentally (citations from Tesch, 1975; Blaxter, 1970), twenty-eight exhibited a critical dependence on visual cues for food location and capture. Vision is often important for some phases of food search even in fishes with poor acuity and discrimination abilities. Hobson (1963) observed that in the field the final approach to motionless prey by sharks is guided by vision. On the basis of sensory deprivation experiments, Gilbert (1963) concluded that at distances greater than 17 m olfaction is more important than vision in guiding lemon sharks (*Negaprion brevirostris*) to prey, but at very close range (≤ 3 m) vision is the primary sense. With some notable exceptions, a light intensity of 10^{-1} mc (meter-candle), corresponding to late dusk, is the lower threshold for effective visual location of food by fishes (Blaxter, 1970).

2. ELECTRODETECTION

Orientation by means of electric stimuli has been convincingly demonstrated in a few fishes, and several hundred species (elasmobranch, chondrostean, siluroid, gymnarchid, mormyrid, and gymnotid fishes) may be capable of receiving and interpreting these stimuli (Bullock, 1973; Protasov, 1973). In principle, electrosensory systems can provide information about the location, size, shape, and quality of objects in the predator's immediate vicinity (Kalmijn, 1974). In fishes that possess well-developed eyes and electrosensory systems, the latter may compensate to allow food location under poor visual conditions. In a series of simple but elegant experiments, Kalmijn (1971) demonstrated that sharks (*Scyliorhinus canicula*) and rays (*Raja clavata*) could locate prey without recourse to optical, chemical, or mechanical stimuli and that the effective stimulus was the electric field generated by the living prey (*Pleuronectes platessa*). The critical evidence was that sharks and rays would attack an electrode buried under sand when a current, similar to that generated by a living plaice, was provided. In additional tests, Kalmijn observed that the predators ignored a piece of fresh fish exposed on the sand surface, but responded to a current from electrodes buried nearby. After digging at the electrodes, the predators swam away without responding to the exposed bait. Kalmijn surmised that electric fields of prey act as a stronger directive force for some fishes, during food search, than do visual or chemical stimuli. Similar experiments (summarized in Kalmijn, 1974) confirmed that species from at least six families are capable of locating prey through passive detection of only the electric fields prey produce.

3. MECHANORECEPTION

There is considerable evidence to indicate that fishes use stimulus cues of sound and turbulence to locate food. Winn (1964) noted that recorded swimming sounds of small forage-fish, *Anchiovella*, produced a general increase in activity of the predator *Caranx latus*. Busnel (cited in Protasov, 1973) commented that fishermen of Senegal and Nigeria attract predatory fish in turbid rivers by using lures which, according to acoustic analyses, produce unique copies of the feeding sounds of the herbivorous fish that serve as prey. The most rigorous studies to date concern the responses of sharks to sound. Field observations suggested that sharks utilize sound to locate struggling fish. Recordings of low-frequency, pulsed sounds, similar to those emitted by struggling fish, attracted large sharks (Carcharhinidae, Sphyrnidae) during playbacks in the natural environment (Nelson and Gruber, 1963). Banner (1972) tested the response of lemon sharks (*N. breviros-tris*) in the field to playbacks of sounds associated with jumping, feeding, and vocalization of fishes and invertebrates normally included in the sharks' diet. The experimental results confirm that sound alone is potentially useful in alerting sharks to the presence and location of prey.

At limited distances, water turbulence produced by prey serves as potential stimuli for the mechanoreceptors of the lateral line system. Wunder (1936) observed that blinded pike (*Esox lucius*) could accurately snap up moving "prey" at distances of less than 10 cm and he suggested that prey turbulence stimulated the canal organs of the head region, thus allowing accurate orientation. Recent studies (Schwartz and Hasler, 1966a,b; Schwartz, 1971), although limited to surface oriented fishes, indicate that at least thirteen species belonging to four families respond spontaneously to locate precisely the source of surface disturbances similar to those created by trapped insects.

4. CHEMORECEPTION

Hara (1971) has reviewed various anatomical and physiological aspects of chemoreception in fishes. A number of laboratory studies have shown that extracts of appropriate invertebrate, vertebrate, or plant food items are effective in stimulating increased exploratory behavior by fishes as diverse as sharks, salmon and cods (Brawn, 1969;

McBride *et al.*, 1962; Tester, 1963). Laboratory and field studies indicate that by following chemical stimuli, fishes may orient precisely to locate food items. Carr *et al.* (1976) observed that prey extracts, delivered into a tank of pinfish (*Lagodon rhomboides*) through a perforated rubber ball, stimulated the fish to increase swimming movements, orient to the ball, and grasp it with their snouts. Catfish (*Ictalurus* spp.) can locate distant chemical cues by taste alone since a facial taste system operates in accurate localization of stimuli by bilaterally steering the trunk musculature (Atema, 1971). In field trials, a number of natural baits and pure compounds attracted winter flounder (*Pseudopleuronectes* sp.), mummichog (*Fundulus* sp.), and Atlantic silversides (*Menidia* sp.) to a release zone (Sutterlin, 1975). Sutterlin suggested that odors in extracts of organisms are similar to those released by intact organisms and enable fish to locate either individual prey or large concentrations of prey. Finally there is evidence that chemical cues may function as either primary or essential stimuli during food search. Moray eels (*Gymnothorax* spp.) respond to chemical extracts of fish (Bardach *et al.*, 1959) and visually impaired morays locate food as quickly as fish with sight intact. By contrast, morays with plugged nares take much longer to locate food. Dogfish (*Mustilus canis*) fail to recognize and locate food substances such as crabmeat when olfactory capsules are occluded with cotton (Sheldon, cited in Hara, 1971).

B. The Nature of Effective Stimuli

Location and identification of a stimulus in space are two basic processes required to produce orientation of predators to prey. The probability of detection and the process of identification are dependent on specific stimulus features of food objects (e.g., movement, size, shape). Once a food item has been detected and identified, its specific features determine the subsequent motor responses of the predator, which can either be to approach, ignore, or actively avoid a particular object (Ewert, 1970, 1974). Every sensory system is adapted to respond to certain kinds of stimuli, thus differential exploitation of foods by fishes will often be an outcome of the process of sensory discrimination, that is, fishes will "selectively" exploit some food items if they react differently to emitted stimuli. For example, the plaice, which is primarily a mollusk eater, is apparently unsuited to hunting for shrimp because it does not respond to the stimuli they present; however, flounder and dab are highly responsive to stimuli generated by shrimp and exploit large numbers of these prey (De Groot, 1969). To understand these differences and to gain further in-

sight into the problems fish face during food acquisition, it is essential
to analyze the key stimuli they use to detect and identify food.

Of all the sensory modes possessed by fishes, we currently understand the nature of effective stimuli best with respect to visual
analysis. Most vertebrates can distinguish five visual properties of an
object: size, form, contrast, motion, and color.

1. STIMULUS SIZE (STIMULUS AMPLITUDE)

The size of biologically important objects influences the probability of detection, by increasing the distance at which fish respond.
Reactive distance should vary as the square root of the prey surface
area (or stimulus emitting region) because light in water is reduced in
proportion to the inverse square of the distance from the source. Many
workers have confirmed that reactive distance of fish to prey increases
in a predictable fashion with prey size (Protasov, 1970; Ware, 1973;
Confer and Blades, 1975; Hyatt, 1978) and this will have significant
implications for patterns of prey exploitation (Section IV,D). For any
predator that uses vision in the search for prey, there will be an upper
and lower limit on the sizes of prey the predator will detect or if
detected, respond to positively (Ewert, 1970, 1974). Braum (1967)
noted that the predatory responses of early larval fish are triggered by
objects up to a maximum size of 1 mm², but the best information to
date concerning orientation by fishes to optimal stimulus sizes comes
from the studies of Protasov (1970). By using models of prey, he not
only demonstrated an optimal stimulus size to elicit approach by fish,
but also revealed that significant differences exist between species
(Fig. 1). The lower size limit of discovery and attraction to prey may

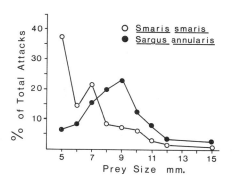

Fig. 1. The relationship between the proportion of attacks initiated during a 10 min
interval by 10–15 cm pickerel (*Smaris smaris*) and by 7–9 cm gilthead (*Sargus annularis*) on prey models of variable size. (Adapted from Protasov, 1970.)

often be determined by the visual acuity of predators while the upper size limit is more likely related to the positive identification of an object as potential food. It is most likely that differences between fish species are innate and that they are directly related to other characteristics of the predators, such as their capacity to capture and ingest prey of particular sizes.

There are fewer quantitative studies documenting the effects of stimulus magnitude on effective search and identification of prey by fishes that rely primarily on nonvisual sensory modes. However, studies by Banner (1972) demonstrate that the reactive distance of sharks is greater to natural auditory stimuli of large amplitude than those of lesser amplitude. Carr *et al.* (1976) presented good evidence that the detection and attack frequency initiated by pinfish on a rubber bulb, diffusing prey extract, was related in a linear fashion to the concentration of the extract.

2. Movement (Stimulus Oscillation)

To visual predators, movement is often a particularly critical stimulus for detection and recognition of prey (Ewert, 1974; Smith, 1976). Because of the poor image transmitting properties of water, high visual acuity can probably only be achieved by fish in a relatively close field; thus, movement perception will be one of the main functions of the visual systems in a large number of fishes (Woodhead, 1966). For some species, movement is essential for successful prey detection. Larval whitefish (*Coregonus wartmani*) will attack stationary or moving prey but pike (*Esox lucius*) will attack only moving prey (Meesters, 1940; Braum, 1967). Confer and Blades (1975) reported that although reactive distance of sunfish varies with size of zooplankton prey, there are significant differences due to movement between species of the same size. For example, *Mesocyclops edax* is detected at a distance of approximately 30 cm, but *Diaptomus sicilis* adults of about the same size, color, and contrast, are detected at a distance of only 12 cm. Ware (1973) observed that trout approach moving targets, regardless of size, from further away than stationary objects with identical properties; he concluded that the contribution of motion to prey detection is additive and a constant. Other studies reveal that movement as a stimulus does not act simply as a constant. Experiments by Protasov (1970) indicate that piscivorous fish react best to baits of high velocity (greater than 5 cm/sec) and that invertebrate eaters react best to baits of intermediate velocity (5 cm/sec). Oscillatory and rotatory target movements are very poor stimuli for goldfish, but unidirectional motion is a powerfully attractive stimulus (Volkmann, 1975). My own studies (unpublished results) on the reactive distance of kokanee (*On-corhynchus nerka*) confirm that the quality of movement is important in detecting prey. Tests with a variety of moving, natural, prey indicate that kokanee respond at the greatest distance to intermediate sized prey (6–10 mm) which exhibit a combination of high linear velocity with pronounced vertical and lateral displacements.

The importance of movement as a stimulus is apparently not restricted to visual predators. In the electrosensitive *Torpedo marmorata*, detection and attacks on prey occur even when the eyes are covered; however, attacks occur only if the prey or lure is moving (Belbenoit and Bauer, 1972). There are no studies of the effect on fish of moving sound stimuli during location and detection of prey, but pulsating auditory stimuli (compared with monotonous ones) act in an analogous fashion to moving and stationary visual stimuli by attracting more fish (Banner, 1972).

3. SHAPE

Many fishes are capable of precise discrimination of a wide range of shapes (Herter, cited in Blaxter, 1970) and some data indicate that this ability will be critical in food exploitation by certain fishes. Shape appears to be an important factor in the recognition of prey versus conspecifics by the piscivorous piranha (*Serrasalmus nattereri*). Markl's (1972) studies suggest that piranhas recognize other piranhas by entirely visual cues and particularly by the characteristically egg-shaped body. Models with a length–height ratio between approximately 1.5 and 2.5 are treated like piranhas and are not attacked. A dark eye and protruding fins make the models even more effective in limiting attacks by piranhas.

4. COLOR AND CONTRAST

Light reflected from the surface of food exhibits a wide range of colors (the property of reflecting light waves of a particular length) and contrast (the difference between the wavelengths reflected from the food items compared to those reflected by the background). Color and contrast differences between otherwise identical food items will frequently alter their probability of consumption by fishes.

Smallmouth bass (*Micropterus* sp.), freshly captured from the field, were attracted in order of decreasing effect to red, yellow, white, green, blue, and black targets (Brown, 1937). Rainbow trout (*Salmo* sp.) which were offered combinations of trout eggs in two colors, against a pale greenish background, consumed blue, red, black, orange, brown, yellow, and green eggs in order of decreasing amounts (Ginetz and Larkin, 1973). This order largely reflected the degree of contrast with background color; but in addition there were true "preferences" for certain colors. For example, consumption of blue eggs on a matched blue background was higher than the consumption of either yellow or red eggs on a blue background. Other studies (Protasov, 1970) have established that significant differences in selection of baits, distinguished only by color, exist among a variety of fishes. Protasov claimed that color selection by various species is a reflection of the colors of natural foods, but close examination of data from a range of studies indicates that the association is not very precise. These studies do indicate that species-specific preferences for color exist and that color stimuli may often play a significant role in biasing the food exploitation patterns.

Given that natural foods are often cryptically colored with respect to background, contrast differences may be particularly important in influencing the probability of fish in detecting food items. Fish are much more effective at locating and exploiting high contrast as compared to low contrast food items (Sumner, 1934; Popham, 1942; Ware, 1973). Greze (1964), using a scale based on the transparency of organisms, observed that a 50% transparent (high contrast) planktonic prey was subject to eight times greater predation effort than one that was 100% transparent (low contrast). Finally, the reactive distances, and therefore the probability of prey detection by rainbow trout, exhibited larger values for high contrast than low contrast, but otherwise identical food targets (Ware, 1973).

Differences of all stimuli emitted by food items compared to those emitted by the background will be an important aspect of effective isolation and identification of food by fishes. For instance, experiments reported by Hawkins and Chapman (1975) confirm that for cod (*Gadus morhua*) a pure tone auditory stimulus is most difficult to detect when components of the background noise are centered at the same or immediately adjacent frequencies as the test tone. The difficulty of detection declines with increasing separation between signal and noise.

5. FURTHER COMMENTS ON EFFECTIVE STIMULI

Many aspects of effective stimuli associated with nonvisual sensory modes are poorly understood but clearly play a significant role in determining response. Of seventeen amino acids tested for attraction of winter flounder (*Pseudopleuronectes americanus*) to a field release location, glycine was by far the most effective. Amines and amino alcohols were ineffective stimuli (Sutterlin, 1975). Unlike flounder, other fishes (*Fundulus* sp. and *Menidia* sp.) were attracted to the release center more effectively by L-alanine or L-histidine than by glycine. The responses of pinfish (*Lagodon rhomboides*) to extracts prepared in identical fashion from various potential prey revealed that considerable differences in potency exist; the concentration of pink shrimp extract required to elicit a set level of 150 attacks from identical groups of pinfish was less than one-tenth the concentration of extract required to elicit the same number of attacks on a target by using extracts of clam, oyster, sea urchin, or whelk (Carr *et al.*, 1976). The major stimulants in five extracts were substances of less than 10,000 molecular weight. Other studies indicate that only α-amino acids are highly stimulatory, L-isomers are always more stimulatory than D-isomers, and stimulatory effectiveness is not directly related to the essential amino acids. Hara *et al.* (1973) suggest it is likely that certain free amino acids or their appropriate mixtures play a fundamental role in differential exploitation of prey; however, with few exceptions (see summary from Johannes and Webb, cited in Bardach, 1975), we currently have little idea of the concentration or identity of amino acids that may be given off by potential foods of fish.

We also know little about the limits of performance by mechanoreceptors or electroreceptors in discrimination of stimulus properties such as size, shape, or movement; however, it is known that the effectiveness of electrical and auditory stimuli is variable (Banner, 1972).

C. Search Procedures and Prey Identity

Predators not only attend to a limited portion of the total stimulus spectrum during food search, but also possess a limited repertoire of search procedures. De Ruiter (1967) classifies search behavior into locomotion, scanning via the sense organs, and special search movements (e.g., turning over leaves by blackbirds). It is reasonable to assume that search behavior, like other aspects of feeding behavior, is adaptive in character, that species-specific search procedures are integrated with particular sensory capacities, and that these have responded over evolutionary time to the identity, distribution, and abundance of prey. Quantitative studies of search patterns used by fishes to locate prey have not reached the sophisticated level of analysis attained in studies of food search by insects and birds (Smith, 1974a,b). However, a number of inferences about species-specific rules for scanning and locomotion procedures of fishes are available from a variety of descriptive studies.

Species-specific search patterns may be initiated upon reception of particular stimuli. Hodgson and Mathewson (1971) conducted a series of experiments on nurse shark (*Ginglymostoma cirratum*) and lemon shark (*Negaprion brevirostris*) responses to chemical stimuli. The experiments indicate that nurse sharks can rely completely upon true gradient searching to locate the source of chemical stimuli; as soon as the stimulus reaches the position of a resting shark, it begins a to-and-fro movement of the head and follows an S-shaped path toward the point of stimulus introduction. By contrast, lemon sharks, after receiving chemical stimuli, immediately orient to the strongest local current and swim rapidly upstream. Lemon sharks are pursuit predators that live largely on a diet of fish, and this response would favor successful location of mobile prey that may change their positions quickly. Nurse sharks are scavengers (Hiatt and Strasburg, 1960), often feeding on stationary or slow moving prey located on the bottom; thus, the slow process of gradient search integrates reasonably well with their food requirements. Therefore, there appears to be a match between the kinds of prey exploited and specific search procedures by fish.

Tyler (1972), in a study of food resource division among thirteen species of demersal fishes, commented that prey species were generally divided among predator groups according to whether they could be caught over, on, or within bottom sediments. Particular techniques of search may well serve as the mechanism driving these patterns of food resource division. In a field study, Hyatt (1978) discovered consistent species-specific differences in the search positions that rainbow trout (*S. gairdneri*) and kokanee (*O. nerka*) maintain in relation to the bottom sediments (Fig. 2). Rainbow trout consistently searched for benthic prey from positions that were further away from the sediment surface than kokanee. Laboratory experiments confirmed that these predators respond at the same distance to various size classes of prey, and because of the relationship between reactive distance and prey size, this means that on average kokanee will detect benthic prey of considerably smaller sizes than trout will. The second basic difference in search techniques was that trout only respond to prey (or preylike objects) that are exposed on the sediment surface but kokanee, moving along at the sediment interface, continuously grab and expel mouthfuls of debris. Undoubtedly the latter technique has some potential for discovery of concealed prey types that trout will seldom detect. Similar observations of behavioral differences in how fishes search for prey on a variety of substrates have been compiled by various authors (Hiatt and Strasburg, 1960; Johannes and Larkin, 1961; Keast and Webb, 1966; Sutterlin, 1975). Johannes and Larkin ob-

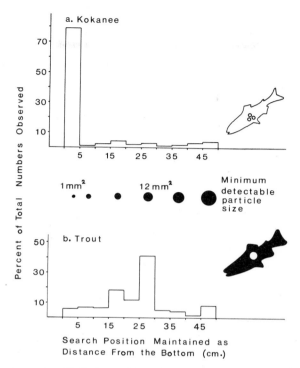

Fig. 2. The proportion of (a) kokanee (*O. nerka*) and (b) trout (*S. gairdneri*) that maintain specific positions while visually searching the bottom sediments for prey. The sequence of solid spheres indicates the relative size of the smallest prey that trout or kokanee can detect from a particular position. The distributions of search positions are based upon direct observations conducted on numerous occasions during the summer months. The estimates of minimum detectable size are based upon laboratory measurements of reactive distance for both trout and kokanee. (From Hyatt, 1978.)

served that shiners (*Richardsonius balteatus*) pursue prey deeper into weed beds and graze one area more thoroughly than trout, that is, like kokanee, shiners are area intensive searchers; trout, by comparison, are area extensive searchers. Grass carp (*C. idella*) in aquaria will locate invertebrates if they are exposed but will never disturb stones under which these prey take refuge (Edwards, 1973). Even for species that do search for prey by disturbing bottom substrates, substantial differences in search procedure may exist. Nikolsky (1963) summarized the depth of substrate penetration achieved by different fishes during prey search (Fig. 3a). Because of the regular patterns in the vertical distribution of various prey types in sediments, these species-specific abilities will lead to substantial differences in the kinds of

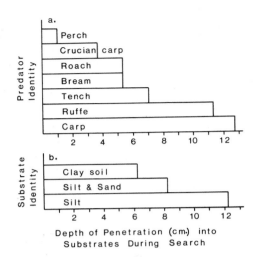

Fig. 3. (a) The depth of penetration into a layer of silt by a variety of fishes engaged in searching for benthic invertebrates. (b) The depth of penetration by carp (*Cyprinus carpio*) into a variety of substrates during the search for food. (Redrawn from Nikolsky, 1963.)

food obtained by different fishes. The type of substrate will also influence the depth of penetration achieved by any single species such as carp (Fig. 3b). Under some conditions a major food resource may go unexploited because of the absence of a species of fish with an appropriate search technique. Munro (1967) observed that the oligochaete, *Branchiura sowerbyi*, forms over 50% of the benthic biomass in Lake McIlwaine, Rhodesia but is not utilized as a source of food by any of the numerous species of fish that are present. The most likely explanation is that none of these species can effectively penetrate the benthic substrates to exploit this food source.

D. Approach, Pursuit, and Attack

The relative vulnerability of a single food item to different fish species or of a variety of food items to a single species is often determined by the match or mismatch of predator and prey characteristics during the approach, pursuit, and attack phase.

Swimming abilities of fishes integrate with morphological characteristics (Section III,E) to determine the probability of successful capture of prey. Fish can cruise at two to three body lengths per second for long periods of time but because the time spent in attack is very short, it is more likely that burst speeds will influence attack success. Burst speeds do appear to be related to the acquisition of particular classes of foods by predatory fishes. With some notable exceptions, fishes capable of high burst speeds regularly capture highly mobile prey such as other fishes while those that possess much lower burst speeds make up the bulk of their diets with less mobile prey (Table II).

Given the variety of body form and accompanying hydrodynamic properties, no general rules emerge with respect to body size and attack procedures by fishes. Different forms adopt different procedures. Fryer and Iles (1972) discussed the body forms and dietary habits of two lake-dwelling cichlids. *Tilapia galilaea* is an inshore dwelling species which feeds mostly on microscopic plants and detritus. *Rhamphochromis longiceps* is an open water piscivore. The collection of small particles by *T. galilaea* requires delicate adjustment of position and can be performed without recourse to high speeds. Consequently *T. galilaea* has a deep body plus long dorsal and anal fins which enable it to maintain stability while collecting small particles from the bottom. The capture of fish calls for marked swimming abilities unless stealth or camouflage are used. *R. longiceps* is torpedolike in form with small dorsal and anal fins set well back "like the feathers of an arrow." Keast and Webb (1966) provide remarkably similar reflections on the body form of the plankton-feeding golden shiner (*Notemigonus crysoleucas*) compared with the highly piscivorous pike (*E. lucius*). Thus different body form and approach techniques will promote divergence of dietary habits while similar body form and approach techniques may promote convergence of dietary habits.

Table II
Food Habits and Swimming Capabilities of Some Fishes[a]

Predator identity	Major food organisms	Burst speeds (body lengths per second)
Tuna (*Euthynnus* spp.)	Small fish, squid, and crustacea	6–21
Salmonids (*Salmo* sp., *Oncorhynchus* sp.)	Small fish, squid, and crustacea	7–13
Herring (*Clupea harengus*)	Zooplankton	6–7
Flatfish (*Pleuronectes* sp., *Platichthys* sp., *Pseudopleuronectes* sp.)	Crustacea, mollusks, and other benthic invertebrates	3–5
Carp (*Cyprinus carpio*)	Omnivorous but eats mainly benthic invertebrates	12

[a] From various sources summarized in Blaxter (1969).

COMMENTS ON PREY VULNERABILITY TO APPROACH AND ATTACK

Prey have many characteristics that serve as adaptations to uncouple or disintegrate an effective approach and attack by predators. Many invertebrates reduce their effective size by crouching close to the substrate as a predacious fish approaches. Others become immobile after an attack and rejection by predators, thus reducing the probability of a subsequent and often lethal attack (Hyatt, unpublished observations). Predacious fish frequently become disoriented during attacks on fish in schools by the unpredictable nature of the stimuli created by many fishes simultaneously moving in different directions (Neil and Cullen, 1974).

Differences in antipredator responses both between and within prey species alter their relative probabilities for inclusion in the diets of specific predators. Escape responses differ dramatically between different species of fish (Nursall, 1973) and zooplankton (Singarajah, 1975) and will account for some differences in the relative vulnerability of these prey to fish. Genetically determined and predator specific behaviors result in different vulnerabilities for single prey species from different populations (McPhail, 1969; Seghers, 1973).

E. Morphology, Handling, and Ingestion

Just as fishes cannot successfully pursue and capture every prey they detect, they will not successfully handle and ingest every prey they capture. Differences in the size, shape, position, and mechanics of jaws, mouth, branchial arches, and dentition will have important consequences for patterns of food acquisition.

1. MOUTH AND JAWS

Because a large mouth appears to be an excellent way to increase the effective density of available prey, we may question why all fish do not have large mouths. One answer is that the selective pressures acting to favor particular shapes, positions, or sizes of mouth often act in an antagonistic fashion.

To process a wide range of food sizes, mouths should evolve toward a maximum gape; however, the head of the fish serves as the bow of a frequently streamlined body and will be under selection for a shape that is hydrodynamically acceptable. In some cases one selective pressure has been much more important than the others resulting in examples of mutually exclusive design that have important implications for food acquisition. The sculpins (Cottidae) of the north temperate zone represent one extreme and have apparently been selected for the ability to process a very wide range of prey sizes since they possess an extremely wide mouth gape. The gape of the Pacific staghorn sculpin (*Leptocottus armatus*) is almost equal to one-fifth of its standard length (Fig. 4). In close association with this, sculpins have adopted a benthic mode of life, a "sit and wait" search procedure, and an approach to prey that often employs stealth. Their diet usually consists of other fishes and a wide range of benthic invertebrates. By way of contrast, the freshwater form of the sockeye salmon (*Oncorhynchus nerka*) is a relatively specialized zooplankton feeder with a mouth size closer to one-twentieth standard length and a reasonably streamlined body (Fig. 4). The "trade-offs" between handling a diversity of prey types in one location and searching for uniformly small and often patchily distributed prey types in many locations are apparent.

Most fish species fall between these extremes in terms of mouth gape. All fourteen of the freshwater species examined by Keast and Webb (1966) displayed mouth dimensions well within the range ex-

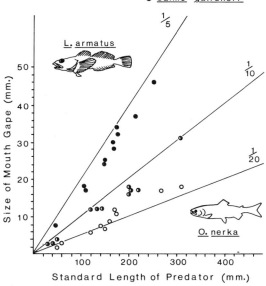

Fig. 4. The relationship between the standard length and the width of mouth gape in Pacific sculpin (*Leptocottus armatus*), rainbow trout (*Salmo gairdneri*), and kokanee (*Oncorhynchus nerka*). Mouth gape is taken as the horizontal distance between the posterior tips of the premaxillaries. The three oblique lines indicate the positions of mouth gapes equal to one-fifth, one-tenth, and one-twentieth of the standard lengths. (From Hyatt, unpublished observations.)

hibited by *Oncorhynchus* and *Leptocottus,* and even the plankton feeding whale shark (*Rhineodon typus*) has a gape of only 10–20% of its length (Gudger, 1941).

Keast and Webb suggest that the mouth of fishes should be studied because its structure dictates the size and type of prey that can be handled. The effect of mouth size and structure on food intake will vary between species of predators as well as between species of prey consumed by one predator. Hartman (1958) compared the maximum dimensions in cross section of various prey accepted by juvenile rainbow trout (*S. gairdneri*) with different sizes of gapes. Rainbow trout would accept as prey, fish with a head depth slightly less than the dimensions of the predators gape; however, the maximum width of stonefly accepted as prey was considerably less than the predator's jaw width (Fig. 5a). One of the critical differences between these prey was

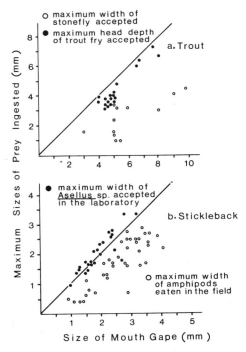

Fig. 5. (a) The relationship between the size of mouth gape of juvenile rainbow trout (*Salmo gairdneri*) and measures of the maximum sizes of prey that they will voluntarily ingest. (Redrawn from Hartman, 1958.) (b) The relationship between the size of mouth gape of threespine sticklebacks (*Gasterosteus aculeatus*) and measures of the maximum sizes of prey (isopods and amphipods) they will ingest under laboratory and field conditions. Measurements of maximum prey width were taken at the fifth thoracic segment. (Redrawn from Burko, 1975.)

that stoneflies used their legs to latch onto the edges of the predators jaws, thus making ingestion extremely difficult. There will also be differences between the maximum prey sizes that fishes are capable of ingesting and the sizes that they will usually ingest under natural conditions. Laboratory experiments (Burko, 1975) indicate that the threespine stickleback (*G. aculeatus*) can consume isopods (*Asellus* sp.) whose maximum body width is very nearly equal to the predators jaw width, but in the field, where a wide range of isopod and amphipod sizes (*Asellus* sp. and *Hyalella* sp.) was available, sticklebacks generally ingested prey that were significantly smaller than the maximum sizes they were capable of handling (Fig. 5b). This may well be due to the relation between prey size and handling time necessary for sticklebacks to ingest prey successfully. Burko demonstrated that at prey sizes above 0.8 of the maximum, handling time increased from less than a minute to times as long as 6–7 min. Given a choice of prey sizes in the field, sticklebacks probably give up more frequently on prey that take long intervals to process.

Mouth shape and position have the potential to influence the quantity and quality of prey that may be ingested but there are few studies to define how important these differences are for different species. The bluntnose minnow (*Pimephales notatus*) has a tubular ventroterminal mouth, and aquarium observations indicate that it is quite unable to get its mouth into the position necessary to pick up food items from the surface (Keast and Webb, 1966).

2. DENTITION

Fishes exhibit a spectacular diversity of dentition and although it is usually assumed that tooth structure and arrangement are intimately related to the nature of the diet, enquiries into the functional aspects of dentition and food processing have been limited in scope and largely descriptive.

At the most fundamental level, it is generally true that the more active predators have strong jaws with sharp teeth, and that teeth on the edge of the jaws serve to bite and catch prey while those on the walls of the pharynx either prevent escape or aid in additional processing during food ingestion. Often a wide range of dental development is associated in a predictable way with the types of foods eaten. In the snapper family (Lutjanidae) the piscivores have well-developed canine dentition, mollusk and crustacean feeders possess short, heavy dentition, and zooplankton feeders have the least developed dentition (Starck and Schroeder, 1970).

Inferences about the critical roles that dentition plays in food processing may be based upon: associations of unique dental morphologies and highly specialized diets, convergence of dental form and dietary habits, developmental changes in dentition accompanied by dietary shifts.

The uniqueness of some dental patterns in association with a highly specialized diet is often the best evidence available to indicate the critical role of particular types of dentition in feeding. All members of the scale-eating cichlids of Lake Tanganyika, Africa possess variously expanded jaw teeth arranged in a single row, remarkably recurved, broadened, coiled, or obtusely pointed in a fashion that doubtless facilitates the removal of scales from other fishes. Similarly specialized tooth shapes have not been encountered in any other known cichlid species (Liem and Stewart, 1976). This does not mean that the same kind of dentition will inevitably arise as the solution to a single problem. Three distinct evolutionary trends in dentition among families of reef fish have developed in association with a diet of live coral (Hiatt and Strasburg, 1960). Butterfly fish (Chaetodontidae) have very small terminal mouths containing fine, incisiform, protruding teeth for biting off individual polyps. Parrot-fish (Scaridae) have a strong, protruding beak of fused teeth with which they can take polyps by heavily scraping coral heads, and the trigger fish (Balistidae) possess very strong, heavy, protruding teeth to break off and ingest the ends of ramose and cespitose coral heads.

Morphological, behavioral, or physiological characteristics that exhibit strong convergence between species in dissimilar habitats may illuminate constraints imposed by common factors in different environments. In the absence of good experimental evidence, convergent evolution in which different groups of distantly related fishes exhibit strikingly similar dental patterns and diets, is reasonably strong evidence indicating critical function related to food handling. The pharyngeal dentition found in different groups of fishes is a particularly good example. Although food is collected by the mouth and is sometimes dealt with by the jaws and their teeth, the real processing, in some cases amounting to mastication, usually takes place in the throat or pharynx where flattened pharyngeal bones fulfill the purpose. The position of the upper and lower pharyngeal bones at the back of the throat is such that all food must pass between them "as between a pair of millstones" (Fryer and Iles, 1972). The number, size, and structure of teeth planted on the surface of these bones differ according to the type of food most commonly processed.

The upper movable surface of the pharyngeal plate of algal eaters is covered with rows of fine recurved teeth (Fig. 6a). Similar teeth occur on the upper bones and when the two sets slide together, parti-

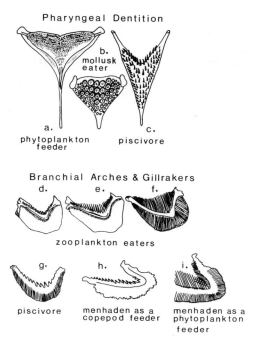

Pharyngeal Dentition

b.
mollusk
eater

a.
phytoplankton
feeder

c.
piscivore

Branchial Arches & Gillrakers
d.   e.   f.

zooplankton eaters

g.   h.   i.

piscivore   menhaden as a   menhaden as a
              copepod feeder   phytoplankton
                                 feeder

Fig. 6. Some characteristic forms of pharyngeal dentition and modified branchial arches in fishes that have a variety of feeding habits. (a, b, and c) Pharyngeal dentition of three African cichlids. (Redrawn from Fryer and Iles, 1972.) (d, e, and f) Anterior gill-arches showing gill-rakers of round whitefish, *Prosopium cylindraceum*, lake whitefish, *Coregonus clupeaformis*, and cisco, *Leucichthys nigripinnis*, respectively. (Redrawn from Koelz, 1929.) (h and i) Anterior gill-arches showing gill-rakers of Atlantic menhaden (*Brevoortia tyrannus*) during the early copepod-eating stages and during the adult phytoplankton-eating stage, respectively. (Redrawn from June and Carlson, 1971.)

cles are raked or combed backward. In fish eaters the pharyngeal bones are considerably different. The lower bone is armed with fewer, sharply pointed and stout teeth that point backward (Fig. 6c). Two sets of these teeth in apposition appear well suited to gripping and forcing backward fishes held in the mouth. Mollusk eaters possess massive pharyngeal bones that have a small number of large, flat-crowned teeth (Fig. 6b) which presumably function to crack the shells of prey, thus exposing the digestible contents. The potential function of these three types of pharyngeal systems is obvious, but there is little information to confirm the advantage these dental adaptations provide compared with a wide range of other species that do not possess the same dental patterns but do consume quantities of algae, fish, and mollusks. Therefore, it is reassuring to observe that these same three extremes of pharyngeal dentition occur among not only the African cichlids but also among the various species of North American Centrarchidae and generally in association with planktivorous, piscivorous, or mollusk eating habits.

Extensive changes in dentition during the developmental history of individual fish species constitutes the final type of evidence in identifying the functional aspects of dentition in handling prey. Copepod-eating, larval menhaden (*Brevoortia tyrannus*) have well-developed recurved dentary and maxillary teeth which, in association with a terminal mouth, appear to be adaptations for the capture and handling of zooplankton. In support of this is the phenomenon that as larger (35–45 mm) Atlantic menhaden shift over to an exclusive phytoplankton diet these teeth are lost (June and Carlson, 1971). On a similar note, Greenwood (1965) found that the heavy mollusk-crushing dentition of the pharyngeal mill in an African cichlid (*Astatoreochromis* sp.) failed to develop under laboratory conditions if fed exclusively on a soft diet. He concluded that the development of massive pharyngeal teeth was induced in part by experience in crushing hard bodied prey at an early age. Nikolsky (1963) noted that similar developmental changes to pharyngeal dentition occur in carp and bream and that such changes are related to developmental shifts in diet.

3. GILL-RAKERS

Convergence of general form and dietary habits, uniqueness of morphological detail, and changes of morphology during development also serve as evidence for modified branchial arches as food processing devices in fishes.

a. Qualitative Assessments. Suyehiro (1942) attributed to Zander (1903 and 1906, cited in Suyehiro, 1942) the often cited position that fish with closely spaced gill-rakers are plankton feeders and those with coarsely spaced rakers are not. So many authors (see Kliewer, 1970, for references) have commented on this association that simply the presence of numerous, long, thin, and closely spaced rakers often results in the designation of a species as a plankton feeder while the absence of well-developed rakers is usually interpreted to mean that a fish must consume larger food items. Certainly all of the fishes that share the unique characteristics of attaining very large body size on a diet of phytoplankton or zooplankton have very elaborate branchial structures which appear to function during food ingestion (e.g., paddlefish, *Polyodon spatula*, Weed, 1925; whale shark, *Rhineodon typus*, Gudger, 1941). The sudden appearance of elaborate gill-raker modifications during development and accompanying dietary shifts in single fish species, is added circumstantial support for an important role of the branchial arches in the collection of specific kinds of food. June and Carlson (1971) discovered that the change from a zooplankton to a phytoplankton diet in young Atlantic menhaden (*B. tyrannus*) is accompanied by an increase in the number, length, and complexity of gill-rakers (compare Fig. 6h to 6i). Similarly the initiation of phytoplankton feeding by the anchovy (*Engraulis anchoita*) is correlated with the formation of elaborate gill-rakers (De Ciechomski, 1967).

However, it is not true that plankton feeding fishes always possess well-developed gill-rakers. Suyehiro (1942) noted that a number of families contain planktivorous species that possess very coarse gill-rakers. Syngnathidae and Fistulariidae have totally degenerated gill-rakers, yet their chief foods are plankton and benthic diatoms. Fryer and Iles (1972) pointed out that the development of a long, tubular mouth by a variety of zooplankton-feeding cichlids (the "utaka" of Lake Malawi, Africa) appears to serve as an adequate substitute for well-developed gill-rakers. *Tilapia melanotheron* feeds, without the benefit of well-developed gill-rakers, on quantities of 50- to 100-μm detritus particles (Pauly, 1976). This may be accomplished with the aid of highly modified pharyngeal dentition which according to Ebeling (1957) is of considerable importance in phytoplankton handling by members of the family Mugilidae.

b. Quantitative Assessments. Although there is undeniably a connection between gill-raker modifications and the trophic status of fishes (compare gill-rakers of plankton feeders to a piscivore, Fig. 6), it is not at all clear to what extent food acquisition is limited by particular morphological arrangements, and we currently understand much less about the quantitative aspects of these "filtering" devices in fishes than in other animal groups such as zooplankton. Many authors (Sverdrup *et al.*, 1942; Young, 1962; Walters, 1966) have assumed that the gill-rakers perform exactly like a sieve in straining food items from the water and thus expect precise relationships to exist between gill-raker numbers, spaces between gill-rakers, gill-raker lengths, and the proportion or size classes of various food types in the diet.

Koelz (1929) observed that various species of whitefishes (Coregonidae) in the Great Lakes of North America exhibit a graded series with respect to number and development of gill-rakers on the branchial arches (Fig. 6d, e, and f); however, the proportion and size ranges of small plankton in the diets of these species are not obviously related to the degree of gill-raker development. Scott and Crossman (1973) report that all three of these whitefish species utilize a variety of invertebrates as food and do not inevitably concentrate on small zooplankton (*Cladocera* and *Copepoda*) or larger invertebrates (mysid shrimp, aquatic insecta) according to the degree of gill-raker development.

In spite of a lack of critical evidence many authors continue to conclude that there will be a precise relation between gill-raker numbers, length, spacing, food type, and food size (Nakamura, 1972; Hutchinson, 1971; Magnuson and Heitz, 1971). Kliewer (1970) conducted a study on a number of lake whitefish (*Coregonus clupeaformis*) populations to assess these aspects. His results indicate that whitefish with shorter gill-rakers ate significantly greater amounts of benthic prey than fish with longer rakers and, somewhat of a surprise, that fish with larger numbers of gill-rakers also ate significantly more benthic food (Fig. 7a). The size of spaces between gill-rakers was not significantly related to the type of food eaten (i.e., benthic or planktonic), but was significantly related to the size of food eaten. Fishes with relatively small spaces between gill-rakers appeared to eat a higher proportion of small food (Fig. 7b). This result must be interpreted cautiously because the correlation was only significant if gill-raker spacing was expressed as a proportion of predator body length. When actual gill-raker space measurements were used there was no significant correlation between food sizes exploited and the sizes of gill-raker spaces. If these fishes handled zooplankton by simply straining them through the gill-rakers, a significant correlation should have occurred between the actual sizes of gill-raker spaces and the absolute sizes of food items; thus, Kliewer concluded that food handling was not a function of simply passive straining by fishes. Magnuson and Heitz (1971) also examined the hypothesis that the proportions of different prey size classes found in the diet of different species of pelagic fishes is related to differences in gill-raker spacing patterns. They found a significant correlation between the proportion of small prey eaten (i.e., Crustacea compared with larger fish and squid) and the mean gill-raker gap, ranked for sixteen size classes of fish (Fig. 7c). These results too must be considered cautiously since the significance of a rank correlation to interpretations of the relation between food size and gill-raker spaces is not immediately apparent, and replotting their data for volume of Crustacea exploited versus the actual mean values of gill-raker spaces indicates that there is no clear relationship. Finally, the conclusion that no precise relation exists between the absolute sizes of gill-raker spaces and sizes of prey ingested

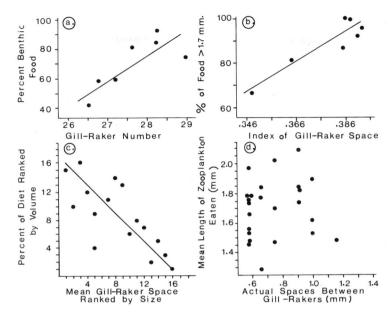

Fig. 7. (a) The relationship between gill-raker number and the proportion of benthic foods eaten by whitefish (*Coregonus clupeaformis*) from seven lakes. (b) The relationship between relative gill-raker space and the proportion of small prey organisms consumed by whitefish from seven lakes. Note that gill-raker spaces are expressed as a percentage of the predators fork length. (Redrawn from Kliewer, 1970.) (c) The relationship between the proportion of the diet (as percentage by volume) made up of small crustaceans (as compared with larger fish and squid) and the mean gill-raker spaces ranked in order for sixteen size classes of Scombrid and Coryphaenid fishes; Kendall rank correlation, $P < 0.001$. (Redrawn from Magnuson and Heitz, 1971.) (d) The relationship between the actual size of spaces between gill-rakers of whitefish (*C. clupeaformis*) and the mean size of zooplankton (*Daphnia* sp.) eaten from a population represented by a diversity of size classes. (Redrawn from Seghers, 1975.)

receives support from the laboratory experiments of Seghers (1975) in which lake whitefish (*C. clupeaformis*) were presented with a mixture of zooplankton (*Daphnia* sp.) sizes. Results from these experiments demonstrated that whitefish are size-selective predators on *Daphnia* sp. but that the selection is not a simple function of gill-rakers acting as a mechanical sieve (Fig. 7d).

The failure to find a precise relationship between gill-raker dimensions and prey size may be because the rakers function as rather "leaky sieves" (see Boyd, 1976, for discussion) or it may be that the "selection" of prey is based on perceptual processes during search, detection, and attack (Galbraith, 1967; Kliewer, 1970; Seghers, 1975). On the basis of studies with both northern anchovy (Leong and O'Connell, 1969) and Pacific mackerel (O'Connell and Zweifel, 1972), it is certain that some predacious fishes have the behavioral flexibility to exploit plankton one at a time or, depending on the density and sizes of prey, to switch over to a mode of operation in which many prey appear to be "filtered" from the water simultaneously.

4. SPECIFIC HANDLING PROCEDURES

Because the majority of fishes engulf and swallow their prey whole, they are not known for complex handling procedures, but some exceptions do occur. Foxx (1972) found that piranha (*Serrasalmus nattereri*) have an attack preference for areas of the body that if damaged will reduce the mobility and increase the vulnerability of prey. In 88% of his trials using piranha and goldfish, the predators attacked the tail area first, often severing the caudal peduncle completely. A similar attack procedure to facilitate handling of large prey is apparently practiced by the tiger fish (*Hydrocyon* spp.) of Africa (Lewis, 1974).

The most elaborate food handling procedures exhibited by fishes are usually observed in association with the ingestion of armored prey. Fricke (1973) summarized his fascinating studies on three families of fishes (Balistidae, Labridae, and Lethrinidae) which each contain species that employ a variety of special techniques to handle and ingest sea urchins, in spite of their spiny armor. The large triggerfish, *Balistes fuscus*, produces a stream of water from its mouth and directs it under a sea urchin to turn it over. It then kills the urchin by biting the oral disc where only a few spines are present. A smaller triggerfish, *Balistopus undulatus*, cannot produce such a strong stream of water with its mouth and consequently it processes urchins by slowly nibbling down the spines until only stumps remain. It then lifts the urchin off the bottom, swims upward with it, drops it and while the urchin is sinking delivers a killing bite to the exposed oral disc. The labrids, *Cheilinus trilobatus* and *Coris angulata*, are particularly remarkable. They push sea urchins over with their heads, bite into the body and then carry the prey to a nearby stone on which they thrash and batter the sea urchin into easily ingestible pieces. These procedures are not only applied to sea urchins. *Balistes fuscus* also uses a water jet to blow crustaceans, marine snails, and mollusks out of the sand.

5. INTEGRATED ASPECTS OF PREDATOR MORPHOLOGY, BEHAVIOR, AND FOOD HANDLING

It is important to remember that a successful food handling procedure for each fish species relies on the integrated action of dentition, structural support elements, and behavior. Small differences between species in any one of these aspects are capable of producing significant differences in the kinds of food processed. For example, the primitive galeoid sand shark (*Carcharias taurus*) feeds only on fishes and invertebrates small enough to swallow whole, while other large galeoid sharks (e.g., *N. brevirostris*) regularly attack and process larger prey. Although these species have similar dentition (long spike-shaped teeth spaced irregularly) for grabbing and holding prey, it appears that the sand shark can not effectively bite and then rotate its body to tear chunks from large prey because it lacks the strong cartilaginous processes that characterize the more advanced jaws of most common galeoids. This cartilage is presumably sufficient to resist the lateral shear acting on the dentition of galeoids that bite and then rotate the body to rip chunks from prey (Springer, 1961). Grass carp (*Ctenopharyngodon idella*) rip pieces from vascular plants by gripping them between the pharyngeal teeth and then following through with violent lateral movements of the body (Cross, 1969). Many other plant-eating fishes exhibit no such behavior and as in the case for sharks, these differences will often bias the animals to process particular kinds of food.

6. COMMENTS ON PREY VULNERABILITY TO HANDLING AND INGESTION

Many prey possess either morphological or behavioral characteristics which increase the probability of rejection during handling by predators. The possession of spines commonly decreases the relative vulnerability of fishes (Hoogland *et al.*, 1957; Mauck and Coble, 1971) and invertebrates (Edmunds, 1974) as compared to similar but spineless counterparts. The ingestion of hard-bodied prey must also present considerable difficulties during the handling stage. Even fishes with specialized dentition for processing mollusks are likely limited in both the quantity and quality of prey that they can obtain since many mollusk species exhibit elaborate morphological characteristics (strong external shell sculpture, elongate or dentate apertures, low spines) that appear to function as antipredator adaptations (Vermeij, 1974). Escape of prey either before or during ingestion is an important problem for fish that feed with the aid of well-developed gill-rakers (Moriarty *et al.*, 1973). Finally palatability of foods must be assessed by fishes during the handling phase of food exploitation. Reliable data concerning the extent to which food palatability is a problem in processing natural foods by fishes are largely lacking. It is clear that many aquatic arthropods and mollusks possess defensive secretions that repel fish at dosages well within the secretory abilities of the prey. A number of experiments indicate that for many fishes chemoreception during the handling stage will play a significant role in determining whether a particular food item is ingested (Sutterlin and Sutterlin, 1970; Atema, 1971).

IV. AN APPRAISAL OF SOME APPROACHES TO ASSESSING PATTERNS OF FOOD EXPLOITATION IN THE NATURAL ENVIRONMENT

A. Qualitative Assessments

The sampling methodology used to reveal patterns of food exploitation by fishes and the process of arriving at legitimate and meaningful interpretations of the biological significance of observed patterns require continuous critical appraisal. The patterns observed and inferences drawn on the basis of such observations are often simply artifacts of the failure to obtain representative data. Sampling procedures that are statistically sound have not been generally adopted to ensure that data truly reflect natural patterns of food acquisition by fishes. In the absence of such procedures, conclusions derived on the basis of extensive but qualitative data are often incorrect. For example, major reference texts (Bigelow and Schroeder, 1953; Scott and Crossman, 1973; Hart, 1973) that serve as sources of information on fish dietary habits often include conclusions that some species of fish are completely opportunistic while others are highly selective. Bigelow and Schroeder concluded that bottom feeding haddock devour all kinds of invertebrates indiscriminately but Zenkevitch (1963) reported that haddock prefer echinoderms (brittle stars and sea urchins) and *Sipunculids,* while feeding to a much lesser extent on mollusks and polychaetes. These authors have arrived at opposite conclusions by relying on data collected with varying degrees of precision. Undoubtedly haddock do consume a wide array of prey. The studies cited by Bigelow and Schroeder report at least 68 species of mollusks, plus unusual fare such as brittle stars and sea cucumbers, as part of the diet. Similarly, Zenkevitch reported that haddock eat a "little bit of everything" including at least 200 forms of benthos, substantial amounts of seaweed, and detritus. Thus, Bigelow and Schroeder carelessly labeled the haddock as an indiscriminate feeder. In drawing his conclusion, Zenkevitch relied on data comparing the abundance of food items that were apparently available in the environment with the abundance of items actually present in the diet of haddock. Because echinoderms made up less than 10% of the bottom fauna but constituted almost 60% of haddock diet, Zenkevitch inferred that echinoderms are the preferred food of haddock. In fact the only legitimate conclusions that the data will support are that the fish exploit a wide variety of food types and that the process of exploitation appears to be density-independent. Many additional examples of weak inference and contradictory conclusions permeate the literature concerning fish feeding habits. In describing the food habits of the goldeye (*Hiodon alosoides*), Scott and Crossman state that "the food is of a great variety, consisting of almost any organism encountered" and that "whatever is most available dominates and there is no indication of any strong food preference." However, later in describing the food of young goldeye, they state the food "is mainly microcrustaceans (Cladocerans and copepods) with minor amounts of insects and other invertebrates." These same authors summarized the dietary habits of northern pike by stating that *E. lucius* "can be classed best simply as omnivorous carnivores in that they eat virtually any living vertebrate available to them within the size range they can engulf." If taken at face value such a conclusion makes it difficult to explain the finding that the proportion of suckers (*Catostomus* sp.) in the diet of pike from Heming Lake, Canada displayed no real increase over an 11-year period in spite of a tenfold increase in the apparent availability of suckers during this same period (Lawler, 1965). Thus, conclusions based on qualitative data must give way to those based on quantitative data.

B. Quantitative Assessments: Indices of Density-Independent Exploitation

In recent years it has become fashionable to conduct studies that compare the abundance of food types that are apparently available in the habitat of a fish to the abundance of items actually present in the diet. These studies have revealed that under most natural circumstances fish will exhibit some degree of density-independent exploitation of foods. A number of authors have suggested indices to express these observations in a quantitative fashion. The "forage ratio" and Ivlev's "electivity index" are the two most commonly encountered in the literature.

1. Forage Ratio

Shorygin (1931) first proposed this as the index of "selection" and defined it as the ratio of the percentage of a given organism in the diet to its percentage in the environment. Hess and Swartz (1941) and Allen (1942) also proposed the use of this index but named it the "forage ratio" and the "availability factor," respectively. I prefer the name forage ratio over the other terms because they incorrectly imply that the mechanisms producing the density-independent patterns of exploitation are known. Therefore

$$\text{forage ratio} = \text{index of selection} = \text{availability factor} = r/p$$

where r, the proportion (as a percentage) of a food item in the diet; p, the proportion of a food item in the environment.

Coche (1967) and Novak and Estes (1974) have recently used this index to characterize the density-independent exploitation of prey by trout (*S. gairdneri*) and sculpins (*Cottus bailieye*), respectively. Values of the ratio can vary between 0 and ∞. The chief disadvantage of the index is that values indicating overexploitation or underexploitation (in a statistical sense) of food groups are distributed asymmetrically with values from 1+ to infinity indicating overexploitation, and values from less than 1 to zero indicating underexploitation of food items. In this respect the "electivity index" of Ivlev (1961) is more useful and has been adopted in the majority of recent studies.

2. IVLEV'S ELECTIVITY INDEX

Ivlev's electivity index (E) is defined as

$$E = \frac{r - p}{r + p}$$

where r, proportion of a food item in the diet; p, proportion of a food item in the environment.

The advantage of the index is that values range from -1 to $+1$, with all values greater than zero indicating overrepresentation of a food item in the diet and all values less than zero indicating underrepresentation of a food item in the diet, compared with its abundance in the environment.

A number of the so-called "overlap indices" are also potentially useful to compare differences in the patterns of food exploitation by fishes with the abundance of foods in the environment or to compare the pattern of food exploitation by one fish species with the pattern exhibited by another (Windell, 1967; Schoener, 1970; Frame, 1974).

C. Problems of Obtaining Biologically Meaningful Data

Many workers have either ignored or failed to identify a number of problems related to the demonstration and interpretation of patterns of food exploitation. These problems are most often mentioned with respect to Ivlev's electivity index but are equally troublesome to the interpretation of any other index that may be used. First, differences in the apparent availability of foods compared with those included in the diet may result from the failure to obtain samples of food from habitats that are representative of those occupied by the predators. This will be especially troublesome in situations where the distribution and abundance of prey types differs dramatically over distances of a few meters; the spatial distribution of the fishes is poorly known and few samples of either fish or prey are taken. This situation frequently occurs for studies of the dietary habits of planktivorous fishes (O'Brien and Vinyard, 1974). Second, differences between the apparent and realized food availabilities may often be no more than a reflection of differences in the probability of "capture" of particular food items by sampling gear or by the fish. Therefore, in order to be certain that a particular species exploits food in a density-independent fashion it is necessary to be confident that the sampling gear used to collect prey will "exploit" the food items in a reasonably density-dependent fashion. Third, different food items may be digested at different rates (Windell, 1967) and thus favor overestimation or underestimation of some types of food. Because many food types possess parts that are highly resistant to digestion, this problem can be partially overcome by careful analysis. In cases where the diet includes a mixture of soft textured foods that are totally digestible and foods that possess digestion-resistant parts, experimental studies of food passage rates should be conducted to provide appropriate correction factors. A fourth and especially serious problem is that patterns of food exploitation are occasionally a function of an arbitrary decision concerning the range of food types which constitute the food complex that is apparently available to a given species. Merrett and Roe (1974) arbitrarily designated the food supply that was apparently available to some mesopelagic fish to consist of only the four groups of prey (calanoid copepods, ostracods, amphipods, and euphasiids) most commonly observed in the diet, but they excluded some abundant groups such as chaetognaths and siphonophores. Similarly, Hutchinson (1971) based electivity values for a planktivorous fish (A. pseudoharengus) upon the abundance of crustacean zooplankton alone because rotifers and phytoplankton were rare in the stomachs of the alewives that he sampled. In these and in many other studies the decision to exclude a potential food type which does not appear in the diet of a single sample of fish does not eliminate the possibility that the predator will exploit the food. For example, Houde (1967) found only a single rotifer in the stomachs of walleye fry (*Stizostedion vitreum vitreum*) even though rotifers numerically dominated the zooplankton of the lake. However, Smith and Moyle (1945) reported that rotifers were an important constituent in the diet of walleye fry from rearing ponds. Similarly, Lam (1974) reported that rabbitfishes (Siganidae), in the field, exploit a relatively narrow range of benthic algae and that they are considered to be primarily herbivorous. However, in captivity *Siganus canaliculatus* feeds readily on all types of food offered and when nonplant foods are present, the fish show no interest in any algae or other plant materials placed in the tank. Thus, siganids are potentially omnivorous even though they often consume a narrow range of plants in nature.

D. Patterns of Food Exploitation in the Natural Environment

In spite of the many problems that may plague efforts to obtain meaningful data, field studies in which many of these problems have been accounted for or eliminated indicate the following.

1. Single species of fish usually exhibit pronounced patterns of density-independent exploitation of foods from the total complex of foods that are apparently available (Fig. 8a, b, d, and e).
2. For each species of fish, the pattern of food exploitation is usually unique even when the predators have been collected simultaneously from the same habitats and potentially have access to an identical set of foods (compare Fig. 8a to b and d to e).
3. Consideration of only limited food groups (e.g., gastropod mollusks), within the total food complex, reveals that density-independent exploitation persists with respect to prey types that have many features in common (Fig. 8c).
4. At very fine levels of discrimination (genus or species), differential exploitation is frequently related to differences in morphology (e.g., lateral plate numbers in the threespine stickleback, Fig. 8f) and especially to differences in the sizes of prey (Fig. 8g, h, and i).
5. Fishes that exploit zooplankton appear to overexploit (in a statistical sense) the largest size classes of prey both within and between species (Fig. 8g).

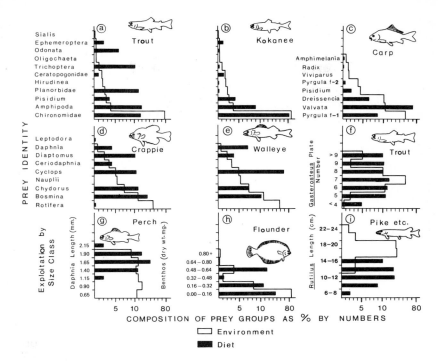

Fig. 8. Patterns of density-independent prey exploitation observed in a variety of field studies. Note the changing scale along the ordinate. (a and b) Trout (*Salmo gairdneri*) and kokanee (*Oncorhynchus nerka*) feeding on a range of benthic invertebrates. (From Hyatt, 1978.) (c) Carp (*Cyprinus carpio*) feeding primarily on gastropod mollusks. (From Stein *et al.*, 1975.) (d and e) Juvenile white crappie (*Pomoxis annularis*) and yellow walleye (*Stizostedion vitreum vitreum*) feeding on crustacean zooplankton. (From Costa and Cummins, 1972.) (f) Rainbow trout (*S. gairdneri*) feeding on threespine stickleback (*Gasterosteus aculeatus*) possessing different numbers of lateral plates. (From Hagen and Gilbertson, 1973.) (g) Yellow perch (*Perca flavescens*) feeding on different size classes of zooplankton, *Daphnia* sp. (From Galbraith, 1967.) (h) Flounder (*Pseudopleuronectes americanus*) feeding on different size classes of benthic invertebrates. (From Levings, 1972.) (i) Pike (*Esox lucius*), zander (*Lucioperca lucioperca*), and sheatfish (*Siluris glanis*) feeding on various size classes of roach (*Rutilus* sp.). [From Popova, 1967, *in* "The Biological Basis of Freshwater Fish Production" (S. D. Gerking, ed.), pp. 359–376. Copyright 1967 by Blackwell.]

6. Piscivores frequently appear to overexploit small or intermediate size classes of forage fishes (Fig. 8i).

E. Problems of Interpretation

Having demonstrated a significant density-independent pattern of food exploitation in a field or laboratory study, an investigator is faced with the problem of interpreting its significance. For example, Costa and Cummins (1972) used their data on the relative abundance of various species of zooplankton and the proportions of these species in the diet of some fish species (Fig. 8d and e) to calculate values of Ivlev's electivity index. On the basis of these values, they concluded that the fishes selected *Daphnia* and *Cyclops* but generally avoided rotifers and copepod nauplii. These authors, along with many others (Houde, 1967; Siefert, 1968; Hutchinson, 1971; Burbidge, 1974; O'Brien and Vinyard, 1974; Repsys *et al.*, 1976), are potentially mistaken in concluding that a large positive value of E indicates "a high degree of selection of a specific item" and that a large negative value indicates "complete avoidance by fish." Confusion is created here by considering the purely statistical phenomenon of density-independent exploitation as critical evidence for the action of only one or two of the biological mechanisms (i.e., behavioral selection or avoidance) which may be operating to produce the pattern. It should be clear from the preceding sections of this article that patterns of "selective" exploitation are a consequence of a large number of predator and prey characteristics that interact at each step of the behavioral chain, from search and detection to manipulation and ingestion of foods. Ivlev (1961) was well aware of this problem and stated that "the selectivity observed will be the result of preference shown by the animal for this or that kind of food only if all the food components are in a state in which the difficulties encountered by the consumer animal in procuring them are absolutely equal." Under both field and laboratory conditions this is virtually never true, thus the phenomenon of density-independent exploitation (electivity) should be considered as a function of several potential mechanisms operating simultaneously. Inferences about the role of these mechanisms may usefully serve as hypotheses which should then be tested by conducting the appropriate experiments. The density-independent exploitation of some food items by size will serve as an example.

A variety of fishes that feed on zooplankton (Lindstrom, 1955; Hrbacek, 1962; Brooks and Dodson, 1965; Galbraith, 1967), phytoplankton (Moriarty *et al.*, 1973), and benthic invertebrates (MacDonald, 1956; Morgan, 1956; Ware, 1973) display an overrepresentation of large size classes of prey in the diet. Although most of these fishes employ vision during the search for food, similar patterns of size-selective, prey exploitation emerge from studies on fishes which have poorly developed vision and which likely use other sensory modes during the search for food (see Section III,A). Repsys *et al.* (1976) observed that catfish (*Ictalurus* sp.) overexploited large planktonic crustacea even though these fish are normally nocturnal and locate prey via chemoreception. Similarly, the elephant snout fish (*Mormyrus kannume*) of Lake Victoria exploits mostly the last instar stages (i.e., the largest sizes) of chironomid larvae. These fish also have poorly developed vision and probably locate prey via electrodetection or chemoreception. Therefore, in spite of a common pattern of exploitation, it is unlikely that a single explanation is suitable for such diverse circumstances. The potential mechanisms producing a common pattern include the following.

1. An increased probability of successful detection and identification of large prey compared with small ones; in cases where the predators do not locate food by visual means, it is likely that prey size is correlated with the magnitude of other kinds of stimuli such as electric fields or auditory stimuli generated by prey (Section III,B)

2. A decreased probability of successful pursuit and attack of small prey due to their more effective escape responses (Section III,D)

3. A decreased probability of successful ingestion of small prey due to limits set by gill-rakers or pharyngeal dentition (Sections III,E,2 and III,E,3)

OTHER EXAMPLES

To some, it may seem a tedious process to eliminate or identify the biological mechanisms actually responsible for the patterns of food exploitation, but unless we understand the underlying mechanisms it will be impossible to predict patterns in many instances or to understand them in others. By way of contrast with the frequently observed overexploitation of large zooplankters, Pieczynski and Prejs (1970) observed that three species of fish (tench, *Tinca tinca;* carp, *Cyprinus carpio;* and Crucian carp, *Carassius carassius*) in Lake Warniak, Poland, exploited a disproportionate number of small water mites (*Hydracarina*). Eighty percent of the mites found in the diets of the fish were less than 1 mm in size although these made up only 40% of the mites present in the environment. Less than 5% of the mites found in the diets were greater than 1.5 mm although these made up more than 25% of the mites present in the environment. The most promising explanation is that all the mites are unpalatable due to the production of noxious chemicals and that fish effectively avoid the big, brightly colored species because they are easy to recognize. Small mites may be included more often in the diet because they are more difficult to recognize and are confused with other prey, they do not produce enough noxious substances to prevent ingestion (Section III,E,6), or the predators ingest them incidentally along with other desirable foods. The latter possibility could depend on whether these fishes feed on planktonic organisms by picking them one at a time or by gulping many of them simultaneously (Section III,E,4). Other studies in which body size was not a successful indicator of zooplankton vulnerability to fish (Zaret, 1972; Nordlie, 1976) emphasize the need to develop precise, experimentally based assessments of the biological mechanisms that control food exploitation by fishes. In many instances experimental assessments will be the only way to obtain a biologically meaningful interpretation. The pattern of stickleback (*G. aculeatus*) exploitation by rainbow trout (*S. gairdneri*) serves as a case in point. Hagen and Gilbertson (1973) observed that trout appeared to over-exploit sticklebacks that had more than eight and less than six lateral plates (Fig. 8f). The pattern was puzzling because if the vulnerability of the prey was determined by body armor at the handling stage, animals with more lateral plates should have represented a steadily decreasing proportion of the fish in the diet of trout. Moodie *et al.* (1973) independently examined predation on sticklebacks with different numbers of plates under laboratory conditions. Their findings indicated that seven-plated sticklebacks were less subject to capture by piscivorous fishes because of particular behavioral responses that enabled them to evade detection and capture by predators. The lateral plates appeared to play little if any role in reducing stickleback vulnerability.

V. SUMMARY

1. Fishes exhibit enormous diversity in the foodstuffs they consume; however, it is possible to classify them broadly into a limited number of feeding types.

2. Classification reveals that in tropical ecosystems there is a more equitable distribution of fishes with respect to the basic types of available forage.

3. Further comparisons of fish communities from tropical and temperate ecosystems indicate that fewer species from temperate ecosystems exhibit the extremes of dietary specialization or generalization attained by many species of tropical fishes.

4. As with most vertebrate predators, the majority of fishes eat mixed diets.

5. Fishes of different ecological groups can have different functional sets of receptors. These are employed in an integrated fashion and there will be a hierarchy of sensory processing which leads to ever finer levels of discrimination during the process of search, detection, identification, and pursuit of prey.

6. Sensory systems of fishes are limited in their capacity to receive information and are biased toward the reception of particular stimuli (effective stimuli) during the search for food. Effective stimuli for visual predators may be defined in terms of size, movement, shape, color, and contrast, and significant differences exist between fish species in the stimulus dimensions of any one of these characteristics that inform fishes of the location or identity of food.

7. The exact nature of effective stimuli associated with nonvisual search for food is poorly understood but selective attention to different stimulus characteristics appears equally important.

8. Differences of response to stimuli will often account for the exploitation of different foods by various fish species.

9. Different stimulus properties of food items will alter their relative probabilities of ingestion by any single fish species.

10. Species-specific search and attack procedures by fishes increase the probability of successfully exploiting particular kinds of prey and at the same time reduce the probability of exploiting others which have different characteristics (e.g., densities, distributions, behaviors).

11. Diverse structural modifications of jaws, mouth, gill-rakers, and dentition in fishes are dramatic testimony to the variety of problems they face in handling and ingesting foods.

12. Members of the families Cottidae and Salmonidae exist at different ends of a range for trophic adaptation. In *Leptocottus armatus* the evolutionary process has favored the development of morphological and behavioral characteristics favoring consumption of a wide range of prey sizes. In *Oncorhynchus nerka* a combination of streamlined form, a small mouth, and the exploitation of a narrow range of prey types has evolved.

13. The critical dimensions of structures controlling the maximum size of food object ingested by fishes varies between species of predator (mouth, esophagus, cleithrum) and between species of prey (head width, body depth, leg span, spine length); however, under natural conditions fishes usually consume food items that are substantially smaller than the maximum sizes they are capable of ingesting.

14. In the absence of adequate experimental results, convergence of general form and dietary habits, uniqueness of morphological detail,

and changes of morphology during development, serve as evidence for the critical involvement of diverse dental patterns and gill-rakers in specific aspects of food processing.

15. Specialized plankton feeders usually possess well-developed gill-rakers; however, substitutes such as long tubular mouths or modified pharyngeal apparatus do allow species without refined gill-rakers to specialize on plankton. Although there is undeniably a connection between gill-raker modifications and the trophic status of fishes, no precise relationships have been demonstrated to occur with respect to food type, food sizes, gill-raker lengths, numbers, or spacing.

16. The majority of fishes do not exhibit complex procedures for handling food and the cases which serve as exceptions usually involve handling of large food items by piscivores or handling prey that are well armored.

17. Although food processing procedures are simple, they rely on the integrated action of dentition, structural support elements, and behavior. Small differences between fishes at this stage can produce large differences in diet.

18. Antipredator adaptations at the handling stage are geared to maximize the probability of food rejection by fishes.

19. Under most natural conditions fishes exhibit some degree of density-independent exploitation of foods. The forage ratio and Ivlev's electivity index have frequently been used to present quantitative expressions of this phenomenon.

20. It is necessary to exercise great care to obtain data that reflect real patterns of food exploitation by fishes. Patterns of density-independent exploitation may be exaggerated or concealed by a failure to obtain representative samples of predators and their food from the field, by different rates of digestion of various food items, and by arbitrary definition of the range of food types that are apparently available.

21. Reliable laboratory and field studies reveal a diversity of patterns of density-independent food exploitation by fishes.

22. Many authors have mistaken the statistical phenomenon of density-independent exploitation for critical evidence of the action of only one or two biological mechanisms (i.e., behavioral selection and avoidance) which may operate to produce the patterns.

23. The example of size-selective prey exploitation is used to suggest that a variety of mechanisms, which may or may not operate in a mutually exclusive fashion, are potentially involved in generating each pattern of food exploitation.

24. Field descriptions of patterns of food exploitation by fishes are most valuable as sources of hypotheses about the biological mechanisms that are involved in determining the limits of food exploitation. However, experimental studies are the major source of critical evidence to test the merit of particular explanations or interpretations of a pattern.

REFERENCES

Ali, M. A. (1959). The ocular structure, retinomotor and photo-behavioural response of juvenile Pacific salmon. *Can. J. Zool.* **37**, 965–996.

Allen, K. R. (1942). Comparisons of bottom faunas as sources of available fish food. *Trans. Am. Fish. Soc.* **71**, 275–283.

Atema, J. (1971). Structures and functions of the sense of taste in catfish (*Ictalurus natalis*). *Brain Behav. Evol.* **4**, 273–294.

Banner, A. (1972). Use of sound in predation by young lemon shark, *Negaprion brevirostris*. *Bull. Mar. Sci.* **22**, 251–283.

Bardach, J. E. (1975). Chemoreception of aquatic animals. *In* "Olfaction and Taste" (D. A. Denton and J. P. Coghlan, eds.), Vol. 5, pp. 121–132. Academic Press, New York.

Bardach, J. E., Winn, H. E., and Menzel, D. W. (1959). The role of senses in the feeding of the nocturnal reef predators *Gymnothorax moringa* and *G. vicinus. Copeia* No. 2, pp. 133–139.

Belbenoit, P., and Bauer, R. (1972). Video recordings of prey capture behaviour and associated electric organ discharge of *Torpedo marmorata* (Chondrichthyes). *Mar. Biol.* **17**, 93–99.

Bigelow, H. B., and Schroeder, W. C. (1953). Fishes of the Gulf of Maine. *U.S. Fish Wildl. Serv., Fish. Bull.* **74**, 577 pp.

Blaxter, J. H. S. (1969). Swimming speeds of fish. *In* "Proceedings of the FAO Conference on Fish Behaviour in Relation to Fishing Techniques and Tactics" (A. Ben-Tuvia and W. Dickson, eds.), FAO Fish. Rep., Vol. 2, No. 62, pp. 69–100. Food and Agricultural Organization, Rome.

Blaxter, J. H. S. (1970). Light. *In* "Marine Ecology" (O. Kinne, ed.), Vol. 1, Part 1, pp. 213–320. Wiley, New York.

Boyd, C. M. (1976). Selection of particle sizes by filter feeding copepods: A plea for reason. *Limnol. Oceanogr.* **21**, 175–180.

Braum, E. (1967). The survival of fish larvae with reference to their feeding behaviour and the food supply. *In* "The Biological Basis of Freshwater Fish Production" (S. D. Gerking, ed.), pp. 113–131. Blackwell, Oxford.

Brawn, V. M. (1969). Feeding behaviour of cod, *Gadus morhua. J. Fish. Res. Board Can.* **26**, 583–596.

Brooks, J. L., and Dodson, S. I. (1965). Predation, body size and composition of plankton. *Science* **150**, 28–35.

Brown, F. A., Jr. (1937). Responses of the large-mouth black bass to colors. *Ill. Nat. Hist. Surv., Bull.* **21**, 37–55.

Bullock, T. H. (1973). Seeing the world through a new sense: Electroreception in fish. *Am. Sci.* **61**, 316–325.

Burbidge, R. G. (1974). Distribution, growth, selective feeding and energy transformations of young of the year blueback herring, *Alosa aestivalis* (Mitchell) in the James River, Virginia. *Trans. Am. Fish. Soc.* 103, 297–311.

Burko, T. (1975). Size-selective predation by the threespined stickleback. M.S. Thesis, Univ. of British Columbia, Vancouver.

Carr, W. E. S., Gondeck, A. R., and Delanoy, R. L. (1976). Chemical stimulation of feeding behaviour in the pinfish, *Lagodon rhomboides:* A new approach to an old problem. *Comp. Biochem. Physiol. A* 54, 161–166.

Coche, A. G. (1967). Production of juvenile steelhead trout in a freshwater impoundment. *Ecol. Monogr.* 37, 201–228.

Confer, J. L., and Blades, P. I. (1975). Omnivorous zooplankton and planktivorous fish. *Limnol. Oceanogr.* 20, 571–579.

Corbet, P. S. (1961). The food of noncichlid fishes in the Lake Victoria basin, with remarks on their evolution and adaptation to lacustrine conditions. *Proc. Zool. Soc. London* 136, 1–101.

Costa, R. R., and Cummins, K. W. (1972). The contribution of *Leptodora* and other zooplankton to the diet of various fish. *Am. Midl. Nat.* 87, 559–564.

Cross, C. G. (1969). Aquatic weed control using grass carp. *J. Fish Biol.* 1, 27–30.

Darnell, R. M. (1961). Trophic spectrum of an estuarine community based on studies of Lake Pontchartrain, Louisiana. *Ecology* 42, 553–568.

De Ciechomski, J. D. (1967). Investigations of food and feeding habits of larvae and juveniles of the Argentine anchovy, *Engraulis anchoita. Calif. Coop. Oceanic Fish. Invest. Rep.* 11, 72—81.

De Groot, S. J. (1969). Digestive system and sensorial factors in relation to feeding behaviour of flatfishes (*Pleuronectiformes*). *J. Cons., Cons. Perm. Int. Explor. Mer.* 32, 385–394.

De Ruiter, L. (1967). The feeding behavior of vertebrates in the natural environment. *In* "Handbook of Physiology" (J. Field, ed.), Vol. 1, Sect. 6, pp. 97–116. Am. Physiol. Soc., Washington, D.C.

Ebeling, A. W. (1957). The dentition of eastern Pacific mullets, with special reference to adaptation and taxonomy. *Copeia* No. 3, pp. 173–185.

Edmunds, M. (1974). "Defense in Animals: A Survey of Antipredator Defences." Longmans, Green, New York.

Edwards, D. J. (1973). Aquarium studies on the consumption of small animals by O-group grass carp, *Ctenopharyngodon idella. J. Fish Biol.* 5, 599–605.

Ewert, J. P. (1970). Neural mechanisms of prey-catching and avoidance behaviour in the toad (*Bufo bufo* L.). *Brain Behav. Evol.* 3, 36–56.

Ewert, J. P. (1974). The neural basis of visually guided behavior. *Sci. Am.* 230, 34–42.

Foxx, R. M. (1972). Attack preferences of the red-bellied piranha (*Serrasalmus nattereri*). *Anim. Behav.* 20, 280–283.

Frame, D. W. (1974). Feeding habits of young winter flounder (*Pseudopleuronectes americanus*): Prey availability and diversity. *Trans. Am. Fish. Soc.* 103, 261–269.

Fricke, H. W. (1973). Behaviour as part of ecological adaptation: *In situ* studies in the coral reef. *Helgol. Wiss. Meeresunters.* 24, 120–144.

Fryer, G. (1959). The trophic interrelationships and ecology of some littoral communities of Lake Nyasa with special reference to the fishes, and a discussion of the evolution of a group of rock-frequenting Cichlidae. *Proc. Zool. Soc. London* 132, 153–281.

Fryer, G., and Iles, T. D. (1972). "The Cichlid Fishes of the Great Lakes of Africa." Oliver & Boyd, Edinburgh.

Galbraith, M. G. (1967). Size selective predation on *Daphnia* by rainbow trout and yellow perch. *Trans. Am. Fish. Soc.* 96, 1–10.

Gilbert, P. W. (1963). The visual apparatus of sharks. *In* "Sharks and Survival" (P. W. Gilbert, ed.), pp. 283–326. Heath, Indianapolis, Indiana.

Ginetz, R. M., and Larkin, P. A. (1973). Choice of colors of food items by rainbow trout (*Salmo gairdneri*). *J. Fish. Res. Board Can.* **30**, 229–234.

Girsa, I. I. (1961). Availability of food animals to some fishes at different conditions of illumination (Russ.). *Tr. Soveshch. Ikhtiol. Kom. Akad. Nauk SSSR* **13**, 355–359.

Greenwood, P. H. (1965). Environmental effects on the pharyngeal mill of a cichlid fish, *Astatoreochromis alluaudi*, and their taxonomic implications. *Proc. Linn. Soc. London* **176**, 1–10.

Greze, V. N. (1964). The determination of transparency among planktonic organisms and its protective significance. *Dokl. Biol. Sci.* **151**, 956–958.

Gudger, E. W. (1941). The food and feeding habits of the whale shark, *Rhineodon typus*. *J. Elisha Mitchell Soc.* **57**, 57–72.

Hagen, D. W., and Gilbertson, L. G. (1973). Selective predation and the intensity of selection acting upon the lateral plates of the threespine stickleback. *Heredity* **30**, 273–287.

Hara, T. J. (1971). Chemoreception. *In* "Fish Physiology" (W. S. Hoar and D. J. Randall, eds.), Vol. 5, pp. 79–120. Academic Press, New York.

Hara, T. J., Carolina Law, Y. M., and Hobden, B. R. (1973). Comparisons of the olfactory response to amino acids in rainbow trout, brook trout and whitefish. *Comp. Biochem. Physiol. A* **45**, 969–977.

Hart, J. L. (1973). Pacific fishes of Canada. *Fish. Res. Board Can., Bull.* No. 180.

Hartley, P. H. T. (1948). Food and feeding relationships in a community of freshwater fishes. *J. Anim. Ecol.* **17**, 1–14.

Hartman, G. F. (1958). Mouth size and food size in young rainbow trout, *Salmo gairdneri*. *Copeia* No. 3, pp. 233–234.

Hawkins, A. D., and Chapman, C. J. (1975). Masked auditory thresholds in the cod, *Gadus morhua* L. *J. Comp. Physiol.* **103**, 209–226.

Hess, A. D., and Swartz, A. (1941). The forage ratio and its use in determining the food grade of streams. *Trans. North Am. Wildl. Conf., 5th* pp. 162–164.

Hiatt, R. W., and Strasburg, D. W. (1960). Ecological relationships of the fish fauna on coral reefs of the Marshall Islands. *Ecol. Monogr.* **30**, 65–127.

Hobson, E. S. (1963). Feeding behaviour in three species of sharks. *Pac. Sci.* **17**, 171–194.

Hodgson, E. S., and Mathewson, R. (1971). Chemosensory orientation in sharks. *Ann. N.Y. Acad. Sci.* **188**, 175–182.

Hoogland, R., Morris, D., and Tinbergen, N. (1957). The spines of sticklebacks (*Gasterosteus* and *Pygosteus*) as means of defense against predators (*Perca* and *Esox*). *Behaviour* **10**, 205–236.

Houde, E. D. (1967). Food of pelagic young of the walleye, *Stizostedion vitreum vitreum*, in Oneida Lake, New York. *Trans. Am. Fish. Soc.* **96**, 17–24.

Hrbacek, J. (1962). Species composition and the amount of zooplankton in relation to fish stock. *Rozpr. Cesk. Akad. Ved, Rada Mat. Prir. Ved* **72**, 1–116.

Hutchinson, B. P. (1971). The effect of fish predation on the zooplankton of ten Adirondack lakes with particular reference to the alewife, *Alosa pseudoharengus*. *Trans. Am. Fish. Soc.* **100**, 323–335.

Hyatt, K. D. (1978). Factors affecting the patterns of prey acquisition by rainbow trout (*Salmo gairdneri*) and kokanee (*Oncorhynchus nerka*) in Marion Lake, British Columbia. Ph.D. Thesis, Univ. of British Columbia, Vancouver.

Ivlev, V. S. (1961). "Experimental Ecology of the Feeding of Fishes." Yale Univ. Press, New Haven, Connecticut.

Johannes, R. E., and Larkin, P. A. (1961). Competition for food between redside shiners (*Richardsonius balteatus*) and rainbow trout (*Salmo gairdneri*) in two B.C. lakes. *J. Fish. Res. Board Can.* **18**, 203–220.

June, F. C., and Carlson, F. T. (1971). Food of young Atlantic menhaden (*Brevoortia tyrannus*) in relation to metamorphosis. *U.S. Fish Wildl. Serv., Fish. Bull.* **68**, 493–512.

Kalmijn, A. J. (1971). The electric sense of sharks and rays. *J. Exp. Biol.* **55**, 371–383.

Kalmijn, A. J. (1974). The detection of electric fields from inanimate and animate sources other than electric organs. *In* "Handbook of Sensory Physiology, III, Electroreceptors and Other Specialized Receptors in Lower Vertebrates" (A. Fessard, ed.), pp. 147–200. Springer-Verlag, Berlin and New York.

Keast, A., and Webb, D. (1966). Mouth and body form relative to the feeding ecology in the fish fauna of a small lake, Lake Opinicon, Ontario. *J. Fish. Res. Board Can.* **23**, 1845–1867.

Kliewer, E. V. (1970). Gill-raker variation and diet in lake whitefish, *Coregonus clupeaformis*, in Northern Manitoba. *In* "Biology of Coregonid Fishes" (C. C. Lindsey and C. S. Woods, eds.), pp. 165–197. Univ. of Manitoba Press, Winnipeg.

Knöppel, H. A. (1970). Food of Central Amazonian fishes. *Amazoniana* **2**, 257–352.

Koelz, W. (1929). Coregonid fishes of the Great Lakes. *Bull. U.S. Fish. Bur.* **43**, 297–643.

Lam, T. J. (1974). Siganids: Their biology and mariculture potential. *Aquaculture* **3**, 325–354.

Larkin, P. A. (1956). Interspecific competition and population control in freshwater fish. *J. Fish. Res. Board Can.* **13**, 327–342.

Lawler, G. H. (1965). The food of pike, *Esox lucius*, in Hemming Lake, Manitoba. *J. Fish. Res. Board Can.* **22**, 1357–1377.

Leong, R. J. H., and O'Connell, C. P. (1969). A laboratory study of particulate and filter feeding of the northern anchovy (*Engraulis mordax*). *J. Fish. Res. Board Can.* **26**, 557–582.

Levings, C. D. (1972). A study of temporal change in a marine benthic community with particular reference to predation by *Pseudopleuronectes americanus*. Ph.D. Thesis, Dalhousie Univ., Halifax, Nova Scotia.

Lewis, D. S. C. (1974). The food and feeding habits of *Hydrocynus forskahlii*, Cuvier and *H. brevis*, Gunther in Lake Kainji, Nigeria. *J. Fish Biol.* **6**, 349–363.

Liem, K. F., and Stewart, D. J. (1976). Evolution of the scale eating fishes of Lake Tanganyika: A generic revision with a description of a new species. *Bull. Mus. Comp. Zool., Harvard Univ.* **147**, 319–350.

Lindstrom, T. (1955). On the relation of fish-size, food-size. *Rep. Inst. Freshwater Res. Drottningholm* **36**, 133–147.

Losey, G. S. (1972). Predation protection in the poison-fang blenny, *Meiacanthus atrodorsalis*, and its mimics *Escenius bicolor* and *Runula laudandus* (Bleniidae). *Pac. Sci.* **26**, 129–139.

Lowe-McConnell, R. H. (1975). "Fish Communities in Tropical Freshwaters." Longmans, Green, New York.

McBride, J. R., Idler, D. R., Jones, E. E., and Tomlinson, M. (1962). Olfactory perception in juvenile salmon. I. Observations on response of juvenile sockeye to extracts of food. *J. Fish. Res. Board Can.* **19**, 327–334.

MacDonald, W. W. (1956). Observations on the biology of chaoborids and chironomids in Lake Victoria and on the feeding habits of the "elephant snout fish" (*Mormyrus kannume*, Forsk.). *J. Anim. Ecol.* **25**, 36–53.

McPhail, J. D. (1969). Predation and evolution of a stickleback (*Gasterosteus*). *J. Fish. Res. Board Can.* **26**, 3183–3208.

Magnuson, J. J., and Heitz, J. G. (1971). Gill-raker apparatus and food selectivity among mackerels, tunas, and dolphins. *U.S. Fish Wildl. Serv., Fish. Bull.* **69**, 361–370.

Markl, V. H. (1972). Aggression and prey capture in piranhas (Serrasalminae, Characidae). *Z. Tierpsychol.* **30**, 190–216.

Marshall, N. B. (1966). "The Life of Fishes." World, New York.

Marshall, N. B. (1971). "Explorations in the Life of Fishes." Harvard Univ. Press, Cambridge, Massachusetts.

Mauck, W. L., and Coble, D. W. (1971). Vulnerability of some fishes to northern pike (*Esox lucius*) predation. *J. Fish. Res. Board Can.* **28**, 957–969.

Meesters, A. (1940). Uber die Organisation des Gesichtsfeldes der Fische. *Z. Tierpsychol.* **4**, 84–149.

Merrett, N. R., and Roe, H. S. J. (1974). Patterns and selectivity in the feeding of certain mesopelagic fishes. *Mar. Biol.* **28**, 115–126.

Moodie, G. E. E., and Lindsey, C. C. (1972). Life history of a unique cyprinid fish, the chiselmouth chub (*Arocheilus alutaceus*) in British Columbia. *Syesis* **5**, 55–61.

Moodie, G. E. E., McPhail, J. D., and Hagen, D. W. (1973). Experimental demonstration of selective predation in *Gasterosteus aculeatus*. *Behaviour* **47**, 95–105.

Morgan, N. C. (1956). The biology of *Leptocerus aterrimus* Steph. with reference to its availability as a food for trout. *J. Anim. Ecol.* **25**, 349–365.

Moriarty, D. J. W., Darlington, J. P. E. C., Dunn, I. G., and Tevlin, M. P. (1973). Feeding and grazing in Lake George, Uganda. *Proc. R. Soc., Ser. B* **184**, 299–319.

Munro, J. L. (1967). The food of a community of East African freshwater fishes. *J. Zool.* **151**, 389–415.

Nakamura, E. L. (1972). Development and use of facilities for studying tuna behaviour. *In* "Behaviour of Marine Animals" (H. E. Winn and B. L. Olla, eds.), Vol. 2, pp. 245–277. Plenum, New York.

Neil, S. R., and Cullen, J. M. (1974). Experiments on whether schooling by their prey affects the hunting behaviour of cephalopods and fish predators. *J. Zool.* **72**, 549–569.

Nelson, D. R., and Gruber, S. H. (1963). Sharks: Attraction by low frequency sounds. *Science* **142**, 975–977.

Nikolsky, G. V. (1963). "The Ecology of Fishes." Academic Press, New York.

Nordlie, F. G. (1976). Plankton communities of three central Florida lakes. *Hydrobiologia* **48**, 65–78.

Norman, J. R., and Greenwood, P. H. (1963). "A History of Fishes." Benn, London.

Novak, J. K., and Estes, R. D. (1974). Summer food habits of the black sculpin (*Cottus baileyi*) in the upper south fork Holston River drainage. *Trans. Am. Fish. Soc.* **103**, 270–276.

Nursall, J. R. (1973). Some behavioural interactions between spottail shiners (*Notropis hudsonius*), yellow perch (*Perca flavescens*), and northern pike (*Esox lucius*). *J. Fish. Res. Board Can.* **30**, 1161–1178.

O'Brien, W. J., and Vinyard, G. L. (1974). Comment on the use of Ivlev's electivity index with planktivorous fish. *J. Fish. Res. Board Can.* **31**, 1427–1429.

O'Connell, C. P., and Zweifel, J. R. (1972). A laboratory study of particulate and filter feeding of the Pacific mackerel, *Scomber japonicus*. *U.S. Fish Wildl. Serv., Fish. Bull.* **70**, 973–978.

Odum, W. (1970). Utilization of the direct grazing and plant detritus food chains by the striped mullet, *Mugil cephalus. In* "Marine Food Chains" (J. H. Steele, ed.), pp. 222–239. Univ. of California Press, Berkeley.

Paloheimo, J. E., and Dickie, L. M. (1966). Food and growth of fishes. III. Relations among food, body size and growth efficiency. *J. Fish. Res. Board Can.* **23**, 1209–1248.

Pauly, D. (1976). The biology, fishery, and potential for aquaculture of *Tilapia melanotheron* in a small West African lagoon. *Aquaculture* **7**, 33–49.

Pieczynski, E., and Prejs, A. (1970). The share of water mites (*Hydracarina*) in the food of three species of fish in Lake Warniak. *Ekol. Pol.* **18**, 445–452.

Popham, E. J. (1942). Further experimental studies of the selective action of predators. *Proc. Zool. Soc. London* **112**, 105–117.

Popova, O. A. (1967). "The predator prey" relationship among fishes (A survey of Soviet papers). *In* "The Biological Basis of Freshwater Fish Production" (S. D. Gerking, ed.), pp. 359–376. Blackwell, Oxford.

Protasov, V. R. (1970). "Vision and Near Orientation of Fish." Isr. Program Sci. Transl., Jerusalem.

Protasov, V. R. (1973). "Electric and Acoustic Fields of Fishes." Natl. Tech. Inf. Serv., U.S. Dep. Commer., Springfield, Virginia.

Repsys, A. J., Applegate, R. L., and Hales, D. C. (1976). Food and food selectivity of the black bullhead (*Ictalurus melas*) in Lake Poinsett, South Dakota. *J. Fish. Res. Board Can.* **33**, 768–775.

Roberts, T. R. (1970). Scale eating American characoid fishes, with especial reference to *Prolodus heterostomus. Proc. Calif. Acad. Sci.* **38**, 383–390.

Schoener, T. W. (1970). Nonsynchronous spatial overlap of lizards in patchy habitats. *Ecology* **51**, 408–418.

Schwartz, E. (1971). Die Ortung von Wasserwellen durch Oberflachenfische. *Z. Vgl. Physiol.* **74**, 64–80.

Schwartz, E., and Hasler, A. D. (1966a). Perception of surface waves by the blackstripe topminnow, *Fundulus notatus. J. Fish. Res. Board Can.* **23**, 1331–1352.

Schwartz, E., and Hasler, A. D. (1966b). Superficial lateral line sense organs of the mudminnow (*Umbra limi*). *Z. Vgl. Physiol.* **53**, 317–327.

Scott, W. B., and Crossman, E. J. (1973). The freshwater fishes of Canada. *Fish. Res. Board Can., Bull.* No. 184.

Seghers, B. H. (1973). An analysis of geographic variation in the antipredator adaptations of the guppy (*Poecilia reticulata*). Ph.D. Thesis, Univ. of British Columbia, Vancouver.

Seghers, B. H. (1975). Role of gill-rakers in size selective predation by lake whitefish, *Coregonus clupeaformis* (Mitchill). *Verh. Int. Ver. Limnol.* **19**, 2401–2405.

Shorygin, A. A. (1931). Foods, selective capacity, and food interrelationships of certain Gobiidae of the Caspian Sea. *Zool. Zh.* **18**, 27–53.

Siefert, R. E. (1968). Reproductive behavior, incubation and mortality of eggs, and postlarval food selection in the white crappie. *Trans. Am. Fish. Soc.* **97**, 252–259.

Singarajah, K. V. (1975). Escape reactions of zooplankton: Effects of light and turbulence. *J. Mar. Biol. Assoc. U.K.* **55**, 627–639.

Smith, J. M. N. (1974a). The food searching behaviour of two European thrushes. I. Description and analysis of search paths. *Behaviour* **48**, 276–302.

Smith, J. M. N. (1974b). The food searching behaviour of two European thrushes. II. The adaptiveness of the search patterns. *Behaviour* **49**, 1–61.

Smith, L. L., and Moyle, J. B. (1945). Factors influencing production of yellow pike perch, *Stizostedion vitreum vitreum*, in Minnesota rearing ponds. *Trans. Am. Fish. Soc.* **73**, 243–261.

Smith, S. M. (1976). Predatory behaviour of young turquoise-browed Motmots, *Eumomota superciliosa. Behaviour* **56**, 309–320.

Springer, S. (1961). Dynamics of the feeding mechanism of large galeoid sharks. *Am. Zool.* **1**, 183–185.

Starck, W. A., II, and Schroeder, R. E. (1970). "Investigations of the Grey Snapper," Studies in Tropical Oceanography, No. 10. Univ. of Miami Press, Miami.

Stein, R. A., Kitchell, J. F., and Knezevic, B. (1975). Selective predation by carp (*Cyprinus carpio*) in Skadar Lake, Yugoslavia. *J. Fish Biol.* **7**, 391–399.

Sumner, F. B. (1934). Does "protective coloration" protect? Results of some experiments with fishes and birds. *Proc. Natl. Acad. Sci. U.S.A.* **20**, 559–564.

Sutterlin, A. M. (1975). Chemical attraction of some marine fish in their natural habitat. *J. Fish. Res. Board Can.* **32**, 729–738.

Sutterlin, A. M., and Sutterlin, N. (1970). Taste responses in Atlantic salmon (*Salmo salar*). *J. Fish. Res. Board Can.* **27**, 1927–1942.

Suyehiro, Y. (1942). A study on the digestive system and feeding habits of fish. *Jpn. J. Zool.* **10**, 1–303.

Sverdrup, H. U., Johnson, M. W., and Fleming, R. H. (1942). "The Oceans." Prentice-Hall, Englewood Cliffs, New Jersey.

Tesch, F. W. (1975). Orientation in space. *In* "Marine Ecology" (O. Kinne, ed.), Vol. 2, pp. 657–707. Wiley, New York.

Tester, A. L. (1963). The role of olfaction in shark predation. *Pac. Sci.* **17**, 145–170.

Tyler, A. V. (1972). Food resource division among northern marine demersal fishes. *J. Fish. Res. Board Can.* **29**, 997–1003.

Vermeij, G. J. (1974). Marine faunal dominance and molluscan shell form. *Evolution* **28**, 656–664.

Volkmann, F. C. (1975). Behavioral studies of the discrimination of visual orientation and motion by the goldfish. *In* "Vision in Fishes: New Approaches in Research" (M. A. Ali, ed.), NATO Advanced Study Institute Series, pp. 731–741. Plenum, New York.

Walters, V. (1966). On the dynamics of filter feeding by the wavyback skipjack (*Euthynnus affinis*). *Bull. Mar. Sci. Gulf Caribb.* **16**, 209–221.

Ware, D. M. (1973). Risk of epibenthic prey to predation by rainbow trout (*Salmo gairdneri*). *J. Fish. Res. Board Can.* **30**, 787–797.

Weed, A. C. (1925). Feeding the paddlefish. *Copeia* No. 146, pp. 67–68.

Windell, J. T. (1967). Food analysis and rate of digestion. *In* "Methods for Assessment of Fish Production in Fresh Waters" (W. E. Ricker, ed.), I.B.P. Handbook No. 3, pp. 215–226. Blackwell, Oxford.

Winn, H. E. (1964). The biological significance of fish sounds. *In* "Marine Bio-Acoustics" (W. N. Tavolga, ed.), pp. 213–231. Pergamon, New York.

Woodhead, P. M. J. (1966). The behaviour of fish in relation to light in the sea. *Oceanogr. Mar. Biol. Annu. Rev.* **4**, 337–340.

Wunder, W. (1936). Physiologieder Subwasserfische. *Handb. Binnenfisch. Mitteleur.* **IIb**, 1–340.

Young, J. Z. (1962). "The Life of Vertebrates." Oxford Univ. Press, London and New York.

Zaret, T. M. (1972). Predators, invisible prey and the nature of polymorphism in the Cladocera (Class, Crustacea). *Limnol. Oceanogr.* **17**, 171–184.

Zenkevitch, L. (1963). "The Biology of the Seas of the U.S.S.R." Allen & Unwin, London.

3

THE BRAIN AND FEEDING BEHAVIOR*

RICHARD E. PETER

I. INTRODUCTION

Is the food intake by a fish another example of regulated homeostasis? Or is food intake dictated by the opportunism presented by the environment? Certainly food has to be available before it can be consumed. Also, it would seem to be a truism that adaptation of a fish to its environment must also involve adaptation of the life cycle to the normal cycles of food availability. However, assuming that the fish is adapted to its environment in this respect, does part of this adaptation involve also the regulation of food intake? The evidence concerning the regulation of food intake is sparse, which makes it difficult to answer the above questions. Fortunately the regulation of food intake in mammals has been studied extensively, and ideas developed from research on mammals may be applicable to fishes, or at least serve as the framework for formulation of questions to ask concerning fishes. Thus, the following presentation is largely speculative, and is given

* The unpublished results reported were supported by a grant from the National Research Council of Canada to the author.

121

with the hope that it will serve to stimulate thinking and research on this topic in fish.

II. THE RELATIONSHIP BETWEEN FOOD INTAKE AND BODY WEIGHT

1. MAMMALS

The idea that the adult mammal controls its food intake in such a way that body weight is regulated to some "set-point" level is well established (Hoebel, 1971; Hoebel and Teitebaum, 1966; Mogenson, 1974; Myers, 1974; Panksepp, 1974, 1975). To illustrate, if food intake is restricted to something less than normal, body weight will decrease as reserves are utilized to maintain metabolism. After this period of starvation, if food is made freely available again the animal will become hyperphagic, compared to the normal prestarvation level of food intake (the normophagic level), and the body weight will be quickly restored to about the normal value. On the other hand, if the animal is force fed so that it becomes obese, once it is given free choice of food intake again it will be hypophagic for a time until the body weight decreases to about the starting value. After regaining the starting body weight, normophagia again occurs so that the normal body weight is maintained. Basically, food intake regulation is a regulation of caloric intake so that body weight is maintained about some set-point level.

Many mammals have annual cycles of body weight, suggesting that the set-point may be adjusted according to some species-specific pattern during the cycle. Mrosovsky and Fisher (1970), Barnes and Mrosovsky (1974), and Mrosovsky (1974) have demonstrated the relationships between food intake and body weight in the golden-mantled ground squirrel, a hibernator with such an annual cycle of body weight. As only one part of the cycle, preparation for hibernation, involves a period of hyperphagia resulting in rapid fattening and weight gain. If this period of hyperphagia is interrupted by food withdrawal for a short period, loss of body weight occurs. When food is made freely available again, intense hyperphagia occurs so that the deficit in body weight is rapidly made up to re-establish the animal at the proper phase of the normal weight gain cycle. Thus, the body weight changes seem to follow a changing set-point, with food, or caloric, intake being the variable controlled to accomplish the body weight modifications.

In addition to annual cycles, a number of other factors can influence the body weight set-point and food intake in mammals. One of these is palatability, or "tastiness," of the diet (e.g., Barnes and Mrosovsky, 1974; Corbit and Stellar, 1964; Hamilton, 1964; Mendelson, 1970). On a more palatable diet a higher level of body weight is generally maintained, providing the caloric value of the diet is not reduced by nonnutritive filler. Satiation is apparently reached less quickly on a highly palatable diet. However, whether palatability itself can alter food intake and subsequently body weight, or whether the more palatable diet contains foodstuffs that cause adjustment in the level of food intake is not known. The central actions of various food substrates on food intake will be discussed below (Section III,A).

2. TELEOSTS

The ideas of set-point regulation of body weight and regulation of food intake to accomplish the body weight set-point have not previously been applied to fishes. For fishes, and from here the discussion will be restricted to teleosts as this is the only group for which a reasonable body of information is available, body weight changes have to be viewed in the perspective of changes due to growth, as well as changes due to the variation in energy reserves (lipids in most teleosts). Growth rates of fish in an experimental situation have been shown to be limitable by ration availability, whether the restriction is by absolute amount of ration or the times of day of feeding (for review, see Chapter 10). Growth rates can also be altered by a variety of other factors, each with a certain degree of species specificity, such as intraspecific social interactions (Barlow, 1973; Barlow et al., 1975; Brown, 1957; Magnuson, 1962), interspecific social interactions (Stein et al., 1972), increasing age (e.g., Brett, 1971a, see also Chapter 10), increasing size (e.g., Brett, 1971a; Brett and Shelbourn, 1975; Johnson, 1966a,b; Paloheimo and Dickie, 1966), migration (e.g., Nikolsky, 1963; Shul'man, 1974), spawning (e.g., Nikolsky, 1963; Shul'man, 1974), temperature changes (e.g., Brett et al., 1969; Kudrinskaya, 1970; Shelbourn et al., 1973), seasonal changes (Johnson, 1966a; Nikolsky, 1963; Saunders and Henderson, 1970), salinity changes (Otto, 1971; Saunders and Henderson, 1970), and high and low dissolved oxygen levels (Stewart et al., 1967). Although each of these factors is associated with a syndrome of physiological changes (some of which are discussed below), in each case it was specifically noted that food intake was also changed, although food supply was not restricted. Some of these examples will be examined in detail below. The independence of food intake and growth rates from the simple restriction of food supply in these various examples suggests that both food intake and body weight, or growth, are regulated relative to some set-point, similar to mammals. Whether the set-point is moving up, as in rapid phases of growth, or whether it is relatively static, as when growth is stopped or slowed, is dependent on various conditions. Also, whether a change in food intake is the initiator of a change in body weight set-point, or whether the set-point is altered first and then food intake changed in a supportive role is not known. Evidence suggestive of both possibilities will be discussed below (Section III). In addition to the above listed factors that may normally serve to alter the growth rates and food intake of fish, other factors such as disease, or build up of insecticides and other pollutants, can also have an effect, particularly a suppressive one. In these cases the metabolism of the fish is altered and the changes in growth and food intake in part reflect the "unhealthy" condition of the animal. Possible mechanisms for altering food intake under these conditions will not be discussed below.

Evidence for self-regulation of food intake by a teleost has been presented by Rozin and Mayer (1961). They trained goldfish to barpress for food and let the individual fish regulate its own intake. Although there was marked variation between individuals, each fish established its own cycle of feeding activity and level of intake (Fig. 1). Further experiments demonstrated that in a free-choice situation the fish regulated their food intake such that a certain amount of nutrient was taken in. For example, when food pellets of two different sizes were provided for alternate 3-day periods, the ratio of the larger to smaller pellets eaten by the fish was such that three out of five of the fish very accurately compensated in order to maintain the same level of nutrient intake (Table I). In other experiments food pellets were made up with kaolin, a nonnutritive filler, added to them so as to decrease the overall nutritive value of each pellet (1.8 kaolin-containing pellets = 1 regular pellet in nutritive value). Goldfish that

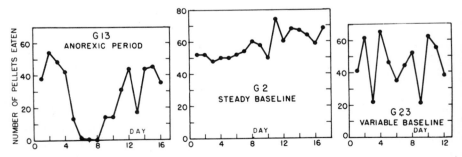

Fig. 1. The daily food intake patterns of three goldfish trained to bar-press for food and allowed to self-regulate their intake. Each fish followed its own individual pattern of feeding and level of food intake. (From Rozin and Mayer, 1961.)

Table I

Daily Food Intake of Five Goldfish Trained to Bar-Press for Food and Allowed to Self-Regulate Their Intakes[a]

| Fish | Mean number of pellets eaten per day (3 days) | | Ratio[b] |
	20 mg	45 mg	
G11	106	35	0.33
G13	33	23	0.70
G25	62	28	0.45
G2	39	18	0.46
G14	48	21	0.44

[a] The fish were fed pellets of different sizes for alternate 3-day periods. From Rozin and Mayer (1961).

[b] Ratio of 45-mg pellets to 20-mg pellets. If nutrient content was controlling lever pressing behavior, the ratio should equal 0.44.

were self-regulating their food intake by bar-pressing were then given the regular pellets or the kaolin-containing pellets for alternate 4-day periods for a total of 32 days. The ratio of kaolin-containing pellets to regular pellets eaten was 1.9, 1.8, and 1.6 for three different goldfish (Fig. 2), indicating accurate compensation for the difference in nutritive value of the two kinds of food. The authors also found that the level of food intake with the regular diet was influenced by temperature. With a 10°C decrease in temperature (25° → 15°C) the fish ate less, and when temperature was increased again, food intake also increased. Q_{10} values for the food eaten at the two temperatures were 2.0, 1.9, and 3.2 for three different fish. Thus, food intake was adjusted approximately the same amount that the metabolic rate was affected by the temperature shifts. This illustrates that caloric or food intake is such as to support the metabolic rate.

In feeding experiments done with pike, *Esox lucius*, Johnson (1966a) found evidence for the regulation of both food intake and growth on a seasonal basis, as well as seasonal changes in food intake to provide the body maintenance requirement. For the experiments pike were held individually in outside tanks and fed live minnows of known weights. The tank temperatures followed seasonal ambient temperature changes, and were similar to the changes of a nearby lake. As shown in Fig. 3, the amount of food required for body maintenance varied on a seasonal basis. Of significance, the "maintenance food" requirement varied independent of temperature to some extent. The peak requirement was in June, followed by a fall over the next

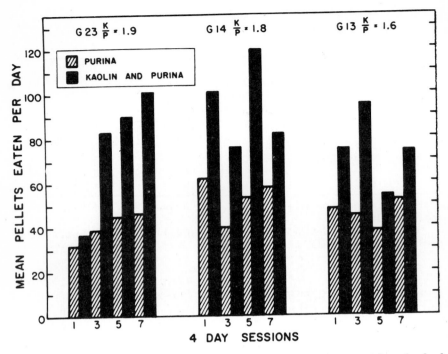

Fig. 2. The effects of dilution of the diet with a nonnutritive material on the food intake per 4-day session for three goldfish. The fish were trained to bar-press for food and were allowed to self-regulate their intake. The diet provided was either the standard 20 mg Purina pellet or adulterated Purina pellets containing 11.3 mg kaolin and 11.3 mg Purina. In terms of nutritive value 1.8 kaolin-containing pellets are equal to 1 regular Purina pellet. (From Rozin and Mayer, 1961.)

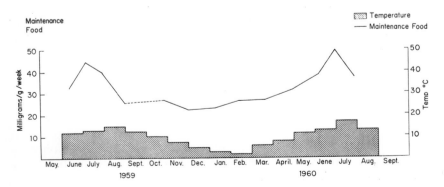

Fig. 3. The changes in food required (maintenance food) to maintain pike at a constant body weight over an extended period. The pike were fed live minnows. Temperatures given represent mean monthly values. (From Johnson, 1966a. Reproduced by permission of *J. Fish. Res. Board Can.*)

month or so to a base level that was constant until the next spring. The peak requirement occurred at the same time of year as the maximum growth rate and maximum food intake of fish fed *ad libitum* (Fig. 4). For the fish fed *ad libitum,* food intake decreased to zero and some weight was lost during the winter months of January and February. The peak in food intake and growth was in June, before the maximum in temperature occurred (refer to Fig. 3 for temperatures). In July 1960, when the temperature was the highest, food intake and the growth rate were decreasing very markedly. Thus, in the pike, food intake and growth are apparently regulated on an annual basis, somewhat independent of temperature changes. These findings may seem contrary to those of Rozin and Mayer (1961) who found that food intake of goldfish was adjusted closely with changes in temperature. However, in the work by Rozin and Mayer, the growth rates of the goldfish were apparently not measured, making it difficult to compare the work with that of Johnson on pike. Perhaps food intake and the growth rate of goldfish are more closely correlated with temperature changes, whereas in pike they are regulated on a seasonal basis somewhat independent of temperature changes.

A number of investigators have noted that fish feed more intensely and that food intake is increased by various species following periods of starvation or deprivation (e.g., Bitterman *et al.,* 1958; Ivlev, 1961; Rozin and Mayer, 1964; Shul'man, 1974; Tugendhat, 1960a,b; Ware, 1972). Thus, deficits in body reserves are made up as quickly as possible when provided with the opportunity, as in mammals. However, when deprivation proceeds beyond a certain point, compensation by increased food intake is no longer possible. This is illustrated by the

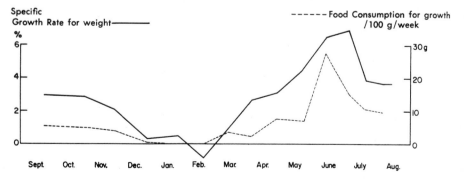

Fig. 4. The changes in growth rate and food intake of individual pike over an extended period. The pike were fed live minnows *ad libitum.* Temperature changes for this period are as in Fig. 3. (From Johnson, 1966a. Reproduced by permission of *J. Fish. Res. Board Can.*)

starvation and refeeding experiments done with sockeye salmon fry by Bilton and Robins (1973). Fry that had been starved for up to 3 weeks and then fed *ad libitum* for 8 weeks were at or near the body weight and length of control animals that were continuously fed *ad libitum*. Fry starved for 4 weeks or longer were not able to overcome the deficits and grow rapidly upon refeeding. Mortality was high in these latter fish, indicating that the more prolonged starvation had caused some irreversible effects. Unfortunately the food intake by the fry was not noted during the experiment, but it is likely that those that were starved for up to 3 weeks were feeding very intensely during the "catch-up" period.

A set of experiments that further demonstrate the effects of deprivation on feeding and growth were done by Tyler and Dunn (1976) on the winter flounder, *Pseudopleuronectes americanus*. They found that the flounder was able to compensate to some extent for decreased meal frequency by increasing meal size. However, once the meal frequency was less than a certain rate, weight loss was evident, indicating starvation. Specifically, either 32, 16, 8, 4 or 2 meals were given per 32-day period. With 16, 8 or 4 meals during the 32-day period, intake per meal was increased such that, although growth did not occur, the fish were able to maintain their body weight. With only 2 meals per 32 days weight loss was evident. When the flounders were given daily meals (32 meals per 32 days) growth occurred.

What is the relationship between apparent food availability and body size in the natural environment? Since in laboratory culture experiments it has been shown that ration level can restrict the growth rate of the whole group of fish in the culture (e.g., Chapter 10; see also Brett *et al.*, 1969; Brett and Shelbourn, 1975; Shelbourn *et al.*, 1973), it would be expected that wild populations with different levels of food availability may also be of proportionately different body size. Contrary to this it has been found in a study on yellow perch that population size and not body size varied with apparent food availability in different regions of a lake (Nakashima and Leggett, 1975). However, more generally it has been observed in wild populations of fish that the rate of growth and the size of depots of energy reserves can be correlated with the availability and nutritional content of the food (e.g., Haines, 1973; Hall *et al.*, 1970; Ivlev, 1961; Martin, 1970; McCaig and Mullan, 1960; Shul'man, 1974), and only over longer term periods can population densities be correlated with food availability. Thus, there can be some restriction of food intake, with a consequent reduction in growth rates and storage of metabolic reserves, without endangering survival of the population. In terms of survival of the species, the set-point level of body weight can be viewed as being regulated between some maximum and a minimum level. The minimum can be viewed as the level at which the animal can still contribute to reproduction of the population. This is chosen here as the minimum because if the animal can still reproduce it must have enough energy intake to divert an adequate amount to ensure maturation of the gonads and successful spawning, as well as the maintenance of the individual. Decreased fecundity of fish under conditions of food shortage have frequently been noted in wild populations. This definition of the minimum level of food intake is different from the "maintenance ration" generally cited in the literature, which is the amount of food just adequate to maintain the body weight at a constant. Although some fish can regulate their food intake to this level, as in the deprivation experiments discussed above, this is considered here as subminimal regulation since this level of food intake will not likely provide enough energy for continuous reproductive success by the population.

Shul'man (1974) has reviewed in detail the annual cycles of fattening observed in a wide variety of teleosts. He has defined six different cycles, based on the variations in total amount of body fat, and the times of year at which the maximum and minimum levels of body fat are observed. The fattening cycle of each species is adapted such that the reserves are used as the energy source for either overwintering, migration, gonadal recrudescence, or spawning, or a combination of these. Interestingly, those fish of a particular species that have longer migratory distances have larger fat stores. Also, each species has a characteristic range of maximum and minimum fatness, and within a species the older fish generally accumulate relatively larger stores. These observations are all suggestive of a set-point regulation of the cycles of metabolic reserves, apparently similar to certain of the mammals. The period of fattening in fish, as in mammals, is associated with more intense feeding compared to other times of year, and, although the accrued maximum level of fatness is dependent to some extent on food availability, there is not necessarily a direct relationship between availability of a specific food item and the time of year of intense feeding. This is also suggestive that food intake is a regulated function.

Several workers have related composition of synthetic diets to growth rates of fish (e.g., Brett, 1971b; Rumsey and Ketola, 1975). However, little or no attention has been paid to the possible effects of palatability of diet on food intake, although palatability could be part of the explanation for increased or decreased consumption and subsequent differences in growth rates of fish on different diets. Fish have been shown to be able to discriminate between foodstuffs on the basis of taste (e.g., Sutterlin and Sutterlin, 1970).

Growth of fish can be enhanced by administration of a variety of hormones (Pickford and Atz, 1957; see also Chapter 9). In such experiments changes in food intake have generally not been commented on. However, food intake was increased following growth hormone administration to hypophysectomized killifish, *Fundulus heteroclitus* (Pickford and Atz, 1957), and intact coho salmon, *Oncorhynchus kisutch* (Higgs *et al.*, 1975). In experiments on plaice, *Pleuronectes platessa,* in which growth was enhanced by treatment with a specific dose of diethylstilbestrol, food intake and efficiency of conversion of food were both increased (Cowey *et al.*, 1973). Similarly, doses of methyltestosterone that enhanced growth of goldfish also caused increased food intake (Yamazaki, 1976). Thus, an effect on food intake is likely a part of the explanation for growth enhancement by hormones. However, whether this is a direct effect of the various hormones is not known. Since increased physiological levels of growth hormone were observed in goldfish in an experiment in which both food intake and growth rate were decreased (Peter *et al.*, 1976), it is likely that for growth hormone at least, food intake is altered by some secondary effect on metabolism in some circumstances.

In summary, a wide number of observations indicate that there is both body weight regulation to a set-point level and food intake regulation in teleosts. Each of these are regulated within the confines of food availability and both of these are affected by a wide number of factors. However, the interdependence of these two regulated phenomena is not clear. The physiological mechanisms for influencing particularly food intake are discussed below.

III. BRAIN REGULATION OF FOOD INTAKE AND FACTORS INFLUENCING INTAKE

A. Brain Regions Involved

1. MAMMALS

The neural control of food intake in mammals has been reviewed frequently in recent years (e.g., Andersson, 1972; Balagura, 1970; Hoebel, 1971; Mogenson, 1974; Myers, 1974; Panksepp, 1974; Stevenson, 1969), and has been the subject of a recent symposium

(Wayner and Oomura, 1975). The hypothalamus is the focal brain region for control of the ingestive behaviors and the vegetative functions of the body. Most of the experimental work on hypothalamic control of food intake in mammals has been done on the laboratory rat. This work and the concepts stemming from it will be briefly reviewed here to serve as a model with which to view the teleosts.

In rats and other mammals, lesions bilaterally in the ventromedial nucleus (VMN) of the hypothalamus, or more generally in the ventromedial hypothalamic (VMH) region, cause hyperphagia leading to obesity. The amount of weight gained is usually up to some level, following which food intake is decreased to some fairly steady level such that the increased body weight is maintained. This effect of VMH lesions is not independent of others, including decreased motivation for performing tasks to acquire food, a general decrease in activity, increased aggressiveness, and irregularity of estrous cycles. Lesions bilaterally in the lateral hypothalamus (LH), on the other hand, cause aphagia or hypophagia. If the animal does not die from inanition, or if it is force fed to prevent death from starvation, food intake generally recommences at some rate less than the normal level, and a less than normal body weight is subsequently maintained. However, the effects of VMH and LH lesioning are dependent on the starting condition of the animal. In the case of the VMH lesions, if the animal is very fat initially due to force feeding, the food intake postlesioning may actually decrease for a short time until the body weight reaches its apparent plateau. However, the plateau will be at some level above the normal body weight. In the case of the LH lesions, if the animal is starved prelesioning, the food intake postlesioning may increase initially until the body weight reaches its plateau again. The plateau will, however, be at some level less than the normal body weight. Thus, by VMH and LH lesioning, the set-point level for body weight seems to be adjusted up or down, respectively, and food intake is modified appropriately to reach the new level.

Other data show that the VMH and LH are directly involved in food intake regulation. Electrical stimulation of the LH causes increases in both the rate and quantity of food intake. This is apparently due to a motivational change, as the feeding can even be initiated in a satiated animal. However, it is not a forced behavior as there can be a short delay after stimulation before feeding commences, and the animal will stop feeding if a distraction impinges. On the other hand, electrical stimulation of the VMH causes cessation of feeding, even in a starved animal.

Based on information such as the above, the concept was advanced

that food intake is regulated by a VMH satiety center and a LH feeding center, with the satiety center able to inhibit the activity of the
feeding center. More specifically, the role of the VMH in feeding,
based on behavioral studies of VMH-lesioned rats by Becker and Kissileff (1974), is to terminate feeding once the meal reaches a certain
size, and to maintain inhibition over the onset of eating between meals
when food is continuously available.

The neural pathways from the VMN to the LH for inhibition of
feeding have been the subject of much controversy. When cuts were
made with a modified Halasz brain knife lateral to the VMN of the rat,
hyperphagia and obesity were observed (e.g., Albert and Storlien,
1969; Palka et al., 1971; Paxinos and Bindra, 1973; Storlien and Albert,
1972) similar to the results of bilateral VMN lesions. These results
were interpreted to show a pathway coursing directly from the VMN
to LH. Fiber connections between these two hypothalamic regions
were also demonstrated (Arees and Mayer, 1967). However, the latter
results have been questioned (Rabin, 1972) and cannot be taken as a
firm demonstration. Rabin (1972), after a thorough review of the literature, concluded that in fact much of the evidence failed to support the
idea of a direct inhibitory connection of VMN with the LH. Also, he
concluded that the available evidence did not clearly support the idea
of a VMN satiety center, in that this region serves to organize and
control the phenomenon of satiation. More recently, a number of
studies in which cuts were made between the VMN and the LH also
failed to support a direct inhibitory connection between the two (e.g.,
Gold, 1973; Sclafani et al., 1975; Sclafani and Maul, 1974). What is
apparent from the lesioning experiments by one technique or another
is that lesions anterolateral to the VMN, and posterior to both the VMN
and LH can cause hyperphagia. Thus, it was proposed by Kapatos and
Gold (1973) and Gold (1973) on the basis of electrolytic and knife-cut
lesioning studies, and by Ahlskog and Hoebel (1973) on the basis of
neurochemically-induced and electrolytic lesioning studies, that a
ventrally located noradrenergic pathway ascending to the forebrain
through the VMN region was the source of the inhibitory effect of
VMH lesions on feeding. In support of this it has been shown that
lesions specifically in the VMN failed to block the occurrence of post-
prandial inhibition of feeding (Panksepp, 1971). However, in spite of
all these doubts there is firm evidence for a role of the VMH in satiation (see also Section III,B). Ahlskog et al. (1975) have shown that
destruction or damage of the noradrenergic pathway through the
VMH, and electrolytic lesions of the VMH that leave the noradrenergic pathways intact, each cause hyperphagia and obesity, with the

combined operations generally causing an even greater effect. Also, recent electrophysiological studies show that there are both excitatory and inhibitory connections between the VMH and LH (Sutin *et al.*, 1975). Thus, a simple inhibitory effect of the VMH on the LH is no longer tenable to explain cessation of feeding.

How then do the VMH and the pathways ascending through this region, both of which apparently have some function in satiation, interact with the LH feeding center? Mogenson (1974) strongly implicates the limbic forebrain in the regulation of feeding. A simplified version of his presentation is as follows: He suggests that the ascending noradrenergic pathways end in the limbic forebrain, and that the limbic forebrain in turn has output to both the LH and VMH. Output from the VMH may inhibit feeding by either an effect on midbrain centers or by projecting more or less directly to the LH. Output from the limbic forebrain and the LH, primarily by the medial forebrain bundles, may both serve to stimulate midbrain regions and various other lower brainstem regions associated with motor aspects of feeding. More recent reviews of forebrain circuitry and VMH–LH connections related to food intake confirm these views (Ban, 1975; Sutin *et al.*, 1975).

In summary, from the many and varied studies it seems tenable that the LH is involved in the organization and activation of feeding behavior and food intake regulation in mammals. It also seems true that the VMH has involvement in regulation of food intake, in part associated with satiation. The VMH is also involved with detection and signaling of deprivation (see Section III,B) to cause LH activation of feeding. The limbic forebrain also plays an important role in feeding behavior, both in activation of feeding and in satiation, as it has important interactions with both the VMH and the LH.

2. FISHES

There are relatively few studies dealing with the neural control of feeding in teleosts. However, the LH has been strongly implicated as a feeding center, similar to mammals.

Demski and Knigge (1971) electrically stimulated a number of forebrain regions of the bluegill sunfish, *Lepomis macrochirus*, via chronically implanted electrodes and made observations on the evoked behaviors. Feeding-like behaviors such as bottom searching, surface searching, snapping up and spitting out of gravel or debris, and snapping up, chewing and swallowing prey were caused by stimulation of particularly the region around the lateral recess of the third ventricle [=nucleus recessus lateralis, NRL, in Peter and Gill (1975) system of nomenclature] in the inferior lobe of the hypothalamus (nucleus diffusus lobi inferioris, NDLI). Feeding and aggressive after-responses were evoked by stimulation more dorsally in the nucleus rotundus [=nucleus glomerulosus, NG of Peter and Gill (1975)]. Since this nucleus has strong fiber connections with the NDLI (see Demski and Knigge, 1971, for literature), the latter result is not unexpected.

The above results were confirmed in a similar study by Demski (1973) on the cichlid *Tilapia heudelotti macrocephala*. However, in addition to being able to evoke feeding behavior by stimulation in the NRL region of the inferior lobe of the hypothalamus, feeding was also evoked by stimulation anteriorly along the midline at about the thalamic–hypothalamic junction. Demski interpreted this result as being due to activation of fibers in the medial forebrain bundle (MFB), a pathway interconnecting olfactory regions in the telencephalon with the preoptic region, and posteriorly with the subglomerulosal region of the inferior lobe of the hypothalamus (see Demski, 1973, for literature). This, of course, implicates the telencephalon in feeding behavior, and raises the possibility that lateral hypothalamic stimulation evokes feeding by activation of ascending MFB fibers.

Grimm (1960) working on goldfish, and Fiedler (1965, 1968) working on wrasse and sunfish, found that feeding behavior could be evoked by stimulation of locations in the telencephalon. Grimm (1960) also found that plugging the olfactory pits of goldfish would block the occurrence of arousal and bottom searching for food after the introduction of a food odor. This suggests that olfactory input and then telencephalon output can serve to activate or arouse feeding behavior. The telencephalon of teleosts has been implicated in arousal of a number of other behaviors in addition to feeding (Aronson, 1963, 1970; Aronson and Kaplan, 1968). The output from the telencephalon with regard to feeding behavior would presumably be via the MFB to the inferior lobe of the hypothalamus. However, the telencephalon is not an essential part of the neural apparatus for feeding in teleosts because it can be almost completely, or completely, ablated and feeding still occur (e.g., Aronson, 1963, 1970; Hale, 1956; Savage, 1969; Segaar, 1965). Furthermore, the food intake of telencephalon-ablated goldfish is apparently similar to that of normal fish (Savage, 1969). But in turn it must be noted that the feeding behavior of telencephalon-ablated fish is not necessarily normal in all respects. The ablated fish have a poor ability for operant conditioning and difficulty in learning maze situations for food reward. Also, the ability to detect food by olfaction is lost, although taste and vision can still function as means for food detection.

Involvement of the LH in feeding behavior has recently been confirmed in a study on goldfish by Savage and Roberts (1975). They observed the evoked behaviors of goldfish that were electrically stimulated, via chronically implanted electrodes, in various hypothalamic, thalamic, and midbrain regions. Figure 5 gives two outline drawings of transverse sections through the diencephalon and mesencephalon with various nuclei named according to their system of nomenclature. Although a wide number of behaviors were recorded, a full feeding response according to their notation involved orientation towards food, assumption of a feeding posture, taking food into the mouth, and ingesting the food. Since the experiments were done on previously satiated fish, the full feeding responses that occurred in response to stimulation were due to induced motivational changes. Figure 6 shows the stimulation sites and incidence of a full feeding response at each site with a threshold stimulus of 20 μA or less. Stimulation of the region lateral or dorsal to the lateral recess of the third ventricle, posterior to the connection of the lateral recess with the central and main part of the third ventricle, was quite effective in evoking feeding responses. Also, stimulation more dorsally in the nucleus subpreglomerulosus region [=nucleus glomerulosus and part of nucleus cerebellosus hypothalami of Peter and Gill (1975)],

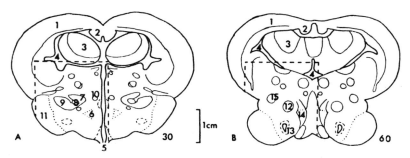

Fig. 5. Drawings of transverse sections through the diencephalon and mesencephalon of the goldfish. A is at 30% of the tectal length, B at 60%. The scale is correct for a 9.5 cm long fish. The dotted lines show the position of the portions of brain drawn in Fig. 6 to show the stimulation sites. 1, Optic tectum; 2, torus longitudinalis; 3, valvula cerebelli; 4, ventricle; 5, pituitary stalk; 6, nucleus anterior tuberis; 7, nucleus preglomerulosus medialis; 8, lateral forebrain bundle; 9, nucleus preglomerulosus lateralis; 10, medial forebrain bundle; 11, nucleus diffusus lobi lateralis; 12, nucleus subpreglomerulosus; 13, posterior recess of third ventricle; 14, corpus mamillare; 15, tractus lobo-cerebellares. (From Savage and Roberts, 1975. Reproduced from *Brain Behav. Evol.* **12**, 42–56, by permission of S. Karger AG, Basel.)

Fig. 6. Sites from which a full feeding response were elicited by electrical stimulation. ●, Sites showing full feeding response in more than 75% of the trials, with thresholds of less than 20 μA; ▲, sites showing a full feeding response in 60–75% of the trials, with thresholds of less than 20 μA; △, sites showing a full feeding response with lower frequencies or with higher thresholds; ○, sites from which a full feeding response could not be elicited. The drawings represent transverse sections at 10% of the tectal length. (From Savage and Roberts, 1975. Reproduced from *Brain Behav. Evol.* **12,** 42–56, by permission of S. Karger AG, Basel.)

and the corpus interpeduncularis, both were quite effective in eliciting feeding behavior. With the addition of the midbrain region, these results are similar to those of Demski and Knigge (1971) and Demski (1973), but add some refinement to the localization of the areas involved in evoking feeding behavior.

Demski (1977) found that feeding responses, represented by the biting and mouthing of food, were consistently evoked by electrical stimulation of the inferior lobe of the hypothalamus of the nurse shark *Ginglymostoma cirratum*. Biting and/or mouthing of food were inconsistently evoked by stimulation of the dorsal hypothalamus and basal regions of the telencephalon. Thus, although these are the only observations available on sharks, the results suggest that homologous hypothalamic regions may be involved in feeding behavior in sharks and teleosts.

There are no published studies dealing with the effects of lesioning the VMH and the LH on food intake in teleosts. However, in a preliminary study (Peter, unpublished observations) lesions of the anterior half of the NRL interrupted the growth of goldfish fed *ad libitum*. For this experiment the fish were held in a 293-liter continuous flow aquarium at 16°–18°C, and on a 16-hr light–8-hr dark photoperiod. Lesions were made with a radiofrequency current generator using the stereotaxic procedure of Peter (1970), as modified by Peter and Gill (1975). Electrodes were No. 00 insect pins insulated to within 0.5–0.6 mm of the tip. The body weight of each fish was taken weekly preoperatively for 4 weeks to assure that each fish was growing and postoperatively for 4 weeks to determine effects of the lesions. Growth changes postoperatively (Fig. 7) are expressed here as a percentage of the weight at the time of operation (weight at operation = 100%). Bilaterally placed lesions in the anterior NRL (Fig. 8) caused an initial loss of weight postoperatively. However, there was weight gain between weeks 1 and 3, indicating there was some recovery from the effects of the lesion. This recovery was perhaps only temporary as there was weight loss again between weeks 3 and 4. Fish with lesions in other locations had growth rates similar to the controls (Fig. 7), and are lumped together to simplify presentation. (Other lesion locations were as follows: dorsomedial nucleus anterior tuberis (NAT)–n. anterioris periventricularis–n. posterioris periventricularis region; anterior NAT–n. lateral tuberis pars anterioris region; bilaterally in n. anterioris hypothalami–anterior n. preglomerulosus pars lateralis (NPGl) region; bilaterally in posterior NPGl–n. glomerulosus region; bilaterally in posterior NPGl–posterior n. ventromedialis thalami region; and bilaterally in posterior NRL region.) The loss of weight and subsequent erratic growth after lesioning the anterior half of the NRL is predictable in view of the results from the electrical stimulation experiments by Demski and Knigge (1971), Demski (1973), and Savage and Roberts (1975). Unfortunately, since food intake was not measured in the present experiment, it cannot be concluded that it was decreased, causing the growth changes. However, Roberts and Savage (unpublished observations, cited in Savage and Roberts, 1975) report aphagia in goldfish after lateral hypothalamic lesions, which does confirm this interpretation. Notably, there were no indications of excessively rapid weight gain, due to hyperphagia, after lesions in the VMH in the present experiment.

The NRL, as well as the nucleus recessus posterioris and a small nucleus in the ventral preoptic region, of teleosts are strongly fluorescent with the Falck–Hillarp method, indicating that they are aminergic (e.g., Baumgarten and Braak, 1967; Ekengren, 1975; L'Hermite and Lefranc, 1972; Weiss, 1970). By the type of fluorescence observed

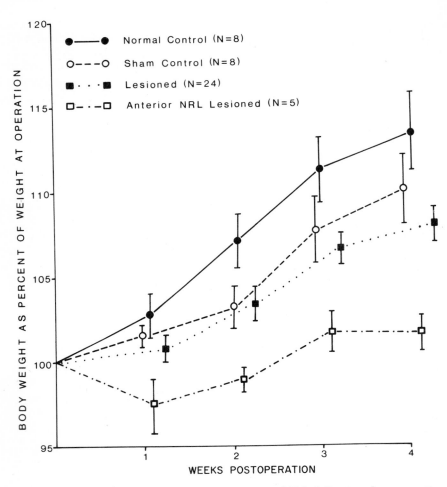

Fig. 7. Growth changes of normal goldfish, or goldfish following sham operation (O---O), lesioning in various parts of the hypothalamus or thalamus (■·····■), or lesioning of the anterior nucleus recessus lateralis (NRL) (□–·–·–·□). Growth changes are expressed as percentages of the body weight at the time of operation (operation weight for each fish = 100%). Values indicated are mean ± SE.

it is likely that the NRL is noradrenergic. These investigators also described a number of aminergic pathways within and traveling through the hypothalamus. It is possible that some of these aminergic fibers are part of the MFB, although it is difficult to be sure from the diagrams and figures given. L'Hermite and Lefranc (1972) described specifically aminergic pathways between the preoptic region and the midbrain, and the preoptic region and the inferior lobe of the

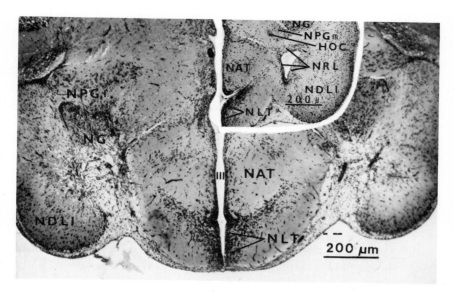

Fig. 8. A cross section through the forebrain of a goldfish showing bilaterally placed lesions (arrows) in the anterior nucleus recessus lateralis. The inset shows the NRL region in an intact fish for comparison. Third ventricle, III; horizontal commissure, HOC; nucleus anterior tuberis, NAT; nucleus diffusus lobi inferioris, NDLI; nucleus glomerulosus, NG; nucleus lateral tuberis, NLT; nucleus preglomerulosus pars lateralis, NPGl; nucleus preglomerulosus pars medialis, NPGm.

hypothalamus in the eel, *Anguilla vulgaris*. Thus, it is possible that there is a homologous neural substrate available for involvement in the regulation of food intake in the teleosts, as compared to mammals.

Electrical stimulation of the optic lobes in wrasse and sunfish has been shown to evoke feeding behaviors (Fiedler, 1965, 1968). Thus, input from the optic tectum may also arouse feeding mechanisms in the hypothalamus.

Aspects of brain involvement in feeding behavior not discussed here, but which cannot be divorced from the topic, are the various neural pathways within the brain for the input and integration of sensory modalities associated with food detection and feeding, and the pathways controlling motor output to accomplish feeding behavior. Discussion of these topics would involve detailed analysis of brain circuitry (for review, see Ariens Kappers *et al.*, 1936) and is beyond the scope of this review.

In summary, the above data demonstrate that the LH has a role as the center for organization and control of feeding behavior in teleosts. The role of the LH in the overall balance of food intake is not clear, however. There is currently no evidence available for or against a satiation function within the hypothalamus or any other brain region, although fish obviously do become satiated. The role of the telencephalon in feeding behavior is unclear, but it likely plays a role in arousal of feeding, probably along with some midbrain regions. The telencephalon probably does not play a role in the overall balance of food intake regulation. Finally, none of the studies to date have provided evidence for regulation of the body weight set-point level by some brain center, although the lesions of the LH are suggestive that it may have some involvement in this regard.

B. Physiological Factors

1. Mammals

Blood glucose levels have long been purported to influence food intake in mammals (Hoebel, 1971; Myers, 1974; Panksepp, 1974, 1975; Rabin, 1972). The original "glucostat" theory proposed that glucose receptors, or glucoreceptors, located in the VMH increase their rate of firing in response to above normal blood glucose levels, thus acting as a satiation signal to the LH. The evidence for such glucoreceptors is based on a number of observations. Serving as some of the original evidence were the findings that gold thioglucose (GTG) injected into mice caused a lesion to develop in the VMH (the supposed satiety center), with subsequent hyperphagia and obesity. The lesioning effect of GTG is apparently related to the glucose moiety of the compound, as its effects in mice are insulin dependent, and other gold salts or gold thiocarbohydrates do not have a similar action. Once GTG is taken up in the VMH by the glucoreceptors, the toxic effect of gold then causes the death of the cells. However, by this same mechanism of action, GTG is also very toxic to many other tissues, especially those that are most active in concentrating glucose. As a result mortality is generally high following intraperitoneal injection of GTG. With the rat, those animals that survive intraperitoneal injection generally develop VMH lesions and obesity, but similar results have not been observed in other mammals or in birds (for review, see Peter *et al.*, 1976). Since VMH glucoreceptors should have a wider distribution within mammals if they are an important component of the food intake regulatory system, either GTG is not an effective means to demonstrate them or the glucoreceptors are, in fact, very restricted in species distribution. Many of the generalized toxic effects of GTG in rats can be avoided if the drug is directly implanted into the brain. Since brain implantation of GTG has only been done in rats, its failure to cause VMH lesions following intraperitoneal injection in mammals other than mice and rats may be more related to failure of the drug to reach the VMH than to its inactivity.

Other evidence for VMH glucoreceptors comes from studies in which glucose and 2-deoxy-D-glucose (2-DG), an analogue that blocks glucose utilization, were infused into the VMH of the rat. Units that are excited by glucose have been found in the VMH. 2-DG infusions, on the other hand, reduce the activity of some VMH units. However, there is much evidence from such studies for similar glucoreceptors in the LH as well.

Recent studies confirm the involvement of hypothalamic and perhaps other brain glucoreceptors in the regulation of food intake (Müller et al., 1974; Panksepp, 1974, 1975), but the function of such receptors is not agreed upon. The suggested scheme by Müller and his colleagues is that glucoprivation acts via the VMH and LH, and possibly other brain sites as well, to cause disinhibition of the LH to initiate feeding. Along with this the VMH causes inhibition of insulin secretion and stimulation of catecholamine secretion from the adrenal medulla, resulting in an increase in blood glucose to alleviate the immediate symptom. The VMH has also been shown to stimulate glucagon secretion (Frohman and Bernardis, 1971). The LH stimulates insulin secretion (Ban, 1975; Gerich et al., 1976). Activation of the LH for feeding will then presumably also result in increased insulin secretion as the meal progresses or following it, thereby accomplishing storage of the foodstuffs. According to this scheme then, the role of the hypothalamic glucoreceptors may be more for the short-term initiation of feeding and prevention of the effects of low blood glucose. In the model advanced by Panksepp (1974, 1975) the short-term termination of feeding is controlled by the LH, perhaps in response to high blood glucose. However, it is likely that hypothalamic glucoreceptors do not have an important function in the termination of feeding as high levels of blood glucose induced by intragastric or intraperitoneal loading do not cause termination of feeding (Wilson and Heller, 1975). On a longer term basis high blood glucose, in the presence of insulin, may have an effect on total intake, because glucose infusion in the VMH has been shown to cause a reduction in food intake over the following day or so in the rat (Myers, 1974; Panksepp, 1974, 1975). This fits with Panksepp's model in which the VMH is involved over the long term in control of food intake.

Glucose receptors have also been suggested to exist in the hepatic portal system. Novin et al. (1973) found that infusion of 2-DG into the hepatic portal vein caused hyperphagia in rabbits. They interpreted this as evidence for glucose receptors within the portal system. The afferent pathway for this glucoprivation signal is via the vagus, as vagotomy delayed the response to 2-DG infusion. On the other hand these receptors are not likely to be important in satiation as gastric filling in the rat with various glucose solutions does not cause cessation of feeding due to an effect specific to glucose (Wilson and Heller, 1975).

Amino acids apparently also play a role in food intake regulation (Myers, 1974, 1975; Panksepp, 1974, 1975). Evidence has been found for both the balance of certain essential amino acids and the total level of amino acids in the blood as having influences on food intake. In general, increased blood levels of amino acids suppress feeding. Some amino acids stimulate neuronal activity in the LH and at least one other forebrain region (zona incerta), whereas others have no action, or inhibit activity (Wayner et al., 1975). Although the mechanisms by which amino acids may influence food intake are not understood, the results available point to the importance of having a balanced diet of protein, as excess of one or another amino acid may have central actions to unbalance food intake regulation.

Distension of the gut seems to be an essential element for satiation. Rats with a gastric fistula installed in such a way that ingested food can be drained from the stomach do not display satiety (Smith et al., 1974). When the fistula is closed the animals have normal satiation. In other experiments by Smith and his colleagues it was demonstrated that distension of the small intestine was specifically involved in satiation. Since an otherwise normal rat still has satiety after vagotomy, the afferent mechanism for the satiety signal may, at least in part, be a blood-borne factor. Cholecystokinin, a hormone released from the intestinal wall in response to stretch, has been shown to cause satiety under various experimental conditions (Smith et al., 1974; Smith and Gibbs, 1975). However, whether it is an important normal physiological mechanism in satiety is not known.

Vagal output is likely to play a role in the altered set-point for body weight and the increased food intake that occurs after VMH lesions. The obesity normally caused by such lesions does not occur after subdiaphragmatic vagotomy (Powley and Opsahl, 1974). This suggests that obesity in this instance is related to some alteration in visceral metabolism. Since the VMH has been shown to stimulate glucagon secretion (Frohman and Bernardis, 1971; Gerich et al., 1976) and catecholamine release from the adrenal medulla (Gerich et al., 1976), decreased release of these hormones could in part contribute to the changes in metabolism leading to obesity. Glucagon levels are, in fact, less than normal in some obese humans (Wise *et al.*, 1972). There is also an increase in blood glucose (Steffens *et al.*, 1972) and insulin (Gerich *et al.*, 1976) levels following VMH lesions. Thus, in obesity induced by VMH lesions part of the apparent alteration in body weight set-point is due to altered endocrine frunction.

On a long-term basis food intake is thought to be influenced by some sort of feedback from adipose tissue and on a shorter term by free fatty acid (FFA) levels in the blood (Myers, 1974, 1975; Panksepp, 1974, 1975). Increased blood levels of FFA may inhibit feeding, and low levels have been correlated with feeding. Also, certain prostaglandins reduce food intake when infused into the hypothalamus. However, the receptors for monitoring these various elements and what is being monitored remain as open questions.

Various hormones, in addition to the insulin functions already mentioned, have effects on food intake. Female rats, and many other female mammals, have cyclic fluctuations of food intake and body weight in correlation with stages of the estrous cycle (for review, see Bray, 1974; Nance and Gorski, 1975). Also, the pattern of diurnal distribution of feeding changes during the estrous cycle. These variations are lost after ovariectomy. Intrahypothalamic implants of estrogen indicate that the hormone suppresses food intake by an action on the VMN. Thus, the cyclic secretion of estrogen on a daily basis and during the estrous cycle could contribute to the food intake changes. Androgens, on the other hand, stimulate food intake in rats, leading to increased body weight. The site of action of the androgens is not known, but evidence suggests that it is also the VMH.

Growth hormone influences food intake in ways dependent on the experimental model being utilized (Bray, 1974). Treatment of hypophysectomized rats with growth hormone causes increased food intake. On the other hand, it suppresses food intake in hyperphagic VMH-lesioned animals. Prolactin has also been shown to increase food intake in hypophysectomized animals. In general, there is also a correlation of body weight with activity of the adrenal gland. However, specific influences of the corticoids on food intake have not been documented, although stress-induced hyperphagia and obesity has recently been demonstrated in the rat (Rowland and Antelman, 1976). Of course a general malaise can also be interpreted as a stress situation, but under such conditions food intake is usually depressed.

Finally, Myers (1974, 1975) and colleagues have shown that both satiation and the tendency to feed can be transferred from one animal to another by taking cerebrospinal fluid, or fluid drawn from parts of the hypothalamus, from one animal and infusing it into the hypothalamus of the other. Thus, the phenomena of satiation and hunger are reflected in the composition of the cerebrospinal fluid and the extracellular fluid of certain parts of the brain. Determining the factors responsible for the behaviors may then be possible by analysis of these fluids. Changes in the concentration of sodium and calcium may apparently be part of the answer. Specifically, feeding can be induced in the satiated rat by infusion into the hypothalamus of fluids with excess sodium or calcium. In addition to the balance of various ions, Myers (1975) also points out that feeding and the balance between food intake and nutritive stores will be influenced by the different neurotransmitters within the brain, and the multitude of other factors (e.g., glucose, FFA, amino acids, hormones) that also influence food intake. An integrated view of all these factors will therefore be necessary to understand food intake regulation. Such a view is obviously only in a formative phase for mammals.

2. Fishes

The effects of blood glucose levels on food intake in teleosts have not been studied per se. However, Peter *et al.* (1976) have done a series of experiments to determine the effects of GTG on goldfish. Contrary to the situation in mice and rats, GTG administered to goldfish by intraperitoneal injection or intraventricular brain injection caused dose-dependent hypophagia and slowing of growth (Fig. 9, Table II). Blood glucose and growth hormone concentrations were, in general, not affected by GTG treatment. Lesions were not observed in the VMH following GTG injection by either route. After brain injection of GTG there was a dose-dependent hypertrophy of the forebrain ependyma. In animals that received a brain injection large enough to affect growth rates, there was disruption of the ependymal lining of the lateral–dorsal aspect of the lateral recess of the third ventricle, accompanied by some hypertrophy of the adjacent NRL. This suggests that the LH may be the site of action of GTG in the goldfish. Similar histological changes were not observed after intraperitoneal injection of GTG, suggesting that other locations may also be involved. In further work, however, the hypothalamus was confirmed as the general site of action. Pellet implants of GTG in the third ventricle at the point where the lateral recesses join the main ventricle were more effective in suppressing growth than the various other brain locations tested (Peter and Monckton, 1975; Peter, unpublished observations). However, it is not likely that the actions of GTG in these various experiments were due to specific effects on glucoreceptors. Gold

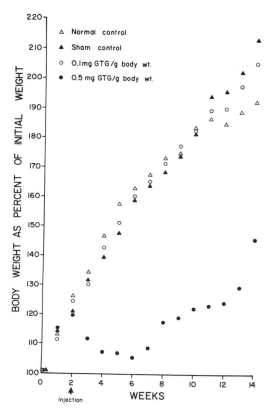

Fig. 9. The mean changes in body weight expressed as a percentage of the initial weight of goldfish after intraperitoneal injection (at week 2) of different dosages of gold thioglucose. The larger dose initially caused an overall loss of body weight. Over the 12 weeks postinjection (weeks 2–14) some fish continued to lose weight and eventually died, others remained stable in body weight, and still others recommenced growth after a number of weeks. These various elements account for the tendency for recovery of growth toward the end of the experiment. For details concerning the number of fish in each group and the variability about each mean, see Peter *et al.* (1976).

thiomalate had an effect on growth in goldfish similar to GTG after intraperitoneal injection (Peter *et al.*, 1976). This suggests that the effects of GTG were due to the general toxicity of gold as it was liberated from the compound during degradation, and not related to the concentration of it in the hypothalamus by glucoreceptors. On this indirect evidence, it may then be that glucoreceptors are not concentrated in some region of the hypothalamus and that blood glucose levels are not an important signal in the regulation of food intake in teleosts. More direct data are obviously needed to solve this problem.

Table II

Food Intake of Goldfish Before and After Injection with Physiological Solution (Sham Control) or 0.5 mg Gold Thioglucose/g Body Weight (GTG Treated)[a]

Group and animal	Daily food intake pre-injection (g ± SE)	Food intake postinjection[b]			
		Week 1	Week 2	Week 3	Week 4
Sham control					
A2	0.75 ± 0.04	–	↓	–	—[c]
B2	0.78 ± 0.09	–	–	↓	–
C1	1.90 ± 0.17	↑↑	↑	↑↑↑	↑↑
D1	1.18 ± 0.11	–	–	–	–
E2	1.01 ± 0.08	–	↓↓↓	–	↓
F2	1.13 ± 0.05	↑↑	↑	–	–
GTG treated					
A1	1.29 ± 0.09	–	↓↓↓	–	—[c]
B1	0.65 ± 0.04	↓↓↓	–	–	–
C2	1.27 ± 0.08	↓↓↓	↓↓↓	↓↓	–
D2	2.13 ± 0.09	↓↓↓	–	↓↓↓	↓↓↓
E1	2.03 ± 0.13	↓↓↓	↓↓↓	↓↓↓	–
F1	0.72 ± 0.04	↓↓↓	↓	↓↓↓	–

[a] There were some significant differences in food intake in the sham control group, but no overall trend was evident. Following GTG treatment food intake was significantly reduced in 15 of the 23 postinjection observation weeks shown in the table. Although food intake was not significantly reduced in 8 of the 23 postinjection observation weeks following GTG, the average daily food intake for these periods (values not given) were always less than the preinjection values. From Peter et al. (1976).

[b] ↑, significant increase compared to preinjection; ↓, significant decrease compared to preinjection, –, no change from preinjection. Single arrow, $P < 0.05$; double arrow, $P < 0.01$; triple arrow, $P < 0.001$.

[c] Data omitted for the fourth week due to sickness of the animals.

The role that various amino acids may play in food intake regulation in fish is unexplored. For aquaculture purposes this topic should be investigated as diet composition may influence food intake and subsequently also growth rates.

Effects of FFA or body fat stores on food intake in fish have not been investigated. Since various species do have annual cycles of fattening that are predictable in terms of timing and size of deposits, and are also correlated with cycles of feeding (see Section II above), it seems apparent that lipid stores and/or FFA must play a role in food intake regulation.

In experiments by Meier (1969, 1972) and his colleagues, fat deposition has been shown to occur in F. grandis and F. chrysotus if injections of cortisol, or thyroxine, and prolactin are given repeatedly at specific times of the day. In animals on a 24-hr light photoperiod it has been found that the cortisol injections can precede the prolactin injections (i.e., cortisol injection on days 1 to 3, prolactin injection on days 5 to 11) and fattening will occur if the prolactin injections are timed appropriately relative to when the cortisol injections were given. Thus, cortisol and thyroxine are able to entrain the physiology of the animal so that fattening can occur in response to prolactin. Photoperiod is also an entraining agent in that prolactin injections repeatedly at a specific time after the onset of light (8 hr) will cause fattening. The latter results have been confirmed in other species (Mehrle and Fleming, 1970; deVlaming and Sage, 1972). Meier has hypothesized that under natural conditions fattening occurs when the daily high blood levels of prolactin and cortisol, or perhaps also thyroxine, coincide with each other in an appropriate way. The daily cycles of secretion of these various hormones are influenced by changes in environmental conditions (e.g., temperature and photoperiod), so that at certain times of year the high concentrations come into phase with each other to induce fattening. Study of the endogenous daily cycles of secretion of these various hormones and how environmental factors may influence their daily cycles are being carried out in a number of laboratories, and hopefully will soon be defined well enough that the normal physiological triggers to induce fattening will be understood. Unfortunately, the changes in food intake that may occur during induced fattening have not been measured.

The changes in metabolism during fattening induced by hormone injections and possible changes in food intake are independent events. Meier (1969) found that the relative amount of lipid in the body increased even in starved fish after injections of prolactin at the time of day appropriate to cause the fattening response. These fish also tended to lose more body weight than starved fish injected with prolactin at a time of day inappropriate to cause fattening. Fed fish that had prolactin-induced fattening generally have a very marked increase in body weight. Clearly the changes in metabolism associated with fat deposition are only one part of the response. Changes in food intake are likely to be either dependent on the changes in metabolism or perhaps a more direct response to the changes in the hormonal regime.

When prolactin is injected early in the daily photoperiod of teleosts on a long photoperiod, loss of fat reserves are generally observed (Mehrle and Fleming, 1970; Meier, 1969, 1972; deVlaming et al., 1975; deVlaming and Sage, 1972). Melatonin injection causes depletion of lipid reserves in *F. similis* and *Cyprinodon variegatus*, although *F. similis* has been found to have an increase in lipids in response to melatonin at one particular stage in its annual cycle (de-Vlaming *et al.*, 1974). Growth hormone and prolactin may both cause lipolysis in kokanee salmon (McKeown *et al.*, 1975). In addition, other hormones may also be lipolytic in teleosts (McKeown *et al.*, 1975). The changes in food intake that may occur during a phase in which lipid reserves are being mobilized have not been studied. Presumably high blood levels of FFA may suppress food intake, as in mammals.

Growth hormone treatment of teleosts has been correlated with increased feeding activity (see Section II). This is likely due to some indirect effects, such as alteration in levels of metabolic substrates, as increased physiological levels of growth hormone and hypophagia can occur together in goldfish (Peter *et al.*, 1976). The effects of growth hormone on feeding may thus be dependent on the condition of the fish, similar to mammals.

Treatment with appropriate doses of diethylstilbestrol and methyltestosterone can each cause enhanced growth rates and greater food intake (see Section II). There is insufficient information available on the metabolic effects of the sex steroids and their analogues to suggest whether some alterations in metabolism or a direct effect of the hormones on the brain may be involved in the stimulation of food intake by these compounds. In some species there are marked differences in body size between the sexes, with the female usually being the larger. Some of this sexual dimorphism may be due to the effects of the sex steroids on food intake, and not just genetically programmed differences in size. Changes in food intake during spawning (see Section II) may also be related to the levels of sex steroids.

The effects of stress-induced increases of cortisol and other corticoids on food intake in fish have not been studied. However, it seems reasonable to assume that food intake would be reduced in a stressful situation. Examples of such a situation may be when a fish is the "loser" in intraspecific (see Section III,C) or interspecific (see Section III,D) competition. However, all stresses are not likely to cause decreased food intake. Meier *et al.* (1973) found that repeated handling at specific times of day could cause either a gain, a loss, or no change in body weight of *F. chrysotus*. This suggests that the stress of handling, and perhaps other stresses as well, may have different effects on the physiology of the animal, including its food intake, depending on the time of day when the event occurs.

The role that stomach and intestinal filling and emptying may have on feeding in fish has not been well studied. The size of the stomach and the speed of emptying can obviously influence meal sizes and intervals between meals. Wide variation has been noted in the amount of stomach filling in various species at a given time of year and on a seasonal basis (e.g., Hellawell, 1972; Hunt and Jones, 1972; Keast, 1968; Miller, 1967). From these studies it can be seen that the fullness of the stomach is not a function of food availability, as there are seasonal variations in food intake and stomach filling that are not necessarily correlated with changes in food availability or environmental temperature.

The causal effects of increasing age or size to decrease growth rates and food intake (see Section II) can only be pondered at this point. These factors are undoubtedly related to changes in the endocrine and metabolic functions of the animal. The effects of aging of fish need to be investigated to help understand the problem.

Obviously, a summary statement of how various physiological factors influence food intake in teleosts is not possible given the current state of our knowledge.

C. Intraspecific Factors

TELEOSTS

The intraspecific social interactions that influence food intake must ultimately relate to various physiological factors within the fish. The function of this section is to review the various intraspecific factors known to influence food intake and to suggest a possible physiological mechanism for the influence.

Brown (1957) reviewed the early work in this area, the most substantive of which was her own research on brown trout fry. She coined the phrase "size hierarchy effect" to describe the common observation in trout hatcheries of the different rates of growth of fry in a population, with the larger fish growing more rapidly than the smaller ones, all the fish having started at about the same size. With time this size difference becomes even greater, unless the larger fish are removed, at which time the smaller ones recommence rapid growth once again. However, the cycle is repeated as the differences in size will recur again after a time, accompanied also by the differences in growth rates. Brown hypothesized that the size hierarchy effect was related to the dominant–subordinate relationships observed to occur among salmon fry, the dominant fish being those that grow fastest. Interestingly, Brown found that the growth of the smaller fish in a group of brown trout fry would not accelerate even under conditions in which an adequate supply of food was available to support rapid growth by all.

Magnuson (1962) related the role of aggressive behavior to food intake and growth of the medaka, *Oryzias latipes*, in a laboratory study. Growth rates were used to reflect the success of fish in different competitive situations, the assumption being that assimilation was the same in all the fish regardless of size or social position. Aggression was found to be related to food availability. In the case of a limited food supply, the larger fish were socially dominant and territorial in the semiconfined area where the food was introduced. The dominants were able to exclude the subordinates from the food area by aggressive interactions, resulting in a relatively greater food intake and a relatively faster growth rate for the dominants. However, in a high density population the dominants were not able to defend the food concentrations, and in this case the amount of aggression and the growth rate by the dominants were less. When the possibilities for isolation were decreased by decreased partitioning of the environment, aggressive interactions increased and growth of all fish was slower. In low densities, when food was supplied in excess, the large fish no longer had a competitive advantage over the small fish in terms of access to food, and all grew rapidly. Also, a dominant and a subordinate in semi-isolation from each other had similar growth rates if food was evenly distributed in the system.

In the Midas cichlid, Barlow (1973) described that, with fish of equal size, the gold-colored morphs were generally dominant in contests for food with the normal cryptic (gray)-colored morphs. This results in a relatively greater food intake by the gold-colored animals and a faster growth rate. With a homogeneous "gold" or "normal" population, all grew at about the same rate. In subsequent work Barlow *et al.* (1975) found that dominance over the food source was established by the largest fish in a small group, regardless of color, and that the subordinate fish had some learning of the color of the dominant. This learning is apparently applied by the fish in subsequent trials to determine whether they should attack to try and gain access to the food source.

In feeding experiments on landlocked Atlantic salmon, Fenderson *et al.* (1968) found that in hatchery fish a strong social hierarchy quickly developed within a small group, with the dominant fish having the greatest food intake from the limited supply provided. When the situation was changed by removal of the dominant, the food consumption by the previously subordinate animals would increase. Unfortunately, the food intake by the dominants and subordinates was not tested in a situation in which there was an unlimited supply. A restricted food supply causes strengthening of the social hierarchy within a small group of Atlantic salmon fry (Symons, 1968). The functional purpose of this behavior system is to secure a feeding territory for some of the members of the population (Keenleyside and Yamamoto, 1962), assuring that at least some will have an adequate food supply for survival. A similar relationship between food availability and level of aggression has also been found in the rainbow trout (Slaney and Northcote, 1974), and other species as well.

From the above observations it is clear that, when the food supply is limited, the establishment of a social hierarchy can result in an "adequate" food supply and rapid growth for the dominant fish, and a "less than adequate" food supply and slow growth for subordinate fish. These differences in food intake can be the result of competitive exclusion of the subordinate from the food supply by the dominant. However, in some situations reported above, the food supply was not limited, in terms of access or availability, but yet the subordinate fish still did not grow rapidly. The explanation for this must lie in the physiological syndrome associated with being subordinate. Being subordinate may be a stress effect, symptomized by high blood levels of the adrenocorticol steroid hormones, particularly cortisol. Brown (1957) also hypothesized that the small subordinate fish were stressed. Unfortunately, the physiological differences between dominant and subordinate fish have not been investigated.

Social facilitation plays a role in the food intake and feeding behavior of some teleosts. In the striped mullet, *Mugil cephalus*, Olla and Samet (1974) found that isolated animals able to periodically view a nonfeeding group of their own species took a relatively long time to initiate feeding, whereas isolates that viewed a feeding group initiated feeding very quickly, and isolates that did not view another group took a time intermediate between the other two. From earlier work, Welty (1934) concluded that social facilitation was reflected in increased food intake in fish. In order for this to obtain, the interactions by the fish in a group likely have to be unencumbered by a social hierarchy. Thus, the role of social facilitation on feeding will be dependent on the social behavior system of the species. To better understand the effects of social facilitation on food intake, studies need to be done in which the intake by the animals in different social situations is carefully recorded.

D. Interspecific Factors

The effects on feeding and food intake of social interactions between fish of different species has received relatively little attention. A number of studies have shown that changes in food selection may occur as a result of interspecific competition (e.g., Schutz and North-cote, 1972; Werner and Hall, 1976). This sort of interaction may not affect the level of food intake. However, if one species assumes social dominance over another, and the subordinate species does not change its ecological niche to adopt another food supply, then the food intake by the subordinate species will likely be decreased, similar to the situation with intraspecific dominant–subordinate interactions. Using laboratory experiments Stein *et al.* (1972) have shown that this is probably the reason for the slower growth of fall chinook salmon when in sympatry with coho salmon in the wild. To understand whether the effects on the subordinate fish are different in interspecific interactions as compared to intraspecific interactions, further experimental study is needed to assess the effects on food intake by the subordinate species in situations in which food is restricted versus when it is unlimited.

E. Environmental Factors

A wide variety of environmental factors, including temperature, dissolved oxygen levels, and salinity (see Section II above), can influence growth rates and food intake in teleosts. Little would be accomplished by citing in detail the specific examples in which each environmental factor was noted to influence growth and food intake, as we obviously do not understand the underlying physiological mechanisms for such changes. Also, the possible mechanisms for the temperature and photoperiod effects have already been dealt with above. Obviously, each of the environmental factors must relate in some way to physiological changes within the animal.

To understand the underlying mechanisms, careful study of the influence of various environmental factors on endocrine function and metabolism need to be undertaken, coupled with studies on the influence of various hormones and metabolic substrates on the control of feeding and food intake.

IV. CONCLUDING REMARKS

Presentation of a model to serve as a working hypothesis for the control of feeding behavior and food intake in teleosts is premature at this stage. Obviously, many avenues for research are open in this field. Approaches need to be made from the environmental, behavioral, metabolic, endocrine, neurophysiological, and neuroanatomical viewpoint to the problem, although all of these elements could not be incorporated in any one study. However, in the design of experiments, and in theory subsequently proposed, the strategy of the animal's control system must be kept in mind: the regulatory system encompasses mechanisms for the short-term onset and termination of feeding, and the long-term balance between body nutrient depletion and repletion, i.e., the set-point level of body weight. Finally, it must also be kept in mind that the mechanisms for control of food intake are an adaptive feature of the life cycle of the fish.

REFERENCES

Ahlskog, J. E., and Hoebel, B. G. (1973). Overeating and obesity from damage to a noradrenergic system in the brain. *Science* **182**, 166–169.

Ahlskog, J. E., Randall, P. K., and Hoebel, B. G. (1975). Hypothalamic hyperphagia: Dissociation from hyperphagia following destruction of noradrenergic neurons. *Science* **190**, 399–401.

Albert, D. J., and Storlien, L. H. (1969). Hyperphagia in rats with cuts between the ventromedial and lateral hypothalamus. *Science* **165**, 599–600.

Andersson, B. (1972). Receptors subserving hunger and thirst. *In* "Handbook of Sensory Physiology. III/1: Enteroceptors" (E. Neil, ed.), pp. 187–216. Springer-Verlag, Berlin and New York.

Arees, E. A., and Mayer, J. (1967). Anatomical connections between medial and lateral regions of the hypothalamus concerned with food intake. *Science* **157**, 1574–1575.

Ariens Kappers, C. U., Huber, G. C., and Crosby, E. C. (1936). "The Comparative Anatomy of the Nervous System of Vertebrates, Including Man," Vols. 1 and 2. Macmillan, New York.

Aronson, L. R. (1963). The central nervous system of sharks and bony fishes with special reference to sensory and integrative mechanisms. *In* "Sharks and Survival" (P. W. Gilbert, ed.), pp. 165–241. Heath, Boston, Massachusetts.

Aronson, L. R. (1970). Functional evolution of the forebrain in lower vertebrates. *In* "Development and Evolution of Behavior" (L. R. Aronson, E. Tobach, D. S. Lehrman, and J. S. Rosenblatt, eds.), pp. 75–107. Freeman, San Francisco, California.

Aronson, L. R., and Kaplan, H. (1968). Function of the teleostean forebrain. *In* "The Central Nervous System and Fish Behavior" (D. Ingle, ed.), pp. 107–125. Univ. of Chicago Press, Chicago, Illinois.

Balagura, S. (1970). Neurochemical regulation of food intake. *In* "The Hypothalamus" (L. Martini, M. Motta, and F. Fraschini, eds.), pp. 181–193. Academic Press, New York.

Ban, T. (1975). Fiber connections in the hypothalamus and some autonomic functions. *Pharmacol., Biochem. Behav.* **3**, Suppl. 1, 3–13.

Barlow, G. W. (1973). Competition between color morphs of the polychromatic Midas cichlid. *Science* **179**, 806–807.

Barlow, G. W., Bauer, D. H., and McKaye, K. R. (1975). A comparison of feeding, spacing, and aggression in color morphs of the midas cichlid. I. Food continuously present. *Behaviour* **54**, 72–96.

Barnes, D. S., and Mrosovsky, N. (1974). Body weight regulation in ground squirrels and hypothalamically lesioned rats: Slow and sudden set-point changes. *Physiol. Behav.* **12**, 251–258.

Baumgarten, H. G., and Braak, H. (1967). Catecholamine im Hypothalamus vom Goldfisch (*Carassius auratus*). *Z. Zellforsch. Mikrosk. Anat.* **80**, 246–263.

Becker, E. E., and Kissileff, H. R. (1974). Inhibitory controls of feeding by the ventromedial hypothalamus. *Am. J. Physiol.* **226**, 383–396.

Bilton, H. T., and Robins, G. L. (1973). The effects of starvation and subsequent feeding on survival and growth of Fulton Channel sockeye salmon fry (*Oncorhynchus nerka*). *J. Fish. Res. Board Can.* **30**, 1–5.

Bitterman, M. E., Wodinsky, J., and Candland, D. (1958). Some comparative psychology. *Am. J. Psychol.* **71**, 94–110.

Bray, G. A. (1974). Endocrine factors in the control of food intake. *Fed. Proc., Fed. Am. Soc. Exp. Biol.* **33**, 1140–1145.

Brett, J. R. (1971a). Satiation time, appetite, and maximum food intake of sockeye salmon (*Oncorhynchus nerka*). *J. Fish. Res. Board Can.* **28**, 409–415.

Brett, J. R. (1971b). Growth responses of young sockeye salmon (*Oncorhynchus nerka*) to different diets and planes of nutrition. *J. Fish. Res. Board Can.* **28**, 1635–1643.

Brett, J. R., and Shelbourn, J. E. (1975). Growth rate of young sockeye salmon, *Oncorhynchus nerka*, in relation to fish size and ration level. *J. Fish. Res. Board Can.* **32**, 2103–2110.

Brett, J. R., Shelbourn, J. E., and Shoop, C. T. (1969). Growth rate and body composition of fingerling sockeye salmon, *Oncorhynchus nerka*, in relation to temperature and ration size. *J. Fish. Res. Board Can.* **26**, 2363–2394.

Brown, M. E. (1957). Experimental studies on growth. *In* "The Physiology of Fishes" (M. E. Brown, ed.), Vol. 1, pp. 361–400. Academic Press, New York.

Corbit, J. D., and Stellar, D. (1964). Palatability, food intake, and obesity in normal and hyperphagic rats. *J. Comp. Physiol. Psychol.* **58**, 63–67.

Cowey, C. B., Pope, J. A., Adron, J. W., and Blain, A. (1973). Studies on the nutrition of marine flatfish. The effect of oral administration of diethylstilboestrol and cyproheptadine on the growth of *Pleuronectes platessa*. *Mar. Biol.* **19**, 1–6.

Demski, L. S. (1973). Feeding and aggressive behaviour evoked by hypothalamic stimulation in a cichlid fish. *Comp. Biochem. Physiol. A* **44**, 685–692.

Demski, L. S. (1977). Electrical stimulation of the shark brain. *Am. Zool.* **17**, 487–500.

Demski, L. S., and Knigge, K. M. (1971). The telencephalon and hypothalamus of the bluegill (*Lepomis macrochirus*): Evoked feeding, aggressive and reproductive behaviour with representative frontal sections. *J. Comp. Neurol.* **143**, 1–16.

deVlaming, V. L., and Sage, M. (1972). Diurnal variation in fattening response to prolactin treatment in two cyprinodontid fishes, *Cyprinodon variegatus* and *Fundulus similis*. *Contrib. Mar. Sci.* **16**, 59–63.

de Vlaming, V. L., Sage, M., Charlton, C. B., and Tiegs, R. (1974). The effects of melatonin on lipid deposition in cyprinodontid fishes and on pituitary prolactin activity in *Fundulus similis*. *J. Comp. Physiol.* **94**, 309–319.

deVlaming, V. L., Sage, M., and Tiegs, R. (1975). A diurnal rhythm of pituitary prolactin activity with diurnal effects of mammalian and teleostean prolactin on total body lipid deposition and liver lipid metabolism in teleost fishes. *J. Fish Biol.* **7**, 717–726.

Ekengren, B. (1975). Aminergic nuclei in the hypothalamus of the roach *Leuciscus rutilus*. *Cell Tissue Res.* **159**, 493–502.

Fenderson, O. C., Evenhart, W. H., and Muth, K. M. (1968). Comparative agonistic and feeding behaviour of hatchery-reared and wild salmon in aquaria. *J. Fish. Res. Board Can.* **25**, 1–14.

Fiedler, K. (1965). Versuche zur Neuroethologie von Lippfischen und Sonnenbarschen. *Verh. Dtsch. Zool. Ges. Kiel, Zool. Anz.* **28**, Suppl. (1965), 569–580.

Fiedler, K. (1968). Verhaltenswirksame Strukturen im Fischgehirn. *Verh. Dtsch. Zool. Ges. Heidleberg, Zool. Anz.* **31**, Suppl. (1968), 602–616.

Frohman, L. A., and Bernardis, L. L. (1971). Effect of hypothalamic stimulation on plasma glucose, insulin, and glucagon levels. *Am. J. Physiol.* **221**, 1596–1603.

Gerich, J. E., Charles, M. A., and Grodsky, G. M. (1976). Regulation of pancreatic insulin and glucagon secretion. *Annu. Rev. Physiol.* **38**, 353–388.

Gold, R. M. (1973). Hypothalamic obesity: The myth of the ventromedial nucleus. *Science* **182**, 488–490.

Grimm, R. J. (1960). Feeding behavior and electrical stimulation of the brain of *Carassius auratus*. *Science* **131**, 162–163.

Haines, T. A. (1973). Effects of nutrient enrichment and a rough fish population (carp) on a game fish population (smallmouth bass). *Trans. Am. Fish. Soc.* **102**, 346–354.

Hale, E. B. (1956). Social facilitation and forebrain function in maze performance of green sunfish, *Lepomis cyanellus*. *Physiol. Zool.* **29**, 93–107.

Hall, D. J., Cooper, W. E., and Werner, E. E. (1970). An experimental approach to the production dynamics and structure of freshwater animal communities. *Limnol. Oceanogr.* **15**, 839–928.

Hamilton, C. L. (1964). Rat's preference for high fat diets. *J. Comp. Physiol. Psychol.* **58**, 459–460.

Hellawell, J. M. (1972). The growth, reproduction and food of the roach *Rutilus rutilus* (L)., of the River Lugg, Herefordshire. *J. Fish. Biol.* **4**, 469–486.

Higgs, D. A., Donaldson, E. M., Dye, H. M., and McBride, J. R. (1975). A preliminary investigation of the effect of bovine growth hormone on growth and muscle composition of coho salmon (*Oncorhynchus kisutch*). *Gen. Comp. Endocrinol.* **27**, 240–253.

Hoebel, B. G. (1971). Feeding: Neural control of intake. *Annu. Rev. Physiol.* **33**, 533–568.

Hoebel, B. G., and Teitebaum, P. (1966). Weight regulation in normal and hypothalamic hyperphagic rats. *J. Comp. Physiol. Psychol.* **61**, 189–193.

Hunt, P. C., and Jones, J. W. (1972). The food of brown trout in Llyn Alaw, Anglesey, North Wales. *J. Fish Biol.* **4**, 333–352.

Ivlev, V. S. (1961). "Experimental Ecology of the Feeding of Fishes." Yale Univ. Press, New Haven, Connecticut. (Transl. from Russ.)

Johnson, L. (1966a). Experimental determination of food consumption of pike, *Esox lucius*, for growth and maintenance. *J. Fish. Res. Board Can.* **23**, 1495–1505.

Johnson, L. (1966b). Consumption of food by the resident population of pike, *Esox lucius*, in Lake Windermere. *J. Fish. Res. Board Can.* **23**, 1523–1535.

Kapatos, G., and Gold, R. M. (1973). Evidence for ascending noradrenergic mediation of hypothalamic hyperphagia. *Pharmacol., Biochem. Behav.* **1**, 81–87.

Keast, A. (1968). Feeding of some Great Lakes fishes at low temperatures. *J. Fish. Res. Board Can.* **25**, 1199–1218.

Keenleyside, M. H., and Yamamoto, F. T. (1962). Territorial behaviour of juvenile Atlantic salmon (*Salmo salar* L.). *Behaviour* **19**, 139–169.

Kudrinskaya, O. M. (1970). Food and temperature as factors affecting the growth, development and survival of pike-perch and perch larvae. *J. Ichthyol.* **10**, 779–788. (Transl. from Russ.)

L'Hermite, A., and Lefranc, G. (1972). Recherches sur les voies monoaminergiques de l'encéphale d'*Anguilla vulgaris*. *Arch. Anat. Microsc. Morphol. Exp.* **61**, 139–152.

McCaig, R. S., and Mullan, J. S. (1960). Growth of eight species of fishes in Quabbin Reservoir, Massachusetts, in relation to age of reservoir and introduction of smelt. *Trans. Am. Fish. Soc.* **89**, 27–31.

McKeown, B. A., Leatherland, J. F., and John, T. M. (1975). The effect of growth hormone and prolactin on the mobilization of free fatty acids and glucose in the kokanee salmon, *Oncorhynchus nerka. Comp. Biochem. Physiol. B* **50**, 425–430.

Magnuson, J. J. (1962). An analysis of aggressive behaviour, growth, and competition for food and space in medaka (*Oryzias latipes (Pisces, Cyprinodontidae)*). *Can. J. Zool.* **40**, 313–363.

Martin, N. V. (1970). Long-term effects of diet on the biology of lake trout and the fishery in Lake Opeongo, Ontario. *J. Fish. Res. Board Can.* **27**, 125–146.

Mehrle, P. M., and Fleming, W. R. (1970). The effect of early and midday prolactin injection on the lipid content of *Fundulus kansae* held on a constant photoperiod. *Comp. Biochem. Physiol.* **36**, 597–603.

Meier, A. H. (1969). Diurnal variations of metabolic responses to prolactin in lower vertebrates. *Gen. Comp. Endocrinol., Suppl.* **2**, 55–62.

Meier, A. H. (1972). Temporal synergism of prolactin and adrenal steroids. *Gen. Comp. Endocrinol., Suppl.* **3**, 499–508.

Meier, A. H., Trobec, T. N., Haymaker, H. G., MacGregor, R., III, and Russo, A. C. (1973). Daily variations in the effects of handling on fat storage and testicular weights in several vertebrates. *J. Exp. Zool.* **184**, 281–288.

Mendelson, J. (1970). Palatability, satiation and thresholds for stimulus-bound drinking. *Physiol. Behav.* **5**, 1295–1297.

Miller, B. S. (1967). Stomach contents of adult starry flounder and sand sole in East Sound, Orcas Island, Washington. *J. Fish. Res. Board Can.* **24**, 2515–2526.

Mogenson, G. J. (1974). Changing views of the role of the hypothalamus in the control of ingestive behaviours. *In* "Recent Studies of Hypothalamic Function" (K. Lederis and K. E. Cooper, eds.), pp. 268–293. Karger, Basel.

Mrosovsky, N. (1974). Natural and experimental hypothalamic changes in hibernators. *In* "Recent Studies of Hypothalamic Function" (K. Lederis and K. E. Cooper, eds.), pp. 251–267. Karger, Basel.

Mrosovsky, N., and Fisher, K. C. (1970). Sliding set-points for body weight in ground squirrels during the hibernation season. *Can. J. Zool.* **48**, 241–247.

Müller, E. E., Pecile, A., Cocchi, D., and Olgiati, V. R. (1974). Hyperglycemic or feeding response to glucoprivation and hypothalamic glucoreceptors. *Am. J. Physiol.* **226**, 1100–1109.

Myers, R. D. (1974). "Handbook of Drug and Chemical Stimulation of the Brain," pp. 302–346. Van Nostrand-Reinhold, New York.

Myers, R. D. (1975). Brain mechanisms in the control of feeding: A new neurochemical profile theory. *Pharmacol., Biochem. Behav.* **3**, Suppl. 1, 75–83.

Nakashima, B. S., and Leggett, W. C. (1975). Yellow perch (*Perca flavescens*) biomass responses to different levels of phytoplankton and benthic biomass in Lake Memphremagog, Quebec–Vermont. *J. Fish. Res. Board Can.* **32**, 1785–1797.

Nance, D. M., and Gorski, R. A. (1975). Neurohormonal determinants of sex differences in the hypothalamic regulation of feeding behaviour and body weight in the rat. *Pharmacol., Biochem. Behav.* **3**, Suppl. 1, 155–162.

Nikolsky, G. V. (1963). "The Ecology of Fishes." Academic Press, New York. (Transl. from Russ.)

Novin, D., Vander Weele, D. A., and Rezek, M. (1973). Infusion of 2-deoxy-D-glucose into the hepatic portal system causes eating: Evidence for peripheral gluco-receptors. *Science* **181**, 858–860.

Olla, B. L., and Samet, C. (1974). Fish-to-fish attraction and the facilitation of feeding behaviour as mediated by visual stimuli in striped mullet, *Mugil cephalus. J. Fish. Res. Board Can.* **31**, 1621–1630.

Otto, R. G. (1971). Effects of salinity on the survival and growth of presmolt coho salmon (*Oncorhynchus kisutch*). *J. Fish. Res. Board Can.* **28**, 343–349.

Palka, Y., Liebelt, R. A., and Critchlow, V. (1971). Obesity and increased growth following partial or complete isolation of ventromedial hypothalamus. *Physiol. Behav.* **7**, 187–194.

Paloheimo, J. E., and Dickie, L. M. (1966). Food and growth of fishes. II. Effects of food and temperature on the relation between metabolism and body weight. *J. Fish. Res. Board Can.* **23**, 869–908.

Panksepp, J. (1971). Is satiety mediated by the ventromedial hypothalamus? *Physiol. Behav.* **7**, 381–384.

Panksepp, J. (1974). Hypothalamic regulation of energy balance and feeding behavior. *Fed. Proc., Fed. Am. Soc. Exp. Biol.* **33**, 1150–1165.

Panksepp, J. (1975). Central metabolic and humoral factors involved in the neural regulation of feeding. *Pharmacol., Biochem. Behav.* **3**, Suppl. 1, 107–119.

Paxinos, G., and Bindra, D. (1973). Hypothalamic and midbrain neural pathways involved in eating, drinking, irritability, aggression, and copulation in rats. *J. Comp. Physiol. Psychol.* **82**, 1–14.

Peter, R. E. (1970). Hypothalamic control of thyroid gland activity and gonadal activity in the goldfish, *Carassius auratus. Gen. Comp. Endocrinol.* **14**, 334–356.

Peter, R. E., and Gill, V. E. (1975). A stereotaxic atlas and technique for forebrain nuclei of the goldfish, *Carassius auratus. J. Comp. Neurol.* **159**, 69–102.

Peter, R. E., and Monckton, E. A. (1975). Induction of hypophagia by gold thioglucose in goldfish. *Proc. Can. Fed. Biol. Soc.* **18**, 105. (Abstr.)

Peter, R. E., Monckton, E. A., and McKeown, B. A. (1976). The effects of gold thioglucose on food intake, growth, and forebrain histology in goldfish, *Carassius auratus. Physiol. Behav.* **17**, 303–312.

Pickford, G. E., and Atz, J. W. (1957). "The Physiology of the Pituitary Gland of Fishes." N.Y. Zool. Soc., New York.

Powley, T. L., and Opsahl, C. A. (1974). Ventromedial hypothalamic obesity abolished by subdiaphragmatic vagotomy. *Am. J. Physiol.* **226**, 25–33.

Rabin, B. M. (1972). Ventromedial hypothalamic control of food intake and satiety: A reappraisal. *Brain Res.* **43**, 317–342.

Rowland, N. E., and Antelman, S. M. (1976). Stress-induced hyperphagia and obesity in rats: A possible model for understanding human obesity. *Science* **191**, 310–312.

Rozin, P., and Mayer, J. (1961). Regulation of food intake in the goldfish. *Am. J. Physiol.* **201**, 968–974.

Rozin, P., and Mayer, J. (1964). Some factors influencing short-term food intake of the goldfish. *Am. J. Physiol.* **206**, 1430–1436.

Rumsey, G. L., and Ketola, H. G. (1975). Amino acid supplementation of casein in diets of Atlantic salmon (*Salmo salar*) fry and of soybean meal for rainbow trout (*Salmo gairdneri*) fingerlings. *J. Fish. Res. Board Can.* **32**, 422–426.

Saunders, R. L., and Henderson, E. B. (1970). Influence of photoperiod on smolt development and growth of Atlantic salmon (*Salmo salar*). *J. Fish. Res. Board Can.* **27**, 1295–1311.

Savage, G. E. (1969). Some preliminary observations on the role of the telencephalon in food-reinforced behaviour in the goldfish, *Carassius auratus. Anim. Behav.* **17**, 760–772.

Savage, G. E., and Roberts, M. G. (1975). Behavioural effects of electrical stimulation of the hypothalamus of the goldfish (*Carassius auratus*). *Brain Behav. Evol.* **12**, 42–56.

Schultz, D. C., and Northcote, T. G. (1972). An experimental study of feeding behaviour and interaction of coastal cutthroat trout (*Salmo clarki clarki*) and dolly varden (*Salvelinus malma*). *J. Fish. Res. Board Can.* **29**, 555–565.

Sclafani, A., and Maul, G. (1974). Does the ventromedial hypothalamus inhibit the lateral hypothalamus? *Physiol. Behav.* **12**, 157–162.

Sclafani, A., Berner, C. N., and Maul, G. (1975). Multiple knife cuts between the medial and lateral hypothalamus in the rat: A reevaluation of hypothalamic feeding circuitry. *J. Comp. Physiol. Psychol.* **88**, 210–217.

Segaar, J. (1965). Behavioural aspects of degeneration and regeneration in fish brain: A comparison with higher vertebrates. *Prog. Brain Res.* **14**, 143–231.

Shelbourn, J. E., Brett, J. R., and Shirahata, S. (1973). Effect of temperature and feeding regieme on the specific growth rate of sockeye salmon fry (*Oncorhynchus nerka*), with a consideration of size effect. *J. Fish. Res. Board Can.* **30**, 1191–1194.

Shul'man, G. E. (1974). "Life Cycles of Fish." Wiley, New York. (Transl. from Russ.)

Slaney, P. A., and Northcote, T. G. (1974). Effects of prey abundance on density and territorial behaviour of young rainbow trout (*Salmo gairdneri*) in laboratory stream channels. *J. Fish. Res. Board Can.* **31**, 1201–1209.

Smith, G. P., and Gibbs, J. (1975). Cholecystokinin: A putative satiety signal. *Pharmacol. Biochem. Behav.* **3** (Suppl. 1), 135–138.

Smith, G. P., Gibbs, J., and Young, R. C. (1974). Cholecystokinin and intestinal satiety in the rat. *Fed. Am. Soc. Exp. Biol.* **33**, 1146–1149.

Steffens, A. B., Mogenson, G. J., and Stevenson, J. A. F. (1972). Blood glucose, insulin, and free fatty acids after stimulation and lesions of the hypothalamus. *Am. J. Physiol.* **222**, 1446–1452.

Stein, R. A., Reimers, P. E., and Hall, J. D. (1972). Social interaction between juvenile coho (*Oncorhynchus kisutch*) and fall chinook salmon (*O. tshawytscha*) in Sixes River, Oregon. *J. Fish. Res. Board Can.* **29**, 1737–1748.

Stevenson, J. A. F. (1969). Neural control of food and water intake. *In* "The Hypothalamus" (W. Haymaker, E. Anderson, and W. J. H. Nauta, eds.), pp. 524–621. Charles C Thomas, Springfield, Illinois.

Stewart, N. E., Shumway, D. L., and Doudoroff, P. (1967). Influence of oxygen concentration on the growth of juvenile largemouth bass. *J. Fish. Res. Board Can.* **24**, 475–494.

Storlien, L. H., and Albert, D. J. (1972). The effect of VMH lesions, lateral cuts and anterior cuts on food intake, activity level, food motivation, and reactivity to taste. *Physiol. Behav.* **9**, 191–197.

Sutin, J., McBride, R. L., Thalman, R. H., and Van Atta, E. L. (1975). Organization of some brainstem and limbic connections of the hypothalamus. *Pharmacol., Biochem. Behav.* **3**, Suppl. 1, 49–59.

Sutterlin, A. M., and Sutterlin, N. (1970). Taste responses in Atlantic salmon (*Salmo salar*) parr. *J. Fish. Res. Board Can.* **27**, 1927–1942.

Symons, P. E. K. (1968). Increase in aggression and in strength of the social hierarchy among juvenile Atlantic salmon deprived of food. *J. Fish. Res. Board Can.* **25**, 2387–2401.

Tugendhat, B. (1960a). The normal feeding behaviour of the three-spined stickleback (*Gasterosteus aculeatus* L.). *Behaviour* 15, 284–318.

Tugendhat, B. (1960b). The disturbed feeding behaviour of the three-spined stickleback: I. Electric shock is administered in the food area. *Behaviour* 16, 159–187.

Tyler, A. V., and Dunn, R. S. (1976). Ration, growth, and measures of somatic and organ condition in relation to meal frequency in winter flounder, *Pseudopleuronectes americanus*, with hypotheses regarding population homeostasis. *J. Fish. Res. Board Can.* 33, 63–75.

Ware, D. M. (1972). Predation by rainbow trout (*Salmo gairdneri*): The influence of hunger, prey density, and prey size. *J. Fish. Res. Board Can.* 29, 1193–1201.

Wayner, M. J., and Oomura, Y. (1975). Central neural control of eating and obesity. *Pharmacol., Biochem. Behav.* 3, Suppl. 1, 1–173.

Wayner, M. J., Ono, T., DeYoung, A., and Barone, F. C. (1975). Effects of essential amino acids on central neurons. *Pharmacol., Biochem. Behav.* 3, Suppl. 1, 85–90.

Weiss, J. (1970). Saisonale Veränderungen des Enzymmusters und des Neurosekretgehaltes sorvie die Innervation des Nucleus praeopticus der Bachforelle (*Salmo trutta fario*) unter besonderer Berücksichtigung der hypothalamischen Hydrencephalokrinie. *Morphol. Jahrb.* 115, 444–486.

Welty, J. C. (1934). Experiments in group behaviour of fishes. *Physiol. Zool.* 7, 85–128.

Werner, E. E., and Hall, D. J. (1976). Niche shifts in sunfishes: Experimental evidence and significance. *Science* 191, 404–406.

Wilson, W. H., and Heller, H. C. (1975). Elevated blood glucose levels and satiety in the rat. *Physiol. Behav.* 15, 137–143.

Wise, J. K., Hendler, R., and Felig, P. (1972). Obesity: Evidence of decreased secretion of glucagon. *Science* 178, 513–514.

Yamazaki, F. (1976). Application of hormones in fish culture. *J. Fish. Res. Board Can.* 33, 948–958.

4

DIGESTION

RAGNAR FÄNGE and DAVID GROVE

FISH PHYSIOLOGY, VOL. VIII
Copyright © 1979 by Academic Press, Inc.
All rights of reproduction in any form reserved.
ISBN 0-12-350408-2

I. INTRODUCTION

Fishes are the dominating vertebrate group as far as number of species is concerned, and in their immense variety have adopted many nutritional habits. One can distinguish piscivores, insectivores, molluscivores, large plant feeders (herbivores), phyto- and zooplanktivores, mud feeders (detritivores), cleaner fish, and, especially in the primitive Cyclostomata, parasites and feeders on carcasses. Some species are extremely specialized in their feeding habits while others are omnivorous. As described elsewhere (Chapter 3) fishes have specific amino acid, lipid, carbohydrate, vitamin, inorganic ion, and water requirements. A wide variety of structural and physiological adaptations permit fishes to capture, digest and absorb these requirements from their food. Several previous surveys deal with digestion and digestive organs in fishes: Biedermann (1911), anatomy, digestive physiology; Pernkopf and Lehner (1937), anatomy of the intestine; Jacobshagen (1937), anatomy of the stomach; Suyehiro (1942), anatomy, feeding habits; Al-Hussaini (1949a,b), functional anatomy; Bernard (1952), digestion; Barrington (1957, 1962), digestion; Bertin (1958), anatomy; Creac'h (1963), proteolytic enzymes; Smit (1968), gastric digestion in lower vertebrates; Barnard (1973), comparative biochemistry; Prosser (1973), comparative physiology; Kapoor *et al.* (1975a,b), digestion, gustatory system.

Our aim in this article is to review the current information on the structure and physiology of the fish alimentary canal but particularly to stress the mechanisms controlling the movement and digestion of food. The rate at which fish digest their food is of primary importance in determining the rates of feeding and growth (see Chapters 3 and 11).

II. FEEDING MECHANISMS

Feeding mechanisms in the jawless Cyclostomata are different from those of other vertebrates. The round suctorial mouth of the adult *Petromyzon* and *Lampetra* is armed with horny teeth. Anticoagulant secretions allow tissue fluids and blood to be ingested and passed directly to the intestine, since cyclostomes possess no stomach. Juvenile (ammocoete) lampreys survive in freshwater for several years by microphagous feeding. At metamorphosis pouchlike gills are formed inside the gill arches which generate a tidal water flow quite independent of the mouth to facilitate respiration while the fish is attached to the host. Hagfish, which possess barbels, have horny teeth on the palate and tongue.

In the gnathostomata, the anterior visceral arches have formed jaws which are relatively simple in the Chondrichthyes. Ectodermal folds inside the jaws produce a series of teeth which move upward to replace those that are lost. The teeth may be homodont (*Raja*) or heterodont (*Heterodontus*) in relation to the diet (Reif, 1976).

In the osteichthyes, the membrane and cartilage bones which form the jaws have a more complex structure provided with an equally complex arrangement of muscles, nerves, and ligaments. Jaw movements in both groups are associated with respiration, biting, scraping, chewing, and rejection of particles. Ballentijn *et al.* (1972) gave a detailed description of the architecture of the jaws of the carp, showing how the movements of the premaxilla and maxilla allow flexible protrusor movements of the mouth to change its shape and position under the influence of the adductor mandibulae 1 alpha and beta, 2 and 3. The mouth can be turned ventrally for feeding and ejection of particles but can be closed without compressing the buccal cavity when full of food. Similar analyses of jaw movements are given in Alexander (1970) and Osse (1969). Keast and Webb (1966) made a detailed comparison of the mouth and body structure of fifteen species of teleosts of one Canadian lake; Hatanaka *et al.* (1954) made a similar study of flatfish, while Hobson (1974) undertook a survey of the feeding relationships among more than 100 species of marine teleosts of the Coral Reefs at Hawaii. The latter author concluded that, in this marine community, the carnivorous habit is central to teleostean evolution. The relatively unspecialized carnivores have limited prey species which are vulnerable to attack. These are mainly nocturnal in activity as are the fish predators. After the final development of modern lithothamnion–scleractinian coral reefs some 50 million years ago, an explosive radiation of acanthopterygian teleosts occurred allowing them to become diurnal carnivores and planktivores, coral eaters, benthos foragers on large echinoids and mollusks or cleaner fish. In contrast to freshwater evolution, herbivorous species probably appeared relatively late. Recognizable adaptations to the new feeding niches are the following.

1. New positions of the paired fins for increased maneuverability
2. Reorganization of the premaxilla/maxilla for greater flexibility and protrusibility

3. Changes in jaw and snout shape such as elongation for snipping off coral polyps (*Chaetodon*) or other sessile invertebrates (*Forcipeger*) leading to the evolution of cleaner fishes (*Labroides phthiriphagus*)
4. Changes in tooth shape such as the delicate incisors of *Chaetodon*, the crushing teeth of blennies, or even the fused plates of the tetraodontiformes
5. Development of accessory structures such as the barbels of *Mullus* to detect prey buried in the substratum
6. Behavioral changes allowing *Coris* to roll over stones to detect small prey or *Sufflamen* to expose buried prey with water jets from fins or gills.

These are only a few examples of adaptations of feeding mechanisms. Lips may be present (*Catostomus, Mugil*) or completely absent (*Sparus*). Accessory external organs bearing taste buds and used for detection and location of food are found as barbels on the snout or lower jaw (Cyprinidae, Siluridae, Gadidae, Mullidae, and others) or as sensory fin areas (Gadidae, Triglidae) (Kapoor *et al.*, 1975a). Teeth may be found not only on the jaws (premaxilla, maxilla, mandible) but also on the prevomer, vomer, palatine, sphenoid, tongue, and the dorsal and ventral regions of the pharynx. Among the teleosts, a bewildering array of tooth and gill raker adaptations are encountered which allow successful ingestion of the preferred food. The teeth may be absent from the jaws (Cyprinidae), minute (planktivorous clupeids), flat and molariform (*Raja, Brama*), incisiform (*Blennius*), pointed or serrate (*Sphyraena*, sharks) or fused into crushing plates (*Tetraodon*). Gill rakers may form blunt "teeth" (most carnivores) or a filter basket (*Dorosoma, Alosa, Polyodon, Labeo*) (see, e.g., Suyehiro, 1942; Weisel, 1973). A tongue is not always present (*Labeo*). It is rarely freely movable, yet in *Dorosoma* it is protrusible and in *Plecoglossus* it forms flaps producing mucus to entrap algal particles scraped off by the comblike teeth. In the pharynx, dorsal and ventral pharyngeal pads may be developed to crush the food, or to compress the algal or detrital ingesta before swallowing (Cyprinidae, Catostomidae, Cobitidae). Pharyngeal and epibranchial organs, with a lumen entering the esophagus, are believed to consolidate food particles and are found in several genera among the Osteoglossiformes, Cypriniformes, Gonorhychiformes, and Clupeiformes (Nelson, 1967). There are no multicellular salivary glands in fish, but solitary mucus-producing gland cells (goblet cells) lubricate the food to facilitate swallowing.

III. ANATOMY AND HISTOLOGY OF THE ALIMENTARY CANAL

The major divisions of the vertebrate alimentary canal are mouth, buccal cavity, pharynx, esophagus, stomach, intestine, rectum, and related organs. In some fishes the digestive canal constitutes a straight tube from the mouth to the anus. More often, however, the canal makes loops and is structurally divided into functionally different parts. Thus one can usually distinguish esophagus, stomach, and intestine and often subdivisions of these. Valves or sphincters often separate different parts of the digestive canal. The principal layers of the gut wall are the mucosa (inner epithelium and adjacent tissues), submucosa, muscularis (usually double layered), and serosa. Associated with the canal are two glands, the liver and the pancreas, which deliver their secretions into the intestinal lumen through special ducts.

It is generally agreed that the structure of the regions of the alimentary canal in a given species is related to its diet but that modifications are superimposed on the basic gut plan of the group to which the species belongs. An example of this is given by Weisel (1962) who examined the cyprinid *Ptychocheilus oregonense,* which preys on young salmon, but which has inherited the toothless and stomachless condition from ancestors assumed to be catostomid suctorial feeders on fine particles.

A number of comprehensive articles have appeared in which the morphology, histology, and cytology of the fish alimentary canal have been described. There are many studies in addition to the reviews mentioned in the Introduction (Ishida, 1935; Kirtisinghe, 1940; Girgis, 1952; Burnstock, 1959a,b; Weisel, 1962; Mohsin, 1962; Hale, 1965; Bishop and Odense, 1966; Keast and Webb, 1966; Bullock, 1963, 1967; Chaichara and Bullock, 1967; Schmitz and Baker, 1969; Frantsusova, 1971; De Groot, 1971; Bucke, 1971; Vegas-Velez, 1972; Chakrabarte *et al.,* 1973; Kayanja *et al.,* 1975). Tanaka (1973) has investigated the structure and function of the digestive system of teleost larvae.

A. Esophagus

Posteriorly, the pharynx passes into a short, wide, muscular esophagus. In elasmobranchs, the esophageal mucosa is often provided with cone-shaped or branched papillae directed backward. Without marked boundaries the esophagus merges caudally with the stomach. In many species the submucosa contains voluminous masses of lymphomyeloid tissue ("organ of Leydig"). In teleosts the mucosa is dominated by characteristic large mucous cells (goblet cells) which may give the epithelium a "frothy" appearance in histological sections. The mucosal epithelium is said to be typically stratified (Kapoor *et al.*, 1975b), although Vegas-Velez (1972) found it to be simple in the species he examined. The mucosa, including the basement membrane, and the stratum compactum are usually thrown into folds which allow distension during swallowing. The muscular coat is typically of striated muscle. If a circular layer is present (*Gadus, Labeo*) it lies outside the longitudinal coat; *Gasterosteus* has only the circular coat. The muscles are innervated by the Xth (vagus) nerve. Glands similar to gastric glands have been observed in the caudal esophagus of some species such as *Mugil capito* (Ghazzawi, 1935), *Cottus gobio* and (*Par*)*enophrys bubalis* (Western, 1969), and *Dorosoma cepedianum* (Schmitz and Baker, 1969). In a variety of fishes examined by Isokawa *et al.* (1965) and Khanna and Mehrotra (1970), esophageal sacs are reported with (*Ariomma, Pampus*) or without (*Tetragonus, Iticus*) teeth. The esophagus may terminate in a cardiac sphincter or valve (*Labeo*) although such demarcation is not invariable (Odense and Bishop, 1966; Schmitz and Baker, 1969). In stomachless fishes the esophagus enters the intestine directly.

B. Stomach

Within the gnathostomata, the stomach is claimed to be a concomitant development with the jaws to receive and store newly ingested food, which may be of large size, and to initiate digestion with pepsin in an acid medium. In elasmobranchs, the stomach is present as a J-shaped organ consisting of a descending *pars cardiaca* and an ascending *pars pylorica* (Fig. 1). The inner epithelium of the mucosa is simple and consists of cylindrical cells which stain with the periodic acid Schiff (PAS) reagent and which probably secrete mucin. Tubular multicellular glands running perpendicular to the luminal border open into mucosal foveolae. The gland cells are of one type only containing acidophilic granules. They may be designated oxyntic cells (Hogben, 1967a,b). A muscularis mucosa is present in the cardiac region only. The muscularis (externa) consists of an inner circular and an outer longitudinal layer of smooth muscles. The circular muscles are strongly developed in the pyloric region to form the pyloric sphincter (Petersen, 1908–1909; Oppel, 1896–1900). In some elasmobranchs, a chamberlike enlargement, the *bursa entiana,* is formed in the pyloric

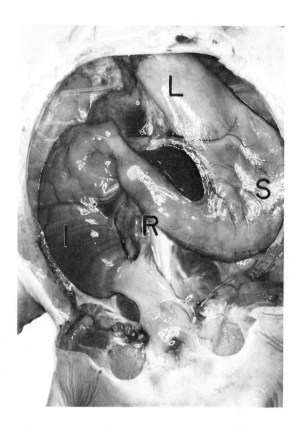

Fig. 1. *Raja radiata.* Abdominal viscera after removal of the liver. S, stomach; I, spiral intestine; L, esophagus with the organ of Leydig; R, rectal gland.

part of the stomach. Doughnut-shaped sphincters are found around veins in the gastric wall of rays (*Raja*) (Weinland, 1901), but similar sphincters occur also elsewhere in the body of rays.

In the osteichthyes the stomach is usually well developed, although in some forms it is reduced or absent (e.g., Cyprinidae, Cyprinodontidae: Rauther, 1940). Tanaka (1969) reported that the stomach is absent from larvae of young fish such as *Cyprinus* and *Salmo*, so the stomach develops (if at all) late in the ontogeny. The loss of stomach may therefore be neotenous and is frequently associated with microphagous habits. The stomachless fishes are stated to have no gastric glands and no pyloric sphincter. However, stomachless predators (labrids, *Ptychocheilus, Scomberesox*) as well as planktivores (*Tilapia, Syngnathus*) prove exceptions to the above generalization. Bertin

(1958) and Vegas-Velez (1972) point out that earlier reviews, tabulating families all of whose members are "stomachless," may be in error since histological examinations have demonstrated the presence of gastric glands in some species (e.g., *Syngnathus, Mugil*). Gupta (1971) observed cells resembling gastric gland cells in the digestive canal of the stomachless carnivorous fish, *Xenentodon cancila*.

Many of the stomachless fishes are provided with pharyngeal chewing devices which permit the ingested food to reach the intestine in a fragmented condition (Rauther, 1940; Bertin, 1958).

Where present, the stomach of teleosts varies greatly in shape. A straight tubelike stomach is found in Gobiidae, Gasterosteiformes, and Symbranchi (*Pomatoschistus, Gasterosteus, Spinachia, Symbranchus*). In flatfish and esocids (*Limanda, Pleuronectes, Esox*) it is curved, whereas in most fishes the stomach may be shaped like one of the letters U, J, or Y (Suyehiro, 1942). In cottids (*Myoxocephalus, Enophrys*) and *Tilapia* the stomach is saclike while in elopids, clupeids, *Gymnarchus, Ophiocephalus, Anguilla* and many others a gastric cecum extends caudally. This cecum sometimes is very long (*Regalecus, Stomias,* Rauther, 1940).

The teleost stomach typically has a lining of columnar epithelial cells, without a striated border. Goblet cells are scattered through the epithelium. Tubular glands, sometimes in groups, are found in the cardiac and fundic regions. The tubules may be simple or occasionally branched, their presence thickening the depth of the mucosa. They open into foveolae. Mucus-producing neck cells may be distinguished. As in elasmobranchs (and in fact in nonmammalian vertebrates generally) the main part of the gastric glands are made up of one type of cell only. This contains abundant secretory granules, probably pepsinogen (Tan and Teh, 1974), but the cells are believed at the same time to be producers of HCl (Barrington, 1957; Iro, 1967). According to Weinreb and Bilstad (1955), gastric gland cells in the rainbow trout (*Salmo gairdneri*) structurally resemble chief cells in other animals. Changes in the microscopic structure of the gastric glands during secretion were reported by Arcangeli (1908). The ultrastructure of gastric gland cells in teleosts has been studied by Ling and Tan (1975) in the coral fish, *Chelmon rostratus,* and by Noaillac-Depeyre and Gas (1978) in the perch, *Perca fluviatilis*. The gastric gland cells in their apical region contain a compact system of tubules. The tubules show resemblances to tubular structures of oxyntic cells of amphibians and higher vertebrates and to structures in the chloride cells of cyclostomes (lamprey). At their bases the cells contain zymogenlike secretory granules and a rich rough endoplasmic reticulum. The secretory granules are released apically by a process of exocytosis. The ultra-structural features of the gastric gland cells are consistent with the hypothesis that they are active both in acid production and in the synthesis of pepsinogen.

Supporting the gastric mucosa is a submucosa often containing a stratum compactum and smooth muscle fibers (Rauther, 1940; Burnstock, 1959a). The muscularis consists of an inner circular and an outer longitudinal layer of smooth muscles, but striated esophageal muscles may extend into the cardiac portion of the stomach (*Perca, Centropristes, Zeus, Solea,* Blake, 1930; Rauther, 1940). In the stomiatid *Cyclothone* the stomach wall contains two layers of diagonally-crossing striated muscle fibers (Nusbaum, 1923) and in the stomachless cyprinid *Tinca* a layer of inner circular and outer longitudinal striated muscle fibers surround the "normal" (smooth muscle) muscularis (Kilarski and Bigäi, 1971). The inner circular layer of smooth muscles of the fish stomach is usually two to three times thicker than the longitudinal coat, and this is accentuated in fishes provided with a gizzard. The pyloric sphincter consists of thickening of the circular smooth muscle layer. In stomachless fishes the sphincter may be absent or replaced by an esophageo–intestinal valve formed by a fold of the mucosa and submucosa.

The capacity of the stomach in relation to the body weight varies between species and is reflected in the size of the meal that can be taken voluntarily. The flatfish *Limanda* for example has a gastric volume of 8 ml/100 g and can ingest up to 10% of its body weight in a meal. The stomachless *Leuciscus rutilus* can consume 15% of its body weight of chironomid larvae, and *Carassius carassius* 21% of its body weight. Sculpins may ingest 30–50% of their body weight at a single feeding.

Several teleosts possess a gizzardlike enlargement of the *pars pylorica* (*Mugil* spp., *Coregonus, Osphromenus, Chanos, Sardinella, Chatoessus, Citharinus, Mormyrus, Notopterus*). In *Dorosoma* the muscularis of the gizzardlike part of the stomach consists of three strongly developed layers of smooth muscles and has a thick mucous cuticula lining the lumen. Fish with gizzards usually are micropha-gous, detritivores, or herbivores. *Mugil cephalus* ingests microalgae and plant detritus together with mineral particles which act as a grinding paste (Rauther, 1940; Schmitz and Baker, 1969; Odum, 1970).

On of us (D.G.) has observed that when two species of small shore fish (*Blennius pholis* and *Ciliata mustela*) are offered barnacles (*Balanus balanoides*) as food, the fish ingest them readily although only the blenny takes this species in the wild. The blenny, however, has no stomach and as the meal is digested the barnacle's calcareous plates are readily transferred to the intestine and defecated within 24 hr at 18°C. C. *mustela,* on the other hand, retains the plates long after the organic part of the meal is evacuated from the stomach and cannot transfer them to the intestine. Clearly the retention of the pyloric sphincter can limit the utilization of available food species.

C. Intestine

In elasmobranchs and holocephalans, the intestinal wall consists of the usual layers of mucosa, submucosa, muscularis, and serosa. In development, independent twisting of the mucosa and supporting tissues leads to the formation of a spiral valve which increases intestinal surface area (Fig. 1). The number of turns in the spiral may be as low as two or three (*Chimaera monstrosa*) or as many as fifty (*Cetorhinus maximus*) and reflects the diet of the species (Rauther, 1940; Bertin, 1958). Spiral valves are also found in the intestine of dipnoi, polypterids, holosteans, acipenseroids, and in the coelacanth, *Latimeria chalumnae*.

In teleosts the intestine may be short and straight or thrown into folds or loops. Its length varies from one-fifth to twenty times the body length and it is longest in microphagous and herbivorous fish (Bryan, 1975). This trend is supported by tabulated data on characinoid and cyprinoid species (Kapoor *et al.,* 1975b). de Groot (1971) reviewed the morphological variations in flatfish guts in relation to diet for 133 species and, in more detailed analysis of 31 species, found that the relative length is greatest in Soleidae (which ingest smaller polychaetes, mollusks, and crustaceans) and least in Psettodidae and Bothidae (which eat fish and larger invertebrates). A similar trend was described by Hatanaka *et al.* (1954). The surface area of the intestine in carnivorous teleosts is usually increased by folds of the mucosa but the intestine length is less than the body length. Al-Hussaini (1947) adopted the term "mucosal coefficient" to describe the relative surface area of the intestine in fishes to allow both for intestinal length and for mucosal foldings. Lange (1962) found that gut length increases with age in species of *Rutilus* as they graduate from yolk, through zooplankton to a diet of larger invertebrates. Angelescu and Gneri (1949) found that starvation of *Prochilodus* leads to intestinal shortening by 30–45%.

Many teleosts have blind tubes connected with the anterior end of the intestine. These intestinal ceca (often termed pyloric ceca) vary in number between 1 and more than 1000 (Suyehiro, 1942) (Figs. 2, 3). They may be relatively free and short (Pleuronectidae) or bound to-

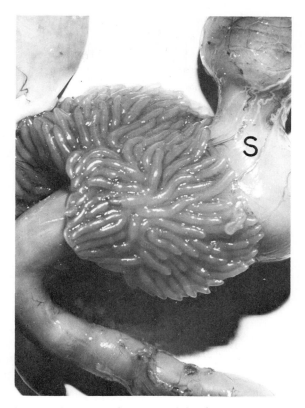

Fig. 2. *Cyclopterus lumpus.* Pyloric part of the digestive system in a freshly dissected specimen. In the middle of the figure are numerous intestinal (pyloric) ceca. A powerful ring-shaped muscle contraction is observed in the stomach (S).

gether to form a compact, glandlike mass (Thunnidae, Xiphidae). Very well developed ceca are found in certain malacopterygians (clupeids, salmonids), gadids, and coryphaenoids. Their presence is not related to the relative gut length nor to the general feeding niche of the species, yet within a chosen group of fishes (e.g., Heterosomata), it seems possible to relate the number of ceca to the diet (de Groot, 1971).

The intestinal ceca closely resemble the intestine in structure, with a well-developed muscularis consisting mainly of circular muscle fibers. The inner epithelium contains goblet cells, but light and electron microscopic studies indicate that they lack cells that secrete digestive enzymes (Jansson and Olsson, 1960; Luppa, 1966; Vegas-Velez, 1972). Between the ceca highly basophilic pancreatic exocrine gland cells may occur in the connective tissue. The function of the ceca is not clear. Since they originate from that region of the intestine

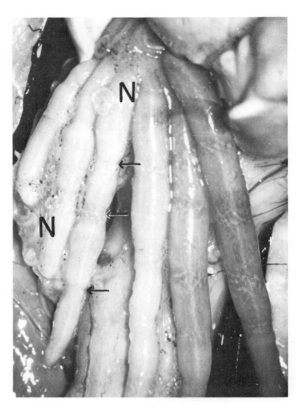

Fig. 3. *Micromesistius poutassou*. Intestinal (pyloric) ceca in a freshly dissected preparation. The ceca to the right in the figure are dilated with semitransparent walls. The ceca to the left show constrictions (arrows) due to contraction of circular smooth muscles. (N, parasitic nematodes in the connective tissue surrounding the ceca).

where bile and pancreatic juices are released, the ceca may form digestive compartments active in resorption of certain nutrients. It has been supposed that they are of especial importance for the resorption of fat and waxes (Greene, 1914; Janson and Olsson, 1960; Benson and Lee, 1975). The ceca often harbor remarkably large numbers of parasites. Thus in the ceca of *Coregonus lavaretus* Reichenbach-Klinke and Reichenbach-Klinke (1970) observed up to 300 individual cestodes.

Stomachless fishes have no intestinal ceca. Other fishes without ceca include *Amia, Chirocentrus, Symbranchus, Anguilla, Esox, Gasterosteus, Agriopus, Megalops, Batrachus*, silurids, and loricarids (Jacobshagen, 1915; Pernkopf and Lehner, 1937).

The mucosa of the intestine is lined by a simple columnar epithelium which possesses the brush border of microvilli (apical plate) typical of absorptive tissues (Odense and Bishop, 1966; Yamamoto, 1966; Krementz and Chapman, 1975). Cartier and Buclon (1973) estimated the number of microvilli per mm^2 as 34×10^6 in cyprinids. Mucus-secreting goblet cells occur scattered among the epithelial cells and show a positive PAS reaction (Gammon *et al.*, 1972). Multicellular glands are not usually found. A stratum compactum is frequently present, but the teleostean intestine seems to lack a muscularis mucosa. The submucosa is very thin and contains scattered collagen and elastic fibers, blood vessels, and nerves. The muscularis consists of inner circular and outer longitudinal smooth muscles. In cyprinids, cobitids, and *Syngnathus*, striped muscles may be found in the muscularis (Rauther, 1940). In the tench (*Tinca tinca*) the striated muscle layer extends over the whole intestine (Ohnesorge and Rauch, 1968; Kilarski and Bigai, 1971).

Ciliated epithelium has been reported in the intestine of a few teleosts (*Syngnathus*) and some primitive fishes (*Polypterus senegalis*, *Polyodon spathula*) (Ishida, 1935; Bucke, 1971; Weisel, 1973; Magid, 1975) although it may be difficult to differentiate cilia from long microvilli on the basis of structure alone. Observations should always be made on living epithelial cells (Odense and Bishop, 1966). Where present, cilia are probably a primitive character since they are found in *Amphioxus* and in ammocoete larvae. In the cyclostome *Myxine glutinosa* the intestinal epithelium produces a peritrophic membrane (Adam, 1960).

Rectum. The terminal part of the intestine is frequently differentiated as a wider "rectum," often demarcated from the intestine by an ileo–rectal valve largely formed by smooth muscles (*Gadus, Gambusia*), but in many species no valve is present (Mohsin, 1962). The mucosa is typically richly endowed with goblet cells. The mucosal epithelium may be provided with microvilli indicating strong absorptive properties (Bullock, 1967). In elasmobranchs the epithelium may be stratified. The rectal gland of elasmobranchs (Fig. 1) is now believed to secrete Na and Cl ions into the lumen as part of the osmoregulatory mechanisms. The muscularis of the rectum is similar to that of the intestine, but striated muscles occur near the anus.

D. Liver and Gallbladder

The liver lies anteriorly in the body cavity and receives blood from the hepatic artery and from one or more portal veins which drain the gastric and intestinal mucosa, the swimbladder (gas gland), the spleen, and the pancreas. Embryologically, the liver develops as an epithelial outgrowth of the digestive tube, and in its simplest form is a branched tubular gland as in the cyclostome *Myxine glutinosa*. In some teleosts the liver is claimed to have a tubular structure (*Anguilla anguilla*, *Muraena, Pleuronectes*). However, in most vertebrates, including fishes, the liver is a complicated structure consisting of anastomosing epithelial lamellae separating blood sinusoids (Elias and Sherrick, 1969). This "muralium" has been found in fishes as diverse as lampreys (Bengelsdorf and Elias, 1950), *Salmo gairdnerii* (Scarpelli *et al.*, 1963), *Poecilia* (Hale, 1965), *Ictalurus* (Hinton and Pool, 1976), and *Micropterus* (Hinton *et al.*, 1972). As in other vertebrates, hepatocytes contain numerous mitochondria, rough endoplasmic reticulum (microsomes), Golgi apparatus, lysosomes, peroxisomes, lipid and glycogen deposits. Intracellular bile canaliculi are present, with the single exception to date of *Ictalurus* (Hinton and Pool, 1976), which link up with intercellular canaliculi. The lining epithelial cells of the canaliculi possess numerous microvilli (David, 1961; Yamamoto, 1965a,b). According to Welsch and Storch (1973) the teleostean liver contains two categories of hepatocytes, lipid-rich and glycogen-rich. In some species (*Tetraodon, Limanda*) lipid-rich cells predominate while in others (*Chromis, Corydoras, Amphiprion*) glycogen-rich cells are more common. Hinton and Pool (1976) concluded that, in *Ictalurus*, large protruding cells, which differ from hepatocytes lining the sinusoids, are Kupffer cells of the reticulo-endothelial system. Lipids may be present in large quantities in some fish livers. In elasmobranchs, squalene and other hydrocarbons accumulate in the liver, helping the fish to gain buoyancy (Corner *et al.*, 1969).

Due to the presence of a large fat vacuole the hepatocytes of elasmobranchs are larger than those of teleosts. In most fishes the exocrine secretion of the liver, the bile, is stored in a gallbladder. The bile duct opens into the anterior intestine or into the intestinal ceca (Western, 1969). The wall of the gallbladder consists of columnar or cuboidal epithelium, a thin submucosa and a muscularis of smooth muscle cells. The similarly constructed *ductus choledochus* is associated with the pancreatic duct in many fish, and its entrance to the intestine is guarded by the smooth muscle sphincter Oddi.

E. Pancreas

In the cyclostome *Myxine glutinosa* (hagfish) the pancreas consists of exocrine gland cells, containing zymogen granules, in the intestinal mucosa. The homologue of the endocrine pancreatic tissue of higher vertebrates is concentrated around the bile duct. In elasmobranchs and holocephalans, the exocrine pancreas forms a discrete organ with a duct emptying into the anterior intestine (duodenum). The microscopic structure may be much like that of the mammalian pancreas [e.g., in *Chimaera monstrosa* (Fujita, 1962)] and numerous scattered islets of endocrine cells are present. In elasmobranchs the spleen and the pancreas are often closely associated. This is also the case in *Protopterus*, the African lungfish, in which the pancreas is a large black-pigmented organ at the posterior end of the spleen and, like the latter, embedded in the gut wall. In teleosts the exocrine pancreas is usually diffuse. It consists of ramified tubules or acini scattered in the connective tissue of the intestinal surface, the mesenteries, between the intestinal ceca or within the liver or spleen (Nagase, 1964; Bishop and Odense, 1966; Western, 1969; Gammon *et al.*, 1972; Hinton and Pool, 1976). In the mesenteries the pancreas forms sheaths around blood vessels. Several small pancreatic ducts open into the intestine or intestinal ceca, although in some cases a single duct may join the bile duct. In cyprinids (*Cyprinus carpio*), labrids (*Crenilabrus melops*), and certain other forms, strands of pancreatic tissue are found within the portal canals of the liver, the pancreatic ducts probably opening into the bile ducts. The combined liver and pancreatic tissue has been termed "hepatopancreas" but must not be confused with so-called hepatopancreas of invertebrates (Ito *et al.*, 1962; Elias and Sherrick, 1969). The exocrine cells of the teleostean pancreas are strongly basophilic and contain zymogen granules. In a minority of teleosts the pancreas forms a distinct organ which is embedded in fat and connective tissue. The compact pancreas accompanies portal vessels on the surface of the intestine (*Esox, Silurus, Parasilurus, Anguilla, Conger, Acerina, Pleuronectes*, Oppel, 1896–1900; Hill, 1926; Chesley, 1934; Suyehiro, 1942; Kukla, 1954, 1958). The secretory cells form acini with narrow lumina, and a common pancreatic duct opens into the anterior intestine. Blood reaches the pancreas from three arteries and drains into the v. portae (Kukla, 1958). For a discussion of the endocrine elements associated with the pancreas, see Section III,F.

F. Endocrine Cells and Hormones

The vertebrate alimentary canal and certain other organs contain large numbers of granulated endocrine cells. Two lines of these cells have been termed enterochromaffin (argentaffin) and enterochromaffinlike (argyrophil) cells. When supplied with amine precursors these cells produce and store amines. For that reason they have been termed APUD cells [abbreviation for Amine Precursor Uptake and Decarboxylation (Pearse, 1977)]. Another term is gastro–entero–pancreatic (GEP) cells (Solcia *et al.*, 1978). A classification on ultra-structural grounds of mammalian GEP cells has been suggested (Solcia *et al.*, 1978). Several investigators have demonstrated GEP cells in nonmammalian vertebrates such as cyclostomes and true fishes (Uggeri, 1938; Read and Burnstock, 1968; Bucke, 1971; Gabe and Matoja, 1971, 1972; Ling and Tan, 1975; Tue, 1975; Östberg, 1976; Falkmer *et al.*, 1978). These endocrine elements are also found in protochordates (amphioxus: Van Noorden and Pearse, 1976).

The GEP cells are assumed to synthesize low molecular weight polypeptides which act as hormones. The various polypeptides show structural resemblances to each other and the hypothesis has been forwarded that different polypeptide hormones (glucagon, secretin, gastrin) evolved together with insulin from an ancestral proinsulinlike molecule (Track, 1973). Several gastrointestinal polypeptide hormones occur in fish (Table I). Nilsson (1970, 1973) obtained evidence for the presence of secretin- and cholecystokininlike substances in the

Table I

Hormones and Hormonelike Substances in Fish Gut and Pancreas

Substance	Tissue	Species	Reference
Insulin	Islets	All	
Gastrin	Islets	*Rhinobatus productus*	Hansen (1975)
Secretin	Intestine	*Myxine glutinosa*	Nilsson (1970, 1973)
Secretin	Intestine	Salmon, dogfish, ray	Bayliss and Starling (1903)
Secretin	Intestine	*Esox, Gadus*	Dockray (1975)
Cholecystokinin (CCK)	Intestine	*Myxine glutinosa*	Nilsson (1973)
CCK	Intestine	*Chimaera monstrosa*	Nilsson (1970)
Cholecystokininlike substance	Intestine	*Lampetra, Petromyzon*	Barrington (1972)
Cholecystokininlike substance	Intestine	*Esox, Anguilla*	Barrington and Dockray (1972)
Ceruleinlike substance	Stomach	*Gadus morhua*	Larsson and Rehfeld (1978)
Substance P	Intestine	*Gadus morhua, Squalus acanthias*	Dahlstedt *et al.* (1959)
Histamine	Stomach	*Salmo gairdnerii, Gadus morhua*	Reite (1969)
Histamine	Stomach	*Esox lucius*	Lorentz *et al.* (1973)
Histamine	Intestine	*Myxine glutinosa*	Reite (1965)

intestine of the hagfish (*Myxine glutinosa*) and the holocephalan *Chimaera monstrosa*. Gabe and Matoja (1972) demonstrated cells structurally resembling mammalian gastrin cells in the stomach mucosa of the teleost, *Mugil auratus*. Östberg *et al.* (1976) showed, in immunofluorescence studies, that the intestinal mucosa of cyclostomes contains cells that react with antisera against mammalian glucagon and gastrin. Larsson and Rehfeld (1978) found that the gastric mucosa of the cod (*Gadus morhua*) contains a rich population of cells demonstrable with antisera against mammalian gastrin and cholecystokinin (CCK), but the immunogenic reactivity was due to a ceruleinlike polypeptide. The authors concluded that, in teleosts, ceruleinlike molecules possibly function as gastrin.

The endocrine pancreas belongs to the GEP system. Islets of endocrine cells are found in the compact pancreas of the eel (*Anguilla anguilla:* Kukla, 1958) and the pike (*Esox lucius:* Bucke, 1971). In teleosts with a diffuse pancreas (*Gadus morhua, Lophius piscatorius, Myoxocephalus scorpius*) aggregations of endocrine cells, separate from exocrine pancreas tissue, form the Brockmann bodies. Within the teleostean islet tissue Falkmer and Olsson (1962) and Bishop and Odense (1966) detected inner B cells (insulin secreting) and more peripheral A cells (glucagon secreting) together with other cell types. According to Brinn (1973) the islets of some fish species contain, in addition to A and B cells, D cells (somatostatin secreting) and a fourth argyrophilic type of granulated cell.

The pancreatic islets are innervated by autonomic nerves, and nervous influence may be a factor in the regulation of endocrine pancreatic secretion. The polypeptide hormones produced by the pancreatic islets are regulators of nutrient homeostasis. The rates at which glucose, amino acids, and free fatty acids enter and leave the extracellular space are influenced by insulin, glucagon, and probably somatostatin (Unger *et al.*, 1978). However, most studies have been done on mammals and very little is known about fishes in these respects.

The peptide-producing endocrine cells of the vertebrate digestive system have been investigated by various methods: immunohistochemistry, silver staining, lead–hematoxylin staining, formaldehyde-induced fluorescence, radioactive labeling, and enzyme studies (Dawson, 1970).

Besides gastrointestinal polypeptides, other hormones or hormonelike substances occur in the digestive system of fishes, for example, histamine. Holstein (1975a) found diamine oxidase, a histamine-deaminating enzyme, in the intestine of teleosts. GEP cells are thought to be the source of the intrinsic factor in the mammalian stomach. It is not known if this factor, needed for the intestinal absorption of iron, occurs in fishes.

G. Blood and Lymph Vessels

In the cyclostome, *Myxine glutinosa*, the intestine is supplied by segmental arteries and the venous drainage goes to the hepatic portal system. In elasmobranchs (sharks, rays) and dipnoans (lungfish) the intestine receives three or four arteries, whereas in teleosts one single celiaco–mesenteric artery supplies the digestive canal. In teleosts the arterial branches to the intestine are often closely followed by veins (Grodzinski, 1938), an arrangement reminiscent of the rete mirabile in the fish swimbladder. Well developed retia mirabilia were further described in the portal, hepatic, intestinal and splenic veins of tuna fish and the elasmobranch, *Alopias vulpes*, by Müller as early as 1840.

Lymphatic vessels and sinuses are found in the digestive system of both cyclostomes and true fishes. In the hagfish, *Myxine glutinosa*, a superficial lymphatic plexus is found in the intestine (Fänge, 1973). In some teleost species the intestine has a system of chyle vessels similar to those of mammals. The chyle vessels converge into a large vessel, the *vas lymphaticum intestinale*. In the marine catfish (*Anarrhichas lupus*) this intestinal lymph vessel is strongly developed and situated to the right of the esophagus (Glaser, 1933). It is not known whether intestinal lymph vessels in fishes are of importance for the transport of absorbed fat from the intestine.

H. Lymphoid Tissue

In most vertebrates lymphoid or hematopoietic tissue is associated with the digestive canal. Thus in the hagfish, *Myxine glutinosa*, the intestinal submucosa contains an extensive granulocyte-producing tissue, assumed to constitute a primitive spleen (Tomonaga *et al.*, 1973). Granulocyte-producing tissue is further found in the submucosa of the esophagus of elasmobranchs (organ of Leydig, Petersen, 1908–1909; Fänge, 1968). Lymphocytes and granulocytes have been observed to invade certain layers of the intestine of fishes, such as membrana propria and epithelium. This infiltration of the gut by leukocytes probably plays a role in a defense system against microbes and parasites. In elasmobranchs the granulocyte-producing tissue of the esophagus is rich in lysozyme (bacteriolytic enzyme) and chitinolytic enzymes. The latter perhaps give some protection against chitin-containing parasites (Fänge et al., 1978). Whether leukocytes of the gut are of any direct importance for the digestion is not known.

IV. DIGESTIVE FLUIDS AND ENZYMES

A. Gastric Secretion

Production of an acid gastric fluid probably occurs in most fishes (e.g., Norris et al., 1973) except those which have no stomach (myxinoids, chimaeroids, many teleosts). In these neither HCl nor pepsin is formed in the gut.

1. HYDROCHLORIC ACID

Large amounts of gastric fluid, usually distinctly acid, are found in elasmobranchs. Weinland (1901) used a glass cannula to obtain 50 ml gastric fluid from a nonfed Scyllium, and Yung (1899) took 500 ml from the large shark, Lamna cornubica. In the gastric fluid of Scyllium stellare pH 1.69 was measured (van Herwerden and Ringer, 1911), but the acidity of the fluid varies. Maximal acidity is observed a few hours after food intake, whereas in the absence of food the gastric fluid may be weakly acid or neutral (Bernard, 1952). Weinland (1901) found both acid and alkaline secretion in the ray, Raja asterias. Babkin et al. (1935a,b) obtained small quantities of gastric fluid with a pH of 3–3.8 from nonfed Raja sp.

Acid gastric fluid has regularly been found in teleosts (e.g., Western and Jennings, 1970). pH values of 3.0–5.0 were measured in the stomach of Micropterus salmoides, Perca fluviatilis, and Tilapia mossambica (Sarbahi, 1951; Fish, 1960). The acidity increases after food intake (MacKay, 1929). Western (1971) noted that in Cottus scorpio and Enophrys (Cottus) bubalis the gastric content was neutral in the absence of food, but 30 hr after ingestion of food a pH of 2.0 was measured. A pH of 2.0 is reached in the stomach of Tilapia a few hours after daily feeding begins (Moriarty, 1973).

2. ENZYMES

The fact that optimal proteolytic activity has been reported at different pH's (2.0, 3.0, 5.0, 8.5) (Creac'h, 1963) indicates that the gastric fluid of fishes contains several types of protease. Pepsin undoubtedly is the major acid protease. It is secreted by the gastric gland cells as a zymogen called pepsinogen, which is inactive. Conversion of pepsinogen into active pepsin is brought about by pepsin in an acid environment. During the activation process amino acids are split off from the NH_2-terminal end of the molecule as a mixture of peptides (Lehninger, 1971). Pepsin is an endopeptidase which cleaves peptide linkages formed by amino groups of aromatic and acidic amino acids. It attacks most proteins but not mucins, spongin, conchiolin, keratin, or low molecular weight peptides (Sumner and Somers, 1947).

Merret et al. (1969) isolated four different pepsinogens from the gastric mucosa of a dogfish (Mustelus canis). Crystalline pepsin has been prepared from the gastric mucosa of teleosts, the Pacific King salmon (Oncorhynchus tschawitscha) (Norris and Elam, 1940) and the halibut (Hippoglossus hippoglossus) (Eriksen, 1945). In these investigations the pepsinogen was extracted by alkaline solutions and then transformed into pepsin by the addition of acid. Extraction with an acid solution was used by Norris and Mathies (1953) in the isolation of pepsin from the tuna fish (Thynnus albacores). The purified tuna fish pepsin differed crystallographically and in its amino acid composition from porcine pepsin.

A few nonproteolytic digestive enzymes may occur in the gastric fluid of fishes. Amylase has been found in the stomach of some fishes (Clupea harengus, Battle, 1935; Dorosoma cepedianum, Bodola, 1966). Lipase is reported from the stomach of Tilapia sp. (Al-Hussaini and Kholy, 1953; Nagase, 1964) and Dorosoma cepedianum (Bodola, 1966), and esterases with pH optima at 5.3 and 8.0 in the stomach of the rainbow trout (Salmo gairdnerii, Kitamikado and Tachino, 1960). Chitinase has been demonstrated in the gastric mucosa of elasmobranchs, insect-feeding teleosts, and the dipnoan Polypterus (Okutani, 1966; Micha et al., 1973). Remarkably strong chitinolytic activity was found in the gastric mucosa of the marine deep water teleost, Coryphaenoides rupestris, and the elasmobranchs, Etmopterus spinax and Raja radiata (Fänge et al., 1978). These three species largely feed on crustaceans and other invertebrates with a chitinous integument.

Hyaluronidase, a mucopolysaccharidase which cleaves beta-N-acetyl-glucosaminidic bonds of hyaluronic acid, has been prepared from the gastric mucosa of the Japanese mackerel, Scomber japonicus (Yamamoto and Kitamikado, 1971).

Among 148 fish species, a few estuarine fishes and one freshwater fish (Ictalurus punctatus) but no offshore marine species were found to contain cellulase in their stomach or anterior intestine. Because no cellulase activity was detected in fishes exposed to streptomycin, it was concluded that the cellulase probably was produced by microorganisms of the gut content (Stickney and Shumway, 1974).

B. Pancreatic Secretion

Due to difficulties in collecting pure pancreatic juice in most fishes the chemical composition of this fluid in representative fish species is not known. Undoubtedly it is rich in enzymes (mostly as zymogens) which serve in digestion of proteins, carbohydrates, fat, and nucleotides. Probably, as in higher vertebrates, the pancreatic juice contains bicarbonates that neutralize hydrochloric acid entering the intestine. Extracts containing pancreatic enzymes have been obtained from the isolated compact pancreas of elasmobranchs and holocephalans, from tissues containing scattered pancreatic acini (intestinal ceca, mesenteries, the "hepatopancreas" of certain fishes), and from the intestinal content.

1. PROTEASES

Trypsin, chymotrypsin, carboxypeptidase, and elastase are stored in the pancreatic cells as inactive zymogen (proenzyme) granules. When arriving in the intestinal lumen, trypsinogen is transformed into trypsin by proteases produced by intestinal mucosal cells (enterokinases). Other pancreatic zymogens are activated by trypsin.

Trypsin is formed by removal of a hexapeptide from the trypsinogen molecule as a result of the hydrolysis of a lysine–isoleucine bond. Trypsin is an endopeptidase with optimal action at a pH of about 7. It cleaves peptide linkages whose carbonyl groups come from arginine or lysine (Lehninger, 1971). Appropriate substrates for estimation of trypsin activity are synthetic peptides such as benzoyl-L-arginine ethyl ester (BAEE) and p-toluenesulfonyl-L-arginine methyl ester (TAME). Trypsinogen, or trypsin, has been demonstrated in the intestinal wall of the cyclostome *Myxine glutinosa* (Nilsson and Fänge, 1970), the pancreas of the holocephalan *Chimaera monstrosa* (A. Nilsson and Fänge, 1969), elasmobranchs (*Ginglymostoma cirratum, Squalus suckleyi*, Zendzian and Barnard, 1967), and the African lungfish (*Protopterus aethiopicus*, Reeck et al., 1970; Reeck and Neurath, 1972), and from intestinal ceca of various teleosts (Creac'h, 1963; *Oncorhynchus*, Croston, 1960, 1965; *Gadus morhua*, Overnell, 1973; *Dicentrarchus (Morone) labrax*, Alliot et al., 1974). The lungfish (*Protopterus aethiopicus*) pancreas also contains a trypsin inhibitor (Reeck et al., 1970). Jany (1976) studied digestive endopeptidases in the

stomachless teleost *Carassius auratus gibelio.* Trypsin and chymotrypsin of pancreatic origin were found but no pepsin, no elastase, and
no collagenase.

Chymotrypsin is formed by the action of trypsin on chymotrypsinogen. It is an endopeptidase which attacks peptide bonds with
carbonyl from aromatic side chains (tyrosine, tryptophan,
phenylalanine). Benzoyl-L-tyrosine-ethylester (BTEE) is an example
of a synthetic peptide useful as a substrate in assays of chymotrypsin
activity. Chymotrypsin has been found in most fishes in which trypsin
has been found.

Elastase is formed when the zymogen, proelastase, is activated by
trypsin. This enzyme is especially active on peptide bonds in the
protein elastin and may be assayed on purified elastin or specific ester
substrates (De Haën and Gertler, 1974). Elastase probably does not
occur in the cyclostome *Myxine glutinosa* (A. Nilsson and Fänge,
1970) but has been found in the pancreas of the holocephalan *Chimaera monstrosa* (A. Nilsson and Fänge, 1969), the elasmobranch
Dasyatis americana, and the teleosts, tuna (*Thynnus secundodorsalis*)
(Zendzian and Barnard, 1967) and the angler, *Lophius piscatorius*
(Lansing *et al.,* 1953). A new type of pancreatic elastase, proelastase A,
has been reported to occur in a dipnoan, the African lungfish (Walsh,
1970).

Carboxypeptidases are exopeptidases which hydrolyze the terminal peptide bonds of their substrates. Carboxypeptidases A and B,
differing in their specificities, are formed by activation of procarboxypeptidases by trypsin. Mammalian carboxypeptidase A is a zinc-
containing enzyme (Vallee, 1955). Carboxypeptidase A, but not
carboxypeptidase B, was found in intestinal extracts of the cyclostome
Myxine glutinosa (A. Nilsson and Fänge, 1970). Carboxypeptidases
have further been found in an elasmobranch (*Squalus acanthias,* Prahl
and Neurath, 1966a,b) and in the teleost, *Dicentrarchus labrax* (Alliot
et al., 1974). Zendzian and Barnard (1967) showed carboxypeptidase
B-like activity in the tuna fish, and Ooshiro (1968, 1971) found
carboxypeptidase A-like activity in the intestinal ceca of another teleost, the mackerel (*Scomber japonicus*). In the Japanese mackerel the
carboxypeptidase A-like activity appeared to depend on Co^{2+} ions
rather then Zn^{2+} ions.

Evolution of pancreatic proteases. Vertebrate trypsin, chymotrypsin, and elastase are structurally related to each other and, together
with a few other proteolytic enzymes, are called serine proteases as
they contain serine in the active site of the molecule. On the basis of
analyses of amino acid sequences of purified trypsinogen from differ-

ent vertebrates, Reeck and Neurath (1972) suggested a possible scheme for the evolutionary changes in the structure of the activation peptide of trypsinogen. The lungfish (*Protopterus*) trypsinogen at the molecular level shows some resemblance with invertebrate trypsins. It has been speculated that different serine proteases have evolved from a common ancestor molecule (Walsh, 1970; Stryer, 1974).

2. Amylase

In some plant-feeding teleosts such as *Tilapia*, amylase has been found in all parts of the digestive tract (Nagase, 1964; Fish, 1960). In a carnivore, the perch (*Perca fluviatilis*), on the other hand, digestive amylase is confined to the diffuse pancreas of the connective tissues surrounding the intestine. According to Nagase (1964) the *Tilapia* amylase has a pH optimum at 6.71, and the pH of the stomach is too low for the enzyme to show any appreciable activity. Pancreatic and intestinal amylase probably are more important than gastric amylase in carbohydrate digestion in *Tilapia*.

3. Chitinase

Chitinase occurs in the digestive system of many fishes and other vertebrates (e.g., Yoshida and Sera, 1970), notably in forms feeding on insects (Micha *et al.*, 1973; Dandrifosse, 1975) or crustaceans. Exceptionally high chitinase activity was found in extracts of the pancreas of the stomachless holocephalan *Chimaera monstrosa*, a fish that feeds largely on shrimps (Fänge *et al.*, 1976, 1978). Little or no pancreatic chitinase was detected in some other marine fish species with similar diets, but these instead possessed strongly active chitinase in the gastric mucosa. The *Chimaera* pancreatic chitinase has a strong optimum around pH 8–10, while gastric chitinases from other species show pH optima at 1.25–3.6. Chitinase splits chitin into dimers and trimers of N-acetyl-D-glucosamine (NAG), which may be further broken down by glucosaminidase (NAGase). This enzyme occurs together with chitinase in the digestive tract. NAG, the end product of the chitinolytic process, probably is of nutritive value since it is resorbed faster than glucose by the intestine (Alliot, 1967). Although chitinase may be produced by intestinal bacteria, most chitinases are synthesized by gastric or pancreatic gland cells.

4. Lipases

Lipases are esterases which split ester bonds. Triglyceride fats, phospholipids, and wax esters are hydrolyzed by lipase. Although

lipase activity has been demonstrated in various parts of the fish digestive system the pancreas is probably the major source (Barrington,
1957; Kapoor *et al.*, 1975b). According to Chesley (1934) lipase is more
abundant in fishes with a compact pancreas than in those with a diffuse pancreas. Brockerhoff (1966) by *in vivo* experiments found lipolytic activity in the intestinal content of cod (*Gadus morhua*) but
Brockerhoff (1966) and Overnell (1973) both failed to find any lipase
activity in extracts of the pyloric ceca or adjacent tissues. Leger (1972)
partly purified a lipase from the rainbow trout (*Salmo gairdnerii*).
Patton *et al.* (1975) found lipase activity in the bile of two marine
fishes, the anchovy (*Engraulis mordax*) and the jack mackerel
(*Trachurus symmetricus*). These authors suggested that in fishes not
pancreatic lipase (EC 3.1.1.3), but another enzyme may function as the
major fat-digesting enzyme.

5. OTHER ENZYMES

Alkaline RNase and phosphodiesterase have been purified from
the intestinal ceca of rainbow trout (*Salmo gairdnerii*, Imura,
1974a,b). An increased activity of carbonic anhydrase found in the gut
of coral fishes is supposed to be an adaptation to ingestion of calcium
carbonate (Smith, 1975).

C. Bile

Production of a detergent-containing fluid, bile, by the liver is
found in all vertebrates. Usually the bile is stored in a gallbladder
with contractile walls. By contraction of the smooth muscles of the
gallbladder the bile is ejected into the lumen of the intestine. During
its storage in the gallbladder the bile becomes more concentrated.
Mammalian bile contains bile salts, cholesterol, phospholipids, bile
pigments, organic anions, glycoproteins, and inorganic ions. Fish bile
has a similar composition. It is weakly alkaline and has a high sodium
and a low chloride concentration (Hunn, 1972). Bile salts are special
types of steroids which are synthesized in the liver from cholesterol.
In fishes (carp, *Cyprinus carpio*) as in other vertebrates, administration of ^{14}C-labeled cholesterol results in production of radioactive bile
salts. In *Myxine glutinosa* (cyclostome), *Chimaera . monstrosa*
(holocephalan), elasmobranchs, dipnoans, and *Latimeria* (coelacanth)
the bile contains bile alcohol sulfate esters (as sodium salts), but in
teleosts as in higher vertebrates the bile contains salts formed with
taurine conjugates of bile acids. However, in one group of teleosts, the

cyprinids (carp fishes) a bile alcohol sulfate is the principal bile salt (Haslewood, 1968). Bile salt molecules have hydrophilic and hydrophobic groups and in solution at a critical concentration they form aggregates called micelles.

In some teleosts the bile contains trypsin, lipase, amylase, or other enzymes from the intrahepatic pancreatic tissue (Babkin and Bowie, 1928). Bile which contains lipase is free from phospholipids (Patton *et al.*, 1975). In fishes as in mammals a large proportion of the secreted bile salts are presumably resorbed from the intestine into the blood and to a large extent returned to the liver. This so-called enterohepatic circulation concerns both bile salts and other bile components.

D. Intestinal Enzymes

Digestive enzymes produced by intestinal cells are located mainly in the brush border of the epithelium (Ugolev and Kooshuck, 1966; Matthews, 1975). However, enzymes in intestinal extracts may partly be pancreatic, as pancreatic enzymes have the tendency in the intestine to adsorb to the glycocalyx of the epithelial cells. Enzyme activities of the fluid of the intestinal lumen, with the exception of the anterior part of the intestine where pancreatic juice is delivered, are low. Cells or fragments of cells continuously released from the intestinal epithelium, extracellular enzymes from the stomach and the pancreas, and enzymes of the ingested food may contribute to enzyme activities in the lumen.

Enzymes thought to be produced by the intestinal mucosa include aminopeptidases, di- and tripeptidases (formerly termed erepsin), alkaline and acid nucleosidases (which split nucleosides), polynucleotidases (which split nucleic acids), lecithinase (which splits phospholipids into glycerol, fatty acids, phosphoric acid, and choline), lipase and other esterases, and various carbohydrate-digesting enzymes: amylases, maltase, isomaltase, sucrase, lactase, trehalase, and laminarinase. The knowledge of intestinal enzymes in fishes is fragmentary. Piavaux (1973) found laminarinase (β-D-1,3-glucan glucohydrolase, EC 3.2.1.58) in intestinal extracts of *Tilapia macrochira*, an African freshwater teleost, which feeds on plankton and plant detritus. Amylase activity is considerably higher in the intestine of herbivorous species such as the carp (*Cyprinus carpio*), than in the intestine and intestinal ceca of more carnivorous forms such as salmon (*Salmo*), cod (*Gadus morhua*), and flounder (*Pleuronectes*) (Phillips, 1969; Kapoor *et al.*, 1975b). Dipeptidase activity was investigated in the intestine of the white grunt (*Haemulon plumieri*) using synthetic dipeptides as substrate. The highest activity was measured in the anterior half of the intestine.

E. Regulation of Secretory Activities

1. GASTRIC ACID SECRETION

a. Elasmobranchs. Continuous secretion of very small amounts of gastric acid in fasting rays (*Raja* sp.) was not influenced by vagotomy or atropine but was inhibited by adrenaline or sympathetic nerve stimulation. Spinal destruction, probably due to elimination of tonic sympathetic nerve influence resulted in a "paralytic secretion" of gastric juice (Babkin *et al.*, 1935a). An influence of vascular sphincters on the composition of the gastric juice was assumed by Weinland (1901) who observed that in the ray, *Raja asterias*, treatment with an ergot preparation caused contraction of the vascular sphincters and alkalinity of the gastric juice. Ungar (1935), working with the isolated perfused stomach of *Torpedo, Squalus* and *Scyliorhinus*, found gastric secretion to be stimulated by acetylcholine and histamine. Hogben (1967a,b) found that the isolated gastric mucosa of the spiny dogfish (*Squalus acanthias*) secretes an acid juice spontaneously. Both histamine and the cholinergic agent carbachol increase the rate by 100–150%, but carbachol was 200 times more effective than histamine. No effect was obtained with a preparation of porcine gastrin, a peptide which causes secretion of gastric fluid in higher vertebrates.

b. Teleosts. In the living fish distension of the stomach serves as a powerful stimulus for gastric secretion (Smit, 1968; Norris *et al.*, 1973). Probably the effect is due to reflex activation of vagal cholinergic fibers. The secretion of gastric acid appears to be intermittent; that is, acid is produced only in connection with digestion or when otherwise stimulated (Gzgzyan *et al.*, 1968).

Histamine is an effective stimulus for gastric acid secretion in the European catfish (*Silurus glanis*: Gzgzyan *et al.*, 1968) and in the cod (*Gadus morhua*: Holstein, 1975b). The facts that exogenous histamine causes acid secretion and that the fish gastric mucosa contains considerable amounts of non–mast-cell histamine (Reite, 1969; Lorentz *et al.*, 1973) indirectly indicate that histamine has a physiologic function in the regulation of acid secretion.

In studies of gastric acid secretion in the cod (*Gadus morhua*) Holstein (1975b, 1976, 1977) found only a low basal output of acid

(5–10 μmol/kg hr). However, the method used to collect gastric juice involved ligating the pylorus, which seriously interferes with the water balance. In spite of their dehydrated condition, the fishes were able to secrete considerable amounts of gastric acid when injected with histamine or carbacholine. Both effects were blocked by the H_2-receptor antagonist metiamide. This provides evidence that histamine is physiologically important as a regulator of acid secretion in fish. Further experiments (Holstein, 1978 personal communication) showed that fishes kept in water balance either by perfusion of the intestine with 33% seawater or by intramuscular injection of hypotonic saline show a relatively high "basal" secretion of acid (50–100 μmol of H^+/kg hr). The intense secretory response after carbacholine is accompanied by vasodilation. The response is blocked by atropine (Holstein, 1977).

The question whether gastric secretion in fishes is influenced by gastrin or other GEP hormones is undecided. Holstein (1975b) found no stimulatory effect of pentagastrin in the cod (*Gadus morhua*), but the experiments were made on fishes not in water balance. Larsson and Rehfeld (1977) suggested that in nonmammalian vertebrates ceruleinlike peptides may serve as "gastrin." In isolated frog stomach (*Rana*) cerulein stimulates gastric secretion more powerfully than gastrin (Negri and Erspamer, 1973).

2. Pepsinogen Secretion

In mammals the secretion of pepsinogen is induced by vagal impulses (Hirschowitz, 1975). The stomach of fishes is richly supplied by vagal fibers, but it is not known whether these have any influence on the pepsinogen producing cells.

3. Pancreatic Secretion

In mammals the pancreatic juice is secreted as the result of stimulation of the exocrine pancreas cells by peptide hormones produced by cells in the anterior intestine and the stomach. The pancreas-stimulating principle of the anterior intestine was termed "secretin" by Bayliss and Starling (1903), but later investigations have shown the existence of more than one hormone which stimulate the pancreas. Thus, in mammals secretin produces a thin watery pancreatic fluid rich in bicarbonate, whereas cholecystokinin (CCK) stimulates a secretion rich in enzymes. Other hormones such as gastrin (regulates gastric secretion) and cholinergic agents (acetylcholine, carbachol, mecholyl) also stimulate the mammalian pancreas.

Babkin (1929, 1933) was able to stimulate pancreatic secretion in a ray (Raja) by the introduction of hydrochloric acid into the anterior intestine. It seems plausible that in elasmobranchs as in mammals, introduction of acid stomach content into the intestine causes release of pancreas-stimulating humoral substances.

4. Secretion and Release of Bile

Release of bile into the intestine is produced by contraction of the smooth muscles of the muscularis of the gallbladder. In the hagfish (Myxine glutinosa) this contraction is probably brought about by cholinergic vagal influences (Fänge and Johnels, 1958). CCK-PZ, which causes contraction of the mammalian gallbladder is probably present in some fishes (see Table I). In a ray, Raja erinacea, the production of bile by the liver seems to take place continuously but in linear relation to the portal vein pressure. The isolated perfused liver continues to produce bile (Boyer et al., 1974).

F. Intestinal Microorganisms

It has been suggested that in certain species of fish the decomposition of food components by microorganisms may be of importance for digestion. Nitrogen-metabolizing bacteria may explain the capacity to utilize urea in the food by the mullet (Mugil auratus) (Albertini-Berhaut and Vallet, 1971). Okutani (1966) found chitinolytic bacteria in the intestine of the marine teleost, Lateolabrax. These bacteria were gram-negative motile rods with a polar flagellum (Vibrio). The bacterial chitinase showed a pH optimum at 7.0, while the gastric mucosa produced a chitinase which was optimally active at a considerably lower pH. Probably in fishes chitinolytic bacteria play a negligible role for digestion in comparison with chitinases produced by the gut mucosa or the pancreas. Cellulase-producing microorganisms were found in the intestine of some estuarine fishes (Stickney and Shumway, 1974). From the little that is known about intestinal microbiology of fishes one is inclined to conclude that microorganisms are less important for the decomposition of food elements than they are in many mammals, especially ruminants. On the other hand, bacteria and other microorganisms are quantitatively an important food component in detritus-feeding fishes such as the mullet (Mugil cephalus) (Moriarty, 1976).

V. DIGESTION AND ABSORPTION

A. Digestion

By the action of enzymes of the digestive fluids and gut epithelial cell proteins, polysaccharides, lipids, and nucleic acids are degraded into smaller molecules, which can be absorbed and assimilated. Some proteins and polysaccharides, however, resist degradation.

Protein. Digestion of protein begins in the stomach in species which possess this structure. The endopeptidase activity of the gastric juice renders proteins soluble and more readily digested by pancreatic and intestinal proteases. In the intestinal digestion of proteins, trypsin and chymotrypsin from the pancreas are of major importance. Polypeptides formed by their interaction are further split by pancreatic carboxypeptidases and by intestinal peptidases. Enzymes such as elastase and collagenase may attack special proteins. The protein digestion leads to a mixture in the intestinal lumen of low molecular peptides and amino acids. Van Slyke and White (1911) found that in the dogfish (*Squalus acanthias*) during digestion of protein di- and tripeptides appear in the intestine.

Fat. Triglycerides, which are highly concentrated stores of metabolic energy, are important components of the food of many fishes. However, in addition to neutral fats, wax esters are a very abundant type of lipid in marine organisms such as certain crustaceans and fishes (Patton *et al.,* 1975). Lipases hydrolyze neutral fat (triglycerides) into diglycerides, monoglycerides, glycerol, and free fatty acids. Brockerhoff (1966) found that in the cod (*Gadus morhua*) after 2 days ingested triglycerides had transformed into the above mentioned decomposition products. Even phospholipids and wax esters are attacked by lipases, but fishes that consume waxes from marine organisms hydrolyze triglycerides four times faster than wax esters. Rahn *et al.* (1973) found that in the freshwater gourami (*Trichogaster cosby*) intestinal hydrolysis of wax esters is followed by oxidation of released alcohols to fatty acids. In the vertebrate gut, products from lipolysis are solubilized by bile salts, which form micelles with these products. Cholesterol and highly nonpolar lipids (hydrocarbons, sterols, fat-soluble vitamins) are particularly dependent on the presence of bile salts for micellar solubilization and subsequent absorption (Borgström, 1974).

Carbohydrates. Carbohydrate-digesting enzymes from the pancreas and in the intestinal epithelium transform oligo- and polysaccharides into hexoses and pentoses. Cellulose is probably utilized only to a small extent and in rather few fish species (cellulose degrading intestinal bacteria, see Stickney and Shumway, 1974), but the presence of high activity of chitinase in the digestive system of many fishes indicates that in some species chitin in the food is broken down to N-acetylamino sugar. Muramic acid, the polysaccharide of bacterial cell walls, is split by lysozyme. This enzyme has a wide distribution in nature, but it is not known if it plays any digestive role in fishes. Undoubtedly an enzyme, which dissolves bacterial cell walls, would be useful in detritus-feeding fishes. For example in the detritus-feeding mullet (*Mugil cephalus*) bacteria make up to 15–30% of the organic material in ingested food (Moriarty, 1976).

B. Absorption

The general problem of intestinal absorption is treated by Davson (1970), and from a comparative physiological point of view by Prosser (1973). In the main the mechanisms of intestinal absorption in fish appear similar to those of mammals. Absorption of the products of digestion takes place by diffusion and by active transport. Everted and noneverted segments of the intestine or *in vivo* techniques have been used to study the transport of different substances through the intestinal epithelium of fishes (Farmanfarmaian *et al.*, 1972).

Protein. The degradation products of protein are absorbed from the intestinal content as amino acids or peptides (Matthews, 1975). Individual amino acids are readily absorbed against concentration gradients and their absorption appears to be coupled to transport of inorganic ions (Smith and Lane, 1971; Farmanfarmaian *et al.*, 1972). Proteins and peptides in the intestinal content are probably also taken up to some extent, without previous degradation, by pinocytosis or related processes. Thus in the goldfish intestine, administered protein (horseradish peroxidase) was found to be absorbed in the distal region, in which the epithelial cells seemed to be specialized for the uptake of large molecules.

Fat. In fishes lipids seem to be absorbed mainly by the epithelial cells of the anterior part of the intestine (Gauther and Landis, 1972). Jansson and Olsson (1960) found that in the perch (*Perca fluviatilis*) the mucosal epithelial cells of the intestinal ceca are strongly sudanophilic, indicating that fat absorption is a function of these cells. In mammals the absorbed fat is transported from the intestine mainly by lymph in chyle vessels (lacteals). Within the blood the lipids form chylomicrons (i.e., particulate complexes of proteins, triglycerides, phospholipids, and cholesterol) or occur as albumin-bound free fatty acids (FFA). Lymph vessels resembling chyle vessels have been found in teleostean intestine (*Anarrhichas lupus,* Glaser, 1933) but chylomicrons have not been found in the blood plasma of fishes (Bilinski, 1974). Malins and Wekell (1970) suggested that in the spiny dogfish, *Squalus acanthias,* the absorbed fat is probably transported by the blood vascular system. The cyclostome *Myxine glutinosa* and certain elasmobranchs contain waxes in their blood plasma (Benson and Lee, 1975).

Carbohydrates. Absorption of glucose by the intestinal epithelium occurs by an active mechanism and can take place against considerable concentration gradients. At low temperature the transport of glucose diminishes. The transport of glucose is associated with electric potentials. Thus in the goldfish, *Carassius auratus,* the serosal side of the intestinal mucosa is positive in relation to the mucosal side. Addition of glucose produces a rise of potential, which is inhibited by phlorhizin (Smith, 1966) Farmanfarmaian *et al.* (1972) investigated the absorption of sugar *in vivo* in the toadfish, *Opsanus tau.* Glucose absorption, which occurs primarily in the anterior intestine, is linear with time and is blocked by phlorhizin.

Salt and water. While freshwater teleosts drink little water, marine teleosts continuously drink water. This was first demonstrated by Smith (1930), who added phenol red to the aquarium water. The quantity of seawater swallowed by marine teleosts per day varies from about 5 to 12% of the body weight (flounder, *Platichthys flesus,* sea perch, *Serranus scriba,* Motais *et al.,* 1969). The absorption of water by the fish intestine is secondary to active transport of sodium (House and Green, 1963). Water and salt movements in the eel (*Anguilla*) have been investigated with the use of isolated sacs of the intestine (Sharratt *et al.,* 1964).

Calcium. In mammals and probably other vertebrates the hormonelike substance 1,25-dihydroxycalciferol, a transformation product of vitamin D, is needed to stimulate the active uptake of calcium in the intestine. However, the physiological role of vitamin D in fishes is little known (Hay and Watson, 1976).

Iron. In mammals absorption of iron is promoted by low pH and by the presence of the reducing agent ascorbic acid. It has been suggested that facilitation of the absorption of iron is an important function of the vertebrate gastric glands, which secrete hydrochloric acid (Granick, 1953). The absorption of iron is an energy-requiring process and involves an iron acceptor in the brush border of the intestinal epithelial cells (Linder *et al.,* 1975).

Total assimilation. Assimilation efficiencies of the various nutrients discussed above are a fundamental part of dietary formulation (see Chapter 1). Assimilation efficiency can be studied by incorporating an inert reference material into the diet which can be readily measured in the subsequent feces. Thus the ratio of nutrient under study to chromic oxide in the food and in the feces is used to calculate the efficiency of assimilation:

$$\text{Assimilation efficiency } (\%) = 100 \times 1 - \frac{(Cr_2O_3: \text{Nutrient}) \text{ in food}}{(Cr_2O_3: \text{Nutrient}) \text{ in feces}}$$

This method has been successfully applied to the study of assimilation efficiency in rainbow trout (Nose and Mamiya, 1963; Nose and Toyama, 1966; Singh and Nose, 1967) and other aquatic animals (Nose, 1967).

VI. MOVEMENT OF FOOD THROUGH THE ALIMENTARY CANAL

A. Methods Used to Measure the Time for Gastric Emptying

The sequence of steps which leads to emptying of food from the stomach of fish has been examined by a number of workers using a variety of techniques. Fish have been given food and killed later to determine the extent of stomach emptying. Construction of the shape of the emptying curve with time requires large numbers of fish but this method has been extensively used (Method 1 in Table II). The degree of breakdown of the food can be measured visually (Fortunatova, 1955; Darnell and Meierotto 1962), by volume (Hunt, 1960), wet or dry weight (Daan, 1973), dry weight of contents after subtracting ash and chitin (Windell, 1966; Windell and Norris, 1969a,b) or calorific and biochemical analysis of the residue (Beamish, 1972; Gerald, 1973). Other workers have avoided killing the fish by causing the fish to vomit (Markus, 1932) or by using a stomach pump (Seaburg, 1956) to collect the residue (Method 2). X-radiography of the fish during its gastric phase of digestion has been successfully used by Molnár and Tölg (1962a,b), Molnár et al. (1967), and Edwards (1971, 1973) (Method 3). Hirao et al. (1960) incorporated ammonium phosphomolybdate containing ^{32}P into eel and trout diets, Kevern (1966) employed cesium isotope while Peters and Hoss (1974) preferred the poorly assimilated [144]Ce to monitor food translocation (Method 4). Other techniques include incorporation of dye in the food (Laurence,

1971), a change to distinguishable items (Blaxter, 1963) or even direct observation of food in the gut of transparent larvae (Rosenthal and Hempel, 1970). Various workers have estimated digestion rates in fish by observing the time between feeding and the production of feces. This technique is relatively straightforward in fish such as *Rutilus* and *Misgurnus* where a small meal subsequently appears as a single dropping (Scheuring, 1928; Bokova, 1938). However, most fish larvae and many microphagous adults feed continuously and it is helpful to label a "quantum" of the diet in such a way that this portion of the meal is detectable in the feces. Much of the work on larval and juvenile fish employs this technique (Lane and Jackson, 1969; Blaxter, 1963; Laurence, 1971). Predacious fish which consume large items of food, however, gradually erode the outer layers of the bolus and continuously pass food into the intestine for further digestion and assimilation. Feces from a given meal start to appear while part of the original meal is still in the gastric phase of digestion. Magnuson (1969) points out that some feces are extruded within 1–2 hr of feeding skipjack tuna. Rozin and Mayer (1961) observed that the first feces containing carmine appeared ca. 7 hr after ingestion by the goldfish but that most of the feces from the meal appeared between 8 and 24 hr after feeding. De Groot (1971) showed that in the turbot the completion of gastric emptying almost coincided with the complete voidance of the meal from the gut. Moriarty and Moriarty (1973a,b) studied the passage of food through the alimentary canal of *Tilapia* and *Haplochromis* which feed continuously on phytoplankton through much of the day. Only part of the ingested plant population is retained in the acid stomach and much of the meal, especially that taken in early morning, passes straight into the intestine and is poorly assimilated. Since the intestine in these species is almost completely emptied of the previous days' food, they were able to follow gastrointestinal motility by the weight of food in different gut segments in serial samples through the day.

As a result of this type of study, more detailed information on gastric emptying has been obtained. After the intake of a meal, there may be a delay of variable extent before the weight of the stomach contents begins to decrease. This delay is usually temperature dependent. Jones (1974) found that small pieces of *Pollachius* fed to gadoids remained unaltered for 3 hr at 6°C but for only 1.5 hr at 12°C. The duration of the delay also depends on the digestibility of the food item. The same author found that *Mytilus* meat starts to decrease in weight almost immediately whereas whole *Centronotus* or crustacea such as *Crangon, Nephrops* or *Carcinus* may require almost a day before disintegration is clearly initiated. Similarly, near the end of gastric emp-

Table II

Emptying Time of the Stomach or Intestinal Bulb in Fishes (Simplified from Various Authors)

Fishes	Temperature (°C)	Time to 100% evacuation (hr)	Fish size (cm or g)	Meal type	Method[a]	Reference	Natural diet[b]
Elasmobranchii							
Squalus acanthias	15	>48		Chopped beef	1c	van Slyke and White (1911)	C/F
Holostei							
Lepisosteus osseus	23–26	25	70–132 g	0.5–1.3 g Gambusia or Molliensia	2a	Hunt (1960)	F
Lepisosteus platyrhynchus	23–26	42	70–132	2.5 g Gambusia, Molliensia	2a	Hunt (1960)	
	26	24		6.7% Gambusia, Molliensia		Netsch and Witt (1962)	
Chaenobryttus gulosus	23–26	26	93 g	2.7% Gambusia, Molliensia	2a	Hunt (1960)	C/F
Amia calva	3	100–190	—	—		Riddle (1909)	C/F
	21	32	11–33	4.9% Gambusia, Molliensia	1a	Herting and Witt (1968)	
Teleostei							
Clupeiformes							
Clupea harengus	7	27 }				Blaxter and Holliday (1963)	MC
	20	10					
Engraulis encrasicolus	18	26 }				Okul (1941)	MC/MH
Clupeonella delicatula	9	4					
Engraulis mordax	16–19	1–4	2.7–18.5	Artemia nauplii	1a	Leong and O'Connell (1969)	MC/MH
Megalops cyprinoides	28	14–18	52 g	2% Gambusia, Metapenaeus	1b	Pandian (1967)	C/F
Esox lucius	18–23	50	40 cm		5	Seaburg and Moyle (1964)	F/HV
Salmo trutta	0	35					
	2–4	12–18 }	7–15 cm	Gammarus pulex	1b	Otto (1976)	
	6–8	10					
	12–15	3					
	5.2	49					
	7.6	37					
	9.8	29	90 g	1% Gammarus	2c	Elliott (1972)[c]	I/C/M/FF
	12	22					
	15	16					
Salmo gairdnerii	8	27 }					
	11	24	30 g	1% Lumbricus	3a	Grove et al. (1976)	I/C/M/FF
	15	22					

194

	8.5	26.5 ⎫	75–85 g	1% Paste	Grove et al. (1978)	3a,b	
	13.5	18.2 ⎬					
	18	15 ⎭					
	0.5	43 ⎫			Reimers (1957)	1a	
	2	26 ⎬	60–80 g	0.5 g *Gammarus*	Windell et al. (1969, 1972)	1c	
	7	18 ⎬	30 g	1.7% Pellet			
	10	13 ⎭	90 g	0.7–1% Capsules			
	12	30		2% *O. nerka* fry			
	15	36	95–213 g	*Chaoborus, Chironomus,*	Armstrong and Blackett (1966)	1a	I/C/M/FF
Salvelinus malma	13	24		*Hydropsyche* or *Acroneuria*	Hess and Rainwater (1939)c	1a	I/C/N/F
Salvelinus fontinalis	5	20–24	9–11 cm	(increasing exoskeleton)			
	7.5	30–30					
	11.5	95					
Oncorhynchus nerka	3.1	147 ⎫	30–40 g	1.5–2.7% Pellet of canned	Brett and Higgs (1970)	1c	I/C/F
	5.5	79 ⎬		salmon			
	9.9	38 ⎬					
	14.9	23 ⎬					
	20.1	18 ⎭					
Cypriniformes							
Barbus liberiensis	22–25	3–5	3–10 cm	Green algae 3.19 g Juvenile	Payne (1975)	1b	I/D
Ptychocheilus oregonensis	6	111	230 g	*S. gairdnerii*	Steigenberger and Larkin	2a	I/FF
	10	38			(1974)		
	15	14					
	20	10					
	24	8					
Misgurnus anguillicaudatus	20–27	20–24 ⎫		*Viviparus* or *Penaeopsis*	Tanaka (1955)	1b	C/MI
Silurus glanis	5	206 ⎭					
	10	87		1 *Acerina*	Fábián et al. (1963)	3a	F/HV
	15	49					
	20	28					
	25	20					
Ictalurus melas	24	6	4–6 cm	1% *Hyalella*	Darnell and Meierotto (1962)	1b	C/M/I
Ictalurus punctatus	10	24 ⎫	380 g	3 g Pellets	Shrable et al. (1969)	1a	C/M/I
	16	24 ⎬					
	22	7–10 ⎬					
	27	3–4 ⎭					
Anguilliformes							
Anguilla japonica	20	9.5		Minced sardine	Hirao et al. (1960)	4	I/M/FF

(Continued)

195

Table II—Continued

Fishes	Temperature (°C)	Time to 100% evacuation (hr)	Fish size (cm or g)	Meal type	Method[a]	Reference	Natural diet[b]
Mugiliformes							
Crenimugil labrosus	8–15	4–8	30–150 g	2–3% Paste	3b	Grove *et al.* (1976)	D/MH
Atherina pontica	26	4				Okul (1941)	MC/MH
Gadiformes							
Gadus morhua	2	72					
	5	58					
	10	25	150–375 g	0.45–0.64 g *Pandalus* tails	1b	Tyler (1970)	C/M/F
	15	20					
	19	20					
	12	72	1240 g	46 g *Clupea*	1b	Daan (1973)	
	8–10	48–130	50–55 cm	11–25% *Gammarus* or *Clupea*	1b	Karpevitch and Bokova (1936, 1937)	
Gadus morhua	6	12–45	18–527 g	0.5–2.5 g *Crangon*, fish, meat, polychaetes	1b	Jones (1974)	C/M/A/F
Melanogrammus aeglef.	10	12–26					
Merlangus merlangus	12	11–16					
Lota lota	1	288			1b	Gomazkov (1959)	C/F
	10	168					
Pleuronectiformes							
Pleuronectes platessa	1	36					
	5	25					
	9	16	280–320 g	1.3–1.5 g *Arenicola*	3a	Edwards (1971)	M/A/C
	14	12					
	20	10					
Platichthys flesus	10	24			1b	de Groot (1971)	M/A/C
	17–18	16					
Solea solea	10	24		5% *Arenicola*	1b	de Groot (1971)	A/C
	14–17	6					
Limanda limanda	8.5	18	100 g	1% Paste	3a	Jobling *et al.* (1977)	M/A/C/E
	16.5	12					
Scophthalmus maximus	10	96–100		5% *Sprattus*	1b	de Groot (1971)	F

196

Species	Temp	Value	Size	Food	Code	Reference	Category
Cyprinodontiformes							
Fundulus heteroclitus	6	27	7–10 cm	0.14 g Clam	2b	Nicholls (1931)	MC
	10	12					
	15	9					
	20	7					
	25	5					
	30	3					
Gambusia affinic	20	5–6		5–10 Larvae *Anopheles*, *Daphnia*	1b	Sokolov and Chvavliova (1936)	I/MC
	30	3–4					
Perciformes							
Perca fluviatilis	5	115					
Perca flavescens	10	63					
	15	49		1 *Alburnus*	3a	Fábián et al. (1963)	I/C/F
	20	27					
	25	21					
	21	20		Young *Perca*	1b	Manteifel et al. (1965)	
	24	15					
Perca flavescens	15	6–12	6 cm	*Daphnia*		Nobel (1973)[c]	
	22	1.5–6.5	3–4 cm				
Stizostedion lucioperca	5	257					
	10	157					
	15	83		1 *Alburnus*	3a	Fábián et al. (1963)	I/C/F
	20	45					
	25	28					
Stizostedion vitreum	14.5	16–25	18–38 cm	0.1–2.2% *Pimephales*	2b	Swenson and Smith (1973)	I/C/F
	20	12–15					
Stizostedion canadense	10	125				Kariya (1969); Fortunatova (1955);	C/F
Sebastes inermis	15	75					
	11	71					
	19	23					
	20	13–22		7%		Manteifel et al. (1965)	
	24	16–20					
Scorpaena porcus	7–13	96–144		Fish	1b	Fortunatova (1950)	C/F
	14–20	68–96					
	20–23	68–96					
	22–25	40–48					

(Continued)

197

Table II—Continued

Fishes	Temperature (°C)	Time to 100% evacuation (hr)	Fish size (cm or g)	Meal type	Method[a]	Reference	Natural diet[b]
Lutianus apodus	28–30	5–10	100–280 g	6% Jenkinsia	1a	Reshetnikov et al. (1972)	C/F
Lutianus cyanopterus							
Lutianus jocu							
Lutianus griseus		20–27		6% Harengula			
Lutianus synagris							
Tilapia nilotica	25	4–10	15–22 cm	Green algae	1b	Moriarty (1973)	MH
Haplochromis nigripennis							
Brachyistius frenatus	23–27	5–8	4–23 cm		5	Bray and Ebeling (1975)	MC/D
Phanerodon furcatus							
Oxyjulis californica							
Gobius minutus	5	18–20	3–5 cm	6 mg Corophium	1b	Healey (1971)	MC
	10	16–18					
	15	14					
Mullus barbatus	15	25		Polychaetes	1a	Lipskaya (1959)	M/A/C/FF
	20	14					
	25	8					
Dicentrarchus (Morone) labrax	15.5	18–28	5–8 g	0.2 g Arenicola, Carcinus	3b	Grove et al. (1976)	A/C/F
	19	10–17					
	24	6–11					
Lepomis macrochirus	17–21	18–21	34–64 g	0.6–1.8% Mixed invertebrates	1c	Windell (1966)	I/C
Lepomis gibbosus	22	20	12–16 cm	250 mg insects	5	Seaburg and Moyle (1964)	I/C/M
Lepomis macrochirus	19–22	14	21 g		1c	Kitchell and Windell (1968)	I/C
Pomoxis nigromaculatus	22	19–28	10–20 cm		5	Seaburg and Moyle (1964)	I/C/FF
Scomber japonicus	22	19	15–27		5	Seaburg and Moyle (1964)	C/M/F
	20	8–21	15 cm			Kariya (1956)	
Katsuwonus pelamis	23–26	12	1600 g	32–295 g Osmerid fish	1b	Magnuson (1969)	C/F
	22–24	25	99–146 g	5 g	1b	Aoyama (1958)	C/F
Trachurus japonicus	22–25	22–26	8–42 g	1.8–4.7 g Chopped fish	1b	Hotta and Nakashima (1968)	C/F

Cottus scorpius	9	Up to 168		50%	Karpevitch and Bokova (1936, 1937)	1b	C/F
Cottus gobio	10	72	5–6 cm	12% *Tubifex*	Western (1971)	1b	I/C
Enophrys bubalis	10	68–100	5–6 cm	32% *Tubifex*, insect larvae	Western (1971)	1b	C/F
Ophiocephalus punctatus	20	48					
	28	24	7 g	8% *Lepidocephalichthys*	Gerald (1973)	1c	C/F
	33	20					
Haemulon plumieri	24	25	19–20	3.9 g *Anchoviella*	Pierce (1936)	1b	F
Ocyurus chrysurus	24	30	19–20 cm	6.5 g *Anchoviella*	Pierce (1936)	1b	F
Micropterus salmoides	5	110					
	10	50					
	15	37	200–700 g	1 *Alburnus*	Molnár and Tölg (1962a,b)	3a	C/F
	20	24					
	25	19					
	20	12–27	91 g	2–8% *Notropis*	Beamish (1972)	1c	
	22	14–16	176 g	3% *Hyporhynchus*	Markus (1932)	2	
	24–25	17–18	89 g	2.7% *Gambusia, Molliensia*	Hunt (1960)	2a	
Prosopium williamsoni	6–11	8–10	22–28 cm	0.4–1.3 g *O. nerka* alevins	McKone (1971)	1a	I/FF
Pungitius pungitius	5	20	0.3–0.6 g	2% *Daphnia*	Cameron *et al.* (1973)	1b	I/MC
	15	7					

[a] Methods: (1) The fish is killed and the stomach contents removed by dissection for weighing or similar treatment. (2) Stomach emptied without killing the fish. (3) X-radiography. (4) Retention of radioactive isotope. (5) A number of fish captured from the wild are then killed at intervals (as in 1). Suffix: (a) force fed; (b) voluntarily fed; (c) digestible organic weight, biochemical or calorific values determined.

[b] Symbols indicate the preferred diet of adult fish: D, detritus; MH, microalgae; MC, microcrustacea and zooplankton; I, insects and larvae; FF, larval and juvenile fish; C, medium to large crustacea; M, molluscs; A, annelids; E, echinoderms; F, fish; HV, other vertebrates.

[c] Authors who have presented more extensive data on the effects of fish size and meal size on gastric evacuation time.

tying less digestible remains such as chitin may form the bulk of the residuum (Karpevitch and Bokova, 1936; Kionka and Windell, 1972).

Once digestion has been initiated, parts of the meal are transferred along the gut. The weight of the residuum decreases with time but the shape of this curve varies in the reports which have been published. In *Gadus* and related species (Jones, 1974; Daan, 1973), *Lepisosteus* (Hunt, 1960), *Stizostedion* (Swenson and Smith, 1973), *Katsuwonus* (Magnuson, 1969), *Megalops* (Pandian, 1967), and other species, the curve is adequately represented by a straight line. On the other hand, many workers have found that the decrease in contents is best described by an exponential curve. A relationship of this kind suggests that the rate of emptying (in grams per hour) is proportional to the instantaneous bulk of food in the stomach and can be expressed as

$$w_t = w_o e^{-b(t-a)}$$

where w_t, content of the stomach at time t after ingestion; w_o, size of the meal; a, delay before disintegration begins; b, instantaneous rate of digestion. It must be remembered that a and b will vary, as described below, with temperature, meal size, food type, fish size, method of feeding, and feeding history of the fish. Exponential rates of emptying have been described for *Oncorhynchus* (Brett and Higgs, 1970), young cod (*Gadus morhua*) (Tyler, 1970), *Salmo trutta* (Elliott, 1972), *Ptychocheilus* (Steigenberger and Larkin, 1974) and others. In his report on gastric evacuation in *Micropterus*, Beamish (1972) points out that emptying at the early and late phases of gastric digestion is faster than predicted by the exponential equation. Hunt and Knox (1968) showed that in mammals a linear relationship exists between the square root of gastric volume and time, related theoretically to the radial distension of the stomach as by a cylinder of changing volume. Kariya *et al.* (1969) were able to show that stomach diameter increases linearly with the square root of meal size in the mackerel while Jobling (personal communication) has shown that in *Pleuronectes* the gastric evacuation curve is linear when the square root of the residuum (dry weight) is plotted against time.

B. Effect of Temperature

The speed with which food moves through the fish alimentary canal has been investigated by many workers, a simplified account of their results being presented in Tables II, III, and IV. The environmental temperature significantly affects the speed with which food is

Table III
Total Emptying Time for Meals from the Digestive Tract of Fish

Species	Temperature (°C)	Time to 100% empty (hr)	Reference
Clupeidae			
Sardinops caerulea	18	12	Lasker (1970)
Salmonidae			
Salmo gairdneri	8	49–51	
	11	46	
	13.5	35	Grove et al. (1978)
	15	40	
	18	30.5	
Salvelinus namaycush	12	60–108	Lane and Jackson (1969)[a]
Esocidae			
Esox lucius	12	72	Lane and Jackson (1969)
Cyprinidae			
Ctenopharyngodon sp.	9	7	Hickling (1966)
Barbus liberiensis	22–25	6–8	Payne (1975)
Rutilus rutilus	3	30	
	6	31	
	10	21–22	Karzinkin (1932)
	14	22	Bokova (1938)
	17	12	
	20	9–10	
	25	6–8	
Pimephales promelas	12	36	Lane and Jackson (1969)
	20	12–24	
Carassius auratus	12	36–48	Lane and Jackson (1969)
	20	60–72	
	25	60	
	25	8–24	Rozin and Mayer (1964)
Cyprinus carpio	12	60	Lane and Jackson (1969)
	23	48	
	12.5	22–50	Kevern (1966)
	25	16–25	
	10	18	Maltzan (cited in Hickling, 1970)
	26	4–5	
Leuciscus baicalensis	0.5	130	
	15	40	Pegel and Popov (1937)
	25	15	
Notemigonus chrysoleucas	20	36	Lane and Jackson (1969)
Catla catla	28–30	18–54	Renade and Kewalramani (1967)
Cirrhina mrigala	28–30	18–60	Renade and Kewalramani (1967)

(Continued)

Table III—*Continued*

Species	Temperature (°C)	Time to 100% empty (hr)	Reference
Catostomidae			
Ictiobus cyprinellus	20	24	Lane and Jackson (1969)
Catostomus commersoni	12	60	Lane and Jackson (1969)
Ameiuridae			
Ictalurus melas	12	84	Lane and Jackson (1969)
Ictalurus punctatus	12	24–36	Lane and Jackson (1969)
Ictalurus catus	20	48	Lane and Jackson (1969)
Ictalurus nebulosus	20	60	Lane and Jackson (1969)
Ictalurus natalis	24	72	Lane and Jackson (1969)
Cobitidae			
Misgurnus fossilis	10	40	Scheuring (1928)
	15	14	
	20	10	
Embiotocidae			
Brachyistius frenatus	23–26	10–12	Bray and Ebeling (1975)
Phanerodon furcatus	23–26	10–12	Bray and Ebeling (1975)
Coridae			
Oxyjulis californica	23–26	10–12	Bray and Ebeling (1975)
Sparidae			
Labeo rohita	28–30	24–54	Renade and Kewalramani (1967)
Percidae			
Perca flavescens	12	36–60	Lane and Jackson (1969)
Stizostedion vitreum	12	60	Lane and Jackson (1969)
Centrarchidae			
Micropterus salmoides	12	48–84	Lane and Jackson (1969)
	20	36–48	Beamish (1972)
	20	60	Lane and Jackson (1969)
Micropterus dolomieui	12	48–72	Lane and Jackson (1969)
Pomoxis annularis	20	60	Lane and Jackson (1969)
Lepomis cyanellus	12	60 ⎫	Lane and Jackson (1969)
	20	48 ⎭	
Lepomis gibbosus	12	84	Lane and Jackson (1969)
Lepomis megalotis	12	72	Lane and Jackson (1969)
Lepomis macrochirus	12	36–84 ⎫	
	17	48	
	20	36–60 ⎬	Lane and Jackson (1969)
	22	36	
	25	36 ⎭	
Cottidae			
Cottus gobio	10	100	Western (1971)
Enophrys bubalis	10	100	Western (1971)

Table III—*Continued*

Species	Temperature (°C)	Time to 100% empty (hr)	Reference
Moronidae			
Dicentrarchus (Morone) labrax	16	36–74	Grove et al. (1976)
	19	20	
Carangidae	24	16	
Katsuwonus pelamis	23–36	14	Magnuson (1969)
Mugilidae			
Crenimugil labrosus	8	6–10 ⎱	Grove et al. (1976)
	18–19	17–18 ⎰	
Mugil cephalus	20–26	4–5	Odum (1970)
Labridae			
Tautogolabrus adspersus	10–15	10–14	Chao (1973)
Cichlidae			
Tilapia nilotica	25	7–15	Moriarty (1973)
	27	15–27	Moriarty (1973)
Blenniidae			
Blennius pholis	8	49 ⎤	
	11	45 ⎥	Grove et al. (1976)
	16	21 ⎥	
	19	17.5 ⎦	
Pleuronectidae			
Pleuronectes platessa	1	158 ⎤	
	5	53 ⎥	
	9	37 ⎬	Edwards (1971)
	14	24 ⎥	
	20	20 ⎦	
	10	72	de Groot (1971)
Patichthys flesus	10	72 ⎱	de Groot (1971)
	17.4	54 ⎰	
Limanda limanda	17	19–24	Jobling et al. (1977)
Soleidae			
Solea solea	10	72	de Groot (1971)

[a] The study by Lane and Jackson (1969) (*Salvelinus namaycush, Esox lucius, Pimephales promelas, Carassius auratus, Notemigonus chrysoleucas, Ictiobus cyprinellus, Catostomus commersoni,* etc.) was limited to young fish (2.5–9 cm) which were allowed a long period of voluntary feeding before isolation and feces collection at 12 hr intervals. Tanaka (1955) noted that the time for gastric evacuation is close to 50% of that for the whole canal to be cleared.

processed (Table II). This effect is illustrated in Fig. 4, in which the time for a meal to be emptied from the stomach or intestinal bulb is plotted in relation to temperature. Jones (1974) found that the evacuation rate of the stomach changes in proportion to $10^{0.035(\Delta t)}$, where Δt is

Table IV

Time for Fish Larvae to Empty Food from the Alimentary Canal

Species	Temperature (°C)	Time (hr)	Remarks	Reference
Clupeidae				
Clupea	7	6–9	Smaller larvae	Various authors, cited
harengus	9	5–7.5	Smaller larvae	in Blaxter (1963)
	11	4–5	Smaller larvae	
	15	4	Smaller larvae	
	12	24–30	Larger larvae	
Clupea pallasi	9	12–19	Duration increased with meal size	Kurata (1959)
Brevoortia	16	5	Copepod diet,	Kjeldson *et al.* (1975)
tyrranus	15	5.1–7.4	*Artemia*	
Sardinops		3–11	Time increases with	Arthur (1956)
caerulea			fish size	
Coregonidae				
Coregonus	14.4	24–48	Time increases with	Hoagman (1974)
clupeaformis			meal or fish size	
Cyprinidae				
Abramis	14	3.5		Panov and Sorokin
brama	20	1.75		(1962)
Cyprinus	18–29	1–8	Continuous feeding	Chiba (1961)
carpio		20	Single meal	
Belonidae				
Belone	22	2.7–4.8	38–45 mm larvae	Rosenthal and Paffen-
belone		2.9–5.3	50–55 mm larvae	höfer (1972)
Sparidae				
Lagodon	16–17	4.7–6.4	Copepod diet	Kjeldson *et al.* (1975)
rhomboides	16	5.15	*Artemia*	
Sciaenidae				
Leiostomus	17	6.1	Copepod diet	Kjeldson *et al.* (1975)
anthurus	16	9.6–10	*Artemia*	
Centrarchidae				
Micropterus	17	3	Continuous feeding	Laurence (1971)
salmoides		5.5	Single meal	
	23	2	Continuous feeding	
		3.7	Single meal	
	20	8	Darkness	
Percidae				
Perca	21	2.1–2.9	Total evacuation for	Nobel (1973)
flavescens			17–19.5 mm fish	
Pleuronectidae				
Pleuronectes	10	6		Ryland (1964)
platessa				

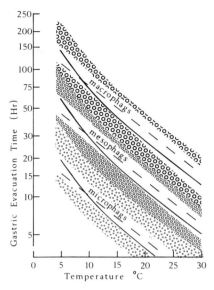

Fig. 4. Relationships between gastric or foregut emptying times and temperature in fishes. Data are taken from Table II and grouped into microphagous (D, MH, MC, I, FF), mesophagous (M, A, E, C), and macrophagous (C, F, HV) based on their preferred natural diet. Note that the time scale is on a logarithmic axis. The effect of temperature on digestion rate in each group is represented by a Q_{10} of 2.6 (Backiel, 1971; continuous lines) or in proportion to $10^{0.035t}$ (Jones, 1974; broken lines).

the temperature change in degrees centigrade. This trend is indicated in the figure by straight lines. Since the time to empty the stomach in many fish deviated from a straight line when log evacuation time is plotted against temperature, it may be more appropriate to represent the effect of temperature on rate in the form of Krogh's curve with a Q_{10} of ca. 2.6 as used by Backiel (1971). Fábián et al. (1963) plotted the log reciprocal of evacuation time against the reciprocal of temperature to demonstrate that their data on predatory fish follow the Arrhenius equation. At temperatures near the upper physiological limits of the species in question, the trend may cease or even reverse (Tyler, 1970; Gomazkov, 1959).

C. Feeding Niches

The data presented in Table II are derived from experiments in which a number of variables which affect gastrointestinal motility are confounded. Within the table however there is evidence that fish which have dissimilar food in nature have dissimilar digestion rates. For example, *Prosopium* in nature feeds on insects and fish fry. When fed salmon fry at a meal size of 1% body weight (8–10°C) the stomach empties within 6–11 hr. The flatfish *Limanda*, which normally ingests marine bivalve mollusks, crustacea, and other invertebrates, requires fully 18 hr to empty the stomach when fed a 1% body weight diet of easily digestible flatfish plaice diet at the same temperature. The piscivorous *Lepisosteus* requires more than 24 hr to digest a meal of fish, also as a meal of 1% (wet weight) of body weight, even though the test was carried out at the higher temperature of 23–26°C. We suggest that the inherent rate of gastrointestinal motility and digestion has evolved to suit the natural diet of fish and that this factor is borne out in Tables II and III. With this in mind, Fig. 4 has been constructed in such a way that the data in Table II have been grouped into microphagous (planktivores, insectivores, and similar), mesophagous (eating larger invertebrates such as mollusks, annelids, and shrimps) and macrophagous (taking crabs, fish, or other vertebrates) species. In many cases the species fall into the appropriate group, related to their natural diet, even when the laboratory data has been obtained using unnatural food items. An interesting example of this pre-adaptation to the feeding niche has been found in our own studies. We fed mullet (*Crenimugil labrosus*) or the blenny (*Blennius pholis*) on the same artificial diet of fishmeal and additives used in flatfish rearing in Aberdeen. The mullet has a long convoluted intestine and normally ingests particulate detritus whereas the littoral blenny predates on barnacles and other invertebrates and has a relatively short intestine with no true stomach. When similar sized animals were fed the same meal size at the same temperature, the mullet completely emptied the gastrointestinal tract three to four times more rapidly than the blenny. The most notable exception to the pattern presented in Fig. 4 is the description by Moore and Beamish (1973) that the ammocoete larva of *Petromyzon marinus,* which feeds on unicellular algae using ciliary mechanisms, digests its food more slowly than teleostean microherbivores.

D. Other Factors That Influence Gastric Motility

It is important, when discussing the emptying rate of meals from the fish stomach, to remark that many workers use the terms "digestion," "evacuation," "emptying," "elimination," and "clearance" rate to describe the *total time* taken for the stomach to empty after a given meal. Others use the same terms to describe the disappearance of food (measured as grams per hour or percentage of the meal per hour) from the stomach. Barrington (1957) stated in his review that, on the available evidence, "a smaller meal may be expected to be digested more readily than a larger one." This statement is based on the relatively larger surface area presented to digestive enzymes by smaller food items. In apparent contradiction to this, Windell (1966) points out that, in his studies, the rate of digestion of a larger meal (measured as the amount disappearing per unit time) is greater than that of a smaller meal.

Appropriate analysis of the gastric emptying process shows that both statements are true. Since enzymes typically attack the food bolus at the surface, the rate of digestion will obviously be proportional to the surface area, namely,

$$dV/dt = - aV^{2/3} \tag{1}$$

where V is the volume (or weight) of the food in the stomach, and a is a constant which depends on such factors as the species and size of the individual fish, temperature, and food type. Accordingly, it is to be expected, as pointed out by Windell, that a larger meal will have a faster digestion rate (grams per hour) and that a plot of log digestion rate (grams per hour) against log meal size will have a slope of 0.67 (for a given sized fish at a stated temperature using a standard food). If the above differential equation (1) is integrated and arranged for any meal size (V_s) to be reduced to $V = 0$, it is seen that the time required to empty the stomach (t_λ) is given by

$$t_\lambda = a'V_s{}^{1/3} \tag{2}$$

and Barrington's (1957) statement remains true: A smaller meal will be digested sooner. It follows that a plot of $\log t_\lambda$ against $\log V_s$ should produce a straight line of slope 0.33, again provided that the experiments are carried out with fish of the same size, at the same temperature, and using the same food.

This relationship may also hold when different sized fish of the same species are examined. If such fish are fed a stated percentage of their body weight, the food bolus will be larger in the bigger fish and will present a relatively smaller surface area to gastrointestinal secretions. Larger fish will digest, for example, a 1% body weight meal at a greater rate (g/hr) than a smaller fish but the time required to complete digestion will be prolonged.

1. MEAL SIZE

In published accounts, it has been reported that a larger meal fed to a given size of fish does take longer to digest (Hunt, 1960; Beamish,

1972; Elliott, 1972; Swenson and Smith, 1973; Jobling *et al.*, 1977). Beamish (1972) found that a four-fold increase in meal size fed to *Micropterus* only doubled the time required to empty the stomach. Using X-radiography, Jobling *et al.* (1977) found that increasing meal size from 1 to 5% (100-g *Limanda* at 16°C) leads only to a fourfold increase in stomach emptying time. Clearly both the time to complete the meal and the rate of digestion increased as predicted by the model. Analysis of the published records for various fish shows that digestion rate, expressed in grams per hour, varies as the following exponents of the wet weight of the meal: 0.46 (gadoids, Jones, 1974), 0.5 (*Limanda*, Jobling *et al.*, 1977), 0.6 (*Prosopium*, McKone, 1971; (*Megalops*, Pandian, 1967), 0.7 (*Micropterus*, Beamish, 1972), and 0.75 (*Salmo gairdnerii*, Windell *et al.*, 1969). In a few fish digestion rate has been found to increase as (meal size)$^{1.0}$, such that overall evacuation time in a given fish is the same for different sized meals (*Lepomis*, Windell, 1966; *Oncorhynchus*, Brett and Higgs, 1970; *Blennius*, Crawford and Grove, unpublished observations). The arrival of larger meals in the fish stomach or foregut will initiate or accelerate peristalsis (Section VII) thereby raising the emptying rate above the level predicted by the simplistic model. Only Steigenberger and Larkin (1974), in a careful study of digestion in *Ptychocheilus*, have found that digestion rate decreases with increased meal size. Clearly more species must be examined before this result can be held anomalous. Elliott (1972) found that the instantaneous rate of gastric evacuation in *Salmo trutta* is not affected by increases in meal size but, since his evacuation curves are exponential, the early digestion rate (g/hr) increases with meal size. A similar conclusion applies to the data for *Oncorhynchus nerka* (Brett and Higgs, 1970).

2. Fish Size

Several workers have reported the digestion rate of fish of different sizes in a form which can be compared with the predictions of the model for gastric emptying time (t_λ). Jobling *et al.* (1977) (Figs. 5 and 6) fed *Limanda* of different body weights to the same percentage of their weight with a flatfish paste diet. The time required to empty the stomach varied as (fish weight)$^{0.386}$. Since the stomach volume is proportional to body weight in this species ($V = 0.081 W - 0.39$ where the units are ml and g, respectively) a stated % body weight meal will fill the stomach to the same extent in different sized fish. A similar analysis of Pandian's (1967) data for *Megalops* fed on prawns shows that t_λ varies as (fish weight)$^{0.41}$.

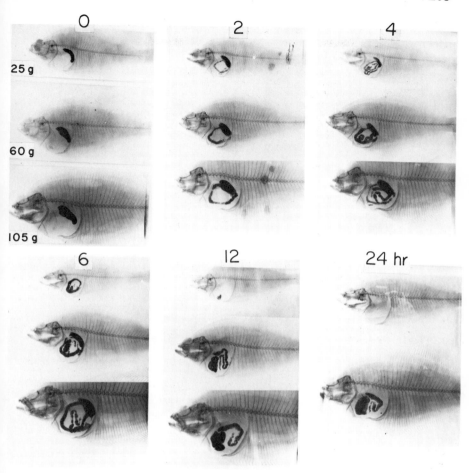

Fig. 5. X-Ray photographs of *Limanda limanda* fed a 1% body weight meal at 16°C. The fish weighed 25, 60, and 105 g. Exposures were taken at 0, 2, 4, 6, 12, and 24 hr after feeding. The delay in stomach and intestinal clearance with increasing size is clearly seen. (From Jobling *et al.*, 1977, *J. Fish Biol.* **10**, 291–298.)

The above reports not only offer reasonable support for the relationships proposed in Eqs. (1) and (2) but offer physiological explanations for the pattern of feeding by fish in their natural environment. It has been suggested (Kariya, 1969; Kariya and Takahashi, 1969; Brett, 1971; Ware, 1972; Elliott, 1972, 1975) that the appetite of fish returns as the stomach empties. It would be natural to expect that, for a given species, fish of different sizes which show the same feeding periodicity must voluntarily regulate their meal size so that digestion

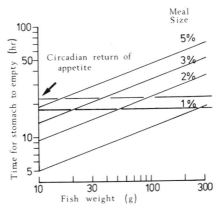

Fig. 6. Relationship between fish weight, meal size (as percentage of body weight) and gastric evacuation time for *Limanda limanda* at 16°C. The decrease in relative digestion rate with size shows that, if feeding is to return at regular intervals (e.g., daily) larger fish must necessarily ingest relatively smaller meals. Voluntary intake in grams will vary as (body weight)$^{0.5}$. (From Jobling *et al.*, 1977, *J. Fish Biol.* **10**, 291–298.)

is completed in time for the next feeding bout (Fig. 6). The stomach contents of freshly captured fish show a relative decrease in amount for bigger fish (Daan, 1973; Steigenberger and Larkin, 1974) and voluntary intake of food has been shown to decrease relative to body weight by many workers (Table V). It must be recognized, in design-

Table V

Magnitude of the Exponent in the Relationship Meal Size = a(Fish Weight)^b
for Voluntarily Feeding Fish

Species	b	Reference
Petromyzon marinus	0.72 (summer)	Moore and Beamish (1973)
(ammocoete)	0.84 (winter)	
Salvelinus fontinalis	0.91	Baldwin (1956)
Salmo trutta	0.75	Elliott (1975)
Oncorhynchus nerka	0.75 (single meal)	Brett (1971)
	0.65 (multiple meal)	
Megalops cyprinoides	0.71	Pandian (1967)
Micropterus salmoides	0.47	Niimi and Beamish (1975)
Navodon modestus	0.52	Suzuki (1976)
Ophiocephalus punctatus	0.76	Gerald (1973)
Blennius pholis	0.38	Wallace (1973)
	1.00	Crawford and Grove (unpublished observations)
Trachurus japonicus	0.56	Hotta and Nakashima (1968)

ing experiments, that the amount taken at a meal by an individual fish will depend on the previous feeding regime (Kariya and Takahashi, 1969; Ishiwata, 1969; Brett, 1971). For example, young *Scomber japonicus* will eat 23% of their body weight of anchovy after several days of deprivation but, on a regular feeding regime, will ingest only 16%.

3. TYPE OF FOOD AND ITS PRESENTATION

A further variable with significant effects on the gastric emptying rate is the type of food ingested by the fish. Jones (1974) found that gadoids digested the following foods (after an appropriate delay) at the stated rate at 12°C: *Nereis/Nephthys*, 0.31 g/hr; *Pollachius* muscle, 0.26 g/hr; *Crangon*, 0.19 g/hr. Elliott (1972) found that brown trout completed 90% elimination of gammarids or oligochaetes in 22 hr at 12°C whereas *Protonemura* (26 hr), *Hydropsyche* (30 hr), and *Tenebrio* (49.5 hr) took longer. A similar observation was made by Reimers (1957) on rainbow trout, where *Helodrilus* (12 hr) was eliminated faster than *Gammarus* (13 hr) or *Arctopsyche* (16 hr) at 10°C when given to 12.4–20 cm fish in 0.5 g meals. Other workers who have detected decreased evacuation rates with less digestible foodstuffs are Pandian (1967) (*Megalops* fed *Gambusia* or *Metapenaeus*), Karpevitch and Bokova (1936, 1937) (*Gadus* fed *Clupea* or *Gammarus*), Western (1971) (*Cottus, Enophrys* fed on *Tubifex, Calliphora* or semifluid meals), Reshetnikov *et al.* (1972) (*Lutianus* fed on *Jenkinsia* or *Harengula*), Renade and Kewalramani (1967) (*Labeo, Cirrhina* or *Catla* fed on various algae, plant detritus, and zooplankton), and Kionka and Windell (1972) (*Salmo* fed various diets). The digestibility of the food will not only affect the emptying rate from the stomach, expressed as g/hr, but may also determine the time after ingestion before weight decrease of the meal can occur. Jones (1974) found that *Merlangus* or *Melanogrammus* start to digest shell-less *Mytilus* almost immediately but that meals of *Ophiopholis*, large crustacea or *Centronotus* require up to 10, 20, and 25 hr, respectively, before weight loss begins. Although in many cases the decrease in digestibility can be attributed to thick or inert casings (e.g., caddis larvae) Windell (1967) suggested that the presence of fat in the food may delay gastric emptying, possibly by a release from the intestinal wall of a hormone similar to enterogastrone which in mammals inhibits gastric motility (Hunt and Knox, 1968). Diets with increased fat levels clearly decrease gastric evacuation rate in the rainbow trout (Windell *et al.*, 1969). Finally, Swenson and Smith (1973) found that when *Stizostedion* were fed the same meal size at the same temperature but with smaller or larger food items, the evacuation rate was faster with the smaller items (they used *Pimephales*).

A further complication to the measurement of gastric emptying rates in fish are the observations by Windell (1966) and by Swenson and Smith (1973) that force-feeding the test animal usually decreases the rate of evacuation of the meal from the stomach when compared with fish which consume the food voluntarily. Of even greater concern, however, are the reports from Kariya and Takahashi (1969), Ishiwata (1969), Tyler (1970), Brett (1971), Jones (1974), and, for larval fish, Laurence (1971), Blaxter and Holliday (1963), Rosenthal and Paffenhöfer (1972), and Nobel (1973) that fish which have been deprived of food for a time prior to feeding show a slower gastric emptying rate than fish tested under continuous feeding. Digestion rates measured in previously deprived fish are usually only 50–68% of those found in actively-feeding fish. In their study on feeding rate of *Tilapia* ingesting blue–green algae, Moriarty and Moriarty (1973a) found that the rate of stomach emptying during the feeding period was faster than that measured after feeding had ceased. This factor is a clear restriction when using laboratory digestion studies to estimate feeding rate in the wild (Healey, 1971).

E. Conclusions

If experiments on gastric evacuation are undertaken (whether to compute daily food intake in nature, to determine the relationship between stomach emptying and the return of appetite for fish culture, or to investigate the physiological control of gastric motility *in vivo*) the following procedures must be controlled and reported with the results.

1. The temperature at which observations are made, together with the previous acclimation history of the fish
2. The frequency of feeding, or the previous duration of food deprivation, since the fish will consume a meal size related to this regime if voluntary feeding is allowed
3. The meal size, as absolute wet and dry weight and as percentage of the body weight
4. The biochemical composition of the food, including the calorific value per unit dry weight of digestible food, both at ingestion and at later stages as stomach emptying proceeds

5. The method of feeding, especially whether the animals were force fed or ingested voluntarily; mention should be made if the fish were handled at intervals during the experiments (e.g., if several X-ray photographs were taken)
6. The length and weight of the fish used in the study

It is likely that other factors, not discussed in the present account, might affect gastric emptying. Reproductive cycles, photoperiod of long or short daylength, presence of parasties and/or the stocking density may all modify gastric motility. Many of the results listed in simplified form in Table II are taken from reports where factors 1–6 were not fully controlled. More recent studies (e.g., Elliott, 1972; Steigenberger and Larkin, 1974) have used factorial designs to test some of the variables likely to affect gastric emptying.

It has often been stated that the return of appetite is closely related to the time taken to reduce the stomach contents. Ware (1972), for example, defined hunger in newly caught fish in relation to the degree of stomach fullness. Magnuson (1969), Brett (1971), and Elliott (1975) all report that hunger returns before the stomach is completely empty. However, the observations of Tugendhat (1960) and Beukema (1968) suggest that stomach emptiness alone does not control the amount ingested. *Gasterosteus* with empty stomachs ingested more food at a faster rate if they had a longer time of deprivation before food was made available. Elliott (1975) found that a trout of stated size ate more in a meal if the temperature is raised. Rozin and Mayer (1961, 1964) found that the goldfish (*Carassius*) feeds continuously, but consumes a greater amount of food per day if the food is diluted with kaolin. A similar conclusion is reported for *S. gairdneri* by Lee and Putnam (1973). In an attempt to clarify this latter phenomenon of "calorie counting," we carried out a study on *Salmo gairdneri* (Grove *et al.*, 1978). As shown in Fig. 7, rainbow trout kept in continuous light exhibit regular periods of feeding activity similar to those described by Adron *et al.* (1973). The records from the demand feeders were subjected to periodogram analysis and showed that, as the food is diluted with kaolin, so the period between meals is progressively shortened and, as a consequence, the food intake per day is increased. Adron *et al.* (1973) suggested that the periodic feeding is determined by the rate of gastric emptying, so X-radiography tests were undertaken to see if dilution of the food with kaolin shortens the time required for the stomach to empty. The gastric evacuation times presented in Table VI clearly show that this is the case. The mechanisms responsible for the increased gastric motility are unknown but may

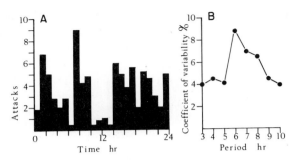

Fig. 7. (A) Activation of a "demand" feeder by a group of 60 g *Salmo gairdnerii* during a 24 hr period under continuous illumination at 18°C. (B) Periodogram analysis of (A) (Williams and Naylor, 1967) shows that appetite returns at 6-hr intervals. (From Grove *et al.*, 1978, *J. Fish Biol.* **12**, 507–516.)

either depend on a decreased production of enterogastrone from the intestine or a change in activity of extrinsic nerves which supply the stomach (see Section VII).

Digestion Efficiency: Pappas *et al.* (1973) added 0.37% chromic oxide to diets fed to channel catfish and observed the disappearance of nutrients (fat, protein, and carbohydrate) relative to the inert marker while the food remained in the stomach. Nearly 50% of these dietary components were removed before the food entered the intestine, al-

Table VI

"Calorie Counting" in *Salmo gairdnerii:* Control of Appetite by Changes in Gastric Motility[a]

Test food	Food (cal/g)	Mean daily intake (cal/100 g/day)	Mean feeding intervals (hr)	Mean gastric evacuation time (hr)
Salmon pellet	4.78	11.3	6	15
75% food/25% kaolin	4.20	15.9	5	14.5
67% food/33% kaolin	3.23	15.0	5	11.8
50% food/50% kaolin	2.18	13.1	4	10

[a] The fish obtained food from a "demand feeder" under continuous illumination at 18°C. Each week the control pellets (Cooper's salmon starter) were replaced by reconstituted pellets diluted with kaolin, and feeding activity was subjected to periodogram analysis to determine the time for return of appetite during the succeeding 7 days. Separate groups of fish were fed the same diets and X-rayed to determine the time for complete gastric evacuation. Note that the stomach empties more quickly when the food is diluted, that appetite returns more quickly and, if the evacuation curve is exponential, this occurs when 70–90% of the previous meal has left the stomach. From Grove *et al.* (1978, *J. Fish Biol.* **12**, 507–516).

though no conclusion was reached whether they undergo gastric absorption or are passed in solution into the intestine for assimilation. Beamish (1972) showed that protein was solubilized more extensively in *Micropterus* stomach than fats, which required intestinal digestion before assimilation occurred. Overall protein digestion is efficient in fish, more than 85% assimilation being the rule unless oxidation of oils occurs in the food employed (Nose and Toyama, 1966). Smith and Lane (1971) reported that protein assimilation of salmonids was impaired if α-cellulose was incorporated in the diet. Increase in the proportion of carbohydrates in the diet, or absolute decrease in the protein content of the diet, has been reported to decrease protein assimilation without explanation of the mechanism involved (see references in Nose, 1967). Nose demonstrated that this apparent decrease in protein assimilation arises from failure of previous workers to allow for the release of metabolic nitrogen into the feces, which becomes significant in biochemical determinations only at low levels of protein intake. A true decrease in protein assimilation was detected by Nose when rainbow trout were given egg albumin, which contains a trypsin inhibitor. The early work of Inaba *et al.* (1962) clearly demonstrates that the protein in whalemeat or soya bean residues is much less digestible than that in white fish meal, but the basis for this observation requires further study. A more extensive account for formulation of fish diets is given in Chapter 1.

VII. PHYSIOLOGICAL STUDIES ON FISH GASTROINTESTINAL MOTILITY

A. Introduction

1. Excitation of Gastrointestinal Smooth Muscles

Our understanding of muscular activity of the gastrointestinal tract in various fish is based on comparisons with higher vertebrates. The major contractile elements in the muscularis are hexagonally packed groups of smooth muscle cells which act as a contractile unit (Bennett and Burnstock, 1968). The cells affect each other electrotonically so that coordinated myogenic contractions, independent of the nervous system, can develop in response to stretch as a result of Ca^{2+} flux across their membrane. In the *in situ* gut, rhythmic myogenic contractions of the circular muscles temporarily divide the intestine into a series of segments to facilitate mixing of the luminal contents.

Superimposed on this activity is the peristaltic process, dependent on cell bodies in the nerve plexuses of the gut wall, usually seen as waves of relaxation and contraction which progress anally along the gut. The combination of these two processes, segmentation and peristalsis, may cause the segments of the intestine to swing slowly from side to side in pendular activity.

2. THE SITES OF DRUG ACTION

In view of the difficulty in separating the roles of muscle and nerve cells, most workers have investigated the neurogenic components by using drugs which are believed to act at the receptor sites on tissues with which neurotransmitters are known to act. This approach is not without problems since few drugs are specific, and may only be considered *selective* within certain dose ranges (Daniel, 1968) which must be determined for each new tissue. It has become general to refer to nerve cells releasing a stated neurotransmitter (e.g., acetylcholine, noradrenaline) as cholinergic, adrenergic, and the receptor they occupy as cholinoceptor, adrenoceptor. When the transmitter, or a similar molecule (agonist), occupies the receptors of a tissue, a response is evoked in proportion to the agonist concentration. A dose–response curve can be constructed and the dose producing 50% of maximum response is a measure of the affinity of the agonist at that site (Ariens and Van Rossum 1957). Similar molecules which occupy the site without eliciting a response may combine with the agonist reversibly, or irreversibly by binding to the receptor chemically. The blocking action of reversible antagonists can be overcome by increasing the agonist concentration, and may be visualized as a parallel shift of the agonist dose–response curve to the right. A measure of the affinity of the antagonist is the observed shift of the dose–response curve in the presence of a stated antagonist concentration. Noncompetitive and irreversible antagonism is usually detected as an insurmountable blockade, accompanied by a depression of the dose–response curve (e.g., Guimaraes, 1969). Examples of these relationships are given in Fig. 8. In Table VII, the drugs which have been commonly used in studies of fish gastrointestinal motility are shown, together with the receptors that they occupy.

3. INTRINSIC NEURONS

Two main techniques for activating neurons in the gut wall are commonly used. Various workers have adapted the technique of Trendelenburg (1917) in which distension of the gut wall *in vitro* initiates

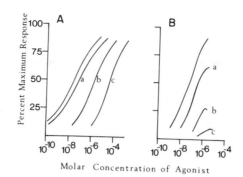

Fig. 8. Analysis of drug effects on *in vitro* Magnus preparations of plaice gastric longitudinal muscle. (A) Dose-related contractions induced by acetylcholine alone. Control $ED_{50} = 2 \times 10^{-8} M$ (and $pD_2 = 7.7$). The addition of atropine to the bath in doses of $3.5 \times 10^{-x} M$ causes a parallel shift of the dose–response curve to the right in relation to the dose: curve a, $x = 9$; curve b, $x = 8$; curve c, $x = 7$. The concentration of atropine for which a control dose of acetylcholine must be *doubled* to obtain the control response is $1.6 \times 10^{-9} M$ (and $pA_2 = 8.8$). (B) A similar study using 5-hydroxytryptamine as stimulant gives a control $ED_{50} = 4 \times 10^{-8} M$ (whence $pD_2 = 7.4$). Atropine here acts noncompetitively when the doses used above are employed. The concentration of atropine required to depress the control maximum by half is $1.2 \times 10^{-8} M$ (whence $pD_2' = 7.9$). (From Edwards, 1972b, *Comp. Gen. Pharmacol.* 3, 345–358.)

activity. To study stimulant effects on inactive preparations, the gut segment is simply suspended in a bath under minimum stretch (Magnus preparation). The second method, which may be used on Trendelenburg or Magnus preparations, is to stimulate the intrinsic neurons by electrodes in the bath placed near (field stimulation) or on either side (transmural stimulation) of the gut wall. The frequency of stimulation may be between 1 and 50 Hz and the pulse duration of 1 msec or less to avoid direct activation of the muscle cells. In addition to recording the muscle responses mechanically (levers, strain gauges, pressure transducers), changes in muscle membrane potential can be monitored using the sucrose gap technique (Campbell and Burnstock, 1968; Ito and Kuriyama, 1971). The use of the latter technique has shown that, in low tone preparations, transmural stimulation may excite inhibitory neurons in the gut wall which hyperpolarize the muscle cell membranes without a clear relaxation being recorded. After cessation of the stimulus, the cells repolarize and may overshoot the resting potential leading to spontaneous electrical activity and "rebound contractions." This phenomenon led Burnstock (1969) and Campbell and Burnstock (1968) to reinterpret the conclusions reached in earlier studies of fish gastrointestinal motility; the vagus nerve was earlier claimed to excite the fish stomach whereas it is now believed to

Table VII

Sites of Action and Names and Abbreviations of Drugs Used to Analyze
Gastrointestinal Motility in Fishes

Receptor/effect	Agonist	Antagonist
1. Muscarinic; cholinoceptor on smooth muscles	Acetylcholine (ACh) Pilocarpine Carbachol Methacol	Atropine (Atr) Hyoscine
2. "Nicotinic"; cholinoceptor on neurons	Acetylcholine Carbachol 1,1-Dimethyl-4-phenyl-piperazonium (DMPP)	Nicotine Hexamethonium (C_6) d-Tubocurarine (dTC) Mecamylamine (Mec)
3. "Nicotinic"; cholinoceptor on striated muscle	Acetylcholine	d-Tubocurarine
4. Adrenoceptors		
α	Phenylephrine Adrenaline (A) Noradrenaline (NA)	Phenoxybenzamine (PBZ) Phentolamine (Phent) Dihydroergotamine (DHE) Piperoxane (Pip) Yohimbine
β	Isoprenaline (Iso) Adrenaline (A) Noradrenaline (NA)	Propranolol (Prop) Butoxamine
5. Tryptaminoceptors "dibenamine type" D	5-Hydroxytryptamine (5-HT)	Methysergide 2-Bromo-lysergic acid diethylamide (BOL) Phenoxybenzamine (PBZ) Ergotamine
6. Purinoceptor	Adenosine triphosphate (ATP)	
7. Histamine receptor (H_1)	Histamine	Mepyramine
8. Cholinesterase inhibition	Eserine	
9. Neuron blocking		
All nerves	Tetrodotoxin (TTX), procaine	
Cholinergic nerves	Morphine, hemi-cholinium	
Adrenergic nerves	Reserpine, bretylium, guanethidine	

inhibit it in the majority of fishes. It has been proposed that intramural neurons exist in vertebrate viscera which excite or inhibit the smooth muscles by releasing a purinergic transmitter, possibly adenosine triphosphate (Burnstock, 1972; Burnstock *et al.*, 1970, 1972).

4. EXTRINSIC NEURONS

The extrinsic nerves to the fish gut originate from the autonomic nervous system. The simplistic account of this system in mammals, attributed to Langley (1921), is that "parasympathetic" nerve tracts from the craniosacral region innervate the intrinsic neurons of the gut wall by way of vagus and pelvic nerves. A "sympathetic" outflow in the thoraco-lumbar region innervates ganglion cells in the sympathetic chain. In both divisions, preganglionic medullated cholinergic nerves leave the CNS. The postganglionic parasympathetic neurons are cholinergic. Activity in the parasympathetic division enhances gastro-intestinal motility whereas sympathetic activity is inhibitory.

It is now acknowledged that adherence to this model is likely to cause misconceptions, and many mammalian exceptions have been reported. Vagal fibers may activate either cholinergic or purinergic intrinsic neurons. The vagus may be joined by postganglionic axons (adrenergic or cholinergic) from the sympathetic division, or itself send preganglionic cholinergic fibers along the nearest convenient "sympathetic" tract. The postganglionic adrenergic nerves of the "sympathetic" system may not innervate gut muscle directly, but en-mesh ganglion cells in the gut wall (Campbell, 1970a,b). These conclusions, reached for higher vertebrates, are equally likely to complicate studies on the less understood fish groups.

B. Cyclostomes

Relatively little information is available on the physiology of intestinal motility in hagfish and lampreys. The intestinal muscles of both groups are poorly developed and food movements may depend on contractions of the body wall. Fänge (1962) and Burnstock (1969) in their reviews report that the intestine in hagfish is contracted by acetylcholine (10^{-8} to 10^{-5} g/ml) but relaxed by adrenaline (10^{-7} to 10^{-5} g/ml), and by atropine. Peristalsis is induced in *Lampetra* rectum by 10^{-4} g/ml 5-hydroxytryptamine.

In the hagfish *Myxine glutinosa*, the two intestinal rami of the vagus nerve unite dorsally to the intestine forming an unpaired nerve-like structure (*ramus impar*) which contains numerous ganglion cells. Cholinergic and adrenergic neurons are present (Fänge and Johnels, 1958; Hallbäck, 1973; Fänge *et al.*, 1963). Patterson and Fair (1933) found that vagal stimulation slightly relaxes the hagfish gut. von Euler and Fänge (1961) detected catecholamines in the gut of *Myxine* and Honma (1970) used the fluorescent histological technique of Falck and Owman (1965) to show that adrenergic neurons and terminals are present in the enteric plexus of *Lampetra japonica* and *L. planeri*, which presumably mediate inhibition. Nonadrenergic nerves in the gut wall of hagfish and lampreys are likely to be cholinergic and possibly also purinergic, coordinating peristalsis, and appear to be innervated by fibers of vagal origin.

More recently, Baumgarten *et al.* (1973) have carried out an extensive study of monoaminergic neurons in the gut wall of *Lampetra fluviatilis* using the fluorescent histochemical technique together with microspectrofluorimetry, chromatography, and electron microscopy. Green fluorescent nerve cells were detected in the submucosa of the fore- and midgut which may inhibit the intestine, and which contain primary catecholamines such as dopamine or noradrenaline. These authors detected an extensive submucosal plexus of tryptaminergic axons and cell bodies which connected with a subserous plexus containing nerve terminals. They propose that the 5-hydroxytryptamine transmitter released by these endings elicits peristalsis but point out that similar intramural neurons have not been described in the gut wall of true vertebrates.

C. Elasmobranchs

In comparison with actinopterygii, relatively little work has been done on elasmobranch gastrointestinal motility. The early reports of Lutz (1931), Nicholls (1933), and Dreyer (1949) state that all parts of the gut are contracted by acetylcholine and pilocarpine. von Euler and Östlund (1957) confirmed the observation of Nicholls that atropine prevents the action of acetylcholine. Adrenaline sometimes causes contraction of the gut (Babkin *et al.*, 1935a,b; Dreyer, 1949; Young, 1933) but according to Nicholls (1933) this is only at lower doses. Inhibitory effects of adrenaline have also been recorded on the dogfish rectum (Lutz, 1931) and ray intestine (von Euler and Östlund, 1957). Without more detailed studies it is impossible to conclude whether adrenaline has different actions in different parts of the gut of different species. Recently Moore and Hiatt (1967) reinvestigated the problem of the action of adrenaline in elasmobranchs. They elected to work with *Squalus acanthias*, using intragastric balloons *in vivo* to record contractions. Injection of adrenaline produced a clear contraction, which they believed would be due to occupation either of α- or β-adrenoceptors. However, they found that neither α (phenoxybenzamine) nor β (propranolol) adrenoceptor blocking agents prevented the action of adrenaline. To extend their study, they tested other sympathomimetic amines which have differential potencies on α- and β-receptors in other vertebrates. Surprisingly, isoprenaline, dopamine, or noradrenaline failed to contract the stomach in control animals. Phenylephrine was effective, but phenoxybenzamine, which in most vertebrates is a powerful phenylephrine antagonist at α-adrenoceptors, failed to antagonize the response. In further studies, aminophylline was found to abolish the adrenaline-induced contraction and the effect of adrenaline decreased rapidly with successive tests in control animals. They concluded that adrenaline had no inhibitory effect on *Squalus* gastrointestinal smooth muscle. The stimulatory action of a first test dose of adrenaline depends not on the usual cell membrane adrenoceptors but on increased glycogenolysis presumably following an increased intracellular build-up of cyclic AMP. However, once under way, this intracellular activity hyperpolarizes the muscle membranes, thus suppressing responses to subsequent adrenaline treatment. Clearly this abnormal action of adrenaline should be investigated in other species. It should not be concluded that adrenoceptors are necessarily aberrant in elasmobranchs since Nilsson *et al.* (1975) have shown α-adrenoceptors in spleen and artery preparations of *Squalus* and *Scyliorhinus* with properties similar to those in teleosts and higher vertebrates.

Gzgzyan and Kuzina (1973) investigated spontaneous peristaltic contractions of the stomach of the ray, *Dasyatis pastinaca*. This species does not exhibit alternation of periods of activity and rest typical of empty mammalian stomachs. The contractions were enhanced when cholinomimetics (carbachol, pilocarpine) were injected but atropine was not a powerful inhibitor of the spontaneous contractions. There is evidence here that cholinergic nerves may play a coordinating role in gastrointestinal peristalsis but, in common with teleosts (see later), myogenic contractions may play a significant part in rhythmic contractions of the gut wall. Myenteric and submucosal plexuses have been described in the stomach and intestine of elasmobranchs (Nicol, 1952).

The digestive canal of elasmobranchs is supplied by vagal and splanchnic nerves. The vagal fibers are believed to be adrenergic (Fänge and Hanson, 1973). Campbell and Burnstock (1968) examined the published records from earlier workers of the effect of vagal stimulation on stomach activity. They concluded that the influence of the vagus is limited to the stomach and proximal intestine and that it is inhibitory. Earlier conclusions that the vagus is excitatory were explained by the long latency of the contraction, reminiscent of the "rebound contractions" described earlier. Campbell (1975) found that the regular spontaneous activity of *Scyliorhinus canicula* stomach, recorded by intragastric balloon, was abolished when the intracranial roots of the vagus nerve were stimulated electrically. He also observed a stimulatory effect of the vagus on esophageal striped muscles. By inference from other vertebrates, Burnstock (1969) suggests that at least part of the inhibitory response mediated by the vagus depends on the presence of inhibitory, nonadrenergic neurons in the stomach wall. The role of sympathetic nerves on gastrointestinal motility is as yet unclear. Burnstock (1969) concluded that the records in earlier publications show evidence of both cholinergic excitatory and adrenergic inhibitory fibers supplying both stomach and intestine in elasmobranchs.

D. Teleosts

Gastrointestinal motility, and its control, has been more extensively studied in teleosts. The sympathetic system, like that of higher vertebrates, consists of two ganglionated chains which extend into the head as far as the trigeminal nerve (V). The esophagus and stomach have a rich vagal innervation, while in many stomachless fish including the cyprinidae the vagal influence extends to the intestine. In the stomachless pleuronectids *Rhombosolea* and *Ammotretis*, however, the vagal influence does not extend beyond the esophagus (Grove and Campbell, unpublished observations). The stomach and intestine are also innervated by fibers from the splanchnic nerves which originate from sympathetic ganglia in the first few spinal segments. The gut of the tench (*Tinca*) contains striated and smooth muscles innervated by the vagus nerve (Ohnesorge and Rehberg, 1963). Only the smooth muscle elements participate in the peristaltic movements. Results of pharmacological analysis indicate that the peristalsis is due to a reflex involving intramural neurons (Ohnesorge and Rauch, 1968). A posterior sympathetic nerve to the rectum has been described in the trout (*Salmo trutta*) by Burnstock (1958a,b). The ramus intestinalis of the vagus nerve probably receives sympathetic fibers from the ganglionated chains and may therefore more properly be termed a vago-sympathetic trunk (Young, 1931).

1. ANALYSIS OF GASTROINTESTINAL RECEPTORS

Few studies have been made to estimate the affinity for specific receptors on gastrointestinal muscle and nerve cells of agonist molecules by dose–response curve analysis. Table VIII summarizes much of the recently published data for a number of teleost fish. In general, cholinesters and 5-hydroxytryptamine (5-HT) contract the muscle coats whereas sympathomimetic amines relax the gut or inhibit spontaneous activity. The problems facing the researcher lie mainly in the complexity of the gut wall, since the site of action of a drug must be determined, and the specificity of the drugs, which have usually been assumed to act in fish as in mammals.

Acetylcholine is probably a direct stimulant of muscarinic receptors on many parts of the gut musculature since atropine is the most potent inhibitor of its action (*Salmo trutta*, Burnstock, 1958a,b; *Gadus morhua*, S. Nilsson and Fänge, 1969; *Carassius auratus*, Ito and Kuriyama, 1971; Saito, 1973; *Pleuronectes platessa*, Edwards, 1972b; Goddard, 1975; *Blennius, Myoxocephalus*, Grove *et al.*, 1974). Acetylcholine also exerts effects on nicotinic receptors of a variety of neurons in the enteric plexus, and the high, potentially nonselective doses of atropine used in the *Gadus* and *Carassius* studies may block this action as well. Where dose–response analysis has been undertaken, the pA_2 of atropine has been found to lie between 8.2 and 9.4. Concentrations of atropine at ca. 10^{-7} g/ml or less should be more than adequate to abolish the muscarinic action of acetylcholine without significant side effects (Edwards, 1972b; Goddard, 1975; Grove *et al.*, 1974). There is no doubt that acetylcholine effects in *Carassius* involve the intramural ganglion cells. Saito (1973) believed that all mechanical effects exerted by the cholinester (10^{-6} g/ml) on the intestine were indirect since tetrodotoxin (10^{-8} g/ml), which prevents axonal conduction by blocking sodium channels in the neuronal membrane, abolished all its actions. He recognized a brief contraction of striated muscle fibers when the drug was applied, followed by a slower smooth muscle contraction. After the administration of atropine (10^{-6} g/ml) to impair the response of cholinergic excitatory cells, he recorded relaxation of the preparation. Since this effect is tetrodotoxin-sensitive, inhibitory neurons activated by preganglionic cholinergic fibers are also present in the intestine. Dimethylphenylpiperazinium (DMPP) (10^{-6} g/ml) which also briefly excites ganglion cells in mammals mimicked acetylcholine. In contrast, the earlier study by Ito and Kuriyama (1971) on silver carp showed that acetylcholine excitation persisted after tetrodotoxin. They proposed that at least part of the stimulus must be directly on the muscle cells. d-Tubocurarine (10^{-6} g/ml) abolishes the effect of acetylcholine on striated muscle in *Carassius* (Saito, 1973) and in *Tinca* (Mahn, 1898). Hexamethonium, which in mammals is a relatively potent nicotinic

Table VIII
Effects of Agonist Drugs on Resting Gastrointestinal Muscle Preparations of Teleosts[a]

Species and organ	Cholinergic stimulants	5-Hydroxytryptamine	Sympathomimetic amines	Purines
Salmo trutta				
a. Stomach	a,b,c. ACh + 10^{-8}–10^{-4} (1)	a,b,c. + 10^{-9} (1)	(1) a. NA + 10^{-6} (1,2)	
b. Intestine	Pil + 10^{-5} (1)		A + 10^{-6} (1,2)	
c. Rectum			ISO + 2×10^{-5} (1,2)	
			b,c. A − 10^{-8}–10^{-5} (1)	
Carassius auratus				
Intestine	ACh (1st) + 10^{-6} (3,4)		NA − 10^{-6}–10^{-5} (3,5)	ATP + 10^{-5} M (5)
(Cholinesters	ACh (2nd) + 10^{-6} (3,4)		A − 10^{-6} (3)	
cause	ACh (3rd) − 10^{-6} (3,4)		ISO − 10^{-6} (3)	
multiple	DMPP (1st) + 10^{-5} (3)			
contractions)	DMPP (2nd) + 10^{-5} (3)			
Tinca tinca				
Intestine	ACh (1st) + 5×10^{-8}–10^{-5} (6,7)	+ 10^{-8} (6,7)	(8) A − 10^{-8}–10^{-7} (7)	
(Cholinesters	ACh (2nd) + 5×10^{-8}–10^{-5} (6,7)		NA − 10^{-8}–10^{-7} (7)	
cause	DMPP + 10^{-8}–10^{-6} (7)	(7)		
multiple				
contractions)				
Anguilla anguilla				
a. Esophagus	a,b,c. ACh + 10^{-5} (9)	(9)	a,c. A − 10^{-5} (9)	
b. Stomach			b. A + then − 10^{-5} (9)	
c. Intestine			Tyr + then − 10^{-5} (9)	
			ISO + then − 10^{-5} (9)	

Species / Tissue							
Gadus morhua Stomach	ACh + 10^{-8}–10^{-5}	(10)			NA + 10^{-7}–10^{-5}	(10)	
	CCh + 10^{-7} – 10^{-5}	(10)			A + 10^{-7}–10^{-5}	(10)	
	Pil + 10^{-7}–10^{-5}	(10)			Pheny + 10^{-7}–10^{-5}	(10)	
					ISO – 10^{-6}–10^{-5}	(10)	
Pleuronectes platessa							
a. Stomach	a. ACh + *3 × 10^{-8} M	(11,14)	a. + 4 × 10^{-8} M	(11,12,14)	a. NA + 10^{-7}– 10^{-5} M	(11)	a. ATP + 10^{-5} M (14)
b. Intestine	MCh + *8 × 10^{-8} M	(11,14)			A 0 10^{-8}– 10^{-3} M	(11)	
	CCh + *2 × 10^{-8} M	(11,14)					
	b. ACh + *3 × 10^{-7} M	(13)	b. – 10^{-6}–10^{-5} M	(13)	b. A – 10^{-10}–10^{-5} M	(13)	b. ATP – 10^{-6} (13)
	MCh + *5 × 10^{-5} M	(13)			DA – 10^{-6}	(13)	
	CCh + *7 × 10^{-6} M	(13)			ISO – 10^{-7}	(13)	
Blennius pholis Intestine	ACh + *8 × 10^{-7} M	(16)	NA – *3 × 10^{-6} M	(16)			
	MCh + *5 × 10^{-8} M	(16)					
Myoxocephalus scorpius Stomach	ACh + *5 × 10^{-7} M	(16)					

a Doses are given as g/ml, except where the author used molar concentrations (M), and ED_{50} (*) values are given where known. Key: +, excitation; −, relaxation; 0, no effect.

b Abbreviations: A, adrenaline; ACh, acetylcholine; ATP, adenosine triphosphate; CCh, carbachol; DA, dopamine; DMPP, dimethylphenylpiperazinium; dTc, d-tubocurarine; 5HT, 5-hydroxytryptamine; ISO, isopropylnoradrenaline; MCh, methacholine; NA, noradrenaline; Pheny, phenylephrine; Pil, pilocarpine; Tyr, tyramine; Vag, vagostimine.

c Numbers in parentheses refer to sources: (1) Burnstock (1958a); (2) Campbell and Gannon (1976); (3) Saito (1973); (4) Ito and Kuriyama (1971); (5) Burnstock et al. (1972); (6) Mahn (1898); (7) Ohnesorge and Rauch (1968); (8) Baumgarten (1967); (9) Nilsson and Fänge (1967); (10) Nilsson and Fänge (1969); (11) Edwards (1972b); (12) Grove et al. (1974); (13) Goddard (1975); (14) Stevenson and Grove (unpublished observations); (15) Gzgzyan et al. (1973); (16) Grove et al. (1976).

blocker, has less specificity in fish although its use has been extensive. Edwards (1972b) found that it noncompetitively antagonized acetylcholine and other cholinesters (pD'_2 = 3.1–4.4) and also 5-HT (pD'_2 = 3.9) stimulation of the plaice stomach. Recently Stevenson and Grove (1977) have found that, after tetrodotoxin (10^{-7} g/ml) the direct muscarinic actions of cholinesters are blocked by hexamethonium (10^{-4}– 10^{-3} g/ml) but not by mecamylamine (10^{-3} g/ml). It is unfortunate that hexamethonium has been the drug of choice for many analyses of teleost gut function. In some of the studies on peristalsis and extrinsic nerves to be reviewed below, it is not unusual to find doses of hexamethonium as high as 10^{-3} g/ml being used as a "selective" nicotinic atagonist. At present the bulk of the evidence suggests that exogenous acetylcholine acts at smooth muscle muscarinic sites (*Salmo, Gadus, Pleuronectes, Blennius, Myoxocephalus*) whereas in the stomachless *Tinca* and *Carassius* much of the action involves ganglion cells of the enteric plexus.

The actions of adrenoceptor stimulants are equally complex. α-Receptors, mediating inhibition of the gut by adrenaline and noradrenaline, have been postulated in *Carassius* intestine (Saito, 1973) and *Salmo* intestine (Burnstock, 1958a,b). The inhibition was antagonized by phentolamine, phenoxybenzamine, or dihydroergotamine in relatively high concentrations (ca. 10^{-5} g/ml). In teleost and elasmobranch cardiovascular systems (S. Nilsson and Grove, 1974; S. Nilsson *et al.*, 1975) these drugs have pA_2 values of the order 6.5–7.5, suggesting that lower doses than used by Saito and Burnstock are sufficient to block α-receptors without side effects. Young (1931) found that ergotoxine (10^{-6} g/ml) was sufficient to block the inhibitory action of adrenaline. β-Adrenoceptors mediating inhibition have been demonstrated in the intestine of *Carassius* (Saito, 1973) and *Pleuronectes* stomach and intestine (Goddard, 1975; Stevenson and Grove, unpublished observations). In *Pleuronectes* the inhibition is reversibly blocked by butoxamine whereas propranolol at doses of 10^{-5} g/ml and above had nonspecific depressant effects on spontaneous activity and acetylcholine responses. In *Rhombosolea* or *Ammotretis*, however, the inhibitory actions of adrenaline, which are mediated by β-adrenoceptors, are competitively antagonized by propranolol (pA_2 = 7.9; Grove and Campbell, unpublished observations). Several anomalous effects of adrenergic agonists have been detected. In *Anguilla* gastric cecum (S. Nilsson and Fänge, 1967), *Gadus* stomach (S. Nilsson and Fänge, 1969), *Salmo* stomach (Burnstock, 1958b; Campbell and Gannon, 1976; and occasionally in *Pleuronectes* stomach (Edwards, 1972b) and intestine (Goddard, 1975) and *Blen-*

nius intestine (Grove *et al.*, 1976), treatment with catecholamines other than isoprenaline causes excitation. In *Gadus* these were antagonized by high doses (10^{-5} g/ml) of α-adrenoceptor antagonists but in *Pleuronectes* were resistant to both α and β-blockade. These effects recall those described for the stomach of elasmobranchs (Section VII,C) and remain similarly unexplained.

5-Hydroxytryptamine is a powerful stimulant of the gastrointestinal tract of various fish (Table VIII) (Fänge, 1962). Few studies have been made to determine the site of action of this amine. Edwards (1972b) showed that atropine ($pD_2' = 7.5$) and hexamethonium ($pD_2' = 3.9$) were noncompetitive antagonists of 5-HT and suggested that it acted in fish, as in mammals (Gershon, 1967), on cholinergic enteric neurons. Grove *et al.* (1974) extended the study and found that the excitatory action persisted after tetrodotoxin, morphine, ergotamine, methysergide, or 2-bromo-lysergic acid diethylamide (each at 10^{-6} and 10^{-5} g/ml). Since hemicholinium impaired the effect, they concluded that much of the action of 5-hydroxytryptamine is indirect, displacing acetylcholine from cholinergic terminals at a site peripheral to that blocked by tetrodotoxin or morphine. In the relaxed *Pleuronectes* intestine, 5-HT stimulates the longitudinal, but not the circular, muscle layer ($pD_2 = 5.2$) and is not affected by attempted blockade with morphine or 2-bromo-LSD. In the active isolated preparation however, the amine inhibits peristalsis when applied to the serosal (but not the mucosal) surface (Goddard, 1975).

Burnstock *et al.* (1970, 1972) have proposed that a new class of neurons releasing a purine as neurotransmitter exists in vertebrates. Adenosine triphosphate has been found to stimulate goldfish intestine (Burnstock *et al.*, 1972) and *Pleuronectes* stomach (Stevenson and Grove, unpublished observations) but to relax *Pleuronectes* intestine (Goddard, 1975).

2. EFFECTS OF DRUGS ON ACTIVE PREPARATIONS OF THE TELEOST GUT

Despite the inadequate knowledge of the sites of action of the agonist and antagonist drugs described above, many workers have used them in attempts to analyze the role of the enteric plexus (Kirtisinghe, 1940) in coordinating peristalsis in the isolated gut. Furthermore, it has become clear that myogenic contractions as well as neurogenic peristalsis may develop when an isolated segment is distended. True peristalsis was described by Burnstock (1958a,b) in the isolated trout gut after distension. At 8°–14°C, a powerful longitudinal contraction precedes each peristaltic wave which then moves anally at ca. 2 cm/min, such waves originating in the cardia, antrum, or about one-third of the way along the intestine. Similar preparations have been made *in vitro* for *Tinca* intestine (Ohnesorge and Rauch, 1968), *Pleuronectes* stomach (Stevenson and Grove, unpublished observations) and intestine (Goddard, 1975), and *in vivo* using intragastric balloons for *Scorpaena* (Gzgzyan *et al.*, 1973), *Salmo* and *Conger* (Campbell, 1975). Typically, an increase in intraluminal pressure of 3–5 cm H_2O initiates rhythmic contractions of the gut preparation. The effects of drugs and the doses used on this activity in various fish are listed in Table IX.

There is general agreement that cholinesters stimulate spontaneous contractions when the gut is distended whereas atropine inhibits them. Eserine has been found to increase peristaltic rate in *Salmo* and *Tinca*. On the assumption that these effects depend on the activity of ganglion cells in the enteric nerve plexus, drugs which block the effect of acetylcholine on ganglion cells (nicotine, DMPP, mecamylamine, or hexamethonium) or which prevent axonal conduction (tetrodotoxin, procaine) have been tested and generally found to diminish or abolish peristalsis. Frequently, the reports have noted that after blockade with nicotinic and muscarinic blocking drugs, or with tetrodotoxin, the pattern of rhythmic contraction changes but is not abolished. Ito and Kuriyama (1971) recorded the mechanical activity of *Carassius* intestine using a strain gauge while simultaneously recording the electrical activity with the sucrose-gap technique. They detected contractions after neuronal blockade which were abolished by exposure to manganous ions, and concluded that these were myogenic contractions based on calcium fluxes across the muscle membranes. Similar activity independent of the enteric nervous system has been described for the gut of *Salmo* (Burnstock, 1958a,b), intestine of *Tinca* (Ohnesorge and Rauch, 1968), and stomach and intestine of *Pleuronectes* (Goddard, 1975; Stevenson and Grove, 1977). The question arises of the role of myogenic contractions in teleost "peristalsis." A major difference exists in the onset of rhythmic activity in isolated, distended fish gut segments when compared with mammalian preparations. In *Pleuronectes*, a delay of several minutes follows distension of the stomach or intestine before peristalsis begins and rings of contraction pass anally (Fig. 9A). In *Tinca* and *Pleuronectes* the longitudinal and circular muscle coats contract synchronously, there being little sign of the "preparatory" and "expulsion" phases typical of mammals (Ohnesorge and Rauch, 1968; Goddard, 1975). The enteric nerve cells appear essential for the initiation of peristalsis, since pretreatment of

Table IX

Effects of Drugs on Active Preparations of Teleost Gastrointestinal Muscles[a]

Drug	Dose (g/ml)	Source	Salmo trutta Stomach and intestine (a.)	(b.)	Carassius auratus intestine	Gadus morhua Stomach	Tinca tinca Intestine	Pleuronectes platessa Stomach and intestine (a.)	(b.)	Scorpaena porcus Stomach
ACh	$10^{-8}-10^{-4}$	(1,3,4,6,7,10,13,14,15)	+(M)		+	+	+	+	+	+
CCH	$10^{-8}-10^{-5}$	(10,13,14)				+	+	+	+	
Eserine or vag	$10^{-6}-10^{-5}$	(1,7)	+							
Atropine	$10^{-7}-10^{-5}$	(1,3,4,7,10,13,14,15)	−(M)		−(M)	−	−	0/(−)		−
dTc	$10^{-7}-10^{-4}$	(7,14)					−		−	
DMPP	$10^{-8}-10^{-6}$	(7)	+				+/−(M)			
Nicotine	$10^{-7}-10^{-4}$	(7)					+/−(M)			
Mecamylamine	$10^{-6}-10^{-4}$	(13,14)								
Hexamethonium	$10^{-7}-10^{-3}$	(1,7,13,14)	−(M)				0(M)	a. 0	b. −	
NA	$10^{-7}-10^{-5}$	(3,4,7,10,13,14)	a. +/−	b. −	−	+	−	a. 0	b. 0/−	
A	$10^{-10}-10^{-5}$	(1,3,4,7,10,13,14)			−	+	−	a. +/−	b. −	
Pheny	$10^{-7}-10^{-5}$	(10,13,14)				+				
ISO	$10^{-8}-10^{-5}$	(3,10,13,14)			−	−			−	
5-HT	$10^{-9}-10^{-5}$	(13,14)	+		+	−		a. +	b. −	
ATP	$10^{-6}-10^{-3}$	(5,13,14)						a. +	b. −	
Tetrodotoxin	$10^{-7}-10^{-6}$	(3,4,13,14)			−(M)			−(M)		
Procaine	$10^{-6}-10^{-4}$	(7,13,14)					−	−(M)		

[a] Drug abbreviations and sources (numbers in parentheses) are given in footnote to Table VIII. Key: +, enhancement of peristalsis; 0, no effect; −, inhibition; (M), unmasking of myogenic rhythms; /, followed by (e.g., +/−, excitation followed by inhibition).

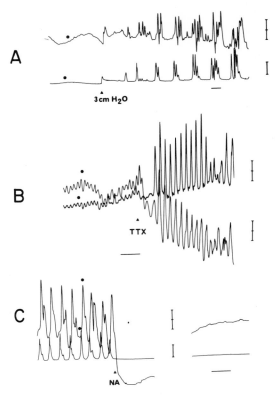

Fig. 9. Trendelenburg preparations of *Pleuronectes platessa* intestine. (A) Peristalsis induced by distension with 3 cm H_2O intraluminal pressure. (B) Change from peristaltic to myogenic activity following 7×10^{-7} g/ml tetrodotoxin (TTX). (C) Inhibition of myogenic activity, in the presence of TTX, by noradrenaline (NA) (10^{-5} g/ml). Upper trace in each figure is longitudinal muscle tension (scale = 1 g) and the lower is intramural pressure (scale = 5 cm H_2O). Time scale = 1 min. (From Goddard, 1975.)

the stomach or intestine of *Pleuronectes* with 10^{-7} g/ml tetrodotoxin prevents the onset of activity. Atropine, even in high concentrations (up to 10^{-5} g/ml) does not prevent the onset, suggesting that non-cholinergic excitatory neurons are involved. However, once activity has been generated, attempts to impair the neurogenic control of activity unmask the myogenic rhythms of many fish preparations as shown in Fig. 9B and Table IX. The amplitude, but not the frequency, of the myogenic activity of *Pleuronectes* and *Limanda* stomach increases with distension (Grove *et al.*, 1976) thereby suggesting a mechanism whereby the digestion rate of larger meals (Section VI) can exceed the predicted exponent of meal size. Ohnesorge and Rauch (1968) discuss the possibility that myogenic activity is controlled by acetylcholine released from muscle cells themselves when stretched, as described in mammals (Ikeda *et al.*, 1958). On this basis, the potentiation by cholinesters or eserine and the depression by atropine, need not involve cholinergic neurons in the maintenance of fish peristalsis.

There is also general agreement that the intestine, and usually the stomach, is inhibited by sympathomimetic amines (Fig. 9C), although in *Salmo* and *Gadus* a paradoxical stimulation of gastric activity is well established. In *Gadus* the excitation is prevented by high doses (10^{-5} g/ml) of α-adrenoceptor blockers (S. Nilsson and Fänge, 1969). Stevenson and Grove (unpublished observations) examined the inhibition of gastric peristalsis in Trendelenburg preparations by sympathomimetic amines. It is likely that the receptors are of the beta variety since the order of potency of agonists is isoprenaline, adrenaline, phenylephrine (pD_2 = 7.8, 7.3, and 5, respectively). Of a wide variety of α- and β-adrenoceptor antagonists tested, only the β_2 blocker butoxamine (10^{-5} g/ml) clearly antagonized inhibition by these amines. The β receptors are likely to reside on the smooth muscle cells rather than enteric neurons, since in tetrodotoxin-treated stomachs exhibiting myogenic activity adrenaline abolished activity.

Other effective agents on active gut preparations include 5-hydroxytryptamine and adenosine triphosphate. 5-HT stimulates *Pleuronectes* stomach smooth muscle and the intestinal longitudinal coat. When applied to a nonactive stomach, whether before or after tetrodotoxin, activity is initiated apparently by a direct stimulation of muscle cells. 5-HT also enhances ongoing gastric peristalsis but inhibits intestinal peristalsis when presented from the serosal side only (Goddard, 1975). Adenosine triphosphate excites *Pleuronectes* stomach (Stevenson and Grove, unpublished observations) and *Carassius* intestine (Burnstock *et al.*, 1972) but inhibits intestinal peristalsis in *Pleuronectes* (Goddard, 1975).

3. Studies on Intrinsic and Extrinsic Nervous Control of the Teleost Gut

A number of workers have stimulated the gut wall electrically to analyze the role of enteric neurons in the control of activity. Saito (1973) and Ito and Kuriyama (1971) described a polyphasic reaction to transmural stimulation in *Carassius*. Saito observed a preliminary twitch (blocked by tubocurarine or atropine), a subsequent relaxation (unaffected by adrenergic or cholinergic blocking agents), and a final slow contraction (partly blocked by atropine but possibly involving a "rebound contraction"). All phases were abolished by tetrodotoxin (10^{-8} g/ml) and presumably represent the activity of several types of nerve cells. Ito and Kuriyama found that field stimulation induced:

1. A fast twitch at higher frequencies (ca. 30 Hz), blocked by tetrodotoxin or d-tubocurarine and apparently caused by striated muscle fibers
2. A subsequent relaxation (3 Hz and above), blocked by tetrodotoxin (10^{-6} g/ml) but not affected by combined α- and β-adrenoceptor blockade with phentolamine and propranolol, each at 10^{-5} g/ml
3. A slow phasic contraction following phase 2
4. A delayed contraction following phase 3 and often lasting several minutes

Phases 3 and 4 were prevented by pretreatment with tetrodotoxin (10^{-6} g/ml) or atropine (10^{-6} g/ml) leading these authors to believe that they depend on cholinergic neurons and are not "rebound" contractions, following activity in the nonadrenergic inhibitory fibers responsible for phase 2.

Edwards (1972a) and Grove et al. (1974) found that transmural excitation of Pleuronectes stomach at 20–30 Hz caused a large contraction within a few seconds of the onset of stimulation and which is readily blocked by low doses of atropine (10^{-8} g/ml). Treatment with tetrodotoxin (10^{-7} g/ml) abolished this action, which is believed to represent the action of cholinergic motor nerves on longitudinal muscle. In contrast, when the stomach is prepared for recording from the circular coat as well (Trendelenburg preparation) a triphasic pattern of mechanical activity is recorded (Fig. 10A). A brief twitch (a) caused by striated muscle elements is followed by an atropine-sensitive primary contraction (b, cf. Fig. 10B) and a subsequent rebound contraction (c). The rebound is closely related to the cessation of stimulation when the period of excitation is varied between 5 and 60 secs, and is resistant to α and β-adrenoceptor blockers or to atropine, but is abolished by tetrodotoxin. Stimulation of the plaice vagus nerve (Edwards, 1972a) causes a powerful contraction of the longitudinal muscles within the period of stimulation which is strongly antagonized by atropine (10^{-9} g/ml) and less effectively by hexamethonium (10^{-6} g/ml) which has proven muscarinic blocking actions on this organ. A brief relaxation followed the primary contraction. When responses were recorded in situ using an intragastric balloon, the biphasic contractions of the smooth muscle coats (Fig. 11A) persisted when stimulation of the vagal roots was carried out intracranially. Lesions in the spinal cord

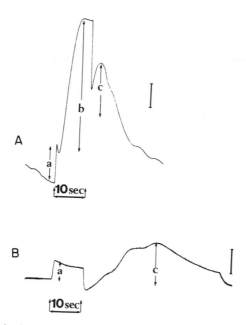

Fig. 10. Trendelenberg preparation of *Pleuronectes platessa* stomach. The fish was injected with reserpine (2 × 3 mg/kg, 72 and 24 hr previously) to abolish adrenergic nerve function. (A) Vagal stimulation at 6 V, 50 Hz, 1 msec pulse-width for 10 sec. (B) The same after 10^{-5} *M* atropine added to the bath (final conc.). Muscarinic blockade abolishes the "primary" cholinergic contraction (b), but leaves the striated muscle twitch (a) and rebound contraction (c) unimpaired. Vertical scale: 2 cm H_2O pressure; temp., 8°C. (From Stevenson and Grove, 1977, *Comp. Biochem. Physiol. C* **58**, 143–151.)

and medulla showed that this mixed innervation derives from the medulla, and not by way of a spinal/sympathetic pathway. Injection of the fish with atropine (1 mg/kg) (Fig. 11A), or bathing the vago-sympathetic ganglion with mecamylamine or hexamethonium (10^{-4} g/ml) abolished the primary contraction often unmasking an inhibitory phase. The plaice vagus apparently carries both excitatory and inhibitory tracts directly from the medulla. In the intact fish, the nerve exerts some inhibitory tone since Edwards (1973) found an acceleration in gastric emptying after vagotomy. Decrease in vagal tone may underly the change in gastric emptying rate when fish are fed low calorie diets (Section VI).

A separate extrinsic nerve supply has been found controlling plaice gastric peristalsis (Stevenson and Grove, 1978). In Trendelenburg preparations, stimulation of the splanchnic nerve at 15–30 Hz produces a biphasic response (Fig. 11B). The excitatory phase is

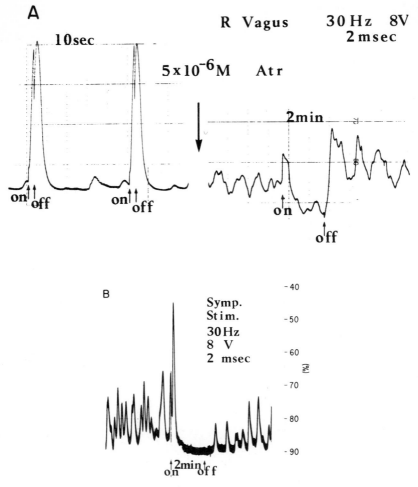

Fig. 11. Effects of stimulation of the extrinsic nerves on *Pleuronectes platessa* stomach activity, recorded by intragastric balloon. (A) Biphasic contraction after stimulation of the vagus nerve. Both primary and secondary ("rebound") contractions are seen. After atropine, excitation is abolished and vagal stimulation causes inhibition. (B) Biphasic response to stimulation of the splanchnic nerve, which carries both cholinergic excitatory and adrenergic inhibitory fibers. (From Stevenson and Grove, 1977, 1978.)

abolished by atropine (10^{-7} g/ml) and the inhibitory phase by reserpine pretreatment (5 mg/kg) or in the presence of butoxamine (3×10^{-5} g/ml). Adrenergic nerve terminals are present in the flatfish enteric plexus and circular muscle coat (Fig. 12) and extracts of the plaice stomach contain 0.05 μg/g adrenaline and 0.04 μg/g noradrenaline when measured by the method of Häggendal (1963) (Grove and S.

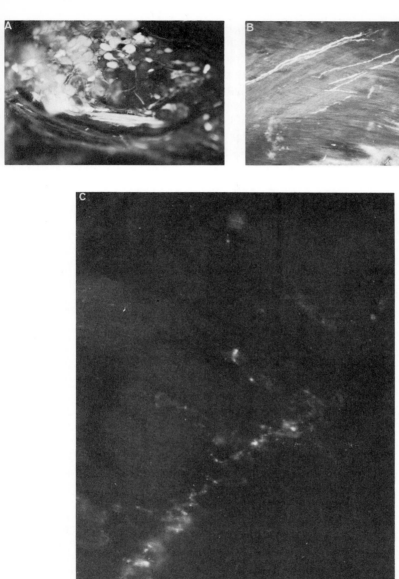

Fig. 12. Adrenergic innervation of flatfish gut. Photomicrographs of tissues prepared using the fluorescent histochemical technique of Falck and Owman (1965). (A) Coeliac ganglion of *Ammotretis* containing fluorescent (adrenergic) ganglion cells. (B) Intestine of *Ammotretis* with adrenergic fibers running in the circular muscle layer. (C) Stomach of *Platichthys* in which adrenergic terminals envelop nonfluorescent ganglion cells in Auerbach's plexus. (From S. Nilsson and Grove, unpublished observations.)

Nilsson, unpublished observations). Extensions of the splanchnic nerves supply the plaice intestine and mediate inhibition (Goddard, 1975). The effect is mimicked by adrenaline (10^{-10} g/ml) and other catecholamines. The intestine contains 0.06 μg/g adrenaline and 0.02 μg/g noradrenaline. However Goddard found a further nonadrenergic inhibitory system in the wall of the plaice intestine. Transmural stimulation relaxes the intestine (Fig. 13) but atropine, ganglion blockers, or α- and β-blockers failed to antagonize the response. Adenosine triphosphate (10^{-6} g/ml) both relaxes the intestine and desensitizes the receptors to transmural stimulation. The response is also blocked by tetrodotoxin, but not reserpine pretreatment, suggesting that intramural inhibitory purinergic neurons are involved.

As a measure of the complexity of the nervous coordination of gastrointestinal activity of teleosts, a comparison with *Salmo trutta* and *S. gairdneri* is revealing. In this more primitive teleost, Burnstock (1969) reinterpreted his early results on vagal stimulation, concluding that the vagus sends cholinergic excitatory nerves to the esophageal striated muscle but inhibits the stomach. The latter response is often accompanied by a strong "rebound" contraction, is unaffected by at-

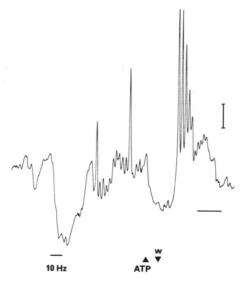

Fig. 13. Magnus preparation of *Pleuronectes platessa* intestine. The inhibitory effects of transmural stimulation (10 V, 10 Hz, 1 msec pulses for 60 sec) are compared with the response to 10^{-6} g/ml adenosine triphosphate. The time course of each relaxation is similar and each is followed by spontaneous "rebound" contractions. Vertical scale, 2 g; horizontal scale, 120 sec. (From Goddard, 1975.)

ropine at high doses (10^{-5} g/ml) and is only impaired by high doses of nicotine (10^{-4} g/ml) and hexamethonium (5×10^{-3} g/ml) which are probably nonselective. His claim that the response is dependent on preganglionic cholinergic fibers depends on the specificity of the above concentrations of ganglion blocking agents. Campbell (1975) showed that spontaneous stomach contractions in *Salmo gairdnerii* were inhibited by intracranial stimulation of the vagal roots (0.5–8 Hz), often accompanied by a large "rebound" contraction at the cessation of stimulation. More recently, Campbell and Gannon (1976) have shown that stimulation of the splanchnic nerve at 5–10 Hz contracts *Salmo* stomach, confirming the studies of Burnstock (1958a). The response was not affected by hyoscine (10^{-7} g/ml) but was abolished by bretylium ($1–3 \times 10^{-6}$ g/ml) and reduced by d-tubocurarine (3×10^{-5} g/ml). Adrenergic fibers were detected histochemically in the splanchnic nerve which proceed to form a plexus in the stomach wall and, in view of the excitatory effect of catecholamines described earlier, are believed to mediate the excitation. Excitation of the posterior splanchnic nerve (Burnstock, 1958a) inhibits or excites the rectum depending on the frequency of stimulation. He proposed that both cholinergic and adrenergic nerves are involved, and the latter have been detected histochemically (Read and Burnstock, 1968a,b, 1969).

Clearly the roles of the extrinsic nerves to the stomach and intestine of *Pleuronectes* and *Salmo* are very different. It is not yet possible to distinguish the roles of intramural neurons which may be linked with, or independent of, extrinsic nerve tracts. The control of peristalsis in *Tinca* is also different in that the vagus in this stomachless fish extends its influence to the intestine (Mahn, 1898) and the gut wall consists of the usual smooth muscle coats together with two layers of striped muscle (circular and longitudinal) with several nerve cell plexuses (Baumgarten, 1965, 1967). In addition to the left and right *rami vagi intestinales*, which exert cholinergic excitatory influences on all muscle coats, a splanchnic nerve originating from the right sympathetic chain innervates the intestine. Baumgarten (1967) also showed that this sympathetic supply carries fluorescent nerves which innervate the muscle layers and which apparently contain dopamine, as judged by the thionylchloride test of Corrodi and Jonsson (1967), and which inhibit the intestine. He also obtained fluorescent histochemical evidence for the presence of 5-hydroxytryptamine in some nerves and pointed out that this amine contracts smooth muscle of the tench intestine. Saito (1973), using the fluorescent technique in *Carassius* intestine, also detected adrenergic nerves which he proposed inhibit spontaneous activity.

4. In Vivo STUDIES

Very few attempts have been made to analyze the control of peristalsis in living, feeding fish. Burnstock (1957) observed in situ peristalsis of Salmo trutta by implanting an abdominal window. Edwards (1973) observed that carbachol dramatically accelerated the gastric emptying of a 1% meal of Arenicola by Pleuronectes whereas atropine delayed gastric emptying. Goddard (1974) found that intestinal transit time was also shortened by carbachol and extended by atropine when these were injected at the appropriate time, but that this region of the gut was less sensitive than the stomach to these drugs. In a separate study, Goddard (1973) demonstrated similar actions of these agents in the stomachless Blennius pholis. After 7 days food deprivation, Pleuronectes (300 g at 15°C) did not show impaired efficiency in transporting food through the gastrointestinal tract but during the subsequent weeks the transit time slowly increased as the physiological effects of starvation developed.

Gzgzyan et al. (1973) examined in vivo gastric contractions in Scorpaena porcus using a balloon to record activity. In addition to concluding that the stomach was under excitatory cholinergic control, they showed that extracts of the pituitary gland of Scorpaena contained a powerful gastric inhibitor, reminiscent of the inhibitory action of oxytocin in mammals. In the dog, it has been found that injected insulin at first depresses gastric activity but then, as hypoglycemia develops, vagal centers in the medulla activate gastric activity. In Scorpaena, insulin injections depressed stomach activity but no compensatory increase accompanied the ensuing hypoglycemia.

E. Summary

Figure 14 represents a generalized account of the mechanisms controlling teleost gastrointestinal motility. Longitudinal and circular coats of smooth muscle fibers develop spontaneous, myogenic activity when the gut is distended. In Pleuronectes, noncholinergic excitatory nerves are necessary to induce this response. Peristalsis depends on intramural neurons (cholinergic and noncholinergic excitatory; noncholinergic/nonadrenergic inhibitory) in synaptic connection with sensory and other neurons in Auerbach's plexus. In flatfish, but to a lesser extent in salmonids and cyprinids, peristalsis in isolated preparations is mainly myogenic, similar to that described by Yung et al. (1965) for the tortoise Amyda japonica. Localized release of 5-hydroxytryptamine generally enhances activity. Cranial vagal fibers

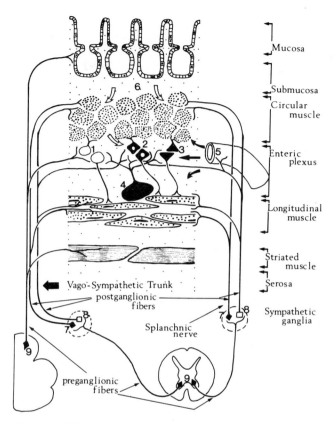

Fig. 14. Diagram of the intrinsic and extrinsic nerve fibers believed to control the muscularis of the teleost stomach, together with sites of action of circulating catecholamines. (1) Nonadrenergic, noncholinergic inhibitory neurons; (2) cholinergic excitatory neurons; (3) nonadrenergic, noncholinergic excitatory neurons; (4) stretch receptors; (5) circulating adrenaline; (6) endogenous 5-hydroxytryptamine; (7) cholinergic excitatory neurons of postganglionic, sympathetic origin; (8) adrenergic inhibitory neurons of postganglionic, sympathetic origin; (9) preganglionic cholinergic neurons from central nervous system. Notes: (a) striated muscle is not always present in the stomach and intestine; (b) in *Salmo*, element 8 to the stomach is excitatory; (c) in *Pleuronectes* an additional element 9 in the vagus synapses with excitatory cholinergic neurons in the jugular ganglion; (d) a vagal element is suggested which controls gastric gland secretion. The intestine is controlled by elements 1–8 but, in stomachless fish, vagal influences may extend to this region.

in fish primarily activate noncholinergic intramural inhibitory fibers. Postganglionic cholinergic excitatory fibers reach the stomach in the splanchnic nerve and, in some acanthopterygians, in the vagal pathway to innervate muscle cells directly. Adrenergic neurons reach the stomach and intestine through the splanchnic route and enmesh ganglion cells in Auerbach's plexus as well as entering the muscularis. Catecholamines from the head kidney (e.g., Grove *et al.*, 1972; Nilsson *et al.*, 1976) generally inhibit activity, even in denervated preparations, but the clear gastric excitation which occurs in elasmobranchs and some teleosts requires further analysis.

It will be obvious from the account given in this section that the present knowledge of gastrointestinal motility and its control in fish is sparse. The onus is on the comparative physiologist to undertake better designed studies if the complexity of the system is to be unraveled. The present authors believe that, for a variety of fish species, careful characterization of tissue receptors and the affinity of agonists and antagonists is long overdue. The doses of antagonists which are *selective* against candidate neurotransmitter substances must be determined. Only then can drugs be used as tools in the study of coordinated peristalsis. The researcher must recognize that both neurogenic and myogenic rhythmic contractions may be involved in the transport of ingested food. Mechanical recordings should be accompanied by electrical measurements of the activity of elements in the gut wall. Detailed study of the ganglion cells of the enteric plexus, including their neurotransmitters, and the origin and distribution of extrinsic nerves should be undertaken. No mention has been made here of the sensory connections of the gut to the central nervous system, although it appears from Section VI that such nerves play a significant part in appetite of fish. Windell and Norris (1969a,b) proposed that the production of enterogastrone in response to high energy food may delay the gastric emptying phase and yet the available data on this and other local gastrointestinal hormones is almost nonexistent. Although motor nerves of various origins impinge on the gastrointestinal tract, and hormones of the adrenal medulla, pituitary, and other endocrines are also likely to affect motility and secretion, very few attempts to investigate their significance to the fish by *in vivo* experiments have been undertaken. Kerkut (1976) has mentioned some of the important contributions to physiology made by comparative pharmacologists; a fuller understanding of the processing of food by the gastrointestinal tract is likely to contribute significantly to the husbandry of fish.

VIII. PERSPECTIVES

In Section VI it is apparent that digestion rate is related to the natural diet. The return of appetite following a meal is to a large extent controlled by the rate of stomach or foregut emptying (Table V). The physiological data presented in Section VII suggest that, when a fish in culture is presented with artificial foodstuffs, the rate of gastrointestinal activity can be changed by including a stimulant or inhibiting drug in the diet. In this way a faster or more efficient rate of food utilization could be obtained, supplemented by further additives (such as steroids) which promote anabolism. Appropriate balance of nutrients, such as amino acids or oligosaccharides, which compete for the active transport sites in the gastrointestinal epithelium, may increase assimilation efficiency. Incorporation in the artificial diets of exogenous enzymes which become active when the gastrointestinal juices are absorbed may allow the fish to overcome the "surface area" limitation to digestion rate. Environmental conditions which minimize the circulation of adrenaline or discharge of the adrenergic nerves ("stress") will allow peristaltic rate and visceral blood flow to be at a maximum. There is a case for incorporating anti-adrenaline agents or sedatives to the diet during "thinning-out" operations on the fish farm, to allow rapid adjustment to the new environment and early return of feeding. This latter suggestion may be most appropriate for marine fish which frequently tolerate crowding to a lesser extent than freshwater fish. These, and similar factors which may emerge, will be additional to improvements in fish production brought about by genetic selection, control of disease, diet formulation, and control of environmental variables such as temperature and photoperiod.

ACKNOWLEDGMENTS

We are indebted to our research colleagues with whom we have discussed and tested our ideas and also to Mr. W. Rowntree for helping prepare the photographs. The British Council (Academic links scheme) aided our cooperative venture. R. F. wrote mainly Sections III, IV, and V and D. G. mainly Sections II, VI, VII, and VIII. Part of the work was done at the Kristineberg Marine Biological Station at Fiskebäckskil, Sweden.

REFERENCES

Adam, H. (1960). Über das Vorkommen einer peritrophischen Membran im Darm von *Myxine glutinosa* L. (Cyclostomata). *Naturwissenschaften* **14**, 333–334.
Adron, J. W., Grant, P. T., and Cowey, C. B. (1973). A system for the quantitative study of the learning capacity of rainbow trout and its application to the study of food preferences and behaviour. *J. Fish Biol.* **5**, 625–636.
Albertini-Berhaut, J., and Vallet, F. (1971). Utilization alimentaire de l'urée chez les muges. *Tethys* **3**, 677–680.

Alexander, R. McN. (1970). Mechanics of the feeding action of various teleost fishes. *J. Zool.* **162**, 145–156.

Al-Hussaini, A. H. (1947). The feeding habits and the morphology of the alimentary tract of some teleosts living in the neighbourhood of the Marine Biological Station, Ghardaqa, Red Sea. *Publ. Mar. Biol. St., Ghardaqa, Red Sea* **5**, 4–61.

Al-Hussaini, A. H. (1949a). On the functional morphology of the alimentary tract of some fish in relation to differences in their feeding habits: Anatomy and histology. *Q. J. Microsc. Sci.* **90**, 109–139.

Al-Hussaini, A. H. (1949b). On the functional morphology of the alimentary tract of some fish in relation to differences in their feeding habits: Cytology and physiology. *Q. J. Microsc. Sci.* **90**, 323–354.

Al-Hussaini, A. H., and Kholy, A. A. (1953). On the functional morphology of the alimentary tract of some omnivorous teleost fish. *Proc. Egypt. Acad. Sci.* **4**, 17–39.

Alliot, E. (1967). Absorption intestinale de l'N-acetyl-glucosamine chez la petite Rousette. *C. R. Soc. Biol. (Paris)* **161**, 2544–2546.

Alliot, E., Febvre, A., and Métailler, R. (1974). Les protéases digestives chez un téléostéen carnivore *Dicentrus labrax*. *Ann. Biol. Anim., Biochim., Biophys.* **14**, 229–237.

Angelescu, V., and Gneri, F. S. (1949). Adaptaciones del aparto digestivo al regimen alimenticio en algunos pesces del Rio Uruguay y del Rio de la Plata. *Rev. Inst. Nac. Invest. Cienc. Nat. (Buenos Aires)* **1**, 161–272.

Aoyama, T. (1958). On the discharge of stomach contents concerning jack mackerel (*Trachurus japonicus*). *Bull. Seikai Reg. Fish Res. Lab. (Seikai-Ku Suisan Kenkyusho Kenkyu Hokoku)* **15**, 29–32.

Arcangeli, A. (1908). Contributo alla conoscenze della strutture minuta dello stomaco del *Box salpa* L. secondo la stato funziale. *Arch. Zool. Ital.* **3**, 261–346.

Ariens, E. J., and Van Rossum, J. M. (1957). pD_x, pA_x and pD'_x values in the analysis of pharmacodynamics. *Arch. Intern. Pharmacodyn. Ther.* **110**, 275–299.

Armstrong, R. H., and Blackett, R. F. (1966). Digestion rate of the Dolly Varden. *Trans. Am. Fish. Soc.* **95**, 429–430.

Arthur, D. K. (1956). Cited in Blaxter (1963).

Babkin, B. P. (1929). Studies on the pancreatic secretion in skates. *Biol. Bull. (Woods Hole, Mass.)* **57**, 272–291.

Babkin, B. P. (1933). Further studies on the pancreatic secretion in the skate. *Contrib. Can. Biol. Fish.* **7**, 1–9.

Babkin, B. P., and Bowie, D. (1928). The digestive system and its function in *Fundulus heteroclitus*. *Biol. Bull. (Woods Hole, Mass.)* **54**, 254–277.

Babkin, B. P., Friedman, M. H. F., and MacKay-Sawyer, M. E. (1935a). Vagal and sympathetic innervation of the skate. *J. Biol. Board Can.* **1**, 239–250.

Babkin, B. P., Chaisson, A. F., and Friedman, M. H. F. (1935b). Factors determining the course of the gastric secretion in elasmobranchs. *J. Biol. Board Can.* **1**, 251–259.

Backiel, T. (1971). Production and food consumption of predatory fish in the Vistula River. *J. Fish Biol.* **3**, 369–405.

Baldwin, N. S. (1956). Food consumption and growth of brook trout at different temperatures. *Trans. Am. Fish. Soc.* **86**, 323–328.

Ballentijn, C. M., van den Burg, A., and Egberink, B. P. (1972). An electromyographic study of the adductor mandibulae complex of a free-swimming carp (*Cyprinus carpio*) during feeding. *J. Exp. Biol.* **57**(1), 261–283.

Barnard, E. A. (1973). Comparative biochemistry and physiology of the digestion. *In* "Comparative Animal Physiology" (C. L. Prosser, ed.), 3th Ed., pp. 133–152. Saunders, Philadelphia, Pennsylvania.

Barrington, E. J. W. (1957). The alimentary canal and digestion. *In* "The Physiology of Fishes" (M. E. Brown, ed.), Vol. 1, pp. 109–161. Academic Press, New York.

Barrington, E. J. W. (1962). Digestive enzymes. *Adv. Comp. Physiol. Biochem.* **1**, 1–65.

Barrington, E. J. W. (1972). The pancreas and intestine. *In* "The Biology of Lampreys" (M. W. Hardisty and I. Potter, eds.), Vol. 2, pp. 135–169. Academic Press, New York.

Barrington, E. J. W., and Dockray, G. J. (1972). Cholecystokinin–pancreozyminlike activity in the eel (*Anguilla anguilla* L.). *Gen. Comp. Endocrinol.* **19**, 80–87.

Battle, H. I. (1935). Digestion and digestive enzymes in the herring (*Clupea harengus* L.) *J. Biol. Board Can.* **1**, 145–157.

Baumgarten, H. G. (1965). Über die Muskulatur und die Nerven in der Darmwand der Schleie (*Tinca vulgaris* Cuv.). *Z. Zellforsch. Mikrosk. Anat.* **68**, 116–137.

Baumgarten, H. G. (1967). Verkommen und Verteilung adrenerger Nervenfasern in Darm der Schleie (*Tinca vulgaris* Cuv.). *Z. Zellforsch. Mikrosk. Anat.* **76**, 248–259.

Baumgarten, H. G., Björkland, A., Lachenmayer, L., Nobin, A., and Rosengren, E. (1973). Evidence for the existence of serotonin-, dopamine- and noradrenaline-containing neurons in the gut of *Lampetra fluviatilis*. *Z. Zellforsch. Mikrosk. Anat.* **141**, 33–54.

Bayliss, W. M., and Starling, E. H. (1903). On the uniformity of the pancreatic mechanism in vertebrata. *J. Physiol. (London)* **29**, 174–180.

Beamish, F. W. H. (1972). Ration size and digestion in largemouth bass, *Micropterus salmoides* Lacépède. *Can. J. Zool.* **50**, 153–164.

Bengelsdorf, H., and Elias, H. (1950). The structure of the liver of cyclostomata. *Chicago Med. Sch. Q.* **12**, 7–12.

Bennett, M. R., and Burnstock, G. (1968). Electrophysiology of the innervation of intestinal smooth muscle. *In* "Handbook of Physiology" (C. F. Code, ed.), Vol. IV, Sect. 6, Alimentary Canal, pp. 1709–1732. Am. Physiol. Soc., Washington, D.C.

Benson, A. A., and Lee, R. F. (1975). The role of wax in oceanic food chains. *Sci. Am.* **232**, 76–86.

Bernard, F. (1952). La digestion chez les poissons. *Trav. Lab. Hydrobiol. Piscicult. Univ. Grenoble* **44**, 61–95.

Bertin, L. (1958). Appareil digestif. *In* "Traité de Zoologie" (P. Grassé, ed.), Tome XIII, Fascicule 2, pp. 1248–1302. Masson, Paris.

Beukema, J. J. (1968). Predation by the three-spined stickleback (*Gasterosteus aculeatus*): The influence of hunger and experience. *Behaviour* **31**, 1–126.

Biedermann, W. (1911). Die Ernährung der Fische. *In* "Handbuch der Vergleichen den Physiologie" (H. Winterstein, ed.), Vol. 2, Part 1, pp. 1049–1115. Fischer, Jena.

Bilinski, E. (1974). Biochemical aspects of fish swimming. *In* "Biochemical and Biophysical perspectives in Marine Biology" (D. C. Malins and J. R. Sargent, eds.), Vol. 1, pp. 239–288. Academic Press, New York.

Bishop, C., and Odense, P. H. (1966). Morphology of the digestive tract of the cod, *Gadus morhua. J. Fish. Res. Board Can.* **23**, 1607–1615.

Blake, I. H. (1930). Studies on the comparative histology of the digestive tube of certain fishes. I. A predacious fish, the sea bass *Centropristes striatus. J. Morphol.* **50**, 39–70.

Blaxter, J. H. S. (1963). The feeding of herring larvae and their ecology in relation to feeding. *Calif. Coop. Oceanogr. Fish Invest. Rep.* **10**, 79–88.

Blaxter, J. H. S., and Holliday, F. G. T. (1963). The behaviour and physiology of herring and other clupeoids. *Adv. Mar. Biol.* **1**, 261–393.

Bodola, A. (1966). Life history of the gizzard shad, *Dorosoma cepedianum* (Le Sueur), in western Lake Erie. *U.S. Fish Wildl. Serv., Fish. Bull.* **65**, 391–425. Cited in Schmitz and Baker (1969).

Bokova, E. (1938). Daily food consumption and digestive rate in *Rutilus rutilus. Rybn. Khoz. (Moscow)* No. 6.

Borgström, B. (1974). Bile salts—their physiological function in the gastrointestinal tract. *Acta Med. Scand.* **196**, 1–10.

Boyer, J. L., Schwarz, J., and Smith, N. (1974). Preliminary studies of bile secretion in the isolated perfused liver of the small skate, *Raja erinacea. Bull. Mt. Desert Isl. Biol. Lab.* **14**, 12–13.

Bray, R. N., and Ebeling, A. W. (1975). Food, activity and habitat of three "picker" type microcarnivorous fishes in the kelp forests of Santa Barbara, California. *U.S. Fish Wild. Serv., Fish. Bull.* **73**(4), 815–829.

Brett, J. R. (1971). Starvation time, appetite and maximum food intake of sockeye salmon, *Oncorhynchus nerka. J. Fish. Res. Board Can.* **28**, 409–415.

Brett, J. R., and Higgs, D. A. (1970). Effect of temperature on the rate of gastric digestion in fingerling sockeye salmon, *Oncorhynchus nerka. J. Fish. Res. Board Can.* **27**, 1767–1779.

Brinn, J. E., Jr. (1973). The pancreatic islets of bony fishes. *Am. Zool.* **13**, 653–665.

Brockerhoff, H. (1966). Digestion of fat by cod. *J. Fish. Res. Board Can.* **23**, 1835–1839.

Bryan, P. G. (1975). Food habits, functional digestive morphology, and assimilation efficiency of the rabbitfish *Siganus spinus* (Pisces, Siganidae) on Guam. *Pac. Sci.* **29**, 269–277.

Bucke, D. (1971). The anatomy and histology of the alimentary tract of the carnivorous fish the pike *Esox lucius* L. *J. Fish Biol.* **3**, 421–431.

Bullock, W. L. (1963). Intestinal histology of some salmonid fishes with particular reference to the histopathology of acanthocephalan infections. *J. Morphol.* **112**, 23–44.

Bullock, W. L. (1967). The intestinal histology of the mosquito fish *Gambusia affinis* (Baird and Girard). *Acta Zool. (Stockholm)* **48**, 1–17.

Burnstock, G. (1957). A "window" technique for observing fish viscera *in vivo. Nature (London)* **180**, 1491–1492.

Burnstock, G. (1958a). The effects of drugs on spontaneous motility and on response to stimulation of the extrinsic nerves of the gut of a teleostean fish. *Br. J. Pharmacol.* **13**, 216–226.

Burnstock, G. (1958b). Reversible inactivation of nervous activity of fish. *J. Physiol. (London)* **141**, 35–45.

Burnstock, G. (1959a). The morphology of the gut of the brown trout (*Salmo trutta*). *Q. J. Microsc. Sci.* **100**, 183–198.

Burnstock, G. (1959b). The innervation of the gut of the brown trout (*Salmo trutta*). *Q. J. Microsc. Sci.* **100**, 199–220.

Burnstock, G. (1969). Evolution of the autonomic innervation of visceral and cardiovascular systems in vertebrates. *Pharmacol. Rev.* **21**, 247–324.

Burnstock, G. (1972). Purinergic nerves. *Pharmacol. Rev.* **24**, 509–581.

Burnstock, G., Campbell, G., Satchell, D., and Smythe, A. (1970). Evidence that adenosine triphosphate or a related nucleotide is the transmitter substance released by nonadrenergic inhibitory nerves in the gut. *Br. J. Pharmacol.* **40**, 668–689.

Burnstock, G., Satchell, D., and Smythe, A. (1972). A comparison of the excitatory and inhibitory effects on nonadrenergic, noncholinergic nerve stimulation and exogenously-applied ATP on a variety of smooth muscle preparations from different vertebrate species. *Br. J. Pharmacol.* **46**, 234–242.

Cameron, J. N., Kostoris, J., and Penhale, P. A. (1973). Preliminary energy budget of the ninespine stickleback (*Pungitius pungitius*) in an Arctic lake. *J. Fish. Res. Board Can.* **30**, 1179–1189.

Campbell, G. (1970a). Autonomic nervous systems. *In* "Fish Physiology" (W. S. Hoar and D. J. Randall, eds.), Vol. 4, pp. 109–132. Academic Press, New York.

Campbell, G. (1970b). Autonomic nervous supply to effector tissues. *In* "Smooth Muscle" (E. Bülbring, A. Brading, A. Jones, and T. Tomita, eds.), pp. 109–132. Arnold, London.

Campbell, G. (1975). Inhibitory vagal innervation of the stomach in fish. *Comp. Biochem. Physiol. C* **50**, 169–170.

Campbell, G., and Burnstock, G. (1968). Comparative physiology of gastrointestinal motility. *In* "Handbook of Physiology." Sect. 6: "Alimentary Canal" (C. F. Code, ed.), Vol. IV, pp. 2213–2266. Am. Physiol. Soc., Washington, D.C.

Campbell, G., and Gannon, B. J. (1976). The splanchnic nerve supply to the stomach of the trout, *Salmo trutta* and *S. gairdneri. Comp. Biochem. Physiol. C* **55**, 51–53.

Cartier, M., and Buclon, M. (1973). Ultrastructure comparée de la bordure en brosse de l'entérocyte en fonction de l'état nutritionnel, chez trois cyprinides: La tanche (*Tinca tinca* L.), la carpe (*Cyprinus carpio* L.) et le poisson rouge (*Carassius auratus* L.). *Ann. Nutr. Aliment.* **27**, 181–189.

Chaichara, A., and Bullock, W. L. (1967). The histopathology of acanthocephalan infections in suckers with observations on the intestinal histology of two species of catastomid fishes. *Acta Zool. (Stockholm)* **48**, 19–42.

Chakrabarte, J., Saharya, R., and Belsare, D. K. (1973). Structure of the gallbladder in some freshwater teleosts. *Z. Mikrosk.-Anat. Forsch.* **87**(1), 23–32.

Chao, L. N. (1973). Digestive system and feeding habits of the cunner, *Tautogolabrus adspersus*, a stomachless fish. *U.S. Fish Wildl. Serv., Fish. Bull.* **71**, 565–586.

Chesley, L. C. (1934). The concentration of proteases, amylase, and lipase in certain marine fishes. *Biol. Bull. (Woods Hole, Mass.)* **60**, 133–144.

Chiba, K. (1961). The basic study on the production of fish seedling under possible control. I. The effect of food in quality and quantity on the survival and growth of common carp fry. *Tansuiku Suisan Kenkyusho Kenkyu Hokoku* **11**, 105–132.

Chujyo, N. (1953). Site of acetylcholine production in the wall of intestine. *Am. J. Physiol.* **174**, 196–198.

Corner, E. D. S., Denton, E. J., and Förster, G. R. (1969). On the buoyancy of some deep sea sharks. *Proc. R. Soc., Ser. B* **171**, 415–429.

Corrodi, H., and Jonsson, G. (1967). Formaldehyde fluorescence method for the histochemical demonstration of biogenic monoamines. A review on the methodology. *J. Histochem. Cytochem.* **15**, 65–78.

Creac'h, P. V. (1963). Les enzymes protéolytiques des poissons. *Ann. Nutr. Aliment.* **17**, A375–A471.

Croston, C. B. (1960). Tryptic enzymes of chinook salmon. *Arch. Biochem. Biophys.* **89**, 202–206.

Croston, C. B. (1965). Endopeptidases of salmon ceca: Chromatographic separation and some properties. *Arch. Biochem. Biophys.* **112**, 218–223.

Daan, N. (1973). A quantitative analysis of the food intake of North Sea cod, *Gadus morhua. Neth. J. Sea Res.* **6**, 479–517.

Dahlstedt, E., von Euler, U. S., Lishajko, F., and Östlund, E. (1959). Observations on the distribution and action of substance P in marine animals. *Acta Physiol. Scand.* **47**, 124–130.

Dandrifosse, G. (1975). Purification of chitinase contained in pancreas of gastric mucosa of frog. *Biochimie* **57**, 829–831.

Daniel, E. E. (1968). Pharmacology of the gastrointestinal tract. *In* "Handbook of Physiology" (C. F. Code, ed.), Part 6, Vol. 4, pp. 2267–2324. Am. Physiol. Soc., Washington, D.C.

Darnell, R. M., and Meierotto, R. R. (1962). Determination of feeding chronology in fishes. *Trans. Am. Fish. Soc.* 91, 313–320.

David, H. (1961). Zur submikroskopischen Morphologie intrazellularer Gallenkapillaren. *Acta Anat.* 47, 216–224.

Davson, H. (1970). "A Textbook of General Physiology." Churchill, London.

Dawson, I. (1970). The endocrine cells of the gastrointestinal tract. *Histochem. J.* 2, 527–549.

De Groot, S. J. (1971). On the interrelationship between morphology of the alimentary tract, food and feeding behavior in flatfishes (Pisces, Pleuronectidae). *Neth. J. Sea Res.* 5, 121–196.

De Haën, C., and Gertler, A. (1974). Isolation and amino-terminal sequence analysis of two dissimilar pancreatic proelastases from the African lungfish. *Biochemistry* 13, 2673–2677.

Dockray, G. J. (1975). Comparative studies on secretin. *Gen. Comp. Endocrinol.* 25, 203–210.

Dreyer, N. B. (1949). The action of autonomic drugs on elasmobranch and teleost involuntary muscle. *Arch. Int. Pharmacodyn. Ther.* 78, 63–66.

Edwards, D. J. (1971). Effect of temperature on rate of passage of food through the alimentary canal of the plaice, *Pleuronectes platessa* L. *J. Fish Biol.* 3, 433–439.

Edwards, D. J. (1972a). Electrical stimulation of isolated vagus nerve–muscle preparations of the stomach of the plaice, *Pleuronectes platessa* L. *Comp. Gen. Pharmacol.* 2, 235–242.

Edwards, D. J. (1972b). Reactions of the isolated plaice stomach to applied drugs. *Comp. Gen. Pharmacol.* 3, 345–358.

Edwards, D. J. (1973). The effects of drugs and nerve section on the rate of passage of food through the gut of the plaice, *Pleuronectes platessa* L. *J. Fish Biol.* 5, 441–446.

Elias, H., and Sherrick, J. C. (1969). "Morphology of the Liver." Academic Press, New York.

Elliott, J. M. (1972). Rates of gastric evacuation in brown trout, *Salmo trutta* L. *Freshwater Biol.* 2, 1–18.

Elliott, J. M. (1975). Weight of food and time required to satiate brown trout, *Salmo trutta* L. *Freshwater Biol.* 5(1), 51–64.

Epple, A. (1969). The endocrine pancreas. *In* "Fish Physiology" (W. S. Hoar and D. J. Randall, eds.), Vol. 2, pp. 275–319. Academic Press, New York.

Eriksen, N. (1945). Purification and properties of halibut pepsin. M.S. Thesis, Univ. of Washington, Seattle. Cited in Jany (1976).

Fábián, G., Molnár, G., and Tölg, I. (1963). Comparative data and enzyme kinetic calculations on changes caused by temperature in the duration of gastric digestion of some predatory fishes. *Acta Biol. Acad. Sci. Hung.* 14, 123–129.

Falck, B., and Owman, C. H. (1965). Detailed methodological description of the fluorescence method for the cellular demonstration of biogenic menoamines. *Acta Univ. Lund.* Sect. II, No. 7, pp. 1–23.

Falkmer, S., and Olsson, R. (1962). Ultrastructure of the pancreatic islet tissue of normal and alloxan-treated *Cottus scorpius*. *Acta Endocrinol. (Copenhagen)* 39, 32–46.

Falkmer, S., Hellman, B., and Voigt, G. E. (1964). On the agranular cells in the pancreatic islet tissue of the marine teleost *Cottus scorpius*. *Acta Pathol. Microbiol. Scand.* 60(1), 47–54.

Falkmer, S., Östberg, Y., and Van Noorden, S. (1978). Entero-insular endocrine systems of cyclostomes. *In* "Gut Hormones" (S. R. Bloom, ed.), pp. 57–63. Churchill Livingstone, New York.

Fänge, R. (1962). Pharmacology of poikilothermic vertebrates and invertebrates. *Pharmacol. Rev.* 14, 281–316.

Fänge, R. (1968). The formation of eosinophilic granulocytes in the oesophageal lymphomyeloid tissue of elasmobranchs. *Acta Zool. (Stockholm)* 49, 155–161.

Fänge, R. (1973). The lymphatic system of *Myxine. Acta Regiae Soc. Sci. Litt. Gothob. Zool.* 8, 57–64.

Fänge, R., and Hanson, A. (1973). Comparative pharmacology of catecholamines. *In* "International Encyclopaedia of Pharmacology" (M. J. Michelson, ed.). Pergamon, New York.

Fänge, R., and Johnels, A. G. (1958). An autonomic nerve plexus control of the gall bladder in *Myxine. Acta Zool. (Stockholm)* 39, 1–8.

Fänge, R., Johnels, A. G., and Enger, P. S. (1963). The autonomic nervous system. *In* "The Biology of *Myxine*" (A. Brodal and R. Fänge, eds.), pp. 124–136. Universitetsforlaget, Oslo.

Fänge, R., Lundblad, G., and Lind, J. (1976). Lysozyme and chitinase in blood and lymphomyeloid tissues of marine fish. *Mar. Biol.* 36, 277–282.

Fänge, R., Lundblad, G., Lind, J., and Wikberg, K. (1978). Chitinolytic enzymes in the digestive system of marine fish. (Submitted for publication.)

Farmanfarmaian, A., Ross, A., and Mazal, D. (1972). *In vivo* intestinal absorption of the sugar in the toadfish (marine teleost *Opsanus tau*). *Biol. Bull. (Woods Hole, Mass.)* 142, 427–445.

Fish, G. R. (1960). The comparative activity of some digestive enzymes in the alimentary canal of *Tilapia* and perch. *Hydrobiologia* 15, 161–179.

Fortunatova, K. R. (1950). Biology of feeding of *Scorpaena porcus* L. *Tr. Sevastop. Biol. Stn., Akad. Nauk Ukr. SSR* 7, 193–235.

Fortunatova, K. R. (1955). Methods of studying the feeding of predatory fishes. *Tr. Sov. Ikhtiol. Kom. Akad. Nauk SSR* 6, 62–84. (In Russ.)

Frantsusova, G. P. (1971). Afferent innervation of the digestive tract in *Stenodus leucichthys. Arkh. Anat., Gist. Embriol.* 60, 39–41. (In Russ.)

Fujita, T. (1962). Über das Inselsystem des Pancreas von *Chimaera monstrosa. Z. Zellforsch. Mikrosk. Anat.* 57, 2272–2277.

Fujita, T. (1974). New aspects on the pancreatic islets. *In* "Recherches Biologiques Contemporaines". (L. Arvy, ed.), pp. 437–440. Imprimerie Wagner, Nancy, France.

Gabe, M., and Martoja, M. (1971). Données histologiques sur les cellules endocrines, gastriques et pancréatiques de *Mugil auratus* (Téléostéen, Mugiliforme). *Arch. Anat. Microsc. Morphol. Exp.* 60, 219–234.

Gammon, R. L., Tiemeier, O. W., and Gier, H. T. (1972). The gross and microanatomy of the digestive tract and pancreas of the channel catfish, *Ictalurus punctatus. Trans. Kans. Acad. Sci.* 75, 141–155.

Gauthier, G. F., and Landis, S. C. (1972). The relationship of ultrastructural and cytochemical features to absorptive activity in the goldfish intestine. *Anat. Rec.* 172, 675–702.

Gerald, V. M. (1973). Rate of digestion in *Ophiocephalus punctatus*, Bloch. *Comp. Biochem. Physiol. A* 46, 195–205.

Gershon, M. D. (1967). Inhibition of gastrointestinal movement by sympathetic nerve stimulation: The site of action. *J. Physiol. (London)* 189, 317–328.

Ghazzawi, F. M. (1935). The pharnyx and intestinal tract of the Egyptian mullets *Mugil cephalus* and *Mugil capito*. II. On the morphology and histology of the alimentary canal in *Mugil capito. Coastal Fish. Serv., Fish. Res. Dir., Notes Mem. (Cairo)* No. 6, pp. 1–31.

Girgis, S. (1952). On the anatomy and histology of the alimentary tract of an herbivorous bottom-feeding cyprinoid fish, *Labeo horie* (Cuv). *J. Morphol.* **90**, 317–362.

Glaser, G. (1933). Beiträge zur Kenntnis des Lymphgefässsystem der Fische. Z. *Gesamte Anat., Abt. 1* **100**, 433–511.

Goddard, J. S. (1973). The effects of cholinergic drugs on the motility of the alimentary canal of *Blennius pholis* L. *Experientia* **29**, 974–975.

Goddard, J. S. (1974). An X-ray investigation of the effects of starvation and drugs on intestinal motility in the plaice, *Pleuronectes platessa* L. *Ichthylolgia (Sarajevo)* **6**(1), 49–58.

Goddard, J. S. (1975). Studies on intestinal control of the plaice, *Pleuronectes platessa* L. Ph.D. Thesis, Univ. of Wales, Cardiff.

Gomazkov, O. A. (1959). Effect of temperature on digestion of the burbot, *Lota Lota*. *Byull. Inst. Biol. Vodokhran., Akad. Nauk SSSR* **5**, 26–28. (In Russ.)

Granick, S. (1953). Investigations in iron metabolism. *Am. Nat.* **87**, 65–75.

Greene, C. W. (1914). The fat-absorbing function of the alimentary tract of the king salmon. *Bull. U.S. Fish. Bur.* **33**, 149–175. Cited in Schmitz and Baker (1969).

Grodzinski, Z. (1938). Das Blutgefässsystem. *In* "Dr. H. G. Bronns Klassen und Ordnungen des Tierreichs." Bd. 6, Abt. 1, Buch 2: "Echte Fische," pp. 1–77. Akad. Verlagsges., Leipzig.

Grove, D. J., Starr, C. R., Allard, D. R., and Davies, W. (1972). Adrenaline storage in the pronephros of the plaice, *Pleuronectes platessa* L. *Comp. Gen. Pharmacol.* **3**, 205–212.

Grove, D. J., O'Neill, J. G., and Spillett, P. B. (1974). The action of 5-hydroxytryptamine on longitudinal gastric smooth muscle of the plaice *Pleuronectes platessa*. *Comp. Gen. Pharmacol.* **5**, 229–238.

Grove, D. J., Goddard, J. S., Tan, S. P., and Wirtz, P. (1976). Unpublished observations from M.Sc. and Ph.D. theses, Univ. of Wales, Cardiff.

Grove, D. J., Lozoides, L., and Nott, J. (1978). Satiation amount, frequency of feeding and gastric emptying rate in *Salmo gairdneri*. *J. Fish Biol.* **12**, 507–516.

Guimaraes, S. (1969). Reversal by pronethalol of dibenamine blockade: A study on the seminal vesicle of the guinea pig. *Br. J. Pharmacol.* **36**, 594–601.

Gzgzyan, D. M., and Kuzina, M. M. (1973). Stomach motor activity in the Black Sea ray, *Dasyatis pastinaca*. *J. Evol. Biochem. Physiol.* (*USSR*) **9**(5), 536–538. (In Russ.)

Gzgzyan, D. M., Zaks, M. G., and Tanasiychuk, O. F. (1968). Effect of histamine, substance "P" and pituitrin on the gastric secretion of the common European catfish (*Silurus glanis* L.). *Probl. Ichthyol.* (*USSR*) **8**, 97–100. (Transl. from *Vopr. Ikhtiol.* 1968, **8**, 126–130.)

Gzgzyan, D. M., Kuzina, M. M., and Tanasiichuk, O. F. (1973). The effects of hypophyseal hormones and insulin on motor activity of the stomach in the scorpion fish *Scorpaena porcus*. *J. Evol. Biochem. Physiol.* (*USSR*) **9**(3), 301–302. (In Russ.)

Häggendal, J. (1963). An improved method for fluorimetric determination of small amounts of adrenaline and noradrenaline in plasma and tissues. *Acta Physiol. Scand.* **59**, 242–254.

Hale, P. (1965). The morphology and histology of the digestive systems of two freshwater teleosts, *Poecilia retculata* and *Gasterosteus aculeatus*. *J. Zool.* **146**, 132–149.

Hallbäck, D. A. (1973). Acetylcholinesterase-containing structures in the intestine of *Myxine glutinosa*. *Acta Regiae Soc. Sci. Litt. Gothob. Zool.* **8**, 24–25.

Hansen, D. (1975). Evidence of a gastrin-like substance in *Rhinobatus productus*. *Comp. Biochem. Physiol. C* **52**, 61–64.

Haslewood, G. A. D. (1968). Evolution and bile salts. *In* "Handbook of Physiology" (C. F. Code, ed.), Sect. 6, Vol. 5, pp. 2375–2390. Am. Physiol. Soc., Washington, D.C.

Hatanaka, M., Kosaka, M., Sato, Y., Yamati, K., and Fukui, K. (1954). Interspecific relations concerning the predaceous habits among the benthic fish. *Tohoku J. Agric. Res.* **5,** 177–189.

Hay, A. W. M., and Watson, G. (1976). The plasma transport of 25-hydroxycalciferol in fish, amphibians, reptiles, and birds. *Comp. Biochem. Physiol. B* **53,** 167–172.

Healey, M. C. (1971). The distribution and abundance of sand gobies, *Gobius minutus* in the Ythan Estuary. *J. Zool.* **163,** 177–229.

Herting, G. E., and Witt, A. (1968). Rate of digestion in the bowfin. *Prog. Fish. Cult.* **30**(1), 26–28.

Hess, A. D., and Rainwater, J. H. (1939). A method for measuring food preference of trout. *Copeia* No. 3, pp. 154–157.

Hickling, C. F. (1966). On the feeding process in the white amur, *Ctenopharyngodon idella. J. Zool.* **148,** 408–418.

Hickling, C. F. (1970). A contribution to the natural history of the English grey mullets (Pisces, Mugolidae). *J. Mar. Biol. Assoc. U.K.* **50,** 609–633.

Hill, W. C. (1926). A comparative study of the pancreas. *Proc. Zool. Soc. London* Part 2, pp. 581–631.

Hinton, D. E., and Pool, C. R. (1976). Ultrastructure of the liver in channel catfish *Ictalurus punctatus* (Rafinesque). *J. Fish Biol.* **8,** 209–219.

Hinton, D. E., Snipes, R. L., and Kendall, M. W. (1972). Morphology and enzyme histochemistry in liver of largemouth bass (*Micropterus salmoides*). *J. Fish. Res. Board Can.* **29,** 531–534.

Hirao, S., Yamada, J., and Kikuchi, R. (1960). On improving the efficiency of feed for fish culture. Transit and digestibility of diet in eel and rainbow trout, observed by the use of ^{32}P. *Tokaiku Suisan Kenkyusho Kenkyu Hokoku* **27,** 67–72.

Hirschowitz, B. I. (1975). Regulation of gastric secretion. *In* "Functions of the Stomach and Intestine" (M. H. F. Friedman, ed.), pp. 145–165. Univ. Park Press, Baltimore, Maryland.

Hoagman, W. J. (1974). Vital activity parameters as related to the early life history of larval and postlarval lake whitefish (*Coregonus clupeaformis*). *In* "The Early Life History of Fish" (J. H. Blaxter, ed.), pp. 547–558. Springer-Verlag, Berlin and New York.

Hobson, E. S. (1974). Feeding relationships of teleostean fishes on coral reefs in Kona, Hawaii. *U.S. Fish Wildl. Serv., Fish. Bull.* **72**(4), 915–1031.

Hogben, C. A. M. (1967a). Response of the isolated dogfish gastric mucosa to histamine. *Proc. Soc. Exp. Biol. Med.* **124,** 890–893.

Hogben, C. A. M. (1967b). Secretion of acid by the dogfish, *Squalus acanthias. In* "Sharks, Skates and Rays" (P. W. Gilbert, R. F. Mathewson, and D. P. Rall, eds.), pp. 299–315. Johns Hopkins Press, Baltimore, Maryland.

Holstein, B. (1975a). Intestinal diamine oxidase of some teleostean fishes. *Comp. Biochem. Physiol. B* **50,** 291–297.

Holstein, B. (1975b). Gastric secretion in a teleostean fish: A method for the continuous collection of gastric effluence from a swimming fish and its response to histamine and pentagastrin. *Acta Physiol. Scand.* **95,** 417–423.

Holstein, B. (1976). Effect of the H_2 receptor antagonist metiamide on carbachol- and histamine-induced gastric acid secretion in the Atlantic cod, *Gadus morhua. Acta Physiol. Scand.* **97,** 189–195.

Holstein, B. (1977). Effect of atropine and SC-15396 on stimulated gastric acid secretion in the Atlantic cod, *Gadus morhua. Acta Physiol. Scand.* **101,** 185–193.

Honma, S. (1970). Presence of monaminergic neurones in the spinal cord and intestine of the lamprey *Lampetra japonica. Arch. Histol. Jpn. (Niigata, Jpn.)* **32,** 383–393.

Hotta, H., and Nakashima, J. (1968). Experimental study on the feeding activity of jack mackerel *Trachurus japonicus*. *Bull. Sekai Reg. Fish. Res. Lab.* (*Seikai-Ku Suisan Kenkyusho Kenkyu Hokoku*) **36**, 75–83.

House, C. R., and Green, K. (1963). Sodium and water transport across isolated intestine of a marine teleost. *Nature* (*London*) **199**, 1293–1294.

Hunn, J. B. (1972). Concentrations of some inorganic constituents in gallbladder bile from some freshwater fishes. *Copeia* No. 4, pp. 860–861.

Hunt, B. P. (1960). Digestion rate and food consumption of Florida gar, warmouth and largemouth bass. *Trans. Am. Fish. Soc.* **89**(2), 206–210.

Hunt, J. N., and Knox, M. T. (1968). Regulation of gastric emptying. *In* "Handbook of Physiology" (C. F. Code, ed.), Sect. 6, Vol. 4, pp. 1917–1935. Am. Physiol. Soc. Washington, D.C.

Ikeda, M., Hayama, T., Chujyo, N., and Hoshi, A. (1958). Acetylcholine content in the stretched wall of intestine. *Am. J. Physiol.* **192**, 27–29.

Imura, K. (1974a). On the alkaline RNase isolated from the pyloric caeca of rainbow trout. *Nippon Suisan Gakkaishi* **40**, 399–408.

Imura, K. (1974b). Purification and properties of the phosphodiesterase in the pyloric caeca of rainbow trout. *Nippon Suisan Gakkaishi* **40**, 807–814.

Inaba, D., Ogino, C., Takamatsu, C., Sugano, S., and Hata, H. (1962). Digestibility of dietary components in fishes. I. Digestibility of dietary proteins in rainbow trout. *Nippon Suisan Gakkaishi* **28**, 367–371.

Iro, S. (1967). Anatomic structure of the gastric mucosa. *In* "Handbook of Physiology" (C. F. Code, ed.), Sect. 6, Vol. 2, pp. 705–741. Am. Physiol. Soc., Washington, D.C.

Ishida, J. (1935). Ciliated intestinal epithelium in a teleost. *Annot. Zool. Jpn.* **15**, 158–160.

Ishiwata, N. (1969). Ecological studies on the feeding of fishes. VII. Frequency of feeding and satiation amount. *Nippon Suisan Gakkaishi* **35**, 979–984.

Isokawa, S., Kubota, K., Kosakai, T., *et al.* (1965). Some contributions to study of esophageal sacs and teeth of fishes. *J. Nihon Univ. Sch. Dent.* **7**, 103–111.

Ito, T., Watanabe, A., and Takahashi, Y. (1962). Histologische und cytologische Untersuchungen der Leber bei Fisch und Cyclostomata nebst Bemerkungen über die Fettspeicherzellen. *Arch. Histol. Jpn.* (*Niigata, Jpn.*) **22**, 429–463.

Ito, Y., and Kuriyama, H. (1971). Nervous control of the motility of the alimentary canal of the silver carp. *J. Exp. Biol.* **55**, 469–487.

Jacobshagen, E. (1915). Ueber die Appendices pyloricae, nebst Bemerkungen zur Anatomie und Morphologie des Rumpfdarmes. *Jena Z. Naturwiss.* **53**, 445–556.

Jacobshagen, E. (1937). Darmsystem. IV. Mittel und Enddarm. Rumpfdarm. *In* "Handbuch der Vergleichenden Anatomie der Wirbeltiere" (L. Bolk, E. Göppert, E. Kallius, and W. Lubosch, eds.), Vol. 3, pp. 563–724. Urban Schwarzenberg, Berlin.

Jansson, B.-O., and Olsson, R. (1960). The cytology of the caecal epithelial cells of *Perca*. *Acta Zool.* (*Stockholm*) **41**, 267–276.

Jany, K.-D. (1976). Studies on the digestive enzymes of the stomachless bonefish *Carassius auratus gibelio* (Bloch): endopeptidases. *Comp. Biochem. Physiol. B* **53**, 31–38.

Jobling, M., Gwyther, D., and Grove, D. J. (1977). Some effects of temperature, meal size and body weight on gastric evacuation time in the dab, *Limanda limanda* (L.). *J. Fish Biol.* **10**, 291–298.

Jones, R. (1974). The rate of elimination of food from the stomachs of haddock, *Melanogrammus aeglefinus*, cod, *Gadus morhua* and whiting, *Merlangius merlangus*. *J. Cons., Cons. Int. Explor. Mer.* **35**(3), 225–243.

Kapoor, B. G., Evans, H. E., and Pevzner, R. A. (1975a). The gustatory system in fish. *Adv. Mar. Biol.* **13**, 53–108.

Kapoor, B. G., Smit, H., and Verighina, I. A. (1975b). The alimentary canal and digestion in teleosts. *Adv. Mar. Biol.* **13**, 109–239.

Kariya, T. (1956). Cited in Kariya (1969).

Kariya, T. (1969). The relationship of food intake to amount of stomach contents in mebaru *Sebastes inermis. Nippon Suisan Gakkaishi* **35**, 533–536. (Engl. summ.)

Kariya, T., and Takahashi, M. (1969). The relationship of food intake to the stomach contents in the mackerel, *Scomber japonicus. Nippon Suisan Gakkaishi* **35**, 386–390.

Kariya, T., Hotta, H., and Takahashi, M. (1969). Relation between condition of the stomach mucous folds and the stomach content in the mackerel. *Nippon Suisan Gakkaishi* **35**, 441–445.

Karpevitch, A. F. (1936). On changes of the digestive juice reaction in marine fishes. *Fiziol. Z.* **30**, 100–123.

Karpevitch, A. F., and Bokova, E. N. (1936). The rate of digestion in marine fishes. Part 1. *Zool. Z.* **15**, 143–148.

Karpevitch, A. F., and Bokova, E. N. (1937). The rate of digestion in marine fishes. Part 2. *Zool. Z.* **16**, 21–44.

Karzinkin, G. S. (1932). Kizuxheniiu fiziologii pishchevarenia. (Ein Betrag zum studium der Physiologie der Verdauung bei den Fischen.) *Tr. Limnol. St. Kosina* **15**, 85–123.

Kayanja, F. I. B., Maloity, G. M. O., and Reite, O. B. (1975). The fine structure of the intestinal epithelium of *Tilapia grahami. Anat. Anz.* **138**, 451–462.

Keast, A., and Webb, D. (1966). Mouth and body form relative to feeding ecology in the fish fauna of a small lake, Lake Opinicon, Ontario. *J. Fish. Res. Board Can.* **23**, 1845–1874.

Kerkut, G. A. (1976). A commentary on comparative pharmacology. *Comp. Biochem. Physiol. C* **55**, 1–3.

Kevern, N. R. (1966). Feeding rate of carp estimated by a radioisotopic method. *Trans. Am. Fish. Soc.* **95**, 363–371.

Khanna, S. S., and Mehrotra, B. K. (1970). Histomorphology of the buccopharynx in relation to feeding habits in teleosts. *Proc. Natl. Acad. Sci., India, Sect. B* **40**, 61–80.

Kilarski, W., and Bigaj, J. (1971). The fine structure of striated muscle fibres of tunica muscularis of the intestine in some teleosts. *Z. Zellforsch. Mikrosk. Anat.* **113**, 472–489.

Kionka, B. C., and Windell, J. T. (1972). Differential movement of digestible and undigestible food fractions in rainbow trout *Salmo gairdneri. Trans. Am. Fish. Soc.* **101**(1), 112–115.

Kirtisinghe, P. (1940). The myenteric nerve plexus in some lower chordates. *Q. J. Microsc. Sci.* **81**, 521–529.

Kitamikado, M., and Tachino, S. (1960). Studies on the digestive enzymes of rainbow trout. II. Proteases. *Nippon Suisan Gakkaishi* **26**, 685–690.

Kitchell, J. F., and Windell, J. T. (1968). Rate of gastric digestion in pumpkinseed sunfish, *Lepomis gibbosus. Trans. Am. Fish. Soc.* **97**, 489–492.

Kjeldson, M. A., Peters, D. S., Thayer, G. W., and Johnson, G. N. (1975). The general feeding ecology of postlarval fishes in the Newport River estuary. *U.S. Fish Wildl. Serv., Fish. Bull.* **73**, 137–144.

Krementz, A. B., and Chapman, G. B. (1975). Ultrastructure of the posterior half of the intestine of the channel catfish, *Ictalurus punctatus. J. Morphol.* **145**, 441–482.

Kukla, M. (1954). The arrangement of blood vessels on the surface and in the wall of the alimentary canal of the eel (*Anguilla anguilla* L.). *Folia Morphol. (Warsaw)* **5**(13), 27–38.

Kukla, M. (1958). The structure and vacularisation of the pancreas of the eel (*Anguilla anguilla* L.). *Acta Biol. Cracov., Ser. Zool.* **1**, 99–111.

Kurata, K. (1959). Preliminary report on the rearing of the herring larvae. *Suisan-Cho Hokkaido-Ku Suisan Kenkyusho Kenkyu Hokoku* **20**, 117–138.

Lane, T. H., and Jackson, H. M. (1969). Voidance time for 23 species of fish. *Invest. Fish Control* No. 33.

Lange, Ø. N. O. (1962). Development of the intestine in *Rutilus rutilus* (L.). *Vopr. Ikhtiol.* **2**, 336–349. (In Russ.)

Langley, J. N. (1921). "The Autonomic Nervous System," Part 1. W. Heffer & Sons, Cambridge.

Lansing, A. I., Rosenthal, T. B., and Alex, M. (1953). Presence of elastase in teleost islet tissue. *Proc. Soc. Exp. Biol. Med.* **84**, 689–691.

Larsson, L.-I., and Rehfeld, J. F. (1978). Evolution of CCK-like hormones. In "Gut Hormones" (S. R. Bloom, ed.), pp. 68–73. Churchill Livingstone, New York.

Lasker, R. (1970). Utilization of zooplankton energy by a Pacific sardine population in the Californian current. In "Marine Food Chains" (J. H. Steele, ed.), pp. 265–284. Oliver & Boyd, Edinburgh.

Laurence, G. C. (1971). Digestion rate of larval largemouth bass. *N.Y. Fish. Game J.* **18**, 52–56.

Lee, D. J., and Putnam, G. B. (1973). The response of rainbow trout to varying protein/ energy ratios in a test diet. *J. Nutr.* **103**, 916–922.

Lehninger, A. L. (1971). "Biochemistry." Worth, New York.

Leong, R. J. H., and O'Connell, C. P. (1969). A laboratory study of particulate and filter feeding of the northern anchovy (*Engraulis mordax*). *J. Fish. Res. Board Can.* **26**, 557–582.

Linder, M. C., Dunn, V., Isaacs, E., Jones, D., Lim, S., Van Volkom, M., and Munro, H. N. (1975). Ferritin and intestinal iron absorption: Pancreatic enzymes and free iron. *Am. J. Physiol.* **228**, 196–204.

Ling, E. A., and Tan, C. K. (1975). Fine structure of the gastric epithelium of the coral fish, *Chelmon rostratus* Cuvier. *Okajimas Fol. Anat. Jpn.* **51**, 285–310.

Lipskaya, N. Y. (1959). Duration of digestion in the Black Sea surmullet (*Mullus Barbatus ponticus*). (In Russ.) *Biol. Abstr.* **48**(1967), 26852.

Lorentz, W., Matejka, E., Schmal, A., Reimann, H. J., Uhlig, R., and Mann, G. (1973). A phylogenetic study on the occurrence and distribution of histamine in the gastroin- testinal tract and other tissues of man and various animals. *Comp. Gen. Pharmacol.* **4**, 229–250.

Luppa, H. (1966). Ein Beitrag zur Funktion der Appendices pyloricae der Fische. Morphologische, histochemische und elektronenoptische Untersuchungen. *Gagen- saurs Morphol. Jahrb.* **109**, 315–339.

Lutz, B. R. (1931). The innervation of the stomach and the action of adrenaline in elasmobranch fishes. *Biol. Bull. (Woods Hole, Mass.)* **61**, 93–100.

MacKay, M. E. (1929). The digestive system of eel-pout (*Zoarces anguillaris*). *Biol. Bull. (Woods Hole, Mass.)* **56**, 8–23.

McKone, D. (1971). Rate at which Sockeye Salmon alevins are evacuated from the stomach of mountain whitefish (*Prosopium williamsoni*). *J. Fish. Res. Board Can.* **28**(1), 110–111.

Magid, A. M. A. (1975). The epithelium of the gastrointestinal tract of *Polypterus senegalus* (Pisces: Brachiopterygii). *J. Morphol.* **146**, 447–456.

Magnuson, J. J. (1969). Digestion and food consumption by skipjack tuna *Katsuwonus pelamis*. *Trans. Am. Fish. Soc.* **98**, 379–392.

Mahn, R. (1898). Untersuchungen über das physiologische Verhalten des Schleiendarms. *Arch. Gesamte Physiol. Menschen Tiere* **72**, 273–307.

Manteifel, B. P., Girsa, I. I., Lesheva, T. S., and Pavlov, D. S. (1965). Diurnal rhythms of feeding and locomotory activity of some freshwater predacious fishes. *Akad. Nauk SSSR Moscow* pp. 3–81. (In Russ.)

Markus, H. C. (1932). The extent to which temperature changes influence food consumption in largemouth bass (*Huro floridana*). *Trans. Am. Fish. Soc.* **62**, 202–210.

Matthews, D. M. (1975). Intestinal absorption of peptides. *Physiol. Rev.* **55**, 537–608.

Merret, T. G., Bar-Eli, E., and Van Vunakis, H. (1969). Pepsinogens A, C, and D from the smooth dogfish. *Biochemistry* **8**, 3696–3702.

Micha, J. C., Dandrifosse, G., and Jeuniaux, C. (1973). Distribution et localisation tissulaire de la synthèse des chitinases chez les vertébrés inférieurs. *Arch. Int. Physiol. Biochim.* **81**, 439–451.

Mohsin, S. N. (1962). Comparative morphology and histology of the alimentary canals in certain groups of Indian teleosts. *Acta Zool.* (*Stockholm*) **43**, 79–133.

Molnár, G., and Tölg, I. (1962a): Experiments concerning gastric digestion of Pike perch (*Lucioperca lucioperca* L.) in relation to water temperature. *Acta Biol. Acad. Sci. Hung.* **13**, 231–239.

Molnár, G., and Tölg, I. (1962b). Relation between water temperature and gastric digestion of largemouth bass (*Micropterus salmoides* Lacepede). *J. Fish. Res. Board Can.* **19**, 1005–1012.

Molnár, G., Tamássy, E., and Tölg, I. (1967). The gastric digestion of living predatory fish. *In* "The Biological Basis of Freshwater Fish Production" (S. D. Gerking, ed.), pp. 135–149. Blackwell, Oxford.

Moore, A., and Hiatt, R. B. (1967). The action of epinephrine on gastrointestinal motility in the spiny dogfish. *Bull. Mt. Desert Isl. Biol. Lab.* **7**, 32–33.

Moore, J. W., and Beamish, F. W. H. (1973). Food of larval sea lamprey (*Petromyzon marinus*) and American brook lamprey (*Lampetra lamottei*). *J. Fish. Res. Board Can.* **30**, 7–15.

Moriarty, C. M., and Moriarty, D. J. W. (1973a). Quantitative estimation of the daily ingestion of phytoplankton by *Tilapia nilotica* and *Haplochromis nigripennis* in Lake George, Uganda. *J. Zool.* **171**, 15–23.

Moriarty, C. M., and Moriarty, D. J. W. (1973b). The assimilation of carbon from phytoplankton by two herbivorous fishes *Tilapia nilotica* and *Haplochromis nigripennis*. *J. Zool.* **171**, 41–55.

Moriarty, D. J. W. (1973). The physiology of digestion of blue-green algae in the cichlid fish *Tilapia nilotica*. *J. Zool.* **171**, 25–39.

Moriarty, D. J. W. (1976). Quantitative studies on bacteria and algae in the food of the mullet *Mugil cephalus* L. and the prawn *Metapenaeus bennettae* (Racek and Dall). *J. Exp. Mar. Biol. Ecol.* **22**, 131–143.

Motais, R., Rankin, I., and Maetz, J. (1969). Water fluxes in marine fishes. *J. Exp. Biol.* **51**, 529–546.

Nagase, G. (1964). Contribution to the physiology of digestion in *Tilapia mossambica* Peters: Digestive enzymes and the effect of diets on their activity. *Z. Vgl. Physiol.* **49**, 270–284.

Negri, L., and Erspamer, V. (1973). Action of caerulein and caerulein-like peptides on 'short-circuit current' and acid secretion in the isolated gastric mucosa of amphibians. *Naunyn-Schmiedeberg's Arch. Pharmacol.* **277**, 401–412.

Nelson, G. J. (1967). Epibranchial organs in lower teleostean fishes. *J. Zool.* **153**, 71–89.

Netsch, M., and Witt, A. (1962). Cited in Herting and Witt (1968).

Nicholls, J. V. V. (1931). The influence of temperature on digestion in *Fundulus heteroclitus. Contrib. Can. Biol. Fish.* **7**, 45–55.

Nicholls, J. V. V. (1933). The effect of temperature variation and certain drugs on gastric motility of elasmobranch fishes. *Contrib. Can. Biol. Fish.* **7**, 449–463.

Nicol, J. A. C. (1952). Autonomic nervous system in lower chordates. *Biol. Rev. Cambridge Philos. Soc.* **27**, 1–49.

Niimi, A. J., and Beamish, F. W. H. (1975). Bioenergetic and growth of largemouth bass (*Micropterus salmoides*) in relation to body weight and temperature. *Can. J. Zool.* **52**, 447–456.

Nilsson, A. (1970). Gastrointestinal hormones in the holocephalian fish *Chimaera monstrosa* (L.). *Comp. Biochem. Physiol.* **32**, 387–390.

Nilsson, A. (1973). Secretin-like and cholecystokinin-like activity in *Myxine glutinosa* L. *Acta Regiae Soc. Sci. Litt. Gothob. Zool.* **8**, 30–32.

Nilsson, A., and Fänge, R. (1969). Digestive proteases in the holocephalian fish *Chimaera monstrosa* (L.). *Comp. Biochem. Physiol.* **31**, 147–165.

Nilsson, A., and Fänge, R. (1970). Digestive proteases in the cyclostome *Myxine glutinosa* (L.). *Comp. Biochem. Physiol.* **32**, 237–250.

Nilsson, S. (1971). Adrenergic innervation and drug responses of the oval sphincter in the swimbladder of the cod (*Gadus morhua*). *Acta Physiol. Scand.* **83**, 446–453.

Nilsson, S., and Fänge, R. (1967). Adrenergic receptors in the swimbladder and gut of a teleost (*Anguilla anguilla*). *Comp. Biochem. Physiol.* **23**, 661–664.

Nilsson, S., and Fänge, R. (1969). Adrenergic and cholinergic vagal effects on the stomach of a teleost (*Gadus morhua*). *Comp. Biochem. Physiol.* **30**, 691–694.

Nilsson, S., and Grove, D. J. (1974). Adrenergic and cholinergic innervation of the spleen of the cod, *Gadus morhua. Eur. J. Pharmacol.* **28**, 135–143.

Nilsson, S., Holmgren, S., and Grove, D. J. (1975). Effects of drugs and nerve stimulation on the spleen and arteries of two species of dogfish, *Scyliorhinus canicula* and *Squalus acanthias. Acta Physiol. Scand.* **95**, 219–230.

Nilsson, S., Abrahamsson, T., and Grove, D. J. (1976). Sympathetic release of adrenaline from the head kidney of the cod, *Gadus morhua. Comp. Biochem. Physiol. C* **55**, 123–127.

Noaillac-Depeyre, J., and Gas, N. (1978). Ultrastructural and cytochemical study of the gastric epithelium in a fresh water teleostean fish (*Perca fluviatilis*). *Tissue & Cell* **10**, 23–37.

Nobel, R. L. (1973). Evacuation rates of young yellow perch, *Perca flavescens* (Mitchell). *Trans. Am. Fish. Soc.* **102**(4), 759–763.

Norris, E. R., and Elam, D. (1940). Preparation and properties and crystallization of tuna pepsin. *J. Biol. Chem.* **134**, 443–454.

Norris, E. R., and Mathies, J. C. (1953). Preparation and properties of tuna pepsin. *J. Biol. Chem.* **204**, 673–686.

Norris, J. S., Norris, D. O., and Windell, J. T. (1973). Effect of simulated meal size on gastric acid and pepsin secretory rates in bluegill (*Lepomis macrochirus*). *J. Fish. Res. Board Can.* **30**(2), 210–204.

Nose, T. (1967). Recent advance in the study of fish digestion. *Tech. Pap. Eur. Inland Fish. Adv. Comm.* **3**, 83–94.

Nose, T., and Mamiya, H. (1963). Protein digestibility of flatfish meal in rainbow trout. *Tansuiku Suisan Kenkyusho Kenkyu Hokoku* **12**, 1–4.

Nose, T., and Toyama, K. (1966). Protein digestibility of brown fish meal in rainbow trout. *Tansuiki Suisan Kenkyusho Kenkyu Hokoku* **15**, 213–224.

Odense, P. H., and Bishop, C. M. (1966). The ultrastructure of the epithelial border of the ileum, pyloric caeca and rectum of the cod, *Gadus morhua. J. Fish. Res. Board Can.* **23**(12), 1841–1843.

Odum, W. E. (1970). Utilization of the direct grazing and detritus food chains by the striped mullet, *Mugil Cechalus. In* "Marine Food Chains" (J. Steele, ed.), pp. 222–240. Oliver & Boyd, Edinburgh.

Östberg, Y. (1976). The entero-insular endocrine organ in a cyclostome, *Myxine glutinosa. Umea Univ. Med. Diss.* No. 15, pp. 1–41.

Östberg, Y., Van Noorden, S., Everson Pearse, A. G., and Thomas, N. W. (1976). Cytochemical, immunofluorescence, and ultrastructural investigations on polypeptide hormone containing cells in the intestinal mucosa of a cyclostome, *Myxine glutinosa. Gen. Comp. Endocrinol.* **28**, 213–227.

Ohnesorge, F. K., and Rauch, R. (1968). Untersuchungen über Peristaltik am Darm der Schleie (*Tinca vulgaris*). *Z. Vgl. Physiol.* **58**, 153–170.

Ohnesorge, F. K., and Rehberg, M. (1963). Der Darm der Schleie (*Tinca vulgaris*) als pharmakologisches Versuchsobject. *Naunyn-Schmiedebergs Arch. Exp. Pathol. Pharmakol.* **246**, 81–82.

Okul, A. V. (1941). The feeding of pelagic fishes in the Azov Sea. *Zool. Zh.* **20**, 587–603. (In Russ.)

Okutani, K. (1966). Studies of chitinolytic systems in the digestive tract of *Lateolabrax japonicus. Bull. Misaki Mar. Biol. Inst., Kyoto Univ.* No. 10, pp. 1–47.

Ooshiro, Z. (1962). Biochemical studies on carboxypeptidase contained in the pyloric caeca of mackerel, *Scomber japonicus. Kagoshima Univ. Fish. Dep. Bull.* **11**, Suppl., 111–151.

Ooshiro, Z. (1968). Biochemical studies on carboxypeptidase contained in the pyloric caeca of mackerel, *Scomber japonicus. Nippon Suisan Gakkaishi* **11**, 111–151.

Ooshiro, Z. (1971). Studies on proteinase in the pyloric caeca of fishes. II. Some properties of proteinases purified from the pyloric caeca of mackerel. *Nippon Suisan Gakkaishi* **37**, 145–158.

Oppel, A. (1896–1900). "Lehrbuch der Vergleichenden Mikroskopischen Anatomie der Wirbeltiere," Vols. 1–3. Fischer, Jena.

Osse, J. W. M. (1969). Functional morphology of the head of the perch (*Perca fluviatilis* L.): An electromyographic study. *Neth. J. Zool.* **19**, 289–392.

Otto, C. (1976). Size, growth, population density and food of brown trout *Salmo trutta* L. in two sections of a south Swedish stream. *J. Fish Biol.* **8**, 477–488.

Overnell, J. (1973). Digestive enzymes of the pyloric cacae and of their associated mesentery in the cod (*Gadus morhua*). *Comp. Biochem. Physiol. B* **46**, 519–531.

Pandian, T. J. (1967). Intake, digestion, absorption and conversion of food in the fishes *Megalops cyprinoides* and *Ophiocephalus striatus. Mar. Biol.* **1**, 16–32.

Panov, D. A., and Sorokin, Y. I. (1962). Speed of digestion in bream larvae. *Byull. Inst. Biol. Vodokhran., Akad. Nauk SSSR* **13**, 24–26. (In Russ.) [Isr. Program Sci. Transl. TT-68-50387 (1968).]

Pappas, C. J., Tiemeier, O. W., and Deyoe, C. W. (1973). Chromic sesquioxide as an indicator in digestion studies in fish. *Progr. Fish Cultunist* **35**, 97–98.

Patterson, T. L., and Fair, E. (1933). The action of the vagus on the stomach-intestine of the hagfish. Comparative studies. VIII. *J. Cell. Comp. Physiol.* **3**, 113–119.

Patton, J. S., Nevenzel, J. C., and Benson, A. A. (1975). Specificity of digestive lipases in hydrolysis of wax esters and triglycerides studied in anchovy and other selected fish. *Lipids* **10**, 575–583.

Payne, A. I. (1975). The reproductive cycle, condition and feeding in *Barbus liberiensis*, a tropical stream-dwelling cyprinid. *J. Zool.* **176**, 247–269.

Pearse, A. G. E. (1977). The diffuse neuroendocrine system and the APUD concept: Related peptides in brain, intestine, pituitary, placenta and anuran cutaneous glands. *Med. Biol.* **55**, 115–125.

Pegel, W. A., and Popov, F. G. (1937). The influence of temperature on digestion of cold-blooded animals. *Tomsk. Univ. Biol. Inst., Tr.* **4**, 59–79. (In Russ.)

Pernkopf, E., and Lehner, J. (1937). Vergleichende Beschreibung des Vorderdarmes bei den einzelnen Klassen der Kranioten. *In* "Handbuch der Vergleichende Anatomie der Wirbeltiere" (L. Bolk, E. Göppert, E. Kallius, and W. Lubosch, eds.), Vol. 3, pp. 349–398. Urban Schwarzenberg, Berlin.

Peters, D. S., and Hoss, D. E. (1974). A radioisotopic method of measuring food evacuation time in fish. *Trans. Am. Fish. Soc.* **103**, 626–629.

Petersen, H. (1908–1909). Beiträge zur kenntnis des Baues und der Entwicklung des Selachierdarmes. *Jena Z. Naturwiss.* **43**, 619–652; **44**, 123–148.

Phillips, A. M., Jr. (1969). Nutrition, digestion and energy. *In* "Fish Physiology" (W. S. Hoar and D. J. Randall, eds.), Vol. 1, pp. 391–432. Academic Press, New York.

Piavaux, A. (1973). Origine non bactérienne de la laminarinase intestinale de *Tilapia macrochir* Boulenger. *Arch. Int. Physiol. Biochim.* **81**, 737–743.

Pierce, E. L. (1936). Rates of digestion in the yellowtail (*Ocyurus chrysurus*) and the white grunt (*Haemulon plumieri*). *Copeia* No. 2, pp. 123–124.

Prahl, J. W., and Neurath, H. (1966a). Pancreatic enzymes of the spiny pacific dogfish. I. Cationic chymotrypsinogen and chymotrypsin. *Biochemistry* **5**, 2131–2146.

Prahl, J. W., and Neurath, H. (1966b). Pancreatic enzymes of the spiny pacific dogfish. II. Procarboxypeptidases B and carboxypeptidase B. *Biochemistry* **5**, 4137–4145.

Prosser, C. L. (1973). "Comparative Animal Physiology," 3rd ed. Saunders, Philadelphia, Pennsylvania.

Rahn, C. H., Sand, D. M., and Schlenk, H. (1973). Wax esters in fish. Metabolism of dietary palmityl palmitate in the gourami (*Trichogaster cosby*). *J. Nutr.* **103**, 1441–1447.

Rauther, M. (1940). Der Intestinaltraktus. *In* "Dr. H. G. Bronns Klassen und Ordnungen des Tierreichs." Teil 1: "Echte Fische," pp. 657–1050. Akad. Verlagsges., Leipzig.

Read, J. B., and Burnstock, G. (1968a). Fluorescent histochemical studies on the mucosa of the vertebrate gastrointestinaltract. *Histochemie* **16**, 324–332.

Read, J. B., and Burnstock, G. (1968b). Comparative histochemical studies of adrenergic nerves in the enteric plexus of vertebrate large intestine. *Comp. Biochem. Physiol.* **27**, 505–517.

Read, J. B., and Burnstock, G. (1969). Adrenergic innervation of gut musculature in vertebrates. *Histochemie* **17**, 263–272.

Reeck, G. R., and Neurath, H. (1972). Pancreatic trypsinogen from the African lungfish. *Biochemistry* **11**, 503–510.

Reeck, G. R., Winter, W. P., and Neurath, H. (1970). Pancreatic enzymes from the African lungfish, *Protopterus aethiopicus*. *Biochemistry* **9**, 1398–1403.

Reichenbach-Klinke, H.-H., and Reichenbach-Klinke, K.-E. (1970). Enzymuntersuchungen an Fischen. II. Trypsin-und Amylase-Inhibitoren. *Arch. Fischereiwiss.* **21**, 67–72.

Reif, W. E. (1976). Morphogenesis, pattern formation and function of the dentition of *Heterodontus* (Selachii). *Zoomorphologie* **83**(1), 1–47.

Reimers, N. (1957). Some aspects of the relation between stream foods and trout survival. *Calif. Fish Game* **43**, 43–69.

Reite, O. B. (1965). A phylogenetical approach to the functional significance of tissue mast cell histamine. *Nature (London)* **206**, 1334–1336.

Reite, O. B. (1969). Phylogenetical persistence of the nonmast cell histamine stores of the digestive tract: A comparison with mast cell histamine. *Experientia* **25**, 276–277.

Renade, S. S., and Kewalramani, H. G. (1967). Studies on the rate of food passage in fish intestines. F.A.O. *Fish. Rep.* **44**, 349–358.

Reshetnikov, Y. S., Claro, R., and Silva, A. (1972). The pattern of feeding and the rate of digestion in some tropical predatory fish. *J. Ichthyol. (USSR)* **12**(5), 818–824.

Riddle, O. (1909). The rate of digestion in cold-blooded vertebrates. The influence of season and temperature. *Am. J. Physiol.* **24**, 447–458.

Rosenthal, H., and Hempel, G. (1970). Experimental studies in feeding and food requirements of herring larvae (*Clupea harengus* L.). *In* "Marine Food Chains" (J. H. Steele, ed.), pp. 344–364. Oliver & Boyd, Edinburgh.

Rosenthal, H., and Paffenhöfer, G. A. (1972). On the digestion rate and calorific content of food and faeces in young gar fish. *Naturwissenschaften* **6**, 274–275.

Rozin, P., and Mayer, J. (1961). Regulation of food intake in the goldfish. *Am. J. Physiol.* **201**, 968–974.

Rozin, P., and Mayer, J. (1964). Some factors affecting short-term food intake in the goldfish. *Am. J. Physiol.* **206**, 1430–1436.

Ryland, J. S. (1964). The feeding of plaice and sandeel larvae in the southern North Sea. *J. Mar. Biol. Assoc. U.K.* **44**, 343–364.

Saito, K. (1973). Nervous control of intestinal motility in goldfish. *Jpn. J. Smooth Muscle Res.* **9**, 79–86.

Sarbahi, D. S. (1951). Studies on the digestive tracts and digestive enzymes of the goldfish *Carassius auratus* L. and the largemouth black bass *Micropterus salmoides* (Lacépède). *Biol. Bull. (Woods Hole, Mass.)* **100**, 244–257.

Scarpelli, D. G., Greider, M. H., and Frajola, W. J. (1963). Observations on hepatic cell hyperplasia, adenoma and hepatoma of rainbow trout (*Salmo gairdneri*). *Cancer Res.* **23**, 848–857.

Scheuring, L. (1928). Beziehungen zwischen Temperatur und Verdauungsgeschwindigkeit bei Fischen. *Z. Fisch. Deren Hilfswiss.* **26**, 231–235.

Schmitz, E. H., and Baker, C. D. (1969). Digestive anatomy of the gizzard shad *Dorosoma Cepedianum* and the threadfin shad *Dorosoma petenense. Trans. Am. Microsc. Soc.* **88**(4), 525–546.

Seaburg, K. G. (1956). A stomach sampler for live fish. *Prog. Fish Cult.* **19**, 137–139.

Seaburg, K. G., and Moyle, J. B. (1964). Feeding habits, digestion rates and growth of some Minnesota warm water fishes. *Trans. Am. Fish. Soc.* **93**, 269–285.

Sharratt, B. M., Bellamy, D., and Chester Jones, I. (1964). Adaptation of the silver eel (*Anguilla anguilla* L.) to sea water and to artificial media together with observations on the role of the gut. *J. Physiol. (London)* **204**, 135–158.

Shrable, J. B., Tiemeier, O. W., and Deyoe, C. W. (1969). Effects of temperature on rate of digestion by channel catfish. *Prog. Fish. Cult.* **31**, 131–138.

Singh, R. P., and Nose, T. (1967). Digestibility of carbohydrates in young rainbow trout. *Tansuiku Suisan Kenkyusho Kenkyu Hokoku* **17**, 21–25.

Smit, H. (1968). Gastric secretion in the lower vertebrates and birds. *In* "Handbook of Physiology." Sect. 6: Alimentary Canal" (C. F. Code, ed.), Vol. V, pp. 2791–2805. Am. Physiol. Soc., Washington, D.C.

Smith, H. W. (1930). The absorption and secretion of water and salts by marine teleosts. *Am. J. Physiol.* **93**, 480–505.

Smith, M. W. (1966). Sodium–glucose interactions in the goldfish intestine. *J. Physiol. (London)* **182**, 559–572.

Smith, R. L. (1975). Carbonic anhydrase in some coral reef fishes: Adaption to carbonate ingestion? *Comp. Biochem. Physiol. A* **50**, 131–134.

Smith, R. L., and Lane, C. E. (1971). Amino acid transport by the fish intestine. *Comp. Biochem. Physiol.* **4**, 93–103.

Solcia, E., and Members of the International Committee on Classification of Gut Endocrine Cells (1978). Lausanne 1977 classification of gastroenteropancreatic endocrine cells. *In* "Gut Hormones" (S. R. Bloom, ed.), pp. 40–48. Churchill Livingstone, New York.

Steigenberger, L. W., and Larkin, P. A. (1974). Feeding activity and rates of digestion of northern squawfish (*Ptychocheilus oregonensis*). *J. Fish. Res. Board Can.* **31**(4), 411–420.

Stevenson, S. V., and Grove, D. J. (1977). The extrinsic innervation of the stomach of the plaice, *Pleuronectes platessa* L. 1. The vagal nerve supply. *Comp. Biochem. Physiol. C* **58**, 143–151.

Stevenson, S. V., and Grove, D. J. (1978). The extrinsic innervation of the stomach of the plaice, *Pleuronectes platessa* L. 2. The splanchnic nerve supply. *Comp. Biochem. Physiol. C* **60**, 45–50.

Stickney, R. R., and Shumway, S. (1974). Occurrence of cellulase activity in the stomach of fishes. *J. Fish Biol.* **6**, 779–790.

Stryer, L. (1974). "Biochemistry." Freeman, San Francisco, California.

Sumner, J. B., and Somers, G. F. (1947). "Chemistry and Methods of Enzymes," 2nd Ed. Academic Press, New York.

Suyehiro, Y. (1942). A study on the digestive system and feeding habits of fish. *Jpn. J. Zool.* **10**, 1–303.

Suzuki, T. (1976). Relations between feeding rates and growth rates of filefish *Navodon modestus* (Günther). *Bull. Jpn. Sea Reg. Fish. Res. Lab.* (*Nihonkai-Ku Suisan Kenkyusho Kenkyu*) **27**, 51–57.

Swenson, W. A., and Smith, L. L. (1973). Gastric digestion, food consumption, feeding periodicity and food conversion efficiency in walleye (*Stizostedion vitreum vitreum*). *J. Fish. Res. Board Can.* **30**(9), 1327–1336.

Tan, C. K., and Teh, S. (1974). The structure of the gut of a coral fish, *Chelmon rostratus* Cuvier. *Okajimos Folia Anat. Jpn.* **51**, 63–79.

Tanaka, K. (1955). Observation of the length of the digestive time of feed by the mud loach *Misgurnus anguillicaudatus. Gyorvigaku Zasshi* **4**, 34–39.

Tanaka, M. (1969). Studies on the structure and function of the digestive system in teleost larvae. II. Characteristics of the digestive system of larvae at the stage of first feeding. *Gyorvigaku Zasshi* **16**(2), 41–49.

Tanaka, M. (1973). Studies on the structure and function of the digestive system of teleost larvae. Ph.D. Thesis, Fac. Agric., Kyoto Univ., Kyoto.

Tomonaga, S., Hirokane, T., Shinohara, H., and Awaya, K. (1973). The primitive spleen of the hagfish. *Dobutsugaku Zasshi* **82**, 215–217.

Track, N. S. (1973). Evolutionary aspects of the gastrointestinal hormones. *Comp. Biochem. Physiol. B* **45**, 291–301.

Trendelenburg, U. (1917). Physiologische und pharmakologische Versuche über die Dünndarmperistaltik. *Arch. Exp. Pathol. Pharmakol.* **81**, 55–129.

Tue, V. T. (1975). Contribution a l'étude histologique des cellules épithéliales glandulaires et endocrines du tube digestif de *Chimaera monstrosa* L. (Pisces, Holocephali). *Vie Milieu, Ser. A* **25**, 41–58.

Tugendhat, B. (1960). The normal feeding behaviour of the three-spined stickleback *Gasterosteus aculeatus. Behaviour* **15**, 284–318.

Tyler, A. V. (1970). Rates of gastric emptying in young cod. *J. Fish. Res. Board Can.* **27**, 1177–1189.

Uggeri, B. (1938). Ricerche sulle cellule enterchromaffini e sulle cellule argentofile dei Pesci. Z. Zellforsch. Mikrosk. Anat. **28**, 648.

Ugolev, A. M., and Kooshuck, R. I. (1966). Hydrolysis of dipeptides in cells of the small intestine. Nature (London) **212**, 859–860.

Ungar, C. (1935). Perfusion de l'estomac des Sélaciens; étude pharmacodynamique de la sécretion gastrique. C. R. Soc. Biol. **119**, 172–173.

Unger, R. H., Dobbs, R. E., and Orci, L. (1978). Insulin, glucagon, and somatostatin secretion in the regulation of metabolism. Annu. Rev. Physiol. **40**, 307–343.

Vallee, B. L. (1955). Zinc and metalloenzymes. Adv. Protein Chem. **10**, 317–384.

van Herwerden, M., and Ringer, W. E. (1911). Die Acidität des Magensaftes von Scyllium stellare. Hoppe-Seyler's Z. Physiol. Chem. **75**, 290.

Van Noorden, S., and Pearse, A. G. E. (1976). The localization of immunoreactivity to insulin, glucagon and gastrin in the gut of Amphioxus (Branchiostoma) lanceolatus. In "The Evolution of Pancreatic Islets" (T. A. I. Grillo, L. Leibson, and A. Epple, eds.), pp. 163–178. Pergamon, Oxford.

Van Slyke, D. D., and White, G. F. (1911). Digestion of protein in the stomach and in intestine of the dogfish. J. Biol. Chem. **9**, 209–217.

Vegas-Velez, M. (1972). La structure histologique due tube digestif des poissons téléostéens. Tethys **4**, 163–174.

von Euler, U. S., and Fänge, R. (1961). Catecholamines in nerves and organs of Myxine glutinosa, Squalus acanthias, and Gadus callarias. Gen. Comp. Endocrinol. **1**, 191–194.

von Euler, U. S., and Östlund, E. (1957). Effects of certain biologically occurring substances in the isolated intestine of fish. Acta Physiol. Scand. **38**, 364–372.

Wallace, J. C. (1973). Observations on the relationship between food consumption and metabolic rate of Blennius pholis L. Comp. Biochem. Physiol. A **45**, 293–305.

Walsh, K. A. (1970). 4. Trypsinogens and trypsins in various species. 5. Chymotrypsinogens–chymotrypsins. In "Proteolytic Enzymes" (G. E. Perlmann and L. Lorand, eds.), Vol. 19, pp. 41–63, 64–108. Academic Press, New York.

Ware, D. M. (1972). Predation by rainbow trout (Salmo gairdneri): the influence of hunger, prey density and prey size. J. Fish. Res. Board Can. **29**, 1193–1201.

Weinland, E. (1901). Zur Magenverdauung der Haifische. Z. Biol. **1**, 35–68; 275–294.

Weinreb, E. L., and Bilstad, N. M. (1955). Histology of the digestive tract and adjacent structures of the rainbow trout, Salmo gairdneri irideus. Copeia **3**, 194–204.

Weisel, G. F. (1962). Comparative study of the digestive tract of a sucker Catostomus catastomus and a predaceous minnow, Ptychocheilus oregonense. Am. Midl. Nat. **68**, 334–346.

Weisel, G. F. (1973). Anatomy and histology of the digestive system of the paddlefish (Polyodon spathula). J. Morphol. **140**, 243–256.

Welsch, U. N., and Storch, V. N. (1973). Enzyme histochemical and ultrastructural observations on the liver of teleost fishes. Arch. Histol. Jpn. (Niigata, Jpn.) **36**, 21–37.

Western, J. R. H. (1969). Studies on the diet, feeding mechanism and alimentary tract in two closely related teleosts, the freshwater Cottus gobio and the marine Parenophrys bubalis (Euphrasen). Acta Zool. (Stockholm) **50**, 185–205.

Western, J. R. H. (1971). Feeding and digestion in two cottid fishes, the freshwater Cottus gobio and the marine Enophrys bubalis. J. Fish Biol. **3**, 225–246.

Western, J. R. H., and Jennings, J. B. (1970). Histochemical demonstration of hydrochloric acid in gastric tubules of teleosts using an in vivo Prussian blue technique. Comp. Biochem. Physiol. **35**, 879–884.

Williams, B. G., and Naylor, E. (1967). Spontaneously induced rhythm of tidal periodicity in laboratory-reared *Carcinus. J. Exp. Biol.* **47**, 229–234.

Windell, J. T. (1966). Rate of digestion in the bluegill sunfish. *Invest. Indiana Lakes Streams* **7**, 185–214.

Windell, J. T. (1967). Rates of digestion in fishes. *In* "The Biological Basis of Freshwater Fish Production" (S. D. Gerking, ed.), pp. 151–173. Blackwell, Oxford.

Windell, J. T., and Norris, D. O. (1969a). Gastric digestion and evacuation in rainbow trout. *Prog. Fish. Cult.* **31**, 20–26.

Windell, J. T., and Norris, D. O. (1969b). Dynamics of gastric evacuation in rainbow trout, *Salmo gairdneri. Am. Zool.* **9**, 584.

Windell, J. T., Norris, D. O., Kitchell, J. F., and Norris, J. S. (1969). Digestive response of rainbow trout, *Salmo gairdneri*, to pellet diets. *J. Fish. Res. Board Can.* **26**, 1801–1812.

Windell, J. T., Hubbard, J. D., and Horak, D. C. (1972). Rate of gastric digestion in rainbow trout, *Salmo gairdneri*, fed three pelleted diets. *Prog. Fish. Cult.* **34**(3), 156–159.

Yamamoto, H., and Kitamikado, M. (1971). Purification of fish gastric hyaluronidase. *Nippon Suisan Gakkaishi* **37**, 621–630.

Yamamoto, T. (1965a). Some observations on the fine structure of the intrahepatic biliary passages in goldfish (*Carassius auratus*). *Z. Zellforsch. Mikrosk. Anat.* **65**, 319–330.

Yamamoto, T. (1965b). Some observations on the fine structure of the epithelium in the intestine of the lamprey (*Lampetra japonica*). *Okajimas Folia. Anat. Jpn.* **40**, 691–713.

Yamamoto, T. (1966). An electron microscope study of the columnar epithelial cell in the intestine of freshwater teleosts: Goldfish (*Carassius auratus*) and rainbow trout (*Salmo irideus*). *Z. Zellforsch. Mikrosk. Anat.* **72**, 66–87.

Yoshida, Y., and Sera, H. (1970). On chitinolytic activities in the digestive tract of several species of fishes and the mastication and digestion of foods by them. *Bull. Nankai Reg. Fish. Res. Lab.* **36**, 751–754. (In Jpn.) Cited in Tanaka (1973).

Young, J. Z. (1931). On the autonomic nervous system of the teleostean fish *Uranoscopus scaber. Q. J. Microsc. Sci.* **74**, 492–535.

Young, J. Z. (1933). The autonomic nervous system of selachians. *Q. J. Microsc. Sci.* **75**, 571–624.

Yung, E. (1899). Recherches sur la digestion des poissons. *Arch. Zool. Exp. Gen.* **7**, 121–201.

Yung, I. K., Woo, K. C., and Doo, W. K. (1965). Peristaltic movement of the tortoise intestine. *Experientia* **21**, 540–541.

Zendzian, E., and Barnard, E. A. (1967). Distribution of pancreatic ribonuclease, chymotrypsin, and trypsin in vertebrates. *Arch. Biochem. Biophys.* **122**, 699–713.

5

METABOLISM AND ENERGY CONVERSION DURING EARLY DEVELOPMENT

CHARLES TERNER

I. INTRODUCTION

In a review published in 1967, Monroy and Tyler (1967) stated that the only egg for which some detailed information on its metabolic condition before and after fertilization was available was the sea urchin egg. This statement is no longer true since during the past 10 years there has been an upsurge of work on mammalian eggs culminating in the *in vitro* fertilization of mammalian eggs and the maintenance of embryos in culture (for review, see Brinster, 1973).

Invertebrate eggs are an attractive material for investigation because they develop rapidly and are available in large numbers during the summer months, affording the researcher an opportunity to transfer his activities to a marine station by the sea shore. Fish eggs have been studied less extensively. They are produced during short spawning seasons, often at inconvenient times of the year, but this disadvantage is outweighed by a prolonged period of embryonic development, es-

261

FISH PHYSIOLOGY, VOL. VIII

pecially in those species which spawn in cold water. As a result, the sequence of stages in embryogenesis is spread over a longer time span, displaying in slow motion the more rapid sequence of events in invertebrate embryogenesis. A number of laboratories which have undertaken the study of fish eggs and embryos have investigated various aspects of current problems: the changes in energy metabolism of the oocyte resulting in maturation and the changes following fertilization; protein and RNA metabolism before and after fertilization; the transfer of genetic information and hormonal control mechanisms. While recognizing that most of the current information is based on experiments using amphibian and invertebrate material, this review will attempt to summarize findings resulting from studies of fish eggs and embryos in papers published since the last review in this series (Blaxter, 1969). A review of related topics appeared in 1974 (Neyfakh and Abramova, 1974).

II. EMBRYOGENESIS

A. Morphological Studies

Chronological tables and illustrations of morphological changes in embryonic and larval development of salmonid fishes are available (Knight, 1963; Vernier, 1969; Ballard, 1973). Various features of the embryonic development of the carp, pike, whitefish, rainbow trout, and brook trout were described by Ignat'eva (1969) and Ignat'eva and Rott (1970), of the herring by Galkina (1970), and of the black prickleback by Wourms and Evans (1974). The morphogenetic development of the salmonid embryo was described by Ballard (1966a,b,c, 1968) and Ballard and Dodes (1968). In a series of papers Wourms (1972a,b,c) examined features of the embryonic development of "annual fishes." In annual fishes, which complete a full life cycle within 1 year, arrest of embryonic development (diapause) occurs repeatedly, but at distinct stages. Fine structural studies of cytodifferentiation of *Fundulus heteroclitus* at various stages of embryonic development have been reported by Trinkaus and Lentz (1967a,b). Lentz and Trinkaus (1971) also described the formation, by means of cell-to-cell attachments, of cohesive monolayers on the surface of the blastoderm of *Fundulus*. The structure of the yolk sac and of the pericardial sac surface of the embryo of *Poecilia reticulata* was described by Dépêche (1973).

The membrane of the unfertilized teleostean fish egg has an opening, the micropyle, through which the fertilizing spermatozoon enters the ovum. After fertilization or exposure of the egg to water the membrane hardens and the micropylar canal is closed. Szöllösi and Billard (1974) have published electron micrographs showing details of the structure of the micropyle of the trout egg before and after closure.

B. RNA and Protein Synthesis

One of the earliest biochemical changes observable in sea urchin eggs during cleavage after fertilization is an increase in the rate of protein synthesis. Mechanical and chemical stimuli can also elicit cleavage and protein synthesis in unfertilized eggs. These stimuli are believed to activate "masked" mRNA stored in the cytoplasm of the oocyte (Spirin, 1966; Monroy and Tyler, 1967). It was shown by Gross and Cousineau (1964) that actinomycin, an inhibitor of the transcription of DNA to mRNA, does not prevent cleavage and blastulation in the sea urchin egg. Similar observations were made in embryos of *Fundulus heteroclitus* (Wilde and Crawford, 1966). The direction by maternal (oogenetic) mRNA is not limited to the earlier stages of embryogenesis. In embryos of the loach *Misgurnus fossilis*, protein synthesis is entirely dependent on oogenetic mRNA up to the midblastula stage (Krigsgaber and Neyfakh, 1972). In addition there is evidence for the continued functional activity of oogenetic mRNA into advanced stages of embryogenesis (Gross, 1968; Davidson, 1968). A specific example is the observation that in the presence of actinomycin, in doses which inhibited the synthesis of high molecular weight mRNA, [^{3}H]leucine continued to be incorporated into the proteins of microtubules of sea urchin embryos (Raff *et al.*, 1971, 1972, 1975). In another study, female toads (*Engystomops pustulosus*) were injected with [^{3}H]uridine to label RNA during oogenesis; the RNA was extracted from the oocytes and hybridized with a DNA fraction prepared from embryos. By this technique maternal mRNA could be detected in embryos up to the tadpole stage (Hough *et al.*, 1973). On the other hand, treatment with actinomycin D or with pactamycin of *Fundulus* eggs within minutes after fertilization resulted in the disruption of morphogenetic processes which normally occur after more than 3 days of embryonic development (Schwartz and Wilde, 1973; Crawford *et al.*, 1973). These findings suggest that some morphogenetic events, while programmed to occur at a later stage, may depend on earlier RNA synthesis. In trout embryos which had been separated from the yolk, the rate of production of ribosomal RNA in early stages of development was retarded, "suggesting the participation of yolk constituents" (Mel'nikova *et al.*, 1972). It is probable that these constituents were oogenetic precursors of ribosomal RNA.

C. Protein Synthesis in Mitochondria, Ribosomes, and Nuclei

The contributions of ribosomes and mitochondria to the total protein synthesis of trout eggs were measured in the reviewer's laboratory (Smith *et al.*, 1970). Unfertilized eggs were kept for 15 days after spawning side by side with the fertilized eggs on hatching trays. After various intervals, the eggs were homogenized and the mitochondria and the ribosomes were isolated and separately incubated with [^{14}C]leucine. Protein synthesis by the mitochondria was inhibited by chloramphenicol (specific for mitochondria), but was resistant to cycloheximide (an inhibitor of cytoplasmic ribosomal protein synthesis). The possibility can therefore be excluded that protein synthesis attributed to mitochondria might have been due to cytoplasmic ribosomes adhering to their surface. In the first few days the mitochondrial contribution to the total protein synthesis was greater than that of the ribosomes. There was a gradual and continuous increase in protein synthesis in the mitochondria from unfertilized as well as fertilized eggs and the levels and rates of increase in both were identical up to day 10. However, the activity of the ribosomes increased more rapidly with age in the embryos than in the unfertilized eggs. Unlike sea urchin eggs, trout eggs did not respond with a burst of protein synthesis to the triggering action of fertilization. Since a gradual acceleration of mitochondrial and ribosomal protein synthesis could also be observed in unfertilized trout eggs, this must be attributed to the progressive activation of preformed components of the system.

According to Krigsgaber and Neyfakh (1968), ribosomal protein synthesis in loach embryos is dependent on preformed ribosomes which presumably had been transferred from the yolk to the blastoderm. Mitochondrial protein synthesis, although observed in sea urchin eggs, had been reported not to exceed 20% of the total protein synthesis, and the mitochondrial contribution was considered too small to support the postfertilization increase (Chamberlain and Metz, 1972). In trout eggs and early embryos, however, a major contribution to protein synthesis is made by mitochondria, as shown by Smith *et al.* (1970). The observation that during the first week the rates of mitochondrial and ribosomal protein synthesis of unfertilized eggs increased at the same rates as in fertilized eggs is more direct evidence that protein synthesis in early development is under the direction of maternal mRNA.

This study was continued by an analysis of mitochondrial RNA and protein synthesis within a short time after fertilization. Unfertilized and fertilized trout eggs were kept at 10°C for 24 hr after spawning. The mitochondria were isolated and incubated with [5-^{3}H]uridine and [^{14}C]leucine. In the mitochondria from both unfertilized and fertilized eggs chloramphenicol (1.5 × 10^{-5} M) strongly inhibited protein synthesis, whereas emetine (10^{-4} M), an inhibitor of cytoplasmic ribosomal protein synthesis, was without effect. Ethidium bromide (10^{-5} M) suppressed the incorporation of radioactive precursors into high molecular weight RNA and protein fractions in mitochondria of unfertilized eggs. In mitochondria of 1-day-old fertilized eggs the rates of incorporation of both precursors were lower and not further depressed by ethidium bromide. Since this inhibitor is believed to interfere with the transcription of mitochondrial, but not of nuclear, DNA, these observations suggest a temporary shutdown of mitochondrial transcription shortly after fertilization (Bourassa and Terner, 1978).

An increase of RNA synthesis in mitochondria of sea urchin eggs within a few minutes after fertilization was reported by Cantatore *et al.* (1974). The presence of an inactive DNA synthesizing system in the mitochondria of unfertilized loach eggs was reported by Gause and Mikhailov (1973). Its activity increased during embryogenesis without observable changes in DNA polymerase activity. Although the mechanism of activation of mitochondrial DNA replication, which occurred within 35 hr after fertilization, was not understood, the possibility of an activation of the generation of primers of DNA polymerase was suggested (Mikhailov and Gause, 1974). Thus, there is evidence from different laboratories pointing to early changes in mitochondrial activities in eggs after fertilization.

The biosynthesis of histones by nuclei of trout embryos was investigated by Trevithick (1969) with the aid of metabolic inhibitors to distinguish between nuclear and cytoplasmic protein synthesis. The incorporation of [^{3}H]arginine into nuclear proteins was strongly inhibited by puromycin, but only partially by cycloheximide.

D. Nucleic Acid Synthesis

The metabolism of nucleic acids has been studied extensively in fish embryos. Their use has been especially advantageous, since their prolonged incubation period permits the spacing of several events of early embryogenesis which in invertebrates occur in a much shorter time span.

Neyfakh (1964) investigated the synthesis of nuclear RNA and the transfer of genetic information to the cytoplasm of embryos of the loach. Irradiation with X rays of embryos at the early blastula stage arrested development before gastrulation. When irradiated at the midblastula stage, embryonic development was arrested at the gastrula stage. When irradiated in the late blastula or early gastrula stages, the arrest occurred before organogenesis. According to the authors, selective radiation damage, which impairs the morphogenetic function of the nucleus, may prevent the transfer of information to the cytoplasm, but allows development to continue under the direction of information which had been transmitted before irradiation. Kafiani et al. (1969) suggested that the nuclei pass through two functional stages: the stage of early cleavage in which the nuclei are not actively transmitting genetic information (the rate of synthesis of DNA-like RNA is low) and the midblastula stage in which the rates of gene transcription and morphogenetic activity are high. At the latter stage, high molecular weight RNA is activated in the nuclei. Following on the observation by Dontsova and Neifakh (1969) of the synthesis of high molecular weight RNA by the isolated blastoderm of loach embryos, Dontsova et al. (1970) took advantage of the uniqueness in the spacing in the trout embryo of developmental events which in other species cannot be separated, for example, in the trout embryo the duration of the mitotic cycle increases before nuclear RNA synthesis can be shown to occur, and the latter event begins before sensitivity to ionizing radiation develops. Measuring the variations in the rate of incorporation of [^{3}H]uridine into RNA up to the time of gastrulation, Dontsova et al. (1970) observed a low rate of cytoplasmic RNA synthesis and an increase associated with the start of synthesis of high molecular weight nuclear RNA. According to Kafiani et al. (1973), the acceleration of RNA synthesis observed in loach embryos between the early blastula and the midblastula stages coincides with the activation of gene transcription and, unlike in amphibians, is accounted for by an increase in the activity of nuclear RNA polymerase.

Neyfakh and Kostomarova (1971) studied the transfer of newly synthesized nuclear DNA-like RNA during mitosis. By incubating isolated loach blastoderm with [^{3}H]uridine and localizing ^{3}H-labeled RNA by radioautography, Neyfakh and Kostomarova showed that RNA, which is synthesized in interphase, moves from the nucleus into the cytoplasm at the onset of mitosis and returns to the nuclei of the daughter cells on completion of cell division. Neyfakh et al. (1972)

continued this study in hybrid embryos hatched from loach eggs fertilized with goldfish sperm. The goldfish chromosomes in a "foreign environment," the loach cytoplasm, contributed to protein synthesis and to morphogenesis. DNA-like RNA was synthesized in the nuclei at the midblastula stage; its release was, however, delayed until the late blastula stage. The hybrids survived through the larval stage, but died at the active feeding stage. The authors consider the gradual transfer of nuclear RNA to the cytoplasm to be the first step of temporal control over the expression of genetic information.

E. Gene Regulation

The studies described above touch upon the general problem of genome control (the switching on and switching off of genes), a striking example of which is seen in the intricate programmed control of differentiation in embryonic development. Earlier work on gene transcription in fish embryos has been reviewed by Kafiani (1970); however, the major concepts have been developed by work with other organisms and a brief summary of current views is given here.

The selective transcription of the genome is explained in terms of the general concept, enunciated by Jacob and Monod (1961), of repression and derepression of individual genes. Bacterial proteins have been isolated which act as "repressors" by binding to DNA at specific sites ("operators"), thereby blocking the transcription of adjacent genes (Gilbert and Müller-Hill, 1966, 1967; Ptashne, 1967). By allowing isolated repressor protein to attach to and protect the operator sites of DNA and by digesting the unprotected sequences with DNase, the operator regions could be isolated and their base sequences determined (Ptashne, 1975; Maniatis *et al.*, 1975). Very readable accounts of this work have been published by Ptashne and Gilbert (1970) and Maniatis and Ptashne (1976).

According to Britten and Davidson (1969) the genomes of animal cells differ from those of bacteria not only by their much larger size and complexity, but also by the arrangement of their DNA sequences. In the genomes of embryos of the sea urchin and of *Xenopus*, long "nonrepetitive" sequences alternate with "repetitive" short sequences (300 nucleotides). The nonrepetitive sequences represent the structural genes which are defined as DNA sequences from which functional polysomic RNA can be transcribed and which accordingly can be detected by their ability to hybridize with polysomal mRNA *in vitro*. In the absence of direct evidence for a function of the repetitive segments it was proposed that they may serve as binding sites for diffusible macromolecules whose attachment would result in the activation and transcription of the adjacent structural gene. It is essential for this theory that the repetitive sequences be contiguous with the structural sequences of the genes, and evidence for this was obtained by Davidson and Britten (1974), Davidson et al. (1975), and Hough et al. (1975).

III. ENERGY METABOLISM

A. Carbohydrates

Metabolic studies of fish eggs are scarce and limited to a few species of teleosts; nevertheless, variations in patterns of metabolism have been reported within this subclass. In confirmation of older observations in invertebrates (Needham, 1942; Brachet, 1950, 1960), embryos of *Fundulus heteroclitus* whose respiration had been arrested by cyanide developed to the blastula stage; the utilization of oxygen was required for further morphogenesis, but was not essential for the immediate survival of postblastular embryos (Wilde and Crawford, 1963; Crawford and Wilde, 1966). By incubating unfertilized and fertilized trout eggs with ^{14}C-labeled substrates, Terner (1968a) showed that the eggs, although supplied with large yolk reserves, metabolize nutrient material taken up from the water. The capacity of the fertilized eggs to oxidize ^{14}C-labeled acetate, pyruvate, and glucose (measured by the production of $^{14}CO_2$) increased continuously during development. The rate of oxidation of glucose appeared to lag behind the rate of oxidation of acetate or pyruvate, but this may have been partly due to dilution of the added [^{14}C]glucose by endogenous glucose. In homogenates of trout eggs (rainbow trout *Salmo gairdneri*, brook trout *Salvelinus fontinalis*, brown trout *Salmo trutta*, and cutthroat trout *Salmo clarki*), glycogen and glucose were metabolized to lactate (Terner, 1968b). Thus, the eggs of at least some teleosts are capable of generating energy by glycolysis. In another teleost, the loach *Misgurnus fossilis* L., however, Yurowitzky and Milman (1972a) found that the hexokinase activity initially present in oocytes is lost during maturation, with the result that there is complete absence of aerobic glycolysis in mature oocytes and in embryos. No such loss of glycolytic capacity was observed in trout embryos (Terner, 1968b). The glycogen reserve of the eggs decreased during embryonic development until, at the eyed-up stage, it was one-half of the level in unfertilized trout eggs. It was calculated that the total carbohydrate reserve could not support the endogenous respiration of the embryos for more than a few hours of the period of development extending over several weeks. Since carbohydrate remains in storage, other substrates, most probably lipids, are the principal energy reserve. In cell-free extracts of trout embryos, the breakdown of glycogen was retarded and the conversion of glucose to lactate was accelerated when high ATP levels were maintained by continuous addition of the nucleotide to compensate for its rapid destruction by ATPase, suggesting that adenine nucleotide control of carbohydrate metabolism is operative in the trout embryo. In embryonated trout eggs, which had been kept at 15°C and were fixed by sudden freezing, the ATP/ADP/AMP concentration ratio was 13 : 1 : 2.4, that is, ATP/(ADP + AMP) = 3.8 : 1. From the observed levels of adenine nucleotides the "energy charge" can be calculated to be 0.83, which is close to the value of 0.85 derived from the steady state concentrations of the components of the adenylate system in living cells (Atkinson, 1968, 1970). The energy charge of the adenylate system is calculated from the concentrations of the adenine nucleotides according to the formula

$$\text{energy charge} = \frac{(ATP) + 0.5(ADP)}{(ATP) + (ADP) + (AMP)}$$

Different findings were reported by Yurowitzky and Milman (1972a). Although the ratio ATP/(ADP + AMP) was 4 : 1 in loach oocytes, it was only 2 : 1 in the embryos. While the mechanism of the decline of the ratio remained unknown, the authors attempted to correlate it with a change in pattern of metabolism from glycogen synthesis and gluconeogenesis during oocyte maturation to glycogenolysis in embryogenesis. From the reviewer's experience the possibility suggests itself that, if the ATPase activity in the loach oocyte were low, but increased in the embryo, the lower nucleotide ratio measured in the embryo may have been an artifact resulting from an experimental technique of fixation that was not rapid enough to prevent the partial breakdown of ATP.

During the maturation of the loach oocyte there is an intensification of glycogen synthesis owing to the increase in activity of glycogen synthetase. The increase of the rate of glycogenolysis after fertilization is believed to be due to an increase in activity of the debranching enzyme (amylo-1,6-glucosidase) rather than to an increase of phosphorylase activity, and only the D-form of glycogen synthetase was found in loach oocytes and embryos (Yurowitzky and Milman, 1973b, 1974, 1975). Changes in [NAD$^+$]/[NADH] and [NADP$^+$]/[NADPH] ratios in the course of maturation of the loach oocyte have been reported (Ermolaeva and Mil'man, 1974; Yermolaeva and Milman, 1974).

Milman and Yurovitsky (1967) and Yurowitzky and Milman (1972a) found mature eggs and embryos of the loach to be devoid of free glucose, although it was present in oocytes before maturation. In contrast, Terner (1968a) noted that trout embryos maintain a pool of free glucose which increases from fertilization to the eyed-up stage and falls at the time of hatching. Two key enzymes of gluconeogenesis, fructose diphosphatase and phosphoenolpyruvate carboxylase, were found to be present and their levels increased with embryonic development. In intact embryos incubated with [^{14}C]pyruvate, radioactive label was incorporated into the pool of free glucose; isolation of the [^{14}C]glucose and its identification by the specific glucose oxidase method demonstrated that gluconeogenesis is operative in trout embryos (Terner, 1968b). Mil'man and Yurovitskii (1970) found that fructose diphosphatase and phosphoenolpyruvate carboxylase are localized mainly in the yolk of the loach embryo. Measuring gluconeogenesis with the aid of ^{14}C-labeled amino acids as precursors and finding the "net flux" of gluconeogenesis in the isolated blastoderm to be negligible, they suggested that fructose diphosphatase in the blastoderm does not take part in the control of gluconeogenesis, but is a component of an "energy wasteful cycle" ("futile cycle"). The activity of fructose diphosphatase was, however, responsive to control by the modifier AMP (Yurowitskii and Mil'man, 1972b, 1973a), as it is known to be in mammalian tissues (Krebs, 1964), whereas in extracts of trout embryos it was not inhibited by 0.15–1.5 mM AMP (Terner, 1968b).

Thus there is ample evidence that in the trout embryo the metabolic pathways of glycolysis and gluconeogenesis are operative, whereas in the loach embryo these reaction sequences may be incomplete or overlapping to the extent that they contribute to a futile cycle. It is obvious that in view of the species differences within a subclass illustrated by the examples given, caution must be exercised in making generalizations.

A comprehensive account of the Soviet scientists' work was published in a monograph by Milman and Yurowitzky (1973).

B. Lipids

The lipid components of the yolk consist mainly of triglycerides and cholesterol. The biosynthesis by the developing trout embryo of structural lipids was studied in the author's laboratory. Because of the large size of the triglyceride pool, attempts to label the triglycerides by incubation of trout eggs with [^{14}C]acetate resulted in ^{14}C-labeled triglycerides of low specific activity. After short incubation of eyed-up embryos with [^{14}C]acetate, radioactivity appeared only in polyglycerophosphatides and in glycolipids. After incubation for several days the radioactive label appeared also in free fatty acids, glycerides, phospholipids, and glycolipids. Of the latter the major component was identified as ceramide galactoside, a cerebroside with nervonic acid as its predominant fatty acid. Its rate of synthesis increased with the age of the embryos and appeared to be related to the development of the nervous system. The pattern of labeling of the various classes of lipids indicates that the trout embryo utilizes the fatty acids of the yolk glyceride reserves as its principal source of metabolic energy and synthesizes *de novo* the fatty acids of its structural lipids from acetyl-CoA units (Terner *et al.*, 1968). The previous belief that "first carbohydrate, then protein and, last of all, fat are used as energy sources during the development of teleost fish" (Deuchar, 1965) can no longer be maintained.

Mounib and Eisan (1973) reported the incorporation of $^{14}CO_2$ into lipids, proteins, and nucleic acids of unfertilized cod eggs. The eggs contained two malic enzymes, distinguished by their specific requirements for the cofactors NAD$^+$ or NADP$^+$.

C. Corticosteroids in Trout Embryos

Gluconeogenesis is known to be under the control of corticosteroids. A search was therefore made in the author's laboratory for the appearance of endogenous corticosteroids in trout embryos. Cortisol and cortisone were found to be present in intact eggs containing embryos at an early stage of development. On incubation of homogenates of isolated embryos with [^{14}C]progesterone, radioactive cortisol and cortisone were produced. As evidence of the functional activity of corticosteroids during embryogenesis, cortisol (10^{-9}–10^{-7} M) was shown to enhance *in vitro* the activity of tyrosine aminotransferase in embryos which had been isolated from trout eggs just before reaching the eyed-up stage. In embryos developing in freshwater, osmoregulation should be functional in early embryogenesis. In a search for mineralocorticoids, it was not possible to detect endogenous aldosterone by standard paper and gas–liquid chromatographic techniques available at the time. However, homogenates of isolated trout embryos could be shown to convert [1,3-^{3}H]corticosterone to aldosterone and 18-hydroxycorticosterone (Pillai *et al.*, 1974). This investigation was followed by the demonstration of a binding protein for cortisol in trout embryos (Pillai and Terner, 1974).

Previous reports of a strong inhibitory effect of cortisol acetate on the development of sea urchin and loach eggs (Chestukhin, 1969) are of uncertain significance, since the hormone was added in very high concentration (10^{-3} M).

The control of steroidogenesis in the fish embryo by the pituitary gland is suggested by cytological studies. An embryonic pituitary gland was found to be present in embryos of the rainbow trout *Salmo gairdneri irideus* at the eyed-up stage. One day after the retinal pigment had become visible, prolactin cells and ACTH cells could be recognized (Nozaki *et al.*, 1974). The early differentiation of the pituitary gland into various cell types of known endocrine function indicates the establishment of control by trophic hormones. The example of corticosteroid metabolism provides evidence of the ability of fish embryos to synthesize hormones and utilize them in the transfer of metabolic information to target cells.

IV. CONCLUDING REMARKS

It may be apparent from the present review that fishes have not been studied enough for the presentation of a complete account of the events related to the control of embryogenesis even of a single species. Although, in order to present a general plan, it is necessary to include observations from other classes, this is a problem not unique to fishes but is consistent with the generally accepted view of the universality of molecular structure and of the principles of genetic regulation and metabolic control. This concept has led researchers to study individual phenomena rather than individual forms of life by gathering and piecing together information derived from a wide range of organisms, often with little concern for their morphological and evolutionary diversity. Thus the major metabolic pathways and gene transcription and translation have been unraveled in microorganisms, whereas the principal models for the study of morphogenesis and differentiation have been echinoderms and chordates. The oviparous fishes offer the same advantages for the study of embryogenesis as sea urchins and amphibians by producing eggs which can be fertilized and allowed to develop under the eyes of the observer. In addition, the prolonged period of embryonic development of many fishes makes it possible to study changes in energy metabolism, hormone production, and sensitivity to hormone action at more clearly defined stages than in more rapidly developing organisms.

Many opportunities for work with fish embryos remain to be exploited. Among these are species differences in great variety, some of which have been discussed in this review. The preferential study by individual research groups of certain species of fish seems to be determined to a large extent by regional programs of fish culture and the availability of material from local fish breeding stations. The use of gonadotropic hormone for the induction of spawning in small fishes kept in the laboratory offers independence from natural breeding seasons. There seems to be a tacit acceptance of a lack of species specificity of the gonadotropic hormones. A well-known example is the response of the male gonad of *Xenopus* to human chorionic gonadotropin and its use in a bioassay and pregnancy test for women. Recent work suggests a greater species specificity of pituitary gonadotropins in fishes than previously assumed (Breton and Billard, 1974). The still controversial view that more than one gonadotropic hormone may be produced in the fish pituitary (Fontaine, 1976) has received support by the recently reported isolation of pituitary gonadotropin of the chum salmon *Oncorhynchus keta* and its separation into two distinct fractions which differed not only in their chromatographic behavior on Sephadex G-25 columns, but also in their differential effectiveness in stimulating cyclic AMP production in male and female immature trout gonads (Idler *et al.*, 1975a,b,c).

In contrast with the concentrated effort directed to the study of steroid and pituitary hormones in adult fishes (see reviews in Idler and Truscott, 1972; Butler, 1973; Fontaine, 1976), work on their production and metabolism in fish embryos is still in its early stages. Apart from the demonstration of the ability of fish embryos to synthesize adrenal corticosteroids, the study of Pillai *et al.* (1974) also illustrates the experimental advantages of investigating embryonic hormone metabolism in oviparous lower vertebrates whose development is uncomplicated by interactions between maternal and fetal tissues.

REFERENCES

Atkinson, D. E. (1968). The energy charge of the adenylate pool as a regulatory parameter. Interaction with feedback modifiers. *Biochemistry* 7, 4030–4034.
Atkinson, D. E. (1970). Enzymes as control elements in metabolic regulation. *In* "The Enzymes" (P. D. Boyer, ed.), Vol. 1, pp. 461–489. Academic Press, New York.
Ballard, W. W. (1966a). The role of the cellular envelope in the morphogenetic movements of teleost embryos. *J. Exp. Zool.* 161, 193–200.

Ballard, W. W. (1966b). Origin of the hypoblast in *Salmo*. I. Does the blastodisc edge turn inward? *J. Exp. Zool.* **161**, 201–210.

Ballard, W. W. (1966c). Origin of the hypoblast in *Salmo*. II. Outward movement of deep central cells. *J. Exp. Zool.* **161**, 211–220.

Ballard, W. W. (1968). History of the hypoblast in *Salmo*. *J. Exp. Zool.* **186**, 257–272.

Ballard, W. W. (1973). Normal embryonic stages for salmonid fishes, based on *Salmo gairdneri* Richardson and *Salvelinus fontinalis* (Mitchill). *J. Exp. Zool.* **184**, 7–26.

Ballard, W. W., and Dodes, L. M. (1968). The morphogenetic movements at the lower surface of the blastodisc in salmonid embryos. *J. Exp. Zool.* **168**, 67–84.

Blaxter, J. H. S. (1969). Development: Eggs and larvae. *In* "Fish Physiology" (W. S. Hoar and D. J. Randall, eds.), Vol. 3, pp. 177–252. Academic Press, New York.

Bourassa, W. L., and Terner, C. (1978). Studies of metabolism in embryonic development. VII. Changes of RNA and protein synthesis in mitochondria of trout eggs after fertilization. Effects of ethidium bromide. *Exp. Cell Res.* (in press).

Brachet, J. (1950). "Chemical Embryology." Wiley (Interscience), New York.

Brachet, J. (1960). "The Biochemistry of Development." Pergamon, New York.

Breton, B., and Billard, R. (1974). Perspectives ouvertes par les données récentes sur l'endocrinologie de la reproduction des poissons pour le contrôle de leur cycle reproducteur. *Colloq. Aquacult., Brest, 1973, Actes Colloq.* No. 1.

Brinster, R. L. (1973). Nutrition and metabolism of the ovum, zygote and blastocyst. *In* "Handbook of Physiology" (R. O. Greep and E. D. Astwood, eds.), pp. 165–185. Williams & Wilkins, Baltimore, Maryland.

Britten, R. J., and Davidson, E. H. (1969). Gene regulation for higher cells. A theory. *Science* **165**, 349–357.

Butler, D. G. (1973). Structure and function of the adrenal gland of fishes. *Am. Zool.* **13**, 839–879.

Cantatore, P., Nicotra, A., Loria, P., and Saccone, C. (1974). RNA synthesis in isolated mitochondria from sea urchin embryos. *Cell Differ.* **3**, 45–53.

Chamberlain, J. P., and Metz, C. B. (1972). Mitochondrial RNA synthesis in sea urchin embryos. *J. Mol. Biol.* **64**, 593–607.

Chestukhin, A. V. (1969). Effect of cortisone on development of the sea urchin *Strongylocentrotus nudus* and loach *Misgurnus fossilis* and biosynthesis of nucleic acids. *Dokl. Biol. Sci.* **187**, 521–524.

Crawford, R. B., and Wilde, C. E., Jr. (1966). Cellular differentiation in the anamniota. IV. Relationship between RNA synthesis and aerobic metabolism in *Fundulus heteroclitus* embryos. *Exp. Cell Res.* **44**, 489–497.

Crawford, R. B., Wilde, C. E., Jr., Heinemann, M. H., and Hendler, F. J. (1973). Morphogenetic disturbances from timed inhibitions of protein synthesis in *Fundulus*. *J. Embryol. Exp. Morphol.* **29**, 363–382.

Davidson, E. H. (1968). "Gene Activity in Early Development." Academic Press, New York.

Davidson, E. H., and Britten, R. J. (1974). Molecular aspects of gene regulation in animal cells. *Cancer Res.* **34**, 2034–2043.

Davidson, E. H., Hough, B. R., Klein, W. H., and Britten, R. J. (1975). Structural genes adjacent to interspersed repetitive DNA sequences. *Cell* **4**, 217–238.

Dépêche, J. (1973). Infrastructure superficielle de la vésicule vitelline et du sac péricardiaque de l'embryon de *Poecilia reticulata* (Poisson Téléostéen). *Z. Zellforsch. Mikrosk. Anat.* **141**, 235–253.

Deuchar, E. M. (1965). Biochemical patterns in early developmental stages of vertebrates. *In* "The Biochemistry of Animal Development" (R. Weber, ed.), Vol. 1, p. 260. Academic Press, New York.

Dontsova, G. V., and Neifakh, A. A. (1969). Synthesis of high-polymer RNA in isolated blastoderms and dissociated cells of the embryonic loach. *Dokl. Biol. Sci.* 1, 25–28.

Dontsova, G. V., Ignatieva, G. M., Rott, N. N., and Tolstorukov, I. I. (1970). Nucleic acids in early embryogenesis of the trout. *Sov. J. Dev. Biol.* 1, 340–353.

Ermolaeva, L. P., and Mil'man, L. S. (1974). The ratio of the oxidized and reduced forms of NAD and NADP in loach oocytes and embryos. *Biochemistry (USSR)* 39, 259–262.

Fontaine, M. (1976). Hormones and the control of reproduction in aquaculture. *J. Fish. Res. Board Can.* 33, 922–939.

Galkina, L. A. (1970). Fertilization and early developmental stages of eggs of *Clupea harengus* n. *maris albi* under freshwater conditions. *Mar. Biol.* 6, 303–311.

Gause, G. G., Jr., and Mikhailov, V. S. (1973). State of the DNA synthesizing system in isolated mitochondria from the mature eggs of the loach (*Misgurnus fossilis*). *Biochim. Biophys. Acta* 324, 189–198.

Gilbert, W., and Müller-Hill, B. (1966). Isolation of the lac repressor. *Proc. Natl. Acad. Sci. U.S.A.* 56, 1891–1898.

Gilbert, W., and Müller-Hill, B. (1967). The lac operator is DNA. *Proc. Natl. Acad. Sci. U.S.A.* 58, 2415–2421.

Gross, P. R. (1968). Biochemistry of differentiation. *Annu. Rev. Biochem.* 37, 631–660.

Gross, P. R., and Cousineau, G. H. (1964). Macromolecule synthesis and the influence of actinomycin on early development. *Exp. Cell Res.* 33, 368–395.

Hough, B. R., Yancey, P. H., and Davidson, E. H. (1973). Persistence of maternal RNA in *Engystomops* embryos. *J. Exp. Zool.* 185, 357–368.

Hough, B. R., Smith, M. J., Britten, R. J., and Davidson, E. H. (1975). Sequence complexity of heterogeneous nuclear RNA in sea urchin embryos. *Cell* 5, 291–299.

Idler, D. R., and Truscott, B. (1972). Corticosteroids in fish. *In* "Steroids in Non-mammalian Vertebrates" (D. R. Idler, ed.), pp. 126–252. Academic Press, New York.

Idler, D. R., Hwang, S. J., and Bazar, L. S. (1975a). Fish gonadotropin(s). I. Bioassay of salmon gonadotropin(s) *in vitro* with immature trout gonads. *Endocr. Res. Commun.* 2, 199–213.

Idler, D. R., Bazar, L. S., and Hwang, S. J. (1975b). Fish gonadotropin(s). II. Isolation of gonadotropin(s) from chum salmon pituitary glands using affinity chromatography. *Endocr. Res. Commun.* 2, 215–235.

Idler, D. R., Bazar, L. S., and Hwang, S. J. (1975c). Fish gonadotropin(s). III. Evidence for more than one gonadotropin in chum salmon pituitary glands. *Endocr. Res. Commun.* 2, 237–249.

Ignat'eva, G. M. (1969). Relative duration of certain processes of early embryogenesis in salmonid fish. *Dokl. Biol. Sci.* 188, 661–663.

Ignat'eva, G. M., and Rott, N. N. (1970). Time relationships between certain processes that occur before the onset of gastrulation in teleosts. *Dokl. Biol. Sci.* 190, 26–29.

Jacob, F., and Monod, J. (1961). Genetic regulatory mechanisms in the synthesis of proteins. *J. Mol. Biol.* 3, 318–356.

Kafiani, C. (1970). Genome transcription in fish development. *In* "Advances in Morphogenesis" (M. Abercrombie, J. Brachet, and T. J. King, eds.), Vol. 8, pp. 209–284. Academic Press, New York.

Kafiani, C. A., Timofeeva, M. J., Neyfakh, A. A., Melnikova, N. L., and Rachkus, J. A. (1969). RNA synthesis in the early embryogenesis of a fish (*Misgurnus fossilis*). *J. Embryol. Exp. Morphol.* 21, 295–308.

Kafiani, C. A., Akhalkatsi, R. G., and Gasaryan, K. G. (1973). Nuclear RNA polymerase activity and template efficiency in developing loach (*Misgurnus fossilis*) embryos. *Biochim. Biophys. Acta* 324, 133–142.

Knight, A. E. (1963). The embryonic and larval development of the rainbow trout. *Trans. Am. Fish. Soc.* **92**, 344–355.

Krebs, H. A. (1964). The Croonian Lecture, 1963. Gluconeogenesis. *Proc. R. Soc., Ser. B* **159**, 545–564.

Krigsgaber, M. R., and Neyfakh, A. A. (1968). Protein synthesis in the blastoderm of the loach embryo. *Dokl. Biol. Sci.* **180**, 178–183.

Krigsgaber, M. R., and Neyfakh, A. A. (1972). Investigation of the mode of nuclear control over protein synthesis in early embryonic development of loach and sea urchins. *J. Embryol. Exp. Morphol.* **28**, 491–509.

Lentz, T. S., and Trinkaus, J. P. (1971). Differentiation of the junctional complex of the surface cells of the developing *Fundulus* blastoderm. *J. Cell Biol.* **48**, 455–472.

Maniatis, T., and Ptashne, M. (1976). A DNA operator–repressor system. *Sci. Am.* **234**(1), 64–76.

Maniatis, T., Ptashne, M., Backman, K., Kleid, D., Flashman S., Jeffrey, A., and Maurer, R. (1975). Recognition sequences and polymerase in the operators of bacteriophage lambda. *Cell* **5**, 109–113.

Mel'nikova, N. L., Timofeeva, M. Y., Rott, N. N., and Ignat'eva, G. M. (1972). Synthesis of ribosomal RNAs in early embryogenesis of the trout. *Sov. J. Dev. Biol.* **3**, 67–79.

Mikhailov, V. S., and Gause, G. G., Jr. (1974). Replication of mitochondrial DNA in early development of *Misgurnus fossilis*. *Dev. Biol.* **41**, 57–71.

Milman, L. S., and Yurovitsky, Y. G. (1967). The control of glycolysis in early embryogenesis. *Biochim. Biophys. Acta* **148**, 362–371.

Mil'man, L. S., and Yurovitskii, Y. G. (1970). Peculiarities of the localization of key enzymes of gluconeogenesis in loach embryos and their role in the regulation of metabolism. *Sov. J. Dev. Biol.* **1**, 232–238.

Milman, L. S., and Yurowitzky, Y. G. (1973). "Regulation of Glycolysis in the Early Development of Fish Embryos," Monograph in Developmental Biology, Vol. 6. Karger, Basel.

Monroy, A., and Tyler, A. (1967). The activation of the egg. *In* "Fertilization" (C. B. Metz and A. Monroy, eds.), Vol. 1, pp. 369–412. Academic Press, New York.

Mounib, M. S., and Eisan, J. S. (1973). Fixation of carbon dioxide and some of the enzymes involved in cod eggs. *Int. J. Biochem.* **4**, 207–212.

Needham, J. (1942). "Biochemistry and Morphogenesis." Cambridge Univ. Press, London.

Neyfakh, A. A. (1964). Radiation investigation of nucleo-cytoplasmic interrelations in morphogenesis and biochemical differentiation. *Nature (London)* **201**, 880–884.

Neyfakh, A. A., and Abramova, N. B. (1974). Biochemical embryology of fishes. *In* "Chemical Zoology" (M. Florkin and B. T. Scheer, eds.), Vol. 8, pp. 261–286. Academic Press, New York.

Neyfakh, A. A., and Kostomarova, A. A. (1971). Migration of newly synthesized RNA during mitosis. 1. Embryonic cells of the loach (*Misgurnus fossilis* L.). *Exp. Cell Res.* **65**, 340–344.

Neyfakh, A. A., Kostomarova, A. A., and Burakova, T. A. (1972). Transfer of RNA from nucleolus to cytoplasm in early development of fish. *Exp. Cell Res.* **72**, 223–232.

Nozaki, M., Tatsumi, Y., and Ichikawa, T. (1974). Histological changes in the prolactin cells of the rainbow trout, *Salmo gairdneri* irideus, at the time of hatching. *Annot. Zool. Jpn.* **47**, 15–21.

Pillai, A. K., and Terner, C. (1974). Studies of metabolism in embryonic development. VI. Cortisol-binding proteins in trout embryos. *Gen. Comp. Endocrinol.* **24**, 162–167.

Pillai, A. K., Salhanick, A. J., and Terner, C. (1974). Studies of metabolism in embryonic development. V. Biosynthesis of corticosteroids by trout embryos. *Gen. Comp. Endocrinol.* **24**, 152–161.

Ptashne, M. (1967). Specific binding of the λ phage repressor to λ DNA. *Nature (London)* **214**, 232–234.

Ptashne, M. (1975). Repressor, operators, and promoters in bacteriophage lambda. *Harvey Lect.* **69**, 143–171.

Ptashne, M., and Gilbert, W. (1970). Genetic repressors. *Sci. Am.* **222**(6), 36–44.

Raff, R. A., Greenhouse, G., Gross, K. W., and Gross, P. (1971). Synthesis and storage of microtubule proteins by sea urchin embryos. *J. Cell Biol.* **50**, 516–527.

Raff, R. A., Colot, S. E., Selvig, S. E., and Gross, P. R. (1972). Oogenetic nature of messenger RNA for embryonic synthesis of microtubule proteins. *Nature (London)* **235**, 211–214.

Raff, R. A., Brandis, J. W., Green, L. H., Kaumeyer, J. F., and Raff, E. C. (1975). Microtubule protein pools in early development. *Ann. N.Y. Acad. Sci.* **253**, 304–317.

Schwartz, R. J., and Wilde, C. E., Jr. (1973). Changes in protein synthesis in the morphogenesis of *Fundulus heteroclitus*. *Nature (London)* **245**, 376–379.

Smith, J. B., MacLaughlin, J., and Terner, C. (1970). Studies of metabolism in embryonic development. IV. Protein synthesis in mitochondria and ribosomes of unfertilized and fertilized trout eggs. *Int. J. Biochem.* **1**, 191–197.

Spirin, A. S. (1966). On "masked" forms of messenger RNA in early embryogenesis and in other differentiating systems. *Curr. Top. Dev. Biol.* **1**, 1–38.

Szöllösi, D., and Billard, R. (1974). The micropyle of trout eggs and its reaction to different incubation media. *J. Microsc. (Paris)* **21**, 55–62.

Terner, C. (1968a). Studies of metabolism in embryonic development. I. The oxidative metabolism of unfertilized and embryonated eggs of the rainbow trout. *Comp. Biochem. Physiol.* **24**, 993–940.

Terner, C. (1968b). Studies of metabolism in embryonic development. III. Glycogenolysis and gluconeogenesis in trout embryos. *Comp. Biochem. Physiol.* **25**, 989–1003.

Terner, C., Kumar, L. A., and Choe, T. S. (1968). Studies of metabolism in embryonic development. II. Biosynthesis of lipids in embryonated trout ova. *Comp. Biochem. Physiol.* **24**, 941–950.

Trevithick, J. R. (1969). Biosynthesis of nuclear proteins in embryos of rainbow trout. *Biochem. Biophys. Res. Commun.* **36**, 728–734.

Trinkaus, J. P., and Lentz, T. L. (1967a). A fine structural study of cytodifferentiation during cleavage, blastula, and gastrula stages of *Fundulus heteroclitus*. *J. Cell Biol.* **32**, 121–128.

Trinkaus, J. P., and Lentz, T. L. (1967b). Surface specializations of *Fundulus* cells and their relation to cell movement during gastrulation. *J. Cell Biol.* **32**, 139–153.

Vernier, J.-M. (1969). Table chronologique du développement embryonnaire de la Truite Arc-en-ciel, *Salmo gairdneri*, Rich. 1836. *Ann. Embryol. Morphog.* **2**, 495–520.

Wilde, C. E., Jr., and Crawford, R. B. (1963). Cell differentiation in the anamniota. I. Initial studies on the aerobic metabolic pathways of differentiation and morphogenesis in teleosts and urodeles. *Dev. Biol.* **7**, 578–594.

Wilde, C. E., Jr., and Crawford, R. B. (1966). Cellular differentiation in the anamniota. III. Effects of actinomycin D and cyanide on the morphogenesis of fundulus. *Exp. Cell Res.* **44**, 471–488.

Wourms, J. P. (1972a). The developmental biology of annual fishes. I. Stages in the normal development of *Austrofundulus myersi* Dahl. *J. Exp. Zool.* **182**, 143–164.

Wourms, J. P. (1972b). The developmental biology of annual fishes. II. Naturally occurring dispersion and reaggregation of blastomers during development of annual fish eggs. *J. Exp. Zool.* **182**, 169–200.

Wourms, J. P. (1972c). The development of annual fishes. III. Pre-embryonic and embryonic diapause of variable duration in the eggs of annual fishes. *J. Exp. Zool.* **182**, 389–414.

Wourms, J. P., and Evans, D. (1974). The embryonic development of the black prickleback, *Xiphister atropurpureus*, a Pacific coast blenninoid fish. *Can. J. Zool.* **52**, 879–887.

Yermolaeva, L. P., and Milman, L. S. (1974). Redox state of nicotinamide adenine nucleotide and phosphorylated state of adenine nucleotide in oocytes and embryos of the loach (*Misgurnus fossilis* L.). *Wilhelm Roux' Arch. Entwicklungsmech. Org.* **174**, 297–301.

Yurowitzky, Y. G., and Milman, L. S. (1972a). Changes in enzyme activity of glycogen and hexose metabolism during oocyte maturation in a teleost, *Misgurnus fossilis* L. *Wilhelm Roux' Arch. Entwicklungsmech. Org.* **171**, 48–54.

Yurowitskii, Y. G., and Mil'man, L. S. (1972b). Role of potassium ions in regulation of fructose diphosphatase activity in oocytes and embryos of the loach *Misgurnus fossilis*. *J. Evol. Biochem. Physiol. (USSR)* **8**, 394–395.

Yurowitskii, Y. G., and Mil'man, L. S. (1973a). Role of an energy-dissipating cycle in the regulation of the hexosephosphate level in the loach embryos and oocytes. *Biochemistry (USSR)* **38**, 253–259.

Yurowitzky, Y. G., and Milman, L. S. (1973b). Factors responsible for glycogenolysis acceleration in early embryogenesis of teleosts. *Wilhelm Roux' Arch. Entwicklungsmech. Org.* **173**, 9–21.

Yurovitskii, Y. G., and Mil'man, L. S. (1974). The glycogen synthetase of loach oocytes and embryos. *Biochemistry (USSR)* **39**, 70–77.

Yurowitzky, Y. G., and Milman, L. S. (1975). Enzymes of glycogen metabolism in developing embryos of a teleost. *Wilhelm Roux' Arch. Entwicklungsmech. Org.* **177**, 81–88.

6

PHYSIOLOGICAL ENERGETICS

J. R. BRETT and T. D. D. GROVES

FISH PHYSIOLOGY, VOL. VIII

I. INTRODUCTION

Physiological energetics, or animal bioenergetics, concerns the rates of energy expenditure, the losses and gains, and the efficiencies of energy transformation, as functional relations of the whole organism. This distinguishes the use of "bioenergetics" as applied to the mechanisms of energy exchange within the cell, but still observing the same physical and chemical laws (e.g., Lehninger, 1965). It also distinguishes the study from "ecological energetics," which involves the transfer of energy from one trophic level to another (e.g., Odum, 1971; Winberg, 1970). All three are closely related.

The majority of such presentations commence with an energy flow diagram indicating the main steps that the energy of food intake follows through the organism, and the paths of energy distribution (e.g., Davies, 1964; Beamish et al., 1975). Each of these steps with their appropriate values is subject to quantitative change, depending on many biotic and abiotic factors. With the thought that the basis of these energy exchanges needs to be elaborated first, it was deemed more fitting to conclude with a quantitatively expressed flow diagram (Fig. 18). It may be of help, however, to keep this format in mind, segments of which constitute the contents of this chapter. For the reader unfamiliar with the various components of expended energy, repeated reference to Fig. 18 is recommended.

The essential concept of biological energetics was captured by Kleiber (1961) with the title "The Fire of Life." Indeed, at the time when coal-fired steam engines were gaining acceptance, research on the comparative fuel requirements and work efficiency of domestic animals was greatly stimulated. Study of the energetic efficiencies of agricultural processes involving the production of such commodities as meat, milk, and eggs (e.g., Brody, 1945) has continued to provide an advanced background of knowledge from which the investigation of fish energetics has drawn insight and inspiration. An understanding of the physical, chemical, and biological basis on which the energetics is built, and the equivalents employed, constitutes the opening section of this chapter. Some necessary distinctions between mammalian and nonmammalian systems are made. An adequately nutritious diet is assumed; the basic source of fuel for the fire of life is solely derived from the food.

Since much of the energetics terminology must be used at the very outset, recurring in later passages where the meaning may be more explicit, a short glossary is included (Section VIII).

II. ENERGY: RELATIONS AND MEASUREMENTS

A. Thermodynamics and Biological Energy Flow

Fish, like other living systems, must conform to the laws of thermodynamics. Matter and energy may be converted but never destroyed. Fish gain matter and energy in food, and they lose absorbed matter and energy as a result of catabolism (which provides energy for maintenance and activity) and the elaboration of reproductive products. Catabolism of substrates results in the production of carbon dioxide and water (which are excreted), heat, and in some cases intermediate partially oxidized products (which may also be excreted).

During aerobic metabolism, 40–50% of the substrate chemical free energy is temporarily trapped in adenosine triphosphate and related labile compounds. These so-called "high energy" compounds provide the immediate driving energy for endergonic processes such as biosynthesis and membrane transport, and are the immediate fuel for conversion to mechanical work in muscle tissues. Chemical free energy is degraded to heat. In the case of homeotherms at temperatures below their thermal neutral range, this metabolic heat contributes to the maintenance of body temperature. With the possible exception of some warm-bodied sharks and tunas (Carey et al., 1971), metabolic heat in the fish represents a complete loss of energy to the animal.

If body mass is to be maintained, absorbed dietary energy (exogenous sources) must equal energy loss for maintenance and activity. When exogenous sources exceed these requirements, growth can occur from the deposition of matter, which for fish is largely protein. Energy is also stored in growth as the chemical energy of covalent bonds in proteins, fats, and carbohydrates. If dietary energy is insufficient to cover catabolism, growth of some organs or body components may occur at the expense of internal (endogenous) sources previously stored in growth. In the absence of any dietary input all energy for maintenance and activity must be provided from endogenous sources.

A good example of the metabolic and growth alternatives available to fish according to the laws of thermodynamics and conservation of matter is provided in comparative data on systems for hatching salmonid eggs and holding alevins. Since the amount of yolk is fixed and hence the amount of matter and energy is fixed, systems that stimulate activity of embryos and alevins result in small "swim-up" fry having limited energy reserves. Systems that minimize such physical activity and hence minimize catabolism result in larger fry (Marr, 1966; Bams, 1970).

In living systems, the fate of absorbed food is more complicated than simple metabolic combustion for energy or deposition of matter in growth. One reason of major importance is that environmental factors strongly influence the biochemical state of the fish (see Chapters 1 and 10). A second concerns particularly the fate of protein. In catabolism, the total heat produced bioenergetically differs from that obtained by direct combustion of food because protein nitrogen is not fully oxidized. Therefore, protein catabolism produces less net energy than expected from the heat of combustion as determined by bomb calorimetry. In growth, only those proteins with an adequate amino acid balance can be utilized. Surplus dietary energy ingested as protein must be stored as fat. Efficient use of body fat at some later time however requires some degradation of carbohydrate and protein.

B. Calorimetry: Metabolic Heat Production

1. DIRECT CALORIMETRY

The basic principles of calorimetry date to the early experiments of Lavoisier and Laplace in 1780 and are summarized by Brody (1945), Kleiber (1961), and Blaxter (1965). In the case of direct calorimetry, heat loss is measured. Lavoisier and Laplace determined the amount of ice melting in a chamber which surrounded a guinea pig. Subsequent direct calorimeters have been based on the measurement of the heat increment in water circulated in the walls of the calorimeter, or of the air passing through the chamber. Another direct technique involves the measurement of heat input by a calibrated electric heater to maintain a given chamber temperature in the presence or absence of an animal. To obtain a more rapid response than that possible with heavy calorimeter chambers having high heat capacity, gradient layer calorimeters have been devised in which the temperature differential across a thin container wall of known thermal conductivity is measured. Calorimeters of this type have been described by Benzinger *et al.* (1958), Pullar (1958), and Mount (1963).

Direct calorimeters have been considered unsatisfactory for studies on fish because of the relatively low heat production by fish and the high heat capacity of the water in the system. Smith (1976), however, argues to the contrary, pointing out that problems of heat loss by vaporization and radiation, which complicate direct calorimetry in terrestrial animals, do not occur in aquatic forms. Using a modified adiabatic calorimeter, Smith (1976) determined the heat production of four species of salmonids. From the data presented, the sensitivity of metabolic response does not appear to be detected as well as by methods depending on measurements of oxygen consumption, due to the high heat buffering capacity of the direct calorimeter used.

2. INDIRECT CALORIMETRY

Indirect calorimetry, which also began with Lavoisier, involves measurement of gas exchange. The calorific equivalents for oxygen uptake and CO_2 production are based on heats of combustion of different energy substrates, measured in a direct bomb calorimeter. The type of physiological fuel involved can be established by determining the RQ, or respiratory quotient (defined as the ratio of moles of CO_2 produced to O_2 utilized), and the urinary or nonfecal nitrogen excretion. Under totally aerobic conditions, the values of RQ vary from 0.7 for fat to 1.0 for carbohydrate. In the classical literature on terrestrial ureotelic animal respirometry, the RQ for protein catabolism is approximately 0.82. In the case of ammonotelic fish, the RQ for protein is nearly 0.9. In either case, if the urinary or nonfecal nitrogen excretion is measured, an estimate can be made of heat production due to amino acid or protein breakdown. This amount, with appropriate adjustment of the oxygen and CO_2 volumes, allows an estimate of heat production from nonprotein energy sources. RQ values of greater than unity have been recorded in cases where there was active synthesis of body fat. In addition, particularly in the case of fish, anaerobic metabolism can result in RQ values of between 1 and 2 (Kutty, 1972). The caloric equivalent of oxygen uptake, Q_{ox}, consequently varies with the physiological state and substrate utilized.

In mammalian studies, the caloric equivalent of respired oxygen during catabolism of a mixed protein (RQ = 0.81) is 4.82 kcal/liter O_2 at standard temperature and pressure (Brody, 1945). The caloric equivalent of oxygen during aerobic carbohydrate catabolism (RQ = 1.0) is 5.04 kcal/liter O_2, and that for mixed fat (RQ = 0.71), 4.69 kcal/liter O_2. At a mean nonprotein RQ of 0.85, the equivalent is 4.86 kcal/liter O_2. Thus, for studies in mammalian metabolism where carbohydrate and fat are the principal energy substrates, a mean oxycalorific value of approximately 4.86 kcal/liter O_2 is appropriate.

For fish, the Q_{ox} for carbohydrate is also taken as 5.04 kcal/liter O_2. Oxycalorific equivalents for fat of 4.66 and 4.69 have been determined from the literature by Brafield and Solomon (1972) and Elliot and Davison (1975), respectively. The same authors record a Q_{ox} of 4.58 kcal/liter O_2 for the ammonotelic catabolism of protein. Since the principal energy sources in carnivorous fish appear to be lipid and protein rather than lipid and carbohydrate, a mean Q_{ox} of 4.63 kcal/liter O_2 may be more appropriate in aerobic, steady state metabolic studies of fish than the frequently applied values of 4.8–5.0 kcal/liter O_2.

Indirect calorimeters for animals have been operated on both the constant volume, open circuit principle, where differential composition of air entering and leaving the chamber is measured, and on the closed circuit principle in which the change of concentration of chamber oxygen is measured after the carbon dioxide produced is absorbed. In studies on fish, tunnel-type respirometers have been operated on a closed circuit system during sampling intervals (Blazka *et al.*, 1960; Brett, 1964; Beamish, 1970). By determining the rates of decrease in dissolved oxygen concentration during closed circuit intervals, precise data on the energy expenditure of fish at different temperatures and at different levels of activity can be obtained. Since dissolved oxygen is more conveniently measured than dissolved carbon dioxide, the RQ of fish is not usually determined. Tunnel respirometers of the type described by Brett (1964) measure the energy exchange of *swimming* fish since movement of water through the apparatus is necessary for both temperature control and representative sampling for oxygen.

Because of some doubt cast by Krueger *et al.* (1968) on the oxycalorific value appropriate for fish respiration, Brett (1973) resorted to testing the basis by a technique first applied to fish by Pentegov *et al.* (1928) and subsequently developed by Idler and Clemens (1959). These authors determined energy expenditure by measuring the change in body composition of migrating salmon, which naturally starve as they proceed from the sea to distant spawning grounds. Thus, by direct determination of the change in energy content (bomb calorimetry) of the flesh and viscera of exercised fish, and by comparison of this with the value deduced from oxygen consumption measurements, an oxycalorific equivalent of 4.8* kcal/liter O_2 was shown to be an acceptable value for fish. The availability of large numbers of fish, providing an opportunity for sacrificing subsamples for determining energy content, provides an opportunity of study not possible with many other vertebrates.

C. Energy Sources

There are some major quantitative differences between the dietary energy sources of fish and land mammals. In contrast to terrestrial

---

* The accuracy of the method did not allow a precise measure, such as 4.63 kcal recommended now.

animals, fish utilize dietary carbohydrate poorly, both at the level of digestion and in their capacity to metabolize absorbed carbohydrate. Raw starches are only 30–40% digestible by salmonids, and digestibility appears to decrease markedly when carbohydrate levels exceed 25% of the ration. More omnivorous or herbivorous species such as catfish and carp have a higher capacity to utilize carbohydrates. A number of species have been shown capable of developing a cellulolytic intestinal microflora (Stickney, 1974). Even in herbivorous fish such as grass carp, the principal digestible energy available in the diet is in the form of protein and simple carbohydrates (e.g., disaccharides, oligosaccharides, and hemicelluloses).

Protein is a major energy source for all fish, and it appears that blood glucose may derive more readily through gluconeogenesis than directly from dietary carbohydrates. This may account for the fact that the optimum protein/calorie ratio in prepared salmonid rations is in the vicinity of 120 mg protein/digestible kcal, as compared to a ratio of approximately 70 mg/kcal in mammalian rations.

Lipids are the principal nonprotein energy source in the natural diets of both carnivorous and omnivorous fish. Provided dietary fats are liquid at ambient temperatures, fats are both highly digestible and readily metabolized.

D. Available Energy

Phillips (1969) assigned caloric values to dietary protein, carbohydrate, and fat that allow an estimate of the available energy of a trout ration based on the proximate composition of the feed. The assigned caloric values take into consideration the average digestibility of the components together with the nonfecal excretory energy losses but not the heat increment, and are therefore an estimate of metabolizable energy rather than net energy (Fig. 18). The values given for dietary fats and carbohydrates are the gross energies (9.45 kcal/g of fat, and 4.0 kcal/g of raw starch) modified only by mean digestibilities of 0.85 for fat and 0.4 for carbohydrate. The resulting metabolizable energy values are 8.0 kcal/g of dietary fat and 1.6 kcal/g of dietary carbohydrate. The latter value is not valid for cooked starch, which has been shown to be approximately 80% digestible by trout (Nose, 1967) resulting in a metabolizable energy closer to 3.3 kcal/g. Phillips' value of 3.9 kcal "available" energy/g of protein was derived by applying an average digestibility factor of 0.9 and by deducting 1.3 kcal/g from the gross energy of digestible protein (5.66 kcal/g).

Brody (1945) refers to a urinary loss of about 1.3 kcal/g of protein catabolized in the form of nitrogenous excretory products, principally urea. This corresponds approximately to the heat of combustion of urea (151.6 kcal/mole—Kleiber, 1961) equivalent to 0.86 kcal/g of protein, plus the energy cost of urea biosynthesis amounting to approximately 0.35 kcal/g of protein (or a total of 1.21 kcal g of protein). The difference between 1.21 and 1.3 kcal/g of protein is due to the other nitrogenous wastes (uric acid and creatinine) which have higher heats of combustion than urea, but which constitute only about 10% of the total urinary nitrogen in mammals. The heat loss due to urea synthesis would be more properly included as part of the heat increment of dietary protein. The 0.35 kcal/g of protein in ureotelic animals does, however, represent a component of energy which is unavailable for purposes other than nitrogen excretion when protein is catabolized.

Since teleost fish excrete primarily ammonia in freshwater, the nonmetabolizable energy fraction of absorbed dietary protein will reflect the heat of combustion of ammonia—5.94 kcal/g of excreted ammonia nitrogen or 0.95 kcal/g of protein (Elliott and Davison, 1975). The metabolizable energy for *absorbed* protein in freshwater teleosts is therefore close to 4.70 kcal/g of protein catabolized. Assuming a mean digestibility of 0.9, the energy available to fish in dietary protein is approximately 4.23 kcal/g rather than the 3.9 of Phillips.

Although ammonia has been shown to contribute up to 75–90% of the nitrogenous excretion of marine teleosts (Wood, 1958; cf. Table V) a small proportion of the nitrogen excreted by marine species is in the form of trimethylamine oxide, some of which may have originated as such in the diet. Excretion of endogenous trimethylamine oxide, synthesized as a result of protein catabolism in the fish, would result in a significant reduction of the metabolizable energy from protein. Methyl carbon and hydrogen excreted as trimethylamine oxide represent a loss of approximately 42.7 kcal/g of trimethylamine nitrogen.

Thus, the amount of energy physiologically available to fish from *synthesized* body carbohydrate, fat, and protein appears to be approximately 4.10, 9.45, and 4.80 kcal/g, respectively. The available energy from protein for marine species excreting a significant amount of trimethylamine oxide would be less. The corresponding mean metabolizable energies of the *dietary* components for salmonids in fresh water, obtained by applying the mean digestibilities used by Phillips (1969), are 1.6, 8.0, and 4.2 kcal/g of raw starch, fat, and protein, respectively. The value for cooked starch consumed by salmonids or for raw starch consumed by herbivorous fish is closer to 3.3 kcal/g. The further energy losses due to the heat increments of the respective dietary components are discussed in Section III,D.

E. Energy Density of Feed

Fish, like other animals, tend to eat to meet their energy requirements (Rozin and Mayer, 1961). Assuming an adequate dietary nutrient balance, the fish can compensate for a low energy density (calories per gram) by eating more of the ration. Compensation of this sort can occur below the limits of the physical capacity of the gut. Up to that point, weight gain may be similar between groups of fish, but the fish on the high energy feed require less feed per unit of gain. However, since fish receiving a ration of high energy density can consume more nutrients at maximum physical intake they are able to grow at a higher rate. These principles are illustrated in data reported for trout by Phillips and Brockway (1959) and Ringrose (1971). Brett (1976a,b) has demonstrated how both the metabolic rates and maximum feed intakes of sockeye salmon receiving isocaloric diets increase with increasing ambient temperature (cf. Figs. 7 and 13). It follows that high energy rations can be important in supporting maximum growth and performance at high temperatures. This has been demonstrated in the experience with chinook fry raised at high ambient temperatures (Robertson Creek Salmon Hatchery, Port Alberni, British Columbia). In this case, the most successful rations were high fat, high energy rations.

F. Storage of Body Substance

The metabolizable energy values discussed in Section D above are the amounts of physiologically useful energy obtained when protein, fat, or carbohydrate is catabolized by the fish. When growth occurs, however, the total body energy increases according to the gross combustion energies of the new tissue components. The minimum energy cost of growth may be estimated from the moles of ATP required per mole of subunit polymerized as protein, fat, or carbohydrate, and by considering the total metabolic energy cost of forming and then breaking a labile phosphate bond. If glucose is the energy source, 1 mole of ATP formed and hydrolyzed represents approximately 18.7 kcal of metabolic heat loss. By this approach, the minimum energy costs of growth are approximately 0.23, 1.24, and 0.56 kcal/g of glycogen, fat, or protein synthesized. The corresponding chemical net energetic efficiencies are approximately 95, 87, and 90%, respectively. The biological net efficiencies are slightly lower than this due to transport and metabolic cycling of precursors prior to synthesis. Brody (1945) reports the net energetic efficiencies of poultry egg production, of milk production, and of prenatal growth in animals to be 77, 61, and 60%, respectively. The gross caloric efficiency of growth of young coho (55%) observed by Averett (1969) indicates a net efficiency of approximately 70% (assuming that the energy cost of maintenance and movement was of the order of 30% of the total energy intake). For practical purposes it is of use to consider that the energy cost of new tissue synthesis is in the range of 15–30% in terms of metabolizable energy of the gross energy of the body components stored. Further discussion of the gross and net efficiencies of growth are given in Section V.

G. Summary of Equivalents

There has been some confusion and inconsistency in the literature on fish energetics according to the caloric equivalents used (mammalian or fish), the substrate being respired, and the amount of energy in question, that is, the absolute caloric content (heat of combustion), the metabolizable content of the food (physiological value), and the metabolically available energy from body resources. A frequent error has been the use of the relatively high caloric equivalents for saturated animal fats rather than the lower caloric content of highly unsaturated fats in fish meal. Table I summarizes the various equivalents, drawn from literature already cited.

III. METABOLISM: RATES OF ENERGY EXPENDITURE

The prime demand for food, before any energy storage or somatic growth can be achieved, is to meet maintenance requirements. Within this need, the highest priority must be ascribed to *basal metabolism*—the minimum rate of energy expenditure to keep the organism alive. Among the various metabolic demands, the determination of this physiological minimum has received most attention (Winberg, 1956) undoubtedly influenced by the considerable documentation of basal metabolism for homeotherms (Brody, 1945; Kleiber, 1961). The case is perhaps not so justified for making comparable measurements on cold-blooded vertebrates, which are characterized by resting metabolic rates that are 10 to 30 times less than mammals, and up to 100 times less than small birds of the same weight (Brett, 1972). It is locomotion that is metabolically costly for fish. Because of such relatively low maintenance requirements fish can withstand long periods of starvation, and frequently cease feeding for many days when experiencing major environmental change (vertical and horizontal migrations) or if by chance they are denied food, as is witnessed by the not infrequent lack of any trace of digested material in the full length of the intestine of captured fish.

Table I
Energy Equivalents Used for Mammalian and Fish Bioenergetics[a]

| Component or substrate | | Energy of food and body resources | | Respiratory energy equivalents Oxycalorific values | | Respiratory quotient |
| | Food | | Body | | | |
	Heat of combustion (kcal/g)	Physiological value (kcal/g)	Metabolic energy (kcal/g)	Q_{ox} (kcal/liter of O_2)	Q_{ox} (cal/mg of O_2)	Ratio, CO_2/O_2
Carbohydrate Mammal	4.10	4.0–3.2	4.10	5.04	3.53	1.0
Fish	4.10	3.3–1.6	4.10	5.04	3.53	1.0
Fat Mammal	9.45	9.0	9.45	4.69	3.28	0.71
Fish	8.66	8.0	9.45–8.66[b]	4.69	3.28	0.70
Protein Mammal	5.65	4.2–3.9	4.70	4.82	3.37	0.81
Fish	5.65	4.5–3.9	4.80	4.58	3.20	0.9
Mixed Mammal	5.95	—	—	4.86	3.40	0.83
Fish	5.89	—	—	4.63	3.25	0.90

[a] The range in physiological values is largely affected by digestibility.
[b] From Beamish et al. (1975).

Because of the natural, ready elevation of the metabolic rate of fish, with concomitant technical difficulties of measuring resting states, the term *standard metabolism* was early applied to describe the minimum rates observed (Krogh, 1914). Cases involving normal, spontaneous activity were called *routine metabolism* (Fry, 1957; Beamish, 1964a). The introduction of activity meters and tunnel respirometers, relating the rate of doing work (power) to swimming speed (performance), allowed estimates of the minimum metabolic rate for complete rest by extrapolation to zero activity—a true basal rate. A functional distinction between basal metabolism (warm-blooded) and standard metabolism (cold-blooded) was no longer necessary. In accordance with custom however, the term standard metabolism will be used throughout for fish. While there is no doubt that this parameter of metabolism offers meaningful insight, particularly with regard to the response to environmental factors, nevertheless the animal's need for foraging and food-processing led Winberg (1956) to make a plea for research to be directed to higher levels of metabolic requirement, in the interest of contributing to problems of bioenergetics involving the feeding, digesting, growing, competing animal.

By measuring *active metabolism,* limits for the maximum rate of energy expenditure at maximum sustained activity could be established experimentally. But, short of knowing what these limits were, and that episodes of chase and escape would undoubtedly invoke active metabolic rates, knowledge of the average daily energy costs for fish under almost any natural conditions was not forthcoming. This led Paloheimo and Dickie (1966) to relate feeding levels to the apparent rate of metabolism that could be derived from energy equations, involving rates of food intake and growth. Assuming certain corrections for the metabolizable fraction of the food, these authors deduced from the literature that maximum food intake would be accompanied by maximum or active metabolic rate. However, ultimate support for this conclusion has not been forthcoming from laboratory experiments on the metabolism of feeding fish (Warren, 1971). It remains a challenge to biologists to devise some remote means of measuring the actual rates of energy expenditure under field conditions. The nearest experimental approach is to determine the metabolic costs associated with the daily processes and activities of normal living, and sum these costs according to closely documented behavior in nature. For captive fish in a hatchery or other aquaculture system (usually fed to near satiation) direct opportunity of metabolic measurement is afforded.

General consideration will be given first to standard metabolic rates, as the minimum energy costs, together with some comparison of maximum rates of energy expenditure (but see also Fry, 1971). The lesser known aspects of metabolic costs associated with feeding, and the apparent patterns of daily energy expenditure, will then be compared with the standard and active metabolic rates.

A. Standard Metabolism

1. LEVEL AND METABOLIC COMPENSATION

Despite the sensitivity of metabolic rate to many factors, particularly excitement, size, and temperature, some idea of the general level for fish can be obtained from the listing in "Biological Data Book" (Altman and Dittmer, 1974). Of the 365 records involving 34 species, 57 entries relate to adequately derived standard metabolism, that is, activity was accounted for and the fish were fully acclimated to the test temperature. These records have a mean of 89 ± 34 (SD) mg of O_2/kg/hr (0.29 ± 0.11 kcal/kg/hr), and an extreme range from 26 to 229 mg of O_2/kg/hr (0.08–0.74 kcal/kg/hr). With few exceptions, the records all apply to fish commonly found in temperate climatic zones. Studies comparing the temperature effect on standard metabolic rates for fish from different climates have been conducted, particularly by Scholander et al. (1953) and Wohlschlag (1960, 1964). Although some of the records appear to border on the high side, possibly because of uncontrolled or unaccounted activity (closer to routine metabolic rates), they clearly demonstrate a compensative, adaptive response of polar species to maintain a higher metabolic rate at low temperatures, above that which would be expected from studies on temperate species (Fig. 1). By contrast, only limited metabolic compensation appears to have taken place among tropical species, which have paced their standard rates downward, mostly in the lower range of their thermal tolerance. These fishes are consequently operating at a higher maintenance level in accordance with their higher environmental temperatures. Applying the same correction factor (×0.56) developed for the temperate species (see Fig. 1), tropical fish would have a minimum energy expenditure of about 0.5 kcal/kg/hr in the midpoint of their temperature range (26°C)—an elevation of 70% over the mean value for temperate species.

Holeton (1973, 1974) has justly questioned the conclusions from earlier studies on metabolic cold adaptation of polar species, pointing out the likelihood of error from prolonged, elevated metabolic rates resulting from capture and handling, and the short acclimation times applied in some cases. However, Holeton's (1974) most extensive

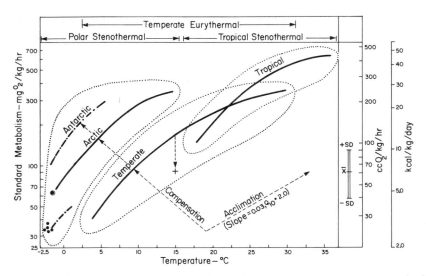

Fig. 1. Schematic representation of relation between temperature and "standard" metabolic rates (log scale) of fish from different climatic zones. Drawn from compilation of Wohlschlag (1964) with selected additions from Holeton (1974) (points for polar species). Dotted lines indicate general range of variability within each zone. Adaptive metabolic compensation is shown for polar species, following the construction line indicated. The effect of acclimation temperature is shown by the average slope (0.03), with a construction line for a temperature increase of 10°C ($Q_{10} = 2.0$). The mean ± 1 SD of the standard values in Altman and Dittmer (1974) (column to right) indicates that the curves are somewhat higher than true standard metabolic rates, the midpoint for the temperate species being about 1.8 times the indicated mean for an average weight of 50–100 g (note the central arrow pointing to the + mark). (Modified from Brett, 1970b; reproduced by permission of Wiley–Interscience.)

records on the Arctic cod, *Boreogadus saida*, were consistent with previously published values for arctic fish, with a mean standard metabolic rate of 70 mg of O_2/kg/hr at $-1.5°C$, a temperature that would be rapidly lethal for most temperate species. Five species of arctic cottids had standard metabolic rates of about 38 mg of O_2/kg/hr at this temperature (shown in Fig. 1), whereas three species of zoarcids and two species of liparids occurred at lower "uncompensated" rates (not shown). It appears that the upper curve for antarctic fish (broken line) in Fig. 1 is unsupported as attributable to metabolic compensation and must represent the upper range for routine metabolic rates. For the balance of polar species investigated by Holeton (1974) there are obviously major exceptions, showing lack of any apparent metabolic compensation to low temperature. No standard rate below 20 mg of O_2/kg/hr has been obtained, which can be taken as the minimum metabolic rate to support life in free-living fish in the 10–100 g size range (approximately 0.07 kcal/kg/hr).

2. TEMPERATURE EFFECT

The foregoing presents the general range of standard metabolism as it varies among species and between climatic zones. The effect of temperature on standard metabolism *within* species has been considered by Fry (1971) in Vol. VI of Fish Physiology, and should be consulted. Some examples occur in the present text, starting with Fig. 1 in which the shape of each curve follows a generalized form, first defined for goldfish, *Carassius auratus*, by Ege and Krogh (1914), and called Krogh's "standard curve." This curvilinear relation was subsequently elaborated upon and supported by Winberg (1956). Since no simple mathematical transformation could be derived, a set of empirical multipliers based on the Q_{10}* values was developed by Winberg (1956) for temperate species. When plotted as the logarithm of standard metabolism against temperature, the curve follows the convex slope in Fig. 1. Variability in level (intercept) and shape of the nontransformed curve can be seen in Figs. 2, 10, and 13. The validity of Krogh's standard curve has been examined by Holeton (1974) who shows that the data points were considerably elevated above those obtained for the same species by Beamish and Mookherji (1964); also that the Q_{10} value for low temperatures (0°–5°C) was undoubtedly in error as a measure of standard metabolism. In general it appears that a simple exponential transformation [log M_S = a + bT (M_S, standard metabolism; T, temperature)] comes close to linearizing the data for goldfish over the temperature range of 10°–30°C, with a mean Q_{10} of 2.3.

Once the level of standard metabolism has been established for a species at the midpoint of its normal, environmental temperature range, then the use of 2.3 as a multiplier would provide a near approximation for temperature effects within the span of ±10°C. This can be shown to apply to brook trout, bullhead, and carp (Beamish, 1964b) and sockeye salmon (Brett, 1964) without introducing an error greater than 20% of the observed value. However, beyond noting the sort of exponential increase, as Fry (1971) comments, "few general statements can yet be made of the relation of standard metabolism to temperature except that different species are adapted to different temperature ranges" (pp. 44–45).

* Q_{10} = the increase in rate for an increase of 10°C.

3. SIZE EFFECT

The effect of size on the standard metabolic rate of animals has been the focus of much research and conjecture (Zeuthen, 1953, 1970; Kleiber, 1961; Gordon, 1972). The general relation is described by the allometric equation, $Y = aX^b$, where Y = rate of oxygen consumption, X = body weight, and a and b are constants characteristic of a given species. The rate per unit weight (Y/X) usually diminishes with increasing size both within and among species such that, for instance, in salmon the standard rate of a 3000-g adult is about one-fifth that of a 1-g fry, or, in the popular elephant–mouse example, the fraction is more like one-fiftieth. Since the exponent b approximates 0.67 (i.e., two-thirds) in many warm-blooded animals it was thought to reflect the simple physical relation of the surface-to-volume law. However, among fishes the value for individual species has rarely been found this low, and would be better placed at 0.86 ± 0.03 (SE)* for the general case (Glass, 1969). This is further supported by the table of values compiled by Kausch (1972) where the mean was also 0.86 (±0.04 (SE), n = 25). Kayser and Heusner (1964) reported an exponent of 0.70 ± 0.02 for four species *grouped* together; the species however are very unevenly distributed in weight dispersion over the full size range, depressing the separate slopes within species. In the interest of accurate assessment the species-specific characteristic is sufficiently unique to justify separate determination for any given species, at the normal temperatures experienced (see table in Glass, 1969).

B. Active Metabolism

Over and above maintenance metabolism, the greatest energy demands for most animals derive from those activities which call on locomotion (attacking, escaping, migrating, jumping). When swimming, a major increase in demand for oxygen arises from the contraction of the large lateral muscles of fish. Burst speeds overtax the capacity of the respiratory–circulatory system to provide sufficient oxygen. Fatigue results. The lesser, marathonlike, sustained speeds are set by the maximum rate of oxygen consumption, defining the upper limit for the *active metabolic rate;* this is equivalent to the "aerobic capacity" of mammals.

* Recalculated from Glass (1969) using estimated rate for the killifish.

1. LEVEL AND TEMPERATURE EFFECT

The effect of temperature on active metabolism has been reviewed by Fry (1957, 1971). A different set of temperature relations than those for standard metabolic rates apply. Four of the better known species are represented in Fig. 2, including their standard rates to provide an indication of the full scope for sustained metabolic rate. Three of the four species are characterized by an optimum temperature above which the active metabolic rate decreases. This phenomenon has not been explained although it has been hypothesized by Jones (1971) that the energy demands of ventilation and associated circulation become excessive beyond critical temperatures (e.g., 15°C for sockeye salmon) restricting increased supply of oxygen to the tissues. There is also provisionary evidence that oxygen becomes a respiratory limiting factor at high temperatures for some salmonids as the oxygen content of air-saturated water decreases with increasing temperature (Brett,

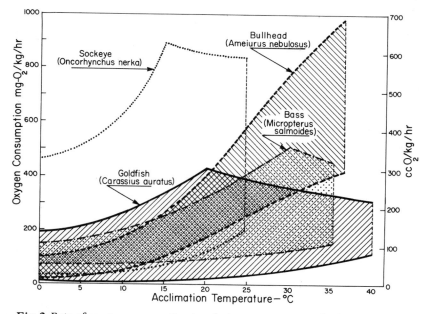

Fig. 2. Rate of oxygen consumption in relation to temperature for four species of fish. Active and standard metabolic rates are indicated in the upper and lower lines for each species, terminated by a vertical line at the upper lethal temperature. Data for sockeye from Brett (1964); for bullhead from Fry (1947); for bass from Beamish (1970); and for goldfish from Fry and Hochachka (1970). (From Brett, 1972; courtesy of North-Holland Publ. Co.)

1964; Fry, 1971). It can be seen (Fig. 2) that the highest active metabolic rates approach 1000 mg of O_2/kg/hr (3.3 kcal/kg/hr). Fast swimming, streamlined fish can readily elevate their metabolic rate from standard levels by a factor of 8 to 10 times.

It would be of interest to have comparable observations on active metabolic rates of sluggish fish to compare with the more active species. As yet such are not available. However, Wohlschlag (1960) made records of the metabolic rate of *Trematomus bernacchii,* a benthic species which lives under a heavy, sea-ice cover in the Antarctic. Although the fish could not be induced to swim continuously, excitement apparently drove the metabolic rate up in a number of cases, rising to 140–180 mg of O_2/kg/hr (0.46–0.58 kcal/kg/hr). This is about one-fifth the active metabolic rate of sockeye salmon. With the exception of hemoglobinless fish (Holeton, 1970), the above records represent the sort of overall range of energy expenditure that may be expected among fishes with regard to their maximum oxygen consumption rates (0.5–3.3 kcal/kg/hr). It cannot of course include the energy expenditure of burst speeds, which are almost entirely anaerobic, relying on subsequent hyperventilation to pay off the accumulated oxygen debt. The extrapolation of power–performance curves reveals that the energy liberated during maximum bursts can reach 100 times the standard rate or 10 times the active rate (Brett, 1972). This may also be demonstrated by computing an "equal energy" curve, such as that in Fig. 3. It underscores the fact that just a few bursts of attack or escape each day would be equivalent to doubling the daily costs incurred from the standard metabolic rate—unlike mammals or birds. There is sufficient quantitative difference in the metabolic rates of cold-blooded vertebrates (ectotherms) compared to warm-blooded vertebrates (endotherms) to warrant one such example: at the same weight (about 2 kg) the highest level of active metabolic rate of a salmon just equals the basal metabolic rate of a rabbit.

2. SIZE EFFECT

Few systematic studies have been conducted on the effect of size on active metabolic rates of fish. This is a very important area of enquiry since it is a common error in most "biological production models" involving energy requirements to incorporate a scaling factor for size according to standard metabolic rates—which is clearly a false premise for growing, active fish. Studies on size effect of sockeye salmon (Brett, 1965) showed a *continuous* change in the weight exponent (W^b), from 0.78 to 0.97, with increasing levels of activity (at 15°C). The mean value of b for all temperatures (5°–20°C) for the active metabolic

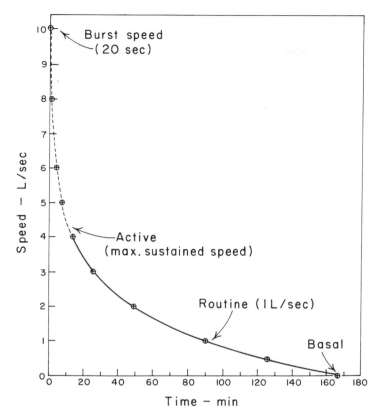

Fig. 3. Time–speed curve for equal energy expenditure of a 50 g, 18 cm sockeye salmon at 15°C. The calculated points are the times to consume 10 mg O_2 (about 35 cal), from a resting (basal) state to swimming at 10 body lengths/sec (L/sec). Routine metabolic rate shown as equal to 1 L/sec. Broken line for extrapolated points. (Derived from data in Brett, 1964.)

rate of this species was 0.98, indicating an almost insignificant effect of weight for most temperature circumstances (Brett and Glass, 1973). Because of the large percentage of muscle in a big fish (increases from about 35% at 10 g to about 65% at 1000 g in salmon) the expected decrease in tissue respiration rate with increasing size is largely offset by the relative increase in mass of "working" tissue.

3. COMBINED EFFECT

A composite graph of the combined effects of weight and temperature on rates of energy expenditure was developed for sockeye salmon by Brett and Glass (1973). It provides a means of approximating

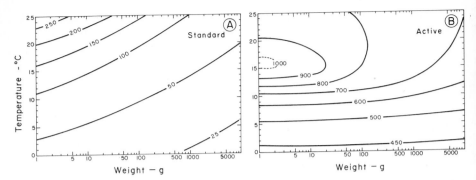

Fig. 4. Developed response surface for standard (A) and active (B) metabolic rates of sockeye salmon in relation to weight (log scale) and temperature. Rates shown in mg of O_2/kg/hr. Dotted isopleth line for 1000 mg of O_2/kg/hr obtained by extrapolation. Equivalents: 100 mg O_2 = 70 cc O_2 = 325 cal. (From Brett and Glass, 1973.)

standard and active metabolism graphically from the plotted isopleths (Fig. 4). The slopes of the lines show how weight has a consistent interacting effect with temperature over the tolerable range, reducing the standard metabolic rate with increasing size at all levels of temperature (Fig. 4A). This contrasts with the circumstance for active metabolism (Fig. 4B) which progresses from almost complete independence of weight below 10°C (isopleth lines are almost horizontal) to increasing dependence for the combined circumstances of high temperature and large size.

C. Feeding Metabolism

The need for a better understanding of the metabolic requirements of *feeding* fish has been recognized by a number of investigators. Warren and associates (Warren and Davis, 1967; Warren, 1971) have repeatedly tried to develop complete energy budgets of growing fish, including measurements of the daily energy expenditure on food processing separate from the requirements of standard metabolism and those generated by the accompanying feeding activity (locomotor requirement and excitement). Such partitioning has its attendant difficulties since the animal does not conveniently separate the components of its total metabolic demands. Some of the metabolic pathways can be recognized if accompanying measurements of CO_2, NH_3, and urea are made. But for the present no such distinctions will be considered—just the increase in oxygen consumption accompanying different levels of ration and temperature.

Using a "mass respirometer" Saunders (1963) was among the first to study the effect of feeding on the metabolic rate of Atlantic cod, *Gadus morhua*, at 10°C. Routine metabolic rates of starving fish (wt = 1 kg) rose from 75 to 112 mg of O_2/kg/hr after feeding and remained at this elevated level for 1 to 2 days, falling gradually back to the fasting, routine rate by the seventh day. More extensive studies on this species were conducted by Edwards *et al.* (1972). Weighed daily rations of plaice fillets were provided to individual cod, and oxygen consumption measured at 12°C in a closed-circuit respirometer. In order to compare temperature relations, a Q_{10} value of 2.5 was applied to convert "basal rates" (determined under light anaesthetic) from 12°C to various levels of seasonal temperature. A maximum increase of 4.7 times the "basal rate" occurred when fed a high ration at 15°C.

The fairly extended metabolic response of cod to feeding was also observed for the subtropical reef fish, aholehole (*Kuhlia sandvicensis*), by Muir and Niimi (1972). When given a single ration of 4.5% body wt, the metabolic rate rose from a routine level of 76 mg of O_2/kg/hr (23°C) to 182 mg of O_2/kg/hr by the fourteenth hr, falling gradually back to the initial level after about 50 hr (Fig. 5A). This contrasts with the excited, more rapid metabolic responses of sockeye salmon, which anticipate a normal feeding time in a daily cycle, raising their metabolic rate from an early morning minimum (0300 hr) of 170 to 370 mg O_2/kg/hr* right at feeding time (0800 hr; Fig. 5B). This peak rate falls as digestion proceeds, reaching a minimum 19 hr later at 15°C.

The above examples serve to illustrate the differences in response between species and the varying approach of investigators to measur-

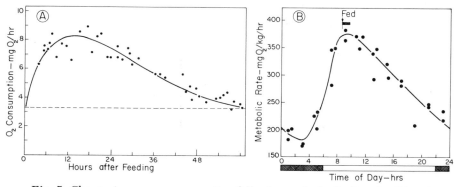

Fig. 5. Change in oxygen consumption following a single feeding for (A) a 44-g aholehole given a 4.7% ration at 23°C (Muir and Niimi, 1972) and (B) a group of sockeye salmon averaging 29 g weight and given a 3% ration at 15°C (Brett and Zala, 1975). Dark period shown cross-hatched in (B).

---

* This relatively high "minimum" was partly due to water velocities of 10–14 cm/sec (1–1.2 length/sec) in the culture tanks.

ing metabolic rates. The computation of daily energy expenditure requires complete information on the hourly metabolic rates over the full 24 hr. To the extent that this was possible to determine, records for eleven species have been compiled in Table II, according to temperature and ration. In some cases only the lowest and highest feeding conditions have been included; not all cases have a true standard metabolic rate determination, and entries under routine metabolic rates are naturally subject to wide variation. Nevertheless some useful generalities can be extracted. Increasing ration has an almost direct, proportional effect on increasing daily metabolic rate. Temperature also increases the feeding metabolic rate both by stepping up the pace and by increasing the daily food intake. The ratio of feeding metabolic rate (F) to standard rate (S) ranges from less than 1.0 for submaintenance rations up to 5.8 for small fish on high rations. The ratio, F/S, for all cases involving a high ration $(n = 8)$ averaged 3.7 ± 1.2 (SD). This may be compared with 1.7 ± 0.4 (SD) for the average ratio of feeding metabolic rate to routine metabolic rate (F/R).

From Winberg's (1956) review, a multiple of 2 times standard rate was given as a near approximation of the metabolic effect of feeding. This is now seen to be understated for cases where accurate estimates of the standard rate have been made, and instead is closer to the ratio for routine rates. However, such early generalities do not take into account the more recent and important finding that the daily metabolic expenditure is highly ration dependent, as may be determined from the entries in Table II. The relation is either a linear function of metabolic rate increasing with ration, up to maximum intake, as has been shown for flatfish, reef fish, and salmon (Fig. 6), or a linear increase at lower levels of ration tapering off at higher rations to form an upper plateau, as in the case of carp (Huisman, 1974).

From a compilation of data available on the feeding metabolic rates of fingerling sockeye salmon a "predictive model"* in the form of isopleths of daily oxygen consumption rate in relation to temperature and ration was computed by Brett (1976b). The response surface (Fig. 7) was developed by use of a set of regression lines, some of which are shown in Fig. 6A. Over the whole surface, from low temperature and low ration to high temperature and high ration, the metabolic rate ranges from 50 to 400 mg of O_2/kg/hr, a factor of 8 times.

The outer encompassing periphery of the response surface is defined by the maximum daily food intake (as measured in Brett *et al.*, 1969), and an inner boundary set by the maintenance ration, below which loss of weight would occur. It applies to sockeye with a mean

---

* No data were available at temperatures above 20°C and very few for 5°C and below. By assuming similar slopes for temperature effects on the *metabolic rate ×
ration* relation, the full response surface was computed (Brett, 1976b).

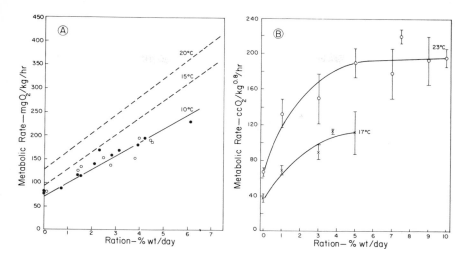

Fig. 6. Response of metabolic rate of three species of fish to different levels of daily ration, at a number of temperatures. (A) Sockeye salmon at 10°C, open circles, and aholehole, *Kuhlia sandvicensis*, at 23°C, solid circles (Muir and Niimi, 1972). Additional lines for sockeye shown for 15° and 20°C (Brett, 1976b). (B) Carp, *Cyprinus carpio*, at 17° and 23°C; variance shown as ±1 SD (Huisman, 1974). Note that metabolic rate in B is expressed as cc $O_2/kg^{0.8}/hr$.

weight of about 20 ± 10 g (range). The strong interaction between temperature and ration is apparent; for a change in ration of 10% of the maximum intake the temperature would have to fall by an average of 3.3°C (13% of the tolerable range) to maintain the same metabolic rate.

Elliot (1969) determined the metabolic rate of hatchery-reared chinook salmon (*Oncorhynchus tshawytscha*) in experimental troughs and raceways. Fingerling fish varied in size from approximately 1 to 25 g and were tested over a temperature range of 6°–16°C. Two conditions were examined: (1) normal hatchery procedure involving a morning cleaning and feeding, followed by an afternoon feeding, and (2) normal activity without feeding, defined as "a state in which oxygen consumption is fairly stable and the fish are in a relatively quiescent state."

The mean metabolic rate for the first condition, over the period of 0800 to 1600 hr, was 277 mg of $O_2/kg/hr$ at 11.7°C, 289 mg of $O_2/kg/hr$ at 12.6°C, and 360 mg of $O_2/kg/hr$ at 15.8°C. This compares quite closely with the daily rate for sockeye of the same weight on maximum ration. Under quiescent, nonfeeding conditions, the mean rate for 6 g and 18 g chinook is also shown in Fig. 7. The rates occur about halfway between the maintenance and maximum values obtained for sockeye. Parallel but less extensive determinations made in raceways were

Table II

Records of Daily Metabolic Rates of Feeding Fish in Relation to Standard and Routine Metabolic Rates[a]

Species	Weight (g)	Temperature (°C)	Ration (% wt)	Metabolic rates (mg of O₂/kg/hr)					Time (hr)	Ratio		Reference and remarks
				Standard	Routine	Peak	Feeding 1	Feeding 2		F/S	F/R	
Gadus morhua												
Atlantic cod	1000	10	(Fed)	—	75	—	112	112	24–48	—	1.5	Saunders (1963); fed on fresh and frozen herring
	1000	15	(Fed)	—	75	—	120	120	24–48	—	1.6	
Cyprinus carpio												
Carp	10 ± 5	10	2	80	—	—	120	116	24	1.5	—	Kausch (1969); standard determined for zero activity; ration determined from *ad libitum* supply
	10 ± 5	15	7	136	—	—	—	230	24	1.7	—	
	10 ± 5	20	6	214	—	—	—	240	24	1.1	—	
Salmo gairdneri												
Rainbow trout	30	15	(High)	—	220	290	264	264	24	—	1.2	Mann (1968); standard rate for sockeye salmon in brackets
	30	15	—	(100)	—	—	—	—	—	(2.6)	—	
Oncorhynchus kisutch												
Coho salmon	3.3	8	11.7	60	240	980	460	350	12	5.8	0.7	Averett (1969) and Warren (1971); ration is percentage fly larvae weight to fish weight; metabolic rates estimated from graphs; seasons were July to September
	3.9	11	5.3	120	220	415	320	270	12	2.2	1.2	
	4.7	14	3.4	170	220	370	350	280	14	1.6	1.3	
	2.6	20	7.7	230	540	1010	900	750	15	3.3	1.4	
Oncorhynchus tshawytscha												
Chinook salmon	18–20	11.7	Ex.	(75)	165	—	277	221	8	(2.9)	1.3	Elliot (1969); ration was excess (Ex.) feeding 2 times/day; standard rates for sockeye salmon; routine rates interpolated; see Fig. 7
	18–20	12.6	Ex.	(80)	180	—	286	228	8	(2.8)	1.3	
	18–20	15.8	Ex.	(96)	220	—	360	287	8	(3.0)	1.3	
	20	12.6	Ex.	(80)	170	366	289	230	8	(2.9)	1.4	

Species												Reference
Cynoglossus sp. (5 species) Sole	9	28	(Low)	105	—	—	(105)	(105)	?	1.0	—	Edwards *et al.* (1971)
	9	28	(Max)	105	—	—	(390)	(390)	?	3.7	—	
Kuhlia sandvicensis Aholehole	71	23	1.4	57	78	118	118	95	72	1.7	1.2	Muir and Niimi (1972); metabolic rate increases proportional to ration; daily feeding rate in first 24 hr approximates 80% of peak rate
	71	23	2.8	57	78	158	158	126	72	2.2	1.6	
	71	23	4.2	57	78	195	195	156	72	2.7	2.0	
	44	23	2.3	62	76	121	121	97	72	1.6	1.3	
	44	23	4.5	62	76	182	182	146	72	2.4	1.9	
Gadus morhua Atlantic cod	500	15 ± 2	(Low)	88	—	—	(88)	(88)	?	1.0	—	Edwards *et al.* (1972); standard = basal under anesthetic
	500	15 ± 2	(Max)	88	—	—	(440)	(440)	?	5.0	—	
Perca fluviatilis Perch	12	14	(Low)	—	175	350	—	146	24	—	0.8	Solomon and Brafield (1972); low ration is approximately maintenance
	12	14	(Max)	—	175	—	—	296	24	—	1.7	
Lepomis macrochirus Bluegill	49	15	Ex.	—	48	93	—	62	24	—	1.3	Pierce and Wissing (1974); ration is excess of mayfly nymphs; nocturnal metabolic rate 26% higher than day
	92	20	Ex.	—	45	62	—	71	24	—	1.6	
	83	25	Ex.	—	89	140	—	120	24	—	1.3	
Cyprinus carpio Carp	31–47	17	10	48	87	—	—	141	24	2.9	1.6	Huisman (1974); standard rate from starving fish
	2–16	23	5	83	156	—	—	243	24	2.9	1.6	
Oncorhynchus nerka Sockeye salmon	10–20	10	4.5	60	100	—	—	190	24	3.2	1.9	Brett (1976b); maximum observed feeding rates for highest rations given; greater rates possible
	10–20	15	6.0	71	125	—	—	315	24	4.4	2.5	
	10–20	20	6.5	120	161	—	—	420	24	3.5	2.6	

a Entries in the column under Feeding 1 have been translated into average *daily* rates in the column under Feeding 2 according to the times during which oxygen consumption rates were measured. This was done by assuming that routine metabolic rates applied for the balance of the 24 hr, if better information was not available in the reference noted. F/S and F/R are the respective ratios of Feeding 2 to standard and routine metabolic rates. Entries in order of year published. Bracketed values were estimated or obtained indirectly.

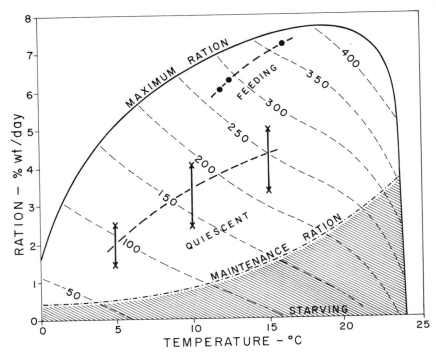

Fig. 7. Isopleths of feeding metabolic rate (mg of O_2/kg/hr) for fingerling sockeye salmon (20 ± 10 g) in relation to ration level and temperature. Ration is expressed in dry weight of food as a percentage of dry weight of fish. The peripheral boundary line of the response surface defines the maximum food intake; the inner maintenance line defines the boundary for positive growth. Confidence limits for any one value are approximately ±20% of the rate depicted. Metabolic rates of juvenile chinook salmon in hatchery troughs are shown (Elliot, 1969). Upper points (●) are for 20 g chinook fed twice a day; lower points (×) relate to the range for fish weighing 6 and 18 g when not fed and considered quiescent under normal activity (routine metabolism). Energy equivalents as in Fig. 4. (From Brett, 1976b.)

about 16% higher than the above trough levels. If this correction is applied to the feeding metabolic rates of the chinook salmon, their 8-hr average falls almost right on the 24-hr average for sockeye.

In the studies on cod (Edwards *et al.*, 1972), evidence for correcting for weight effect by using the approximate power function ($W^{0.8}$) is presented. While this may be applied with some confidence to standard metabolic rates there is no convincing argument developed for applying such an exponent value to the metabolic rates of feeding fish. Two phenomena make the presumption questionable. First, there is nothing to support the physiological assumption that the metabolic cost of digesting a unit of food would be different in small and large fish. Second, maximum food intake (as a proportion of body weight) decreases with increasing size. This association confounds, but does not discredit, the weight exponents for fed fish reported by Saunders (1963); the values for fed fish (0.76–0.83) were lower than those for starved fish (0.79–0.89), which could be a result of the above-noted confounding.

D. Heat Increment

Within the total energy expenditure associated with feeding there is a segment derived from the biochemical transformation of ingested food into a metabolizable, excretable form. A major contribution to this exothermic loss apparently comes from deamination of protein, mostly in the liver (Krebs, 1964; Buttery and Annison, 1973). Rubner (1902) first drew attention to this phenomenon in domestic animals. The nutritional quality of the food affected the magnitude of the resulting *heat increment*. Kleiber (1967) noted that the original German term was wrongly translated as "specific dynamic action" (SDA) instead of "specific dynamic effect." Further, the concept of "specificity," restricting the heat loss to the specific process of deamination, is no longer applicable since similar but smaller energy releases accompany lipid and carbohydrate catabolism. In homeotherms these latter losses can amount to 13% of the caloric content of lipid and 5% for carbohydrate, compared to 30% for protein (Harper, 1971). In consequence, preference is given to the use of the term *heat increment*, as favored by Kleiber (1967).

For fish, the first experimental determinations of metabolic loss ascribed to heat increment were reported by Warren and Davis (1967), presenting previously unpublished data of H. Sethi on cichlids (*Cichlasoma bimaculatum*). Greater elaboration of the phenomenon was provided by Averett (1969), following extensive studies on juvenile coho salmon, *Oncorhynchus kisutch*. Under conditions of varying ration and temperature, heat increments ranging from 4 to 45% of the caloric content of the food were reported, with most values occurring between 9 and 15%. In their reviews, Warren and Davis (1967) and Warren (1971) give special attention to this incremental heat loss. With subsequent insight these authors recognized that the upper values obtained by Averett (1969) were attributable in part to metabolic excitation accompanying feeding. It is absolutely necessary to separate the energy expenditure of excitability and increased activity (occurring in conjunction with food intake) by rigorous experimental technique, otherwise the partitioning is a useless exercise.

Beamish's (1974) carefully conducted study on largemouth bass, *Micropterus salmoides*, is almost a model in this regard of appropriate energy partitioning. In one experiment, conditioned fish were forced to swim in a tunnel respirometer at each of three velocities (1.4, 1.9, and 2.5 body lengths/sec). Continuously monitored oxygen consumption rates ceased to fluctuate after 14–18 hr, as the fish became habituated. A single ration of 4% weight was fed after 24 hr, and monitoring continued until prefeeding metabolic rates were resumed. The diet consisted of freshly thawed shiners. In a second experiment a fixed swimming speed was imposed and four separate rations used (2, 4, 6, and 8% weight). After training the fish to accept food from forceps, a final experiment was conducted to determine the metabolic expenditures associated with the feeding procedure itself. By withholding the food after the fish bit at the forceps, the ancillary excitation of feeding could be assessed and then deducted to obtain an unconfounded estimate of heat increment. This daily energy loss per unit weight did not differ significantly with the weight of the fish (mean, 65 g; range, 9–190 g; temperature, 25°C) but varied as a mean percentage of 14.2% ± 4.2 (SD) of the food energy ingested, or 17.2% of the metabolizable energy (Fig. 8). Peak oxygen consumption rates rose to 80–100% of the active metabolic rate, lasting from 1 to 2 hr. At the highest ration (8% weight) the average total daily metabolic rate can be computed as 339 mg of O_2/kg/hr, partitioned as 35% standard, 41% excitation, and 24% heat increment (standard metabolic rate, from Beamish, 1970).

Three other studies on heat increment have been conducted with sufficient attention to the problem of increased activity and excitability, that bear comparison with the largemouth bass. Aholehole, *Kuhlia sandvicensis*, were fed chopped tuna flesh at two ration levels (2.3 and 4.5% weight) (Muir and Niimi, 1972). Elevated metabolic rates, over baseline estimates for nonfeeding swimming fish, indicated that 76 mg of O_2/g of ration, or approximately 16% of the energy of the food, was appearing as heat increment. Energy expenditure for food utilization in bluegill, *Lepomis macrochirus*, fed to satiation on mayfly nymphs, ranged from 4.8 to 24.4% ($\overline{X} = 12.7 \pm 1.5$%) of the total intake, which was considered by Pierce and Wissing (1974) to be mostly heat increment. Satiation feeding (although with quite variable daily intake) of the sluggish sargassum fish, *Histrio histrio*, resulted in a range of heat increments from 15.2% for 1-g fish to 36.2% for 28-g fish; the overall average was 23.7% (Smith, 1973). This is in agreement with the mean value of 23.3% obtained by Miura *et al.* (1976) for biwamasu salmon, *Oncorhynchus rhodurus*, when fed chopped pieces of fish.

Drawing on the more refined experiments, it would appear that a

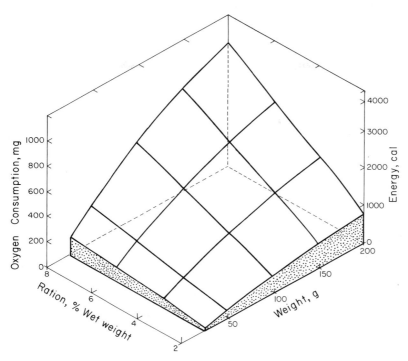

Fig. 8. Heat increment or specific dynamic action of ingested food (minnows) expressed in terms of oxygen consumption and equivalent energy expenditure of largemouth bass, *Micropterus salmoides*, in relation to ration and weight (Beamish, 1974). The results are shown in absolute values. In relative terms, according to the food intake, approximately 14% of the energy of the ration is lost to heat increment, fish weight having no significant effect on this percentage.

loss of about 12–16% of the ingested food energy can be attributed to heat increment. This generalization applies under favorable environmental conditions to ration levels that are well above maintenance. More evidence is necessary for extreme conditions before further conclusions can be drawn. It is safe to say, however, that unless serious attention to the effect of diet composition on heat increment is the aim, it is doubtful if the necessary effort to correctly partition feeding metabolism into its respective components is warranted.

E. Other Metabolic Costs

As was pointed out, determination of energy expenditure of fishes has evolved and expanded in its centers of attention from standard and routine metabolic rates to active and feeding metabolic rates, including partitioning of the heat increment fraction. Earlier generalizations that standard metabolism was about one-quarter active metabolism, and that twice the standard rate was a fair approximation of average daily metabolic costs, are no longer tenable (Ware, 1975). The enquiry now stands at the threshold of breaking the barrier of knowledge on metabolic rates of free-roaming fishes, with all their diversity of behavior and energy-saving strategies. Unlike the captive fish, little of this is open to direct study, with but one notable exception—the spawning migration of nonfeeding fish such as the salmon and shad. Otherwise it is only possible to examine "components" of wild behavior in the walled-off compartments devised by experimenters, not excluding the artificial stream. Although information is most limited and spotty, this area of enquiry is highly deserving of attention. A few cases are presented.

1. AGGRESSION

A chance to record the metabolic rate accompanying aggression in the pumpkinseed, *Lepomis gibbosus,* was afforded while studying group respiratory rates (Brett and Sutherland, 1965). Four fish of similar size had been selected. Within a few hours of being placed in the respirometer, periods of intense contesting for selected areas in the tunnel ensued. Energy expenditure rose to a rate of 180 mg of O_2/kg/hr (0.63 kcal/kg/hr) during peaks of attack and defense, equivalent to about one-half the active metabolic rate of this species.

On another occasion, under very similar circumstances, young sockeye salmon were being exercised as a group (Brett, 1973). Aggression that broke out at the start of one series reached peak metabolic demands of 450 mg of O_2/kg/hr—about one-half the active metabolic rate. Gradually over 6 days this diminished from a daily average of 360 mg of O_2/kg/hr (0.83 kcal/kg/hr) to 180 mg of O_2/kg/hr (0.58 kcal/kg/hr), slightly above the expected rate for 36-hr starved fish forced to swim at 1.25 lengths/sec, the actual imposed velocity in this case.

These two examples serve to illustrate the great energy drain that can occur when unresolved disputes for territory occur, amounting to one-third to one-half the active metabolic rate when contesting is intense.

2. MIGRATION

Observations on the rate of swimming of migrating young sockeye in Babine Lake have been reported by Johnson and Groot (1963), and Groot and Wiley (1965). From a knowledge of the temperature, weight, and speeds of migration at near-surface and subsurface levels the likely energy requirements can be calculated. For the period of about 6 hr a day, when fish are on the move, from 1.3 to 2.1 kcal/kg/hr are expended, that is, average metabolic rates of 380–640 mg of O_2/kg/hr (equivalent rates of 30–50 kcal/kg/day).

Precise records of energy drain exist for adult sockeye salmon (2–3 kg) migrating up the Fraser River, British Columbia (Idler and Clemens, 1959). By means of the method of caloric equivalents described previously, the change in body constituents, from estuary to spawning grounds, was traced by sampling particular races and determining the progressive depletion of body reserves. Over 90% of the fat and, in the female, 50–60% of the total protein may be utilized. This amounts to 2.1 kcal/kg/hr equivalent to 600 mg of O_2/kg/hr, or about three-quarters of the active metabolic rate for fish of this size.

Less precise estimates can be made of the energy expenditure of American shad, *Alosa sapidissima*, which enter the Connecticut River to spawn in the spring (Leggett, 1972). Adults do not normally feed in fresh water. The average weight loss of spent fish ranged from 48 to 55% of the weight prior to entering fresh water. Allowing for the weight of ovaries (13–15%) or testes (8–9%), the average somatic weight loss of females and males was 690 g (45%) and 613 g (48%), respectively, over an 83-day period, that is, 7.8 g/day. Using a mean weight of pre- and postmigrating fish, and allowing for an expected increase in water content of the spent fish, the energy expenditure for this species was about 10 kcal/kg/day.

3. SPAWNING

Although measurements have been made of the metabolic rates of salmon at a stage close to spawning time (Awakura, 1963; Brett, 1965) none are available for the cost of digging and ultimate spawning. The caloric content of the body and gonads, however, has been measured (Brett and Glass, 1973). When these values are applied to the body and ovary weights of female sockeye salmon on the spawning grounds (Stuart Lake race; see Idler and Clemens, 1959) the energy content of the ovary equals about 25% of the remaining body caloric content.

4. DAILY PATTERNS: VERTICAL MIGRATIONS

Many fishes, particularly the mesopelagic fishes of the oceans such as the myctophids or lantern fishes, rise in the evening through several hundred meters of water column, descending at dawn. They feed on plankton near the surface at night, resting during the day at depth.

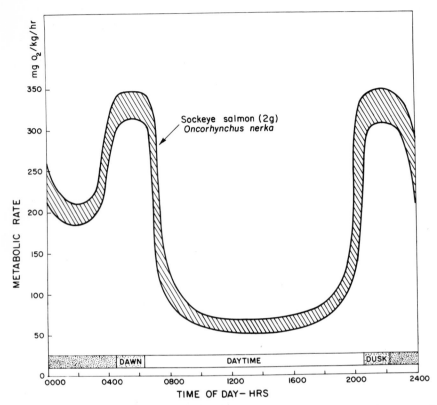

Fig. 9. Schematic representation of daily midsummer metabolic rates of young sockeye salmon in Babine Lake, British Columbia. Feeding occurs almost entirely near the surface at dawn and dusk, with descent from surface temperatures of 16–17°C to deep water temperatures of 5°–6°C (Brett, 1971). The daily mean computes to 183 mg of O_2/kg/hr or 14.2 kcal/kg/day. Energy equivalents as in Fig. 4.

Some clupeoids make similar vertical migrations. Alexander (1972) has analyzed the energetics for fish of different buoyancy, considering the daily energy cost of remaining at depth. Estimates of 30 mg of O_2/kg/hr (over standard metabolism) were calculated for fish with a swimbladder, and 13–20 mg of O_2/kg/hr for fish deriving their buoyancy from a high proportion of lipids. The swimbladder was considered the most economical mechanism for near-surface buoyancy, but not so at considerable depth.

Similar vertical migrations have been reported during the summer for the lake-dwelling stage of young sockeye salmon, involving descent from surface temperatures of 17°C to approximately 5°C at 30 m or deeper. The possibility of an energy-saving device through the mechanism of behavioral thermoregulation, which favorably balances daily metabolic expenditure, was considered by Brett (1971). Reduced metabolic rates at low temperatures would confer an "energy bonus" when food was limiting. Using the information on feeding times and amounts, and applying appropriate metabolic rates according to the temperatures experienced, a daily pattern of energy expenditure was computed (Fig. 9). This shows a remarkable resemblance to the daily metabolic patterns of the little brown bat, *Myotis lucifugus*, which undergoes daily torpor periods apparently to conserve energy (Gordon, 1972, p. 362).

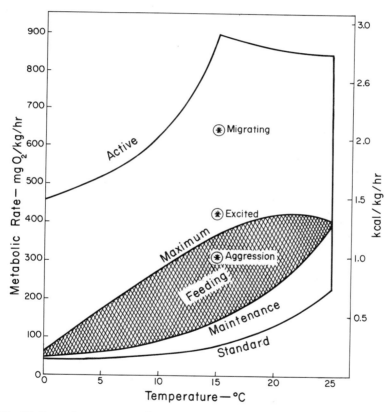

Fig. 10. Rate of energy expenditure of fingerling sockeye salmon showing standard and active metabolic rates (lower and upper lines) in relation to temperature. Metabolic rates associated with feeding (from maintenance to maximum ration) shown cross-hatched. Maximum oxygen consumption rate observed for aggression and excitement, and calculated rate for lake migration shown for a temperature of 15°C.

Table III

Metabolic Rates Associated with Various Activities of Sockeye Salmon[a]

	mg O$_2$/kg/hr	kcal/ kg/hr	kcal/ kg/day	Multiple of standard	Percent-age of active
Standard	75	0.24	5.8	1.0	9.6
Active	790	2.52	60.5	10.5	0.0
Routine	120	0.39	9.3	1.6	15.2
Feeding (maximum)	320	1.03	24.8	4.3	40.5
Feeding (maintenance)	140	0.45	10.8	1.9	17.7
Heat increment (maximum)	110	0.35	8.5	1.5	13.9
Aggression (maximum)	310	1.00	24.0	4.1	39.2
Excitability (maximum)	420	1.36	32.6	5.6	53.2
Migration	640	2.07	49.6	8.5	81.0

[a] Data determined for a 100 g fish at 15°C. See text for basis.

F. Summary

The rate at which energy is expended by fish can be seen to vary greatly according to species, climatic zone, temperature, size, and level of activity. Standard metabolic rates (maintenance metabolism) for fish in the 10- to 100-g size range vary from an average of 0.2 kcal/kg/hr for arctic species at $-1.5°C$ through the temperate species at 0.3 kcal/kg/hr (15°C) to the tropical fishes at 0.5 kcal/kg/hr (26°C). An elevation in temperature of 10°C increases the metabolic rate by a factor of about 2.3 times; this varies considerably between species and over the tolerable temperature range.

Through increased activity, streamlined fast-swimming forms can elevate their metabolic rate on a sustained basis by as much as 10 times, whereas more sluggish, cold-adapted species are confined to a multiple of 2 to 3 times.

A loss of 14% of the energy value of the food occurs as a result of the costs of metabolic processing. This heat increment, along with the activity accompanying feeding, raises the daily metabolic rate of fish that are feeding heavily by a factor of approximately 4 times the standard rate.

The various metabolic rates of sockeye salmon have been compiled to show their relative magnitudes (Fig. 10, Table III). Upstream migration of the spawning adult· salmon is the most costly *sustained* energy expenditure of all the activities recorded.

IV. EXCRETION: RATES OF ENERGY LOSS

A. Composition and Energy Loss in Feces

The nondigestible fraction of a diet, along with sloughed intestinal epithelial cells, mucus, catabolized digestive enzymes, and bacteria constitute the main components of feces. Smith and Thorpe (1976) also include metabolic fecal nitrogen secreted into the gut, which according to Nose (1967) could amount to 5–17% of the total fecal nitrogen. Diet formulation has a primary aim of using selected, highly digestible nutritious components with a balanced energy content such that utilization is maximized and fecal loss reduced (Hastings, 1969). These involve drawing on many available sources of feedstuff that are not necessarily normal to the diet of wild fish. Only the loss from natural foods will be considered here. Various terms are used in the literature to describe the first stage of incorporating food (less feces) into the body, all with the same general meaning: absorption, assimilation, digestibility, and availability. This brings the food to the second step or metabolizable level, a fraction of which is also excreted mostly as soluble nitrogenous wastes. This latter aspect is considered subsequently.

The opportunistic nature of feeding of many fishes presents a wide variety of organisms, which can change with season and alter as the fish grows, affecting fecal loss. An omnivorous stage in early life is frequently followed by a maturing, carnivorous stage. When the zooplankton feeding young ayu, *Plecoglossus altivelis,* leave the sea to enter freshwater they become grazing herbivores utilizing the diatoms and blue–green algae of streams (Kawanabe, 1969). Chitin and cellulose (fiber) are nondigestible fractions of such diets contributing to a more copious fecal production than the highly digestible flesh and bone of a fish diet. Hence the composition of the feces necessarily varies; it also changes significantly from the composition of the food. Since the ash fraction is often 2 to 3 times that of the food (Rosenthal and Paffenhöfer, 1972; Hickling, 1966; Kelso, 1972), the fecal caloric content on a dry weight, unit basis ranges from one-quarter to one-half that of the food (Table IV). Studies on fecal production that only measure dry weight miss this essential energetic assessment, leaving the impression of greater loss than the caloric values reveal.

In many cases the accuracy of fecal measurement leaves much to be desired. Suspended and soluble fractions may be lost. Allowing feces to accumulate over a number of days invites inaccuracy from

Table

Records of the Digestibility of Various Natural Foods in Terms of Percentage

Species	Size (g)	Temperature (°C)	Food — Diet	Food — Calories (kcal/g/dry)	Food — Ration (% weight)
Holacanthus bermudensis Angel fish	200+	19–28	Algae (e.g., *Enteromorpha* sp.)	2.5	Ad lib.
Epinephelus guttatus Red hind	350	19–28	Fish (e.g., *Sardinella* sp.)	—	Excess
Chanos chanos Milkfish	4–80	25–30 (?)	1. Diatoms 2. Algae	— 1.2	20–25 20–25
Cichlasoma bimaculatum Cichlid		20–32 36	Tubifex (*Tubifex* sp.)	5.5 5.5	Ad lib. Ad lib.
Ophiocephalus striatus Ophiocephalus	30	28	Prawn (*Metapenaeus* sp.)	—	3–8
Megalops cyprinoides Megalops	2–150	28	Prawn (*Metapenaeus* sp.)	—	Satiat[n]
Megalops cyprinoides Megalops	50	28	Fish (*Gambusia* sp.)	—	Satiat[n]
Micropterus salmoides Largemouth bass	81	21	Guppy (*Lebistes* sp.)	5.3 — —	0.8 2.7 3.2
Cottus perplexus Reticulate sculpin	1–5	10	Midge larvae (*Chironomus* sp.)	5.3	—
Salmo clarkii Cutthroat trout	2–8	10	Midge larvae (*Chironomus* sp.)	5.3	—
Oncorhynchus kisutch Coho salmon	3	5–17	Fly larvae (*Musca* sp.)	5.6	1–4
Ctenopharyngodon idella Grass carp	20–70	23	Lettuce (*Lactuca* sp.)	3.5	Ad. lib.
Cyprinus carpio Common carp	0.5	25?	Algae (Chlorophyceae)	—	10
Cynoglossus sp. Flatfish	10	28	Polychaete (*Diopatra* sp.)	5.3	Excess
Perca fluviatilis Perch	12	14	Amphipods (*Gammarus* sp.)	4.3	0.3
Gadus morhua Cod	400	14–16	Plaice (*Pleuronectes* sp.)	5.6	Various
Micropterus salmoides Largemouth bass	7–91	25	Shiner (*Notropis* sp.)	6.5	2.8
Stizostedion vitreum Walleye	113–502	12 20	1. Amphipods 2. Crayfish 3. Perch 4. Shiners	4.96	1–5 1–5 1–5 1–5
Blennius pholis Blenny	19	25	Squid	0.97 (wet)	High
Histrio histrio Sargassum fish	1–13 28	21–24	Shrimp (*latreutes* sp.)	—	Satiat[n] Satiat[n]
Ctenopharyngodon idella Grass carp	1100	23	Egeria (*Egeria* sp.)	—	1.2
Salmo gairdneri Rainbow trout	15	5 17 20	Tubifex (*Tubifex* sp.)	— — —	5 5 5
Siganus spinus Rabbitfish	5.4 (cm) 12.9 (cm)	— —	Algae (*Enteromorpha* sp.)	— —	— —
Salmo trutta Brown trout	10–302 10–302	4 20	Amphipod (*Gammarus* sp.)	4.4	Maximum 0.1 Maximum Maximum 0.1 Maximum

[a] Data placed in order of publication. Entries of fecal dry weight for herbivores converted to caloric content (in parentheses) by the ratio obtained by Stanley (1974a).

Loss in the Feces, as either Dry Weight or Caloric Content[a]

Cal-ories (kcal/ g/dry)	Feces Fraction % dry	% kcal	Comments	Reference
—	—	77.7	Feces siphoned daily and water filtered. Food composition was 50–70% carbohydrate	Menzel (1958)
—	—	4.7	Feces siphoned off daily. Fed 3 species of fish	Menzel (1960)
—	50	(42)	Feces occur in compact form. Siphoned without	Tang and Hwang (1966)
	65	(55)	loss of dilution or suspension	
—	—	15	Refers to unpublished results of	Warren and Davis (1967)
		30	H. Sethi	
—	—	9.4	Feces collected after 7–10 days by filtration. Chitin fraction subtracted from food and feces	Pandian (1967a,b)
—	—	8.5	No size effect on absorption	Pandian (1967b)
—	—			
—	—	7.3	Fish fed once per day to satiation	Pandian (1967c)
—	—			
2.8	22	9.8	Three levels of ration used. All feces	Blackburn (1968)
—	19	9.3	filtered and losses accounted for	
—	17	7.3		
—	—	18.1	Food and feces caloric equivalent determined by wet combustion method	Brocksen et al. (1968)
—	—	14.5	Food and feces caloric equivalent determined by wet combustion method	Brocksen et al. (1968)
—	—	15	Nonassimilated food showed no trend with season or temperature	Averett (1969)
—	—	13.2	High fraction of nondigestable cellulose	Fischer (1970)
—	—	15–20	Control fish fed on live plankton—nearly the same as formulated diets	Singh and Bhanot (1970)
1.4	19	5.0	Caloric content of feces only 26% of food	Edwards et al. (1971)
—	—	13	Possibly missed small fraction of feces, collected daily	Solomon and Brafield (1972)
3.1	2.3	1.3	Fed on fillets of plaice. Some feces lost in suspension	Edwards et al. (1972)
—	—	10.4	Feces collected within 30 min	Beamish (1972)
1–3.0	—	17.9	Four diets used. Feces collected within	Kelso (1972)
1–3.0	—	16.5	1 hr by pipette and filtering. Ration level	
1–3.0	—	3.1	had no effect on assimilation; large fish	
1–3.0	—	2.1	showed some decrease	
—	4	4	Two temperatures tested. Feces collected by pipette each morning prior to feeding	Wallace (1973)
—	—	27	Two sizes tested. Water filtered daily. Fecal	Smith (1973)
—	—	18	loss greatest in smaller fish	
—	50	42	Feces siphoned and suspended organics filtered. (Also called white amur)	Stanley (1974a)
—	—	28.2	Three temperatures tested. All organic refuse	Brocksen and Brugge (1974)
—	—	22.1	determined by chemical oxidation	
—	—	15.2		
—	64	(54)	Two sizes tested. Used ^{14}C-labeling of algae to	Bryan (1974)
—	84	(70)	determine gross carbon assimilation	
—	—	15	All particulate feces and dissolved organic	Elliott (1976b)
—	—	29	material analyzed. Six temperatures and	
—	—	11	five levels of ration used, within range	
—	—	21	recorded here	

bacterial action (Iwata, 1970; Smith and Thorpe, 1976), although this would not be of consequence if the total organic content of the water were analyzed (e.g., Davies, 1963, 1964). Blackburn (1968) gently agitated fresh fecal matter for 24 hr and showed that 16.8% of the feces was lost in suspension. Since the caloric content of the feces was low (2.7 kcal/g) this loss represented a 4% unaccounted reduction in original food energy content. It was further demonstrated that different fecal samples from the same fish had significantly different caloric content. A similar sort of loss (18%) in protein nitrogen content of fecal matter of bass was reported by Beamish (1972). Soluble organic material was shown by Elliott (1976a) to amount to only 1–4% of the total fecal energy, depending on ration level (fresh amphipods fed to brown trout).

Various methods of determination exist, beyond simply siphoning off the gross particulate matter and supplementing this with fine-pore filtration. The addition of an inert reference material to the diet (1% chromic oxide) is frequently used in nutritional studies (Furukawa and Tsukahara, 1966). Approximate homogeneity of the indicator chemical in the consumed ration, and a similar rate of passage through the gut, appear to be justifiable assumptions. Bryan (1974), quoting Bakus (1969), supported the method of using ^{14}C-labeled food—in this case taken up by algae through photosynthesis. Gross carbon assimilation was determined in short-term experiments that were not considered to be significantly affected by any respiratory loss of ^{14}C in 6–8 hr from uptake.

For caloric assessment Warren and Davis (1967) substituted and recommended the wet combustion method (Kanzinkin and Tarkovskaya, 1964) to determine the relative energy content of food and feces using powerful oxidants.

In the case of carnivorous fish, feeding on invertebrates with a hard exoskeleton (e.g., amphipods, prawns, midge larvae), the energy loss in the feces was $16.8 \pm 5.9\%$ (SD) $(n = 14;$ Table IV). A diet of soft-bodied invertebrates (e.g., polychaetes, squid—but not tubifex) showed only a 4.5% loss for the two cases recorded. Tubifex, however, was relatively poorly digested, losing $22.1 \pm 7.0\%$ (SD) $(n = 5)$. A piscivorous diet had a mean, nonassimilated fraction of $6.1 \pm 3.4\%$ (SD).

For herbivorous fish, grazing many hours of the day, the digestibility of the selected plant life (algae, grasses) is comparatively low. Among grass carp, *Ctenopharyngodon idella*, Hickling (1966) analyzed the food and fecal ash content, which increased from 6 to 12% following passage through the gut. A fecal loss of 30–40% of the weight of the ingested food was considered normal. Stanley (1974a,b) recorded a 50% dry-weight loss, equivalent to a 42% energy loss for this species (Table IV). Milkfish, *Chanos chanos*, were found to lose from 50 to 65%, depending on the diet (Tang and Hwang, 1967). However, when three species of algae (Chlorophycae) were incorporated in an experimental feed for common carp, the algae *Mougeotia* sp. was highly digestible (Table IV) and provided best growth, with *Sirogonium* sp. being comparatively poor (Singh and Bhanot, 1970).

The above considerations take some account of the dietary quality of the food and the type of fish but not of the other potentially influential factors such as temperature, size of meal, frequency of feeding, weight of fish, and prestarvation. With the exception of Elliott's (1976a) work, none of these factors show significant effects except at extremes. Brocksen and Brugge (1974) reported a significant temperature effect, with higher temperatures conferring a greater efficiency on the assimilation of tubifex by rainbow trout (fecal caloric loss = 28.2% at 5°C, falling to 15.2% at 20°C). However, on a fixed 5% ration, as applied, the *relative* satiation would be greater at 5°C than 20°C possibly influencing the percentage loss. Blackburn (1968) found that the highly efficient digestive process of largemouth bass feeding on guppies improved from 3.2% loss for daily feeding to only 0.8% loss when fed once every 5 days. This is in some contrast to Pandian's (1967b) findings that starvation of 10–40 days had no significant effect on the efficiency of digestion by *Megalops cyprinoides*.

Elliott (1976a) has conducted the most searching analysis to date, determining the separate and combined effects of temperature, ration, and body weight on the fecal energy loss of brown trout feeding on

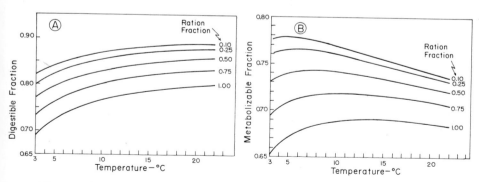

Fig. 11. The fraction of daily energy intake available to brown trout in relation to temperature at different levels of ration (*R*). Ration is expressed as a fraction of the maximum intake (R/R_{max}). (A) is the amount absorbed, the balance being lost in feces; (B) is the amount available after both fecal and nitrogenous excretion losses (ammonia-N and urea-N) have been subtracted. (From Elliott, 1976a.)

Gammarus pulex. When ration was provided as a fixed proportion of the maximum intake (0.1–1.0 R_{max}) for the particular temperature tested (e.g., $10°C$), the percentage energy loss increased exponentially with increasing ration from a mean low of 14% (0.1 R_{max}) to 23% (1.0 R_{max}). Temperature had a significant effect; at a given ration (e.g., R_{max}) fecal loss decreased with increasing temperature, falling from approximately 29% at $4°C$ to 20% at $19°C$ (Fig. 11A). Weight, however, had no significant influence within the range studied (11–250 g).

B. Metabolizable Energy and Nitrogen Excretion Rates

The protein of natural diets (except the hard, cuticle-forming keratins) is usually assimilated to a greater degree than other components of the food. This assumes that the protein is available for assimilation. If the cell wall of plant material is not broken down, then an inability to digest cellulose can obviously prevent access to the protein content. Sunfish, feeding on mealworms, absorbed 96–98% of the protein at all levels of ration (Gerking, 1955). Birkett (1969) obtained an average absorption of 92.3% for plaice, sole, and perch feeding on live invertebrates. Animal proteins in general have been shown to have high assimilation efficiencies, for example, beef heart, 96% (Morgulis, 1918), whitefish meal, 92%, casein, 99% (Nose, 1967), and fish protein concentrate, over 90% (Cowey and Sargent, 1972). This is also true for some plant proteins but with greater variability [e.g., 73–93% for algal diets fed to silver carp (Chiang, 1971) dropping to 54–63% for goldfish (Nose, 1960)]. Converted by enzymatic hydrolysis into their constituent amino acids, dietary protein is usually readily absorbed through the gut wall into the blood stream. However, if protein is in excess of the requirements of the organism, or the constituent amino acids are poorly balanced in relation to growth needs, deamination occurs, with excretion of nitrogen, mainly as ammonia and urea across the gills (Forster and Goldstein, 1969). This dietary or *exogenous* fraction of nitrogenous excretion represents an energy loss from the food, the balance of which is available for metabolic use, either as an energy source or as a growth increment. This balance is the *metabolizable* fraction (see Fig. 18), providing a measure of the *physiological value* of the diet. It can only be determined experimentally. It is not to be confused with the normal loss of energy associated with protein catabolism, which occurs through the excretion of the *energy-containing* end products of ammonia and urea (Krueger *et al.*, 1968). It is this latter, chemically determinable reduction in energy content, that defines a major portion of the remaining physiological value, which for protein is approximately 88% of the gross caloric content (as presented in Section I). The principle of physiological value applies equally well to carbohydrates and lipids as it does to proteins, but the former constituents tend to be fully utilizable once absorbed, mostly as sugars and fatty acids, respectively. There is great importance associated with determining the N excretion rates as a means of assessing the metabolizable protein available to the organism (Smith, 1971). The problem, however, is to distinguish the normal, maintenance waste fraction (endogenous) from the true metabolizable fraction (Iwata, 1970). This can be seen best by summarizing the expressions involved: Consumed N = Assimilated N + Fecal N; Assimilated N = Metabolizable N + Excreted N (mostly exogenous); Metabolizable N = Retained N + Excreted N (mostly endogenous). This follows the sequential steps depicted in Fig. 18 (Birkett, 1969).

Unlike mammals which tend to use a large proportion of carbohydrate and lipid for energy (conserving proteins unless starving), fish normally exploit a portion of their protein stores as an energy source. In the case of the mullet, *Rhinomugil corsula*, Kutty and Mohamed (1975) showed this to amount to 14–15% of the total energy expended (in routine metabolism). The dynamic state of tissue proteins, involving catabolic and anabolic processes in an open system, led to the well-recognized concept of a metabolic pool of amino acids (see discussion in Cowey and Sargent, 1972). Exogenous and endogenous sources are common to this pool. To separate the relative contributions, endogenous nitrogen excretion was defined for domestic animals as "the lowest level of N excretion attained after an empirically defined time interval on a low nitrogen but otherwise complete diet" (Brody, 1945, p. 59). This maintains the normal energy requirements of the animal without involving it in any serious nutritional stress from complete lack of a nitrogen source, (i.e., just maintaining the status quo without sacrificing body protein). In accordance with the need for meeting energy expenditure, Savitz (1969, 1971) force-fed just enough glucose to bluegill, *Lepomis macrochirus*, to meet their estimated maintenance metabolism following the method of Gerking (1955) (see excretion rates in Table V). However, according to Iwata (1970) it is very difficult to determine endogenous nitrogen excretion accurately because it is hard to give an amount of nonprotein food equivalent in calories to the maintenance metabolic requirement of fish.

Records of N excretion of starved animals provide a near approximation of the endogenous fraction, particularly for cold-blooded vertebrates that normally endure prolonged periods of starvation (Stover,

Table V

Daily Rates of *Endogenous* Nitrogen Excretion[a]

Species	Weight, average or range (g)	Temperature (°C)	Salinity (‰)	Starvation (days)	Nitrogen excretion (mg of N/kg/day)			Reference
					Ammonia	Urea	Total	
Lepomis macrochirus	50	26	Fresh	3–8	—	—	154	Gerking (1955)
Lepomis macrochirus	144	26	Fresh	3–8	—	—	95	Gerking (1955)
Leptocottus armatus	165–391	12	30	1	41	14	64	Wood (1958)
Platichthys stellatus	310–335	12	30	1	61	9	73	Wood (1958)
Taeniotoca lateralis	360	12	30	1	14	13	30	Wood (1958)
Salmo gairdneri	129	13	Fresh	6–14	75	35	136	Fromm (1963)
Lepomis macrochirus	10–100	7	Fresh	3	—	—	58	Savitz (1969)
Lepomis macrochirus	10–100	31	Fresh	3	—	—	289	Savitz (1969)
Carassius auratus	1	20	Fresh	7	105	—	135	Iwata (1970)
Carassius auratus	10	20	Fresh	7	75	—	100	Iwata (1970)

Species							Reference
Salmo gairdneri	50–100	13	Fresh	7	—	40	160	Olson and Fromm (1971)
Carassius auratus	1–4	22	Fresh	7	—	71	—	Olson and Fromm (1971)
Lepomis macrochirus	70–90	24	Fresh	3	—	—	114	Savitz (1971)
Lepomis macrochirus	70–90	24	Fresh	7–28	—	—	74	Savitz (1971)
Perca fluviatilis	12	14	Fresh	7+	170	—	—	Solomon and Brafield (1972)
Salmo gairdneri	900	10	Fresh	4	31	—	—	Nightingale (1974)
Salmo gairdneri	900	15	Fresh	4	55	—	—	Nightingale (1974)
Salmo gairdneri	900	20	Fresh	4	85	—	—	Nightingale (1974)
Oncorhynchus nerka	29	15	Fresh	1–22	175	46	221	Brett and Zala (1975)
Salmo gairdneri	30–40	12	Fresh (smolt)	10	—	—	92	Smith and Thorpe (1976)
Salmo gairdneri	30–40	12	Fresh (postsmolt)	10	—	—	60	Smith and Thorpe (1976)
Salmo gairdneri	30–40	12	Salt (smolt)	10	—	—	97	Smith and Thorpe (1976)
Salmo gairdneri	30–40	12	Salt (postsmolt)	10	—	—	93	Smith and Thorpe (1976)
Dicentrarchus labrax	5–235	16–20	Salt	7–10	72	12	84	Guérin-Ancey (1976)

[a] Listed in order of publishing date.

1967). Error may be minimized by extrapolating back to zero time (postdigestive state), where it is obvious that any decay in maintenance metabolism can be considered as insignificant. The best method is undoubtedly the use of ^{15}N-labeled protein, tracing the fraction that is excreted before any new dietary protein has been made available for energy.

Gerking (1955) demonstrated a weight-dependent relation for endogenous N excretion (approx. $W^{0.54}$), following the well-known relation of decreasing rate of metabolism with increasing size. A considerably higher value for the weight exponent ($W^{0.9}$) was reported for starving crucian carp (Iwata, 1970). It is undoubtedly important to take size into consideration when comparing various endogenous rates among fishes; however, some doubt has been cast on the correct exponent (Savitz, 1969; Iwata, 1970), and it can be expected that considerable species variation will occur, as indicated in the compilation of Table V. Davies (1963) determined the metabolizable fraction of live, white worms (*Enchytraeus albidus*) when fed to goldfish, *Carrasius auratus*, at various ration levels. The metabolizable portion of the food varied quite significantly, ranging from 72% at a ration of 1.5% dry weight/day to 86% for a ration of 4.5%.

Few studies on fish have been performed where the exogenous and endogenous fractions of nitrogen metabolism could be separated. By tracing the hourly pattern of ammonia and urea excretion of fingerling sockeye salmon, following a single daily meal (Fig. 12), it was shown that a strong pulse of ammonia excretion occurred, peaking 4–4.5 hr after the start of feeding (Brett and Zala, 1975; see also Durbin, 1976). A formulated diet based on fish meal was used (Oregon moist pellets). From comparison with the endogenous excretion of starving "control" fish, the exogenous fraction could be computed. This amounted to approximately 15 mg of N/kg/day, or 27% of the nitrogenous intake. Assuming that 97% of the dietary protein would normally be absorbed in the gut, the metabolizable fraction would only amount to 70% of the intake, that is, a loss of 3% in feces and 27% excreted. This example serves to illustrate the areas of potential loss not usually determined for the components of fish food. The complication of separating the sources of the excretory products is further involved by posing the question of identifying the metabolic source for the accompanying heat increment, which in fish could be derived in part from protein catabolism.

The daily fluctuations of ammonia and urea excretion for heavily feeding fish have been studied in at least two hatcheries. Quite irregular patterns in the relative proportions of these nitrogenous products were recorded by Burrows (1964) for chinook and coho salmon. Stock-

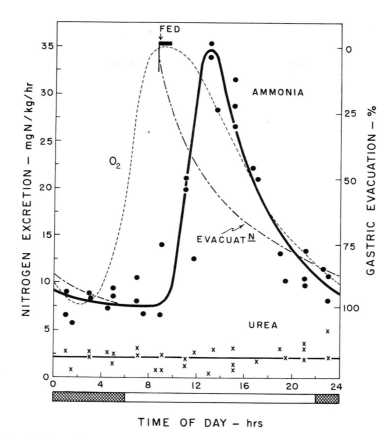

Fig. 12. Diurnal variation in rate of ammonia (●) and urea (×) excretion of fingerling sockeye salmon (*Oncorhynchus nerka*). Ration = 3% weight/day; temperature = 15°C; photoperiod = 16 hr with dark period shown cross-hatched; freshwater pH = 7.0–6.7. For comparison the associated oxygen consumption rate and expected gastric evacuation rate have been included, scaled to the peak and baseline excretion of ammonia. Note that the oxygen consumption rates, which peaked at 370 mg of O_2/kg/hr, are about 10 times the nitrogen excretion rates. (From Brett and Zala, 1975.)

ing density, handling stress, and temperature levels were factors contributing to the variability. When ammonia did predominate, as in the sockeye experiments, it tended to occur over 14 hr of the day, with a peak at 1600 hr. In the studies of McLean and Fraser (1974) on hatchery-reared coho, ammonia nitrogen accounted for over 60% of the total nitrogen excreted, rising to over 90% on some days (at 40 mg of N/kg/hr). With fish of 12–15 g weight, on a dawn to dusk feeding regime, a daily pulse in ammonia excretion occurred, reaching a peak 9 hr after the start of feeding.

When applied to the maintenance ration, Brett (1976a) further attempted to account for the energy losses occurring over the full range of temperature tolerance (approximately 0°–25°C), using Phillips' (1969) physiological values for energy equivalents (Fig. 13). Despite the relatively low caloric value so computed (3.92 kcal/g, i.e., a factor of 0.72 times the total energy content) the appropriate feeding metabolic rate did not equate to the physiological value at temperatures

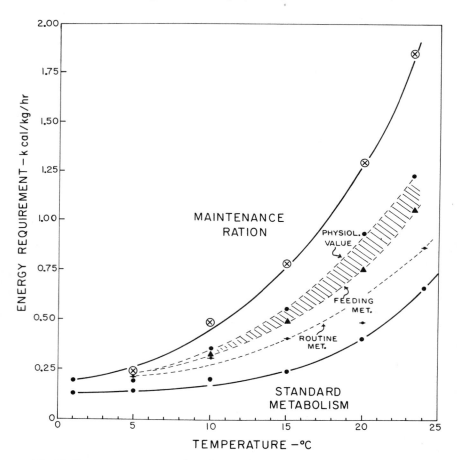

Fig. 13. Energy requirement of yearling sockeye for standard metabolism and maintenance ration, over the tolerable range of temperature. The metabolic rate associated with feeding at the maintenance level (F_{met}) is shown together with the amount normally expended in spontaneous activity (routine metabolism). The net energy available from the maintenance ration is shown as the physiological value. The shaded area is unaccounted energy, probably attributable to an increasing inefficiency of food conversion (i.e., a reduced physiological value at higher temperatures). (From Brett, 1976a.)

above 10°C. The unaccounted caloric difference, amounting to as much as 18% at 20°C, was considered to be the result of an increasing inefficiency of food conversion at higher temperatures, indicating lower physiological values than the ones applied.

While determining the effect of temperature, ration, and size on the energy loss in the feces of brown trout, Elliott (1976a) also examined the daily energy loss from excretion of nitrogenous wastes—ammonia and urea. The proportion of the daily food intake (gammarids) lost from these sources increased as temperature increased and the level of intake decreased (i.e., just the opposite to the relation for fecal loss). Thus, nitrogenous excretory products accounted for about 4–6% loss in energy at 4°C, and from 11 to 15% loss at 20°C. The greater percentage loss at each temperature occurred at the lowest intake ($0.1 R_{max}$). Endogenous excretory levels were not distinguished from exogenous sources, although nitrogen excretion rates were shown to decrease over a 4- to 6-day period. The fractional loss was independent of weight (4- to 300-g range). The combined energy losses (fecal plus soluble nitrogenous) resulted in a range from 22 to 35% on a low ration, and from 26 to 31% on a high ration (Fig. 11B).

By experimenting with different types of natural diets, Elliott (1976a) further showed that although there were some significant differences in the comparative energy losses according to pathway, the total energy losses from each diet were very similar. Consequently, the fraction of energy available for growth and metabolism remained remarkably constant at each ration level and temperature.

C. Summary

From the foregoing it can be seen that there is a wide range in the value of food energy available for metabolism and growth. Fecal energy alone accounted for over 50% in some herbivorous fish, leading to the belief that the metabolizable fraction could easily be as low as 40%. For omnivorous, invertebrate-feeding fish, such as the brown trout and perch, a range of 25–30% loss would apply for most combinations of temperature and ration. Carnivorous fish, feeding on other fish, could be expected to approach a 20% loss under favorable environmental conditions—the only category supporting the former generalization of Winberg (1956) on the average metabolizable fraction of the diet. Least information is available on the energy loss from nitrogenous excretion. An erroneous assumption by Winberg (1956), that the energy value of the chief excretory product (ammonia) was negligible, has contributed in the past to overrating the potential net energy from the food.

These cases have been followed through indicating how gaps in knowledge of fish energetics, frequently riding on assumptions from mammalian physiology, can lead to potential errors.

V. GROWTH: RATES OF ENERGY GAIN AND FOOD CONVERSION

The net energy derived from the food (metabolizable minus heat increment) is that portion available for all additional forms of metabolism and activity, of which the three major components—swimming, maintenance, and growth—make the greatest demands. Growth has the lowest immediate priority, but in the long run growth and reproduction dictate the species survival. The need to collect and convert enough food to meet the growth requirements is almost always pressing in nature. Records of large fish demonstrate the rare capacity afforded to particular members of populations (World Record Board, National Fresh Water Fishing Hall of Fame, Box 33, Hayward, Wisconsin 54843).

By developing the relation between experimentally controlled rations (of high diet quality) and growth rate, the full range of capacity to grown can be defined. The characteristics of this growth–ration relation (the GR curve) and its derivative, the conversion-efficiency relation (the KR curve), are dealt with in Chapter 10 (see Fig. 4 of that chapter). It is sufficient here to present a few of the main findings, and extend these to elaborate on the principles of energy conversion and the effects of environmental factors.

A. Ration and Fish Size Relations

In almost all cases recorded, the higher the ration the greater the growth rate. Only carp, *Cyprinus carpio*, has been shown to experience some rate reduction at maximum ration (Huisman, 1974, 1976). This case serves to illustrate an extreme of the general "glutton effect" of high rations, in which the efficiency of food utilization is decidedly less than at some submaximal point defining the optimum ration (R_{opt}). The position of R_{opt} in the growth–ration curve may range from a relatively low to a relatively high ration according to species and environmental conditions, especially temperature effect (e.g., Elliott, 1976b). Two cases of extreme difference are shown in Fig. 14A. A

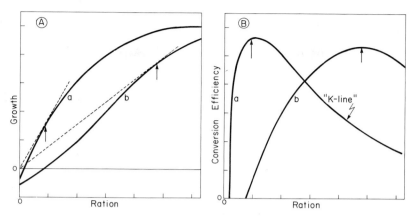

Fig. 14. (A) Variation in form of the relation between growth rate (G) and ration (R), both expressed in terms of % dry body weight/day. Broken lines are the tangents to the curves a and b with arrows indicating the defined position of optimum ration. (B) Corresponding conversion efficiency curves $(G/R \times 100\%)$ with maximum efficiency (K_{max}) indicated by arrows also corresponding to the positions in A for R_{opt}. The segment of curve a in B, from optimum to maximum ration, is the K-line of Paloheimo and Dickie (1966), shown to be linearized by an exponential transformation.

point of major significance is the consequence to the conversion-efficiency relation illustrated in Fig. 14B. In the first case (curve a) a long and well-defined reduction in efficiency accompanies increasing ration beyond R_{opt}, whereas in the second case (curve b) there may be little to no reduction of efficiency at any point as ration increases. This variation of pattern accounts for much of the controversy over how efficiency changes with ration, the so-called "K-line" of Paloheimo and Dickie (1966) being applicable to those cases resembling curves of type a (Fig. 14). K refers to the gross efficiency (K_1) used by Ivlev (1945). K was shown to decrease exponentially with increasing ration.

Relative growth rate* is greatest at the smallest size (see Chapter 10, Figs. 22 and 29). Some exception to this principle may occur at first feeding of larval or fry stages, where effective feeding behavior is being learnt and the digestive system may not be fully differentiated or completely free of yolk. Once maximum feeding is established, growth rates can easily reach 8–10% body weight/day, as in young salmon (see Chapter 10), requiring feeding rates of 25–30% per day. Winberg (1956) compiled tables of average daily gain in weight during initial periods of development, giving examples for bream and carp

* Equivalent to specific or instantaneous growth rate. See Chapter 11.

where the growth rate exceeded 35% per day. Although for very brief periods, it is apparent that some of these early forms would have to consume more than their body weight in a day.

On maximum ration at a young stage the rate at which fish can accumulate body calories exceeds the rate at which metabolism expends calories. Both rates are relatively high initially. With increasing size and age these two relative rates decrease but at different declining slopes such that eventually the rate of metabolic energy expenditure is considerably higher than the associated capacity to deposit energy (Fig. 15). This fundamental circumstance is a basic cause for decreasing conversion efficiency accompanying increasing size, which would approach zero as size reached an upper limit. The relation is apparent in the studies of Kinne (1960) on *Cyprinodon macularius*, of Smith (1973) on *Histrio histrio*, and for salmonids (see Chapter 10, Fig. 22). Huisman (1974) has shown that conversion efficiency declines exponentially with increasing weight in carp (35–210 g). Such a decline with size could easily confound the K-line relation of Paloheimo

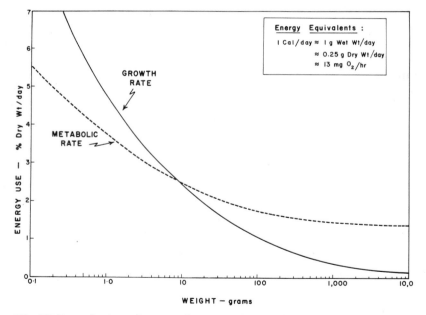

Fig. 15. General relation between the energy deposited in growth and that expended in total metabolism according to size (sockeye salmon). The ratio of the energy involved in the two systems (growth : metabolism) changes from >1.0 at approximately 10 g to <1.0 above 10 g, accounting largely for the decrease in food conversion efficiency accompanying increasing size. (From Brett, 1970a.)

and Dickie (1966) where increased ration bringing decreased efficiency could well be associated with increased fish size.

B. Gross and Net Conversion Efficiencies

Possibly the most meaningful and simplest indicator of adequacy of diet, ration level, state of health, and environmental suitability for an organism is the capacity to convert food into flesh: the gross conversion efficiency,

$$K_1 = (G/R) \times 100\%$$

Growth rate (G) and ration (R) may be expressed in terms of wet weight, dry weight, or caloric content. Wet weight is only suitable when the moisture content of both the food and the fish are approximately the same. Determinations based on dry weight will vary from those in calories according to whether the level of ration (if of constant composition) induces weight gains that increase or decrease the relative fat content of the fish (e.g., Huisman, 1974).

Net conversion efficiency provides a measure of the capacity to convert that fraction of the ration in excess of the maintenance level into flesh,

$$K_2 = (G/R - R_{maint}) \times 100\%$$

Its measurement will depend greatly on the accuracy of determining the maintenance ration—not always accomplished too successfully. Its usefulness is unquestionable where partitioning of energy or nutritional adequacy are under investigation. Otherwise this demanding refinement hardly adds much to the acknowledged value of knowing the gross efficiency.

Many growth studies include determinations of gross conversion efficiencies. For juvenile fish up to maturity, gross efficiencies mainly range between 10 and 25%, depending on size, age, diet, ration, and environmental conditions (Pandian, 1967a,b; Yoshida, 1970; Chesney and Estevez, 1976). Highest efficiencies are associated with conversion of yolk in early development, for example, 65–70% in salmonids (Marr, 1966) and 75–80% in sardine (Lasker, 1962). Maximum efficiencies for post embryonic stages were earlier thought to have an upper ceiling of about 35% (Ivlev, 1945). However, in caloric equivalents considerably higher values have been recorded, for example, 55% for 1–2 gm coho salmon feeding on live fly larvae at 8°–14°C (Averett, 1969), 52% for immature cichlids feeding on oligochaetes at 28°C (Warren and Davis, 1967), and 60% for 3.8 g mackerel feeding on chopped anchovy at 15°C (Hatanaka and Takahashi, 1956). Herbivorous fishes are characteristically on the low side of efficiency (10–20%) because of the large nondigested fraction of their normal diet (Welch, 1968). However, Stanley (1974a,b) has reported that grass carp feeding on green algae reach an efficiency of 40% or more by virtue of a compensating low metabolic rate (see also Stanley and Jones, 1976). Using a radioactive tracer, ^{137}Cs, Kevern (1966) was able to trace the annual efficiency of a population of carp, *Cyprinus carpio*, (150–200 g) feeding on algae and detritus. On a caloric basis, and a mean ration of 3.9% wet weight day, an efficiency of only 6.5% was determined.

C. Nitrogen Retention

Although growth must be considered as a net increase in any of the body constituents, not excluding water, continued elaboration of tissues cannot proceed in the absence of an adequate supply of protein. In consequence nitrogen retention (protein synthesis) rather than carbon or caloric retention has been considered as the fundamental unit of growth (Brody, 1945; Maynard and Lousli, 1962). There is an optimum nitrogen content of the diet, subject in particular to age and temperature effects, ranging from 35 to 55% of the diet composition (see Chapter 1).

Such an approach to growth was adopted by Gerking (1971) in a study of the effects of nitrogen consumption and body weight on the N retention of bluegill sunfish, *Lepomis macrochirus*. His findings are in keeping with the basic principles of growth and conversion efficiency already set forth, with the modifier that protein would be required to a lesser extent than calories as size and age increased, diminishing to a state of nitrogen maintenance for an old fish mainly metabolizing for energy and not for growth. With an increase in nitrogen consumption, the rate of N retention of sunfish increased linearly up to a maximum for the range of rations provided, following a conversion efficiency curve similar to that in Fig. 14B, type b. Maximum protein conversion efficiency decreased from 39% for a 14-g fish to 10% at 85 g.

Iwata's (1970) detailed studies on nitrogen retention in relation to absorption in the carp, *Carassius auratus cuvieri*, showed that efficiency of retention increased during the latter half of a 20-day experiment. The improvement was ascribed to recovery of the fish from the primary stress of handling and effects of anaesthetic. From the data obtained during the tenth to twentieth day of feeding it is apparent that the carp exhibited a relation for gross conversion efficiency of protein in relation to nitrogen absorption that falls in an intermediary position between the a- and b-type curves of Fig. 14A,B. Maximum gross efficiency was approximately 35%. This appears to be in agreement with Birkett's (1969) findings for plaice, sole, and perch which ranged from 27.5 to 49.0% of the "gross efficiency."*

D. Environmental Factors

The manipulation of environmental conditions has been shown to provide favorable combinations that significantly improve both growth and conversion efficiency. Growth rate of 20 g sockeye at an optimum temperature of 15°C in freshwater was increased from 1.4 to 2.4% per day by using isosmotic salinity, an increasing photoperiod, cover from direct light, and low water velocity (Brett and Sutherland, 1970). The salmon not only increased their daily food intake but also showed improved conversion efficiency at any given level of ration. As presented in Chapter 10, abiotic factors can be classified in terms of how they influence any activity, a classification first elaborated by Fry (1947) and subsequently extended (Fry, 1971). Thus, temperature governs the rates of metabolic reactions (Controlling Factor), salinity imposes a metabolic load on internal regulation (Masking Factor), daily light cycles affect endocrine activity (Directive Factor), and oxygen, size, and ration can each restrict growth through one mechanism or another (Limiting Factor). The consequences of these factors will be considered as they relate to conversion efficiency in a manner similar to that developed for the growth–ration curves (see Chapter 10, Figs. 18 and 26).

1. TEMPERATURE

In three species receiving most attention (coho, Averett, 1969; sockeye, Brett et al., 1969; trout, Elliott, 1975a,b), at temperatures below 10°C conversion efficiency rises most rapidly from the base level (maintenance ration) reaching a peak efficiency at an intermediate ration (Fig. 16). This relation changes progressively with rising temperature such that at 17°C and above, the highest efficiency occurs near the maximum ration. The shift in the relation of conversion effi-

---

* Birkett's (1969) paper should be consulted for his use of "gross efficiency," which by definition would be equivalent to net efficiency as used in this text.

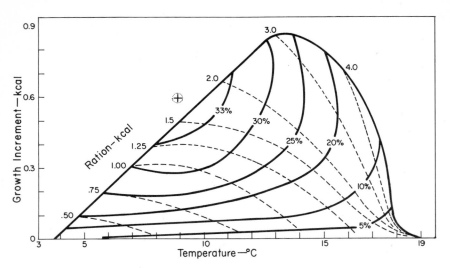

Fig. 16. The relation of temperature to daily growth increment (broken lines) according to fixed ration (shown for each curve), measured in terms of energy values for brown trout of 50 g initial weight. Gross conversion efficiencies (%) drawn as isopleths (solid lines) over the growth curves (Elliott, 1976b). The probable center of the isopleth configuration has been added (circled +) for comparison with Fig. 27B, Chapter 10. The interaction axis of the isopleths would run through this center approximately parallel to the outer diagonal boundary line (set by the maximum food intake).

ciency to ration with increasing temperature follows a pattern of change not unlike that from curve a to curve b depicted in Fig. 14B. Over the complete range of tolerable temperatures maximum efficiency occurred between 5 and 10°C (coho), at 9°C (trout), and at 11°C (sockeye).

In cases where growth–ration curves are established for a wide range of temperatures, the interaction between ration and temperature on growth rate can be determined, from which the full range of conversion efficiencies may be derived. Two such examples are shown, for trout (Fig. 16) and sockeye (Chapter 10, Fig. 27B). Although the central axes of the isopleths are not particularly different, the epicenter for the efficiency isopleths of trout lies just outside the boundary ($R_{\max}$) of the figure, in contrast to that for the sockeye where it lies well inside the configuration (at 11.5°C and 4.0% per day ration). The sockeye were more prone to feed well at low temperatures, accounting for much of the difference. Brown trout did not feed much above the maintenance level when at the lowest (3.8°C) and the highest (19.5°C) temperatures studied.

2. SALINITY

When exposed to increasing salinity the capacity to regulate ionic balance of stenohaline freshwater fish decreases rapidly above the blood isosmotic level of $10 \pm 2\%_{00}$ (see Chapter 10, Fig. 13). The increasing ionic load of this Masking Factor becomes intolerable. This is reflected in the fall of conversion efficiency reported by Shaw *et al.* (1975) for Atlantic salmon *parr*, which decreased from a maximum of 22% in fresh water to 7% in salt water (Chapter 10, Fig. 14B). By contrast the euryhaline pupfish, *Cyprinodon macularius*, showed greatest efficiency at $15\%_{00}$ with a maximum in excess of 30% conversion when on a high ration at temperatures from 17° to 22°C (Kinne, 1960) (see Chapter 10, Fig. 28B).

3. PHOTOPERIOD

The effect of this Directive Factor on conversion efficiency has received relatively little attention. By stimulating the production of growth hormone (Chapter 9) improved efficiency would likely follow the path of change noted earlier for sockeye salmon, following environmental manipulation (Brett and Sutherland, 1970). This conjecture is supported by the findings of Gross *et al.* (1965) for sunfish, *Lepomis cyanellus*, which had greatest consumption and maximum conversion efficiency (48%) after exposure to an increasing photoperiod of 8–16 hr light per day.

4. OXYGEN

Factors such as oxygen concentration have a predictable effect on conversion efficiency by restricting the development of the normal GR curve, as discussed in Chapter 10. Below the critical O_2 level of 5 ppm, growth becomes dependent on oxygen concentration for many fish, for example, largemouth bass, *Micropterus salmoides* (Stewart *et al.*, 1967) (see Chapter 10, Fig. 16). A precipitous decline from an average of 27% efficiency above 5 ppm O_2 occurred among the bass, falling to 0% at approximately 2 ppm of O_2.

E. Summary Configurations

The above environmental relations have been depicted in general form in Fig. 17, and bear comparison with the recapitulation of growth × ration (GR) curves presented in Chapter 10, Fig. 18. Maximum

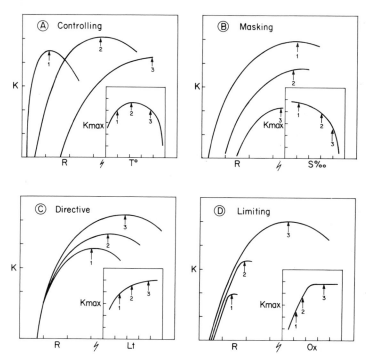

Fig. 17. Recapitulation of gross conversion efficiency × ration curves (*KR*) in relation to abiotic entities illustrating the basic forms for: (A) *temperature* (T°), a Controlling Factor; (B) *salinity* (S‰), a Masking Factor; (C) *light* (L$_t$ = static photoperiods), a Directive Factor; and (D) *oxygen* (O$_x$), a Limiting Factor. Each *KR* curve is shown at three levels of each abiotic entity, with an arrow for the position of maximum efficiency (*K*$_{max}$) associated with optimum ration (*R*$_{opt}$). The insert box shows the path of change of *K*$_{max}$ as the abiotic entity increases over the tolerable range. The intercept on the x-axis is defined by *R*$_{maint}$. *K* is represented in terms of % and *R* as % dry body weight/day. Compare with Chapter 10, Fig. 18.

efficiency (*K*$_{max}$) usually occurs at an intermediate ration (*R*$_{opt}$) below *R*$_{max}$, though not always, as is apparent for extremes of temperature, salinity, and limiting levels of oxygen. The change in efficiency with increasing salinity is depicted for a freshwater fish; the relation would be reversed for a marine species, with maximum efficiency occurring at relatively high salinities, that is, the curve in "B-box" would be reversed. Least information is available for the influence of photoperiod. Limiting effects of reduced oxygen concentration probably impose some increased maintenance ration through increased ventilation; this speculation is shown by the separation of the curves in D, which would otherwise be superimposed.

VI. ENERGY BUDGETS

A. Balanced Equations

Since biological systems conform to the laws of thermodynamics, all the energy ingested (I) by a fish must turn up in one form or another through metabolism (M), growth (G) and excretion (E), where $I = M + G + E$.

As presented in Section III, total metabolism (M) may be divided into a variety of levels, namely, standard metabolism (M_S), routine metabolism (M_R), feeding metabolism (M_F) and active metabolism (M_A). If the *additional* energy demand (scope) explicit in each is considered to be additive, then

$$M = M_S + aM_{R-S} + bM_{F-S} + cM_{A-S}$$

where the constants (a, b, and c) apply to estimates of the fraction of time each day that routine, feeding, and active metabolism occur. Determination of any subcomponent helps in an estimate of M. In a similar manner growth (G, see Section V) may include a subcomponent related to production of gametes (G_G) to add to somatic or general body growth (G_S), where $G = G_G + G_S$. Finally, excretion (E, see Section IV) has subcomponents mainly as feces (E_F), urea and ammonia (E_U), and a small amount of mucus and sloughed epidermal cells from the skin (E_S). When all assembled,

$$I = (M_S + aM_{R-S} + bM_{F-S} + cM_{A-S}) + (G_S + G_G) + (E_F + E_U + E_S)$$

Each of these subcomponents has already been considered (except E_S), including the range of values and variations that can be expected, particularly from the effects of environmental factors. Because they form a balanced equation, determinations (or well-founded estimates) of any three of the main components provide a value for the fourth by difference. However, all errors in assessing the three components obviously become a pooled error in the fourth, unless by chance they cancel out.

B. Carnivores

Following the major review of metabolism and growth of carnivorous fish by Winberg (1956), and greatly influenced by Ivlev's (1939) pioneering studies in fish energetics, a general value of 20% of the food intake was assigned to excretion (E). A simplified form of the general equation $(0.8^*I = M + G)$ was consequently accepted as applicable in many cases. Coupled with measurements of growth (G) and estimating total metabolism† (M), the amount of food consumed (I) was deduced by difference (Mann, 1965); or alternatively by knowing I and G, M could be deduced (Paloheimo and Dickie, 1966). Efforts have been made to measure most of the components and subcomponents directly, notably by Warren and associates (Warren, 1971). In addition, two of the most complete studies are those by Solomon and Brafield (1972) on perch, and by Elliott (1976b) on brown trout. These important contributions along with a selection of others have been assembled in Table VI. Although some degree of arbitrary selection has been made, they all represent advanced studies in the field and can be analyzed for an assessment of energy budgets. It must be noted that these are almost entirely for carnivorous fish, feeding mostly on stream invertebrates and fish (perhaps some distinction should be made between insectivores and carnivores, as in mammals).

While noting the uniqueness of species and their individualistic response to different diets and environmental conditions, some average relations have been determined. Using only those energy budgets where the fish were fed a ration well above maintenance, and where temperatures were not extreme (i.e., cases where conversion efficiency was not less than 20%), and excluding the one herbivorous fish, a mean budget $(n = 15)$ with 95% confidence limits computes to

$$99.5\,I = (43.8 \pm 7.0)M + (28.9 \pm 5.7)G + (26.8 \pm 3.2)E$$

or balances to

$$100\,I = (44 \pm 7)M + (29 \pm 6)G + (27 \pm 3)E$$

This general expression is different from the one developed by Winberg (1956) (Table VI), being reduced in the relative metabolic fraction (-16%) and greater in both the fractions for growth $(+9\%)$ and excretion $(+7\%)$. This does not mean that the absolute metabolic rates were lower; indeed, there is reason to believe these are generally higher (Solomon and Brafield, 1972), which in turn would mean a greater food requirement. As stated previously, Winberg (1956) did not ascribe any caloric value to nonfecal excretion, an oversight that accounts for most of the error in that component, necessitating a com-

---

* Here expressed as a fraction, as used by Winberg (1956), rather than as a percentage (80%) as used in the equations below.

† Frequently by relating established values of standard metabolism (M_S) with estimates of feeding (M_F) and activity effects (M_A), for the weight and temperatures concerned.

pensatory change in the others. Regardless of this change, it should be noted that the variability is least for the excretion fraction, a fact which tends to apply over a wide range of rations and natural diets (Elliott, 1976b). Hence, changes in rations affecting growth rate will likely have greater effect on the relative metabolic fraction. Turning to the growth fraction, a value of $29 \pm 6\%$ is indicative of the conversion efficiency of young, fast-growing, well-fed fish. Accepting the fairly consistent provision of 59% net energy from the food (Fig. 18, derived from Table VI), growth and metabolism are competing components for this remaining energy. Increased metabolism will inevitably result in a relative decrease in growth, particularly when food is limiting or the fish cannot compensate by increasing its food intake. Further, it is

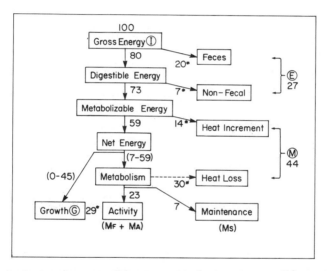

Fig. 18. Average partitioning of dietary energy for a carnivorous fish. A ration of 100 calories is equivalent to about 2% dry weight/day for a 1 kg fish. Nonfecal energy is mostly that excreted as ammonia and urea. Physiological value is equal to metabolizable energy. Circled lettering is according to equations for energy budgets (Table VI): $I = M + G + E$, where I, rate of ingestion; M, metabolic rate; G, growth rate; E, excretion rate. Amounts marked with asterisk total to 100; figures in brackets indicate the possible range of net energy distributed to metabolism and growth. Thus, of the net energy (59 cal) a minimum of 7 cal is required for maintenance of a nonfeeding fish (standard metabolism); alternatively all 59 cal could be required by a very active fish for metabolism alone. A growing fish would require some increase in the metabolic rate over the standard rate, here shown as reducing the upper limit of energy available for growth by an additional 7 cal [i.e., $59 - 14(M_F) = 45$]. Note that Metabolism gives rise to Heat Loss (30 cal) which is the sum of Maintenance (7 cal) and Activity (23 cal), which in turn are coupled with Heat Increment (14 cal) to make up the total metabolism ($M = 44$ cal).

Table

Energy Budgets Listed in Order of Increasing Assumptions

Species	Start-ing weight (g)	Aver-age tem-pera-ture (°C)	Diet	In-gest[a] I (kcal)	Energy budget			
					I (%)	$= M$ (%)	$+ G$ (%)	$+ E$ (%)
Perca fluviatilis	12.0	14	*Gammarus*	14.1	98.7*	54.4	20.5	23.9
Perch	12.8	14	*Gammarus*	9.8	103.3*	59.0	20.4	23.9
	18.7	14	*Gammarus*	6.8	105.8	74.4	6.4	25.0
Salmo trutta	50.1	3.8	*Gammarus*	0.28	100	(62)	5	33
Brown trout	50.7	9.5	*Gammarus*	1.60	100*	(36)	33	31
	50.1	15.0	*Gammarus*	3.84	100*	(49)	20	31
	50.6	17.8	*Gammarus*	5.14	100	(66)	3	31
Salmo gairdnerii	3500	15	Prepared	112	99.2*	25.9	50.1	23.2
Rainbow trout	3500	15	Prepared	102	99.9*	22.5	48.0	29.4
Histro histro	1.0	7	Shrimp	1.2	100*	33.4	34.8	31.8
Sargasum fish	13			9.5	100*	41.9	24.0	34.1
	28			19.7	100*	55.1	15.8	29.1
Oncorhynchus kisutch	1.2	17	Fly larvae	1.5	100	(43)	32	25
Coho salmon	—	—		—	100*	(35)	32	(33)
Ctenopharyngodon idella	1200	23	*Egeria*	96	122	8	74	40
Grass carp					100	8.3	(50)	41.7
Gadus morhua	223	15	Chopped fish	415	100	(76.5)	22.2	1.3
Atlantic cod	—	—		—	100*	(68.5)	22.2	(9.3)
Gasterosteus aculeatus	1.0	15	Tubifex	—	100	46	4.5	(49.4)
Stickleback								
Esox lucius	250	12	*Phoxinus*	—	100*	43	29	(28)
Pike								
Lepomis macrochirus	49	15	Mayfly nymphs	620	100	36	(44)	(20)
Bluegill sunfish								
	—	—		—	100*	36	(36)	(28)
	92	20		1060	100	47	(33)	(20)
	—	—		—	100*	47	(25)	(28)
	63	25		1213	100	50	(30)	(20)
	—	—		—	100*	50	(22)	(28)
Carnivorous fish No. 1	—	—		—	100	60	20	20
Carnivorous fish No. 2	—	—		—	100	44	29	27
Herbivorous fish	—	—	—	—	100	37	20	43

[a] Values determined in calories consumed (I) and expressed as percentage of this amount. I, ingestion; fractions. Growth as a percentage of ingestion is equivalent to conversion efficiency (column G). Only those cases are included. If one or more values were determined by difference, ingestion (I) occurs as 100%; otherwise because different times and expressions of amounts were used. Only "recalculated" values were used for statistics. marked with an asterisk were used in deriving the "general budget" for carnivorous fish, including cases where

VI

(Values in Parentheses) Regarding One or More Components

Excretion			
Fe-cal (%)	Non-fecal (%)	Remarks	Reference
16.3	7.6	All components monitored continuously. Three levels of ration selected, including a near maximum and near maintenance level	Solomon and Brafield (1972)
15.7	8.2		
14.6	10.4		
29	4	Four temperatures selected in relation to a variety of rations near maximum for each temperature. Metabolic rate determined by difference. Weight not significant for range studied (12–258 g)	Elliott (1976b)
24	7		
22	9		
21	10		
15.2	(8.0)	Domesticated stock of hatchery trout tested using a diet with 55% protein and 13% fat. Nonfecal estimated as 8%	Cho (1975)
2.14	(8.0)		
27.7	(4.1)	Use of organic carbon determinations for food and feces. Three size classes tested, with decreasing conversion efficiency as size increased	Smith (1973)
26.1	(4.1)		
17.7	(11.4)		
—	—	Approximated from data in Fig. 29. Feces was 25%. (Recalculated assuming an 8% nonfecal component)	Avertt (1969); Warren (1971)
25	(8)		
—	—	Fixed amount of food used—1 kg wet food per day. (Recalculated, metabolic rate least accurate determination; no net gain in protein)	Stanley (1974a)
—	—		
—	—	Metabolic rate determined by difference. Nonfecal not measured. (Recalculated by assuming an 8% nonfecal component)	Edwards *et al.* (1972)
1.3	(8)		
20	(29.4)	Nonfecal portion assigned the balance, by difference (unusually high)	Walkey and Meakins (1969)
—	—	Metabolism is equivalent to maintenance requirement and is therefore on the low side	Johnson (1966); Welch (1968)
(17)	(3)	Fecal loss assumed as 20% with growth determined by difference	Pierce and Wissing (1974)
(20)	(8)	(Recalculated, assigning 8% to nonfecal loss)	
(17)	(3)	Same as above. Higher temperature	
(20)	(8)	(Recalculated)	
(17)	(3)	Same as above. Highest temperature	
(20)	(8)	(Recalculated)	
20	—	General budget developed from pertinent literature published to 1955. Nonfecal energy considered insignificant	Winberg (1956); Welch (1968)
20	7	General budget for young fish feeding well (selected data, to 1976)	This table
41	2	General budget for young fish feeding well	See text

M, total metabolism; G, growth; and E, total excretion. Excretion is further divided into fecal and nonfecal cases showing positive growth are presented. Where various temperatures, rations, or sizes were tested not all I is the sum of the components. The column for calories consumed is only useful for comparison within species The third to last entry is the general equation for carnivorous fish developed by Winberg (1956). The budgets good growth and/or recalculation were involved.

apparent that for some species efficient food conversion is only possible over the environmental range of temperature where the population is naturally abundant (e.g., for sockeye, Brett, 1971). Thus, given a good estimate of growth of a normal population the likely values for the other components of the energy budget can be determined.

C. Herbivores

The herbivorous fishes are undoubtedly the most important species of food-fish in the world for aquaculture, yet they have received far less attention in physiological studies than the carnivores. In particular, the families Cyprinidae (carps), Cichlidae (tilapias), Mugilidae (mullets), and Chanidae (milkfish) form the basis of much of the world production (Bardach et al., 1972). In the foregoing considerations, they have been shown to be characterized by generally low total digestibility of their plant food (e.g., 60–70%) (Hickling, 1966) (Table IV). High amounts of fiber and ash contribute to this more copious fecal matter, since the digestibility of the protein, fat, and carbohydrate components of some algal feeds may be as high as 92, 97, and 80%, respectively (Singh and Bhanot, 1970). However, Chiang (1971) recorded a wide variability in protein digestibility, ranging from (a) 73–93% for silver carp, Hypophthalmichthys molitrix, feeding on phytoplankton, (b) 54–63% for goldfish (chlorophyceous diet), and (c) 23–69% for milkfish (miscellaneous plant material).

Stanley (1974a) attempted to determine an energy budget for large grass carp or white amur, Ctenopharyngodon idella, feeding on the aquatic macrophyte Egeria densa (Table VI). Although there was no net gain in nitrogen retention, it was determined that from a caloric intake of 96 kcal approximately 40 kcal were excreted, 74 kcal deposited in growth, and 8 kcal expended in metabolism. Within the acknowledged difficulties of measuring all components of the energy budget with equal accuracy, the divergence from a balanced equation appears greatest in growth (a remarkable mean conversion efficiency of 74%), and in metabolism at only 8%. Greatest support can be given for the fraction in excretion (42%) which agrees well with an overall mean rate of fecal production for herbivores of 40.5 ± 8.6 (SE) from Table IV. If a nonfecal total fraction of 2.5% is added to this (since the protein fraction of plant diets is frequently one-fourth to one-third that of meats) then a total of 43% excretion would be considered to characterize herbivores.

Proceeding to conversion efficiency (G in the balanced equation), the generally lower capacity of herbivores would likely not exceed

20% for young fish. This leaves 37% for metabolic rate. Any improvement by reducing the metabolic fraction would turn up as increased growth, and vice versa. Some herbivores are known to be fairly sluggish (carp) while others may become remarkably active (milkfish). Thus, any generalized equation for herbivores will leave considerably greater limits of confidence than that for carnivores. As recorded in Table VI, the general equation may be written: $100 I = 37 M + 20 G + 43 E$. Lack of sufficient data does not permit determining confidence limits for the values given, which would appear to be greater than those for carnivores, judging from the excretion records.

When grass carp were fed only on lettuce (*Lactuca sativa*), Fischer (1972) showed that they were hardly able to maintain their body weight. For fish ranging in body weight from 40 to 120 g an average balanced equation at 22°C worked out to $100 I = 16 M + 3 G + 81 E$. When switched to a diet of tubificidae, considerable improvement in growth occurred, resulting in an average equation of $100 I = 23 M + 17 G + 60 E$.

The omniverous nature of this species (see review in Fischer and Lyakhnovich, 1973) appears to put it part way between the strictly carnivorous and herbivorous forms or stages. Excretion remains exceedingly high (60% of energy intake) even on the tubificid diet. Indeed, there is obviously a general range in functional capacity, developmental stage, and feeding opportunity between carnivore, insectivore, omnivore, and herbivore. Experiments with single diets deny the benefits of mixed diets so essential for the subtle requirements of good growth.

VII. CONCLUDING COMMENTARY

Studies on the growth of fishes have been more concerned with the nutritional adequacy of the diet than the provision of an energy component suited to the life stage, environmental conditions, and routine activity of the species. By studying the metabolic rates of *feeding fish* in relation to *ration level* and *temperature*, energy requirements can be assessed with increased confidence—but only in the laboratory, or hatchery rearing-pond. For the wild fish, energy expenditure still defies direct measurement. Detailed records of daily activities in nature (e.g., Olla *et al.*, 1974) hold promise for assigning likely hourly energy expenditures, calculated from tables of metabolic rates corresponding to type activities. The nearest such table is that for sockeye salmon (Table III) from which the daily metabolic pattern in Fig. 9

was derived. However, the phenomenon of metabolic acclimation under conditions of fluctuating environmental temperature has yet to be investigated with any thoroughness.

Advances in the development and application of direct calorimeters for fish should result in a final conclusion to the refined but possibly still inaccurate oxycalorific equivalent for respired oxygen. Such expression of metabolic energy release for carnivores and herbivores alike needs greater attention. Further, it is possible that in some species the anaerobic fraction or the oxygen debt accumulated at times of active metabolism may not be fully replaced by respiration, leading to underestimation of this important parameter of energetics.

The variability between species in metabolic response to temperature and the effect of size on metabolic rate make any generalization subject to considerable inaccuracy where a particular species is involved. The mean values of weight exponents and temperature Q_{10}'s may offer a useful measure for the broad insights of the ecologist but hardly satisfy the more detailed interests of the physiologist. The assumption that the exponent, approximately 0.8, for weight relations derived for standard metabolism ("metabolic weight") can be applied directly to other levels of metabolism, as well as to digestion, excretion and growth, needs critical examination.

Improved techniques and procedural systems have allowed better assessment of the digestible, metabolizable, and net energy of the diet but all too few cases have been recorded, even among the better known carnivorous fishes. Comparable knowledge and detailed studies on herbivorous fishes are greatly needed, particularly in view of the expanding aquaculture interest for plant-eating species.

Bioenergetic studies are of fundamental importance to the advance of aquaculture. Manipulation of environmental factors in conjunction with improved diets holds great promise that gross conversion efficiencies of 50% and over can be achieved, at least for young fish.

Finally, among families of fishes, the Salmonidae appear to have received greater bioenergetic attention than any other family. It would be a mistake to apply the findings too generally. The diversity of morphology of fishes is undoubtedly matched by a comparable diversity in physiology.

VIII. GLOSSARY

absorption   uptake of digested material across a membrane such as the gut wall
active metabolism   mg of O_2/kg/hr; rate of oxygen consumption during maximum sustained activity

adiabatic calorimeter one in which heat flow through the wall is prevented by a water jacket maintaining similar temperature inside and out

ammonotelic organisms that excrete ammonia as the major end product of nitrogen metabolism

anabolism metabolic processes resulting in increased body substance (=growth)

assimilation pertaining to that part of the food intake which is utilized

availability as applied to diet, that fraction of the food that is capable of being absorbed

basal metabolism mg of O_2/kg/hr; as applied to mammals, the minimum metabolic rate of a resting animal in the postabsorptive state under thermoneutral conditions

catabolism the breakdown of metabolites resulting in loss of energy

calorie the amount of energy required to heat 1 kg of water by 1°C, at 15°C (= 1000 cal)

cellulolytic inducing the hydrolysis of cellulose

digestible energy kcal; energy remaining from food *less* feces

direct calorimetry energy expenditure by an organism through direct measurement of heat loss

ectotherm an organism that does not maintain body temperature by internal heat production (cf. poikilotherm)

endergonic absorbing energy

endogenous arising from body substance (i.e., the metabolic energy source is internal)

endotherm pertaining to organisms that maintain body temperature by internal heat production (cf. homiotherm)

energy the term used to designate all forms of work and heat

energy budget a quantitative accounting of all the energy inputs and losses of an organism

energy density kcal/g; the energy per unit weight of food

exergonic producing or losing energy

exogenous arising from the food (i.e., having an external source from that of body substance)

feeding metabolism mg of O_2/kg/hr; the metabolic rate of an organism during feeding and absorption

gross conversion efficiency K or K_1, %; the percentage of total food intake converted into body substance

gross energy kcal; the total energy of food as determined by bomb calorimetry

gluconeogenesis formation of blood glucose as a result of the breakdown of amino acids

heat increment kcal/kg/hr; the metabolic heat loss from the digestion and transformation of food (=specific dynamic action, or SDA)

indirect calorimetry kcal/liter of O_2; measurement of the respiratory gas exchange of an organism for estimating energy exchange by oxycalorific equivalents

isocaloric the provision of equal amounts of dietary energy to different treatments in a feeding experiment

instantaneous growth rate the relative growth rate, as expressed in specific growth rate (see specific growth rate)

K-line the rate of decrease in gross food conversion efficiency (K) with increasing ration (R) (i.e., the straight line obtained from the equation $\log K = a + bR$)

maintenance metabolism the minimum rate of energy expenditure maintaining an organism at rest (=standard metabolism)

maintenance ration that level of food intake just maintaining a fixed weight (i.e., maintaining energy equilibrium)

metabolic compensation a shift in standard metabolic rate as an adaptation to living under extremes of climatic conditions

metabolizable energy   the energy available from the food less total exogenous excretion (fecal + nonfecal)

net conversion efficiency K_2, %; the percentage of food intake *less* maintenance fraction converted into body substance

net energy   metabolizable energy *less* heat increment (i.e., the food energy available for work and growth)

nonfecal excretion   the excretion of soluble metabolites via the kidney and gills (principally nitrogenous)

oxycalorific equivalent   kcal/liter of O_2; the number of calories released by a substrate per liter of oxygen consumed (also expressed as mg of O_2/cal)

routine metabolism   mg of O_2/kg/hr; the metabolic rate of an organism during normal spontaneous activity

standard metabolism   mg of O_2/kg/hr; minimum rate of oxygen consumption of an organism at rest in the postabsorptive state and thermally acclimated

specific dynamic action   SDA; see heat increment

specific growth rate G, as % wt/day; the percentage increase in body weight (W) per unit time (T), determined as:

$$[(\ln W_2 - \ln W_1) \times 100]/(T_2 - T_1)$$

ureotelic   organisms that excrete urea as the major endproduct of nitrogen metabolism

REFERENCES

Alexander, R. M. (1972). The energetics of vertical migration by fishes. *Symp. Soc. Exp. Biol.* **26**, 273–294.

Altman, P. L., and Dittmer, D. S., eds. (1974). "Biological Data Book," Vol. 3, Part V, "Fishes," 2nd Ed., pp. 1624–1630. Fed. Am. Soc. Exp. Biol., Biol. Handbooks, Bethesda, Maryland.

Averett, R. C. (1969). Influence of temperature on energy and material utilization by juvenile coho salmon. Ph.D. Thesis, Oregon State Univ., Corvallis.

Awakura, T. (1963). Physiological and ecological studies on the fishes at spawning stage. I. On oxygen consumption of spawning chum salmon *Oncorhynchus keta* (Walbaum). *Sci. Rep. Hokkaido Fish Hatch.* **18**, 1–10.

Bakus, G. J. (1969). Energetics and feeding in shallow marine waters. *In* "International Review of General and Experimental Zoology" (W. J. Felts and R. J. Harrison, eds.), pp. 275–369. Academic Press, New York.

Bams, R. A. (1970). Evaluation of a revised hatchery method tested on pink and chum salmon fry. *J. Fish. Res. Board Can.* **27**, 1429–1452.

Bardach, J. E., Ryther, J. H., and McLarney, W. O. (1972). "Aquaculture. The Farming and Husbandry of Freshwater and Marine Organisms." Wiley (Interscience), New York.

Beamish, F. W. H. (1964a). Influence of starvation on standard and routine oxygen consumption. *Trans. Am. Fish. Soc.* **93**, 103–107.

Beamish, F. W. H. (1964b). Respiration of fishes with special emphasis on standard oxygen consumption. II. Influence of weight and temperature on respiration of several species. *Can. J. Zool.* **42**, 177–188.

Beamish, F. W. H. (1970). Oxygen consumption of largemouth bass, *Micropterus salmoides*, in relation to swimming speed and temperature. *Can. J. Zool.* **48**, 1221–1228.

Beamish, F. W. H. (1972). Ration size and digestion in largemouth bass, *Micropterus salmoides* Lacepede. *Can. J. Zool.* 50, 153–164.

Beamish, F. W. H. (1974). Apparent specific dynamic action of largemouth bass, *Micropterus salmoides. J. Fish. Res. Board Can.* 31, 1763–1769.

Beamish, F. W. H., and Mookherji, P. S. (1964). Respiration of fishes with special emphasis on standard oxygen consumption. I. Influence of weight and temperature on respiration of goldfish, *Carassius auratus* L. *Can. J. Zool.* 42, 161–175.

Beamish, F. W. H., Niimi, A. J., and Lett, P. F. K. P. (1975). Bioenergetics of teleost fishes: Environmental influences. In "Comparative Physiology—Functional Aspects of Structural Materials" (L. Bolis, H. P. Maddrell, and K. Schmidt-Nielsen, eds.), pp. 187–209. North-Holland Publ., Amsterdam.

Benzinger, T., Huebscher, R. G., Minard, D., and Kitzinger, C. (1958). Human calorimetry by means of the gradient principle. *J. Appl. Physiol.* 12, S1–S28.

Birkett, L. (1969). The nitrogen balance in plaice, sole and perch. *J. Exp. Biol.* 50, 375–386.

Blackburn, J. M. (1968). Digestive efficiency and growth in largemouth black bass. M.A. Thesis in Zoology, Univ. of California, Davis.

Blaxter, K. L., ed. (1965). "Energy Metabolism." Academic Press, New York.

Blazka, P., Volt, M., and Cepela, M. (1960). A new type of respirometer for the determination of the metabolism of fish in an active state. *Physiol. Bohemoslov.* 9, 553–558.

Brafield, A. E., and Solomon, D. J. (1972). Oxycalorific coefficients for animals respiring nitrogenous substratus. *Comp. Biochem. Physiol. A* 43, 837–841.

Brett, J. R. (1964). The respiratory metabolism and swimming performance of young sockeye salmon. *J. Fish. Res. Board Can.* 21, 1183–1226.

Brett, J. R. (1965). The relation of size to rate of oxygen consumption and sustained swimming speed of sockeye salmon (*Oncorhynchus nerka*). *J. Fish. Res. Board Can.* 22, 1491–1501.

Brett, J. R. (1970a). Fish—The energy cost of living. In "Marine Aquiculture" (W. J. McNeil, ed.), pp. 37–52. Oregon State Univ. Press, Corvallis.

Brett, J. R. (1970b). 3. Temperature. 3.3 Animals. 3.32 Fishes. In "Marine Ecology," Vol. 1, "Environmental Factors" (O. Kinne, ed.), Part 1, pp. 515–560. Wiley (Interscience), New York.

Brett, J. R. (1971). Energetic response of salmon to temperature. A study of some thermal relations in the physiology and freshwater ecology of sockeye salmon (*Oncorhynchus nerka*). *Am. Zool.* 11, 99–113.

Brett, J. R. (1972). The metabolic demand for oxygen in fish, particularly salmonids, and a comparison with other vertebrates. *Respir. Physiol.* 14, 151–170.

Brett, J. R. (1973). Energy expenditure of sockeye salmon, *Oncorhynchus nerka*, during sustained performance. *J. Fish. Res. Board Can.* 30, 1799–1809.

Brett, J. R. (1976a). Scope for metabolism and growth of sockeye salmon, *Oncorhynchus nerka*, and some related energetics. *J. Fish. Res. Board Can.* 33, 307–313.

Brett, J. R. (1976b). Feeding metabolic rates of sockeye salmon, *Oncorhynchus nerka*, in relation to ration level and temperature. *Envir. Can., Fish. Mar. Serv.* Tech. Rep. No. 675, 18 pp.

Brett, J. R., and Glass, N. R. (1973). Metabolic rates and critical swimming speeds of sockeye salmon (*Oncorhynchus nerka*) in relation to size and temperature. *J. Fish. Res. Board Can.* 30, 379–387.

Brett, J. R., and Sutherland, D. B. (1965). Respiratory metabolism of pumpkinseed (*Lepomis gibbosus*) in relation to swimming speed. *J. Fish. Res. Board Can.* 22, 405–409.

Brett, J. R., and Sutherland, D. B. (1970). Improvement in the artificial rearing of sockeye salmon by environmental control. *Fish. Res. Board Can., Gen. Ser. Circ.* No. 89.

Brett, J. R., and Zala, C. A. (1975). Daily pattern of nitrogen excretion and oxygen consumption of sockeye salmon (*Oncorhynchus nerka*) under controlled conditions. *J. Fish. Res. Board Can.* **32,** 2479–2486.

Brett, J. R., Shelbourn, J. E., and Shoop, C. T. (1969). Growth rate and body composition of fingerling sockeye salmon, *Oncorhynchus nerka,* in relation to temperature and ration size. *J. Fish. Res. Board Can.* **26,** 2363–2394.

Brocksen, R. W., and Brugge, J. P. (1974). Preliminary investigations on the influence of temperature on food assimilation by rainbow trout *Salmo gairdneri* Richardson. *J. Fish Biol.* **6,** 93–97.

Brocksen, R. W., Davis, G. E., and Warren, C. E. (1968). Competition, food consumption, and production of sculpins and trout in laboratory stream communities. *J. Wildl. Manage.* **32,** 51–75.

Brody, S. (1945). "Bioenergetics and Growth." Reinhold, New York.

Bryan, P. G. (1974). Food habits, functional digestive morphology, and assimilation efficiency of the rabbitfish *Siganus spinus* (Pisces: Siganidae) on Guam. M.S. Thesis, Univ. of Guam, Agana.

Burrows, R. E. (1964). Effects of accumulated excretory products on hatchery-reared salmonids. *Fish Wildl. Serv. (U.S.), Res. Rep.* **66,** 1–12.

Buttery, P. J., and Annison, E. F. (1973). Considerations of the efficiency of amino acid and protein metabolism in animals. *In* "The Biological Efficiency of Protein Production" (J. G. W. Jones, ed.), pp. 141–171. Cambridge Univ. Press, London and New York.

Carey, F. G., Teal, J. M., Kanwisher, J. W., and Lawson, K. D. (1971). Warm-bodied fish. *Am. Zool.* **11,** 137–145.

Chesney, E. J., Jr., and Estevez, J. I. (1976). Energetics of winter flounder (*Pseudopleuronectes americanus*) fed the polychaete, *Nereis virens,* under experimental conditions. *Trans. Am. Fish. Soc.* **105,** 592–595.

Chiang, W. (1971). Studies on feeding and protein digestibility of Silver carp. *Hypophthalmichthys molitrix* (C. & V.). *Chin.-Am. Jt. Comm. Rural Reconstr., Fish. Serv.* No. 11, pp. 96–114.

Cho, C. Y. (1975). Aquaculture with emphasis on nutrition and diet formulation. Prog. Rep. to D.O.E., Fish. Mar. Serv., from Dep. Nutr., Guelph, Ontario.

Cowey, C. B., and Sargent, J. R. (1972). Fish nutrition. *Adv. Mar. Biol.* **10,** 383–492.

Davies, P. M. C. (1963). Food input and energy extraction efficiency in *Carassius auratus. Nature (London)* **198,** 707.

Davies, P. M. C. (1964). The energy relations of *Carassius auratus* L. I. Food input and energy extraction efficiency at two experimental temperatures. *Comp. Biochem. Physiol.* **12,** 67–79.

Durbin, A. G. (1976). Oxygen consumption and ammonia excretion of adult Atlantic menhaden, *Brevoortia tyrannus.* Grad. School Oceanogr., Univ. Rhode Island, Kingston, 16 pp.

Edwards, R. R. C., Blaxter, J. H. S., Gopalon, U. K., Mathews, C. V., and Finlayson, D. M. (1971). Feeding, metabolism, and growth of tropical flatfish. *J. Exp. Mar. Biol. Ecol.* **6,** 279–300.

Edwards, R. C. C., Finlayson, D. M., and Steele, J. H. (1972). An experimental study of the oxygen consumption, growth, and metabolism of the cod (*Gadus morhua* L.). *J. Exp. Mar. Biol. Ecol.* **8,** 299–309.

Ege, R., and Krogh, A. (1914). On the relation between the temperature and the respiratory exchange in fishes. *Int. Rev. Gesamten Hydrobiol. Hydrogr.* **1,** 48–55.

Elliot, J. E. (1969). The oxygen requirements of chinook salmon. *Prog. Fish Cult.* **31**, 67–73.

Elliott, J. M. (1975a). The growth rate of brown trout, *Salmo trutta* L., fed on maximum rations. *J. Anim. Ecol.* **44**, 805–821.

Elliott, J. M. (1975b). The growth rate of brown trout (*Salmo trutta* L.) fed on reduced rations. *J. Anim. Ecol.* **44**, 823–842.

Elliott, J. M. (1976a). Energy losses in the waste products of brown trout (*Salmo trutta* L.). *J. Anim. Ecol.* **45**, 561–580.

Elliott, J. M. (1976b). The energetics of feeding, metabolism and growth of brown trout (*Salmo trutta* L.) in relation to body weight, water temperature and ration size. *J. Anim. Ecol.* **45**, 923–948.

Elliott, J. M., and Davison, W. (1975). Energy equivalents of oxygen consumption in animal energetics. *Oecologia* **19**, 195–201.

Fischer, Z. (1970). The elements of energy balance in grass carp (*Ctenopharyngodon idella* Val.). Part 1. *Pol. Arch. Hydrobiol.* **17**, 421–434.

Fischer, Z. (1972). The elements of energy balance in grass carp (*Ctenopharyngodon idella* Val.). Part II. Fish fed with animal food. *Pol. Arch. Hydrobiol.* **19**, 65–82.

Fischer, Z., and Lyakhnovich, V. P. (1973). Biology and bioenergetics of grass carp (*Ctenopharyngodon idella* Val.). *Pol. Arch. Hydrobiol.* **20**, 521–557.

Forster, R. P., and Goldstein, L. (1969). Formation of excretory products. *In* "Fish Physiology." Vol. 1. (W. S. Hoar and D. J. Randall, eds.), pp. 313–350. Academic Press, New York and London.

Fromm, P. O. (1963). Studies on renal and extra-renal excretion in the freshwater teleost, *Salmo gairdneri. Comp. Biochem. Physiol.* **10**, 121–128.

Fry, F. E. J. (1947). Effects of the environment on animal activity. *Univ. Toronto Stud., Biol. Ser.* **55**, 1–62.

Fry, F. E. J. (1957). The aquatic respiration of fish. *In* "The Physiology of Fishes" (M. E. Brown, ed.), Vol. 1, pp. 1–63. Academic Press, New York.

Fry, F. E. J. (1971). The effect of environmental factors on the physiology of fish. *In* "Fish Physiology" (W. S. Hoar and D. J. Randall, eds.), Vol. 6, pp. 1–98. Academic Press, New York.

Fry, F. E. J., and Hochachka, P. W. (1970). Fish. *In* "Comparative Physiology of Thermoregulation" (G. C. Whittow, ed.), Vol. 1, pp. 79–134. Academic Press, New York.

Furukawa, A., and Tsukahara, H. (1966). On the acid digestion method for the determination of chromic oxide as an index substance in the study of digestibility of fish feed. *Bull. Jpn. Soc. Sci. Fish.* **32**, 502–506.

Gerking, S. D. (1955). Endogenous nitrogen excretion of bluegill sunfish. *Physiol. Zool.* **28**, 283–289.

Gerking, S. D. (1971). Influence of rate of feeding and body weight on protein metabolism of bluegill sunfish. *Physiol. Zool.* **44**, 9–19.

Glass, N. R. (1969). Discussion of calculation of power function with special reference to respiratory metabolism in fish. *J. Fish. Res. Board Can.* **26**, 2643–2650.

Gordon, M. S. (1972). "Animal Physiology: Principles and Adaptations," 2nd Ed. Macmillan, New York.

Groot, C., and Wiley, W. L. (1965). Time-lapse photography of an ASDIC echo-sounder PPI-scope as a technique for recording fish movements during migration. *J. Fish. Res. Board Can.* **22**, 1025–1034.

Gross, W. L., Roelofs, E. W., and Fromm, P. O. (1965). Influence of photoperiod on growth of green sunfish, *Lepomis cyanellus. J. Fish. Res. Board Can.* **22**, 1379–1386.

Guérin-Ancey, O. (1976). Étude expérimentale de l'excretion azotée du bar (*Dicentrar-

chus labuax) en cours de croissance. II. Effects du jeûne usr l'excretion d'ammoniac et d'urée. *Aquaculture* 9, 187–194.

Harper, H. A. (1971). "Review of Physiological Chemistry," 13th Ed. Lange Med. Publ., Los Altos, California.

Hastings, W. H. (1969). Nutritional score. *In* "Fish in Research" (O. W. Neuhaus and J. E. Halver, eds.), pp. 263–292. Academic Press, New York.

Hatanaka, M., and Takahashi, M. (1956). Utilization of food by mackerel *Pneumatophorus japonicus. Tohoku J. Agric. Res.* 7, 51–57.

Hickling, C. F. (1966). On the feeding process in the white amur, *Ctenopharyngodon idella. Proc. Zool. Soc. London* 148, 408–419.

Holeton, G. F. (1970). Oxygen uptake and circulation by a hemoglobinless Antarctic fish (*Chaenocephalus aceratus* Lonnberg) compared with three red-blooded Antarctic fish. *Comp. Biochem. Physiol.* 34, 457–472.

Holeton, G. F. (1973). Respiration of Arctic char (*Salvelinus alpinus*) from a high arctic lake. *J. Fish. Res. Board Can.* 30, 717–723.

Holeton, G. F. (1974). Metabolic cold adaptation of polar fish: Fact or artifact. *Physiol. Zool.* 47, 137–152.

Huisman, E. A. (1974). A study on optimal rearing conditions for carp (*Cyprinus carpio* L.). Ph. D. Thesis, Agricult. Univ. Wageningen. (Spec. Publ., Organisatie ter Verbetering van de Binnenvisserij, Utrecht.)

Huisman, E. A. (1976). Food conversion efficiencies at maintenance and production levels for carp, *Cyprinus carpio* L., and rainbow trout, *Salmo gairdneri* Richardson. *Aquaculture* 9, 259–273.

Idler, D. R., and Clemens, W. A. (1959). The energy expenditure of Fraser River sockeye salmon during the spawning migration to Chilko and Stuart Lakes. *Int. Pac. Salmon. Fish. Comm., Prog. Rep.* 80 pp.

Ivlev, V. S. (1939). Energy balance of carps. *Zool. Zh.* 18, 303–318.

Ivlev, V. S. (1945). The biological productivity of waters. *Usp. Sovrem. Biol.* 19, 98–120. [Engl. transl., *J. Fish. Res. Board Can.* 23, 1727–1759 (1966).]

Iwata, K. (1970). Relationship between food and growth in young crucian carps, *Carassius auratus cuvieri*, as determined by the nitrogen balance. *Jpn. J. Limnol.* 31, 129–151.

Johnson, L. (1966). Experimental determination of food consumption of pike, *Esox lucius*, for growth and maintenance. *J. Fish. Res. Board Can.* 23, 1495–1505.

Johnson, W. E., and Groot, C. (1963). Observations on the migration of young sockeye salmon (*Oncorhynchus nerka*) through a large, complex lake system. *J. Fish. Res. Board Can.* 20, 919–938.

Jones, D. R. (1971). Theoretical analysis of factors which may limit the maximum oxygen uptake of fish: The oxygen cost of the cardiac and branchial pumps. *J. Theor. Biol.* 32, 341–349.

Kanzinkin, G. S., and Tarkovskaya, O. I. (1964). Determination of caloric values of small samples. *In* "Techniques for the Investigation of Fish Physiology" (E. N. Pavlovskii, ed.), pp. 122–124. IPST, Jerusalem.

Kausch, H. (1969). The influence of spontaneous activity on the metabolic rate of starved and fed young carp (*Cyprinus carpio* L.). *Verh. Ver. Limnol.* 17, 669–679.

Kausch, H. (1972). Stoffwechsel und Ernährung der Fische. *Handb. Tierernäehr., Band II* 8, 690–738. [Fish. Res. Board Can., Transl. Ser. No. 2489 (1973).]

Kawanabe, H. (1969). The significance of social structure in production of the "Ayu," *Plecoglossus altivelis. In* "Symposium on Salmon and Trout in Streams" (T. G.

Northcote, ed.), pp. 243–251. Inst. Fish., Univ. of British Columbia, Vancouver, B.C.

Kayser, C., and Heusner, A. (1964). Étude comparative du métabolisme énergétique dans la série animale. *J. Physiol. (Paris)* **56**, 489–524.

Kelso, J. R. M. (1972). Conversion, maintenance, and assimilation for walleye, *Stizostedion vitreum vitreum*, as affected by size, diet, and temperature. *J. Fish. Res. Board Can.* **29**, 1181–1192.

Kevern, N. R. (1966). Feeding rate of carp estimated by a radioisotopic method. *Trans. Am. Fish. Soc.* **95**, 363–371.

Kinne, O. (1960). Growth, food intake, and food conversion in a euryplastic fish exposed to different temperatures and salinities. *Physiol. Zool.* **33**, 288–317.

Kleiber, M. (1961). "The Fire of Life. An Introduction to Animal Energetics." Wiley, New York. (Rev. Ed., R. E. Krieger Publ. Co., Huntington, New York, 1975.)

Kleiber, M. (1967). An old professor of animal husbandry ruminates. *Annu. Rev. Physiol.* **29**, 1–20.

Krebs, H. A. (1964). The metabolic fate of amino acids. *In* "Mammalian Protein Metabolism" (H. N. Munro and J. B. Allison, eds.), Vol. 1, pp. 125–176. Academic Press, New York.

Krogh, A. (1914). The quantitative relation between temperature and standard metabolism in animals. *Int. Z. Phys.-Chem. Biol.* **1**, 491–508.

Krueger, H. M., Saddler, J. B., Chapman, G. A., Tinsley, I. J., and Lowry, R. R. (1968). Bioenergetics, exercise, and fatty acids of fish. *Am. Zool.* **8**, 119–129.

Kutty, M. N. (1972). Respiratory quotient and ammonia excretion in *Tilapia mossambica*. *Mar. Biol.* **16**, 126–133.

Kutty, M. N., and Mohamed, M. P. (1975). Metabolic adaptations of mullet *Rhinomugil corsula* (Hamilton) with special reference to energy utilization. *Aquaculture* **5**, 253–270.

Lasker, R. (1962). Efficiency and rate of yolk utilization by developing embryos and larvae of the Pacific sardine *Sardinops caerulea* (Girard). *J. Fish. Res. Board Can.* **19**, 867–875.

Leggett, W. C. (1972). Weight loss in American shad (*Alosa sapidissima*, Wilson) during the freshwater migration. *Trans. Am. Fish. Soc.* **101**, 549–552.

Lehninger, A. L. (1965). "Bioenergetics. The Molecular Basis of Biological Energy Transformations." Benjamin, New York.

McLean, W. E., and Fraser, F. J. (1974). Ammonia and urea production of coho salmon under hatchery conditions. *Envir. Prot. Ser. Pac. Region, Surveillance Rep.* **EPS 5-PR-74-5**.

Mann, H. (1968). Der Einfluss der Ernährung auf den Sauerstoffverbranch von Forellen. *Arch. Fischereiwiss.* **19**, 131–133.

Mann, K. H. (1965). Energy transformation by a population of fish in the River Thames. *J. Anim. Ecol.* **34**, 253–257.

Marr, D. H. A. (1966). Influence of temperature on the efficiency of growth of salmonid embryos. *Nature (London)* **212**, 957–959.

Maynard, A. L., and Lousli, K. J. (1962). "Animal Nutrition," 5th Ed. McGraw-Hill, New York.

Menzel, D. W. (1958). Utilization of algae for growth by the angel fish, *Holacanthus bermudensis*. *J. Cons., Cons. Perm. Int. Explor. Mer* **24**, 308–313.

Menzel, D. W. (1960). Utilization of food by a Bermuda reef fish, *Epinephelus guttatus*. *J. Cons., Cons. Perm. Int. Explor. Mer* **25**, 216–222.

Miura, T., Suzuki, N., Nagoshi, M., and Yamamura, K. (1976). The rate of production and food consumption of the Biwamasu, *Oncorhynchus rhodurus,* population in Lake Biwa. *Res. Popul. Ecol. (Kyoto)* **17,** 135–154.

Morgulis, S. (1918). Studies on the nutrition of fish. Experiments on brook trout. *J. Anim. Sci.* **2,** 263–277.

Mount, L. E. (1963). The thermal insulation of the newborn pig. *J. Physiol. (London)* **168,** 698–705.

Muir, B. S., and Niimi, A. J. (1972). Oxygen consumption of the euryhaline fish aholehole (*Kuhlia sandvicensis*) with reference to salinity, swimming, and food consumption. *J. Fish. Res. Board Can.* **29,** 67–77.

Nightingale, J. W. (1974). Bioenergetic responses of nitrogen metabolism and respiration to variable temperature and feeding interval in Donaldson strain rainbow trout. Ph.D. Thesis, Univ. of Washington, Seattle.

Niimi, A. J., and Beamish, F. W. H. (1974). Bioenergetics and growth of largemouth bass (*Micropterus salmoides*) in relation to body weight and temperature. *Can. J. Zool.* **52,** 447–456.

Nose, T. (1960). On the effective value of freshwater green algae, *Chlorella ellipsoidea,* as a nutritive source to goldfish. *Bull. Freshwater Fish. Res. Lab.* **10,** 1–10.

Nose, T. (1967). On the metabolic fecal nitrogen in young rainbow trout. *Bull. Freshwater Fish. Res. Lab.* **17,** 97–106.

Odum, E. P. (1971). "Fundamentals of Ecology," 3rd Ed. Saunders, Philadelphia, Pennsylvania.

Olla, B. L., Bejda, A. J., and Martin, A. D. (1974). Daily activity, movements, feeding, and seasonal occurrence in the tantog, *Tantoga onitis. U.S. Fish Wildl. Serv., Fish. Bull.* **72,** 27–35.

Olson, K. R., and Fromm, P. O. (1971). Excretion of urea by two teleosts exposed to different concentrations of ambient ammonia. *Comp. Biochem. Physiol. A* **40,** 999–1008.

Paloheimo, J. E., and Dickie, L. M. (1966). Food and growth of fishes. III. Relations among food, body size, and growth efficiency. *J. Fish. Res. Board Can.* **23,** 1209–1248.

Pandian, T. J. (1967a). Food intake, absorption and conversion in the fish *Ophiocephalus striatus. Helgol. Wiss. Meeresunters.* **15,** 637–647.

Pandian, T. J. (1967b). Intake, digestion, absorption and conversion of food in the fishes *Megalops cyprinoides* and *Ophiocephalus striatus. Mar. Biol.* **1,** 16–32.

Pandian, T. J. (1967c). Transformation of food in the fish *Megalops cyprinoides.* I. Influence of quality of food. *Mar. Biol.* **1,** 60–64.

Pentegov, B. P., Mentov, Y. N., and Kurnaev, E. F. (1928). Physicochemical characteristics of spawning migration fast of chum salmon. *Bull. Pac. Sci. Fish. Res. Stn. (Vladivostok)* **2,** 47 pp.

Phillips, A. M., Jr. (1969). Nutrition, digestion, and energy utilization. *In* "Fish Physiology" (W. S. Hoar and D. J. Randall, eds.), Vol. 1, pp. 351–432. Academic Press, New York.

Phillips, A. M., Jr., and Brockway, D. R. (1959). Dietary calories and the production of trout in hatcheries. *Progr. Fish. Culturist* **21,** 3–16.

Pierce, R. J., and Wissing, T. E. (1974). Energy cost of food utilization in the bluegill (*Lepomis macrochirus*). *Trans. Am. Fish. Soc.* **103,** 38–45.

Pullar, J. D. (1958). Direct calorimetry of animals by the gradient layer principle. *Proc. Symp. Energy Metab., 1st, Copenhagen* pp. 95–98.

Ringrose, R. C. (1971). Calorie-to-protein ratio for brook trout (*Salvelinus fontinalis*). *J. Fish. Res. Board Can.* **28,** 1113–1117.

Rosenthal, H., and Paffenhöfer, G. A. (1972). On the digestion rate and calorific content of food and feces in young gar fish. *Naturwissenschaften* **59**, 274–275.

Rozin, P., and Mayer, J. (1961). Regulation of food intake in the goldfish. *Am. J. Physiol.* **201**, 968–974.

Rubner, M. (1902). "Die Gesetze des Energieverbranchs bei der Ernahrung." Deuticke, Vienna.

Saunders, R. L. (1963). Respiration of the Atlantic cod. *J. Fish. Res. Board Can.* **20**, 373–386.

Savitz, J. (1969). Effects of temperature and body weight on endogenous nitrogen excretion in the bluegill sunfish (*Lepomis macrochirus*). *J. Fish. Res. Board Can.* **26**, 1813–1821.

Savitz, J. (1971). Nitrogen excretion and protein consumption of the bluegill sunfish (*Lepomis macrochirus*). *J. Fish. Res. Board Can.* **28**, 449–451.

Scholander, P. F., Flagg, W., Walters, V., and Irving, L. (1953). Climatic adaptation in arctic and tropical poikilotherms. *Physiol. Zool.* **26**, 67–92.

Shaw, H. M., Saunders, R. L., Hall, H. C., and Henderson, E. B. (1975). The effect of dietary sodium chloride on growth of Atlantic salmon (*Salmo salar*). *J. Fish. Res. Board Can.* **32**, 1813–1819.

Singh, C. S., and Bhanot, K. K. (1970). Nutritive food values of algal feeds for common carp, *Cyprinus carpio* (Linnaeus). *J. Inland Fish. Soc. India* **2**, 121–127.

Smith, K. L., Jr. (1973). Energy transformations by the Sargassum fish, *Histrio histrio* (L.). *J. Exp. Mar. Biol. Ecol.* **12**, 219–227.

Smith, M. A. K., and Thorpe, A. (1976). Nitrogen metabolism and trophic input in relation to growth in freshwater and saltwater *Salmo gairdneri. Biol. Bull. (Woods Hole, Mass.)* **150**, 139–151.

Smith, R. R. (1971). A method for determining digestibility and metabolizable energy of fish feeds. *Prog. Fish Cult.* **33**, 132–134.

Smith, R. R. (1976). Studies on the energy metabolism of cultured fish. Ph.D. Thesis, Cornell Univ., Ithaca, New York.

Solomon, D. J., and Brafield, A. E. (1972). The energetics of feeding, metabolism and growth of perch (*Perca fluviatilis* L.). *J. Anim. Ecol.* **41**, 699–718.

Stanley, J. G. (1974a). Energy balance of white amur fed *Egeria. Hyacinth Control J.* **12**, 62–66.

Stanley, J. G. (1974b). Nitrogen and phosphorus balance of grass carp, *Ctenopharyngodon idella*, fed elodea, *Egeria densa. Trans. Am. Fish. Soc.* **103**, 587–592.

Stanley, J. G., and Jones, J. B. (1976). Feeding algae to fish. *Aquaculture* **7**, 219–223.

Stewart, N. E., Shumway, D. L., and Doudoroff, P. (1967). Influence of oxygen concentration on the growth of juvenile largemouth bass.

Stickney, R. R. (1974). Occurrence of cellulase activity in the stomachs of fishes. *J. Fish Biol.* **6**, 779–782.

Stover, J. H. (1967). Starvation and the effects of cortisol in the goldfish (*Carassius auratus* L.). *Comp. Biochem. Physiol.* **20**, 939–948.

Tang, Y.-A., and Hwang, T.-L. (1966). Evaluation of the relative suitability of various groups of algae as food of milkfish produced in brackishwater ponds. *Proc. World Symp. Warm-Water Pond Fish Cult., Rome* FAO Fish. Rep. No. 44, Vol. 3, pp. 365–372.

Walkey, M., and Meakins, R. H. (1969). Energy transformation in a host–parasite system. *Parasitology* **59**, 1–26.

Wallace, J. C. (1973). Observations on the relationship between the food consumption and metabolic rate on *Blennius pholis* L. *Comp. Biochem. Physiol. A* **45**, 293–306.

Ware, D. M. (1975). Growth, metabolism, and optimal swimming speed of a pelagic fish. *J. Fish. Res. Board Can.* **32**, 33–41.

Warren, C. E. (1971). "Biology and Water Pollution Control." Saunders, Philadelphia, Pennsylvania.

Warren, C. E., and Davis, G. E. (1967). Laboratory studies on the feeding bioenergetics and growth of fishes. *In* "The Biological Basis of Freshwater Fish Production" (S. D. Gerking, ed.), pp. 175–214. Blackwell, Oxford.

Welch, H. E. (1968). Relationships between assimilation efficiencies and growth efficiencies for aquatic consumers. *Ecology* **49**, 755–759.

Winberg, G. G. (1956). "Rate of Metabolism and Food Requirements of Fishes." Beloruss. State Univ., Minsk. [Fish. Res. Board Can., Transl. Ser. No. 194 (1960).]

Winberg, G. G. (1970). Energy flow in aquatic ecological system. *Pol. Arch. Hydrobiol.* **17**, 11–19.

Wohlschlag, D. E. (1960). Metabolism of the Antarctic fish and the phenomenon of cold adaptation. *Ecology* **41**, 287–292.

Wohlschlag, D. E. (1964). Respiratory metabolism and ecological characteristics of some fishes in McMurdo Sound, Antarctica. *In* "Biology of the Antarctic Seas" (M. O. Lee, ed.), Antarctic Research Series, Vol. 1, pp. 33–62. American Geophysical Union.

Wood, J. D. (1958). Nitrogen excretion in some marine teleosts. *Can. J. Biochem. Physiol.* **36**, 1237–1242.

Yoshida, Y. (1970). Studies on the efficiency of food conversion to fish body growth. III. Total uptake of food and the efficiency of total food conversion. *Bull. Jpn. Soc. Sci. Fish* **36**, 914–916.

Zeuthen, E. (1953). Oxygen uptake as related to body size in organisms. *Q. Rev. Biol.* **28**, 1–12.

Zeuthen, E. (1970). Rate of living as related to body size in organisms. *Pol. Arch. Hydrobiol.* **17**, 21–30.

7

CYTOGENETICS

J. R. GOLD

*The chromosome structural arrangement is
the umbilical cord of the species.*
Verne Grant (1963)

I. INTRODUCTION: CHROMOSOMES AND FISHES

It is estimated that there are between 20,000 and 23,000 living species of fishes belonging to three diverse classes: *Agnatha* (lampreys and hagfishes); *Chondricthyes* (sharks, skates, and chimeras); and *Osteichthyes* (bony fishes). Of all these, the chromosome numbers of only 650–700 species have been reported; complete karyotypes are known for only about 500 species (ca. 2–3%). In contrast, over 30% of the living species of eutherian mammals have been studied cytologically, in some cases extensively.

FISH PHYSIOLOGY, VOL. VIII
Copyright © 1979 by Academic Press, Inc.
All rights of reproduction in any form reserved.
ISBN 0-12-350408-2

At the outset, it would be helpful to distinguish between "cytogenetics" and "cytotaxonomy." The former refers to the study of heredity through the study of chromosomes (the bearers of the genes), and the cytological mechanisms of inheritance. Cytotaxonomy, on the other hand, refers to the study of phenetic and/or phylogenetic relationships among species, based on comparisons of chromosome number and morphology.

In this review, emphasis is placed on cytogenetics of fishes. Readers interested more in cytotaxonomy of fishes are urged to consult the review by Ohno (1974), and the references contained in the checklists of fish chromosome numbers by Roberts (1967), Gyldenholm and Scheel (1971), Chiarelli and Capanna (1973), Denton (1973), and Park (1974).

II. TECHNIQUES AND METHODS OF KARYOTYPING FISHES

Obtaining consistent chromosome "spreads" of good quality is the limiting factor in the study of chromosome cytology in fishes. Early fish cytologists were handicapped by numerous technical difficulties, resulting in several reports of chromosome number and morphology now considered incorrect (Chiarelli and Capanna, 1973; Denton, 1973; Ohno, 1974). A possible exception was the sectioning technique utilized by Nogusa (1960), whose reports on chromosome numbers of several fish species are in agreement with later studies.

With the "revolution" in techniques of mammalian cytology during the last quarter century (German, 1973), several innovative procedures, including pretreatment with mitotic inhibitors and exposure of cells to hypotonic solution, have greatly simplified the preparation of fish chromosomes. Often the choice of method depends on the time and facilities available. Roberts (1967), Denton (1973), and Blaxhall (1975) have reviewed some of the literature on obtaining and preparing chromosomes from fish. A brief review of the sources from which fish chromosomes may be obtained is presented below.

A. Chromosomes Obtained from Live Fish or Embryos

Procedures involving preparations of mitotic chromosomes from actively dividing somatic tissues of live specimens or from embryos have been the most widely used among fish cytologists and have the dual advantages of being rapid and inexpensive. The soft organs (kidney, spleen, and liver) have proved to be good sources of chromosomes (Ohno *et al.*, 1965; Nygren *et al.*, 1968a,b, 1971a,b; Chen, 1969; Davisson *et al.*, 1972; Wilmot, 1974; Gold, 1974; Zenzes and Voiçulescu, 1975). Kidney probably gives the best results since in most fishes the renal intertubular tissue contains the hematopoietic organs (Catton, 1951), and thus provides numerous rapidly proliferating blood cells. Equally good sources are the epithelial cells from gills (McPhail and Jones, 1966; Lieppman and Hubbs, 1969), from fins or scales (Denton and Howell, 1969), and from cornea (Drewry, 1964). The use of epithelial cells instead of soft organ tissue has the advantage that the specimens may be kept alive. Swarup (1959a), Simon (1963), Booke (1968), and Endo and Ingalls (1968) developed techniques to obtain chromosomes from the blastula of early embryos. Several disadvantages to using embryonic material were summarized by Roberts (1967).

Testes are useful for meiotic as well as mitotic chromosome preparations (Roberts, 1964; Nygren *et al.*, 1968a,b; Chen, 1969), but usually can be used only during active spermatogonial division (Roberts, 1967; Blaxhall, 1975). Furthermore, the connective tissue stroma of testes often makes satisfactory spreading of cells difficult (Roberts, 1967). Meiotic chromosomes have also been obtained from ovaries (Ohno *et al.*, 1965; Davisson *et al.*, 1973).

Preparation of the chromosomes from any of the tissues noted above is relatively straightforward, and most of the various procedures are thoroughly outlined in Denton's (1973) book on fish chromosome methodology. Slight variations were suggested by Gold (1974) and Zenzes and Voiçulescu (1975) for soft organ tissues, and by Endo and Ingalls (1968) for embryos.

B. Chromosomes Obtained from Cell Culture

The overwhelming successes of mammalian cytologists in using short- or long-term cell cultures as sources of chromosomes have prompted several fish researchers to initiate studies along this line. Short-term cell cultures using soft organ tissue (Wolf *et al.*, 1960; Wolf and Quimby, 1969; Roberts, 1964; Barker, 1972; Yamamoto and Ojima, 1973; Abe and Muramoto, 1974; Wilmot, 1974), scale epithelium (Ojima *et al.*, 1972), fin explant (Regan *et al.*, 1968; Leuken and Foerster, 1969), gill (Chen and Ebeling, 1975), and gonad (Roberts, 1964, 1968, 1970; Chen, 1970), have all provided good sources of chromosomes from several different freshwater and marine species. The methods involved in initiating cell cultures, including digestion and centrifugation procedures, are well reviewed by Wolf and Quimby (1969). Techniques for harvesting cells and preparing chromosomes are essentially the same as those from tissues of live specimens, and are outlined in Denton (1973).

Long-term cell cultures have also been used as sources of fish chromosomes (Clem *et al.*, 1961; Regan *et al.*, 1968; Wolf and Quimby, 1969; Rio *et al.*, 1973; Hayashi *et al.*, 1976), but are not recommended for general karyotyping (Chen and Ebeling, 1975). In attaining the potential for indefinite subculturing, several cell lines in both fishes (Wolf and Quimby, 1969) and mammals (Nelson-Rees *et al.*, 1967) have been found to be heteroploid. For example, Rio *et al.* (1973) examined the chromosomes of a cell line of the goldfish, *Carassius auratus*, which had been subcultured over 3 years, and found chromosome numbers ranging from 47 to 193, with an indistinct modal number of 94 (22%). The normal diploid number of *C. auratus* is 104 (Chiarelli *et al.*, 1969). In contrast, Regan *et al.* (1968) found the chromosome number of an 8-year cell line derived from fin tissue of the blue striped grunt, *Haemulon sciurus*, to be stable and $2n = 46$. Although the normal complement of *H. sciurus* contains $2n = 48$ chromosomes, the authors point out that their cell line could have been initiated from a population carrying a fixed polymorphism similar to that described by Roberts (1964) in the green sunfish, *Lepomis cyanellus*.

The most promising cell culture technique for obtaining chromosomes from fish is that of using leukocytes, since the tedious digestion and centrifugation procedures need not be employed. Leukocytes do not normally undergo cell division once in the circulating blood; however, Nowell (1960), Moorhead *et al.* (1967), and many others have shown that leukocytes in several eukaryotes may be stimulated to divide *in vitro* in the presence of certain mitogenic chemicals. Unfortunately, leukocyte culture has not enjoyed great success in fishes. Among the few reported successes are the following: Labat *et al.* (1967) and Ojima *et al.* (1970) in carp and goldfish; Heckman and Brubaker (1970) and Heckman *et al.* (1971) in goldfish and trout; Kang and Park (1975) in the Japanese eel, *Anguilla japonica;* Legendre (1975) in *Anguilla anguilla;* and Thorgaard (1976) in the rainbow trout. Heckman *et al.* (1971) noted that rainbow trout leukocytes grew only under increased oxygen tension, and suggested that perhaps alternate techniques might have to be developed for different groups of fishes. Barker (1972) devised a method for obtaining chromosomes from the few immature leukocytes in the circulating blood of the marine fish, *Pleuronectes platessa*, and indicated that the method was applied to other marine species as well.

C. Methods of Staining and Examining Chromosomes

Recently, staining techniques which result in differential banding of somatic metaphase chromosomes and permit the identification of individual chromosome pairs have been developed for a variety of organisms. Although there are few reports of successful "banding" of fish chromosomes (Abe and Muramoto, 1974; Zenzes and Voiçulescu, 1975; Thorgaard, 1976), there is little doubt with the successes in mammals that these techniques will also become widely used in studying fish chromosomes. The traditional methods of staining fish chromosomes using aceto-orecin or Giemsa (e.g., Fig. 1) are adequate for enumerating the chromosome complement of a species, but do not always permit the resolution necessary to ascertain possible chromosomal heteromorphy. Furthermore, there now are suggestions that the banding patterns observed on mammalian chromosomes not only reflect chromosomal phenotypes, but may also indicate functional genetic aspects of a given chromosome or chromosomal segment (Hoehn, 1975). A bibliography of the literature on banding techniques may be found in Nilsson (1973).

Another method which holds great promise for the study of fish chromosomes is the use of a scanning electron microscope. Webb (1974) was the first to use this technique in fishes, and his results are encouraging. The centromeres of each chromosome were readily visible (facilitating arm length determinations), and the three-dimensional surface structure of the chromosome was impressively revealed. According to Webb, the only difficulty seems to be obtaining sufficient numbers of spreads for analysis. This should be overcome by cell culture.

III. SEXUALITY, SEX CHROMOSOMES, AND SEX DETERMINATION

As a group, the fishes display an almost complete range of sexuality from hermaphroditism to unisexuality to bisexuality or gonochorism (Yamamoto, 1969). This diversity is unparalleled among the vertebrates. A few instances of unisexuality are known in the ambystomid salamanders and in a few lizard species (White, 1973a), but hermaphroditism is unknown elsewhere.

In the following, the various modes of sexuality found among fishes are briefly considered, with the emphasis placed on the genetic and/or cytogenetic mechanisms which influence sexuality. Readers more interested in the physiology of sexuality, secondary sex characteristics, or sex differentiation are referred to the reviews by Gordon (1957) and Yamamoto (1969).

A. Hermaphroditism

Hermaphroditism is the normal and functional coexistence in an individual of both maleness and femaleness. In fishes, two basic types of hermaphroditism are recognized (Yamamoto, 1969). Synchronous or balanced hermaphrodites possess both male and female tissues which ripen together and function simultaneously. Histologically, the gonad of a synchronous form consists of an "ovotestes" divided into ovarian and testicular regions (D'Ancona, 1950). In theory, synchronous hermaphrodites have the capability for self-fertilization; this has been substantiated in the Florida serranid, *Serranus subligerus* (Clark, 1959, 1965), and in the oviparous cyprinodontid, *Rivulus marmoratus* (Harrington, 1961). Asynchronous or consecutive hermaphrodites are those that function as one sex when young and then transform to the other sex when aged. Protandrous forms function first as males, and then as females; protogynous forms are first females, and then transform to males. An important basic histological characteristic of asynchronous hermaphrodites, whether protandrous or protogynous, is that juveniles possess both ovarian and testicular tissue (Yamamoto, 1969).

Atz (1964) and Yamamoto (1969) listed the various hermaphroditic fish species which are found in thirteen families belonging to five orders; the majority belong to the order *Perciformes*. The synchronous hermaphrodites include species in several genera of the sea bass family *Serranidae,* the already mentioned cyprinodontid, *Rivulus marmoratus,* and a few species from four families of the order *Myctophiformes.* Atz (1964) also listed one Alaskan population of the stickleback, *Gasterosteus aculeatus,* of the order *Gasterosteiformes* as being a synchronous hermaphrodite.

The asynchronous type of hermaphroditism appears to be more prevalent, although this may reflect a paucity of information on hermaphroditic fishes. Protandrous forms (δ to $\female$) are encountered chiefly among the sea breams and porgies of the family *Sparidae;* a few others are cited in Atz (1964) and Yamamoto (1969). Protogynous forms ($\female$ to δ) are found in at least four perciform families, including both the *Sparidae* and *Serranidae* (Yamamoto, 1969). The swamp eel, *Monopterus albus,* of the order *Synbranchiformes* is also a protogynous hermaphrodite (Liem, 1963; Chan, 1970). It should be pointed out that many families contain both synchronous and asynchronous forms, and that in all families which contain hermaphrodites there are bisexual species.

The physiological and genetic bases of sexuality in the hermaphrodites are not well understood. Different histological patterns of gonadal development in protogynous forms have been described (Smith, 1959; Reinboth, 1967); essentially the process appears to be an orderly transition from one gonadal type to the other. There is evidence, however, that sex transformation in asynchronous forms may not always be completed (Larrañeta, 1964, cited in Yamamoto, 1969), and also that all individuals of a given species may not inverse sex in sequence (Smith, 1959; Reinboth, 1965). Nonetheless, in the asynchronous forms the transformation from one sex to the other may be viewed as ordinary histological differentiation; a decrease in the number of cells of one gonadal type is followed by an increase in the number of cells of the other type. An intervening "intersex" stage may prevail in some species (Yamamoto, 1969). Ohno (1974) has proposed that the switch may result from an antagonism between masculinizing inducers (androgenic steroids) and feminizing inducers (estrogenic steroids). In a protogynous species, the feminizing inducers produced by ovarian cells would suppress testicularization until a time— perhaps at ovulation—when the amount of feminizing inducer decreases and testicularization commences. Obviously, no such antagonism exists in synchronous hermaphrodites.

The above suggests that sexuality in a hermaphroditic species is a process of sex *differentiation*, rather than sex *determination*. Thus, one would not expect to find evidence of genetically or morphologically differentiated sex chromosomes (heterosomes) in either synchronous or asynchronous hermaphrodites. The chromosomes of ovarian and testicular tissue should differ to the same extent as they would between any other tissues of the same individual, for example, between liver and spleen. Vestigial heterosomes might possibly be found in a hermaphroditic species recently derived from a bisexual species which had possessed well-developed heterogamety, but even this seems unlikely. On the other hand, Ohno (1974) has pointed out the very puzzling fact that of all the bisexual species with well-differentiated (heteromorphic) heterosomes, several belong to orders which are known or thought to contain hermaphrodites (including the *Perciformes*, Nogusa, 1960; *Myctophiformes*, Chen, 1969; *Gasterosteiformes*, Chen and Reisman, 1970). One would expect those orders with hermaphroditic species to have maintained their chromosomes in the "least committed state" (Ohno, 1974). Since none of the hermaphrodites have yet been examined cytologically, no answer can be given for the paradox.

The experimental evidence of self-fertilization in R. *marmoratus* (Harrington, 1961) and S. *subligerus* (Clark, 1959) raises the question as to the extent of self-fertilization in natural populations of synchronous hermaphrodites. Genetically, complete selfing is the most intense form of inbreeding. Genetic variability within a lineage would be reduced by one-half every generation, and in the effective absence of genetic recombination the only new source of genetic variation would be mutation. On the other hand selfing could permit the selection for highly adapted genotypes, and prevent the breakup of co-adapted gene complexes which would normally occur with outcrossing. Theoretical aspects of this advantage are more thoroughly discussed by Crow and Kimura (1965). One also might expect that self-fertilization would be advantageous in colonizing a new habitat in areas of low population density, or in any situation where finding a mate is difficult. A single hermaphroditic individual capable of self-fertilization could successfully colonize an uninhabited area.

Few studies exist on the frequency of self-fertilization in natural populations of synchronous hermaphrodites. Clark (1959, 1965) found that isolated individuals of S. *subligerus* can fertilize their own eggs in captivity, but form spawning pairs when placed in proximity with other conspecifics. Harrington and Kallman (1968) demonstrated that the laboratory-reared offspring of wild-caught specimens of R. *marmoratus* were not only isogenic, but also homozygous for several histocompatibility loci. Whether the wild-caught individuals self-fertilized as a direct result of transplantation to the laboratory could not be determined. However, in a subsequent study, it was found that exposure to low temperature could transform both juvenile and adult R. *marmoratus* hermaphrodites into males (Harrington, 1971). Apparently, R. *marmoratus* has options for future changes in its mode of reproduction (Ohno, 1974).

B. Unisexuality

Unisexuality in fishes can be defined broadly as those modes of reproduction in which individual females produce female offspring exclusively. The term "thelytoky," defined by White (1973a) as unisexual reproduction by a process not involving fertilization (i.e., fusion of male and female pronuclei) is inadequate here. Some unisexual fishes undergo gametic fusion, and the paternal genome is expressed phenotypically (*hybridogenesis*). Others merely use the sperm of closely related, usually sympatric, congeners to stimulate divisions of the egg nucleus; the male pronucleus degenerates and makes no genetic contribution to the developing embryo (*gynogenesis* or *pseudogamy*). In either case, the unisexual form is dependent on males of a bisexual species for sperm.

The first "all-female" species described was the Amazon molly, *Poecilia* (=*Mollienisia*) *formosa* (Hubbs and Hubbs, 1932), a small, viviparous toothcarp belonging to the family *Poeciliidae* (order *Atheriniformes*). Throughout its range from southern Texas to northeastern Mexico (Darnell and Abramoff, 1968), *P. formosa* is sympatric with one or both of the related species *Poecilia mexicana* and *Poecilia latipinna*. Hubbs and Hubbs (1932, 1946) initially suspected that *P. formosa* was an interspecific *P. mexicana* × *P. latipinna* hybrid which had lost its ability to produce bisexual offspring. Wild *P. formosa* females were phenotypically intermediate between *P. mexicana* and *P. latipinna*, and when crossed to males of either suspected parental species produced only female progeny of *formosa* phenotype. The hybrid origin of *P. formosa* was later substantiated beyond reasonable doubt by the genetic studies of Abramoff *et al.* (1968) using plasma protein markers, and by Prehn and Rasch (1969) using chromosomal markers. Further studies (Hubbs and Hubbs, 1946; Hubbs, 1955) revealed that matings of *P. formosa* with males from several different species invariably produced all-female progeny with strictly matroclinous inheritance; paternal characters were not apparently transmitted to the offspring (see also Meyer, 1938; Haskins *et al.*, 1960; Hubbs, 1964). Tissue transplant experiments gave similar results; female offspring were genetically identical to both their mothers and sisters (Kallman, 1962, 1970a). These findings led to the conclusion that unisexual reproduction in *P. formosa* was the result of gynogenesis.

Chromosome studies have shown that *P. formosa*, like its progenitors, *P. mexicana* and *P. latipinna*, is a diploid species with $2n = 46$ chromosomes (Drewry, 1964; Schultz and Kallman, 1968; Prehn and Rasch, 1969). A few triploid ($3n = 69$) individuals were identified among laboratory stocks of *P. formosa* perpetuated by crosses to males of *Poecilia sphenops* or *Poecilia vitatta* (Rasch *et al.*, 1965; Schultz and Kallman, 1968). Subsequently, Prehn and Rasch (1969) and Rasch *et al.* (1970) discovered *P. formosa*-like triploid "clones" in nature that bore a close phenotypic resemblance to *P. mexicana*, and which apparently had arisen from the addition of a functional haploid *P. mexicana* genome to the diploid one of *P. formosa* (Balsano *et al.*,

1972; Menzel and Darnell, 1973). Unlike the laboratory-produced triploids, the ones from nature were fertile and produced all-female triploid broods (Rasch and Balsano, 1973; Strommen et al., 1975). Balsano et al. (1972) found that triploid clones comprise a significant, but variable fraction of several isolated P. formosa populations. Although not substantiated, the triploids are thought to reproduce by gynogenesis (Strommen et al., 1975).

Production of diploid (or triploid) offspring by gynogenesis requires that the correct ploidy level be maintained in the developing embryo without the contribution of the paternal genome. This may occur cytologically by apomixis, where meiosis is abortive; or by automixis, where meiosis is normal, but correct ploidy is maintained by events occurring prior to or following the meiotic divisions (White, 1973a). In diploid or triploid P. formosa the exact cytological mechanisms of all-female production are not known. However, in another poeciliid genus, Poeciliopsis, found along the northwestern coast of Mexico (Moore et al., 1970; Schultz, 1971), several unisexual forms have been discovered, and, in a few, the cytological features of all-female production have been verified.

Like P. formosa, the unisexuals of Poeciliopsis consist of both diploid and triploid forms, and apparently arose through interspecific hybridization (see review in Schultz, 1971). Initially a number of different Poeciliopsis unisexuals were identified. Some forms, known as P. Cx and P. Cz, were diploid with $2n = 48$ chromosomes; others, for example, P. Cy, were triploid with $3n = 72$ chromosomes (Miller and Schultz, 1959; Schultz, 1961, 1966, 1967, 1969, 1971). The Poeciliopsis diploid unisexuals maintain the all-female characteristic from generation to generation by a unique mechanism called hybridogenesis (Schultz, 1969), which prevents independent assortment of maternally and paternally derived chromosomes. Although both maternal and paternal genomes are expressed phenotypically, only the haploid female genome is transmitted to the ovum; diploidy is restored via fertilization with males from sympatric bisexual species. The cytological mechanism is essentially automictic, but not strictly so. Cimino (1972b) found that during the mitotic oogonial divisions preceding meiosis, a unipolar spindle is formed which attracts one set of chromosomes (maternal); the other set (paternal) is lost. The meiotic events, if any, are unknown but the ova produced are haploid and matroclinous (Schultz, 1973).

Additional important features of hybridogenesis were revealed by studies on the unisexual P. Cx. This form, which inhabits the Rio Mocorito, resembled another hybridogen, Poeciliopsis monacha–

lucida, in several morphological features, but differed slightly in attributes characteristic of the bisexual species, *Poeciliopsis viriosa* (Schultz, 1961, 1966). From biochemical and morphological studies (Schultz, 1969; Vrijenhoek, 1972), it was evident that *P. monacha–lucida* had arisen from interspecific hybridization between *Poeciliopsis monacha* and *Poeciliopsis lucida,* some 200 km to the north in the Rio Fuerte. Since *P. monacha* was unknown in the Rio Mocorito, and was closely related to *P. viriosa,* it was tentatively suggested that *P. Cx* arose in a similar fashion to *P. monacha–lucida,* but from interspecific hybridization between *P. viriosa* and *P. lucida.* In an extended series of studies, Vrijenhoek and Schultz (1974) demonstrated that *P. Cx* was in fact the *P. monacha–lucida* of the Rio Fuerte, but that *P. viriosa* genes had become introgressed into the *monacha* (maternal) genome. Laboratory crosses of genetically marked *P. monacha–lucida* × *P. viriosa* hybrids (which are chromosomally *monacha–viriosa* since the *monacha* genome is maternal in *P. monacha–lucida*) × *P. viriosa* males showed independent assortment both for the genetic markers used and for sex. Apparently, neither the cytological features of hybridogenesis nor the all-female character are irreversible, and when a hybridogen enters unfamiliar territory bisexuality and gene exchange with the endemic species may be favored by natural selection. Vrijenhoek and Shultz's (1974) study further revealed that it is the *monacha* genome which is invariably maternal and that "the *monacha–lucida* unisexuals have played a central role in the origin of other unisexual 'species' of *Poeciliopsis*" (p. 317). The dramatic spread of *P. monacha–lucida* via hybridogenetic combinations with the paternal genomes of *Poeciliopsis latidens* and *Poeciliopsis occidentalis* is now well documented (Schultz, 1961, 1966, 1969, 1971; Moore *et al.,* 1970; Vrijenhoek and Schultz, 1974). The adaptive advantages and evolutionary implications of hybridogenesis (and gynogenesis) are too lengthy to be considered here. Readers are referred to the reviews on the subject by Schultz (1971) and Maslin (1971).

The triploid *Poeciliopsis* unisexuals reproduce by gynogenesis. Schultz (1967) mated males from several different bisexual *Poeciliopsis* species to triploid *P. Cy,* and in each instance no evidence of paternally derived traits was found. The cytological mechanisms here are truly automictic. Prior to meiosis, the triploid oogonia undergo an endomitotic replication which raises the chromosome number to hexaploid (Cimino, 1972a). Meiosis then proceeds, and the ova produced are triploid with genetic complements identical to that of the mother. Chromosome segregation from the hexaploid meiocyte, however, is not random (Cimino and Schultz, 1970; Cimino, 1972a).

Other instances in natural populations of all-female unisexual fishes have been reported, based on observations of highly dispropor-tionate sex ratios. Schultz (1971) cautioned, however, that skewed sex ratios could result from several causes (e.g., differential mortality) and should be viewed skeptically. Certain populations of the silver cru-cian carp, *Carassius auratus gibelio* (order *Cypriniformes*), are all-female producing, triploid with $3n = 141$* chromosomes, and gynogenetic (Cherfas, 1966, 1972). In nature, these triploid unisexuals can apparently utilize the sperm from related cyprinid species. Cher-fas (1966, 1972) demonstrated that the cytological mechanism is apomictic. During late prophase and early metaphase of meiosis I, the chromosomes are unpaired (univalent). Multipolar figures appear which then culminate in a tripolar spindle that eventually disinte-grates, resulting in an aborted reductional division. Although an equational division apparently occurs following ovulation, the ova are triploid and matroclinous. Most interestingly, Cherfas (1972) observed that the breakdown of the tripolar spindle first involved a transition to a bipolar spindle where, at least in some cells, the chromosomes were oriented in a $1:2$ ratio. If, as suggested by Cherfas, this ratio corre-sponded to a haploid : diploid arrangement, then a similar mechanism might account for the rare diploid segregants found from *Poeciliopsis* triploids (Cimino and Schultz, 1970). Several *C. auratus* populations in Japan are also triploid, and probably gynogenetic (Kobayasi *et al.*, 1970; Kobayasi, 1971; Muramoto, 1975; Ojima *et al.*, 1975).

C. Bisexuality (Gonochorism)

The great majority of fishes reproduce bisexually, and have sepa-rate sexes which in nature are regularly encountered in an approxi-mate $1:1$ ratio. Because of this, it is frequently assumed that sex de-termination depends to a large extent on genes which reside on a single pair of "sex" chromosomes or *heterosomes*. In the highly evolved eutherian mammals, this is strictly the case. Males possess a pair of genetically nonhomologous heterosomes (X and Y) and produce both X- and Y-bearing sperm (*heterogamety*); females possess two X chromosomes and produce only X-bearing ova (*homogamety*). The two heterosomes, X and Y, are morphologically differentiated (*heteromorphic*) in size and shape and are easily identified cytologi-cally. Insofar as sex determination in mammals is concerned, a single Y

---

* The high triploid chromosome number of *C. a. gibelio* reflects the apparent tetrap-loid origin of *Carassius auratus* [see Ohno and Atkin (1966) and Section IV,D of this review]. Reported diploid ($2n$) chromosome numbers of *C. auratus* range from 94 to 104 (see list in Chiarelli and Capanna, 1973).

chromosome is sufficient to determine maleness. Among the fishes, no such generalizations can be made.

1. HETEROGAMETY

Cytological evidence of heterogamety (heteromorphy) was claimed for several fishes by the early cytologists (for references, see Chen, 1969), but since has been questioned in view of the technical difficulties which then prevailed. Ebeling and Chen (1970) listed three criteria by which cytological heterogamety may be established: (1) the invariant occurrence in mitotic cells of a heteromorphic chromosome pair in one but not the other sex, (2) the atypical behavior— usually an end-to-end association—of a single bivalent at meiosis I, and (3) the presence of two different haploid karyotypes at meiosis II, each possessing one of the heteromorphic chromosome pairs. Although all three criteria are rarely fulfilled, relatively reliable evidence of sex chromosome heteromorphy has been reported for over twenty-five species of fish.

Chen (1969) described cytological heterogamety in twelve species from three orders of deep-sea fishes. In the mesopelagic deep-sea smelts of the family *Bathylagidae* (order *Salmoniformes*), four species, including *Bathylagus wesethi*, *B. stilbius*, *B. ochotensis*, and *B. milleri*, were classified as male heterogametic (XX:XY). In each species, the presumed X was the largest chromosome of the complement, and varied only slightly among the species in size (smaller in *B. ochotensis*) and in centromere position (submetacentric in *B. stilbius*, but metacentric in the others). The presumed Y also varied interspecifically, being the second largest chromosome in *B. stilbius* and *B. ochotensis*, and the smallest in *B. wesethi*. In *B. milleri*, the Y was indistinguishable from a number of very small chromosomes. Meiotic preparations were observed only in *B. wesethi*. At the first meiotic metaphase, a single "sex" bivalent was observed in an end-to-end configuration; at meiosis II, two morphotypes were found, one with the presumed X and the other without (see also Chen and Ebeling, 1966).

In another salmoniform, *Sternoptyx diaphana*, of the hatchetfish family *Sternoptychidae*, Chen (1969) found evidence of male heterogamety of the XX:XO type. This form of heterogamety is not at all rare in animal groups, and usually arises either from loss of the Y chromosome, or from fusion of the Y with an autosome or with the X (White, 1973a). Spermatogonial metaphases of *S. diaphana* contained $2n = 35$ chromosomes, the presumed X being the largest among five acrocentric chromosomes. In meiotic II metaphases, two morphotypes were observed, one with $n = 18$ and one with $n = 17$.

Of the remaining deep-sea fishes studied by Chen (1969), three lantern fish [one neoscopelid, *Scopelengys tristis*, and two myctophids, *Lampanyctus ritteri* and *Lampanyctus* (=*Parvilux*) *ingens*, of the order *Myctophiformes*] were apparently of the XX:XO type; while one myctophid, *Symbolophorus californiensis*, and three prepercoid melamphids, *Melamphaes parvus*, *Scopeloberyx robustus*, and *Scopelogadus mizolepis bispinosus* (order *Beryciformes*), were all of the XX:XY type. Full details of the cytological observations are given in Chen (1969) and Ebeling and Chen (1970). It should be noted, however, that females were available for analysis only in *L. ritteri*.

Among the shallow-water fishes, male heterogamety of the XX:XY type from cytological evidence is reported for the stickleback, *Gasterosteus wheatlandi* (Chen and Reisman, 1970; Ebeling and Chen, 1970); the gobiid, *Mogrunda obscura*, and the cottid, *Cottis pollux* (Nogusa, 1960); and two species of killifish, *Fundulus diaphanus* and *Fundulus parvipinnis* (Chen and Ruddle, 1970). In both fundulines, the heterosomes were identified as the fourth largest pair in the diploid complement. Interestingly, the presumed Y was metacentric in *F. diaphanus*, but acrocentric in *F. parvipinnis* (Ebeling and Chen, 1970; Chen and Ruddle, 1970). LeGrande (1975) found evidence of an XX:XO system in the flatfish, *Symphurus plagiusa* (*Pleuronectiformes*). Females of *S. plagiusa* contained $2n = 46$ chromosomes, whereas males had $2n = 45$. The missing element was a small metacentric.

Female heterogamety of the WZ:ZZ type (WZ = ♀, ZZ = ♂) also has been found cytologically. Chen and Ebeling (1968) discovered that karyotypes from several tissues of female mosquitofish, *Gambusia affinis*, invariably contained a large, unpaired metacentric chromosome (W) not found in males. The stickleback, *Apeltes quadracus*, is also female heterogametic (Chen and Reisman, 1970), but the presumed W is acrocentric.

Uyeno and Miller (1971) reported multiple sex chromosomes (X_1X_2Y) in an undescribed cyprinodontid killifish, related to the genus *Cyprinodon*. Mitotic karyotypes revealed that females ($2n = 48$) possessed five pairs of acrocentric chromosomes, whereas males ($2n = 47$) possessed only four pairs of acrocentrics, plus a single, outsized metacentric. In late spermatogonial prophase I, the long metacentric appeared in a trivalent configuration with two small acrocentrics. Based on similar cases described in other animals, Uyeno and Miller suggested that the large metacentric arose in the male karyotype through the fusion of a Y chromosome with an autosome. Since by definition the homologue of the fused autosome is considered a "sex" chromosome (X_2), the species is said to possess multiple sex chromosomes, that is, $♀ = X_1X_1X_2X_2$ and $♂ = X_1X_2Y$ where X_1 was the initial heterosome. Subsequently, Uyeno and Miller (1972) reported an apparently identical case of multiple sex chromosomes in an undescribed goodeid species.

Other fishes for which cytological heteromorphy has been described are the cyprinid, *Scardinus erythrophthalmus*, and the European eel, *Anguilla anguilla* (Chiarelli *et al.*, 1969). Since the sex of the individuals which were karyotyped was unspecified, it is not known whether the heteromorphic pair of chromosomes observed in both species were sex chromosomes.

In several species, heterogamety has been adduced from genetic rather than cytological evidence. In some instances, sex chromosomes have been identified by following the inheritance patterns of nonallelic sex-linked marker genes which affect morphological traits. Thus, male heterogamety of the XX:XY type has been demonstrated for several of the small, live-bearing poeciliid fishes, including two species in the genus *Poecilia* (Winge, 1922; Breider, 1935), and several species in the genus *Xiphophorus* (Kosswig, 1935, 1959; Bellamy, 1936; Gordon and Smith, 1938; Gordon, 1946, 1947; Kallman, 1965a), and for the cyprinodontid, *Oryzias latipes* (Aida, 1921). With one possible exception (see below), the heterosomes, X and Y, are not heteromorphic in these species (Winge, 1922; Iriki, 1932, cited in Denton, 1973; Friedman and Gordon, 1934; Wickbom, 1943), nor are they genetically nonhomologous throughout their length. Various authors, including Aida (1921), Winge and Ditlevsen (1947), Yamamoto (1961), and Kallman (1965b), have shown that genetic crossing over between the X and Y occurs at an appreciable frequency, although regions of nonhomology evidently exist since some genes behave as essentially X- or Y-linked. Genetic evidence of XX:XY male heterogamety also has been reported for the anabantid, *Betta splendens* (Kaiser and Schmidt, 1951), and for cultivated stocks of the cyprinid, *Carassius auratus* (Yamamoto and Kajishima, 1969).

The platyfish, *Xiphophorus maculatus*, is both male and female heterogametic. Initially, Bellamy (1922) and Gordon (1927) found that "domesticated" stocks of *X. maculatus* of unknown origin were female heterogametic and male homogametic (WZ:ZZ). When other xiphophorine species (e.g., *Xiphophorus variatus*) were then found to be XX:XY male heterogametic (Kosswig, 1935; Bellamy, 1936), it was questioned as to how or why two different modes of sex determination would arise in such closely related forms. This odd situation became confounded further by Gordon's (1946, 1947) discovery that *X. maculatus* populations in Mexico were male heterogametic. From a series of crosses between "domesticated" ♀ heterogametic and "wild" ♂ heterogametic forms of X. *maculatus*, Gordon (1946, 1947) concluded that the Z of the ♀ heterogametic forms was equivalent to the Y of the ♂ heterogametic forms. Since he could find no evidence that W was equivalent to X, he suggested that the use of WY:YY (WY = ♀, YY = ♂) was more appropriate than WZ:ZZ. Later, Gordon (1951) found naturally occurring populations of ♀ heterogametic X. *maculatus* from the Belize River in the former British Honduras.

It now appears that Gordon's appreciation of the sex chromosomes in X. *maculatus* was correct, although at the time (Gordon, 1952) he believed that X. *maculatus* was separated into two major, isolated populations or races—one to the west in Mexico which was ♂ heterogametic (XX:XY), and one to the east in British Honduras which was ♀ heterogametic (WY:YY). Kallman (1965b, 1970b, 1973) sampled X. *maculatus* extensively throughout its native range from near Veracruz, Mexico, southeast to British Honduras, and found that the species is polymorphic for three sex chromosomes, W, X, and Y. Of the six possible zygotic combinations, four (WW, WX, WY, and XX) normally differentiate into females; the remaining two (XY and YY) normally differentiate into males. Since the Y is ubiquitous, the mode of heterogamety is dependent on the frequency of the W or X chromosome. For example, in the Belize River where Gordon (1951) first discovered ♀ heterogametic populations, the frequency of the X is low (ca. 0.045) and the dominant mode is ♀ heterogamety (WY:YY). To the northwest in the Rio Jampa, Mexico, the frequency of the W is apparently negligible, and the populations are ♂ heterogametic (XX:XY). A geographic cline, however, is not indicated. Both the W and X are widespread, and WX ♀ are not infrequent. In fact, the only trend noted by Kallman (1973) was that the X is possibly more prevalent in populations at the periphery of the platyfish distribution, suggesting that the W chromosome was a secondary modification of an already existant sex chromosome which arose in the center of the species' range. The existance of three different heterosomes leading to both ♀ and ♂ heterogamety is best documented in X. *maculatus*, but also may obtain in another poeciliid, *Poecilia sphenops* (Schröder, 1964, cited in Kallman, 1973), and in the cichlid, *Tilapia mossambica* (Hickling, 1960).

2. Sex Determination

As pointed out by Yamamoto (1969), sex in bisexual fishes is determined much in the same manner as demonstrated by Bridges

(1925) in *Drosophila*. That is, "a given property, the *sex included*, depends upon all the chromosomes, some of which pull in one direction and others in the other direction, some strongly and others faintly or not demonstrably at all" (Bridges, 1939).

The early genetic experiments by Winge (1922) on the guppy, *Poecilia reticulata*, indicated that the species was ♂ heterogametic (XX : XY). Occasionally, however, he found exceptional individuals which were heterosomally of one sex, but phenotypically and functionally of the other sex. These "exceptions" proved fertile in crosses to "normal" individuals of the same sex chromosome constitution, but of the opposite sex. Exceptional XX ♂ crossed to normal XX ♀ produced all XX (♀) progeny, and exceptional XY ♀ crossed to normal XY ♂ produced male (2XY, 1YY) and female (1XX) offspring in a 3 : 1 ratio. Winge (1934) and Winge and Ditlevsen (1947, 1948) interpreted these results as indicating that minor male (M) and female (F) determining genes were situated throughout the genome. Normally, these minor autosomal genes were hypostatic to the heterosomal sex-determining genes; the sex of an individual was a function of its sex chromosome constitution. However, through chance genetic or chromosomal recombinations, the sum of the autosomal male- or female-potency could override the usually epistatic sex chromosome genes, and thus produce the "exceptional" individuals.

Similar explanations have been proposed by several authors to account for the sporadic appearance in nature and in the laboratory of these so-called "sex reversals" among the heterogametic xiphophorines (Kallman, 1968), and for hormone-induced sex reversals of *O. latipes* (Yamamoto, 1963). In the latter species, Aida (1936) established an XX : XX bisexual strain by selective breedings of exceptional XX ♂, but suggested that the XX ♂ could have stemmed from a lowering of the female-potency of X chromosome genes. Although Aida's suggestion may have partial validity, the general consensus is that sex-determination (at least in a number of poeciliids and *O. latipes*) is polyfactorial, with epistatic sex genes located on the sex chromosomes. Kosswig (1964) has discussed this mode of sex determination in some detail, and Yamamoto (1969) has presented a simple, but useful, model based on three overlapping normal distribution curves.

The number, location, and mode of interaction of the autosomal M and F genes are unknown; the gene action, however, is not strictly additive. Kallman (1968) found evidence of specific sex transformer genes (♀ → ♂) in *X. maculatus*. In this instance, a fortuitous combination of autosomal genes, derived from crosses of two specific strains, was apparently sufficient to override the strong female-potency of the W chromosome.

In several species, sex determination appears to be completely "polygenic," there being no genetic or cytological evidence of sex chromosome heterogamety. The most thoroughly studied example is the swordtail, *Xiphophorus helleri*. Over the years, Kosswig and his collaborators have found that sex ratios vary considerably among and within stocks of *X. helleri* and have proposed that "polygenes in their manifold recombinations decide about the sex of a specimen" (Kosswig, 1964, p. 195). A single pair of sex-indifferent autosomes, designated xx, are considered homologous to the sex chromosomes found in the heterogametic xiphophorines.

Interspecific hybridization studies between polygenic *X. helleri* and heterogametic *X. maculatus* have indicated that the M and F autosomal genes of *X. helleri* may, in certain combinations, be epistatic to sex-determining heterosomes (Kosswig, 1964). In the F_1 of crosses between *X. helleri* ♀♀ and *X. maculatus* ♂♂ (XY and YY), both male and female offspring were found among the chromosomal classes Xx and Yx. Sengün (1941, cited in Yamamoto, 1969), however, observed that Wx individuals from crosses of *X. helleri* ♂♂ to *X. maculatus* ♀♀ (WY) were all female, and that Xx individuals from crosses of *X. helleri* ♂♂ to *X. maculatus* ♀♀ (XX) were of both sexes. Presumably, this not only indicates that the W heterosome of *X. maculatus* has greater female-potency than the X, but also that the W itself has a very strong feminizing tendency—a fact substantiated by the somewhat infrequent occurrence of exceptional WY ♂♂ in natural populations of *X. maculatus* (Kallman, 1973). Based on his discovery that the sex transformer genes of *X. maculatus* may cause fluctuations in sex ratios, Kallman (1968) has suggested the interesting possibility that similar sex transformer genes may be prevalent in *X. helleri*.

Other species for which there is evidence of a polygenic mode of sex determination include two Caribbean poeciliids, *Poecilia caudofasciata* and *Poecilia vittata* (Breider, 1935, 1936), and possibly one anabantid, *Macropodus concolor* (references in Yamamoto, 1969).

3. EVOLUTION OF SEX CHROMOSOMES

In those species for which there is genic evidence of heterogamety, the sex chromosomes do not appear morphologically differentiated.* Furthermore, the presumed sex chromosomes invariably show exten-

---

* A possible exception is one stock of *X. maculatus* which may have heteromorphic X and Y chromosomes (Anders *et al.*, 1969).

sive genetic homology. For example, in the guppy, *P. reticulata*, Winge (1934) and Winge and Ditlevsen (1947) have shown that during male meiosis a number of sex-linked genes cross over freely between the morphologically similar X and Y chromosomes. In the platyfish, *X. maculatus*, crossing over between the W and Y occurs about as frequently as between the X and Y (Gordon, 1937; Kallman, 1965b). These results suggest that the sex chromosomes in these species may be viewed simply as a homologous pair of autosomes which acquired one or more sex-determining genes. Further evidence that extensive genetic homology still exists between the sex chromosomes is demonstrated by the occurrence of viable WW (in *X. maculatus*) and YY individuals (many species) among laboratory or natural populations (Yamamoto, 1969; Kallman, 1973).

The number of sex-determining genes is unknown. Ohno (1974) has suggested that the key to sex determination in fishes may reside in the regulation of a single enzyme which converts androstenedione to testosterone. If the enzyme is present, testosterone is produced and testicular development is stimulated. In the functional absence of the enzyme, estrogens are produced and ovarian development results. Thus sex determination could depend upon allelic relationships at a single regulatory locus (Ohno, 1974). Such a hypothesis is not incompatible with the results from numerous studies on the physiology of sex determination in fishes (Yamamoto, 1969). Theoretically, crossing over between the sex chromosomes should have no consequence; an "X" of one generation would merely be the "Y" of the next. A single locus system determining sex, however, is suspect for several reasons, not the least of which is its susceptibility to mutation. It is reasonable to assume, then, that at least a few nonallelic, closely linked genes determine sex. The finding of essentially X- and Y-linked Mendelian genes in *P. reticulata* (Winge and Ditlevsen, 1947) indicates that stretches of nonhomology exist on the X and Y, and that crossovers within the region containing the sex determining loci are extremely rare (if they occur at all).

The position of the sex determining loci on the chromosome is of importance, since crossing over within the region containing the sex genes (or between this region and the centromere) might prove disruptive. Anders *et al.* (1973) have shown that the male and female determining regions of the *X. maculatus* and *X. variatus* sex chromosomes· X and Y map adjacent to the centromere, and proximal to all known sex-linked Mendelian genes. Since the chromosomes of *X. maculatus* and *X. variatus* are acrocentric (Friedman and Gordon, 1934), this location is ideal. Chances of crossing over between the "sex" region and the centromere would be small, and should be reduced further by an expected nonrandom distribution of crossovers along the length of the chromosome. In other experimentally tractable animals (e.g., *Drosophila*) there is ample evidence that crossing over itself is reduced in regions proximal to the centromere. Once situated, each sex-determining region might then undergo paracentric inversion(s) to further reduce any chance of recombination (Ohno, 1967).

Along these lines, Yamamoto's (1961) study on *O. latipes* is of interest. Using two alleles at a sex-linked locus for body color (R and r, where R– = orange red and rr = white), he constructed a stock in which the R allele was Y-linked, and the r allele X-linked. In normal ♂ ♂ (X^rY^R), crossing over between the R locus and the sex-determining loci occurred at a low frequency (ca. 0.2%). But when sex-reversed X^rY^R ♀ ♀ were examined, it was found that crossing over had increased more than fivefold. Apparently, there may also be physiological regulation of crossing over between the X and Y, which normally prevents the disruption of the specific "sex" genes.

The isolation of a "differential segment" containing the sex-determining genes in a position adjacent to the centromere may represent the initial stages leading to morphological differentiation or heteromorphy of the sex chromosomes. A subsequent pericentric inversion involving the sex region of one heterosome might further isolate the sex-determining genes, and result in detectable heteromorphy. In time, unequal crossing over, or perhaps nonreciprocal translocation, could bring about extreme heteromorphy with most if not all of the Mendelian genes eliminated from one heterosome. From studies on different families of snakes, Ohno (1967) has diagrammed this progression to heteromorphy, beginning with the cytologically detectable pericentric inversion.

Whether the chromosomal location of the sex-determiners in *X. maculatus* represents the incipient stages of heteromorphy is unknown. Anders *et al.* (1969) reported heteromorphic X and Y chromosomes in one stock of *X. maculatus*, which may be the result of unequal exchange or translocation (Anders *et al.*, 1973). Other *X. maculatus* populations, however, are apparently homomorphic, as are *O. latipes* and other closely related poeciliids (references in Denton, 1973; Chiarelli and Capanna, 1973). As an alternate hypothesis, Ebeling and Chen (1970) suggested that homomorphy may be adaptive since it could ensure a certain lability of sexual expression.

Among the deep-sea bathylagids studied by Chen (1969), the cytological progression to increasing heteromorphy (as outlined in Ohno, 1967) is apparently observed in the four species of the genus

Bathylagus which possess heteromorphic sex chromosomes (Ebeling and Chen, 1970). The same may also be true for the two fundulines, *F. diaphanus* and *F. parvipinnis*, even though the centromere position of the heteromorphic Y differs in these two species (Chen and Ruddle, 1970; Ebeling and Chen, 1970). However, in the remaining species where heteromorphic sex chromosomes have been reported, there is no indication as to whether they represent isolated exceptions or the initial stages of incipient heteromorphy. In short, the above examples notwithstanding, the sex chromosomes in most fishes have remained in a relatively undifferentiated state.

IV. CHROMOSOME NUMBERS, CHROMOSOME MORPHOLOGY, AND GENOME SIZES: THE EVOLUTION OF FISH KARYOTYPES

> *The karyotype, as far as its morphological features are concerned, is also part of the phenotype, as it is the result of an evolution whose course may have been varied.*
>
> Mario Benazzi (1973)

Considering the vast number of living fish species, their diversity in morphology, and the antiquity of the group as a whole, one might expect to find a corresponding wealth of karyotypic diversity. Surprisingly, this does not appear to be the case. Many orders are relatively uniform in karyotype, although they may differ in evolutionary age by tens of millions of years. For example, the haploid (n) karyotype of 24 acrocentric chromosomes is found throughout several diverse orders of the subclass *Teleostei* (class *Osteichthyes*) and appears to be the predominant karyotype in the recently evolved *Perciformes* (Roberts, 1964, 1967; Denton, 1973; Chiarelli and Capanna, 1973). This has led to the suggestion that the 24 acrocentric chromosome complement may be ancestral to all modern fishes, and perhaps was possessed by the primordial teleost (*Leptolepis*) over 100 million years ago (Ohno, 1974). An even more entertaining possibility is that this chromosome configuration may have been ancestral to all vertebrates (Ohno *et al.*, 1968; Ohno, 1974). Nogusa (1960) and Taylor (1967) found this karyotype in two hagfish species of the primitive order *Cyclostomata*. These hagfish species represent the primitive jawless *Agnatha* which separated from the main line of vertebrate evolution more than 300 million years ago. But other primitive species from the classes *Agnatha*, *Chondrichthyes*, and *Osteichthyes*, do not have a 24 acrocentric karyotype (Denton, 1973; Chiarelli and Capanna, 1973).

A. Chromosome Numbers and Genome Sizes

Reported chromosome numbers in fishes range from a low of $n = 8$ in the cyprinodontid, *Notobranchius rachovii* (Post, 1965), and the anabantid, *Sphaerichthys osphromonoides* (Calton and Denton, 1974), to a high of $n = 84$ in the petromyzontiform lamprey, *Petromyzon marinus* (Potter and Rothwell, 1970). The wide range in chromosome number is misleading. About 35–40% of almost 500 assayed fish species from 76 families in 26 orders have $n = 24$ chromosomes (Denton, 1973; Chiarelli and Capanna, 1973). The distribution is strongly leptokurtic; almost 70% of species have chromosome numbers in the range $n = 22$–26, and ca. 80% of species fall in the range $n = 20$–28. Two minor peaks are found in the ranges $n = 40$–52 (ca. 5–6%) and $n = 82$–84 (ca. 0.8%). The former peak contains species from the teleost families *Salmonidae*, *Cyprinidae*, *Catostomidae*, and *Cobitidae*, a few petromyzontiform lampreys, one chondrostean, and a few skates of the order *Rajiformes*. The latter peak contains four species of petromyzontiform lampreys.

Chromosome numbers and variabilities in chromosome number distinguish certain major taxonomic groupings of fishes. For example, salmoniform species have higher chromosome numbers (median ≈ 36 haploid chromosomes) than cypriniform species (median ≈ 25 haploid chromosomes), and they also are more variable in chromosome number. Within the *Salmoniformes*, chromosome numbers are in the range $n = 11$–51 and are distributed platykurtically; within the *Cypriniformes*, chromosome numbers are in the range $n = 18$–52 but are distributed leptokurtically. Trends also are observed at lower taxonomic levels. Briefly, groups with chromosome numbers in the range $n = 22$–26 tend to be relatively invariant in chromosome number, whereas groups with higher or lower chromosome numbers tend to be more variable.

There are at least three cytological mechanisms which may bring about changes in chromosome number: (1) polyploidization, where the chromosome number is increased to an exact multiple of the basic chromosome set, (2) Robertsonian rearrangements, where centric fusion of two nonhomologous acrocentric chromosomes produces a single metacentric, or where centric dissociation of a single metacentric produces two nonhomologous acrocentrics (Robertson, 1916), and

(3) aneuploidy, where nondisjunction or endoreduplication results in gain or loss of individual chromosomes.

Genome size, or the amount of DNA per nucleus, also shows wide variation among fishes. Haploid DNA contents range from 0.4 pg $(10^{-12}$ g) per nucleus in tetraodontiform puffers (Hinegardner and Rosen, 1972) to 124 pg per nucleus in the lungfish, *Lepidosiren paradoxa* (Markert, 1968). Bachmann *et al.* (1972), however, found that DNA contents of 195 fish species, including representatives from all three classes, exhibited a unimodal distribution, skewed towards higher DNA values. When log transformed, the distribution was normal around a strong mode at 1.7 pg (diploid amount). Among the teleost fishes, the distribution is similar but the estimated mode (haploid amount) is 1.0 pg (Hinegardner, 1968; Hinegardner and Rosen, 1972).

The evolutionary implications of changes in genome size in fishes have been studied by Hinegardner (1968), Hinegardner and Rosen (1972), Bachmann *et al.* (1972), and Ohno (1970, 1974), and only their broad conclusions are summarized here. There is usually homogeneity of DNA amounts within families and lower taxonomic categories, and genome sizes tend to be relatively stable despite changes in morphology and/or physiology. Notable exceptions occur in the families *Cyprinidae, Cyprinodontidae,* and *Callichthyidae,* where species may differ in DNA content by more than twofold. A greater variation in DNA amounts is observed among certain orders. For example, the average DNA content in 104 perciform species is 1.04 pg, whereas 32 siluriform species average 1.78 pg (Hinegardner and Rosen, 1972). The increased DNA content of the *Siluriformes* stems primarily from the high DNA contents of the families *Callichthyidae* and *Loricariidae* (averaging 2.68 pg per species).

Decreases in DNA content often are associated with increasing specialization in body form and design. More specialized species have less DNA per cell than do more generalized forms.* This inverse relationship between genome size and degree of specialization holds for fishes as a group, and also within certain taxa (Mirsky and Ris, 1951; Hinegardner, 1968; Hinegardner and Rosen, 1972). A rationale for this pattern of DNA loss has been developed by Ohno *et al.* (1968), Ohno (1970, 1974), and Bachmann *et al.* (1972). In their view, increases in genome size, particularly when provided by polyploidization, result

---

* The terms "specialized" and "generalized" as used here follow the definitions of Hinegardner and Rosen (1972). Specialized groups (or species) are those which share few features in common with related members of the same taxon; generalized species share many features in common with phylogenetic relatives.

in major adaptive shifts; following such a shift, loss of "excess" DNA accompanies specialization.

Within taxonomic families, a significant correlation exists between genome size and variation in genome size (Hinegardner and Rosen, 1972). Specialized families with small genome sizes tend to be less variable in genome size, and almost all families with very low average DNA contents per species (0.4–0.6 pg) have very little variation in DNA content among species.

Exceptions to the trend of decreasing genome size with increasing specialization are found among the catfishes of the order *Siluriformes*. Species in the specialized families *Callichthyidae* and *Loricariidae* have, on the average, two- to threefold more DNA per nucleus than does the average species in other siluriform families (Hinegardner and Rosen, 1972). Other exceptions include the specialized families *Scaridae* and *Gobiidae* of the order *Perciformes,* which have more DNA than 36 other perciform families (Hinegardner and Rosen, 1972). The biological significance of these exceptions is unknown.

The most poignant exceptions to the trend are the dipnoan lung fishes (order *Lepidosireniformes*) of the subclass *Crossopterygii.* These ancient, but specialized, fishes have genome sizes some 80–100 times as large as the average fish and at least 25–40 times as large as their closest living relative, *Latimeria chalumnae* (Mirsky and Ris, 1951; Ohno and Atkin, 1966; Cimino and Bahr, 1974). The enormous genome size of the dipnoans is a characteristic shared only with the Urodele *Amphibia* (Morescalchi, 1973).

Cytological mechanisms which could lead to increases in genome size include polyploidy (Ohno *et al.,* 1968; Ohno, 1970), "lateral increases" through differential polynemy (Rothfels *et al.,* 1966), "longitudinal increases" through accidental DNA doubling (Sparrow and Nauman, 1974), unequal crossing over (Spofford, 1972; Ohno, 1974), and regional disturbances in DNA replication (Keyl, 1965; Price, 1976). Mechanisms which might lead to decreases in genome size include unequal crossing over, regional disturbances in DNA replication, or misrepair of chromosome "breaks" (Bachmann *et al.,* 1972; Spofford, 1972; Sparrow *et al.,* 1972; Price, 1976).

Based on the log normal distribution of DNA content observed among fishes, Bachmann *et al.* (1972) suggested that changes in genome sizes were small, numerous, and cumulative, and most likely stemmed from successive duplications and/or deficiencies. Large changes such as implied by polyploidy were exceptional. Goin and Goin (1968) and Bachmann *et al.* (1972) view the decreases in genome size as due to loss of unnecessary and/or redundant DNA. Whether these DNA losses occur subsequent to loss of gene function, or whether the losses themselves cause loss of gene function, is problematic (Hinegardner and Rosen, 1972).

Although there is little direct evidence to indicate the cytological processes responsible for the decreases in genome size in fishes, one line of reasoning suggests that much of the reduction in genome size occurs during chromosomal rearrangements which produce changes in chromosome number. Among diploid teleost fishes there is a highly significant, positive correlation between chromosome number and genome size (Hinegardner and Rosen, 1972). This correlation holds when species with probable polyploid ancestry (see Section IV,D) are excluded from the calculation. Species or species groups with higher chromosome numbers tend to have larger genome sizes. Although exceptions exist (Hinegardner and Rosen, 1972), the clear implication is that reduction in chromosome number is accompanied by reduction in genome size, and hence may be viewed as another process correlated with increasing specialization and advancement.

Even though the overall picture of fish karyotype evolution is as yet unclear, a few salient features are apparent. The trend in fish karyotype evolution is toward smaller genome size. This presumably is accomplished in part by chromosomal rearrangements which reduce chromosome number. DNA loss may in itself be adaptive by altering certain biophysical parameters related to genome size (Bennett, 1972; Price, 1976), and then too, reduction in chromosome number may be adaptive through tightening of linkage (Mather, 1953; Stebbins, 1958). Nikolsky (1976) has suggested that reduction in chromosome number (and also in genome size) in fishes may be associated with increasing habitat stability and effectiveness of food resource utilization.

One may further speculate that once a species or species group reaches a small genome size, chromosome structural changes which result in further DNA loss should no longer be easily tolerated. If this is true, highly specialized taxa should be relatively invariant in genome size and in karyotype. In general, this is the case. In the highly specialized order *Perciformes*, the average species (excluding those from the families *Scaridae* and *Gobiidae*) has ca. 0.97 pg of DNA per haploid nucleus (Hinegardner and Rosen, 1972). This estimate is relatively low when compared with most other teleostean orders. The *Perciformes* are also relatively homogeneous in genome size and chromosome number (Hinegardner and Rosen, 1972; Denton, 1973). In contrast, species from the less specialized order *Cypriniformes* [excluding two cyprinids, *Carassius auratus* and *Cyprinus carpio*, and the family *Catostomidae*, which are apparently polyploid (see Section IV,D)] have on the average ca. 1.33 pg of DNA per haploid nucleus, and also are more variable in genome size and chromosome number (Hinegardner and Rosen, 1972; Denton, 1973). The same trend holds for comparisons within orders. Many species in the families *Gobiidae* (*Perciformes*), *Callichthyidae* and *Loricariidae* (*Siluriformes*) each have average DNA contents higher than the average species in their respective orders. They also appear to be more heterogeneous in genome size and chromosome number (Hinegardner and Rosen, 1972; Scheel *et al.*, 1972; Denton, 1973). The indication is that taxa with high DNA contents may have greater flexibility in terms of chromosomal rearrangement.

B. Chromosome Morphology and Polymorphism

1. MORPHOLOGY

Among diploid species, each pair of homologous chromosomes is assumed to differ genetically from all other chromosome pairs in the same cell. Outward manifestations of some of these differences comprise the morphological "phenotype" or karyotype and include differences between chromosome pairs in relative size, shape, and centromere position. Karyotypic differences among species or taxa may be used to determine phenetic similarities and phylogenetic relationships.

The concept of "symmetrical" versus "asymmetrical" karyotypes has been developed by several authors (Stebbins, 1958, 1971; White, 1973a) to indicate the apparent degree of chromosomal heterogeneity within a karyotype. In a perfectly symmetrical karyotype all chromosomes are approximately the same size and shape and have medially located centromeres. The trout (family *Salmonidae*) karyotype (Fig. 1) is quite symmetric; the metacentrics essentially comprise one size group, and the acrocentrics a second. By contrast, the minnow (family *Cyprinidae*) karyotype (Fig. 2) is highly asymmetric; this is typical of most *Cyprinidae*. Not only are there apparent differences in chromosome size, there are also obvious differences in centromere position even within groups of chromosomes of approximate size.

Degrees of asymmetry in karyotype among the fishes are broadly taxon specific. As noted above, the *Cyprinidae* predominantly have highly asymmetrical karyotypes, the *Salmonidae* much less so. Examples also exist within orders, for example, the highly asymmetrical karyotypes of the salmoniform family *Bathylagidae* contrast sharply with the symmetry of the *Salmonidae* (Chen, 1969).

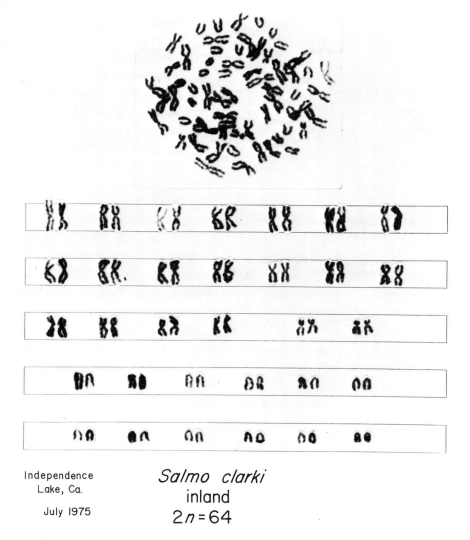

Independence Lake, Ca.

July 1975

Salmo clarki

inland

$2n = 64$

Fig. 1. Somatic metaphase karyotype from kidney cell of inland cutthroat trout, *Salmo clarki* (family *Salmonidae*), $2n = 64$.

Generally, fish chromosomes are smaller in size than chromosomes in most vertebrates. The length of the "average" fish chromosome is between 2 and 5 μm. Many species possess numerous small chromosomes of 2 μm or less, but which are nonetheless easily seen through the light microscope. Very large chromosomes of 15–30 μm in length, such as those found in the lungfish, *Lepidosiren paradoxa* (Ohno and

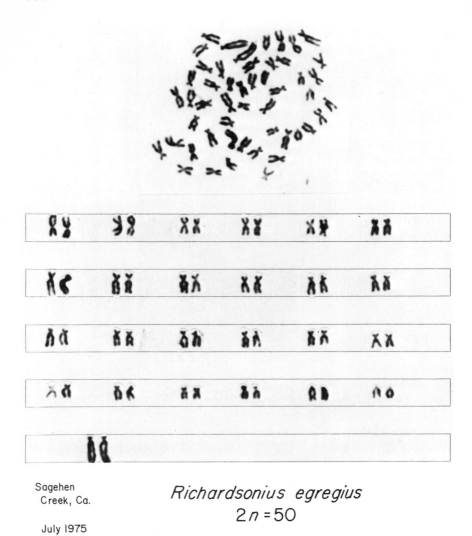

Sagehen
Creek, Ca.

July 1975

Richardsonius egregius
$2n = 50$

Fig. 2. Somatic metaphase karyotype from kidney cell of Lahontan redside, *Richardsonius ergregius* (family *Cyprinidae*), 2n = 50.

Atkin, 1966), or the extremely bizzare SM (*supermacro*) chromosomes found by Post (1973) in two forms of the family *Diretmidae* (*Beryciformes*), are rare. In the past the overall small size of fish chromosomes often has precluded the accurate determination of centromere position, but improved cytological techniques now available should remedy this situation. Most workers follow the standard nomenclature of Levan *et al.* (1964) in identifying particular chromosome shapes.

Extremely small chromosomes (microchromosomes) have been reported in a few species (Ohno and Atkin, 1966; Ohno *et al.*, 1969b; Chen, 1969). Ohno *et al.* (1969b) found between 26 and 48 microchromosomes in the karyotypes of three very primitive species, *Hydrolagus colliei* (ratfish), *Scaphirhynchus platorhynchus* (sturgeon), and *Lepisosteus productus* (gar). Elsewhere among the vertebrates microchromosomes are found in the birds (Ray-Chaudhuri, 1973) and in certain reptiles (Ohno, 1967; Gorman, 1973). Ohno *et al.* (1969b) pointed out that although microchromosomes could have arisen independently at the reptilian stage of vertebrate evolution, their presence in these relic fishes suggests that the evolution of terrestial vertebrates from the fishes was polyphyletic, and furthermore, that the fish ancestral to all birds may already have possessed microchromosomes.

2. POLYMORPHISM

A chromosomal rearrangement (inversion or translocation) arises in a single individual and may be passed on to its progeny. If adaptive, it may increase in frequency in the population. The "new" and "old" gene arrangements are inherited essentially as if they were Mendelian alleles at a single locus. Thus, when more than one chromosomal *morph* is present in a population, the population is said to be polymorphic and individuals are considered as homozygous or heterozygous in the traditional genetic sense.

The importance of chromosomal polymorphism to evolution is stated by White (1973a): "Where a parallelism exists . . . between intraspecific chromosomal rearrangements and interspecific ones we may legitimately conclude that balanced chromosomal polymorphism has given rise to cytotaxonomic differences between species through one chromosome sequence undergoing fixation in one incipient species while an alternative sequence undergoes fixation in a second incipient species" (p. 764). This correlation between fixation of alternate chromosomal sequences and incipient speciation is well documented in both plant and animal species, and has been considered as evidence that some chromosomal rearrangements are adaptive and strongly influenced by natural selection (Grant, 1963; Dobzhansky, 1951, 1970; White, 1973a,b). Occasionally, a chromosomal polymorphism may persist in a species in a "balanced" condition, presumably through a mode of selection which maintains heterozygosity. This too is well documented in both plants and animals (see above references).

In contrast to other animal species, there are few reports of chromosomal polymorphisms among the fishes, although they certainly must occur given the diversity of karyotypes observed in fishes. Why chromosomal polymorphism appears low in fishes is unknown. One possibility is that chromosomal rearrangements in fishes tend to be fixed rapidly [perhaps as a result of small effective population sizes (see Wright, 1943; Wilson *et al.*, 1975)]. A more plausible alternative is that the present techniques used in fish cytology provide insufficient resolution of chromosome structure to discern most polymorphisms. Further, the number of individuals examined per population usually is small, and existing polymorphisms may go unnoticed.

The most widely known instances of chromosomal polymorphism in fishes occur among certain genera of the family *Salmonidae*, notably in the genus *Salmo*. Thorgaard (1976) has recently presented convincing evidence of an intrapopulation chromosomal polymorphism in the rainbow trout (*Salmo gairdneri*). In the most thoroughly studied species, the Atlantic salmon, *Salmo salar*, reported chromosome numbers range from $2n = 54$ to $2n = 60$. European populations of *S. salar* range in chromosome number from $2n = 58–60$ (Prokofieva, 1934; Svärdson, 1945; Rees, 1964, 1967; Nygren *et al.*, 1968a, 1972), whereas North American populations range from $2n = 54–57$ (Boothroyd, 1959; Roberts, 1968, 1970).

Changes in chromosome number without change in the number of chromosome arms (*nombre fondamental*) constitute *primae facie* evidence for Robertsonian fusions or dissociations (Matthey, 1945, 1973). Arm number estimates in North American *S. salar* reveal that most of the polymorphism is Robertsonian. Boothroyd (1959) and Roberts (1968, 1970) found estimates of 72 arms in seven North American populations, despite variations in chromosome number from $2n = 54–57$. Among European *S. salar* there is evidence of variation in arm number as well as in chromosome number. Prokofieva (1934) and Svärdson (1945) reported karyotypes of *S. salar* consisting of $2n = 60$ chromosomes; in both, six chromosome pairs were metacentric giving an arm number estimate of 72. Rees (1964) also found a *S. salar* karyotype of $2n = 60$ from a hatchery population in Wales. After a 2-year study of the same population Rees (1967) acknowledged a karyotype of $2n = 58$ chromosomes containing 74 chromosome arms as the correct one. This same karyotype ($2n = 58$ with 74 arms) has been found among several Swedish *S. salar* populations (Nygren *et al.*, 1968a, 1972).

The karyotype data from *S. salar* (North American and European) suggest moderate chromosomal restructuring within the species involving both Robertsonian and non-Robertsonian rearrangement. Centric fusions and/or dissociations could account for the observed variation in chromosome number, while uneven translocation or pericentric inversion could account for the discrepancies in arm number. The real situation is more complex. Several investigators have commented on the apparent absence in S. *salar* of a fixed karyotype *within* individuals; cells examined from the same or different tissue of a single specimen showed extensive intraindividual variation (or polymorphism) in chromosome number, sometimes by as much as 16 chromosomes (Nygren *et al.*, 1968a). Rees (1967) and Boothroyd (1959) felt that much of the observed variation was more apparent than real and was due to counting error and/or artifacts caused by chromosome preparation technique. It was also suggested (Svärdson, 1945; Boothroyd, 1959) that some of the variation could stem from genuine aneuploidy, without phenotypic alteration.

Ohno *et al.* (1965) observed similar intraindividual polymorphism in the rainbow trout, *Salmo gairdneri,* but found that much of the variation followed a distinct Robertsonian pattern. By counting both chromosome and chromosome arm numbers in several cells from various tissues of the same individual, they identified seven distinct karyotypes ranging from $2n = 58$–65, each of which possessed 104 chromosome arms. Since the modal karyotype (35% of counts) of $2n = 60$ with 104 arms was consistent with other published karyotypes of *S. gairdneri* (Wright, 1955; Bungenberg de Jong, 1955; Simon and Dollar, 1963), Ohno *et al.* (1965) concluded that they were "witnessing for the first time an example of Robertsonian polymorphism within single individuals" (p. 118). Similar patterns of intraindividual Robertsonian polymorphisms have subsequently been described in several salmonid species, including *S. salar* (Roberts, 1968, 1970; see also Ohno *et al.*, 1969a; Davisson *et al.*, 1973; Gold and Gall, 1975; Zenzes and Voiçulescu, 1975).

The cytological mechanisms that produce intraindividual polymorphism are not well understood. Beçak *et al.* (1966a) suggested two possible alternatives: (1) During zygotic development, certain chromosomes undergo Robertsonian exchange (fusion and dissociation) without apparent harm to cell viability, or (2) zygotes begin development structurally heterozygous for several Robertsonian rearrangements, then during cell division undergo a segregation pattern called *somatic segregation* which tends to restore structurally homozygous cell types. The first alternative predicts that all individuals, whether structurally heterozygous or homozygous for any number of Robertsonian rearrangements, should display the intraindividual polymorphism pattern, at least to some extent; the second predicts that individuals should exist which as zygotes were structurally homozygous for all the Robertsonian rearrangements segregating in the population, and hence should display a fixed karyotype in all cells. Whether either or both of the two alternatives is operative in salmonid species which show this polymorphic pattern is yet unknown (Beçak et al., 1966a; Ohno et al., 1969a; Gold and Gall, 1975; Roberts, unpublished observations).

C. Chromosome Changes and Speciation

The idea that chromosomal rearrangement is involved in the process of speciation is an old one in evolutionary biology, and is thoroughly discussed in the literature (Dobzhansky, 1951, 1970; Mayr, 1973; White, 1973a,b). White (1968) notes that "it is a matter of empirical observation that . . . even the most closely related [higher animal] species are usually found to differ in karyotype. . . . The only sure exceptions to this generalization seem to be certain (homosequential) species complexes in the genus Drosophila" (p. 1065).

There are at least two reasons why chromosomal rearrangements may be important to speciation. The first is by providing a postmating reproductive isolation mechanism that renders F_1 hybrids partially or completely sterile (White, 1973a,b). Individuals heterozygous for one or more structural rearrangements would be expected to produce duplication-deficiency gametes due to chromosome pairing and segregation irregularities at meiosis.

The second reason is that chromosomal restructuring, rather than point mutation, may effect significant changes in the patterns of gene regulation (Wallace, 1963; Stebbins, 1969; Wilson et al., 1974a,b). Insofar as speciation is concerned, it has been proposed that "genetic revolutions" may occur by changes in gene arrangement rather than by accumulated changes in structural genes. The evidence for this is that rates of organismal evolution (e.g., anatomy or way of life) in certain groups are correlated with rates of chromosomal evolution, but not with rates of protein evolution (Wilson et al., 1974a,b, 1975; Prager and Wilson, 1975).

Among the fishes, several taxa show evidence of extensive chromosomal rearrangement, much of which presumably is associated with speciation. Several cyprinodontids in the genera Aplocheilus, Aphyosemion, Epiplatys, and Fundulus differ markedly in karyotype. These differences apparently stem from both Robertsonian rearrangement and uneven translocation or pericentric inversion (Post, 1965; Scheel, 1968, 1972; Chen, 1971; see also Gyldenholm and Scheel, 1971; Denton, 1973). Similar examples are found among species or taxa of certain pleuronectiforms (LeGrande, 1975), the perciform *Gobiidae* (Nogusa, 1960; Chen and Ebeling, 1971), and neotropical cichlids (Thompson, 1976).

Among the *Salmonidae*, several genera reflect rather extensive chromosomal rearrangement. Booke (1968, 1970, 1974) examined eleven North American species from *Prosopium* and *Coregonus* and found evidence of both Robertsonian rearrangement and uneven translocation or pericentric inversion; six species of *Prosopium* had the same arm number (100) but different chromosome numbers (from 64 to 82), and five species of *Coregonus* had the same chromosome number (80) but different arm numbers (from 98 to 106). European species of *Coregonus* may differ in chromosome number (references in Booke, 1968). Simon (1963) found that five species of *Oncorhynchus* ranged in chromosome number from $2n = 52$–74, and in arm number from 102 to 112. Perhaps the most karyotypically variable taxa is the genus *Salmo;* reported chromosome numbers (seven species) range from $2n = 54$–80, and arm numbers range from 72 to 106 (Svärdson, 1945; Wright, 1955; Boothroyd, 1959; Simon and Dollar, 1963; Rees, 1967; Roberts, 1967, 1968, 1970; Nygren *et al.*, 1968a, 1972; Miller, 1972; Gold and Gall, 1975).

It is well known, however, that many species and genera within the *Salmonidae* hybridize in culture and in nature (Hubbs, 1955; Buss and Wright, 1956; see list of references in Dangel, 1973). A case in point are the trouts (*Salmo*) endemic to the Pacific Northwest. The chromosome numbers (arm numbers) reported for these trouts are *S. clarki clarki* (coastal subspecies), $2n = 68$ (104); *S. clarki henshawi* (inland subspecies), $2n = 64$ (104); *S. gairdneri*, $2n = 60$ (104); *S. aguabonita*, $2n = 58$ (104); and *S. apache*, $2n = 56$ (106) (Wright, 1955; Ohno *et al.*, 1965; Miller, 1972; Wilmot, 1974; Gold and Gall, 1975; Gold *et al.*, 1977; Thorgaard, 1976). These species apparently comprise a Robertsonian series with the predominant rearrangement being centric fusion (Gold *et al.*, 1977). And yet, hybridization between several of these species occurs freely, and often at a very high frequency (Gould, 1966; Behnke, 1970, 1972; Dangel, 1973; Gold and Gall, 1975; Gold *et al.*, 1976). Moreover, in several cases the hybrids and their offspring are apparently fertile (Behnke, 1972). The overall indication is that among related species in the *Salmonidae*, partial or complete sterility due to structural heterozygosity may not contribute to reproductive isolation.

The same appears to be true among fish taxa which by present cytological resolution are extremely conservative karyotypically. Roberts (1964) examined twenty of the thirty extant species of the North American perciform family *Centrarchidae* and found that fifteen had virtually indistinguishable karyotypes of $n = 24$ acrocentric chromosomes. All fifteen are known to produce fertile hybrids (references in Avise and Gold, 1977). The implication is again that sterility due to structural heterozygosity may not contribute to reproductive isolation.

In Section IV,A, it was noted that fishes are remarkably conservative in karyotype. About 35–40% of all species karyotyped have $n = 24$ chromosomes, and about 70% fall in the range $n = 22-26$. Of these, many show little or no variation in arm number (Denton, 1973). This is more pronounced in certain taxa. In the order *Perciformes*, 70–75% of the species examined have $n = 24$ chromosomes, and with three exceptions, all have karyotypes in the range $n = 22-26$. Furthermore, of those perciforms (excluding the gobiids) for which arm number data are available (Denton, 1973), ca. 70% have arm numbers of 48. In view of the large number of extant fish species and the relatively high frequency of hybridization which generally characterizes fish (Hubbs, 1955), the overall conservative nature of fish karyotypes suggests that chromosomal restructuring in fishes may not be a prime contributor to reproductive isolation. It also suggests that some speciation events may not necessarily be accompanied by (observable) chromosomal changes.

Wilson *et al.* (1975) recently published estimates of rates of chromosomal change in fifteen vertebrate groups. Their estimate for teleost fishes (23 genera) was 1.5 changes in arm number and 1.1 changes in chromosome number per lineage per 100 million years. This may possibly be an overestimate. Of the 23 teleost genera (one-hundred eight species) sampled, 6 (forty-seven species) were from genera which are extremely variable in karyotype (G. L. Bush, personal communication). However, since Wilson *et al.* (1975) restricted their sample only to genera known to occur as fossils, the estimate for teleosts may be regarded as reasonable, although perhaps somewhat elevated. When compared to other vertebrates the rates of chromosomal evolution in teleosts were over threefold less than in placental mammals, about the same as in snakes and lizards, higher than in frogs, and much higher than in turtles, crocodiles, and salamanders (Wilson *et al.*, 1975). Since the number of extant fish species is easily double that of most other vertebrate groups, the indication again is that many speciations in fishes may not be accompanied by gross chromosomal change.

In a recent study, Avise and Gold (1977) compared the karyotypes of several North American cyprinids (*Leuciscinae*) with those of the North American sunfish genus *Lepomis* (see also Gold and Avise, 1977). The subfamily *Leuciscinae* is a highly speciose taxon (ca. two-hundred fifty species), whereas the genus *Lepomis* is species poor (eleven species); both are thought to be of approximate evolutionary age. At the level of gross chromosomal organization, we found little evidence of greater chromosomal evolution among the speciose *Leuciscinae* than among the species-poor *Lepomis*. This suggests that the rates of regulatory evolution, as reflected in gross chromosomal rearrangement, do not appear more rapid among the speciose *Leuciscinae*.

The foregoing discussion indicates that many speciations in fishes may occur in the absence of chromosomal rearrangement. The data, however, should only be treated as suggestive, since important chromosome structural changes could have occurred beyond the resolution of present cytological techniques. Certainly, the application of higher resolution methodology as discussed in Section II,C is to be encouraged.

D. Polyploidy and Aneuploidy

1. POLYPLOIDY

Incipient polyploidy among bisexual vertebrates is extremely rare, and has been verified only recently among certain *Amphibia* (Beçak *et al.*, 1966b, 1967, 1970). In contrast, polyploidy is common among higher plants, and apparently has played a major role in speciation and evolution (Stebbins, 1971). White (1973a) has listed a few reasons why polyploidy should be rare in bisexual species, including Muller's (1925) suggestion that heterogametic sex-determining mechanisms might be disturbed.

In fishes, the all-female triploid unisexuals (Section III,B) are the only substantiated instances in nature of polyploidy at the population level. There are reports of tetraploid ($4n$) individuals among Japanese populations of *Carassius auratus* (Kobayasi *et al.*, 1970), and moreover that some of these populations contain triploids of both sexes (Muramoto, 1975). However, since unisexuality occurs in *C. auratus* these populations may be gynogenetic (Kobayasi, 1971). Isolated and very rare triploid individuals have been identified cytologically in the rainbow trout, *Salmo gairdneri* (Cuellar and Uyeno, 1972), and the western roach, *Hesperoleucus symmetricus* (Gold and Avise, 1976), but they are expected to be sterile.

There is, however, circumstantial evidence that a few extant groups or species of fish are ancestral polyploids. Ohno and his colleagues have published in depth on the subject. In their view these "polyploids" are likely the result of nature's experimentation with gene duplication (Ohno and Atkin, 1966; Atkin and Ohno, 1967; Muramoto *et al.*, 1968; Ohno *et al.*, 1967, 1968, 1969a; Ohno, 1970, 1974). White (1946, 1973a) and others have questioned whether polyploidy could have occurred at all in a bisexual species and have noted that the evidence "has never been of a conclusive kind."

Actually, there are only a few living fishes for which ancestral polyploidy of this sort has been suspected. These include the Northern Hemisphere genera of the lamprey family *Petromyzontidae* (Ohno *et al.*, 1968; Howell and Denton, 1969; Potter and Robinson, 1973; but see Robinson *et al.*, 1975), the cypriniform family *Catostomidae* (Uyeno and Smith, 1972), three species in the *Cyprinidae* (*C. auratus*, *Cyprinus carpio*, and *Barbus barbus*, Ohno and Atkin, 1966; Ohno *et al.*, 1967, 1968; Ohno, 1974), the loach, *Misgurnus fossilis* (Raicu and Taisescu, 1972), the family *Salmonidae* (see below), and one form of the beryciform genus *Diretmus* (Post, 1973). The usual evidence for ancestral polyploidy is that the species or groups in question have both genome sizes and chromosome numbers which are approximately twofold greater than those of closely related taxa.

Some caution is advised, however, before one considers a species as ancestrally polyploid since there are instances where either genome size or chromosome number (but not both) appear to have been increased substantially. For example, the clown loach, *Botia macracantha*, of the cypriniform family *Cobitidae* has $n = 49$ chromosomes as compared to the Khulli loach, *Ancanthophthalmus khulli*, which has $n = 25$. Both species have genome sizes of about 1.0 pg per haploid nucleus which is average for most cypriniforms (Muramoto *et al.*, 1968). A similar example is found in the salmoniform *Bathylagidae* (Ebeling *et al.*, 1971).

The reverse situation, substantial increase in genome size without apparent concomitant increase in chromosome number, also has been observed in genera of the siluriform *Callichthyidae* and *Loricariidae* (Muramoto *et al.*, 1968; Hinegardner and Rosen, 1972), the hagfish family *Eptatretidae* (Atkin and Ohno, 1967; Taylor, 1967), and the

Cyprinidae (references in Ohno, 1970, 1974). A most striking example occurs in the salmoniform suborder *Esocoidei* (Beamish *et al.*, 1971).

Much of the literature on polyploidy in fishes has centered on the family *Salmonidae*.* Initially, Svärdson (1945) suggested the group comprised a polyploid series with a haploid set of 10 chromosomes. This hypothesis has since been questioned by several authors, and no longer seems tenable in view of Rees's (1964) critical study. There is accumulated evidence, however, which suggests that the *Salmonidae* are ancestral polyploids which arose from tetraploidization, rather than successive events as suggested by Svärdson. Ohno (1974) recently reviewed most of this evidence and observed that (1) with one possible exception (Ohno *et al.*, 1969a), the chromosome arm numbers and genome sizes in species examined from each salmonid subfamily are approximately double those of related taxa, (2) there are now a number of biochemical–genetic studies which indicate duplication for several nonlinked genes (see also Chapter 8), and (3) in a few salmonid species there are ring and rod multivalents observed during prophase of meiosis I. The latter is an important point since if the salmonids are autotetraploids (as seems to be indicated—see Ohno *et al.*, 1969a) a few tetravalents might be expected to occur.

Unfortunately, in those species where meiotic multivalents are regularly observed, for example, *Salmo salar*, *Salmo gairdneri*, and *Salmo aguabonita* (Ohno *et al.*, 1965; Nygren *et al.*, 1968a; Gold and Gall, 1975), there is ample evidence of extensive Robertsonian rearrangement at both the interspecific and intraindividual levels. This could in itself explain any number or type of multivalent by assuming a karyotype of more than two metacentric chromosomes with partial or heterobrachial homologies (Gropp *et al.*, 1972). Zenzes and Voiçulescu (1975) published a C-banded somatic karyotype of *Salmo trutta*, in which the $2n = 80$ chromosomes were arranged into sets of twos (metacentrics) and fours (acrocentrics). However, since specific identification of all individual chromosomes was not possible, whether the chromosomes in all sets of two or four were homologous could not be determined.

Somewhat different reasoning led Nygren *et al.* (1972) to the conclusion that "reciprocal translocations rather than autopolyploidy cause the occurrence of multivalents in the meiosis of (*S. salar*)," although at the time they were apparently considering Svärdson's model. Nevertheless, even if all the multivalents observed in some

* The family *Salmonidae* is considered here to comprise the three subfamilies *Salmoninae*, *Coregoninae*, and *Thymallinae* (after Norden, 1960).

species were the result of numerous Robertsonian rearrangement between nonhomologues, this could still reflect a process of "diploidization" from a tetraploid state (Ohno *et al.*, 1969a).

2. ANEUPLOIDY

As noted previously, the only substantiated report of an aneuploid fish was a trisomic ($2n = 85$) Eastern brook trout, *Salvelinus fontinalis*, identified from both cytological and biochemical–genetic evidence by Davisson *et al.* (1972). It is not surprising perhaps that the aneuploid should be found in a salmonid species, or that the apparently nondisjoined chromosome was a metacentric, presumably derived from a Robertsonian fusion event. Among strictly diploid species, aneuploids are expected to be extremely rare because of their severe effect on developmental processes.

V. CYTOGENETICS AND FISH CULTURE

The discipline of "cytogenetics," along with its practical application, has yet to be used extensively in fish breeding or fish culture. In contrast, the manipulation of chromosomes and/or chromosome sets has proved a valuable technique in plant breeding, and general karyology has been useful to some extent in animal breeding. Many of these methods should be useful in fish breeding or in other phases of fish culture. A few potentialities for the practical application of cytogenetics to fish culture are briefly discussed below.

A. Manipulation of Chromosome Sets

A serious difficulty in fish breeding programs is the establishment and maintenance of inbred lines, since the amount of time, labor, and expense is very often prohibitive. Purdom and Lincoln (1973), however, have suggested that highly inbred strains could be produced in only a few generations by increasing the rate of inbreeding through artificial parthenogenesis, that is, sperm-stimulated development or gynogenesis of diploidized eggs.

Earlier, several Russian investigators working with carp (*Cyprinus carpio*), loach (*Misgurnus fossilis*), and sturgeon (*Acipenser ruthenus*) found that a low frequency of diploids was recovered from haploid eggs "fertilized" with radiation-inactivated sperm (references in Purdom and Lincoln, 1973). Most of the gynogenetic embryos were monoploid, but the frequency of diploids could be increased by exposing the embryos to temperature shocks subsequent to fertilization. Purdom and Lincoln (1973) extended this work to the plaice, *Pleuronectes platessa,* and by carefully studying the duration of temperature shock they were able to recover substantially high frequencies of diploid "gynogenomes."

The reestablishment of diploidy following temperature shock in these gynogens probably stems from a failure of either the first or second meiotic divisions of the egg, or of the first mitotic division of the embryo. In *P. platessa* and *C. carpio,* the failure is apparently at the second meiotic division (Purdom and Lincoln, 1973; Cherfas, 1975). Regardless, the degree of inbreeding (homozygosity) should be at least 50% (failure of meiosis I), and could be as high as 100% (failure of first mitosis). Failure of meiosis II should result in inbreeding levels between 50 and 100%, depending on the amount of crossing-over occurring during meiosis I (Purdom and Lincoln, 1973). Thus, since each gynogenome is expected to be genetically unique, a different inbred line should result following a second generation.

The survival of the first generation diploid gynogens was low in both *P. platessa* and *C. carpio* (Purdom and Lincoln, 1973; Cherfas, 1975), but such an approach to inbreeding should be considered for other fish species.

The application of temperature shocks along with fertilization by *normal* sperm may also be useful to fish culture since polyploids should be produced. This has been accomplished in the stickleback, *Gasterosteus aculeatus* (Swarup, 1959a,b), in plaice and plaice × flounder (*Platichthys flesus*) hybrids (Purdom, 1973), and the blue tilapia, *Tilapia aurea* (Valenti, 1975). Usually, triploids are the result, although Swarup (1959a,b) recovered several heteroploid *G. aculeatus,* and Valenti (1975) may have recovered a few tetraploid *T. aurea.*

The potential value of the triploids is twofold. First, it is known in animals that nuclear and cell sizes increase in proportion to increases in chromosome number (Fankhauser, 1945; Swarup, 1959b). Thus, if cell number and division time are the same for both diploid and triploid individuals, the triploids should have increased growth rates. Swarup (1959b) found that growth rate and final size of *G. aculeatus* triploids were the same as diploid controls. Since the triploids possessed larger overall cell size, Swarup (1959b) concluded that size regulating mechanisms were operating in such a way that the increased cell size of the triploids was compensated for by a reduction in the number of cells per organ. Purdom (1973), however, found that plaice

× flounder triploids grew considerably faster than expected, and suggested that the effect was due to triploidy per se. The difference between the growth rates of diploid and triploid *G. aculeatus* and those of plaice × flounder hybrids may be due to the determinate growth pattern of *G. aculeatus* (Purdom, 1973); in plaice, growth is continuous throughout life and certain size regulating mechanisms may be absent. Valenti (1975) has obtained similar results with *T. aurea;* in all cases, polyploid fish were larger at 14 weeks of age than were diploid siblings.

A second practical use of triploid fish follows from the fact that they are expected to be sterile due to irregular segregation during meiosis. Sterile fish would be useful in stocking programs where the genetic integrity of wild populations may be threatened by hatchery introductions. Other possible advantages of stocking sterile fish were discussed by Purdom and Lincoln (1973).

B. Use of General Karyology

In several animal species, including man, many instances of embryonic rejection may be attributed to chromosomal anomalies, particularly those arising from nondisjunction and polyploidy (see reviews in Carr, 1966, 1970; Fechheimer, 1968, 1972; Bruere, 1974). The role of chromosomal anomalies in reproductive failure is not known, yet several abortuses are apparently either aneuploid, polyploid, or chromosomally mosaic. Possible causes of chromosomal anomalies in animals also are unknown, but may include ionizing radiation, delayed fertilization, or aging of ova. The anomalies are easily identified through general karyology or karyotyping.

Instances of reproductive failure among cultured fish species are common, but these usually are attributed to inadequate management practices which result in physiological stress on the adults, eggs, or embryos. It is also conceivable that chromosomal anomalies play a major role in fish reproductive failure. Adults with altered reproductive ability due to structural heterozygosity for chromosomal rearrangement might be identified prior to their use in breeding programs.

Bruere (1974) has reviewed the evidence for reduced fertility in animals due to structural heterozygosity for Robertsonian rearrangement. In some species such as the tobacco mouse, *Mus poschiavinus,* heterozygosity for Robertsonian rearrangement apparently leads to aneuploid gametes, and hence to reduced fertility (Gropp *et al.,* 1972; Tettenborn and Gropp, 1970). In other species (e.g., sheep) there is little evidence that either aneuploid gametes or aneuploid embryos result from matings of individuals heterozygous for a Robertsonian rearrangement to structurally normal individuals (Bruere, 1974). In fishes, very little is known on this subject, although it is interesting that progeny survival of rainbow trout, a species with apparently numerous Robertsonian chromosomal polymorphisms, may be as high as 85–90% (Gold and Gall, unpublished observations).

A final possibility is that individual chromosomes could potentially serve to identify specific strains or hybrids, much in the same manner as biochemical–genetic markers (see Chapter 8). Cytological techniques presently used in fishes, however, would have to be improved appreciably. To date, only a few species hybrids have been identified cytologically (Prehn and Rasch, 1969; Setzer, 1970; Chen and Ebeling, 1975).

ACKNOWLEDGMENTS

I wish to thank Dr. C. L. Leinweber (Range Science, Texas A & M University), Mr. C. Busack (Animal Science, University of California at Davis), and Drs. H. J. Price and G. E. Hart (Genetics, Texas A & M University) for their many suggestions on various parts of this chapter. I am particularly grateful and indebted to Ms. M. F. Gold (College Station, Texas) and Drs. J. C. Avise (Zoology, University of Georgia) and F. L. Roberts (Zoology, University of Maine) for their numerous, valuable contributions to the entire chapter.

REFERENCES

Abe, S., and Muramoto, J. (1974). Differential staining of chromosomes of two salmonoid species, *Salvelinus leucomaenis* (Pallas) and *Salvelinus malma* (Walbum). *Proc. Jpn. Acad.* **50**, 507–511.

Abramoff, P., Darnell, R. M., and Balsano, J. S. (1968). Electrophoretic demonstration of the hybrid origin of the gynogenetic teleost *Poecilia formosa. Am. Nat.* **102**, 555–558.

Aida, T. (1921). On the inheritance of color in a fresh-water fish *Aplocheilus latipes* Temminck and Schlegel, with special reference to the sex-linked inheritance. *Genetics* **6**, 554–573.

Aida, T. (1936). Sex reversal in *Aplocheilus latipes* and a new explanation of sex differentiation. *Genetics* **21**, 136–153.

Anders, A., Anders, F., and Rase, S. (1969). XY females caused by X-irradiation. *Experientia* **25**, 851.

Anders, A., Anders, F., and Klinke, K. (1973). Regulation of gene expression in the Gordon–Kosswig melanoma system. II. The arrangement of chromatophore determining loci and regulating elements in the sex chromosomes of xiphophorin fish, *Platypoecilus maculatus* and *Platypoecilus variatus. In* "Genetics and Mutagenesis of Fish" (J. H. Schröder, ed.), pp. 53–63. Springer-Verlag, Berlin and New York.

Atkin, N. B., and Ohno, S. (1967). DNA values of four primitive chordates. *Chromosoma* **23**, 10–13.

Atz, J. W. (1964). Intersexuality in fishes. *In* "Intersexuality in Vertebrates Including Man" (C. N. Armstrong and A. J. Marshall, eds.), pp. 145–232. Academic Press, New York.

Avise, J. C., and Gold, J. R. (1977). Chromosomal divergence and rates of speciation in two families of North American fishes. *Evolution* **31**, 1–13.

Bachmann, K., Goin, O. B., and Goin, C. J. (1972). Nuclear DNA amounts in vertebrates. *Brookhaven Symp. Biol.* **23**, 419–450.

Balsano, J. S., Darnell, R. M., and Abramoff, P. (1972). Electrophoretic evidence of triploidy associated with populations of the gynogenetic teleost *Poecilia formosa*. *Copeia* No. 2, pp. 292–297.

Barker, C. J. (1972). A method for the display of chromosomes of plaice, *Pleuronectes platessa*, and other marine fishes. *Copeia* No. 2, pp. 365–368.

Beamish, R. J., Merrilees, M. J., and Crossman, E. J. (1971). Karyotypes and DNA values for members of the suborder *Esocoidei* (Osteichthyes: Salmoniformes). *Chromosoma* **34**, 436–447.

Beçak, W., Beçak, M. L., and Ohno, S. (1966a). Intraindividual chromosomal polymorphism in green sunfish (*Lepomis cyanellus*) as evidence of somatic segregation. *Cytogenetics* **5**, 313–320.

Beçak, M. L., Beçak, W., and Rabello, M. N. (1966b). Cytological evidence of constant tetraploidy in the bisexual South American frog *Odontophrynus americanus*. *Chromosoma* **19**, 188–193.

Beçak, M. L., Beçak, W., and Rabello, M. N. (1967). Further studies on polyploid amphibians (Ceratophrydidae). I. Mitotic and meiotic aspects. *Chromosoma* **22**, 192–201.

Beçak, M. L., Beçak, W., and Vizotto, L. D. (1970). A diploid population of the polyploid amphibian *Odontophrynus americanus* and an artificial intraspecific triploid hybrid. *Experientia* **26**, 545.

Behnke, R. J. (1970). The application of cytogenetic and biochemical systematics to phylogenetic problems in the family Salmonidae. *Trans. Am. Fish. Soc.* **1**, 237–248.

Behnke, R. J. (1972). The systematics of salmonid fishes of recently glaciated lakes. *J. Fish. Res. Board Can.* **29**, 639–671.

Bellamy, A. W. (1922). Sex-linked inheritance in the teleost *Platypoecilus maculatus*. *Anat. Rec.* **24**, 419–420.

Bellamy, A. W. (1936). Interspecific hybrids in *Platypoecilus:* One species ZZ–WZ; the other XY–XX. *Proc. Natl. Acad. Sci. U.S.A.* **22**, 531–535.

Benazzi, M. (1973). Cytotaxonomy and evolution: General remarks. *In* "Cytotaxonomy and Vertebrate Evolution" (A. B. Chiarelli and E. Capanna, eds.), pp. 3–14. Academic Press, New York.

Bennett, M. D. (1972). Nuclear DNA content and minimum generation time in herbaceous plants. *Proc. R. Soc., Ser. B* **181**, 109–135.

Blaxhall, P. C. (1975). Fish chromosome techniques—A review of selected literature. *J. Fish Biol.* **7**, 315–320.

Booke, H. E. (1968). Cytotaxonomic studies of the coregonine fishes of the Great Lakes, USA: DNA and karyotype analysis. *J. Fish Res. Board Can.* **25**, 1667–1687.

Booke, H. E. (1970). Speciation parameters in coregonine fishes: I. Egg-size. II. Karyotype. *In* "Biology of Coregonid Fishes" (C. C. Lindsey and C. S. Woods, eds.), pp. 61–66. Univ. of Manitoba Press, Winnipeg.

Booke, H. E. (1974). A cytotaxonomic study of the round whitefishes, genus *Prosopium*. *Copeia* No. 1, pp. 115–119.

Boothroyd, E. R. (1959). Chromosome studies on three Canadian populations of Atlantic salmon, *Salmo salar* L. *Can. J. Genet. Cytol.* 1, 161–172.

Breider, H. (1935). Geschlechtsbestimmung und-differenzierung bei *Limia nigrofascida, Caudofasciata, vittata* und deren Artbastarden. *Z. Indukt. Abstamm.- Vererbungsl.* 68, 265–299.

Breider, H. (1936). Weiteres über die Geschlechtsbestimmung und Geschlechts-differenzierung bei *Limia vittata* Poy. *Zool. Anz.* 114, 113–119.

Bridges, C. B. (1925). Sex in relation to chromosomes and genes. *Am. Nat.* 59, 127–137.

Bridges, C. B. (1939). Cytological and genetic basis of sex. *In* "Sex and Internal Secretions" (E. Allen, ed.), 2nd Ed., pp. 15–63. Williams & Wilkins, Baltimore, Maryland.

Bruere, A. N. (1974). The discovery and biological consequences of some important chromosome anomalies in populations of domestic animals. *World Congr. Genet. Appl. Livestock Prod., 1st* pp. 151–175.

Bungenberg de Jong, C. M. (1955). Cytological studies in *Salmo irideus*. *Genetics* 27, 472–485.

Buss, K., and Wright, J. E. (1956). Results of species hybridization within the family Salmonidae. *Prog. Fish Cult.* 18, 149–158.

Calton, M. S., and Denton, T. E. (1974). Chromosomes of the chocolate gourami: a cytogenetic anomaly. *Science* 185, 618–619.

Carr, D. H. (1966). Cytogenetics and abortions. *In* "Comparative Aspects of Reproductive Failure" (K. Benirschke, ed.), pp. 96–117. Springer-Verlag, Berlin and New York.

Carr, D. H. (1970). Chromosome abnormalities and spontaneous abortions. *In* "Human Population Cytogenetics" (P. A. Jacobs, W. H. Price, and P. Law, eds.), Pfizer Medical Monographs, No. 5, pp. 103–118. Univ. of Edinburgh, Edinburgh.

Catton, W. T. (1951). Blood cell formation in certain teleost fishes. *Blood* 6, 39–60.

Chan, S. T. H. (1970). Natural sex reversal in vertebrates. *Philos. Trans. R. Soc. London, Ser. B* 259, 59–71.

Chen, T. R. (1969). Karyological heterogamety of deep-sea fishes. *Postilla* 130, 1–29.

Chen, T. R. (1970). Fish chromosome preparation: Air-dried displays of cultured ovarian cells in two killifishes (*Fundulus*). *J. Fish. Res. Board Can.* 271, 158–161.

Chen, T. R. (1971). A comparative study of twenty killifish species of the genus *Fundulus* (Teleostei: Cyprinodontidae). *Chromosoma* 32, 436–453.

Chen, T. R., and Ebeling, A. W. (1966). Probable male heterogamety in the deep-sea fish *Bathylagus wesethi* (Teleosti: Bathylagidae). *Chromosoma* 18, 88–96.

Chen, T. R., and Ebeling, A. W. (1968). Karyological evidence of female heterogamety in the mosquitofish *Gambusia affinis* (Baird and Girard). *Copeia* No. 1, pp. 70–75.

Chen, T. R., and Ebeling, A. W. (1971). Chromosomes of the Goby fishes in the genus *Gillichthys*. *Copeia* No. 1, pp. 171–174.

Chen, T. R., and Ebeling, A. W. (1975). Karyotypes from short- and long-term cultures of hybrid killifish and platyfish tissues. *Copeia* No. 1, pp. 178–181.

Chen, T. R., and Reisman, H. M. (1970). A comparative chromosome study of the North American species of sticklebacks (Teleostei: Gasterosteidae). *Cytogenetics* 9, 321–332.

Chen, T. R., and Ruddle, F. H. (1970). A chromosome study of four species and a hybrid of the killifish genus *Fundulus* (Cyprinodontidae). *Chromosoma* 29, 255–267.

Cherfas, N. B. (1966). Natural triploidy in females of the unisexual forms of the goldfish (*Carassius auratus gibelio* Bloch). *Sov. Genet.* **2**(5), 9–13.

Cherfas, N. B. (1972). Results of a cytological analysis of unisexual and bisexual forms of silver crucian carp. *In* "Genetics, Selection, and Hybridization of Fish," pp. 79–90. Isr. Program Sci. Transl., Jerusalem.

Cherfas, N. B. (1975). Investigation of radiation-induced diploid gynogenesis in the carp (*Cyprinus carpio* L.). *Genetika* **11**, 78–86.

Chiarelli, A. B., and Capanna, E. (1973). Checklist of fish chromosomes. *In* "Cytotaxonomy and Vertebrate Evolution" (A. B. Chiarelli and E. Capanna, eds.), pp. 206–232. Academic Press, New York.

Chiarelli, B., Ferrantelli, O., and Cucchi, C. (1969). The caryotype of some Teleostea fish obtained by tissue culture *in vitro*. *Experientia* **25**, 426–427.

Cimino, M. C. (1972a). Meiosis in triploid all-female fish (*Poeciliopsis*, Poeciliidae). *Science* **175**, 1484–1486.

Cimino, M. C. (1972b). Egg production, polyploidization and evolution in a diploid all-female fish of the genus *Poeciliopsis*. *Evolution* **26**, 294–306.

Cimino, M. C., and Bahr, G. F. (1974). The nuclear DNA content and chromatin ultra-structure of the coelacanth *Latimeria chalumnae*. *Exp. Cell Res.* **88**, 263–272.

Cimino, M. C., and Schultz, R. J. (1970). Production of a diploid male offspring by a gynogenetic triploid fish of the genus *Poeciliopsis*. *Copeia* No. 4, pp. 760–763.

Clark, E. (1959). Functional hermaphroditism in a serranid fish. *Science* **129**, 215–216.

Clark, E. (1965). Mating of groupers. *Nat. Hist.* **74**, 22–25.

Clem, W. L., Moewus, L., and Sigel, M. M. (1961). Studies with cells from marine fish in tissue culture. *Proc. Soc. Exp. Biol. Med.* **108**, 762–766.

Crow, J. F., and Kimura, M. (1965). Evolution in sexual and asexual populations. *Am. Nat.* **99**, 439–450.

Cuellar, O., and Uyeno, T. (1972). Triploidy in rainbow trout. *Cytogenetics* **11**, 508–515.

D'Ancona, U. (1950). Détermination e differenciation du sexe chez les poissions. *Arch. Anat. Microscop. Morphol. Exp.* **39**, 274–294.

Dangel, J. R. (1973). An annotated bibliography of interspecific hybridization of Sal-monidae. *FAO (FAO UN) Fish. Circ.* No. 133, pp. 1–32.

Darnell, R. M., and Abramoff, P. (1968). Distribution of the gynogenetic fish, *Poecilia formosa*, with remarks on the evolution of the species. *Copeia* No. 2, pp. 354–361.

Davisson, M. T., Wright, J. E., and Atherton, L. M. (1972). Centric fusion and trisomy for the LDH-B locus in brook trout, *Salvelinus fontinalis*. *Science* **178**, 992–994.

Davisson, M. T, Wright, J. E., and Atherton, L. M. (1973). Cytogenetic analysis of pseudolinkage of LDH loci in the teleost genus *Salvelinus*. *Genetics* **73**, 645–658.

Denton, T. E. (1973). "Fish Chromosome Methodology." Thomas, Springfield, Illinois.

Denton, T. E., and Howell, W. M. (1969). A technique for obtaining chromosomes from the scale epithelium of teleost fishes. *Copeia* No. 2, pp. 392–393.

Dobzhansky, T. (1951). "Genetics and the Origin of Species," 3rd Ed. Columbia Univ. Press, New York.

Dobzhansky, T. (1970). "Genetics of the Evolutionary Process." Columbia Univ. Press, New York.

Drewry, G. (1964). Appendix I, Chromosome number. *Bull. Tex. Mem. Mus.* No. 8, pp. 5–72.

Ebeling, A. W., and Chen, T. R. (1970). Heterogamety in teleostean fishes. *Trans. Am. Fish. Soc.* **1**, 131–138.

Ebeling, A. W., Atkin, N. B., and Setzer, P. Y. (1971). Genome sizes of teleostean fishes: Increases in some deep-sea species. *Am. Nat.* **105**, 549–562.

Endo, A., and Ingalls, T. H. (1968). Chromosomes of the zebra fish. *J. Hered.* **59**, 382–384.

Fankhauser, G. (1945). The effect of change in chromosome numbers on amphibian development. *Q. Rev. Biol.* **20**, 20–78.

Fechheimer, N. S. (1968). Consequences of chromosomal aberrations in mammals. *J. Anim. Sci.* **27**, Suppl. I, 27–50.

Fechheimer, N. S. (1972). Causal basis of chromosome abnormalities. *J. Reprod. Fertil., Suppl.* **15**, 79–98.

Friedman, B., and Gordon, M. (1934). Chromosome numbers in Xiphophorin fishes. *Am. Nat.* **68**, 446–455.

German, J. (1973). Studying human chromosomes today. *Am. Sci.* **58**, 1–20.

Goin, O. B., and Goin, C. J. (1968). DNA and the evolution of vertebrates. *Am. Midl. Nat.* **80**, 289–298.

Gold, J. R. (1974). A fast and easy method for chromosome karyotyping in adult teleosts. *Prog. Fish Cult.* **36**, 169–171.

Gold, J. R., and Avise, J. C. (1976). Spontaneous triploidy in the California roach, *Hesperoleucus symmetricus* (Pisces: Cyprinidae). *Cytogenet. Cell Genet.* **17**, 144–149.

Gold, J. R., and Avise, J. C. (1977). Cytogenetic studies in North American minnows (Cyprinidae). I. Karyology of nine California genera. *Copeia* No. 3, pp. 541–549.

Gold, J. R., and Gall, G. A. E. (1975). Chromosome cytology and polymorphism in the California High Sierra golden trout (*Salmo aguabonita*). *Can. J. Genet. Cytol.* **17**, 41–53.

Gold, J. R., Pipkin, R. E., and Gall, G. A. E. (1976). Artificial hybridization between rainbow (*Salmo gairdneri*) and golden trout (*Salmo aguabonita*). *Copeia* No. 3, pp. 597–598.

Gold, J. R., Avise, J. C., and Gall, G. A. E. (1977). Chromosome cytology in the cutthroat trout, *Salmo clarki* (Salmonidae). *Cytologia* **42**, 377–382.

Gordon, M. (1927). The genetics of a viviparous top-minnow *Platypoecilus;* the inheritance of two kinds of melanophores. *Genetics* **12**, 253–283.

Gordon, M. (1937). Genetics of *Platypoecilus*. III. Inheritance of sex and crossing over of the sex chromosomes in the platyfish. *Genetics* **22**, 376–392.

Gordon, M. (1946). Interchanging genetic mechanisms for sex determination. *J. Hered.* **37**, 307–320.

Gordon, M. (1947). Genetics of *Platypoecilus maculatus*. IV. The sex-determining mechanism in two wild populations of Mexican platyfish. *Genetics* **32**, 8–17.

Gordon, M. (1951). Genetics of *Platypoecilus maculatus*. V. Heterogametic sex-determining mechanisms in females of a domesticated stock originally from British Honduras. *Zoologica (N.Y.)* **36**, 127–134.

Gordon, M. (1952). Sex determination in *Xiphophorus* (*Platypoecilus*) *maculatus*. III. Differentiation of gonads in platyfish from broods having a sex ratio of the three females to one male. *Zoologica (N.Y.)* **37**, 91–100.

Gordon, M. (1957). Physiological genetics of fishes. *In* "The Physiology of Fishes" (M. E. Brown, ed.), Vol. 2, pp. 431–501. Academic Press, New York.

Gordon, M., and Smith, G. M. (1938). The production of a melanotic neoplastic disease in fishes by selective matings. IV. Genetics of geographical species hybrids. *Am. J. Cancer* **34**, 543–565.

Gorman, G. C. (1973). The chromosomes of the Reptilia, a cytotaxonomic interpretation. *In* "Cytotaxonomy and Vertebrate Evolution" (A. B. Chiarelli and E. Capanna, eds.), pp. 349–424. Academic Press, New York.

Gould, W. R. (1966). Cutthroat trout (*Salmo clarkii* Richardson) × golden trout (*Salmo aguabonita* Jordan) hybrids. *Copeia* No. 3, pp. 599–600.

Grant, V. (1963). "The Origin of Adaptations." Columbia Univ. Press, New York.

Gropp, A., Winking, H., Zech, I., and Müller, H. (1972). Robertsonian chromosomal variation and identification of metacentric chromosomes in feral mice. *Chromosoma* 39, 265–288.

Gyldenholm, A. O., and Scheel, J. J. (1971). Chromosome numbers of fishes. I. *J. Fish Biol.* 3, 479–486.

Harrington, R. W., Jr. (1961). Oviparous hermaphroditic fish with internal self-fertilization. *Science* 134, 1749–1750.

Harrington, R. W., Jr. (1971). How ecological and genetic factors interact to determine when self-fertilizing hermaphrodites of *Rivulus marmoratus* change into functional secondary males, with a reappraisal of the modes of intersexuality among fishes. *Copeia* No. 3, pp. 389–432.

Harrington, R. W., Jr., and Kallman, K. D. (1968). The homozygosity of clones of the self-fertilizing hermaphroditic fish *Rivulus marmoratus* Poey (Cyprinodontidae, Atheriniformes). *Am. Nat.* 102, 337–343.

Haskins, C. P., Haskins, E. F., and Hewitt, R. E. (1960). Pseudogamy as an evolutionary factor in the poeciliid fish *Mollienisia formosa. Evolution* 14, 473–483.

Hayashi, M., Ojima, Y., and Asano, N. (1976). A cell line from teleost fish: Establishment and cytogenetical characterization of the cells. *Jpn. J. Genet.* 51, 65–68.

Heckman, J. R., and Brubaker, P. E. (1970). Chromosome preparation from fish blood leukocytes. *Prog. Fish Cult.* 32, 206–208.

Heckman, J. R., Allendorf, F. W., and Wright, J. E. (1971). Trout leukocytes: Growth in oxygenated cultures. *Science* 173, 246–247.

Hickling, C. F. (1960). The Malacca *Tilapia* hybrids. *J. Genet.* 57, 1–10.

Hinegardner, R. (1968). Evolution of cellular DNA content in teleost fishes. *Am. Nat.* 102, 517–523.

Hinegardner, R., and Rosen, D. E. (1972). Cellular DNA content and the evolution of teleostean fishes. *Am. Nat.* 106, 621–644.

Hoehn, H. (1975). Functional implications of differential chromosome banding. *Am. J. Hum. Genet.* 27, 676–686.

Howell, W. M., and Denton, T. E. (1969). Chromosomes of ammocoetes of the Ohio brook lamprey, *Lampetra aepyptera. Copeia* No. 2, pp. 393–395.

Hubbs, C. (1964). Interactions between a bisexual fish species and its gynogenetic sexual parasite. *Bull. Tex. Mem. Mus.* No. 8, pp. 1–72.

Hubbs, C. L. (1955). Hybridization between fish species in nature. *Syst. Zool.* 4, 1–20.

Hubbs, C. L., and Hubbs, L. C. (1932). Apparent parthenogenesis in nature, in a form of fish of hybrid origin. *Science* 76, 628–630.

Hubbs, C. L., and Hubbs, L. C. (1946). Breeding experiments with the invariable female, strictly matroclinous fish, *Mollienisia formosa. Rec. Genet. Soc. Am.* 14, 48.

Kaiser, P., and Schmidt, E. (1951). Vollkommene Geschlechtsumwandlung beim weiblichen siamesischen Kampffish, *Betta splendens. Zool. Anz.* 146, 66–73.

Kallman, K. D. (1962). Gynogenesis in the teleost, *Mollienisia formosa* (Girard), with a discussion of the detection of parthenogenesis in vertebrates by tissue transplantation. *J. Genet.* 58, 7–24.

Kallman, K. D. (1965a). Sex determination in the teleost *Xiphophorus milleri. Am. Zool.* 5, 246–247.

Kallman, K. D. (1965b). Genetics and geography of sex determination in the poeciliid fish, *Xiphophorus maculatus. Zoologica* (N.Y.) 50, 151–190.

Kallman, K. D. (1968). Evidence for the existance of transformer genes for sex in the teleost *Xiphophorus maculatus*. *Genetics* **60**, 811–828.

Kallman, K. D. (1970a). Genetics of tissue transplantation in Teleostei. *Transplant. Proc.* **2**, 263–271.

Kallman, K. D. (1970b). Sex determination and the restriction of sex-linked pigment patterns to the X and Y chromosomes in populations of a poeciliid fish, *Xiphophorus maculatus*, from the Belize and Sibun Rivers of British Honduras. *Zoologica (N.Y.)* **55**, 1–16.

Kallman, K. D. (1973). The sex-determining mechanism of the platyfish, *Xiphophorus maculatus. In* "Genetics and Mutagenesis of Fish" (J. H. Schröder, ed.), pp. 19–28. Springer-Verlag, Berlin and New York.

Kang, Y. S., and Park, E. H. (1975). Leukocyte culture of the eel without autologous serum. *Jpn. J. Genet.* **50**, 159–161.

Keyl, H. G. (1965). Demonstrable local and geometric increase in the chromosomal DNA of *Chironomus. Experientia* **21**, 191–193.

Kobayasi, H. (1971). A cytological study on gynogenesis of the triploid Ginbuna (*Carassius auratus langsdorfi*). *Dobutsugaku Zasshi* **80**, 316–322.

Kobayasi, H., Kawashima, Y., and Takeuchi, N. (1970). Comparative chromosome studies in the genus *Carassius*, especially with a finding of polyploidy in the Ginbuna (*C. auratus langsdorfi*). *Gyoruigaku Zasshi* **17**, 153–160.

Kosswig, C. (1935). Genotypische und Phänotypische Geschlechts bestimmung bei Zahnkarpfen. V. Ein X(Z)-Chromosom als Y-Chromosom in fremden Erbgut. *Wilhelm Roux' Arch. Entwicklungsmech. Org.* **133**, 118–139.

Kosswig, C. (1959). Beiträge zur genetischen Analyse xiphophoriner Zahnkarpfen. *Biol. Zentralbl.* **78**, 711–718.

Kosswig, C. (1964). Polygenic sex determination. *Experientia* **20**, 190–199.

Labat, R., Larrouy, G., and Malaspina, L. (1967). Technique de culture des leucocytes de *Cyprinus carpio* L. *C. R. Acad. Sci., Ser. D* **264**, 2473–2475.

Legendre, P. (1975). A field-trip microtechnique for studying fish leukocyte chromosomes. *Can. J. Zool.* **53**, 1443–1446.

LeGrande, W. H. (1975). Karyology of six species of Louisiana flatfishes (Pleuronectiformes: Osteichthyes). *Copeia* No. 3, pp. 516–522.

Leuken, W., and Foerster, W. (1969). Chromosomenuntersuchungen bei Fischen mit einer vereinfachten Zellkulturtechnik. *Zool. Anz.* **138**, 168–76.

Levan, A., Fredga, K., and Sandberg, A. A. (1964). Nomenclature for centromeric position on chromosomes. *Hereditas* **52**, 201–220.

Liem, K. F. (1963). Sex reversal as a natural process in the *Synbranchiform* fish *Monopterus albus. Copeia* No. 2, pp. 303–312.

Lieppman, M., and Hubbs, C. (1969). A karyological analysis of two cyprinid fishes, *Notemigonus crysoleucas* and *Notropis lutrensis. Tex. Rep. Biol. Med.* **27**, 427–435.

McPhail, J. D., and Jones, R. L. (1966). A simple technique for obtaining chromosomes from teleost fishes. *J. Fish. Res. Board Can.* **23**, 767–768.

Markert, C. L. (1968). Panel discussion: Present status and perspectives in the study of cytodifferentiation at the molecular level. I. Initial remarks. *J. Cell. Physiol.* **72**, Suppl. 1, 213–219.

Maslin, T. P. (1971). Parthenogenesis in reptiles. *Am. Zool.* **11**, 361–380.

Mather, K. (1953). The genetical structure of populations. *Symp. Soc. Exp. Biol.* **7**, 66–93.

Matthey, R. (1945). L'évolution de la formule chromosomiale chez les vertébrés. *Experientia* **1**, 50–56.

Matthey, R. (1973). The chromosome formulae of eutherian mammals. *In* "Cytotaxonomy and Vertebrate Evolution" (A. B. Chiarelli and E. Capanna, eds.), pp. 531–616. Academic Press, New York.

Mayr, E. (1973). "Animal Species and Evolution." Harvard Univ. Press, Cambridge, Massachusetts.

Menzel, B. W., and Darnell, R. M. (1973). Morphology of naturally-occurring triploid fish related to *Poecilia formosa. Copeia* No. 2, pp. 350–352.

Meyer, H. (1938). Investigations concerning the reproductive behavior of *Mollienisia* "*formosa.*" *J. Genet.* **36**, 329–366.

Miller, R. R. (1972). Classification of the native trouts of Arizona with the description of a new species, *Salmo apache. Copeia* No. 3, pp. 401–422.

Miller, R. R., and Schultz, R. J. (1959). All-female strains of the teleost fishes of the genus *Poeciliopsis. Science* **130**, 1656–1657.

Mirsky, A. E., and Ris, H. (1951). The desoxyribonucleic acid content of animal cells and its evolutionary significance. *J. Gen. Physiol.* **34**, 451–462.

Moore, W. S., Miller, R. R., and Schultz, R. J. (1970). Distribution, adaptation and probable origin of an all-female form of *Poeciliopsis* (Pisces: Poeciliidae) in northwestern Mexico. *Evolution* **24**, 806–812.

Moorhead, J. F., Connolly, J. J., and McFarland, W. (1967). Factors affecting the reactivity of human lymphocytes *in vitro. J. Immunol.* **99**, 413–419.

Morescalchi, A. (1973). Amphibia. *In* "Cytotaxonomy and Vertebrate Evolution" (A. B. Chiarelli and E. Capanna, eds.), pp. 233–348. Academic Press, New York.

Muller, H. J. (1925). Why polyploidy is rarer in animals than in plants. *Am. Nat.* **59**, 346–353.

Muramoto, J. (1975). A note on triploidy of the Funa (Cyprinidae, Pisces). *Proc. Jpn. Acad.* **51**, 583–587.

Muramoto, J., Ohno, S., and Atkin, N. B. (1968). On the diploid state of the fish order Ostariophysi. *Chromosoma* **24**, 59–66.

Nelson-Rees, W. A., Kniazeff, A. J., and Darby, W. B., Jr. (1967). Debut and accumulation of centric fusion products: An index to age of certain cell lines. *Cytogenetics* **6**, 436–450.

Nikolsky, G. (1976). The interrelation between variability of characters, effectiveness of energy utilisation, and karyotype structure in fishes. *Evolution* **30**, 180–185.

Nilsson, B. (1973). A bibliography of literature concerning chromosome identification—with special reference to flourescence Giemsa staining techniques. *Hereditas* **73**, 259–270.

Nogusa, S. (1960). A comparative study of the chromosomes in fishes with particular considerations on taxonomy and evolution. *Mem. Hyogo Univ. Agric. (Hyogo Noka Daigaku Kiyo), Biol. Ser.* 3(1), 1–62.

Norden, C. R. (1960). Comparative osteology of representative salmonid fishes, with particular reference to the grayling (*Thymallus articus*) and its phylogeny. *J. Fish. Res. Board Can.* **18**, 679–791.

Nowell, P. C. (1960). Phytohaematagglutinin; an initiator of mitosis in cultures of normal leukocytes. *Cancer Res.* **20**, 462–466.

Nygren, A., Nilsson, B., and Jahnke, M. (1968a). Cytological studies in Atlantic salmon. *Ann. Acad. Regiae Sci. Ups.* **12**, 21–52.

Nygren, A., Edlund, P., Hirsch, U., and Ahsgren, L. (1968b). Cytological studies in perch (*Perca fluviatilis* L.), pike (*Esox lucius* L.), pike-perch (*Lucioperca lucioperca* L.), and ruff (*Acerina cernua* L.). *Hereditas* **59**, 518–524.

Nygren, A., Nilsson, B., and Jahnke, M. (1971a). Cytological studies in *Thymallus thymallus* and *Coregonus albula. Hereditas* **67**, 269–274.

Nygren, A., Nilsson, B., and Jahnke, M. (1971b). Cytological studies in Hypotremata and Pleurotremata (Pisces). *Hereditas* **67**, 275–282.

Nygren, A., Nilsson, B., and Jahnke, M. (1972). Cytological studies in Atlantic salmon from Canada, in hybrids between Atlantic salmon from Canada and Sweden and in hybrids between Atlantic salmon and sea trout. *Hereditas* **70**, 295–306.

Ohno, S. (1967). "Sex Chromosomes and Sex-linked Genes" (A. Labhart, T. Mann, L. T. Samuels, and J. Zander, eds.), Monographs on Endocrinology, Vol. 1. Springer-Verlag, Berlin and New York.

Ohno, S. (1970). "Evolution by Gene Duplication." Springer-Verlag, Berlin and New York.

Ohno, S. (1974). Protochordata, Cyclostomata and Pisces. *In* "Animal Cytogenetics" (B. John, ed.), Vol. 4, Chordata 1, pp. 1–91. Borntraeger, Berlin.

Ohno, S., and Atkin, N. B. (1966). Comparative DNA values and chromosome complements of eight species of fishes. *Chromosoma* **18**, 455–466.

Ohno, S., Stenius, C., Faisst, E., and Zenzes, M. T. (1965). Postzygotic chromosomal rearrangements in rainbow trout (*Salmo irideus* Gibbons). *Cytogenetics* **4**, 117–129.

Ohno, S., Muramoto, J., Christian, L., and Atkin, N. B. (1967). Diploid–tetraploid relationship among old-world members of the fish family *Cyprinidae. Chromosoma* **23**, 1–9.

Ohno, S., Wolf, U., and Atkin, N. B. (1968). Evolution from fish to mammals by gene duplication. *Hereditas* **59**, 169–187.

Ohno, S., Muramoto, J., Klein, J., and Atkin, N. B. (1969a). Diploid–tetraploid relationship in clupeoid and salmonoid fish. *In* "Chromosomes Today" (C. D. Darlington and K. R. Lewis, eds.), Vol. 2, pp. 139–147. Oliver & Boyd, Edinburgh.

Ohno, S., Muramoto, J., Stenius, C., Christian, L., Kittreel, W. A., and Atkin, N. B. (1969b). Microchromosomes in holocephalian, chondrostean and holostean fishes. *Chromosoma* **26**, 35–40.

Ojima, Y., Hitosumachi, S., and Hayashi, M. (1970). A blood culture method for fish chromosomes. *Jpn. J. Genet.* **45**, 161–162.

Ojima, Y., Takayama, S., and Yamamoto, K. (1972). Chromosome preparation from cultured scale epithelium of teleost fish. *Jpn. J. Genet.* **47**, 445–446.

Ojima, Y., Hayashi, M., and Ueno, K. (1975). Triploidy appeared in the backcross offspring from Funa–carp crossing. *Proc. Jpn. Acad.* **51**, 702–706.

Park, E. H. (1974). A list of the chromosome numbers of fishes. *Rev. Coll. Liberal Arts Sci., Seoul Natl. Univ.* **20**, 346–372.

Post, A. (1965). Vergleichende Untersuchungen der Chromosomenzahlen bei Süsswasser-Teleosteern. *Z. Zool. Syst. Evolutionsforsch.* **3**, 47–93.

Post, A. (1973). Chromosomes of two fish species of the genus *Diretmus* (Osteichthyes, Beryciformes: Diretmidae). *In* "Genetics and Mutagenesis of Fish" (J. H. Schröder, ed.), pp. 103–111. Springer-Verlag, Berlin and New York.

Potter, I. C., and Robinson, E. S. (1973). The chromosomes of the cyclostomes. *In* "Cytotaxonomy and Vertebrate Evolution" (A. B. Chiarelli and E. Capanna, eds.), pp. 179–203. Academic Press, New York.

Potter, I. C., and Rothwell, B. (1970). The mitotic chromosomes of the lamprey, *Petromyzon marinus* L. *Experientia* **26**, 429–430.

Prager, E. M., and Wilson, A. C. (1975). Slow evolutionary loss of the potential for interspecific hybridization in birds: A manifestation of slow regulatory evolution. *Proc. Natl. Acad. Sci. U.S.A.* **72**, 200–204.

Prehn, L. M., and Rasch, E. M. (1969). Cytogenetic studies of *Poecilia* (Pisces). I. Chromosome numbers of naturally occurring poeciliid species and their hybrids from eastern Mexico. *Can. J. Genet. Cytol.* **11**, 880–895.

Price, H. J. (1976). Evolution of DNA content in higher plants. *Bot. Rev.* **42**, 27–52.

Prokofieva, A. (1934). On the chromosome morphology of certain pisces. *Cytologia* **5**, 498–506.

Purdom, C. E. (1973). Induced polyploidy in plaice (*Pleuronectes platessa*) and its hybrid with the flounder (*Platichthys flesus*). *Heredity* **29**, 11–24.

Purdom, C. E., and Lincoln, R. F. (1973). Chromosome manipulation in fish. *In* "Genetics and Mutagenesis of Fish" (J. H. Schröder, ed.), pp. 83–89. Springer-Verlag, Berlin and New York.

Raicu, P., and Taisescu, E. (1972). *Misgurnus fossilis*, a tetraploid fish species. *J. Hered.* **63**, 92–94.

Rasch, E. M., and Balsano, J. S. (1973). Cytogenetic studies of *Poecilia* (Pisces). III. Persistence of triploid genomes in the unisexual progeny of triploid females associated with *Poecilia formosa*. *Copeia* No. 4, pp. 810–813.

Rasch, E. M., Darnell, R. M., Kallman, K. D., and Abramoff, P. (1965). Cytophotometric evidence for triploidy in hybrids of the gynogenetic fish, *Poecilia formosa. J. Exp. Zool.* **160**, 155–170.

Rasch, E. M., Prehn, L. M., and Rasch, R. W. (1970). Cytogenetic studies of *Poecilia* (Pisces). II. Triploidy and DNA levels in naturally occurring populations associated with the gynogenetic teleost, *Poecilia formosa* (Girard). *Chromosoma* **31**, 18–40.

Ray-Chaudhuri, R. (1973). Cytotaxonomy and chromosome evolution in birds. *In* "Cytotaxonomy and Vertebrate Evolution" (A. B. Chiarelli and E. Capanna, eds.), pp. 425–483. Academic Press, New York.

Rees, H. (1964). The question of polyploidy in the Salmonidae. *Chromosoma* **15**, 275–279.

Rees, H. (1967). The chromosomes of *Salmo salar*. *Chromosoma* **21**, 472–474.

Regan, J. D., Sigel, M. M., Lee, W. H., Llamas, K. A., and Beasley, A. R. (1968). Chromosomal alterations in marine fish cells *in vitro*. *Can. J. Genet. Cytol.* **10**, 448–453.

Reinboth, R. (1965). Sex reversal in the black sea bass *Centropristes straitus*. *Anat. Rec.* **151**, 403.

Reinboth, R. (1967). Protogynie bei *Chelidoperca hirundinacea* (Cuv. et Val.) (Serranidae)—Ein Diskussionsbeitrag zur Stammesgeschichte amphisexueller Fische. *Annot. Zool. Jpn.* **40**, 181–193.

Rio, G. J., Magnavita, F. J., Rubin, J. A., and Beckert, W. H. (1973). Characteristics of an established goldfish cell line. *J. Fish Biol.* **5**, 315–321.

Roberts, F. L. (1964). A chromosome study of twenty species of Centrarchidae. *J. Morphol.* **115**, 401–418.

Roberts, F. L. (1967). Chromosome cytology of the Osteichthyes. *Prog. Fis Cult.* **29**, 75–83.

Roberts, F. L. (1968). Chromosomal polymorphism in North American landlocked *Salmo salar*. *Can. J. Genet. Cytol.* **10**, 865–875.

Roberts, F. L. (1970). Atlantic salmon (*Salmo salar*) chromosomes and speciation. *Trans. Am. Fish. Soc.* **99**, 105–111.

Robertson, W. R. B. (1916). Chromosome studies. 1. Taxonomic relationships shown in the chromosomes of *Tettigidae* and *Acrididae*. V-shaped chromosomes and their significance in *Acrididae*, *Locustidae*, and *Gryllidae*. Chromosomes and variation. *J. Morphol.* **27**, 179–332.

Robinson, E. S., Potter, I. C., and Atkin, N. B. (1975). The nuclear DNA content of twenty lampreys. *Experientia* **31**, 912–913.

Rothfels, K., Sexsmith, E., Heimburger, M., and Krause, M. O. (1966). Chromosome size and DNA content of species of *Anemone* L. and related genera (Ranunculaceae). *Chromosoma* **20**, 54–74.

Scheel, J. J. (1968). "Rivulins of the Old World." T. F. H. Publications, Jersey City, New Jersey.

Scheel, J. J. (1972). Rivulin karyotypes and their evolution (Rivulinae, Cyprinodontidae, Pisces). *Z. Zool. Syst. Evolutionsforsch.* **10**, 180–209.

Scheel, J. J., Simonsen, V., and Gyldenholm, A. O. (1972). The karyotype and some electrophoretic patterns of fourteen species of the genus *Corydoras*. *Z. Zool. Syst. Evolutionsforsch.* **10**, 144–152.

Schultz, R. J. (1961). Reproductive mechanism of unisexual and bisexual strains of the viviparous fish *Poeciliopsis*. *Evolution* **15**, 302–325.

Schultz, R. J. (1966). Hybridization experiments with an all-female fish of the genus *Poeciliopsis*. *Biol. Bull. (Woods Hole, Mass.)* **130**, 415–429.

Schultz, R. J. (1967). Gynogenesis and triploidy in the viviparous fish *Poeciliopsis*. *Science* **157**, 1564–1567.

Schultz, R. J. (1969). Hybridization, unisexuality, and polyploidy in the teleost *Poeciliopsis* (Poeciliidae) and other vertebrates. *Am. Nat.* **103**, 605–619.

Schultz, R. J. (1971). Special adaptive problems associated with unisexual fishes. *Am. Zool.* **11**, 351–360.

Schultz, R. J. (1973). Origin and synthesis of a unisexual fish. *In* "Genetics and Mutagenesis of Fish" (J. H. Schröder, ed.), pp. 207–211. Springer-Verlag, Berlin and New York.

Schultz, R. J., and Kallman, K. D. (1968). Triploid hybrids between the all-female teleost *Poecilia formosa* and *Poecilia sphenops*. *Nature (London)* **219**, 280–282.

Setzer, P. Y. (1970). An analysis of a natural hybrid swarm by means of chromosome morphology. *Trans. Am. Fish. Soc.* **99**, 139–146.

Simon, R. C. (1963). Chromosome morphology and species evolution in the five North American species of Pacific salmon (*Oncorhynchus*). *J. Morphol.* **112**, 77–97.

Simon, R. C., and Dollar, A. M. (1963). Cytological aspects of speciation in two North American teleosts, *Salmo gairdneri* and *Salmo clarki lewisi*. *Can. J. Genet. Cytol.* **5**, 43–49.

Smith, C. L. (1959). Hermaphroditism in some serranid fishes from Bermuda. *Pap. Mich. Acad. Sci., Arts Lett.* **44**, 111–119.

Sparrow, A. H., and Nauman, A. F. (1974). Evolutionary change in genome and chromosome sizes and in DNA content in the grasses. *Brookhaven Symp. Biol.* **25**, 367–389.

Sparrow, A. H., Price, H. J., and Underbrink, A. G. (1972). A survey of DNA content per cell and per chromosome of prokaryotic and eukaryotic organisms: Some evolutionary considerations. *Brookhaven Symp. Biol.* **23**, 451–494.

Spofford, J. B. (1972). A heterotic model for the evolution of duplications. *Brookhaven Symp. Biol.* **23**, 121–143.

Stebbins, G. L. (1958). Longevity, habitat, and release of genetic variability in the higher plants. *Cold Spring Harbor Symp. Quant. Biol.* **23**, 365–378.

Stebbins, G. L. (1969). "The Basis of Progressive Evolution." Univ. of North Carolina Press, Chapel Hill.

Stebbins, G. L. (1971). "Chromosomal Evolution in Higher Plants." Addison-Wesley, Reading, Massachusetts.

Strommen, C. A., Rasch, E. M., and Balsano, J. S. (1975). Cytogenetic studies of *Poecilia* (Pisces). V. Cytophotometric evidence for the production of fertile offspring by triploids related to *Poecilia formosa*. *J. Fish Biol.* **7**, 667–676.

Svärdson, G. (1945). Chromosome studies on Salmonidae. *Rep. Inst. Freshwater Res. Drottningholm* No. 23, pp. 1–151.

Swarup, H. (1959a). Production of triploidy in *Gasterosteus aculeatus*. *J. Genet.* **56**, 129–142.

Swarup, H. (1959b). Effect of triploidy on the body size, general organization and cellular structure in *Gasterosteus aculeatus* (L.). *J. Genet.* **56**, 143–155.

Taylor, K. M. (1967). The chromosomes of some lower chordates. *Chromosoma* **21**, 181–188.

Tettenborn, U., and Gropp, A. (1970). Meiotic nondisjunction in mice and mouse hybrids. *Cytogenetics* **9**, 272–283.

Thompson, K. W. (1976). Some aspects of chromosomal evolution of the Cichlidae (Teleostei: Perciformes) with emphasis on neotropical forms. Ph.D. Thesis, Univ. of Texas at Austin.

Thorgaard, G. H. (1976). Robertsonian polymorphism and constitutive heterochromatin distribution in chromosomes of the rainbow trout (*Salmo gairdneri*). *Cytogenet. Cell Genet.* **17**, 174–184.

Uyeno, T., and Miller, R. R. (1971). Multiple sex chromosomes in a Mexican cyprinodontid fish. *Nature (London)* **231**, 452–453.

Uyeno, T., and Miller, R. R. (1972). Second discovery of multiple sex chromosomes among fishes. *Experientia* **15**, 223–225.

Uyeno, T., and Smith, G. R. (1972). Tetraploid origin of the karyotype of Catostomid fishes. *Science* **175**, 644–646.

Valenti, R. J. (1975). Induced polyploidy in *Tilapia aurea* (Steindachner) by means of temperature shock treatment. *J. Fish Biol.* **7**, 519–528.

Vrijenhoek, R. C. (1972). Genetic relationships of unisexual hybrid fishes to their progenitors using lactate dehydrogenase isozymes as gene markers (*Poeciliopsis*, Poeciliidae). *Am. Nat.* **106**, 754–766.

Vrijenhoek, R. C., and Schultz, R. J. (1974). Evolution of a trihybrid unisexual fish (*Poeciliopsis*, Poeciliidae). *Evolution* **28**, 306–319.

Wallace, B. (1963). Genetic diversity, genetic uniformity, and heterosis. *Can. J. Genet. Cytol.* **5**, 239–253.

Webb, C. J. (1974). Fish chromosomes: A display by scanning electron microscopy. *J. Fish Biol.* **6**, 99–100.

White, M. J. D. (1946). The evidence against polyploidy in sexually reproducing animals. *Am. Nat.* **80**, 610–619.

White, M. J. D. (1968). Models of speciation. *Science* **159**, 1065–1070.

White, M. J. D. (1973a). "Animal Cytology and Evolution." Cambridge Univ. Press, London.

White, M. J. D. (1973b). Chromosomal rearrangements in mammalian population polymorphism and speciation. *In* "Cytotaxonomy and Vertebrate Evolution" (A. B. Chiarelli and E. Capanna, eds.), pp. 95–128. Academic Press, New York.

Wickbom, T. (1943). Cytological studies on the family Cyprinodontidae. *Hereditas* **29**, 1–24.

Wilmot, R. L. (1974). A genetic study of the red-banded trout. Ph.D. Thesis, Oregon State Univ., Corvallis.

Wilson, A. C., Maxson, L. R., and Sarich, V. M. (1974a). Two types of molecular evolution. Evidence from studies of interspecific hybridization. *Proc. Natl. Acad. Sci. U.S.A.* **71**, 2843–2847.

Wilson, A. C., Sarich, V. M., and Maxson, L. R. (1974b). The importance of gene rearrangement in evolution: evidence from studies on rates of chromosomal, protein, and anatomical evolution. *Proc. Natl. Acad. Sci. U.S.A.* **71**, 3028–3030.

Wilson, A. C., Bush, G. L., Case, S. M., and King, M. C. (1975). Social structuring of mammalian populations and rate of chromosomal evolution. *Proc. Natl. Acad. Sci. U.S.A.* **72**, 5061–5065.

Winge, Ö. (1922). One-sided masculine and sex-linked inheritance in *Lebistes reticulatus. J. Genet.* **12**, 145–162.

Winge, Ö. (1934). The experimental alteration of sex chromosomes into autosomes and vice versa, as illustrated by *Lebistes. C. R. Trav. Lab. Carlsberg, Ser. Physiol.* **21**, 1–49.

Winge, Ö., and Ditlevsen, E. (1947). Colour inheritance and sex determination in *Lebistes. Heredity* **1**, 65–83.

Winge, Ö., and Ditlevsen, E. (1948). Colour inheritance and sex determination in *Lebistes. C. R. Trav. Lab. Carlsberg, Ser. Physiol.* **24**, 227–248.

Wolf, K., and Quimby, M. C. (1969). Fish cell and tissue culture. *In* "Fish Physiology" (W. S. Hoar and D. J. Randall, eds.), Vol. 3, pp. 253–305. Academic Press, New York.

Wolf, K., Quimby, M. C., Pyle, E. A., and Dexter, R. P. (1960). Preparation of monolayer cell cultures from tissues of some lower vertebrates. *Science* **132**, 1890–1891.

Wright, J. E. (1955). Chromosome numbers in trout. *Prog. Fish Cult.* **17**, 172–176.

Wright, S. (1943). Breeding structure of populations in relation to speciation. *Am. Nat.* **74**, 232–248.

Yamamoto, K., and Ojima, Y. (1973). A PHA-culture method for cells from the renal tissue of teleosts. *Jpn. J. Genet.* **48**, 235–238.

Yamamoto, T. (1961). Progenies of induced sex-reversal males in the medaka, *Oryzias latipes. J. Exp. Zool.* **146**, 163–179.

Yamamoto, T. (1963). Induction of reversal in sex differentiation of YY zygotes in the medaka, *Oryzias latipes. Genetics* **48**, 293–306.

Yamamoto, T. (1969). Sex differentiation. *In* "Fish Physiology" (W. S. Hoar and D. J. Randall, eds.), Vol. 3, pp. 117–175. Academic Press, New York.

Yamamoto, T., and Kajishima, T. (1969). Sex-hormonic induction of reversal of sex differentiation in the goldfish and evidence for its male heterogamety. *J. Exp. Zool.* **168**, 215–222.

Zenzes, M. T., and Voiçulescu, I. (1975). C-banding patterns in *Salmo trutta*, a species of tetraploid origin. *Genetica (The Hague)* **45**, 531–536.

8

POPULATION GENETICS*

FRED W. ALLENDORF and FRED M. UTTER

I. INTRODUCTION

The title of this chapter covers a very broad area of study and is perhaps slightly misleading. A more exact description of the topic to be presented is the use of isozymic gene frequency data in the understanding of the genetic structure of fish populations. Almost all of our knowledge about the genetics of natural populations of fish has been gained from electrophoretic data—mostly within the present decade because of the maturation of isozyme methodology within this period.

* The authorship of this chapter is equal and alphabetical.

407

FISH PHYSIOLOGY, VOL. VIII
Copyright © 1979 by Academic Press, Inc.
All rights of reproduction in any form reserved.
ISBN 0-12-350408-2

This recent increase of knowledge is part of a new era in the ease of studying genetic variation in natural populations (Lewontin, 1974). The genetic structure of most economically important fish populations still remains virtually unknown, however; even during the present decade hypothetical gene frequencies had to be postulated for populations of chinook salmon of the Columbia River because of the absence of gene frequency data (Simon, 1972) in spite of the fact that these populations have been the object of intensive fisheries research for many years.

A major area of fish population genetics which is not discussed in the present paper is that of quantitative genetics. For many years, domesticated fish populations have been subjected to massive artificial selection programs with the intent of producing more desirable fish. Until recently these programs have generally lacked an appreciation of the genetic processes underlying such changes in the genetic composition of populations. In recent years this situation has changed with two notable examples being quantitative genetic work with the rainbow trout (Gall, 1974, 1975) and the carp (Moav and Wohlfarth, 1973); it has become evident that the basic principles of quantitative genetics can be readily applied to the culture of fish. The reader interested in this area should examine these works as well as the following pertinent references (Aulstad *et al.*, 1972; Gjedrem, 1975, 1976; Sneed *et al.*, 1971; Cherfas, 1972).

Recognizing that major gaps remain in the understanding of population structures of fish, the methodology is now available for obtaining answers concerning the structure of fish populations that were previously impossible to obtain; a representative number of fish species has been studied electrophoretically for this reason. de Ligny (1969, 1972) reported that variants of an apparent genetic origin had been found in more than 47 species while reviewing electrophoretic studies carried out on fishes (predominantly teleosts) through 1970. Although much of the earlier work was primarily devoted to descriptions of genetic systems (presumably because of the novelty of the ease of detecting single gene variations in natural populations), some more penetrating biological questions were being asked concerning gene–environment interactions, physiological functions of variants, and population structures of species. A recently published book (Altukhov, 1974) discusses theoretical aspects of fish population genetics as well as presenting an extensive list of references of electrophoretic studies of fish populations. In a recent review article Utter *et al.* (1974) discuss what they regard as the most promising direction to be pursued through applications of electrophoretic data in the study of fishes. An extremely useful discussion of the applications of biochemical genetics to aquaculture with emphasis on the lobster has also recently appeared (Hedgecock *et al.*, 1976).

Fish are a uniquely useful group of organisms for studies of population genetics. They are a large and highly variable group comprising nearly 45% of all vertebrate species. Kosswig (1973) has recently reviewed the role and advantages of using fish in research on genetics and evolution from the classical genetics perspective. Fish are especially useful for studies using isozymic techniques for a number of reasons. Fish populations, compared to those of other vertebrates, are generally quite large and collection of individuals is relatively easy. Fish occupy a wide range of aquatic (as well as some nonaquatic) habitats. Fish are poikilothermic animals and thus are more directly sensitive to environmental temperatures than are homeothermic vertebrates. Populations are structured by a variety of life history patterns including such extremes as anadromy—where a series of at least semi-isolated spawning populations spend the majority of their life cycle together in the ocean—to the other extreme of catadromy— where a single ocean spawning population diverges to spend the growing part of the life cycle in separate fish-water streams. Populations are also strongly influenced by geographical barriers to migration such as isolated lakes, streams, and impassable falls. Many fish species are easily cultured, permitting verification of Mendelian variants and experiments simulating environmental variables that may differentially affect particular allelic products. These are just some of the features that illustrate the usefulness of fish in studying evolutionary processes in natural populations.

Many fish populations are presently undergoing drastic fluctuations in size and distribution as a result of man's interest in their harvest and culture. Behnke (1973) has estimated that 99% of the original populations of cutthroat trout (*Salmo clarki*) in the interior of the United States have been lost in the last 100 years. Many other species of negligible direct importance to man, but of enormous importance indirectly through interactions with other species, have also been drastically affected by man's physical and biological manipulation of the environment. Understanding the present structure of fish populations therefore has a significant practical dimension superimposed on the pure scientific interest of these findings. The effects of plantings and the excessive harvest of existing populations can be best evaluated and managed only if the present population structures are known.

The present paper is intended to complement and extend the review of Utter *et al.* (1974). The emphasis is directed toward the use of isozymic genetic data in studies of fish populations with special focus on the application of this data to the culture and management of fish populations. We have directed much of our discussion toward the salmonid fishes because of the importance of salmonids and our own extensive experience with this group, although the same principles are applicable to other organisms. A preliminary methodology section is presented which emphasizes the amount of genetic information available and the importance of confirming the genetic basis of electrophoretic variants. An overview of some of the areas where isozyme studies have been used in the study of fish populations is followed by a more detailed examination of the concept of the amount and distribution of genetic variation in populations of fish. The final section discusses the management applications of this population genetic data in both artificial and natural populations of fish.

II. ISOZYME METHODOLOGY

A. Basic Techniques of Electrophoresis

The major advantages of electrophoretic methodology for genetic studies of populations are (1) the relative ease of application and efficiency of the techniques and (2) the direct relationship between appropriately chosen protein variants and the gene. The basic electrophoretic methodology has been presented in detail elsewhere (see Utter *et al.*, 1974, for gel techniques; Shaw and Prasad, 1970, for staining methods). The present discussion is supplemental to these descriptions and emphasizes ways in which a greater amount of genetic information per individual can be obtained.

Optimal resolution for a given protein system involves a number of variables including buffer system, distance of migration in a given buffer system, tissue in which protein is expressed, and species. As an example, we will consider enzyme systems in rainbow trout that are affected by this kind of variation.

Malate dehydrogenase (MDH) of rainbow trout (*Salmo gairdneri*) occurs in two cytoplasmic forms and at least two mitochondrial forms (reviewed in Clayton *et al.*, 1975). The distribution of these forms in different tissues and their expression on different buffer systems is summarized in Table I. The alkaline buffer systems (1 and 2) resolve clearly only the B form of MDH in muscle extracts while the acidic buffer systems (3 and 4) have a much broader range of expression. Many other protein systems have optimal expression on the alkaline buffer systems and most of the genetic variation of MDH reported in

Table I

Tissue Distribution and Expression of MDH Isozymes of Rainbow Trout in Different Tissues with Different Buffer Systems

Buffer system[a]	Muscle MDH[b]			Liver MDH			Eye MDH		
	A	B	M	A	B	M	A	B	M
1 and 2	0	+++[c]	0	0	0	0	0	0	0
3 and 4	+	++	+	+++	+	0	+++	+	0

[a] (1) Tris–citrate gel buffer, pH 8.5; lithium hydroxide–boric acid tray buffer, pH 8.1 (Ridgway et al., 1970). (2) Tris–boric acid–EDTA continuous system, pH 8.7 (Markert and Faulhaber, 1965). (3) Sodium phosphate, pH 6.5 (Wolf et al., 1970). (4) N-(3-Aminopropyl)-morpholine-citrate, pH 6.1 (Clayton and Tretiak, 1972).

[b] Forms of MDH: (A) cytoplasmic A; (B) cytoplasmic B; (M) mitochondrial.

[c] +, ++, and +++ indicate relative intensities with clear expression; 0 indicates no activity or inadequate resolution.

rainbow trout occurs in the B system; thus, it may seem expedient not to screen for MDH activity on acidic buffer systems. However, genetic variants have been seen for the A and mitochondrial systems as well (Allendorf et al., 1975; Clayton et al., 1975) and data from these systems should be included in population surveys where possible.

Similar variation with buffer systems is seen with other enzymes. For instance, one locus is expressed on the alkaline systems and at least three loci are resolved with the acid systems for α-glycerophosphate dehydrogenase (AGPD) from muscle extracts of rainbow trout. Clear expression for rainbow trout AGPD variants in buffer system (1) is obtained only if marker dye migration has not exceeded 4 cm; AGPD variants in pink salmon are clearly resolved on all of the buffer systems independent of migration distance. Glucose-6-phosphate dehydrogenase (G6PD) expression from liver extracts of rainbow trout is clear only on buffer system (1) and then only if the total anodal migration has not exceeded 4 cm. These examples illustrate the kinds of variability of expression seen for many protein systems with regard to different buffer systems, migration distances, tissues, and species. It is necessary to effectively deal with these variables or risk the loss of genetic information.

Broadening the capability for specific staining of enzyme systems is also necessary for increasing the amount of genetic information that can be obtained. Fluorescent detection of enzyme activity is neither new nor inadequately described in the literature (Coates et al., 1975; Chen et al., 1972) but has not received the universal application of other staining methods. We have recently incorporated fluorescent techniques into our methodology and review them here because (1) they expand the number of genetic systems that can be examined, and (2) they have the potential for giving greater sensitivity and economy to some staining methods.

We use fluorescent staining methods for two major categories of enzymes. A method is available for staining for specific esterases using esters of the fluorescent compound 4-methyl umbelliferone. This procedure has many advantages over the usual azo-dye linked methods. It is both more specific and sensitive which allows detection of loci not previously detectable. In addition, these fluorescent methods are quick, simple, reliable, and inexpensive.

The other enzyme category which can be examined using fluorescent methods is NAD and NADP dependent enzymes. These procedures take advantage of the fluorescence of the reduced cofactor at the site of enzyme activity. The staining procedure is the same as the conventional method except that a tetrazolium salt (either NBT or MTT) and PMS are not used. The advantages of the method are that it is less expensive and more sensitive than the conventional method. A disadvantage is that immediate recording of the data is necessary because of diffusion of the fluorescent product and the eventual loss of fluorescence.

The potential uses of fluorescent staining procedures are not limited to these applications. A variety of other fluorescent procedures are possible; the further exploitation of these precedures should play an important role in increasing the number of loci which can be examined by electrophoresis.

B. Data Interpretation

Electrophoretic variants are useful for genetic analysis of populations only if the data reliably reflect genetic variations. A serious potential pitfall of the inexperienced worker seeking genetic data from electrophoretic patterns is attributing a genetic basis to nongenetic variations. The genetic variants we are concerned with are a stable attribute of the individual and once expressed in the organism's development, remain qualitatively the same throughout the remainder of its life. Variations in protein expression from a given tissue reflecting other than simple Mendelian differences among individuals may arise from a number of causes including (1) ontogenetic changes in gene expression (Shaklee et al., 1974), (2) changes reflecting environmental differences such as temperature, salinity, or disease (Hochachka and Somero, 1973; Amend and Smith, 1974), (3) changes resulting from dissection or extraction procedures, and (4) changes brought about from conditions and length of storage. Nongenetic variation must be ruled out prior to concluding that a particular variant has a simple genetic basis; the use of such data in making genetic comparisons among populations can be grossly misleading and result in erroneous conclusions.

We have observed a considerable amount of electrophoretic variation of both enzymes and nonenzymatic proteins that lack a genetic basis. Two instances that were particularly deceiving were variations in liver esterases of sockeye salmon (*Oncorhynchus nerka*) and the sculpin (*Leptocottus armatus*). Initial extractions in both cases indicated two allele systems of monomeric enzymes with phenotypic frequencies fitting a Hardy–Weinberg distribution. Extractions made a day later from the same livers showed different patterns of variation (e.g., extracts from the same liver would frequently give single-banded phenotypes one day, and double-banded or the alternate single-banded phenotype on the other day and vice versa).

The ultimate test of the genetic basis of a particular variant is actual breeding data where the phenotypic ratios of progeny for a particular protein system are consistent with the known phenotypes of the parents. Family data are impractical or impossible to obtain for many organisms; however, other data are usually sufficient to give assurance that a particular pattern of variation has a genetic basis. Genetic variants are usually evident on the basis of predictable banding patterns of heterozygotes for a given protein system; heterozygotes of a monomer, dimer, and tetramer predictably give rise to one, three, and five-banded phenotypes, respectively (for more detailed discussions of relationships between subunit structures and heterozygous phenotypes, see Shaw, 1964; Utter *et al.*, 1974). The expression of a predictable pattern of electrophoretic variation for a given protein system in a species is usually sufficient evidence for simple genetic variation, provided consistent individual phenotypes are obtained from multiple tests of a tissue.

The strongest criterion, next to family data, for a genetic basis of a particular protein variant is parallel expression of the variant in different tissues of the organism. Such expression is virtual confirmation of a genetic polymorphism. A weakness of this criterion, if used alone, is that the converse situation, when a variant is expressed in only a single tissue, is not positive evidence for artifactual expression because of the possibility of the expression of specific genes in single tissues. This principle is demonstrated in the expression of aspartate amino-transferase (AAT) in muscle and eye extracts of salmonids (Allendorf,

1975; May, 1975; Allendorf and Utter, 1976). The common form in both tissues has identical electrophoretic mobility; but genetic variants have demonstrated that two loci code for muscle AAT while a third locus is expressed in the eye. In spite of this caution, parallel expression in different tissues remains strong positive evidence for genetic polymorphism.

The extensive gene duplication of salmonid fishes, and certain other teleosts having an apparent tetraploid ancestry (Ohno, 1970), complicate interpretations of electrophoretic patterns. Approximately 50% of the structural gene loci detected by electrophoretic methods are duplicated in rainbow trout and Pacific salmon (Allendorf *et al.*, 1975). Complexities of interpretation arise when subunits encoded by different loci form active proteins of similar or identical electrophoretic mobility. This situation is outlined in Fig. 1 for a dimeric protein where patterns of duplicated and nonduplicated loci are compared. Electrophoretic expression is based on four rather than two gene doses, and the number of possible phenotypes is five rather than three for a simple two allele polymorphism. Interpretation of the three phenotypes where both kinds of subunits are expressed becomes quantitative rather than qualitative and is based on the relative intensity of the three bands rather than the presence or absence of particular bands.

Two genetic possibilities exist for this kind of variation, disomic inheritance at two loci or a single tetrasomic locus. It has been shown that the distinction between two polymorphic disomic loci and a single polymorphic tetrasomic locus can be made only through breeding studies (Allendorf *et al.*, 1975). Only disomy has been demonstrated in salmonids where breeding data have been available (Allendorf and Utter, 1973; Allendorf *et al.*, 1975, May *et al.*, 1975).

Disomic inheritance of two polymorphic loci creates ambiguities regarding assignment of genotypes to phenotypes where both kinds of subunits are expressed. Phenotypes expressing 3 : 1 ratios of subunits are homozygous for one locus and heterozygous for the other, but the

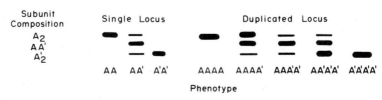

Fig. 1. Phenotypes expected from genetic variation of a dimeric protein expressed at a single and a duplicated locus (A = allele *100*; A¹ = allele 75).

loci cannot be assigned. The phenotype expressing $2:2$ ratios of sub-units may be either heterozygous at both loci or homozygous at both loci for different alleles.

The variability of expression of the common phenotypes of the dimeric enzyme phosphoglucose isomerase (PGI) in different tissues of four salmonid species is useful in demonstrating some of the principles of identification of biochemical genetic variation in a well-studied and somewhat complex system. The molecular basis of PGI isozymes of salmonids was identified by Avise and Kitto (1973) through observation of monomorphic phenotypes and has been confirmed by subsequent observations and breeding studies involving genetic variants of PGI (May, 1975). PGI phenotypes from eye, muscle, heart, and liver of rainbow trout, cutthroat trout, coho salmon (*Oncorhynchus kisutch*), and masu salmon (*O. masou*) (Fig. 2) display a potentially confusing array of patterns. Three loci code for PGI activity in these species, *PGI-1*, *PGI-2*, and *PGI-3*, listed in order of increasing mobility of the common homodimers. PGI subunits appear to unite at random in these species within and between loci, giving rise to six-banded common phenotypes from extracts of skeletal muscle, a tissue in which each locus is expressed at approximately the same intensity.

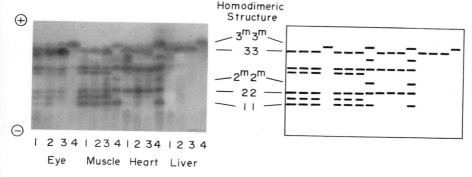

Fig. 2. Expression of PGI from four tissues of salmonids. (1) Rainbow trout; (2) cutthroat trout; (3) coho salmon; (4) masu salmon. Subunits for *PGI-1* produce electrophoretically identical homodimers in each of these species and are indicated by 1. Subunits for *PGI-2* and *PGI-3* in the two trout species and in coho salmon produce electrophoretically identical homodimers and are indicated by 2 and 3; subunits for *PGI-2* and *PGI-3* in masu salmon produce more negatively charged homodimers than the other species and are indicated by 2^m and 3^m. The other bands seen are heterodimers formed by the combination of protein subunits from two different loci. These heterodimers are intermediate in mobility between the homodimers of the two subunits present in the heterodimer.

Expression of the three PGI loci varies in the other tissues. *PGI-3* is strongly expressed in all tissues of all species; in the liver it is the only locus giving rise to clearly distinguishable bands. *PGI-1* and *PGI-2* are expressed at reduced intensities compared to *PGI-3* in the eye; expression of both loci is virtually absent in the eye of the masu salmon. *PGI-2* and *PGI-3* are preferentially expressed in the heart except in the masu salmon where, as the skeletal muscle, all three loci are expressed.

There are some bands of Fig. 2 that are not explained by the above model, for example, a band slightly slower than the 33 band of rainbow trout (quite diffuse in tissues other than the eye), and a pair of bands between the 22 and the 33 bands that are particularly evident in the muscle of coho salmon. Such bands are most likely the result of posttranslational protein changes affecting the charge or shape of the molecule. We have observed a direct relationship between the occurrence of such additional PGI bands and the duration of frozen storage.

Genetic variation of PGI superimposes additional complexities beyond those observed in the homozygous phenotypes of Fig. 2. Additional bands are formed by random interactions of subunits both within and between loci if the variant allele codes for a subunit that is differently charged from any of those involved in the invariant phenotypes.

Often, however, a variant allele of one PGI locus codes for subunits giving rise to enzymes that are electrophoretically identical to the common forms of another locus; we have seen this kind of variation between *PGI-1* and *PGI-2* in a number of salmonid species. Muscle extracts of heterozygous individuals give rise to six-banded phenotypes but the bands have uneven intensities because of the disproportionate synthesis of subunits. The eye PGI phenotype of coho salmon in Fig. 2 is similar to a heterozygous phenotype in which a *PGI-1* variant allele gives rise to a subunit which is electrophoretically identical to the common form of *PGI-2*. However, the even intensity of the bands of the muscle phenotype of coho salmon indicates that uneven intensities of the bands containing the *PGI-1* and the *PGI-2* subunits in the eye reflects differential synthesis of subunits of the two loci in the eye of this individual rather than allelic variation.

The complexity of PGI expression in salmonids is matched (or surpassed) by other systems such as lactate dehydrogenase (LDH) and MDH. A competent worker must acquire an understanding of the genetic significance of qualitative and quantitative variations observed in any complex system if data from that system are to be used in the study of populations.

Null alleles (i.e., alleles which either do not produce a protein or produce a protein with greatly reduced activity) are a phenomenon that can complicate interpretations of electrophoretic data. They are difficult to detect in simple systems (where only a single locus is expressed) except through the absence of activity in the homozygote. Detection becomes even more difficult where multiple loci are expressed. It can be impossible to differentiate between a null allele and an allele giving rise to an active protein having the same mobility as a protein for the same enzyme encoded by another locus. Dosage of heterozygotes in these instances is not a reliable criterion because it is difficult to differentiate between an active heterozygote [where a 3 : 1 ratio of subunits interact (see Fig. 1)] and a null heterozygote (with a 2 : 1 ratio of subunits). Null alleles must be considered in isozyme studies because of the tendency to identify heterozygotes of the null allele as a homozygote for the active allele which is present. This confusion results in distorted estimates of genotype frequencies and a deficiency of heterozygotes observed.

Before moving on we would like to emphasize the importance of both the quality and the quantity of the electrophoretic data collected in the study of populations. Meaningful parameters pertaining to the genetic variability and the identity of a particular taxon can only be estimated with data from many loci. It has been observed that estimates of average heterozygosity in a population obtained from data from 20 to 25 loci are highly correlated with estimates of variability that were derived from morphological criteria (Soule et al., 1973); we regard this number of loci as a minimum for obtaining reliable estimates for average heterozygosity (see Section IV,B). Avise (1974) has emphasized the need for examining large numbers of loci in attempts to derive systematic relationships among conspecific populations (see also Sections III,B and V,B).

Amid the need to collect data from as many loci as possible, it is important to exercise restraint in the distinction of genetic polymorphism and artifact. If interpretation of a particular system is doubtful because of inadequate resolution or inexplicable patterns, it is better to disregard that system rather than to make guesses in the absence of reasonable assurance. If multiple bands are observed in the absence of genetic variation, it is best to conservatively estimate the number of loci involved, and it is almost certainly erroneous to ascribe a separate locus to each band. It is much easier to modify a conservative interpretation later as more data become available than to defend or retract an erroneous conclusion drawn from inappropriately interpreted data.

C. A Uniform System of Nomenclature

Many systems of nomenclature have arisen to describe allelic variation that is detected by electrophoresis. Most of these systems are based either on electrophoretic mobility or frequency of variants. Symbols describing allelic variants (often presented as superscripts following a designated locus) include letters, numbers, and different numbers of apostrophes. Use of one or another system of nomenclature has been rather arbitrary in electrophoretic studies of fishes and has often been based on the personal preferences of the worker who first described a particular electrophoretic variant.

Most systems of nomenclature are unambiguous if only a single locus and two alleles are expressed for a particular biochemical system. The potential for confusion rises rapidly, however, as the number of loci and alleles increase. Wright *et al.* (1975) have summarized a particularly confusing set of systems of nomenclature that have been used by various investigators for five loci coding for LDH in salmonids. This potential for confusion is detrimental to effective communication, and we propose here a uniform system of nomenclature that minimizes ambiguity.

The system that we propose has been used by investigators of *Drosophila* for some time (e.g., Prakash *et al.*, 1969). An abbreviation is chosen to designate each protein; when in italics, these same abbreviations represent the loci coding for these proteins. In the case of multiple forms of the same enzyme, a hyphenated numeral is included; the form with the least anodal migration is designated one, the next two, and so on. Allelic variants are designated according to the relative electrophoretic mobility. One allele (generally the most common one) is arbitrarily designated *100*. This unit distance represents the migration distance of the isozyme coded for by this allele. Other alleles are then assigned a numerical value representing their mobility relative to this unit distance. Thus, an allele of the most cathodal LDH locus coding for an enzyme migrating one-half as far as the common allele would be designated *LDH-1(50)*.

III. MAJOR AREAS OF CURRENT INVESTIGATIONS

The application of the techniques of electrophoresis to the study of fish populations has found a broad base of biological questions of both basic and applied interest. The following section outlines the major categories of these investigations. This discussion is intended both to point out the type of biological questions that can be pursued with fish as an organism and also to raise questions of special importance to the study of fish populations.

A. The Measurement of Natural Selection

The argument raging around the adaptive significance of isozymic polymorphisms has dominated the experimental population genetics literature in recent years. The controversy between the two conflicting groups, as colorfully portrayed by Lewontin (1974), has not subsided. Rather, it appears that polarization is increasing among strong advocates favoring either a selective or a neutral explanation for most of the genetic variation that has been detected by electrophoresis. Much of the most valuable experimental evidence towards resolving this controversy has come from studies of fish populations.

The most ambitious attempt to measure natural selection in natural populations of fish is that undertaken on the eelpout (*Zoarces viviparus*) in the seas surrounding Denmark. This fish is exceptionally well suited for this purpose because of its abundance, ease of capture, and its ovoviviparous life history. The young are large enough when taken from pregnant females to be typed electrophoretically for a series of enzymes which allows the verification of the genetic basis of the variation, including linkages (Hjorth, 1971). More importantly, a statistical procedure has been developed that uses such mother–offspring combinations to provide a much more powerful test for natural selection than is generally possible (Christiansen and Frydenberg, 1973; Christiansen *et al.*, 1973).

A very important result of this investigation is the demonstration of just how difficult it is to measure natural selection. The authors have carried out an intensive study of a single polymorphic esterase locus. In their procedure of partitioning selection into five components, they have calculated that the smallest selection differences that could have been detected with a 50% probability in these five components from a sample of 1282 adult and 782 offspring ranged from 7 to 33%. Since quite low selection differentials on the order of 1% is compatible with polymorphism at a large number of loci within the genome, it can be seen that the task of demonstrating such selection is formidable.

A more extensive report of this study involving the examination of 4206 adult and 2210 fetal young eelpouts has since appeared (Christiansen *et al.*, 1974). The only selection component that deviated significantly from the null hypothesis was that of zygotic selection acting in favor of both homozygotes at the expense of the heterozygotes. This study does provide evidence of natural selection operating within a natural population for a specific isozyme locus. However, the demonstration of selection against the heterozygote does not provide the answer to the important question of how the polymorphism is maintained.

Systematic patterns of geographical variation in natural populations are often interpreted in adaptive terms. However, any geographical pattern of gene frequencies for a single locus can be explained by the random drift of neutral alleles if one is free to construct a suitable breeding structure. Even gene frequency clines can be explained by a model of neutrality (Kimura and Maruyama, 1971). Further evidence in the form of correlations between patterns of genotypic variation and specific characteristics of the environment is necessary. However, temperature, precipitation, and other environmental variables often are so strongly correlated with latitude, longitude, or altitude that it is difficult to relate particular patterns of genetic variation with environmental factors independent of geographical effects (Selander and Johnson, 1973). It is only when specific functional properties of allelic proteins can be related to the distribution of environmental variables that an adaptive interpretation becomes compelling.

A very fruitful area for such studies has been the examination of functional properties of fish isozymes in relation to temperature and their geographical distribution in populations of fish. Temperature, as an environmental variable, has the advantages of being both easy to measure in nature and having a relatively straightforward and determinable effect on enzyme kinetics. Some of the most convincing examples of the adaptive significance of allelic proteins have come from studies along these lines.

Merritt (1972) has demonstrated a functional basis for the maintenance of a cline for a two allele LDH polymorphism in the minnow *Pimephales promelas*. He has found that the homozygous form for the allele that predominates in northern areas possesses significantly higher Michaelis constants (K_m's) for pyruvate at 25°C and above than the isozymes of the other two genotypes. Thus, the pattern of variation of the northern allele in nature fits the distribution expected from *in vitro* studies. Koehn (1969) has also presented evidence of selective maintenance of esterase polymorphisms in two species of fish by finding associations between the temperature-dependent properties of the isozymes *in vitro* and their distribution in natural populations. Further evidence of temperature-dependent properties of isozymes and their distribution in nature was provided by Johnson (1971) in the high cockscomb, *Anoplarchus purpurescens*. He demonstrated differential survival of larval LDH types at high and low temperatures through laboratory experiments. These experiments and field data showed a consistent correlation of allele frequencies with temperature supporting the author's conclusion of an adaptive interpretation of the polymorphism.

We would like at this time to interject a warning concerning a potential confounding factor in studies correlating *in vitro* enzymatic properties with geographical distribution. Electrophoresis is estimated to detect 33% of the possible amino acid substitutions (Lewontin, 1974); this allows a great deal of room for "hidden" genetic variation to exist. Such variation can sometimes cause a serious experimental problem. For example, assume a clinal polymorphism for which the allele frequencies are significantly correlated with temperature. To test our hypothesis of temperature being an important selective factor we propose to isolate the different allelic isozymes and to measure the effect of temperature on their kinetic properties. This isolation procedure is most easily accomplished by using the homozygotes for the particular allele. In doing this, however, there would be a tendency for one allelic product to come from fish from one end of the cline and the other allelic product from the other end of the cline. If we then detect kinetic differences in the isozymes derived in this manner, such differences could actually be due to some hidden allelic variation and not to the allelic differences we can measure with electrophoresis. The clinal distribution of the electrophoretic alleles could possibly be strictly due to geographically determined genetic drift.

Such an example is not totally hypothetical. Guilbert (cited in Utter *et al.*, 1974) studied K_m's of electrophoretically different forms of a two allele LDH system in sockeye salmon. He observed differences among the three phenotypes taken from fish returning to western Alaska, indicating that the heterozygous phenotypes had the greatest NADH binding capacity (i.e., lowest K_m) at 15°C. The common electrophoretic phenotype of western Alaskan fish was the only LDH form observed among sockeye returning to British Columbia and Washington. The K_m's of this phenotype were very different between the two areas, however. The K_m of Washington fish was the same as that of the heterozygous phenotype from western Alaska while the K_m of the common phenotype from western Alaska was much higher. The data suggest that the electrophoretically identical phenotype from two areas represents genetically different enzymes, and that conclusions relating function and distribution must be drawn with caution if data for different genetic types are collected from different areas.

An important present and future area of investigation is the examination of the effect of warm-water effluent on fish populations. The addition of warm-water effluent into the aquatic environment provides a situation where a possible genetic response of fish populations to this environmental change can be studied. Recent reports have examined this situation in three fish species (Mitton and Koehn, 1975; Nyman, 1975; Yardley *et al.*, 1974). Such investigations provide a "natural" experiment whereby possible natural selection on these loci can be sought. In addition, these studies have important practical implications as to possible harmful effects of warm-water effluent on populations of aquatic species.

Another very different approach to detecting natural selection in populations involves taking advantage of unusual life history patterns present in some species. The eel and salmon are fish with strikingly contrasting life histories. These extreme patterns have been used in the American eel and pink salmon to test hypotheses of selection versus neutrality at isozyme loci.

Investigations in the early decades of this century (Schmidt, 1922, 1925) provided evidence for a single mass spawning population of the American eel (*Anguilla rostrata*) in a region northeast of the West Indies. Eels from streams along the east coast of the United States were examined for gene frequencies of several isozyme loci under the assumption that any significant difference in gene frequencies between groups must be due to divergence from a common pool of zygotes (Williams *et al.*, 1973). Significant differences in gene frequencies were found both between groups of eels from different streams and between eels of different ages from the same streams. The authors interpreted this difference as strong evidence for a selective basis of the variation with the simple statement: "The alternative of selective neutrality can be immediately dismissed" (Williams *et al.*, 1973, p. 200).

We feel, however, that such conclusions must be based on a firmer understanding of the population structure of this species. The case for a single mass spawning panmictic population is not conclusive. Another problem that must be considered is the theoretical expectations of the result of this unusual life history where a single panmictic population disperses over such a large range of environments. It seems at least intuitively reasonable that large blocks of coadaptive genes are likely to evolve in such a situation. The effect then measured at a single locus may be the cumulative effect of a large number of genes in nonrandom association rather than a direct effect of the specific locus we are looking at. A more detailed experimental examination of these populations should provide valuable evidence of the evolutionary processes which are actually present.

A study with pink salmon (*O. gorbuscha*) (Aspinwall, 1974) provides a highly interesting contrast to the eel study. Pink salmon mature at exactly 2 years of age (±10 days) and return to their native streams to spawn after spending the great majority of their life cycle in marine waters. The effect of this rigid timing is that there are two genetically isolated populations within many streams in odd and in even years. Confirmation of this rigid 2-year life cycle is the existence in the southern part of the range of large numbers of pink salmon returning in odd years but a complete absence of any even-year populations.

A number of populations from both even and odd years were examined for two polymorphic enzyme systems throughout the range of the pink salmon. The allele frequencies of the two systems revealed uniformity throughout the geographical range within even- or odd-year cycles. However, significant differences were found among some even- and odd-year populations that were sharing a single stream. These populations come as close as possible to sharing the same environment but yet being genetically isolated.

These results thus show uniformity of allele frequencies to even- or odd-year populations inhabiting a wide range of environmental situations, presumably due to a small amount of gene flow between populations of a given cycle. At the same time some populations sharing the same environment do not have similar allele frequencies, as would be expected under a model of selective maintenance of the polymorphisms. The author concludes that a model of selective neutrality combined with random genetic drift is the most plausible explanation of these results.

There are problems, however, in the assumption that similar environments will produce similar gene frequencies. A basic problem in this case is Lewontin's "historicity principle" (Lewontin, 1967). Lewontin shows that populations exposed to identical fluctuating environments, which differ only in the order of the environments, may exhibit vastly different gene frequencies. Thus, even if all biological assumptions are correct concerning these pink salmon populations, random environmental fluctuations from year to year may bring about *selected* gene frequency differences between odd- and even-year populations in the same stream.

These two studies provide striking contrast in both organisms studied and in conclusions drawn. The conclusions of both studies are somewhat preliminary in that they are based on important assumptions of the life cycles involved. Further results from these studies should provide more definite answers to the question being asked. Both of these studies are excellent examples of the kind of well-designed investigations that are needed to answer the current questions of population genetics. The differences point to the danger of extrapolating significantly beyond the experimental organism in question. Further studies simply reporting on genetic variation in additional species are not particularly useful at this time.

The study of the selective components of enzyme polymorphisms has important implications in the management of fish populations as well as being an important area of basic research. When using isozymic allele frequencies to delineate the boundaries of breeding populations we must have some assurance that we are dealing with relatively stable gene frequencies which are not being greatly changed by selection during the length of a generation. Such stability has been observed in salmonid populations that we have studied, as indicated in Section V.

In addition, there has been a widespread effort to detect the selective basis for particular polymorphisms and to turn this knowledge into practical use by artificial selection of a particular genotype. As stated before, a very low selection differential is compatible with the selective maintenance of polymorphisms. The hope of finding a large effect on economic performance of the vast majority of protein variants appears to be largely illusory (Robertson, 1972). This rather pessimistic prospect does not imply that these protein variants are without considerable potential for practical applications, however. We believe that the use of such allelic variants as effectively neutral markers of the genome is an extremely promising and relatively unexplored area of application. These applications are discussed in detail in Section V.

B. Systematics

Protein markers have found increasing use in systematic studies in recent years. Cumulative comparisons among loci between two taxonomic groups can be summarized by a variety of methods into indices of similarity or, conversely, genetic distance. Such data can be arranged in matrix form for comparisons among many taxonomic units, and dendrograms suggesting taxonomic relationships can be constructed from these matrices.

Avise (1974) has reviewed systematic studies using electrophoretic data and has concluded that electrophoretic techniques are an extremely valuable tool for systematics, particularly at the subgeneric level. The data are objective and have a purely genetic basis; a survey of the literature revealed that relationships indicated by electrophoretic data usually corresponded closely to those previously indicated by classical systematic criteria.

Ayala (1975) has compiled electrophoretic data from systematic studies of a broad range of organisms. His data reveal that similarity indices (based on Nei, 1972) in all organisms reported tend to average above 0.90 for populations within a species. The indices in *Drosophila* drop to about 0.80 for semi-species and sibling species, and considerably lower for morphologically distinct species. A similar reduction in genetic similarity as taxonomic distance increased was observed for other organisms.

Isolation of sympatric populations of chars and brown trout in certain Scandinavian lakes was suspected on the basis of different spawning times and locations, and differential growth rates; this genetic isolation was confirmed through electrophoretic studies (Nyman, 1972; Allendorf *et al.*, 1976). The reasons for this isolation of sympatric conspecific populations are unclear and may reflect recent convergence following previous isolation, or perhaps even sympatric speciation. Regardless of cause, it is clear that electrophoretic data are a powerful tool for studying the question of genetic isolation of sympatric populations.

Some electrophoretic studies dealing with systematic relationships among fish species are consistent with the tendency of electrophoretic data to confirm relationships established by other criteria, and the indication that the process of speciation involves considerable genetic reorganization. Studies of Pacific salmon (Utter *et al.*, 1973), sunfish (Avise and Smith, 1974), and rockfish (Johnson *et al.*, 1972) all fall into this category.

There are a number of interesting exceptions to these generalities among electrophoretic studies of fish systematics, however. Some of these studies concern comparisons among populations that have been given different specific, or even generic rank, but whose genetic similarities lie within the range usually attributed to conspecific populations. The specific status of some of the populations examined remains in question in spite of considerable morphological differentiation because of the allopatric nature of the populations and the fertility of hybrids between them. Such is the case with certain pupfish (*Cyprinodon*) populations of the southwest United States (Turner, 1974) and rainbow, golden (*Salmo aguabonita*), and red-banded trout (*Salmo* sp.) populations discussed in Section V of this review. High genetic similarities among morphologically distinct pairs of fish species that occur sympatrically have been reported for rainbow and cutthroat trout (Utter *et al.*, 1973), two native cyprinids of California of different genera—hitch (*Lavinia*) and roach (*Hesperoleucus*) (Avise *et al.*, 1975), and two endemic cichlids of Mexico (Kornfield and Koehn, 1975). Such high genetic similarities among morphologically distinct and, in some instances, reproductively isolated populations are inconsistent with the much larger differences that are observed among most other species pairs and seem to be the result of recent speciation rather than a reflection of such factors as sampling error or convergent evolution (see Avise *et al.*, 1975, for a more detailed discussion of alternative possibilities). These studies indicate that the amount of genetic rearrangement involved in the process of speciation can vary considerably.

The above studies also suggest that loci determining metabolic proteins evolve at different rates in some organisms than those determining morphological characteristics. This observation is not limited to fishes. King and Wilson (1975) observed that man and chimpanzee are biochemically as closely related as subspecies of most other vertebrates studied. Cavalli-Sforza (1974) compared evolutionary trees of major human groups based respectively on purely genetic (i.e., protein and blood group) and anthropometric data and noted significant discrepancies between the two trees. He concluded that single gene data are probably more reliable for such reconstructions than are morphological data because of the limitations of an unknown number of gene loci and short-term environmental effects inherent in the latter method. This conclusion reinforces the potential value of electrophoretic data as a tool for systematic studies of fishes.

IV. GENETIC VARIATION IN POPULATIONS OF FISH

A. The Nature of Genetic Variation

Genetic variation among organisms sharing a common gene pool is necessary before evolutionary change can occur. The importance of genetic variation has long been recognized in the fields of plant and animal breeding. Knowledge of the amount of genetic variation present for a particular trait allows one to predict the amount of change expected from selection on that trait. If all of the loci relevant to the trait lack genetic variation, then no scheme of selection can be successful. The measure of genetic variation important in this situation is defined as the heritability (h^2) of a trait and is expressed as

$$h^2 = \frac{V_A}{V_P}$$

where V_P is the phenotypic variance of the character in the population and V_A is the additive genetic variance of the trait (Falconer, 1960).

The amount of change expected under selection is a function of both the heritability of the trait under selection and the intensity of selection. It should be pointed out that since h^2 is a function of the amount of genetic variation in a population, the value of h^2 will undergo change in the course of a population's evolution. Therefore, this quantity is inadequate for predicting both the long-term progress expected under selection and the eventual limit to the selection process.

The widespread success of selection for traits in agricultural animals (Falconer, 1960; Brewbaker, 1964) is evidence of extensive genetic variation for many traits in a wide range of organisms. Even more impressive are the results of selection experiments with populations of *Drosophila*. There appears to be practically no character that cannot be successfully selected for in *Drosophila* (Lewontin, 1974). The conclusion from these results is that genetic variation exists in natural populations which is relevant to almost every aspect of an organisms's phenotype.

The description of genetic variation in populations is the fundamental observational basis of evolutionary genetics. There has historically been a great deal of controversy as to just how much genetic variation actually exists in natural populations (Lewontin, 1974). This controversy went unresolved in spite of the evidence from artifical selection experiments of a large amount of genetic variation in populations. It was not until the advent of electrophoresis, which has allowed the direct measuring of the genetic variation in natural populations at many loci, that this controversy could be settled. The presence of a high amount of genetic variability in a wide range of organisms has been convincingly demonstrated by isozyme studies over the last 10 years. These studies have been recently reviewed by Powell (1975).

B. Measuring Genetic Variation

It should be pointed out before detailing the measure of genetic variation in fish populations that the amount of genetic variation estimated via electrophoresis is a minimum estimate. There is a large class of amino acid substitutions which is not detectable by electrophoresis because of the dependence of electrophoretic separation on changes in charge or shape of the protein molecule. It has been generally assumed that electrophoresis can detect approximately one-third of all possible amino acid substitutions. However, a detailed study of amino acid sequences in primate hemoglobin suggests that electrophoresis may in fact detect considerably less than one-third of the actual genetic variation present for a particular molecule (Boyer, 1972). New techniques are currently being developed which expand the capability of electrophoresis to detect genetic differences (Singh *et al.*, 1974; Johnson, 1976).

The simplest, most direct, and most informative measure of genetic variation from gene frequency data is the average proportion of heterozygotes per locus (i.e., average heterozygosity). This quantity can be calculated by directly counting the proportion of heterozygotes observed or by using the expected Hardy–Weinberg proportions and calculating the average proportion of heterozygotes using the observed gene frequencies.

In estimating the average heterozygosity (H) from populations there are a number of pitfalls that must be avoided to arrive at a reliable estimate of genetic variation. These considerations have been reviewed by Lewontin (1974). A critical factor is the number of loci examined; Nei and Roychoudhury (1974) have outlined the statistical procedures appropriate for estimating the variance of heterozygosity measures and in doing so have emphasized the importance of examining as many loci as possible.

Another serious problem is the type of loci used in heterozygosity estimates. It may be questioned whether the amount of genetic variation at the many other types of loci throughout the genome is reflected by that detected through isozyme loci. However, we believe that it is a reasonable assumption that the amount of variation at isozyme loci reflects the *relative* amounts of genetic variation found at other loci in the genome. We believe this to be true because many of the processes affecting the amount of genetic variation act uniformly on the genome (e.g., effective population size).

A more serious and avoidable error in estimating H is the predominant use of a single class of isozyme loci. Different classes of isozyme loci have been found to have consistently different amounts of variation over a wide range of organisms (Powell, 1975). The reliance on a single major class of isozyme loci (e.g., esterases) in the estimating of heterozygosity introduces a large bias into the estimate. Therefore, if one's goal is to estimate heterozygosity in a population using isozyme data, one must strive to examine both a large number and wide range of isozymic loci.

In the course of our studies with salmonid populations we have accumulated a large amount of data which is appropriate for estimating average heterozygosity in several salmonid species. These estimates are presented in Table II. The average heterozygosity of these species ranges between 0 and 6%. The reliability of interspecific heterozygosity estimates has been challenged. The potential unreliability of such comparisons can be demonstrated by a simple example. Assume we estimate the average heterozygosity of a species based on some twenty genetic loci. If a twenty-first locus is added, which is polymorphic for two alleles with approximately equal gene frequencies, then our estimate of average heterozygosity will increase by approximately 2.5%. This example, taken into consideration with the

Table II

Average Heterozygosity in Nine Species of Salmonids[a]

Species	Common name	Number of populations	H	Range of H
Oncorhynchus				
O. gorbuscha	Pink salmon	6	0.039	0.032–0.047
O. keta	Chum salmon	5	0.045	0.043–0.048
O. kisutch	Coho salmon	10	0.015	0.000–0.025
O. nerka	Sockeye salmon	10	0.018	0.008–0.024
O. tshawytscha	Chinook salmon	10	0.035	0.024–0.052
Salmo				
S. apache	Apache trout	1	0.000	
S. clarki	Cutthroat trout			
	Coastal form	6	0.063	0.022–0.077
	Interior form	2	0.023	0.021–0.025
S. gairdneri	Rainbow trout	41	0.060	0.020–0.098
S. salar	Atlantic salmon	2	0.024	0.020–0.028

[a] These estimates are based on at least 30 loci in each species. These estimates were made by calculating the average heterozygosity in each individual population using allele frequencies, assuming Hardy–Weinberg proportions, and then averaging these estimates for all populations within a species. The following loci were used: AAT-1, -2, and -3; AGPD-1 and -2; ADH (alcohol dehydrogenase); GMP-1, -2, and -3 (general muscle protein); CPK-1 and -2 (creatine phosphokinase); IDH-1 and -2 (isocitrate dehydrogenase); LDH-1, -2, -3, -4, and -5; MDH-1, -2, -3, and -4; PGI-1, -2, and -3; PMI (phosphomanose isomerase); PGM-1 and -2; 6PGDH (6-phosphoglucose dehydrogenase); SDH-1 and -2 (sorbitol dehydrogenase); SOD (superoxide dismutase); TFN. From 50 to 100 individuals were examined in each population to estimate allele frequencies from that population.

narrow range of heterozygosities usually encountered in studies of natural populations (0–25%), should serve as warning on placing too much reliability on interspecific heterozygosity estimates.

We should point out, however, that the estimates of heterozygosity presented in Table II have been historically consistent and thus appear to be reliable. The sockeye salmon and rainbow trout were initially the most thoroughly examined species in our laboratory. Our initial observation, based on relatively few loci, was that the rainbow trout had a much higher amount of genetic variation than did the sockeye salmon. As additional loci have been added to our techniques over the years, this relationship has been extremely consistent.

Altukhov et al. (1972) have reported the average heterozygosity in chum salmon to be 3.2%; this estimate is very close to our estimate of 4.5%. They proposed that the low value of heterozygosity in chum salmon, relative to other species which have been studied, is the result of the great amount of gene duplication found in the salmonids. A look at H values in other salmonid species in Table II shows that a low amount of genetic variation is not the rule among salmonids. Therefore, the low amount of genetic variation found in some salmonids (e.g., sockeye salmon and coho salmon) cannot be readily explained by the presence of many duplicated genes, since some salmonid species do show a significantly larger amount of genetic variation.

Althukhov and his group have published a large amount of work with Asian salmonid populations which closely parallels our own work. Anyone interested in genetic variation in salmonids should be aware of the work that has been done on these species in the Soviet Union. We have included in the bibliography a selected listing of those papers by Altukhov's group of which we are aware (Altukhov et al., 1975a,b).

The comparison of the amount of genetic variation between populations within a species is more reliable than interspecific comparisons. One reason for this reliability is that such estimates are not as sensitive to the inclusion of additional loci as are estimates for a species as a whole. This principle holds because a locus which is highly polymorphic in a particular population is usually polymorphic in most populations from that species. Therefore, the inclusion of an additional locus will tend to affect all populations similarly.

An interesting observation pertaining to Table II is that estimates of H for different populations within a species are remarkably consistent. To demonstrate this point, Table III shows the estimates of H from forty-one populations of rainbow trout. These populations are from four major sources: (1) natural populations of anadromous rainbow

Table III

Average Heterozygosities in Each of 41 Populations of Rainbow Trout[a]

Location[b]	Heterozygosity
Hatchery anadromous populations	
Big Creek, OR	0.058
Chelan, Columbia River, WA	0.050
Cowlitz River, WA (winter-run)	0.062
Cowlitz River, WA (summer-run)	0.068
Chambers Creek, WA	0.072
Deschutes River, OR	0.049
Clearwater River, ID	0.041
Pahsimeroi River, ID	0.048
Siletz River, OR	0.059
Snake River, ID	0.045
Wells Dam, Columbia River, WA	0.064
Washougal River, WA	0.070
Native anadromous populations	
Arnold Creek, WA	0.068
Deer Creek, WA (upstream)	0.056
Deer Creek, WA (downstream)	0.039
Clallam River, WA	0.071
Falls Creek, WA	0.071
Dickey River, WA	0.080
Gobar Creek, WA	0.086
Hoko River, WA	0.071
Kalama River, WA	0.056
Nooksack River, WA	0.050
Pysht River, WA	0.056
Quinault River, WA	0.062
Sauk River, WA	0.060
Soleduck River, WA (upstream)	0.056
Soleduck River, WA (downstream)	0.067
Stillaguamish River, WA (upstream)	0.053
Stillaguamish River, WA (downstream)	0.056
Twin River, WA	0.066
Wild Horse Creek, WA	0.067
Wishkah River, WA	0.080
Hatchery resident populations	
Chambers Creek, WA	0.060
Dream Lake, WA	0.056
Halle Sø, Denmark	0.098
Hoptrop, Denmark	0.067
Lem, Denmark	0.061
Mors, Denmark	0.075
Puyallup, WA	0.059
University of Washington, WA	0.020
Native resident population	
Chester Morse Lake, WA	0.055

[a] These estimates are based on those loci listed in Table II with the exception of *TFN*, which is polymorphic in rainbow trout but was not included because it was not examined in all populations (Allendorf, 1975).

[b] OR, Oregon; WA, Washington; ID, Idaho.

trout (i.e., steelhead), (2) hatchery populations of steelhead, (3) natural populations of nonanadromous rainbow, and (4) hatchery populations of rainbow trout which are maintained in the hatchery for their complete life cycle. The differences in the amount of genetic variation found in rainbow trout populations are discussed in the next section.

It is interesting to note the wide differences in H among species seen in Table II. There appears to be no simple explanation for these differences. As discussed before, the low H seen in some species cannot be simply explained by the extensive gene duplication in salmonids because of the high H seen in the rainbow trout, which has the same amount of gene duplication as other salmonid species. The low H seen in *Oncorhynchus* species is also not satisfactorily explained by the recent rapid evolutionary divergence of this group because of the similar low H in the Atlantic salmon, an old phylogenetic lineage which has not undergone recent divergence. Likewise, an explanation based on the relative effective population sizes of these species is not tenable. The sockeye salmon with low H is represented by immense populations in major river systems (e.g., the Fraser River and Bristol Bay drainages) while the steelhead (rainbow trout) is restricted to comparatively small population sizes throughout its range.

Whatever the cause of these differences in H among species, they are consistent and appear to be reflected in other ways. The rainbow trout has historically been the most successful salmonid species in adapting to new environmental conditions—whether in a hatchery or when planted in the wild. This greater adaptability of the rainbow trout is very likely due to the greater genetic variation throughout the genome as reflected by the high H values for this species measured with isozyme loci. The possible use of H as an indicator of the potential of a stock for genetic change is discussed in the next section.

C. Importance of Genetic Variation in Fish Culture

As emphasized earlier, genetic variation is required in a population if the attributes of that population are going to be changed via selection. Another factor which must be remembered is that this same genetic variability which makes change possible is decreased when such changes are accomplished. As stated by Gall (1972): "We must make every effort to learn and understand both the biology and genetics of the organism before we attempt to tamper with the essential but perishable resource, genetic variability" (p. 159).

The estimation of the additive genetic variability in a population for a particular trait is difficult and involved. On the other hand, the estimation of the average amount of genetic variation in a population of isozyme loci is simple and straightforward. We would like to point out a possible relationship between these two measures of genetic variation.

We submit that a high amount of genetic variability as measured by isozyme loci is an indicator of high genetic variability throughout the genomes of that population. Therefore, populations with high average heterozygosities should also demonstrate high additive genetic variance for phenotypic traits which are of importance to fish culturists. Although this is currently only an hypothesis, it is one that can and should be tested; both parameters involved can be directly estimated.

The practical importance of such a relationship is great. One would be able to predict the relative success of a particular stock undergoing a program of artificial selection. In addition, one could create a stock with exceptionally high genetic variation by forming a new stock composed of individuals from populations which are initially extremely genetically different. Therefore, using average heterozygosity as an indicator of potentially valuable genetic variation, one could evaluate the potential of a current stock for genetic change, select a stock to use in a selection program which has an initial great deal of genetic variation, or create new stocks with exceptionally high genetic variation.

A word of caution must be interjected at this point. Estimating genetic variation by isozyme frequencies cannot replace the value of estimating the heritability of particular traits. When developing a program of selection one must consider the comparative heritability of particular traits to be selected. However, estimates of variability based on isozyme data allows one to predict the expected success to be obtained using different populations within a single species.

The contention has often been made that the artificial propagation of salmonid populations has led to the reduction of genetic variability in such populations. This contention can now be tested by examining the amount of genetic variation in different populations within a species as measured with isozymes. An examination of natural and hatchery populations of anadromous rainbow trout in Table III shows that there is no indication of loss of genetic variation in these hatchery populations. Thus, the procedures in developing and maintaining these populations in a hatchery during part of their life cycle has not resulted in a loss of genetic variation.

The examination of hatchery populations of rainbow trout which are completely maintained in a hatchery reveals a different situation, however. Eight populations of nonanadromous rainbow trout are shown in this table. The estimates of H in seven of these populations are between 5.6 and 9.8%. The exceptional population is the University of Washington hatchery stock which has an H value of 2.0%. Seven of the hatchery populations show no evidence of a reduction in genetic variation. However, the dramatic reduction in H of the University of Washington stock seems to indicate a significant loss in genetic variation in this stock. This stock has been maintained in the hatchery for a period of 40 years, during which it has been subjected to both a reduced population size and an ongoing program of mass selection for several characters, for example, body size, egg production, and age at maturity (Donaldson and Olson, 1955). Thus, it appears that this program has caused a drastic reduction in the amount of genetic variability in this stock. Such a reduction is potentially deleterious to the production characteristics of a stock. Studies with a variety of cultured organisms have shown that reduction in genetic variability such as this very often results in reduced viability of the population. The recently reported (Hershberger et al., 1976) poor survival of individuals from this stock during the early life of the zygote may be attributable to this loss of genetic variation.

The loss of genetic variation because of hatchery procedures is of concern to all those involved in the maintenance of hatchery populations of fish. An awareness of the factors involved in the loss of genetic variation in such a manner is extremely important. The reader is encouraged to pursue such sources as Falconer (1960) and Crow and Kimura (1970) to acquire such an understanding. We will present a brief review, however, of the important factors involved.

The loss of genetic variation in artifically cultured populations of plants and animals because of inbreeding is well known. However, the actual genetic processes underlying this phenomenon are often misunderstood. "Inbreeding" is a term which is used to describe a variety of different circumstances (Jacquard, 1975).

In the present context, the loss in genetic variation is largely attributable to a limited population size. The magnitude of this loss in a single generation is inversely proportional to the size of that population as expressed by the following equation

$$\Delta F = \frac{1}{2N_e}$$

where F is the inbreeding coefficient, which represents the proportional loss in genetic variation, and N_e is the effective size of the population (Crow and Kimura, 1970).

The effective size of a population is rarely equal to the total number of reproductive individuals in that population. Rather, the effective population size is defined as the size of an ideally behaving population that would have the same decrease in genetic variation as the observed population. A major factor which influences N_e is the relative number of males and females in a population. The effective population size of a population with different numbers of males and females is expressed by the relationship

$$\frac{1}{N_e} = \frac{1}{4N_f} + \frac{1}{4N_m}$$

where N_f and N_m are the numbers of females and males in the population, respectively (Crow and Kimura, 1970). The important consideration is that the value of N_e is strongly influenced by the sex which is the least frequent. For example, a population consisting of 1 male and 1 female has an N_e of 2, while a population with 1 male and 100 females has an N_e of approximately only 4.

Another major factor in the loss of genetic variation is fluctuations in population size from generation to generation. A greatly reduced population size for a single generation can have a drastic effect. Such bottlenecks should be avoided in the propagation of hatchery stocks.

Loss of genetic variation because of limited population size can be avoided given the proper concern and an understanding of the principles involved. Our results have shown that such a loss has occurred in at least one hatchery stock of rainbow trout. The potential estimation of genetic variation in a stock via isozyme examination provides a means to assess any possible loss of genetic variation in a hatchery stock. This capability may prove extremely valuable in estimating the potential value of a particular stock.

V. THE USE OF GENETIC DATA IN FISHERY MANAGEMENT

This section is based on some of our applications of electrophoretic data in the management of fish populations. These examples are drawn from our own experience but the concepts are much more generally applicable. The reader is also referred to the following sources for other discussions of applying genetic principles in the management of fish populations (Calaprice, 1969, 1970, 1976; Rasmuson, 1968; Purdom, 1972, 1976).

The capability of managing a fishery on the basis of its component populations is an objective that has generally eluded salmonid biologists until recently because of the difficulties involved in defining these populations. Tagging and marking studies have provided useful information concerning origins and degrees of straying of fish but have been limited by the need for handling all treated individuals. Natural features such as scale characters and relative mineral composition have also proven actually or potentially useful for identification of areas of origin through reflections of natal environments. None of the above approaches are capable of genetically defining population structures, however, and such definitions are necessary if management is to be based on population structures.

Properly selected electrophoretic data can genetically define populations on the basis of gene frequencies at different loci. Such data permit estimation of relative genetic similarity among populations of a species in the same manner that similarity among species can be estimated. Frequencies of variants in a given population are stable attributes of that population and tend to persist at the same level over many generations in salmonid populations that we have studied; estimates are therefore usually cumulative over time and may be added together to provide greater precision rather than requiring redefinition every year.

As a result of these attributes, electrophoretic studies of salmonid populations of the Pacific Northwest carried out by our unit and other groups have genetically defined these populations in much greater detail and clarity than had previously been possible. The overall picture is far from complete because of the complexities of salmonid populations, but an overview of major population units can be presented for some species on the basis of data that has been collected to date. This section presents such an overview, and discusses some areas of fish culture where we believe that electrophoretic data are particularly applicable.

A. Species Identification

Adult salmonid species are usually unambiguously identifiable on the basis of external morphology, particularly during the time of spawning. However, morphological identification of very young salmonids is often not possible. Such individuals are readily identifiable to species from biochemical genetic data obtained from freshly collected material (Fig. 3). This capability permits precise species assessment in stream surveys, even at very early life history stages; it has proven particularly useful for identification of cutthroat and rainbow trout, which often occur sympatrically and are morphologically very

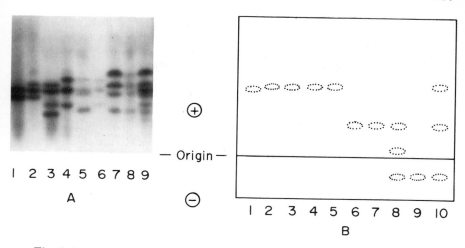

Fig. 3. Isozyme patterns useful in identification of nine salmonid species of two genera. Each species can be unambiguously identified through combined use of these two systems alone. (Note expression of bands from both parent species in hybrids.) (A) Creatine kinase (from muscle): 1, cutthroat trout; 2, rainbow trout; 3, masu salmon (*O. masou*); 4, sockeye salmon; 5, chum salmon; 6, pink salmon; 7, coho salmon; 8, chinook salmon; 9, chinook–masu salmon hybrid. (B) Superoxide dismutase (from liver): 1, chum salmon; 2, sockeye salmon; 3, masu salmon; 4, rainbow trout; 5, cutthroat trout; 6, coho salmon; 7, pink salmon; 8, pink–chinook salmon hybrid; 9, chinook salmon; 10, masu–chinook salmon hybrid.

similar as juveniles. Hybrids among the salmonid species that we have studied are usually clearly identifiable through biochemical genetic markers; we have often assisted other workers by identifying the appropriate ancestry of suspected hybrid individuals.

The absence of qualitative biochemical genetic differences among nominal species is insufficient evidence to demonstrate that these populations are conspecific, because sufficient genetic differences for reproductive isolation may exist at unsampled loci. Such data, of course, do not preclude conspecificity of the populations in question and are supportive of positive evidence derived by other means. A study of twenty biochemical genetic loci in rainbow trout, California golden trout (*S. aguabonita*), and red-banded trout (*Salmo* sp.)—a group of trout indigenous to dessicated basins of eastern Oregon (Behnke, 1965; Wilmot, 1974)—indicated that these three natively allopatric groups shared common alleles at every locus (Allendorf and Utter, 1974). Conspecificity of these three groups of trout is also supported by the tendency of rainbow and golden trout to hybridize readily when placed in the same drainages (Behnke, 1965). The question of species remains unresolved, however, because of the tendency of rainbow trout and cutthroat trout, two valid species, also to hybridize under some conditions (Behnke, 1965). Management of these three groups of trout should emphasize continued separation if their identities are to remain distinct (Behnke, 1965).

B. Identification of the Genetic Structure of Natural Populations

A major goal of our studies is the genetic characterization of salmonid populations of the Pacific Northwest. Fulfillment of this goal can provide major insights into the breeding structure of these species and lead to a sound basis for identification of areas of origin in mixed fisheries. The problem has been complicated by the transplantation of stocks from one area to another, but sufficient variation persists among major groups within species to characterize these groups on the basis of frequency differences of biochemical genetic markers. We summarize here some of our more extensive studies of either natural populations or hatchery populations derived principally from native fish from the area of the hatchery. The data were obtained over a number of years and appear to reflect stable genetic attributes of the populations studied (regardless of whether the variation is maintained by random or selected processes).

An electrophoretic survey of 32 loci of anadromous rainbow trout (i.e., steelhead) populations of the Pacific Northwest (Allendorf, 1975) revealed considerable genetic heterogeneity among them and indicated some relationships that had not previously been known (Fig. 4). A major division occurs among populations at a point coinciding with the crest of the Cascade Mountains (Fig. 5).

A similar east–west division of LDH variants has also been reported for both migratory and nonmigratory rainbow trout on the Fraser River (Huzyk and Tsuyuki, 1974). Two major taxonomic units were proposed for rainbow trout of this area on the basis of these findings, a coastal group and an inland group. The inland group presumably descended from rainbow trout residing in large lakes formed from inland drainages of the Fraser and Columbia Rivers during the last period of glaciation (McKee, 1972). The coastal areas were apparently repopulated by another group—possibly the Asiatic rainbow trout (*S. mykiss*) (Behnke, personal communication)—when the glaciers receded. The evidence indicating that geographic separation is the principle basis for genetic isolation of rainbow trout populations differs from previous conceptions in which anadromy and time of return to freshwater were regarded as the primary indications of genetic differences (Behnke, 1965; Withler, 1966; Millenbach, 1973).

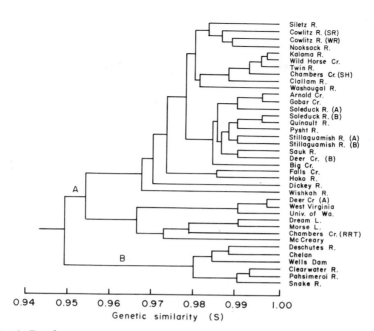

Fig. 4. Dendrogram (unweighted average linkage method) of genetic similarity among 38 populations of rainbow trout based on Rogers (1972) similarity coefficients obtained from gene frequencies at 32 loci. The five populations contained in branch B represent the inland group of rainbow trout found east of the Cascade Crest. All those populations in branch A are representatives of the coastal form. (From Allendorf, 1975.)

Coho salmon of the Pacific Northwest have been examined in sufficient detail to identify some major population units. Populations of both the Columbia and Fraser Rivers and their tributaries are distinguished by high frequencies of the gene coding for the most anodal form (*TFN-103*) of the serum protein transferrin (Utter *et al.*, 1970, 1973; May, 1975). Coho salmon populations returning to other areas between and on either side of these two large river systems are characterized by a predominance of the most cathodal migrating form of transferrin (*TFN-97*) and the presence of a third form of intermediate electrophoretic mobility (Fig. 5). The discontinuous distribution of transferrin alleles among coho salmon populations cannot be directly explained on the basis of glacial events (as could discontinuities of rainbow trout populations of these rivers) because the predominance of the *TFN-103* allele extends the full length of both large rivers. This distribution may be related to environmental factors of the large rivers, a possibility which is currently being investigated.

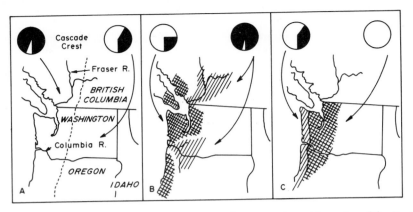

Fig. 5. Major population groups in three anadromous salmonid species of the Pacific Northwest. (A) Average proportion of *LDH-4* (*100*) allele in rainbow trout (represented by shaded areas of circles) in areas east and west of the Cascade Crest in Washington, Oregon, Idaho, and southern British Columbia. (B) Average proportion of *TFN* (*103*) allele of coho salmon (represented by shaded areas of circles) in coastal and Puget Sound drainages of Washington, Oregon, and British Columbia (checked areas), and Columbia River and Fraser River drainages (lined areas). (C) Average frequencies of *PGI-2*(*68*) allele of fall chinook salmon (represented by shaded areas of circles) in Puget Sound and Columbia River drainages (checked areas) and coastal populations (lined areas) of Washington and Oregon.

Populations of chinook salmon returning to fresh water in the fall have been examined from Puget Sound, along the Washington and Oregon coasts, and in the Columbia River (Utter *et al.*, 1973; Kristiansson, 1975; May, 1975). Coastal populations of Washington and Oregon have high frequencies of variants of PGI and also usually have phosphoglucomutase (PGM) variation; both PGI and PGM are invariant in samples taken from Puget Sound and the Columbia River (Fig. 5). These differences indicate considerable genetic isolation of coastal and noncoastal populations of fall-run chinook salmon and are potentially very useful for determining major areas of origin of ocean caught fish.

These biochemical genetic studies of salmon and trout populations have defined major natural population units that had not been previously identified. This knowledge is presently being used in the management of these fisheries. Allocation to one or another major grouping in mixed fisheries can be made using known gene frequencies from the major groups and gene frequencies obtained from sampling of the fisheries. An actual example of this application is outlined in Table IV for coho salmon collected in the Pacific Ocean near Neah Bay, Wash-

Table IV

Estimated Proportions of Columbia and Fraser Rivers Fish in Samples of Coho Salmon Caught in the Pacific Ocean near Neah Bay, Washington[a]

Collection number	Collection date	Sample size	Frequency of TFN-103 in ocean catch (F_m)	Estimated proportion of Columbia and Fraser Rivers fish in sample (p)
1	25 Jul 75	53	0.537	0.39
2	12 Aug 75	106	0.425	0.25
3	8 Sep 75	110	0.464	0.30
4	11 Sep 75	114	0.504	0.35

[a] These estimates were calculated assuming fixation of the TFN-103 allele in the Columbia and Fraser Rivers and an average frequency of the TFN-103 allele of 0.250 estimated from 59 collections from other areas of Washington State. The proportion (p) was calculated by $p = (F_m - F_w)/(F_{fc} - F_w)$, where F_m, F_{fc}, and F_w are the frequencies of TFN-103 in the ocean catch, Columbia and Fraser Rivers, and Washington State coastal and Puget Sound streams, respectively.

ington, during the fall of 1975. The data indicate that a sizable proportion of this fishery is supported by the two large river systems.

These data can also be used to suggest modifications of current management practices. Steelhead returning to the Columbia River in the summer have previously been regarded as a single genetic unit, contrasted with those returning in the winter (Millenbach, 1973). Planting of summer-run steelhead derived from coastal fish in inland areas (where all steelhead are summer-run fish) have been unsuccessful (personal communication, Washington State Department of Game). This failure is understandable now that the major grouping of steelhead appears to be based on inland and coastal populations rather than time of return, and it is not surprising that coastal fish were not successful when planted in inland areas.

Statistical differences also exist among natural populations within major subgroupings of these salmonid species. For instance, coho salmon populations in south Puget Sound are characterized by average gene frequencies of a variant LDH allele of about 0.05; this allele is virtually absent in north Puget Sound populations (May, 1975). Differences among coastal steelhead populations are seen in the frequencies of alleles at many loci. These variations are generally more subtle than those observed between major subgroups; they are useful in defining populations and are potentially applicable in identifying component populations of mixed fisheries. The application in the latter case is complicated by multiple genetic variants and populations beyond the simple instance outlined in Table IV. Analysis is feasible, provided appropriate baseline data are available, by fitting the appropriate data parameters through computerized methods using the techniques of maximum likelihood and minimum χ^2 (Kempthorne, 1957; Krieger et al., 1965). Personnel of our group are presently developing and applying this analytical capability.

It is interesting at this point to consider the quite different population structures of these three closely related sympatric species which have similar life histories. Each species has its own major population groupings which are defined through distinct arrays of polymorphic loci. Although some possible explanations for a particular distributional pattern within a species have been suggested above, possible reasons for such diversity among species remain obscure. It seems likely that many factors have interacted to bring about these differences including sequence of initial entry to the region, differential tendencies of straying from natal streams, and differential adaptive capabilities of a species to a particular region. Regardless of cause, it is apparent that generalizations regarding population structure cannot be safely made even among closely related species; and it follows directly that management practices that are based on population structures must arise from direct evidence rather than inferential data from closely related species.

C. Hatchery Populations

The emphasis to this point has concerned the management applications of protein variants of natural populations. The focus now shifts to hatchery populations.

Determining the effects of plantings of hatchery fish on native salmonids of the same species is a major concern to management biologists. Native fish are a valuable reservoir of genetic variation and provide a useful supplement to the fishery, even in stocks that are largely maintained through hatcheries. Although native fish may be more adapted to a particular area than hatchery fish, they are potentially endangered through hatchery plantings by factors including (1) competition for spawning and rearing grounds resulting from large hatchery releases, (2) possible earlier hatching of progeny of hatchery fish resulting in a competitive advantage, and (3) hybridization of native and hatchery fish resulting in disruption of adaptive gene pools.

Biochemical genetic markers are very useful for studying the effects of hatchery plantings on native fish, provided there are differences in gene frequencies between the two groups. Genetically marked hatchery fish require no special handling prior to release, and long-term effects of plantings can be measured because genetic markers are passed on to subsequent generations.

We are presently collaborating with the Washington State Department of Game in studying the effects of plantings of two stocks of hatchery steelhead maintained by the department which have been introduced in certain rivers of Washington State. The Kalama River, a tributary of the lower Columbia River, has been heavily planted in recent years with summer-run steelhead from the department's hatchery on the Washougal River, which enters the Columbia River 30 miles upstream from the Kalama. Native fish from the Kalama River and winter-run hatchery fish that have been planted in the river both lack genetic variants for AGPD while the Washougal hatchery stock has a variant form of the enzyme with a gene frequency of about 0.15. This difference has been useful in tracing the effects of plantings of Washougal hatchery fish in the Kalama River and other river systems of western Washington.

Data from these preliminary studies indicate some interesting interactions of Washougal hatchery fish with other steelhead stocks of the Kalama River. Hatchery fish planted in the main stream of the Kalama River tend to enter tributaries prior to their seaward migration. Adult fish from hatchery plantings return near the point of release and many of them spawn successfully. Descendants of these fish apparently hatch earlier than those of other stocks based on their larger size in a given sampling area (although existing data cannot exclude other factors such as faster growth rates). Almost all of the residualized steelhead (i.e., fish that remain in the river rather than migrating to sea) appear to be from the Washougal hatchery. These data indicate considerable long-term competition of Washougal hatchery fish with the native steelhead stocks of the Kalama River. Plans for more detailed studies of this competition are outlined below.

D. Genetic Marking of Stocks

The potential value of a genetic marker for the identification of populations increases as the differences in its frequency increases between populations. The sample size needed to demonstrate differences between two populations decreases to the point where individual fish can be identified if different alleles for a particular protein are fixed in the two populations. Such a situation rarely occurs naturally within a species, particularly among populations where gene flow is possible, but can be created through artificial propagation.

We are presently working with the Washington State Department of Game to create genetically marked stocks for maximizing genetic differences between these stocks and native fish in areas where the stocks are to be planted. One such stock is being bred from Washougal hatchery fish for introduction into previously unplanted tributaries of the Kalama River. Breeding is based only on the AGPD variation. Fish are selected for breeding by a screening process involving muscle biopsy, tagging, and electrophoresis of muscle samples. The breeding scheme is outlined in Table V. Initial selection based on homozygous males [AGPD (140/140)] bred with randomly selected females will be repeated each year for 4 years (we have presently completed 2 years). After that time the progeny of the first-year crosses will return as adults to the point of release; this point is a previously unstocked pond where only selected progeny have been reared. These fish will be screened for AGPD (140/140) individuals of both sexes to be used exclusively as breeders, and this procedure will also be repeated each year for 4 years. The 140 allele will now be fixed in this stock and spawning fish returning to the site of release can be spawned randomly. Fish returning to the pond should be screened periodically to assure that a significant influx of unmarked fish does not enter the spawning population.

Table V

Breeding Scheme for Fixation of AGDP-140 Allele in a Derivative Stock of Washougal River Hatchery Anadromous Rainbow Trout

	AGPD				
	Genotype frequencies			Allele frequencies	
	100/100	100/140	140/140	100	140
Parental population					
Total population	0.72	0.25	0.02	0.85	0.15
Breeding males	0	0	1.00	0	1.00
Breeding females	0.72	0.25	0.02	0.85	0.15
First generation after selection					
Total population	0	0.85	0.15	0.42	0.58
Breeding males	0	0	1.00	0	1.00
Breeding females	0	0	1.00	0	1.00
Final derivative stock					
Total population	0	0	1.00	0	1.00

There are two potential genetic pitfalls that must be kept in mind in a breeding scheme of this kind. The first of these is the possibility that the variant form of AGPD (or any other protein that might be selected) has a selective disadvantage contrasted with the common form of the enzyme. If such a disadvantage exists, the selected stock would be less genetically fit than the parent stock and conclusions drawn from the selected stock pertaining to the parent stock would be biased. It is therefore important to select a marker that is not obviously associated with any negative characteristic, and to carry out controlled tests on selected and parent stocks to be reasonably sure that both stocks are comparable for measurable variables other than the selected marker. This danger is especially present when selecting for an allele which is very rare in the original population. An allele present at an original frequency of less than 0.01 is much more likely to have a potential harmful effect than an allele which is present at a frequency of 0.15 and is therefore already present in 25% of the fish in the original population (assuming Hardy–Weinberg proportions). We have not detected any differences among the three AGPD phenotypes of the Washougal stock, and relative attributes of the parent and selected stocks are being monitored.

The other genetic potential pitfall that must be remembered is inbreeding (see Section IV,C). Inbreeding depression is a loss of vigor that occurs in most sexually reproducing organisms as a result of the reduction of genetic variation accompanying breeding of closely related individuals. This depression reflects factors including the expression of deleterious recessive genes, and the reduction of beneficial interactions both within and between loci. Inbreeding coefficients above 0.10 (this represents a loss of 10% of the genetic variation present in the original population) are often sufficient to result in detectable loss of vitality. The inbreeding coefficients induced in founder populations originating from 100 females and 1–10 males are plotted in Fig. 6. It is apparent that use of a single male would create a potential danger to the stock from inbreeding, and that this danger is reduced dramatically as up to 6 additional males are used—where a leveling off point is reached. Initial inbreeding is not significantly increased in subsequent generations provided selection of the parents is randomized and adequate numbers of individuals, say 50 or more of each sex, are bred. Both of these potential sources of genetic weakness must be anticipated in planning a genetic marking program but neither represents a serious obstacle if caution is exercised.

We foresee artificial genetic marking of hatchery stocks becoming a very useful management tool. The breeding scheme outlined in Table

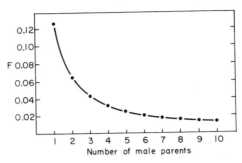

Fig. 6. Coefficients of inbreeding (F) after one generation of mating based on 100 female parents and varying numbers of male parents. $F = (N_m + N_f)/8N_mN_f$, where N_m, number of male parents, and N_f, number of female parents, equal 100.

V was a special case of an anadromous stock being brought to fixation with males only being selected in the first generation. The process can be simplified in hatcheries where brood stock are maintained because the same individuals can be used on successive years. If fixation of a marker gene is not required, useful gene frequency changes can be achieved in a single generation even if only males are selected. Change of gene frequency can obviously be accelerated if initial selection for both sexes is feasible.

Genetically marked hatchery stocks have many management applications in addition to evaluating the effect of hatchery plantings on native fish populations. Any management program based on differential harvest of a particular anadromous fishery would be much easier to implement if specific segments of the fishery were genetically marked. It has been indicated above that sufficient genetic differences already exist among major population units to permit some allocation of component stocks of mixed fisheries. Hatchery marking would significantly enhance this capability by creating additional differences where none presently exist.

Genetic marking of hatchery fish also has a significant potential for application in nonanadromous fisheries. Sufficient biochemical genetic variation appears to be present in most trout species to permit differential marking of all hatchery strains that may be planted in a particular region. The results of particular plantings can be closely followed in closed stream or lake systems and the performance of different hatchery strains can be precisely evaluated in the same environment. Measurements of such factors as relative survival, growth, reproduction, and dispersal can be readily obtained.

The above discussion has mentioned a number of applications of genetically marked hatchery stocks. Many additional uses of the method certainly exist and, of course, the concept can be generally applied to any cultured species. We anticipate that genetic marking will ultimately become a routine aquacultural procedure and urge management and hatchery personnel to seriously consider the process in their future planning.

VI. SUMMARY

We have reviewed genetic studies of fish populations emphasizing electrophoretic studies because the overwhelming majority of knowledge pertaining to genetics of fish populations has come from this source. Fish are especially appropriate as a group for studies of genetics of natural populations because of their extreme diversity, large population sizes, and their poikilothermic physiology which increases their susceptibility to thermal effects of the environment. Fish are also widely cultured, and there are many aspects of population genetics that are directly applicable to cultured populations of fish.

A discussion of isozyme methodology focuses on the genetic interpretation of electrophoretic variation. Use of multiple buffer systems, multiple tissues, and a wide variety of staining procedures are necessary to maximize the amount of genetic data that can be detected in a given species. Much electrophoretic variation is nongenetic and it is essential to identify only genetic variation in order to preclude a faulty data base for an investigation. The best criterion for a genetic basis of a variant is actual breeding data although alternate criteria may be reliably invoked if breeding studies are not feasible. Multiple systems of nomenclature of isozymic variants of fish have led to ambiguities and confusion. A unified system of nomenclature is proposed which would minimize this confusion and which is based on a numerical designation of allelic variants where the most common form is arbitrarily used as a standard.

A section is devoted to major areas of isozyme investigations of fishes. Much effort has been directed towards the measurement of natural selection. Extensive studies of eelpouts have indicated that demonstration of selection is itself very difficult, and even when demonstrated, may still not be informative regarding the method of maintenance of polymorphism. Relationships have been demonstrated in some fish species between geographic patterns of protein variation and the functional properties of a protein. Such correlations are evidence for selection but do not necessarily constitute proof; electrophoretically identical but functionally different proteins must also be considered, as well as the possibility of groups of co-adapted loci. Studies of eels and pink salmon are examples of well-planned investigations that respectively provide evidence for and against selection as the primary method of maintenance of polymorphisms in a species; more studies of this kind are encouraged. Systematic studies based on isozymic data have provided a useful complement to studies of the same taxa where other criteria have been used. Exceptional instances tend to reinforce the value of using protein data in systematic studies.

A discussion of genetic variation in populations of fish puts forth the concept that the relative amount of genetic variation in particular populations within a species is an indicator of the potential of that stock to undergo genetic change in a program of artificial selection. A review of the amount of genetic variation based on our own studies shows that (1) different species of salmonids have significantly different amounts of genetic variation as measured by average heterozygosity, (2) populations within a species have remarkably similar levels of heterozygosity, and (3) hatchery management procedures, in at least one case, have resulted in the loss of genetic variation in artificially cultured stocks. We also emphasize the role of limited population size in causing the loss of genetic variation in hatchery populations of fish.

Protein variations have been applied to define genetically some populations of fishes in much greater detail than had previously been possible. This capability offers some new possibilities for fisheries management. Biochemical identification of species can frequently be made on samples where morphological criteria are not useful. Identification of population groups provides data for management based on the genetic structure of these populations. Proportions of component populations of mixed fisheries can be determined if known differences in frequencies of isozyme variants exist among these populations. Biochemical genetic differences among hatchery populations can be maximized through selection for specific protein variants. Such genetic marketing can be done without affecting performance characteristics of a particular stock. This procedure has considerable potential as a tool in the management of fish populations.

Note. This chapter reviews research on population genetics of fish conducted prior to August 1976 (the date the chapter was written).

ACKNOWLEDGMENTS

F.W.A. worked on the initial stages of this chapter while at the Department of Genetics, Aarhus University, Aarhus, Denmark, and he gratefully acknowledges the support given to him by that department and by the Danish Science Council.

We would like to thank the following agencies for their cooperation and assistance during the course of our studies: the U.S. National Marine Fisheries Service; the Washington Cooperative Fisheries Unit, College of Fisheries, University of Washington; the Washington State Departments of Game and Fisheries; the Alaskan Department of Fish and Game; the U.S. Fish and Wildlife Service; and the Oregon Cooperative Fisheries Unit, Oregon State University.

REFERENCES

Allendorf, F. W. (1975). Genetic variability in a species possessing extensive gene duplication: Genetic interpretation of duplicate loci and examination of genetic variation in populations of rainbow trout. Ph.D. Thesis, Univ. of Washington, Seattle.

Allendorf, F. W., and Utter, F. M. (1973). Gene duplication within the family Salmonidae: Disomic inheritance of two loci reported to be tetrasomic in rainbow trout. *Genetics* **74**, 647–654.

Allendorf, F. W., and Utter, F. M. (1974). Biochemical genetic systematics of the genus Salmo. *Anim. Blood Groups Biochem. Genet.* **5**, Suppl. 1, p. 33. (Abstr.)

Allendorf, F. W., and Utter, F. M. (1976). Gene duplication in the family Salmonidae. III. Linkage between two duplicated loci coding for aspartate aminotransferase in the cutthroat trout (*Salmo clarki*). *Hereditas* **82**, 19–24.

Allendorf, F. W., Utter, F. M., and May, B. P. (1975). Gene duplication within the family Salmonidae. II. Detection and determination of the genetic control of duplicate loci through inheritance studies and the examination of populations. *In* "Isozymes," Vol. 4, "Genetics and Evolution" (C. L. Markert, ed.), pp. 415–432. Academic Press, New York.

Allendorf, F. W., Ryman, N., Stennek, A., and Stahl, G. (1976). Genetic variation in Scandinavian populations of brown trout (*Salmo trutta* L.): Evidence of genetically distinct sympatric populations. *Hereditas* **83**, 73–82.

Altukhov, Y. P. (1974). "Populyatsionnaya Genetika Ryb" (Population Genetics of Fishes). Pishchevaya Promyshlennost' Press, Moscow. [Transl., Fish. Mar. Serv. Transl. Ser. No. 3548 (1975).]

Altukhov, Y. P., Salmenkova, E. A., Omelchenko, V. T., Sachko, G. D., and Slynko, V. I. (1972). The number of mono- and polymorphous loci in the population of the tetraploid salmon species *Oncorhynchus keta* Walb. *Genetika* **8**(2), 67–75.

Altukhov, Y. P., Salmenkova, E. A., Konovalov, S. M., and Padoukin, A. I. (1975a). Stationary distributions of frequencies of lactate dehydrogenase and phosphoglucomutase genes in population system of local fish stock *Onchorhunchus nerka* I. Stability of the stock in generations under simultaneous variability of subpopulations making up their structure. *Genetika* **11**(4), 44–53.

Altukhov, Y. P., Pudovkin, A. I., Salmenkova, E. A., and Konovalov, S. M. (1975b). Stationary distribution of frequencies of lactate dehydrogenase and phosphoglucomutase genes in population system of local fish stock, *Onchorhynchus nerka*.

II. Random genetic drift, migration, and selection as factors of stability. *Genetika* 11(4), 54–62.

Amend, D., and Smith, L. (1974). Pathophysiology of infectious hematopoietic necrosis virus disease in rainbow trout (*Salmo gairdneri*): Early changes in blood and aspects of the immune response after injection of IHN virus. *J. Fish. Res. Board Can.* **31**, 1371–1378.

Aspinwall, N. (1974). Genetic analysis of North American populations of the pink salmon, *Oncorhynchus gorbuscha*, possible evidence for the neutral mutation–random drift hypothesis. *Evolution* **28**, 295–305.

Aulstad, D., Gjedrem, T., and Skjervold, H. (1972). Genetic and environmental sources of variation in length and weight of rainbow trout (*Salmo gairdneri*). *J. Fish. Res. Board Can.* **29**, 237–241.

Avise, J. C. (1974). Systematic value of electrophoretic data. *Syst. Zool.* **23**, 465–481.

Avise, J. C., and Kitto, G. B. (1973). Phosphoglucose isomerase gene duplication in the bony fishes: An evolutionary history. *Biochem. Genet.* **8**, 113–132.

Avise, J. C., and Smith, M. H. (1974). Biochemical genetics of sunfish. II. Genetic similarity between hybridizing species. *Am. Nat.* **108**, 458–472.

Avise, J. C., Smith, J. J., and Ayala, F. J. (1975). Adaptive differentiation with little genic change between two native California minnows. *Evolution* **29**, 427–437.

Ayala, F. J. (1975). Genetic differentiation during the speciation process. *In* "Evolutionary Biology" (T. Dobzhansky, M. K. Hecht, and W. C. Steere, eds.), Vol. 8, pp. 1–78. Plenum, New York.

Behnke, R. J. (1965). A systematic study of the family Salmonidae with special reference to the genus Salmo. Ph.D. Thesis. Univ. of California, Berkeley.

Behnke, R. J. (1973). The rationale of preserving genetic diversity: Examples of the utilization of intraspecific races of salmonid fishes in fisheries management. Colorado State Univ., Ft. Collins.

Boyer, S. H. (1972). Extraordinary incidence of electrophoretically silent genetic polymorphisms. *Nature (London)* **239**, 453–454.

Brewbaker, J. L. (1964). "Agricultural Genetics." Prentice-Hall, Englewood Cliffs, New Jersey.

Calaprice, J. R. (1969). Production and genetic factors in managed salmonid populations. *In* "Symposium on Salmon and Trout in Streams" (T. G. Northcote, ed.), pp. 377–388. Inst. Fish., Univ. of British Columbia, Vancouver.

Calaprice, J. R. (1970). Genetics and mariculture. *Fish. Res. Board Can. Tech. Rep.* No. 222.

Calaprice, J. R. (1976). Mariculture—ecological and genetic aspects of production. *J. Fish. Res. Board Can.* **33**, 1068–1084.

Cavalli-Sforza, L. L. (1974). The genetics of human populations. *Sci. Am.* **231**, 81–89.

Chen, S. H., Fossum, B. L. G., and Giblett, E. R. (1972). Genetic variation of the soluble form of NADP-dependent isocitric dehydrogenase in man. *Am. J. Hum. Genet.* **24**, 325–329.

Cherfas, B. I., ed. (1972). "Genetics, Selection, and Hybridization of Fish." (Transl. from Russ., Isr. Program Sci. Transl., Jerusalem; available from U.S. Dep. Comm., Nat. Tech. Inf. Serv., Springfield, Virginia.

Christiansen, F. B., and Frydenberg, O. (1973). Selection component analysis of natural polymorphisms using population samples including mother–offspring combinations. *Theoret. Pop. Biol.* **4**, 425–445.

Christiansen, F. B., Frydenberg, O., and Simonsen, V. (1973). Genetics of *Zoarces* populations. IV. Selection component analysis of an esterase polymorphism using population samples including mother–offspring combinations. *Hereditas* **73**, 291–304.

Christiansen, F. B., Frydenberg, O., Gyldenholm, A. O., and Simonsen, V. (1974). Genetics of *Zoarces* populations. VI. Further evidence, based on age group samples of a heterozygote deficit in the *EST III* polymorphism. *Hereditas* **77**, 225–236.

Clayton, J. W., and Tretiak, D. N. (1972). Amine citrate buffers for pH control in starch gel electrophoresis. *J. Fish. Res. Board Can.* **29**, 1169–1172.

Clayton, J. W., Tretiak, D. N., Billeck, B. N., and Ihssen, P. (1975). Genetics of multiple supernatant and mitochondrial malate dehydrogenase isozymes in rainbow trout (*Salmo gairdneri*). *In* "Isozymes," Vol. 4, "Genetics and Evolution" (C. L. Markert, ed.), pp. 433–448. Academic Press, New York.

Coates, P. M., Mestriner, M. A., and Hopkinson, D. A. (1975). A preliminary genetic interpretation of the esterase isozymes of human tissues. *Ann Hum. Genet.* **39**, 1–20.

Crow, J. F., and Kimura, M. (1970). "An Introduction to Population Genetics Theory." Harper, New York.

de Ligny, W. (1969). Serological and biochemical studies of fish populations. *Oceanogr. Mar. Biol. Annu. Rev.* **7**, 411–513. .

de Ligny, W. (1972). Blood groups and biochemical polymorphisms in fish. *Eur. Conf. Anim. Blood Groups Biochem. Polymorphism, 12th*, pp. 55–65.

Donaldson, L. R., and Olson, P. (1955). Development of rainbow trout brood stock by selective breeding. *Trans. Am. Fish. Soc.* **15**, 93–101.

Falconer, D. S. (1960). "Introduction to Quantitative Genetics." Ronald Press, New York.

Gall, G. A. E. (1972). Phenotypic and genetic components of body size and spawning performance. *In* "Progress in Fishery and Food Science" (R. W. Moore, ed), University of Washington Publications in Fisheries, New Series, Vol. 5, pp. 159–163. Univ. of Washington, Seattle.

Gall, G. A. E. (1974). Influence of size of eggs and age of female on hatchability and growth in rainbow trout. *Calif. Fish Game* **60**, 26–36.

Gall, G. A. E. (1975). Genetics of reproduction in domesticated rainbow trout. *J. Anim. Sci.* **40**, 19–28.

Gjedrem, T. (1975). Possibilities for genetic gain in salmonids. *Aquaculture* **6**, 23–29.

Gjedrem, T. (1976). Possibilities for genetic improvement in salmonids. *J. Fish. Res. Board Can.* **33**, 1094–1099.

Hedgecock, D., Shleser, R. A., and Nelson, K. (1976). Applications of biochemical genetics to aquaculture. *J. Fish. Res. Board Can.* **33**, 1108–1119.

Hershberger, W. K., Brannon, E. L., Donaldson, L. R., Yokoyama, G. A., and Roley, S. E. (1976). Salmonid aquaculture studies: Selective breeding. *In* "1975 Research in Fisheries," Contrib. No. 444, p. 61. Annu. Rep., Coll. Fish., Univ. of Washington, Seattle.

Hjorth, P. (1971). Genetics of *Zoarces* populations. I. Three loci determining the phosphoglucomutase isoenzymes in brain tissue. *Hereditas* **69**, 233–242.

Hochachka, P., and Somero, G. N. (1973). "Strategies of Biochemical Adaptation." Saunders, Philadelphia, Pennsylvania.

Huzyk, L., and Tsuyuki, H. (1974). Distribution of *LDH-B″* gene in resident and anadromous rainbow trout (*Salmo gairdneri*) from streams in British Columbia. *J. Fish. Res. Board Can.* **31**, 106–108.

Jacquard, A. (1975). Inbreeding: One word, several meanings. *Theor. Popul. Biol.* **7**, 338–363.

Johnson, A. G., Utter, F. M., and Hodgins, H. O. (1972). Electrophoretic investigation of the family Scorpaenidae. *U.S. Fish Wildl. Serv., Fish. Bull.* **70**, 403–413.

Johnson, G. B. (1976). Hidden alleles at the α-glycerophosphate dehydrogenase locus in *Colias* butterflies. *Genetics* **83**, 149–167.

Johnson, M. S. (1971). Adaptive lactate dehydrogenase variation in the crested blenny, *Anoplarchus. Heredity* **27**, 205–226.

Kempthorne, O. (1957). "An Introduction to Genetic Statistics." Wiley, New York.

Kimura, M., and Maruyama, T. (1971). Pattern of neutral polymorphism in a geographically structured population. *Genet. Res.* **18**, 125–131.

King, M. C., and Wilson, A. C. (1975). Evolution at two levels: Molecular similarities and biological differences between humans and chimpanzees. *Science* **188**, 107–116.

Koehn, R. K. (1969). Esterase heterogeneity: Dynamics of polymorphism. *Science* **163**, 943–944.

Kornfield, I. L., and Koehn, R. K. (1975). Genetic variation and speciation in New World cichlids. *Evolution* **29**, 427–437.

Kosswig, C. (1973). The role of fish in research on genetics and evolution. *In* "Genetics and Mutagenesis of Fish" (J. H. Schroder, ed.), pp. 3–16. Springer-Verlag, Berlin and New York.

Krieger, H., Morton, N. W., Mi, M. P., Ezevedo, E., Freire-Maia, A., and Yasuda, N. (1965). Racial admixture in northeastern Brazil. *Ann. Hum. Genet.* **29**, 113–125.

Kristiansson, A. C. (1975). Biochemical genetic variation among selected populations of chinook salmon (*Oncorhynchus tshwytscha*) in Oregon and Washington. M.S. Thesis, Oregon State Univ., Corvallis.

Lewontin, R. C. (1967). The principle of historicity in evolution. "Mathematical Challenges to the Neo-Darwinian Interpretation of Evolution," Wistar Institute Symposium Monograph No. 5, pp. 81–94. Wistar Inst., Philadelphia, Pennsylvania.

Lewontin, R. C. (1974). "The Genetic Basis of Evolutionary Change." Columbia Univ. Press, New York.

McKee, B. (1972). "Cascadia: The Geologic Evolution of the Pacific Northwest." McGraw-Hill, New York.

Markert, C. L., and Faulhaber, I. (1965). Lactate dehydrogenase isozyme patterns of fish. *J. Exp. Zool.* **159**, 319–332.

May, B. P. (1975). Electrophoretic variation in the genus *Oncorhynchus:* The methodology, genetic basis, and practical applications to fisheries research and management. M.S. Thesis, Univ. of Washington, Seattle.

May, B., Utter, F. M., and Allendorf, F. W. (1975). Biochemical genetic variation in pink and chum salmon: Inheritance of intraspecies variation and apparent absence of interspecies introgression following massive hybridization of hatchery stocks. *J. Hered.* **66**, 227–232.

Merritt, R. B. (1972). Geographic distribution and enzymatic properties of lactate dehydrogenase allozymes in the fathead minnow, *Pimephales promelas. Am. Nat.* **196**, 173–184.

Millenbach, C. (1973). Genetic selection of steelhead trout for management purposes. *Int. Atl. Salmon J.* **4**, 253–257.

Mitton, J. B., and Koehn, R. K. (1975). Genetic organization and adaptive response of allozymes to ecological variables in *Fundulus heteroclitus. Genetics* **79**, 97–111.

Moav, R., and Wohlfarth, G. W. (1973). Carp breeding in Israel. *In* "Agricultural Genetics: Selected Topics" (R. Moav, ed.), pp. 293–318. Wiley, New York.

Nei, M. (1972). Genetic distance between populations. *Am. Nat.* **106,** 283–292.

Nei, M., and Roychoudhury, A. K. (1974). Sampling variances of heterozygosity and genetic distance. *Genetics* **76,** 379–390.

Nyman, L. (1972). A new approach to the taxonomy of the "*Salvelinus alpinus* species complex." *Inst. Freshwater Res. Drottningholm, Rep.* No. 52, pp. 103–131.

Nyman, L. (1975). Allelic selection in a fish (*Gymnocephalus cernua* (L.)) subjected to hotwater effluents. *Inst. Freshwater Res. Drottningholm. Rep.* No. 54, pp. 75–82.

Ohno, S. (1970). "Evolution by Gene Duplication." Springer-Verlag, Berlin and New York.

Powell, J. (1975). Protein variation in natural populations of animals. *In* "Evolutionary Biology" (T. Dobzhansky, M. I. Hecht, and W. C. Steere, eds.), Vol. 8, pp. 79–119. Plenum, New York.

Purdom, C. E. (1972). "Genetics and Fish Farming," Lab. Leaflet (New Ser.) No. 25. Min. Agric. Fish. Food, Fish. Lab., Lowestoft, Suffolk, England.

Purdom, C. E. (1976). Genetic techniques in flatfish culture. *J. Fish. Res. Board Can.* **33,** 1088–1093.

Prakash, S., Lewontin, R. C., and Hubby, J. L. (1969). A molecular approach to the study of genic heterozygosity in natural populations. IV. Patterns of genic variation in central, marginal and isolated populations of *Drosophila pseudoobscura. Genetics* **61,** 841–858.

Rasmuson, M. (1968). "Populationsgenetiska Synpunkter pa Laxodlingsverksamheten i Sverige," Rep. LFI Medd. 3/1968. Swed. Salmon Res. Inst., Stockholm.

Ridgway, G. J., Sherburne, S. W., and Lewis, R. D. (1970). Polymorphism in the esterases of Atlantic herring. *Trans. Am. Fish. Soc.* **99,** 147–151.

Robertson, F. W. (1972). Value and limitations of research in protein polymorphism. *Eur. Conf. Anim. Blood Groups Biochem. Polymorphism, 12th* pp. 41–54.

Rogers, J. S. (1972). Measures of genetic similarity and genetic distance. Studies in Genetics VII, Univ. Texas Publ., 7213:145–153.

Schmidt, J. (1922). The breeding places of the eel. *Philos. Trans. R. Soc. London, Ser. B* **211,** 179–208.

Schmidt, J. (1925). The breeding places of the eel. *Smithson. Rep.* 1924, pp. 279–316.

Selander, R. K., and Johnson, W. E. (1973). Genetic variation among vertebrate species. *Annu. Rev. Ecol. Syst.* **4,** 75–91.

Shaklee, J. B., Champion, M. J., and Whitt, G. (1974). Developmental genetics of teleosts: A biochemical analysis of lake chubsucker ontogeny. *Dev. Biol.* **38,** 356–382.

Shaw, C. R. (1964). The use of genetic variation in the analysis of isozyme structure. *Brookhaven Symp. Biol.* No. 17, pp. 117–130.

Shaw, C. R., and Prasad, R. (1970). Starch gel electrophoresis of enzymes—A compilation of recipes. *Biochem. Genet.* **4,** 297–320.

Simon, R. C. (1972). Gene frequency and the stock problem. *In* "The Stock Concept in Pacific Salmon" (R. C. Simon and P. A. Larkin, eds.), pp. 161–169. Inst. Fish., Univ. of British Columbia, Vancouver.

Singh, R. S., Hubby, J. L., and Lewontin, R. C. (1974). Molecular heterosis for heat-sensitive enzyme alleles. *Proc. Natl. Acad. Sci. U.S.A.* **71,** 1808–1810.

Sneed, K. E., Pillay, S. R., and Rabanal, H. R., eds. (1971). "Seminar/Study Tour in the U.S.S.R. on Genetic Selection and Hybridization of Cultivated Fishes, 19 April—29 May, 1968, Lectures," UN Dev. Programme No. TA 2926. FAO *U.N.*, Rome.

Soule, M., Yang, S. Y., Weiler, M. G., and Gorman, G. C. (1973). Island lizards: The genetic–phenetic variation correlation. *Nature (London)* **242**, 190–192.

Turner, B. J. (1974). Genetic divergence of Death Valley pupfish species: Biochemical versus morphological evidence. *Evolution* **28**, 281–294.

Utter, F. M., Ames, W. E., and Hodgins, H. O. (1970). Transferrin polymorphism in coho salmon (*Oncorhynchus kisutch*). *J. Fish. Res. Board Can.* **27**, 2371–2373.

Utter, F. M., Allendorf, F. W., and Hodgins, H. O. (1973). Genetic variability and relationships in Pacific salmon and related trout based on protein variations. *Syst. Zool.* **22**, 257–270.

Utter, F. M., Hodgins, H. O., and Allendorf, F. W. (1974). Biochemical genetic studies of fishes: Potentialities and limitations. *In* "Biochemical and Biophysical Perspectives in Marine Biology" (D. C. Malins and J. R. Sargent, eds.), Vol. 1, pp. 213–238. Academic Press, New York.

Williams, G. C., Koehn, R. K., and Mitton, J. B. (1973). Genetic differentiation without isolation in the American eel, *Anguilla rostrata*. *Evolution* **27**, 192–204.

Wilmot, R. L. (1974). A genetic study of the red-band trout (*Salmo* sp.). Ph.D. Thesis, Oregon State Univ., Corvallis.

Withler, I. L. (1966). Variability in life history characteristics of steelhead trout (*Salmo gairdneri*) along the Pacific coast of North America. *J. Fish Res. Board Can.* **23**, 365–393.

Wolf, U., Engel, W., and Faust, J. (1970). Zum Mechanismus der Diploidisierung in der Wirbeltierevolution: Koexistenz von tetrasomen und disomen Genloci der Isocitrat-Dehydrogenases bei der Regenbogenforelle (*Salmo irideus*). *Humangenetik* **9**, 150–156.

Wright, J. E., Heckman, J. R., and Atherton, L. M. (1975). Genetic and developmental analyses of LDH isozymes in trout. *In* "Isozymes," Vol. 3, "Developmental Biology" (C. L. Markert, ed.), pp. 375–401. Academic Press, New York.

Yardley, D., Avise, J. C., Gibbons, J. W., and Smith, M. H. (1974). Biochemical genetics of sunfish. III. Genetic subdivision of fish populations inhabiting heated waters. *Therm. Ecol., Proc. Symp., Augusta, Ga., USAEC, 1973* pp. 255–263.

9

HORMONAL ENHANCEMENT OF GROWTH

*EDWARD M. DONALDSON, ULF H. M. FAGERLUND,
DAVID A. HIGGS, and J. R. McBRIDE*

FISH PHYSIOLOGY, VOL. VIII

I. INTRODUCTION

The purpose of this chapter is to review the comparative endocrinology of the growth regulating hormones in fish and to describe research aimed at the application of this knowledge to fish culture. While hormonal and nonhormonal growth promoters have been used in the husbandry of mammals and birds, no use has yet been made of these substances for the production of fish. Increased construction and operating costs for salmon hatcheries have led to the search for technology to maximize the yield to the fishery per unit of input at the hatchery level. Three factors which enter into yield per unit cost are amenable to alteration by the use of hormonal growth promoters. First, the time to reach the appropriate size for release can be shortened. Second, the fish can be grown to a larger size prior to release at the normal time, thus increasing survival after release. Third, the food conversion efficiency can be improved, an important factor considering that the cost of food is the largest single item in hatchery operating budgets.

Three types of hormones have been shown to increase growth rates in fish both alone and in combination. These are the pituitary growth hormones, the anabolic steroid hormones, and the thyroid hormones. In addition to these, the insulins are a fourth group of hormones which play a significant metabolic role and may be capable of growth promotion alone or in combination with other hormones.

II. GROWTH HORMONE*

A. Regulation and Characterization

1. ENDOGENOUS SOURCE OF GROWTH HORMONE

Growth hormone is produced by the somatotrops (growth hormone cells or α cells) located in the pars distalis of the pituitary gland. The staining characteristics and functional identification of the somatotrops have been reviewed by Ball and Baker (1969) and Holmes and Ball (1974). The tinctorial characteristics of these classical acidophil cells in the sockeye salmon (*Oncorhynchus nerka*) have been tabulated (cell type 5) by Van Overbeeke and McBride (1967). In ultrastructural studies, granules, 200–350 nm in diameter, have been shown to be elaborated by the Golgi complex of the somatotrops (Nagahama and

* This section was prepared by E. M. D. and D. A. H.

Yamamoto, 1969, 1970). In functional studies, McKeown and Van Overbeeke (1971) were able to demonstrate that fluorescent antibody to mammalian growth hormone is located selectively in the α cells of the sockeye salmon (*Oncorhynchus nerka*). The growth hormone cells are present in particularly large numbers during rapid growth in the Atlantic salmon (*Salmo salar*) smolt (Olivereau, 1954) and during sea growth in the Pacific salmon (*Oncorhynchus*) (Olivereau and Ridgeway, 1962). Recent ultrastructural studies in coho salmon have confirmed that these cells in the freshwater smolt are more active than in the parr; after transfer to saltwater for 2 months, the cells are larger and show more granulation, rough endoplasmic reticulum, and Golgi apparatus (Nagahama *et al.*, 1977). The growth hormone cells of coho salmon which failed to grow after transfer to seawater were highly granular, suggesting that growth hormone in these fish was synthesized but not released. The thyroid tissue, endocrine and exocrine pancreas were less active than in control fish (Clarke and Nagahama, 1977). These changes were probably the cause of the low growth in these fish but may have been correlated with low caloric intake at an earlier stage in ontogeny.

Castration of maturing Pacific salmon results in acceleration of growth (Robertson, 1961; McBride *et al.*, 1963). This growth is associated with hypertrophy and hyperplasia of the growth hormone cells (McBride and Van Overbeeke, 1969).

2. REGULATION OF SECRETION

Hypophysectomy results in cessation of growth in length in the killifish *Fundulus heteroclitus* (Pickford, 1953a) and in *Poecilia formosa* and *P. latipinna* (see review in Ball, 1969) and causes severely reduced or halted growth in the rainbow trout *Salmo gairdnerii* (Donaldson and McBride, 1967). Removal of the pituitary gland also stopped growth in the dogfish, the only elasmobranch which has been examined (Vivien, 1941). Hypophysectomy followed by ectopic homotransplantation of the pituitary resulted in a very low growth rate in *Poecilia formosa* (Ball *et al.*, 1965) which was correlated with a reduction in typical growth hormone cells in the transplant (Olivereau and Ball, 1966). In contrast pituitary autotransplants in *Anguilla* contained somatotrops having moderate activity (Olivereau, 1970) and pituitary homotransplants in *Gasterosteus* contained somatotrops having normal activity after 2 weeks (Leatherland, 1970). These observations suggest that the hypothalamus in *Poecilia* may exert a dominant stimulatory influence over growth hormone secretion while in *An-*

guilla and *Gasterosteus* there may be a greater degree of inhibitory control.

In the mammalia (see review in McCann *et al.*, 1974) and aves (Hall and Chadwick, 1976) growth hormone appears to be under dual hypothalamic control. Growth hormone inhibiting factor (GIF, somatostatin) is a tetradecapeptide which inhibits growth hormone secretion from the pituitary gland of mammals both *in vivo* and *in vitro* (Brazeau *et al.*, 1973, 1974; Grant *et al.*, 1974). Growth hormone releasing factor (GRF) (Schally *et al.*, 1971) is a decapeptide which is identical to the N terminal sequence of the B chain of porcine hemoglobin. It is capable of the release of bioassayable but not immunoassayable growth hormone and may not be the true physiological GRF (see review in McCann *et al.*, 1974). While the role of these or similar factors in the release of growth hormone from the fish pituitary has yet to be reported, somatostatin has been shown to inhibit thyroid stimulating hormone release in the goldfish (Peter and McKeown, 1975) a property which this factor also has in the mammalia.

Stimuli for growth hormone secretion in the mammalia fall into three categories: first, a lowering of energy substrate (e.g., hypoglycemia, fasting, low plasma fatty acids); second, changes in the plasma concentrations of certain amino acids (e.g., arginine); third, stress (e.g., exercise, anesthesia, pyrogen) (see review in Muller, 1974).

The role of brain amines in the release of growth hormone from the mammalian primate pituitary has been reviewed by Wilson (1974) and Muller (1974). Catecholamines, especially norepinephrine or its precursor L-DOPA stimulate growth hormone release. This release can be blocked by phentolamine indicating involvement of α-adrenergic receptors. The β-receptor system is thought to be inhibitory to growth hormone release as the β blocking agent propranolol increases plasma growth hormone in the human (Imura *et al.*, 1968). Some observations in the rat on the relative role of the α and β receptors does not coincide with the primate data (see review in Wilson, 1974). Dopamine inhibited release of growth hormone from incubated *Poecilia latipinna* pituitary glands (Wigham *et al.*, 1975). However, this effect was not reversed by the dopamine antagonist DMPEA or the α- or β-receptor blocking agents phentolamine and propranolol; in fact the latter compound enhanced the effect of dopamine. Nor did these factors alone have any effect on growth hormone release. None of the α- or β-adrenergic antagonists tested had any significant effect on growth hormone release. The significance of these findings to the physiological control of growth hormone release in the teleost remains to be elucidated.

Growth hormone release from *Anguilla* pituitary glands during short-term incubation (Ingleton *et al.*, 1973) or long-term organ culture (Baker and Ingleton, 1975) is greater in a low osmotic pressure (110 mM sodium) medium than in a high osmotic pressure (170 mM sodium) medium; however, transfer from a high to a low osmotic pressure medium did not stimulate secretion. Addition of 1 μg/ml of *dl*-thyroxine to isotonic medium inhibited growth hormone release but not prolactin. Trout (*Salmo gairdneri*) pituitary glands released steadily decreasing amounts of growth hormone during long-term culture. The osmotic pressure of the medium had no effect on rate of release. However, during short-term incubation at 5°C trout pituitary glands released more growth hormone into low than into high osmotic pressure media (Baker and Ingleton, 1975). The most significant fact from the point of view of growth regulation may be the direct inhibitory effect of thyroxine on growth hormone release from the pituitary. However, the dose was pharmacological and is in conflict with data that suggest a stimulatory role of thyroxine on growth hormone release in the intact fish (see Section IV).

3. Purification of Fish Growth Hormone

The earliest successful attempt to purify fish growth hormone was that of Wilhelmi (1955), 10 years after the production of the first purified bovine growth hormone preparation (Li *et al.*, 1945). Fresh frozen pituitaries, largely from pollack (*Pollachius virens*) (preparation F6B) or hake (*Urophycis tenuis*) (preparation F80GH) (Pickford and Atz, 1957, p. 332), were ground and suspended in 0.3 M KCl and adjusted to pH 5.5. After centrifugation, the pH of the supernatant fluid was adjusted to pH 8.5. Ethyl alcohol was added to 10% by volume and the precipitate removed by centrifugation. The ethanol content was then raised to 30% to precipitate the growth hormone fraction as a crude concentrate. This concentrate was further purified by repeated steps in which it was dissolved at alkaline pH and reprecipitated at acidic pH. The growth hormone was finally precipitated at pH 8.5 by addition of ethanol to a concentration of 20%. The second (hake) preparation was more potent than the first (pollack) preparation in the *Fundulus* assay (Pickford, 1954a) but Wilhelmi (1955) does not indicate if there was any difference in procedure between the two preparations.

The growth hormone of the blue shark (*Prionace glauca*) has been isolated by Lewis *et al.* (1972) using disc electrophoresis to monitor the purification procedure. The pars distalis from each shark was homogenized in distilled water prior to storage at $-20°C$. For extraction the homogenate was adjusted to pH 10 with NaOH and made 10^{-3} M in diisopropylphosphofluoridate. After centrifugation the extract was adjusted to pH 8, concentrated on a Diaflow UM-10 membrane, and gel filtered on a Sephadex G-150 column developed with $0.01 M$ NH_4HCO_3. Tubes containing presumptive growth hormone and prolactin were identified by disc electrophoresis, combined, and concentrated. The growth hormone and "prolactin" fractions were separated by ion exchange chromatography on DEAE-cellulose equilibrated with $0.01 M$ NH_4HCO_3. The growth hormone was eluted by a gradient obtained by mixing equal parts of $0.01 M$ and $0.1 M$ NH_4HCO_3. The growth hormone extracted from the blue shark (Lewis *et al.*, 1972) had a higher biological activity than the hake growth hormone (Wilhelmi, 1955) when assayed in the *Fundulus* assay (Table III) (Lewis *et al.*, 1972; Pickford, 1973).

In this laboratory a crude preparation of growth hormone has been purified from chinook salmon (*Oncorhynchus tshawytscha*) pituitary glands (Higgs *et al.*, 1978a). This preparation was obtained by a modification of the procedure described by Papkoff and Li (1958) for purification of growth hormone from the pituitary gland of the humpback whale. The pituitary glands were collected from maturing chinook salmon caught at sea in July and frozen at $-20°C$ prior to extraction. The glands were homogenized in $Ca(OH)_2$. After adjustment to pH 10.0 the homogenate was stirred and then centrifuged. The supernatant fluid was conserved and the residue reextracted in $Ca(OH)_2$ solution. The two extracts were combined, an equal volume of saturated $(NH_4)_2SO_4$ was added, and the pH adjusted to 7. After standing overnight the precipitate was collected by centrifugation, dissolved in cold distilled water, and dialyzed for 23 hr against tap water at $4°C$. The dialysate was then lyophilized and the powder stored at $-40°C$.

The Pacific salmon pituitary extract (oncPE) obtained in this way had an activity of approximately $0.1 \times$ NIH-GH-B-17 when bioassayed for induction of growth in intact coho salmon (*Oncorhynchus kisutch*) (Higgs *et al.*, 1978a).

Recently Farmer *et al.* (1976) have isolated growth hormone from the teleost *Tilapia mosambica* using a side fraction from the purification of *Tilapia* prolactin (Farmer *et al.*, 1975). The yield of growth hormone was 1.4 g/kg wet weight pituitaries. The *Tilapia* growth hormone had a biological activity similar to that of NIH-GH-bovine when assayed for growth promotion in intact coho salmon or in intact juvenile *Tilapia* (Clarke *et al.*, 1977).

4. BIOLOGICAL RELATEDNESS OF PISCINE GROWTH HORMONES

While pituitary materials have been collected from most classes of fish, they have only been tested for growth promoting activity in the teleosts, amphibia, or mammalia. Most of these fish pituitary preparations have been tested in the mammalia, specifically in the rat tibial plate assay (Greenspan et al., 1949) (Table I). Of the various classes of fish, only the teleostei have been used as bioassay recipients for growth hormone preparations from other vertebrates (Table I). It is clear from the range of donor and recipient response relationships shown in Table I that the growth hormone of teleosts is distinct from that of other vertebrates including the other classes of fish. Pituitary preparations from elasmobranchs, dipnoids, chondrosteans, and holosteans are all active in the rat tibia assay; teleostean preparations are inactive in the rat assay with the exception of *Tilapia* GH (Farmer *et al.*, 1976). Elasmobranch pituitary extracts promote growth in both the rat assay and in the teleosts while pituitary preparations from other classes of fish have been tested in the rat but not in the teleost. In addition to being inactive in the mammalian assay, the teleostean pituitary does not promote growth when transplanted into the anuran tadpole (Enemar and von Mecklenburg, 1962) (Table I). This relative lack of biological activity of teleostean growth hormone in nonteleostean bioassays is reversed when growth hormones from other vertebrates are tested in the teleost; both mammalian and elasmobranch growth hormones show good biological activity (Table I). Immunological studies tend to bear out the relationships seen in degree of biological activity (Hayashida, 1970, 1975; Farmer *et al.*, 1976).

5. PHYSICAL CHARACTERIZATION OF FISH GROWTH HORMONES

Observations on the first fish (pollack and hake) growth hormone preparations showed that they were of lower molecular weight than mammalian (bovine and equine) preparations. The sedimentation constant in glycine buffer at pH 9.7 averaged 1.78 (1.43–2.15) for pollack GH and was 1.78 for hake GH (Wilhelmi, 1955). The species are reversed in Wilhelmi's report (Pickford, 1957, p. 332). This sedimentation constant for pollack and hake GH corresponded in the absence of a diffusion constant to a molecular weight of 22,000–26,000 which was approximately half the MW of beef growth hormone (Pickford, 1957). It has since been shown that the molecular weights of growth hormones depend on the pH and nature of the buffer. In the dissociating solvent guanidine hydrochloride (5.6 M) bovine, porcine, and human GH all had molecular weights by sedimentation equilibrium in the

Table I

Growth Hormone Activity of Pituitary Extracts and Growth Hormone Preparations from Various Vertebrate Sources in Various Groups of Vertebrates Excluding the Effects of Mammalian Growth Hormone Preparations in Mammals[a]

Growth bioassay	Source of pituitary extract or growth hormone preparation									
	Agnatha	Elasmobranch	Dipnoi	Chondrostei	Holostei	Teleostei	Amphibia	Reptilia	Aves	Mammalia
Teleostei		+ (12)[b]				+ (7, 8, 11, 27, 28)				+ (7, 9, 10, 25, 28)
Amphibia	− (13)					− (13)	+ (18, 29) ± (13)	+ (13, 18)	+ (13, 18)	+ (18–21)
Reptilia										+ (22, 23)
Aves										+ (24)
Mammalia		− (1) + (16)	+ (1) + (17)	+ (2) + (3)	+ (3)	− (2, 4–6, 17) + (26)	+ (5) + (17, 29)	(14) + (17)	+ (15) + (17)	

[a] +, positive effect; −, negative result.

[b] Numbers in parentheses indicate reference: 1, Geschwind (1967); 2, Hayashida and Lagios (1969); 3, Hayashida (1971); 4, Wilhelmi (1955); 5, Solomon and Greep (1959); 6, Moudgal and Li (1961); 7, Pickford (1954a); 8, Higgs et al. (1976); 9, Higgs et al. (1975); 10, Swift (1954); 11, Swift and Pickford (1965); 12, Lewis et al. (1972); 13, Enemar and von Mecklenburg (1962); 14, Papkoff and Hayashida (1972); 15, Farmer et al. (1974); 16, Hayashida (1973); 17, Hayashida (1970); 18, Nicoll and Licht (1971); 19, Zipser et al. (1969); 20, Cohen et al. (1972); 21, Licht et al. (1972); 22, DiMaggio (1960); 23, Licht and Hoyer (1968); 24, Bates et al. (1962); 25, Pickford et al. (1959); 26, Farmer et al. (1976); 27, Clarke et al. (1977); 28, Higgs et al. (1978a); 29, Farmer et al. (1977).

region of 22,000 (Wilhelmi and Mills, 1969). This suggests that growth hormone preparations with observed molecular weights in the region of 45,000 are dimers.

The molecular weight of the growth hormone of the blue shark was 22,000 when determined by disc electrophoresis (Lewis *et al.*, 1972). The fact that both this elasmobranch growth hormone preparation and the teleostean growth hormone preparations exist as monomers having molecular weights of 22,000 suggests that growth hormone may only exist as a dimer in some of the higher vertebrates. *Tilapia* GH had a molecular weight by gel filtration of 22,000 (which is very similar to that of the human GH monomer) and a sedimentation coefficient, $s_{20,w}$, in the ultracentrifuge of 2.19 (Farmer *et al.*, 1976).

The isoelectric point of the hake growth hormone, pH 6.2–7.2, is similar to that for bovine growth hormone, while the pollack growth hormone had a lower isoelectric point similar to that of porcine or equine growth hormone (Wilhelmi, 1955; Pickford, 1957).

6. CHEMICAL CHARACTERIZATION OF FISH GROWTH HORMONE AND FEASIBILITY OF SYNTHESIS

Growth hormone and prolactin are believed to be derived from a common ancestral hormone by gene duplication followed by evolutionary divergence. Several growth hormones, for which the amino acid sequence is known, contain internal homologies which suggest that the molecule evolved from a smaller molecule by one or more partial duplications prior to the divergence of the growth hormone and prolactin lines (Niall *et al.*, 1973; Wallis, 1975). Furthermore Fellows (1973) has detected homology between growth hormone and the α subunit of luteinizing hormone using computer techniques for the analysis of sequence data.

Lewis *et al.* (1972) presented amino acid composition data for elasmobranch growth hormone (component S) which showed that it contained six half-cystine residues and two tryptophan residues. In these respects it resembles mammalian prolactin rather than mammalian growth hormone which contains four half-cystine residues (two disulfide bridges) and one tryptophan residue at position 96 in a total chain of 191 amino acids (see reviews in Niall *et al.*, 1973; Li, 1975). *Tilapia* GH resembles ovine GH in having two disulfides, a single tryptophan, low methionine and histidine content, and a high glutamic acid and leucine content. On the other hand, *Tilapia* GH differs from ovine GH in having a high aspartic acid and serine content and a low alanine, methionine, and phenylalanine content (Farmer *et al.*, 1976).

Of particular relevance to the use of growth hormone for growth promotion in fish is the possibility of synthesizing a biologically active core as has been accomplished for adrenocorticotropin. This latter hormone, however, consists of a chain of only 48 amino acids and has no secondary or tertiary structure. Growth hormone, on the other hand, with its chain of 191 amino acids and two disulfide bridges is a globular protein with considerable α helix content (Li, 1975). In 1970, Li and Yamashiro synthesized a version of human growth hormone with 188 amino acids and a misplaced tryptophan residue (Li *et al.*, 1969). This protein had 10% of the growth promoting activity of native growth hormone and it has been suggested that this retention of some activity indicates that neither the full 191 amino acids nor the specific three-dimensional structure are essential for activity (Aloj and Edelhock, 1972; Niall, 1971). On the other hand, one could argue that the loss of 90% of the activity which resulted from the relatively minor changes in the molecule suggests that the specific three-dimensional structure is critical for biological activity. Segments of the human growth hormone sequence corresponding to residues 87–123 and 124–155 (Chillemi and Pecile, 1971; Chillemi *et al.*, 1972). and 95–136 (Blake and Li, 1973) have been synthesized and have only slight but measurable growth promoting activity. Bewley and Li (1975) have presented a review of the arguments for and against the active core hypothesis. One proposal is that the growth hormone isolated from the pituitary does not in fact interact with the receptor. Instead an active fragment may be released which interacts with a carrier protein or directly with the receptor (Niall *et al.*, 1973). If this first proposal were correct the appropriate synthetic fragment would possess full activity in the appropriate assay. A second perhaps more likely proposal is that while native hormone may interact directly with the receptor on the basis of their specific compatible three-dimensional structures, it may be possible to synthesize a relatively large fragment of the native molecule which is capable of some interaction with the receptor. This synthetic protein would not be expected to be as potent as the native hormone on a molar basis (Bewley and Li, 1975) but could well be feasible in terms of cost per unit of biological activity depending on the size of the fragment and the complexity of its synthesis.

B. Effect of Exogenous Growth Hormone on Growth

1. MAMMALIAN GROWTH HORMONE

a. Administration Route and Frequency. The first studies to show a positive effect of mammalian pituitary extracts on growth of intact fish, namely, the guppy, *Lebistes reticulatus* (Tuckmann, 1936), swordtail, *Xiphophorus helleri* (Régnier, 1938), and brook trout, *Salvelinus fontinalis* (Cantilo and Regalado, 1942) used the oral route for hormone administration (Table II). The apparent success obtained by these researchers is surprising because growth hormone would be expected to be destroyed by digestive enzymes. However, Pickford (1957) suggested that there may only be partial digestion of orally ingested growth hormone prior to its entry into the blood stream. In support of this idea, enzymatic cleavage of bovine growth hormone (bGH) by trypsin, chymotrypsin, papain, and pepsin can result in fragments which retain some growth promoting activity (literature reviewed in Daughaday *et al.*, 1975). The results of these early studies have not been confirmed to this day. Certainly, the oral route would offer the most convenient and practical approach to administer growth hormone to fish if resource enhancement or aquaculture is the main goal.

Most studies since the late 1930s and early 1940s have utilized injection or pellet implantation procedures. Injection has either been intraperitoneal (ip) or intramuscular (im) (Table II). Injection frequency for growth hormone has ranged from once per day for *Salmo gairdneri* (Chartier-Baraduc, 1959) to once every 14 days for *Oncorhynchus kisutch* (Higgs *et al.*, 1976). In the latter study, im injection of underyearling coho salmon with bGH once every 14 days resulted in a significant increase in weight and length after 4 weeks. But im injections twice per week resulted in significantly greater growth than that noted for the once every 14-day interval (Figs. 1 and 2).

A procedure for implanting bGH cholesterol pellets into coho salmon was modified from that of Robertson and Rinfret (1957) for gonadotropin (Higgs *et al.*, 1975). This was subsequently improved by Higgs *et al.* (1976). The purpose in each case was to obtain slow release of hormone into the circulation so that hormone administration would be more practical and less stressful to the fish. With the improved method it was possible to implant pellets (im) once every 3 weeks into underyearling coho salmon at 10°C without any significant reduction in maximum growth response to bGH (Figs. 1B and 2B).

Pickford (1963) (Table II) was unable to stimulate growth with bGH administered via the water. Recently it has been shown that bGH, administered by a hyperosmotic saline immersion technique (modified from Amend and Fender, 1976), is capable of marginally enhancing growth and improving food conversion efficiency in 1- to 2-g juvenile coho at 10°C (Higgs, Donaldson, McBride, MacQuarrie, and Nichols, 1977 unpublished observations).

Effect of Mammalian Growth Hormone Preparations

Species	Hormone	Administration method			Environmental factors			Nutritional state
		Dose	Route	Frequency	Temperature (°C)	Photoperiod	Salinity	
Salmo trutta (brown trout)	E1[a]							
	bGH	5 µg/g	ip[b]	2×/week		March	FW[c]	Liver and dog food to satiation.
	C (0.6% saline)[e]	0	ip	2×/week				
	C (uninjected)	0						
	E2							
	bGH[f]	5 µg/g	ip	2×/week		July	FW	Liver and dog food to satiation.
	C (0.6% saline)	0	ip	2×/week				
	C (uninjected)	0						
Salmo gairdnerii (rainbow trout)	bGH	2 µg/g	ip	7×/week	12–13		FW	
Salmo irideus (rainbow trout)	bGH	70 µg/fish	im[g]	1×/4 days	16		FW	Pellets (40% protein) fed to excess (i.e., 1 g/fish/day).
	C (0.7% saline)	0	im	1×/4 days				
Salmo salar (Atlantic salmon)	pGH[h]	1 µg/g	ip	1×/2 days	11.5	Reciprocal photoperiod to that of June and July.	FW	Ewos salmon pellet food p4 (Astra-Ewos AB) 2–3×/day to satiation.
	C (0.6% saline)	0						
Salvelinus fontinalis (brook trout)	Bovine anterior pituitary extract.		o[j]	2×/week				
	C	0						
Oncorhynchus kisutch (coho salmon)	bGH (NIH-GH-B17)	3.33 µg/g	ip	3×/week	10	12L : 12D	FW	Excess (8% wet body weight/day) Oregon Moist Pellet salmon diet (OMP-2) fed 2×/day.
	bGH	33.3 µg/g	inj	3×/week				
	C (0.65% alkaline saline)	0	ip	3×/week				
	bGH	20 µg/g	ip pellet	1×/2 weeks				
	bGH	200 µg/g	ip pellet	1×/2 weeks				
	C (cholesterol pellet)	0	ip	1×/2 weeks				
Oncorhynchus kisutch	bGH (NIH-GH-B17)	0.1 µg/g	im	1×/week	10	12L : 12D	FW	Excess (8% wet body weight/day) Oregon Moist Pellets fed 2×/week
	bGH	0.32 µg/g	im	1×/week				
	bGH	1.0 µg/g	im	1×/week				
	bGH	3.2 µg/g	im	1×/week				
	bGH	10 µg/g	im	1×/week				
	C (0.65% alkaline saline)	0	im	1×/week				
	C (uninjected)	0						
	E1							
Oncorhynchus kisutch	bGH (NIH-GH-B17)	5 µg/g	ip	2×/week	10	12L : 12D	FW	Excess (8% wet body weight/day) Oregon Moist
	C (0.65% alkaline saline)	0	ip	2×/week				
	bGH	5 µg/g	im	2×/week				
	C	0	im	2×/week				

and Pituitary Extracts on Growth of Fish

Hy-po-phy-sec-tomy	Comments	Dura-tion (days)	Initial age and/or weight	Weight gain Over con-trol (%)	Weight gain Over initial weight (%)	% day	Ref-erence
							Swift (1954)
No	Low endogenous GH. Noted effect of handling.	28	4 years			0.7[d] 0.3 0.2	
No		28	4 years			1.1 0.4 1.0	
	Weight increased. Decreased condition factor. Increased muscle water.	14					Chartier-Baraduc (1959)
No	Two out of three bGH fish had exophthalmos.	27	4.0–9.3 g		108 (one fish) 38		Enomoto (1964)
No	pGH fish ate more food than controls. Fin tips were blackened, and yellowing occurred around operculae and fins. pGH led to a decline in condition factor and permitted survival when fish transferred to SW.	28	Aged 2+, 11.5 cm (fork length)			L, W[i] 0.46, 1.11 0.19, 0.86 0.41, 1.45 0.22, 1.35 0.59, 1.01 0.23, 0.64 first 2 weeks 3rd week 4th week	Komourdjian et al. (1976a)
No		240	1 month fry 0.22 g	170.9	9368 3395		Cantilo and Regalado (1942)
No	bGH lowered condition factor, increased % water and reduced % protein in flesh and stimulated greater feeding activity.	56	Yearling ranged 9.08–13.86 g in each treatment group			L, W 0.55, 1.51 0.55, 1.59 0.17, 0.57 0.50, 1.30 0.58, 1.60 0.28, 0.77	Higgs et al. (1975)
No	All doses of bGH lowered condition factor. Refer to Fig. 3 for data of % length/day.	70	Yearling; range in treatment group means was 13.4–15.3 g.			W 0.75 0.92 1.01 1.11 1.41 0.67 0.88	Higgs et al. (1978a)
No	bGH lowered condition factor. Decline reversed when hormone treatment withdrawn. bGH	84	Underyearling: ranged from 4.42 to 8.54 g in each treatment group.			L, W 0.78, 2.32 0.14, 0.47 0.81, 2.42 0.17, 0.53	Higgs et al. (1976)

(Continued)

Table II—

Species	Hormone	Administration method			Environmental factors			Nutritional state
		Dose	Route	Frequency	Temperature (°C)	Photoperiod	Salinity	
	bGH	10 μg/g	im					Pellet
	C	0	im	1×/week				salmon diet
	bGH	20 μg/g	im	1×/2 weeks				fed 2×/day.
	C	0	im	1×/2 weeks				
	bGH	30 μg/g	p^k	1×/3 weeks				
	bGH	90 μg/g	p	1×/3 weeks				
	C (cholesterol pellet)	0	p	1×/3 weeks				
	C (uninjected) E2	0						
	bGH (NIH-GH-B17)	20 μg/g	im	1×/2 weeks	6–19	Natural	SW^t	As above (25.6–29.8‰
	C (0.65% alkaline saline)	0	im	1×/2 weeks				
	bGH	20 μg/g	im	1×/2 weeks				
	C	0	im	1×/2 weeks				
Oncorhynchus kisutch	bGH (NIH-GH-B17) C (0.65% alkaline saline) C (uninjected)	10 μg/g	im im	1×/week 1×/week	10	Natural	FW	Excess (8% wet body weight/day Oregon Moist Pellet salmon die fed 2×/da
Oncorhynchus kisutch	bGH (NIH-GH-B17) C (uninjected)	10 μg/g 0	im	1×/week	10	12L:12D		Presented either Oregon Moist Pellet (OMP) or moist diet with poul offal (PO) and white fish meal a main prote sources. Fish fed either 0.8 body weight/da [dry feed (wet fish (g or to satiation.

Hypophysectomy	Comments	Duration (days)	Initial age and/or weight	Weight gain Over control (%)	Weight gain Over initial weight (%)	% day	Reference
	fish had higher % protein and lower % lipid in muscle. bGH treatment stimulated ovarian maturation and thyroid activity, and led to hyperplasia and hypertrophy of pancreatic islets. bGH fish fed more actively than controls. Refer to Figs. 1 and 2.					0.71, 2.09 0.14, 0.42 0.70, 2.00 0.17, 0.51 0.78, 2.24 0.79, 2.29 0.17, 0.59 0.16, 0.50	
No	Control and enhanced fish from E1 used for E2. Coho survived transfer to SW. Low water temp., salinity and/or photoperiod history cited as likely causes of poor growth responses of all treatment groups.	70	Yearling 25.5 g 22.8 g 98.1 g 100.3			$\dfrac{W}{0.35}$ 0.12 0.54 0.28	
No	bGH lowered condition factor, elevated muscle water, increased mean nuclear diameter of interrenal tissue led to hyperplasia of pancreatic islets and greater granulation of β cells and stimulated ovarian maturation.	59	Yearling; ranged from 10.0 to 15.0 g in each treatment group.	Refer to Fig. 20 for data on growth rates.			Higgs *et al.* (1977a)
	Refer to Figs. 6 and 7 for effect of bGH on appetite and feed conversion of OMP-fed fish. bGH treatment did not affect appetite of PO-fed fish but did improve feed conversion. bGH administration lowered condition factor and % protein in muscle, but elevated muscle water.	56	Yearling	Diet PO(0.88%) bGH PO(0.88%)C OMP(0.84%) bGH OMP(0.84%)C PO(Sat.)bGH PO(Sat.)C OMP(Sat.) bGH OMP(Sat.)C		L, W 0.35, 0.86 0.23, 0.65 0.40, 1.00 0.25, 0.75 0.52, 1.40 0.34, 1.04 0.55, 1.53 0.30, 0.91	Markert *et al.* (1977)

(Continued)

Table II—

Species	Hormone	Dose	Route	Frequency	Temperature (°C)	Photoperiod	Salinity	Nutritional state
Oncorhynchus nerka (sockeye salmon)	bGH (NIH-GH-B17)	1 μg/g	ip	1×/2 days	17.5	12L:12D	FW	Fed Oregon Moist Pellet diet at 5% body weight/day (dry weight basis)
	C (saline)	0	ip	1×/2 days				
Carassius carassius (golden crucian carp)	Bovine pituitary extract in saline. Also, acetone powders used.		ip	Every 2–7 days.				
Carassius auratus (goldfish)	Bovine anterior pituitary alkaline extract.	1 ml/fish	ip	1×/day	25			Aquarium fish food
	C (saline)		ip					
Fundulus heteroclitus (killifish)	bGH (Wilhelmi batch 65AS)	10 μg/g 4 μg/g	ip	3×/week	17.4–10.4	8 hr	SW	Excess pablum and liver 2×/day.
	C (saline)	0	ip	3×/week				
Fundulus heteroclitus	C (uninjected)	0						
	bGH (Wilhelmi 429)	10 μg/g	ip	3×/week	16–20			Pablum and liver plus fresh liver.
	bGH (Armour 22K22)	10 μg/g	ip	3×/week	16–20			
	bGH (Wilhelmi 508B)	10 μg/g	ip	3×/week				
	C (saline)	0	ip	3×/week				
	C (mock)	0						
Fundulus heteroclitus	bGH (Wilhelmi 568)	10 μg/g	ip	3×/week	20	8 hr		Liver–pablum and daphnia
	bGH (Wilhelmi 508B)	10 μg/g	ip	3×/week				
	bGH (Organon 66450)	10 μg/g	ip	3×/week				
Fundulus heteroclitus	bGH (Wilhelmi 508B)	10 μg/g	ip	3×/week	10			Liver–pablum and daphnia
	bGH				15			
	bGH				25			
	C (saline)	0	ip	3×/week	20			
	E1							
Fundulus heteroclitus	bGH (NIH-GH-B1)	1 μg/g	ip	3×/week	20			
	bGH	3 μg/g	ip	3×/week				
	bGH	10 μg/g	ip	3×/week				
	bGH	30 μg/g	ip	3×/week				
	E2							
	bGH	1 μg/g	ip	3×/week				
	bGH	3 μg/g	ip	3×/week				
	bGH	10 μg/g	ip	3×/week				
	bGH	30 μg/g	ip	3×/week				
	control							
Fundulus heteroclitus	bGH							
	sGH[n]							
	sGH							
	sGH							
	hGH[o]							
	oGH[p]							
	pGH							

Hy-po-phy-sec-tomy	Comments	Dura-tion (days)	Initial age and/or weight	Weight gain Over con-trol (%)	Weight gain Over initial weight (%)	% day	Ref-erence
No	bGH treatment re-duced blood sodium levels after transfer to seawater, and lowered body lipid and condition factor.	14	Underyearling	—	Refer to Fig. 4 for data on growth rates.		Clarke (1976)
No	March–July control. March–Sept. control.	99 99 176 176	23.0 g 18.1 23.0 g 17.0	— — — —	31.5 62.1 71.3 98.2	— — — —	Nixo-Nicoscio (1940)
No	Extract toxic. After 1 week no increase in length. Weight increase may be osmotic.	5 5	7–14 g		−21 −5		Jampolsky and Hoar (1954); Hoar (1951)
No	All fish lost weight when injected every day.	71	7.0–10.6 g avg/gp of 5.		12.7 3.2 −5.0 6.2		Pickford and Thompson (1948)
Yes	3 GH preps. given in sequence to same fish.	21 14 28			−6.5 8.8 12.5		Pickford (1953b)
...ock Yes	3 GH preps. given in sequence to same fish. No controls.	56 56 56	2 year		−17.5 12.3 39.5		Pickford (1954a)
Yes	See Pickford, 1954a for response at 20°C.	56			9.0 22.4 44.6 −1.2		Pickford (1954b, 1957)
...es	Dosage of 1 $\mu g/g$ not significant. Linear log dose response over range 3–30 $\mu g/g$.	28			1^m 1.5 3.5 5.3		Pickford (1959)
	Dosage of 1 $\mu g/g$ not significant. Linear log dose response over range 3–30 $\mu g/g$.	56			1 3.3 7.5 11.3		
s	Length increase ex-pressed as percent of growth induced by bGH (NIH-GH-B1)				100 15.9 26.0 40.3 33.8 99.9 55.1		Pickford et al. (1959)

(Continued)

Table II—

Species	Hormone	Administration method Dose	Route	Fre quency	Temperature (°C)	Photoperiod	Salinity	Nutritional state
Fundulus heteroclitus	bGH (NIH-GH-B1)	10 mg/liter in aquarium water.		1×/day			SW	Ate well.
Fundulus heteroclitus	bGH (NIH-GH-B1) C (0.6% saline)	20 μg/g	ip ip	3×/week 3×/week	20	8 hr	SW	Aronson's mix ture plus brine shrimp.
Fundulus heteroclitus	bGH (NIH-GH-B6) bGH + oLH saline bGH bGH + bLH saline	10 μg/g 10 + 0.5 μg/g 5 μg/g 5 + 0.02 μg/g	ip ip ip ip ip ip	3×/week 3×/week 3×/week 3×/week 3×/week 3×/week	20 10	8 hr 8 hr	SW SW	
Xiphophorus helleri (swordtail)	Anterior pituitary powder		o					
Lebistes reticulatus (guppy)	Anterior pituitary powder		o					

b. Hormone Dosage and Potency, Environmental Temperature, Ration, and Economics. The report of an effect of a purified growth hormone preparation on growth in intact killifish (Pickford and Thompson, 1948) marked the beginning of 25 years of research in this area by Grace Pickford. Her research has had a marked impact on our understanding of the effects of both mammalian and piscine growth hormones in the teleost. The use of hypophysectomized male *Fundulus* maintained on an adequate diet (Pickford, 1953b, 1954a) at a constant temperature of 20°C (Pickford, 1954a, 1957) permitted the examination of dose response effects and the development of a bioassay (Pickford, 1959). This has proved useful for the comparison of diverse growth hormone preparations from the elasmobranchii, teleostei, and mammalia (Tables II and III). Pickford found, in her 1959 study, that bGH caused a linear log dose-response in length increase of hypophysectomized killifish over a dose range of 9–90 μg bGH/g body weight/week. A dosage as low as 3 μg/g/week did not significantly stimulate growth (Table II). It would appear that to obtain an adequate growth response in hypophysectomized *Fundulus* at 20°C, 30 μg of bGH/g/week is required. Water temperature was found to modify the growth response of *Fundulus*. For example, weight gain

Continued

Hypophysectomy	Comments	Duration (days)	Initial age and/or weight	Weight gain		% day	Reference
				Over control (%)	Over initial weight (%)		
Yes	Not significant.	37			0		Pickford (1963)
Yes		42			17.0		Swift and Pickford (1962a, b, 1965)
					−11.4		
Yes	Growth in bGH *groups significant relative to control.	42			5.76		Pickford et al. (1972)
					10.70		
Yes	Growth in bGH + bLH group significant relative to control.	56			−5.96		
					0.51		
					2.38		
No		60	Young	50%[q]	3.01		Regnier (1938)
No	Increased growth.	60					Tuckmann (1936)

[i] L = length (cm); W = weight (g).
[j] Oral.
[k] Pellet.
[l] Seawater.
[m] Length change over initial length.
[n] Simian growth hormone.
[o] Human growth hormone.
[p] Ovine growth hormone.
[q] Length gain.

relative to initial weight was fivefold greater at 25°C than at 10°C (Pickford, 1957)(Table II). The effect of temperature on the response of intact carp (*Cyprinus carpio*) to bovine GH has been investigated by Adelman (1977). Bovine GH had a greater relative effect above the optimum temperature for growth, suggesting that endogenous GH release was lower at the higher temperature. Photoperiod had no effect on the response of the carp to GH. In the black bullhead (*Ictalurus melas*) linear growth arrested by hypophysectomy was reinitiated by injecting bovine GH (1 μg/g) every second day (Kayes, 1977a). Treatment of hypophysectomized *Ictalurus melas* with bovine GH at 12°C resulted in significant gains in weight but not in length. Growth in weight was maximally stimulated at 18°C, whereas growth in length was maximal at 24°C. The results suggest an effect of temperature on tissue responsiveness to growth hormone (Kayes, 1977b).

Purified mammalian growth hormone was first administered to a salmonid by Swift (1954). In this study, 10 μg of bGH/g/week were administered to intact brown trout, *Salmo trutta*, for 28 days. Significant enhancement of growth relative to both saline and uninjected controls was noted in March. However, in July the growth of treated trout was only slightly better than that of untreated controls,

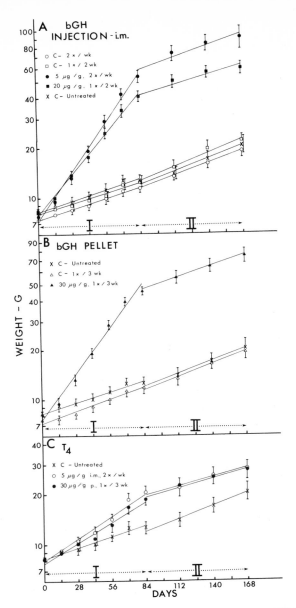

Fig. 1. Influence of bovine growth hormone [bGH; NIH-GH-B17 (0.92 IU/mg)] and Na L-thyroxine pentahydrate (T₄; T₄ doses refer to free acid) on growth in weight of underyearling coho salmon at 10°C. Geometric means for wet weights of bGH-treated (A, B) or T₄-treated (C) fish are shown. Vertical lines through the geometric means denote 95% confidence intervals. Administration of the hormones in phase I (0–84 days) was by dorsal musculature injection or implantation of a hormone : cholesterol pellet. Control fish received alkaline saline or a cholesterol pellet. Hormone treatment was stopped in phase II (84–168 days). Each point is a mean of 40–65 fish in phase I and 16–44 fish in phase II. (From Higgs *et al.*, 1976, *J. Fish. Res. Board Can.* 33, 1585–1603.)

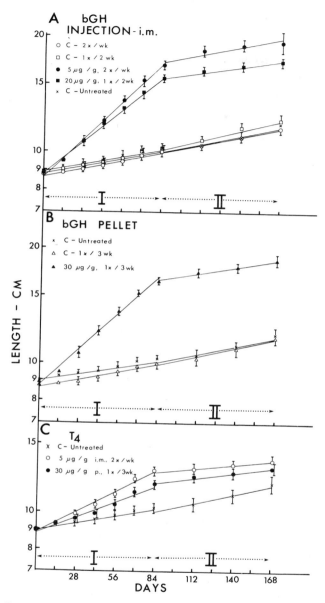

Fig. 2. Influence of bovine growth hormone (bGH) and L-thyroxine (T_4) on growth in length of underyearling coho salmon at 10°C. Geometric means for fork lengths of bGH-treated (A, B) or T_4-treated (C) fish are shown. Vertical lines through the geometric means denote 95% confidence intervals. Refer to Fig. 1 for additional details. (From Higgs *et al.*, 1976, *J. Fish. Res. Board Can.* **33**, 1585–1603.)

Species	Hormone	Administration method			Environmental factors		
		Dose	Route	Frequency	Temperature (°C)	Photoperiod	Salinity
Oncorhynchus kisutch	(*Oncorhynchus tshawytscha*) pituitary extract (oncPE)	10 μg/g 100 μg/g	im[a] im	1×/week-	10	12L : 12D	FW[b]
	C (0.65% alkaline saline)	0	im	1×/week	—	—	—
	C (uninjected)	0	—	—	—	—	—
Carassius auratus	(*Oncorhynchus tshawytscha*) pituitary extract	1 ml/fish	ip	1×/day	25		FW
Fundulus heteroclitus	Fish GH chiefly *Pollachius virens*, (F6B)	10 μg/g 2 μg/g	ip[d] ip	3×/week 3×/week	20 20	8 hr 8 hr	
	Fish GH chiefly *Urophycis tenuis* (F80GH)	10 μg/g	ip	3×/week	20	8 hr	
Fundulus heteroclitus	*Fundulus heteroclitus* brei of whole pituitary	1 mg wet weight/g	ip	1×/2 days	15–17		1/week SW[e] then 3/week FW
Fundulus heteroclitus	*Urophycis tenuis* acetone powder of whole pituitary	300 μg/g	ip	3×/week	20		
Fundulus heteroclitus	*Urophycis tenuis* brei of whole pituitary	2 mg wet weight/g	ip	3×/week	20		
Fundulus heteroclitus	Wilhelmi F80GH C (saline)	10 μg/g	ip	3×/week	20		
Fundulus heteroclitus	Pollack GH		ip				
Fundulus heteroclitus	*Perca fluviatilis* pituitaries lyophilized and homogenized in 0.6% saline C (0.6% saline)	100 μg/g	ip	3×/week	20	8 hr	SW

Preparations on Fish Growth

Nutritional state	Hy-po-phy-sec-tomy	Comments	Dura-tion (days)	Initial age and/or weight	Weight gain		%/day	Reference
					Over control (%)	Over initial weight (%)		
							L, W[c]	
Excess ration (Oregon Moist Pellets) presented twice daily.	No	Activity approximately 0.1 × NIH-GH-B17 (refer to Fig. 3). oncPE treatment decreased condition factor and increased gonadosomatic indices of males relative to those of controls.	70				0.37, 1.03 0.53, 1.53	Higgs et al. (1978a)
—							0.25, 0.67	
—							0.32, 0.88	
Aquarium fish food.	No	Toxic after 1 week, no change in length.		7–14 g		10.8		Jampolsky and Hoar (1954)
Liver–pablum and daphnia	Yes Yes	No controls	56 56	2 years-		8.4 0		Pickford (1954a)
	Yes		56			26.3		
	Yes	Weight loss. In 2 of 4 fish small length increase.				0		Burden (1956)
	Yes	Weight loss, length no change.	30			0		Pickford (unpublished, in Pickford, 1957)
	Yes	Slight weight loss.	30			2.5[f]		Pickford (unpublished, in Pickford, 1957)
	Yes		35			21.5		Pickford (1957)
	Yes	Length increase 55.1% of that induced by bGH NIH-1.				−4.7		Pickford et al. (1959)
ronson's mixture plus brine shrimp.	Yes	Pituitaries collected in February April June July August Sept. Nov.	42			19.0 46.1 41.3 24.1 4.7 30.1 23.9 −11.4		Swift and Pickford (1962a,b, 1965)

(*Continued*)

Table III—

		Administration method			Environmental factors		
					Temper-		
				Fre-	ature	Photo-	
Species	Hormone	Dose	Route	quency	(°C)	period	Salinity
	Hake GH (*Urophycis*)	20 µg/g	ip	3×/week			
	Shark GH (*Prionace*)						

[a] Intramuscular.   [b] Freshwater.   [c] L = length (cm); W = weight (g).

while the handled controls had a lower growth rate (Table II). These data suggest that endogenous growth hormone secretion is greater in the summer months and that stress of handling can have an adverse effect on the growth rate of salmonids. Water temperature was not reported, but it can be assumed that the lower growth rate in both hormone-treated and uninjected trout in March was related to a lower ambient temperature. These results can be correlated with a later study which showed that the growth hormone content of the pituitary of the perch is highest in February and lowest in July. This may signify that active release of pituitary growth hormone occurs during the summer months (Swift and Pickford, 1962a,b, 1965) (Table II).

There have been two studies in which rainbow trout have been recipients of bGH treatment. Enomoto (1964) gave approximately 15 µg hormone/g/week to young *Salmo irideus* at 16°C for 27 days. Growth enhancement was found in one of three fish; the other two fish developed exophthalmos. Chartier-Baraduc (1959) administered approximately 15 µg hormone/g/week to *Salmo gairdneri* maintained between 12° and 13°C. The mean weight for experimental fish was higher than that of control fish after 14 days of treatment.

Recent work on salmonids has been conducted on Atlantic salmon (*Salmo salar*) and on two species of Pacific salmon (Table II). Komourdjian *et al.* (1976a) found that porcine growth hormone (pGH) administered at a dosage of 3.5 µg/g/week (1 µg/g/2day) for 28 days significantly stimulated growth of Atlantic salmon at 11.5°C. A series of studies has assessed the influence of bGH on growth of underyearling and yearling coho salmon held at 10°C. In a preliminary investigation, Higgs *et al.* (1975) demonstrated that doses of 10 and 100 µg hormone/g/week administered to yearling intact coho salmon at 10°C by ip injection or by hormone; cholesterol pellet implantation was

Continued

Nutritional state	Hy-po-phy-sec-tomy	Comments	Dura-tion (days)	Initial age and/or weight	Over control (%)	Weight gain Over initial weight (%)	%/day	Reference
	Yes	Only 2 assay fish. Gain in length estimated from graph.	35			3.7[f]		Lewis *et al.* (1972); Pickford (1973)
	Yes	Shark GH more potent than hake GH.	28			4.9		

[d] Intraperitoneal.   [e] Seawater.   [f] Length change over initial length.

equipotent in enhancing growth (Table II; Fig. 3). Subsequent studies (Higgs *et al.*, 1976, 1978a) showed that 10 μg bGH/g/week was the minimum dosage of hormone necessary to promote maximum growth of coho salmon at 10°C. However, stimulation of growth over that of uninjected controls was obtained with a dosage as low as 0.32 μg/g/week (2.9×10^{-4} IU/g/week) (Fig. 3).

Clarke (1976) and Clarke *et al.* (1977) injected underyearling sockeye salmon, *O. nerka*, held at 17.5°C and fed a restricted ration (5% body weight/day) with 3.5 μg bGH/g/week (1 μg/g/2day) for 14 days. The growth rates (weight and length) of treated fish were significantly higher than those of the controls (Fig. 4). Brett *et al.* (1969) have shown that underyearling sockeye at 17.5°C on a ration of 5% body weight/day (dry weight basis) are capable of rapid growth. This suggests that while endogenous growth hormone secretion may be high at 17.5°C, it is still not optimal. Studies on coho salmon show that underyearlings (Higgs *et al.*, 1976) which should have the greatest potential for rapid growth because of size and/or age are able to respond to bGH treatment better than yearlings (Higgs *et al.*, 1975). In both of these studies of coho salmon, excess food was presented daily. Ration level has been shown by Markert *et al.* (1977) to influence not only the growth rate of bGH-treated yearling coho salmon held at 10°C, but also the cost of bGH treatment. This was prohibitive at present for salmon enhancement and aquaculture programs (Table IV). However, they did obtain a 2.5- to 3-fold reduction in the cost per gram gain in wet weight over control by administering bGH to coho fed to satiation rather than to those on a restricted ration (0.8% body weight/day; dry feed/wet fish). Hence, administration of 1–10 μg bGH/g/week to small rapidly growing salmon, fed a well balanced diet to satiation, should yield the greatest gains in growth at the most economical cost. Additional

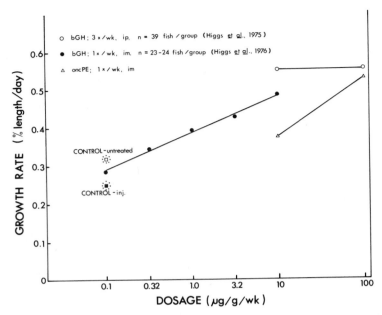

Fig. 3. Influence of dosage (μg/g/week) of bovine growth hormone [bGH; NIH-GH-B17 (0.92 IU/mg)] and of a chinook salmon pituitary extract (oncPE) on the growth rate (length) of yearling coho salmon. Period of hormone treatment was either 56 days (Higgs et al., 1975) or 70 days (Higgs et al., 1978a). Coho were held in 10°C well water and maintained on a 12L:12D photoperiod. Excess ration (Oregon Moist Pellets) was presented twice daily. Control fish received 100 μl of 0.7% NaCl (pH 9.5) or were untreated. Refer to Section II,A,3 of the text for the methods used to prepare oncPE. (Adapted from Higgs et al., 1975, 1978a.)

studies assessing the interactions between temperature, diet, ration level, less purified (expensive) mammalian hormone preparations and doses, and fish age may provide information which will enable growth hormone treatment to be both practical and economical in the future.

It is apparent from the long-term studies conducted by Pickford on *Fundulus* (Table II) and those by Higgs et al. (1975, 1976, 1977a, 1978a) and Markert et al. (1977) on coho salmon that bGH treatment does not result in hormone–antibody complexes with subsequent loss of potency.

While bGH was the first mammalian preparation to be tested in fish, it has been shown that only ovine growth hormone (oGH) has a similar potency to bGH in *Fundulus*. Porcine, simian, and human growth hormone have a lower potency (Pickford, 1959) (Table II). In a recent study by Komourdjian et al. (1978), hypophysectomy of *Salmo gairdneri* resulted in a reduced linear growth rate of 0.08% per day,

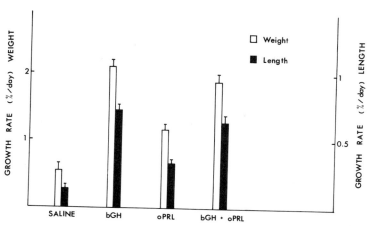

Fig. 4. Influence of bovine growth hormone [bGH; NIH-GH-B-17 (0.92 IU/mg)] and ovine prolactin [oPrl; NIH-PS-11 (26.4 IU/mg)], administered singly and in combination, on the growth rate in weight and length of underyearling sockeye salmon. Dosage of bGH was 1 μg/g body weight, while that for oPrl was 5 μg/g, controls received saline. The fish were held in 17.5°C freshwater and maintained on a 12L:12D photoperiod for 14 days prior to hormone treatment. All fish were individually marked so that they could be distinguished in a common tank. Hormone injections into the peritoneal cavity were administered on alternate days for 14 days. A restricted ration [5% body weight; dry weight of food (g)/dry weight of fish (g)] of Oregon Moist Pellets was fed daily. Vertical lines represent one standard error of mean. Dunnett's test showed that the growth rates (weight or length) of all hormone-treated fish were significantly higher ($P < 0.01$) than those of the controls. (From Clarke, 1976, *Reg. Conf. Comp. Endocrinol.*, Oregon State Univ., *Corvalis* Abstr. No. 38.)

whereas replacement therapy with porcine GH at 3.4 and 6.8 mU/g increased growth rates to 0.29 and 0.44% per day, respectively, at 10°C.

2. Piscine Growth Hormone

There have been two attempts (Jampolsky and Hoar, 1954; Higgs *et al.*, 1978a) to demonstrate the presence of a growth-promoting factor in pituitary extracts from chinook salmon. Jampolsky and Hoar (1954) found that daily injections of alkaline extracts of the whole pituitary gland into goldfish, *Carassius auratus*, initially induced weight gains, but later led to weight loss and high mortality. In contrast, im injection of 10 or 100 μg/g body weight of the partially purified salmon growth hormone (oncPE) prepared by Higgs *et al.* (1978a) once per week into yearling coho salmon held at 10°C led to growth enhancement during the 70-day experimental period (Fig. 3 and Table III). oncPE-treated male, but not female, coho had higher gonadosomatic indices relative to those of controls suggesting the presence of a low level of gonado-

Table IV

Estimated Cost (¢) per Gram Gain in Weight over Control of Administering Bovine Growth Hormone (bGH) to Yearling Coho Salmon

Group	Diet[a]	Ration	bGH	Initial weight (g/100 fish)	Final weight (g/100 fish)	Weight gain (g)	Weight gain over control (g)	Total bGH administered (mg) per 100 fish	Cost ¢[b]/gain (g)
1	PO	0.88%	+	2475.5	3986.2	1510.7	488.0	240.2	9.84
2	PO	0.88%	−	2280.0	3302.7	1022.7	—	—	—
3	OMP	0.84%	+	2563.7	4504.3	1940.6	631.7	258.1	8.17
4	OMP	0.84%	−	2492.7	3801.6	1308.9	—	—	—
7	OMP	Satiation	+	2274.5	5338.9	3064.4	1559.8	253.7	3.25
8	OMP	Satiation	−	2239.6	3744.2	1504.6	—	—	—

[a] The fish were fed a Poultry Offal (PO) or Oregon Moist Pellet (OMP) diet formulation for 56 days either as a restricted ration [0.8%/day; dry weight of food (g)/wet weight of fish (g)] or to satiation. Refer to Fig. 7 for additional information. From Markert et al. (1977, Can. J. Zool. 55, 74–83).
[b] To calculate this result a price of $20.00/100 mg of bGH was used (Research Products Division, Miles Laboratories Inc., Kankakee, Illinois).

tropin. Additional work is required to obtain purified salmon growth hormone preparations for two reasons. First, they could conceivably prove to be more potent and less expensive than the crude or purified mammalian preparations. Second, such preparations would enable the development of a homologous radioimmunoassay for determining plasma concentrations of salmon growth hormone. Until recently it was only possible to monitor relative changes in serum (McKeown and Van Overbeeke, 1972) or plasma (Leatherland et al., 1974) growth hormone levels in Oncorhynchus using a heterologous radioimmunoassay. However, Fryer (1977) has described a radioreceptor assay for teleost GH which involves the displacement of ^{125}I-labeled Tilapia GH from the microsomal membrane fractions of Tilapia liver or kidney. Information on blood levels of growth hormone, fractional turnover rates, and distribution volumes will permit calculation of secretion rates. Data of growth hormone secretion rates for fish in various physiological and environmental states will provide information on the regulation and role of growth hormone in fish and will permit refinement of hormone doses and treatment intervals in growth studies. Preliminary data from a study by Higgs, Donaldson and Dye (1975, unpublished data) show that the fractional turnover rate of ^{125}I-labeled human growth hormone in plasma of yearling coho salmon at 10°C is considerably slower than found in mammals. Other studies assessing the effect of piscine growth hormone preparations on fish growth have employed hypophysectomized male F. heteroclitus as the assay species (Table III). With the Fundulus assay, Pickford has been able to compare the potency of hake and pollack growth hormone, prepared by Wilhelmi (1955), to that of bovine growth hormone. She has also compared the potency of shark growth hormone, prepared by Lewis and co-workers (1972) to that of hake growth hormone (Tables II and III). In the first case, hake growth hormone (preparation F80GH) induced a weight increase of 26.3% over 56 days. Three preparations of bGH (Table II) given in sequence at the same dosage stimulated a greater weight increase, namely, 39.5% (Pickford, 1954a). Pickford (1954a) pointed out that the lesser response to fish growth hormone may have been caused by the presence of an inactive component or could be correlated with the total lack of thyroid stimulation by this preparation when compared to the bGH preparations. In the second case, hake growth hormone stimulated a length increase of approximately 3.7% in 35 days, while the response to shark growth hormone was approximately 4.9% after 28 days in the same individual recipients (Lewis et al., 1972; Pickford, 1973).

C. Morphological and Physiological Effects and Mechanism of Action

1. Condition Factor, External Appearance, and Salinity Transfer of Salmonids

Treatment of salmonids with purified mammalian growth hormone depresses condition factor (Fig. 5); an index of the fatness or leanness of fish (Chartier-Baraduc, 1959; Clarke, 1976; Higgs *et al.*, 1975, 1976, 1977a, 1978a; Komourdjian *et al.*, 1976a; Markert *et al.*, 1977). This also occurs after treatment of coho salmon with a chinook salmon pituitary extract (Higgs *et al.*, 1978a). In addition Komourdjian *et al.* (1976b) noted that untreated Atlantic salmon exposed to longer daylength not only had higher growth rates and lower condition factors than those on shorter daylength but also an increase in the number and apparent activity of pituitary somatotrops. The evidence therefore indicates that increased titers of growth hormone induce greater growth

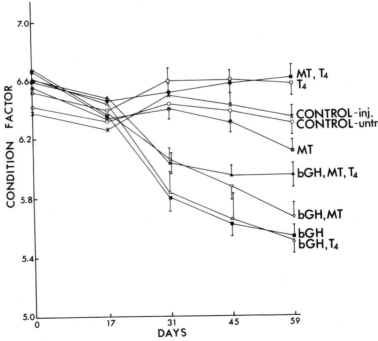

Fig. 5. Mean values ± 2 SE for condition factor of groups of yearling coho salmon at 10°C administered either bovine growth hormone (bGH), L-thyroxine (T₄), 17α-methyltestosterone (MT), combinations of these hormones, or no hormone for 59 days. Refer to Fig. 20 for details. (From Higgs *et al.*, 1977a, *Can. J. Zool.* **55**, 1048–1056.)

in length than weight. In yearling coho salmon this has been observed (Higgs *et al.*, 1978a) at the lowest dosage of bGH administered to date, namely, 0.1 μg (0.92 × 10^{-4} IU)/g/week. The decline, however, is less than that following treatment with 10 μg bGH/g/week. With this latter dosage the condition factor of both underyearling and yearling coho salmon has been observed to progressively decrease until cessation of hormone administration. The trend is then reversed and the condition factor returns to the control range. This occurs much faster in yearling than in underyearling coho salmon (Higgs *et al.*, 1975, 1978a). So far there is no evidence that the decline in condition factor found in yearling coho salmon treated with bGH is influenced by diet formulation or ration level (Markert *et al.*, 1977) (Table II). Recently Komourdjian *et al.* (1978) have shown that porcine growth hormone has no effect on the condition factor of hypophysectomized trout (*Salmo gairdneri*).

In Atlantic salmon, growth hormone administration can evoke both xanthophore and melanophore responses. For example, Komourdjian *et al.* (1976a) found that porcine growth hormone treatment (Table II) for 14 days leads to a blackening of the fin margins. Also, they observed yellowing around the operculae and on the fins of hormone-treated fish. Parr marks remained visible. A period of 14–28 days treatment of coho salmon at 10°C with bGH also results in darkening of the dorsal and caudal fin margins. Higgs *et al.* (1976) found that the parr marks of underyearling coho salmon were visible even after 84 days of bGH treatment.

The changes in condition factor of salmon, and to some extent in pigmentation, after growth hormone treatment and withdrawal are similar to those which occur at the time of parr–smolt transformation. For example, Vanstone and Markert (1968) have observed that the condition factors of untreated presmolt and postsmolt coho salmon are higher than those of smolting coho. More specifically, evidence accumulated over the last 20 years indicates that growth hormone participates, probably in association with other hormones, in the physiological changes which allow juvenile salmonids in freshwater to transfer to, and survive in, seawater. For example, Smith (1956) found that mammalian GH treatment develops salinity tolerance in brown trout. Changes in the salinity preference of young coho salmon have been observed after prolonged mammalian GH treatment (McInerney, unpublished observations, cited in Ball, 1969). There is evidence for increased utilization of growth hormone during both direct and stepwise transfer of juvenile salmon to seawater (McKeown, unpublished observations, cited in Chester Jones *et al.*, 1974). Clarke (1976) has recently found that bGH treatment of underyearling sockeye salmon

(*O. nerka*) significantly reduced their blood sodium levels 24 hr after transfer of the fish to seawater. Komourdjian *et al.* (1976a) showed that administration of pGH to Atlantic salmon prevented mortality when the fish were transferred from freshwater to seawater (30% salinity) over a period of 22 hr. Considerable mortality of untreated fish occurred which apparently was not size related. Finally Higgs *et al.* (1976) demonstrated that prolonged bGH treatment of underyearling coho salmon in freshwater does not impair their ability to survive transfer to seawater or respond to hormone treatment after transfer. Further clarification of the role of growth hormone in osmoregulation in salmonids will have to await information on the degree of interaction between mammalian growth hormone and the receptors for growth and prolactin in fish or the isolation of sufficient piscine growth hormone for physiological studies.

2. Mechanisms by which Growth Hormone Enhances Fish Growth

There is evidence that growth hormone enhances fish growth by stimulating greater voluntary food intake (appetite) and by improving food conversion, calculated as ratio of food intake to weight gain. Changes in both of these parameters after growth hormone treatment reflect actions of growth hormone on metabolism. Other hormones whose secretion is indirectly or directly stimulated by growth hormone may be involved.

a. Appetite. The food intake of hypophysectomized *F. heteroclitus* is increased after bGH treatment (Pickford, 1957) as is that of intact Atlantic salmon parr treated with pGH (Komourdjian *et al.*, 1976a). Higgs *et al.* (1975, 1976, 1977a, 1978a) have observed greater feeding activity in yearlings and underyearling coho salmon treated with bGH. Studies on other poikilotherms have also shown that mammalian growth hormone treatment results in greater food consumption (Licht and Hoyer, 1968; Zipser *et al.*, 1969; Brown *et al.*, 1974). None of the above investigations attempted to relate changes in daily food intake, after growth hormone treatment, to those of body weight. But this has recently been done by Markert *et al.* (1977). In their study on yearling intact coho salmon, appetite was expressed either as mean daily dry food (milligrams) consumed per gram body weight or as mean daily caloric intake per gram. While bGH treatment did not significantly elevate the food consumption of coho, there was nevertheless a consistent increase in the appetite of bGH-treated fish relative to that of the controls for the last 4 weeks of the experiment

(Fig. 6). Markert *et al.* (1977) have outlined two ways in which growth hormone could affect appetite of fish. First, as has been proposed for weanling rats (Wade, 1974), growth hormone may have a direct action on centers in the hypothalamus which influence food intake. Second, growth hormone may induce a number of metabolic changes which feed back on the hypothalamic centers affecting appetite. For example, Brown *et al.* (1974) who studied the effect of bGH on food consumption of the snapping turtle, *Chelydra serpentina*, suggested that metabolites such as glucose, amino acids, or fatty acids may act as signals which could alter the activity of the nerves in the hypothalamic centers. The importance of the hypothalamus as a center influencing

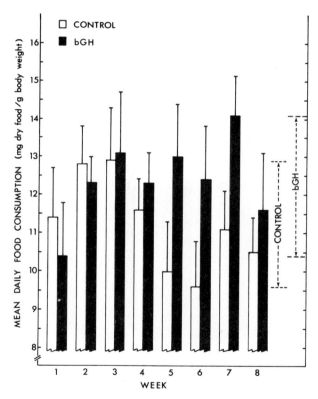

Fig. 6. Influence of bovine growth hormone (bGH) on the voluntary food intake [mean daily dry weight of food (mg) consumed/g body weight] of yearling coho salmon at 10°C for each 7-day period of a 56 day experiment. Control (untreated) and bGH-treated coho were fed Oregon Moist Pellets to satiation twice daily. The range in mean daily food consumption of control and bGH-treated coho during the test period is indicated. Refer to Fig. 7 for additional information. (Modified from Markert *et al.*, 1977.)

feeding responses of fish was shown by Demski (1973; see also Chapter 3).

  b. Food Conversion. Administration of mammalian growth hormone improves food conversion in the pig (Machlin, 1972) and chicken (Myers and Peterson, 1974). Higgs *et al.* (1975, 1976) postulated such an effect for salmon and a recent study by Markert *et al.* (1977) supports this. Markert *et al.* (1977) observed that the amount of dry food or protein fed per gram gain in weight of yearling coho salmon at 10°C was less for bGH-treated fish relative to that of controls at two ration levels (Fig. 7). The greatest differences in food conversion between bGH-treated and control fish occurred when a satiation ration

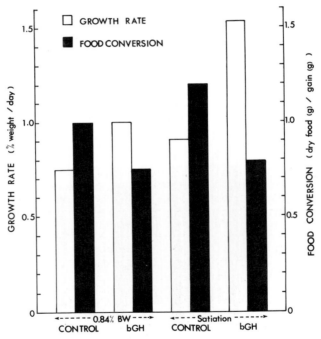

Fig. 7. Influence of bovine growth hormone [bGH; NIH-GH-B-17 (0.92 IU/mg)] on growth rate and food conversion of yearling coho salmon over a period of 56 days. Paired groups of coho were held in 10°C well water and maintained on a 12L:12D photoperiod. The groups were fed twice daily with Oregon Moist Pellets at two ration levels [0.84% body weight/day; dry weight of food (g)/wet weight of fish (g) and satiation]. Coho in one of the groups at each ration level received, by intramuscular injection, 10 μg bGH/g body weight once per week. Coho in the other group were not treated (control). Ration (0.84%) and hormone dosage were adjusted according to the new mean fish weight in each group at every 14-day interval. (Modified from Markert *et al.*, 1977.)

was provided. The mechanism(s) by which growth hormone improves food (or protein) conversion of coho and probably other species of fish is open to some speculation. Markert *et al.* (1977) have outlined some of the possibilities. These include: (1) Stimulation of fat mobilization and oxidation. Percentage of lipid in muscle of coho salmon declines following bGH treatment (Higgs *et al.*, 1975, 1976, 1977a) (Table V; Markert *et al.*, 1977). Also Clarke (1976) found that bGH treatment decreased whole body lipid (expressed as % of dry weight) of under-yearling sockeye salmon. In this case, more of the ingested amino acids would be available for growth since lipid would be preferentially used as an energy source (Frye, 1971; Snyder and Frye, 1972). (2) An action of growth hormone on the rate of protein synthesis and/or breakdown. Higgs *et al.* (1975, 1976) estimated that the mean amount of protein in bGH-treated intact underyearling or yearling coho was substantially higher than that of control fish. Also, their estimated ratios of protein (g) to dry body weight (g) were higher for bGH-treated fish relative to those of control fish. Enomoto (1964) found that the ratios of crude protein to crude ash in the whole body of *Salmo gairdnerii* treated with growth hormone were larger than those of his control fish. Reduction of total plasma protein as well as plasma urea was observed in *Cottus scorpius* after treatment with mammalian

Table V

Proximate Analyses of Muscle from Yearling Coho Salmon at 10°C, Administered Either Bovine Growth Hormone, L-Thyroxine, 17α-Methyltestosterone, Combinations of these Hormones, or No Hormone for 59 Days[a]

Treatment[b]	Protein (%)[c]	Lipid (%)[c]	Moisture (%)[c]
Control, solvent injected	18.0 ± 0.34*	3.05 ± 0.40*,**	76.5 ± 0.80*,**
bGH, 10 μg/g	18.3 ± 0.40*	2.60 ± 0.28*	78.0 ± 0.47***
T₄, 1 μg/g	18.0 ± 0.21*	4.05 ± 0.15†	76.3 ± 0.57*,**
MT, 1 mg/kg diet	17.9 ± 0.57*	3.28 ± 0.30**,***	76.0 ± 0.28*
bGH + T₄	18.3 ± 0.50*	2.64 ± 0.19*	77.6 ± 0.33**,***
bGH + MT	17.6 ± 0.43*	3.69 ± 0.22***,†	76.7 ± 0.70*,**
T₄ + MT	17.2 ± 0.34*	4.23 ± 0.24†	76.5 ± 0.47*,**
bGH + T₄ + MT	17.2 ± 0.70*	2.88 ± 0.32*,**	76.7 ± 0.23*,**

[a] Mean percentages ($n = 8$ fish) ± 2 SE are shown. Refer to Fig. 20 for additional details.

[b] bGH, bovine growth hormone; T₄, L-thyroxine; MT, 17α-methyltestosterone.

[c] An unblanced analysis of variance of group percentages (untransformed and arcsin square root transformed) for muscle lipid and water, using a least squares solution (BMD 10V-general linear hypothesis) indicated $P < 0.001$. Treatment groups with the same asterisk or dagger superscripts form a homogeneous subset (Student Newman–Keuls test, $P = 0.05$).

growth hormone over a period of 15 days (Matty, 1962). Finally, Matty
and Cheema (1976 unpublished data) have demonstrated that both
porcine and shark growth hormone treatment of intact rainbow trout
can stimulate (significant effect obtained only for pGH) *in vivo* incorporation of L-[^{14}C]leucine into skeletal muscle protein (Fig. 8). These
studies indicate that growth hormone treatment creates a positive nitrogen balance in fish because of a stimulatory influence on protein
synthesis. In mammals, growth hormone improves the uptake of amino
acids into cells, influences the rate of RNA synthesis, and has an effect
on translational mechanisms (Talwar *et al.*, 1975). Similar mechanisms
may be operative in fish. (3) A stimulation by direct or indirect means
of insulin synthesis and release. Higgs *et al.* (1976, 1977a) found that
prolonged bGH administration to underyearling or yearling coho salmon increased the number and size (underyearlings only) of the islets
of Langerhans and in particular β-cell granulation within the islets,
relative to that of the controls. They concluded that greater insulin
synthesis in the β cells may have been in response to one or more of
three possible actions of growth hormone. First, bGH may have inhibited peripheral glucose utilization. This would have led to
hyperglycemia necessitating greater insulin synthesis and release to
lower blood glucose. Studies on other fish (Matty, 1962; Enomoto,
1964; McKeown *et al.*, 1975) provide support for this idea. Second,
growth hormone may have acted directly on pancreatic islets to stimulate synthesis and/or release of insulin. There is only evidence from
mammalian studies for this hypothesis (Martin and Gagliardino, 1967;
Curry and Bennett, 1973; Parman, 1975a,b). Third, the effect of
growth hormone on the endocrine component of the pancreas may
have been a consequence of the increased food consumption of the
bGH-treated fish. There is another possible mechanism not considered by Higgs *et al.* (1976). Higgs *et al.* (1977a) found that administration of bGH to yearling coho salmon significantly increased the mean
nuclear diameters of the interrenal tissue (Table VI). Glucocorticoids
have been shown to influence pancreatic islet function in mammals
(Curry and Bennett, 1973; Van Lan *et al.*, 1974) and this may occur in
fish. Increased insulin synthesis and release, irrespective of the mechanism, would, on the basis of studies conducted on fish by Jackim and
LaRoche (1973) and Matty and Cheema (1976, unpublished data) result in greater protein synthesis and hence improvement of protein
conversion. In mammals there is a specific relationship between
growth hormone and insulin in growth regulation. Muscle growth is
accomplished by nuclear replication which is considered to be under
the control of growth hormone, while cytoplasmic growth is under the
control of insulin (Cheek, 1971). A similar situation may exist in fish.

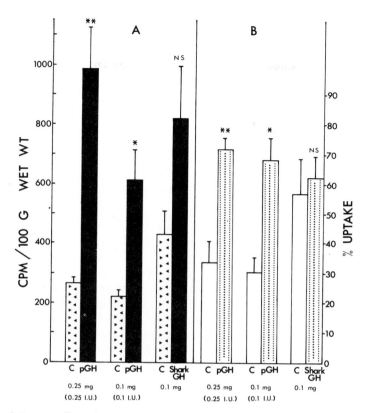

Fig. 8. *In vivo* effect of porcine growth hormone (pGH) and shark growth hormone on incorporation of L-[¹⁴C]leucine into skeletal muscle of rainbow trout (range in mean weight 20–30 g) held at 13–14°C. In one experiment, each fish received an intraperitoneal (ip) injection of 0.25 mg pGH. After 16 hr 5 μCi of L-[¹⁴C]leucine/100 g body weight was administered (ip). Fish (n = 6/treatment) were sacrificed exactly 6 hr later and muscle samples were frozen ($-20°C$) or homogenized immediately. Controls (C) were injected with $0.005\,M$ HCl, which was the solvent used to dissolve pGH. In a second experiment pGH was administered at a dosage of 0.1 mg/fish, while in a third, 0.1 mg of shark growth hormone was administered to each fish. In the latter experiments there were 3 fish in each treatment group. Results were expressed either as incorporation of radioactivity into skeletal muscle protein (A) (cpm/100 g wet weight) or as percentage uptake of radioactivity by the skeletal muscle (B). * or ** indicates that P was less than 0.01 or 0.001, respectively, while NS indicates $P > 0.05$. (From Matty and Cheema, 1976 unpublished observations.)

Markert *et al.* (1977) have postulated that some of the above actions of growth hormone on metabolism may have decreased the specific dynamic action of food on metabolism. Thus more of the energy from ingested food would be available for growth.

Table VI

Mean Nuclear Diameters of Interrenal Tissue from Yearling Coho Salmon at 10°C,
Administered Either Bovine Growth Hormone, 17α-Methyltestosterone,
a Combination of bGH + MT, or No Hormone for 59 Days

Treatment[a]	Number of fish	Nuclear diameter[b] (μm ± 2 SE)
Control, solvent injected	6	6.21 ± 0.061*
bGH, 10 μg/g fish	5	6.55 ± 0.22**
MT, 1 mg/kg diet	5	6.87 ± 0.14***
bGH + MT	5	6.82 ± 0.18***

[a] bGH, bovine growth hormone; MT, 17α-methyltestosterone. Refer to Fig. 20 for additional details.

[b] Treatment groups with the same asterisk superscripts form a homogeneous subset. All other pairs of group means are significantly different (analysis of variance indicated $P < 0.001$; Student Newman–Keuls test, $P = 0.05$).

c. Somatomedin. There is considerable evidence that growth hormone does not have a direct stimulatory influence on essential growth tissues of mammals but induces the formation of a secondary hormonal agent somatomedin (originally called sulfation factor) which acts at the cellular level (Daughaday *et al.*, 1975). The term somatomedin has been defined by Van Wyk *et al.* (1974) as all growth hormone-dependent plasma factors (peptides) that mediate the growth of responsive tissues. For example, the action of growth hormone on cartilage and bone growth is mediated by somatomedin; Talwar *et al.* (1975) consider the liver and kidney as possible sites for somatomedin generation in response to growth hormone. There is no information regarding the presence of somatomedin in fish.

III. STEROID HORMONES AND SUBSTANCES RELATED IN FUNCTION*

A. Biological Function

1. NATURAL STEROIDS

Common names, systematic nomenclature, and structural formulas of compounds mentioned in this section of the chapter are listed in Table VII.

* This section was prepared by U. H. M. F. and J. R. McB.

Table VII

Steroid Hormones and Substances with Related Function

Common name	Systematic nomenclature	Structure
Androgens		
Testosterone: R = H	17β-Hydroxy-androst-4-ene-3-one	
Testosterone propionate: R = $COCH_2$—CH_3	17β-Acetoxy-androst-4-en-3-one	
1-Dehydrotestosterone acetate	17β-Acetoxy-androst-1,4-dien-3-one	
11-Ketotestosterone	17β-Hydroxy-androst-4-en-3,11-dione	
Methenolone acetate	17β-Acetoxy-1β-methyl-androst-4-en-3-one	
17α-Methyltestosterone	17β-Hydroxy-17α-methyl-androst-4-ene-3-one	
Methylandrostenediol, methandriol	17α-Methyl-androst-5-ene-3β,17β-diol	
17α-Ethynyltestosterone, ethisterone, pregneninolone	17β-Hydroxy-17α-ethinyl-androst-4-en-3-one	
Oxymetholone	17β-Hydroxy-2-hydroxy-methylene-17α-methyl-5α-androstan-3-one	

(Continued)

Table VII—*Continued*

Common name	Systematic nomenclature	Structure
4-Chlorotestosterone acetate	4-Chloro-17β-acetoxy-androst-4-en-3-one	
Ethylestrenol	17β-Hydroxy-17α-ethyl-estr-4-en-3-one	
Stanozolol	17β-Hydroxy-17α-methyl-5α-androstano-(3,2-C)-pyrazole	
Dimethazine	2α,17α-Dimethyl-17β-hydroxy-5α-androstan -3,3'-azine	

Estrogens
Estradiol: R = H
Estradiol benzoate:
  R = COC₆H₅
Estradiol monopalmitate:
  R = CO(CH₂)₁₄CH₃
	Estra-1,3,5(10)-triene-3,17β-diol
	Estra-1,3,5(10)-triene-3,17β-diol 3-benzoate
	Estra-1,3,5(10)-triene-3,17β-diol 3-palmitate	

Estrone	3-Hydroxy-1,3,5(10)-triene-17-one	
Estriol	Estra-1,3,5(10)-triene-3,16α,17β-triol	
Diethylstilbestrol	4,4'-(1,2-Diethyl-1,2-ethenediyl) bis-phenol	
Progestogens Progesterone	Pregn-4-en-3,20-dione	

Table VII—*Continued*

Common name	Systematic nomenclature	Structure
Pregnanediol	5β-Pregnane-3α, 20α-diol	
Melengestrol acetate	17α-Acetoxy-6-methyl-16-methylene-pregna-4,6-diene-3,20-dione	
orticosteroids Cortisol	$11\beta,17\alpha,21$-Trihydroxy-pregn-4-ene-3,20-dione	
Deoxycorticosterone acetate	21-Acetoxy-pregn-4-ene-3,20-dione	

The steroid hormones of vertebrates may be separated into four groups on the basis of physiological function. These functions *in mammals* are outlined below. Representatives of all four groups of hormones have been identified in a number of fish species and physiological function in fish is similar to that in mammals with the exception of progestogens, the function of which in fish is poorly understood.

a. *Androgens* or male sex hormones are mainly produced by the testis, particularly in the mature animal. Smaller amounts are also synthesized by the ovary and the adrenal. The main target tissues are those which are involved in the expression of male sex function and character. In addition, many androgens have a considerable anabolic or growth promoting effect either on specific sex related tissues such as prostate or seminal vesicle or under certain conditions on the total organism.

The most extensively studied androgen is testosterone which is widespread in the animal kingdom including Pisces. A number of precursors and metabolites of testosterone present in the animal body also have considerable androgenic potency. A notable one is the 5α-isomer of the reduction product dihydrotestosterone, which is more active than testosterone on target tissues in mammals.

b. *Estrogens* or female sex hormones are produced by the ovary from testosterone via a number of intermediary products (Baggett *et al.*, 1956). They are responsible for female sex function but also have important anabolic effects. There are three important estrogens: estradiol, estrone, and estriol.

c. *Progestogens* are also female sex hormones, produced by a number of tissues of which the corpus luteum of the ovary is the most important. Among functions of progestogens may be mentioned facilitation of implantation of the fertilized ovum and continued maintenance of pregnancy. Progesterone is the most important progestogen. Anabolic activity in higher vertebrates has been ascribed to some progestogens.

d. *Corticosteroids* are mainly produced by the adrenal cortex or its homologue, the interrenal gland of fishes. These compounds, of which cortisol is a notable representative, have a variety of functions relating to the maintenance of normal body processes such as regulating carbohydrate, protein, and lipid metabolism, combating the effects of injury, and maintaining body fluids and electrolytes. Corticosteroids do not have significant anabolic effects.

2. STEROIDS WITH MODIFIED STRUCTURE

The natural steroid hormones outlined above are all derived from the basic hydrocarbon cyclopentanophenanthrane. The divergence in physiological effects of the hormones may be ascribed to differences either in spatial configuration of the molecule or in the nature of unsaturation and functional groups located at certain key positions in the molecule. By identifying these positions and altering the structure of the molecule accordingly it has been possible to enhance certain of the physiological effects while suppressing others. One of the notable successes in this regard has been the manipulation of the ratio of anabolic to androgenic effects. Changes in effectiveness of different routes of administration may also be influenced. Esterification of the 17β-hydroxyl group protects the group from oxidation and enhances activity during oral administration.

In determining the anabolic/androgenic ratio, anabolic activity is expressed as the effectiveness of a substance, measured in percentage of the appropriate reference standard, in increasing the weight of the levator ani muscle of the castrated male rat. Androgenic activity is similarly related to the ability of a reference compound to increase the weight of the seminal vesicle or ventral prostate of the castrated rat. The index derived in this manner is important for comparing the effectiveness of hormones, but may only be taken as an indication of activity, since the anabolic/androgenic ratio for a particular hormone varies from species to species. The validity of this index is also minimized by the questionable assumption that growth of the levator ani equals general muscle growth.

3. COMPOUNDS NOT RELATED TO STEROIDS

A number of compounds are known to mimic the physiological effects of the natural hormones, although not related in chemical structure. A notable example is diethylstilbestrol, which possesses estrogenic and anabolic properties in mammals superior to those of natural estrogens.

In this context, mention should be made of a group of substances of plant origin possessing estrogenic properties. Structurally these phyto-estrogens often contain a sapogenin component which is either triterpenoid or steroid in nature. Estrogen-like activity of several species of clover, caused by the presence of such substances, has been determined on the basis of teat length in wethers feeding on these clovers (Braden *et al.*, 1971).

B. Mode of Action of Anabolic Steroids

The importance of endocrine factors in mammalian growth was realized in ancient times when castration of boys and men resulted in reduced body size. Kochakian *et al.* (1948) identified the responsible testicular agent by reversing weight loss of castrated hamsters through administration of natural androgens. The female of most mammalian species is smaller than the male. Estrogens have been found to decrease growth by inhibiting the secretion of growth hormone from the anterior pituitary of rats.

It thus appears that androgens are anabolic steroids in contrast to the estrogens. In ruminants, however, both androgens and estrogens have anabolic effects, and both groups of hormones have been used as growth promoters. Testosterone was the commonly used androgen but, because of adherent excessive masculinization, which has a detrimental effect on the quality of the meat, its use has been discontinued. Feminization on the other hand may occur with estrogens. To prevent deterioration of meat quality through expression of sexual character, combinations of estrogen and androgen have been used. VanderWal *et al.* (1975), for instance, obtained good growth response in calves with a combination of equal parts of estradiol and testosterone.

The effect of anabolic agents is generally defined as an increase in nitrogen retention. There are two possible sources of the increased nitrogen: increased food intake or improved food utilization. The control of *food intake* is located in the hypothalamus. The principal areas involved in feeding behavior are the ventromedial nucleus which has inhibitory or "satiety" effects on a primary food-seeking area located in the lateral hypothalamus. The exact mechanism by which endocrine factors influence these areas is not known but Bray (1974) postulated the involvement of catecholamines. A concise review of the role of endocrine factors in the control of food intake may be found in the same paper. The subject of feeding as it relates to fish is treated separately in this volume (Chapter 3).

A fully acceptable theory explaining the mechanisms by which anabolic steroids improve *food utilization* has not been advanced. Preston (1975) has summarized such attempts as they relate to the effects of estrogens in ruminants and lists them as follows: (1) increased ACTH production, causing an increased output of androgens by the adrenals; (2) increased production and/or release of growth hormone; (3) increased secretion of insulin, stimulating amino acid incorporation into protein; (4) increased production of thyroid hormone; (5) direct effect on the tissues in enhancing utilization of nonprotein nitrogen. These theories fall short of explaining all the observed phenomena, and contradictions may be found.

In the past there was no indication of a direct effect of anabolic steroids on the cells of skeletal muscle. Cytoplasmic or nuclear localization was not observed (Mainwaring, 1972; Giannopoulos, 1973); neither testosterone nor dihydrotestosterone accumulates in skeletal muscles (Gloyma and Wilson, 1969; Hansson *et al.*, 1972) and a cytoplasmic androgen receptor protein could not be demonstrated (Jung and Baulieu, 1972; Wainman and Shipounoff, 1941). Recently, however, Michel and Baulieu (1974) succeeded in demonstrating the specific binding of testosterone and other anabolic steroids to a cytoplasmic protein from the quadriceps muscle of rats and preliminary evidence of binding of [^{3}H]testosterone to proteins larger than albumin has been reported (Powers and Florini, 1975). In the latter report, the authors also demonstrated that testosterone caused a 25% stimulation of labeling index by autoradiographic analysis of myoblast cultures from thigh muscles of rats incubated with [^{3}H]thymidine. These reports indicate a direct involvement of anabolic steroids in the function of muscle cells.

The meager knowledge about the effect of steroid hormones on appetite and food conversion efficiency in fish is touched upon in Section III,D,5.

C. Growth Promoters Currently in Use for Livestock

In order to lend perspective to the question of possible benefits from the use of steroid growth promoters in the culture of fish, some comments are provided below on the economic importance and current usage of growth promoters in animal husbandry.

The search for compounds with enhanced anabolic activity has produced some drugs which have had a substantial impact on the production efficiency in the beef and poultry industry on the North American continent. The importance of diethylstilbestrol as a growth promoter has been demonstrated by Monfort (1974), who has calculated that lifting of the ban on the use of DES (imposed in 1973 and lifted in 1974 in the United States) is equivalent to adding 3 to 5 million acres to the production of the 1974 corn crop in the United States.

The following hormonal growth promoters are permitted on the North American continent for use on certain livestock in strictly defined amounts (Umberger, 1975; Health and Welfare Canada, Bureau of Veterinary Medicine, 1978 personal communication).

In Canada	In the United States
Feed additives	Feed additives
1. Melengestrol acetate	1. Diethylstilbestrol
(Diethylstilbestrol was withdrawn	2. Melengestrol acetate
effective January 1973)	Implants
Implants	1. Diethylstilbestrol
1. Estradiol benzoate in combination	2. Estradiol benzoate in combination
with progesterone	with progesterone
2. Estradiol benzoate in combination	3. Estradiol benzoate in combination
with testosterone propionate	with testosterone propionate
	4. Estradiol monopalmitate

D. Anabolic Responses in Fish

In this section, only those anabolic responses which affect muscle or bone mass will be considered. Growth of specific tissues normally under control of sex hormones is beyond the scope of this chapter. The effect of anabolic steroids on the mammalian organism has been reviewed exhaustively by Krüskemper (1968).

It is a matter of record that the response to any hormonal agent will be influenced by a number of factors associated with either mode of administration or the internal and external environment of the animal. The studies undertaken which relate to the anabolic effect on fish of steroids and functionally related substances are few and, thus, the effects of these factors remain largely unknown. However, some tentative conclusions can be drawn. Anabolic responses are summarized in Table VIII and side effects, when noted, in Table X. To increase the general usefulness of the tables, many of the compounds will be identified by their trivial or common names.

1. ANDROGENS

a. Species. Reports of anabolic effects of androgens and androgen analogues have been limited to eight salmonid and seven nonsalmonid species. Early studies were mainly concerned with guppy and platyfish, species which are readily adaptable to laboratory conditions. Later studies involved species of greater commercial value reflecting a change in interest from the purely academic question of effects on sexual character and sex reversal to a more practical outlook on the possibility of applying androgens as a means of improving the economics of fish culture. Unfortunately, in a number of the earlier studies the anabolic responses were of secondary interest and were inadequately recorded.

The recorded metabolic effect on the guppy, *Lebistes reticulatus* (Eversole, 1939, 1941; Scott, 1944; Clemens *et al.*, 1966), and the platyfish, *Platypoecilus maculatus* (Cohen, 1945), was generally one of growth suppression of females. In these species the female normally attains a larger size than the male. When treated with androgens the growth of the males was not affected but that of the females was stunted, and the treated females approached in size the normally smaller males. The only exception to this trend was noted by Svärdson (1943), who found that when sexually mature *Lebistes* females were treated with testosterone propionate there was a statistically significant increase in length. As mentioned in Section III,D,3, estrogens enhanced the growth of male guppies and platyfish and caused them to approach normal females in size. The growth inhibiting effect of androgens in *Lebistes* and *Platypoecilus* may thus be regarded as a sex-related response.

A negative growth response to androgens was also reported by Ashby (1957) in brown trout, *Salmo trutta*. This report and those mentioned in the previous paragraph are the only cases of a definite growth inhibition caused by moderately low doses of androgens. In the majority of studies involving the *Salmonidae* the growth response was positive.

b. Choice of Androgen. The anabolic effects of thirteen androgenic steroids have been investigated in fish. The synthetically produced 17α-ethynyltestosterone (pregneninolone) has progestational and weakly estrogenic effects in mammals. However, in three species of fish this hormone induced distinct androgenic effects and has therefore been included with the androgens.

The majority of androgens tested induced anabolic responses. Synthetic steroids, which have been developed with the purpose of maximizing the anabolic and minimizing the androgenic response in mammals have generally shown the best anabolic response of all androgens when tested in fish. To date, the most frequently used androgen has been the synthetic steroid, 17α-methyltestosterone. This steroid enhances biological activity, both anabolic and androgenic, in mammals. When tested in fish, *Tilapia* (Guerrero, 1975), goldfish (Yamazaki, 1976), and all salmonids responded with weight increases. A negative growth response occurred only in the guppy (Clemens *et al.*, 1966).

Two naturally occurring androgens, testosterone and 11-ketotestosterone, which have been identified in salmon plasma (Idler *et al.*, 1960; Schmidt and Idler, 1962; Grajcer and Idler, 1963), were found to have anabolic properties when tested in coho salmon. Not unexpectedly, 11-ketotestosterone, which possesses ten times the androgenic activity of testosterone in *Oryzias latipes* (Arai, 1967), induced greater changes in the testes than testosterone (McBride and Fagerlund, 1976). In the same study administration of the two steroids, when tested against 17α-methyltestosterone, resulted in body weight gains approximately 50% (testosterone) and 65% (11-ketotestosterone) of those attained with 17α-methyltestosterone. In agreement with results obtained with mammals, not only the anabolic, but also the side effects of 17α-methyltestosterone were greater than those of the natural androgens.

c. Steroid Administration. The anabolic effects of the androgens tested have been shown by a number of investigators to be dose-dependent. In the section dealing with side effects, it is noted that all androgens and particularly the synthetic compounds, when adminis-

Species	Steroid	Route	Administration Fre-quency	Administration Dura-tion (days)	Dose
	(A) Androgens				
Salmo salar (Atlantic salmon)	Ethylestrenol	Diet	NR[c]	NR	2.5 mg/kg
S. trutta (brown trout)	Testosterone	Aquarium water	Daily	111	50–60 μg/liter (50 fish)
S. gairdneri (rainbow trout)	Methenolone acetate	Diet (dry)	Daily	48	5 mg/kg
	4-Chloro-testoster-one acetate	im[d]	Every 4 days	30	1 mg 2.5 6.25 12.5
		im	Every 4 days	30	0.5 1 2.5
	Ethylestrenol	Diet	NR	NR	2.5 mg/kg
	Dimethazine	Diet (dry)	Daily	48	5 mg/kg 10 20
S. gairdneri (steelhead trout)	17α-Methyl-testosterone	Diet (moist)	Daily	285	1 mg/kg 2
S. gairdneri irideus (horai masu)	17α-Methyl-testosterone	Diet	Daily	28 49	1 mg/kg
Oncorhynchus nerka (kokanee)	17α-Methyl-testosterone	Diet	Daily	85	1 mg/kg 10 30
O. kisutch (coho salmon)	Testosterone	Diet (moist)	Daily	244	1 mg/kg 10
	11-Ketotestosterone	Diet (moist)	Daily	255	1 mg/kg 10
	17α-Methyl-testosterone	Diet (moist)	Daily	42	1 mg/kg 10 50
		Diet (moist)	Daily	53 35	100 mg/kg 500
		Diet (moist)	Daily	392	0.2 mg/kg 1 10
		Diet (moist)	Daily	269	0.2 mg/kg 1 0.2 1
		Diet (moist)	Daily	371 256	1 mg/kg 1
		Diet (moist)	Daily	221	1 mg/kg
		Diet (moist)	Daily	255	1 mg/kg 10
	Oxymetholone	Diet (moist)	Daily	244	1 mg/kg 10

VIII

and Substances Related in Function

	Gain or loss as % of control					
Final weight	Group increment over initial weight	Statistics[a]	Initial age, weight, or length	Temperature[b] (°C)	Experimental conditions	Reference
Enhanced			NR	NR		Simpson (1976a)
−13.2			Alevins	8		Ashby (1957)
0			NR	NR		Matty (1975)
	12.8	NS	103–125 g	8	Sexually active season.	Hirose and Hibiya (1968b)
	86.4–125.8	S				
	36.3					
	44.4					
	39.2	NS	103–105	10	Sexually inactive season.	
	50.0	NS				
	96.9	S				
Enhanced			NR	NR		Simpson (1976a)
14.7		S	3.5 g	16–18		Matty and Cheema (1976 unpublished)
8.4		NS				
3.0		NS				
47.5			1.1 g	9.5→13.1→5.2		Fagerlund and McBride (1977)
36.5			1.2			Yamazaki (1976)
19.0			3.8 g	8		
23.3			0.5			
18.4			50–51 g	8		
0						
Retarded						
30.1			2.0 g	8→13→5*		McBride and Fagerlund (1976)
41.1			2.0			
31.3			1.4 g	7→12→5**		
77.0			1.4			
18.6			3.7 g	15→11		McBride and Fagerlund (1973)
29.0			3.8			
27.7			3.9			
11.5 (104 days)			0.4 g	8.5→12		Fagerlund and McBride (1975 unpublished)
10.4 (104 days)			0.4			
.2			0.8 g	11.5→5.5→9.8		Fagerlund and McBride (1975)
1.5			0.8			
24.9			0.7			
3.7			1.4 g	11.5***		Fagerlund and McBride (1977)
5.6			1.3			
04.8			1.2	16.5***		
48.8			1.4			
3.7			0.9 g		SW[e] adapted (see Fig. XI)	
0.8			1.0			
6.9			7.9 g	11→2.8→5.5	Hatchery trial	McBride and Fagerlund (1976)
8.0			1.3 g	7→12→5**		
0.0			1.3			
4.3			2.0 g	8→13→5*		
0.3			2.0			

(Continued)

Table VIII—

| Species | Steroid | Route | Administration | | |
			Frequency	Duration (days)	Dose
	4-Chloro-testosterone acetate	Diet (moist)	Daily	255	1 mg/kg 10
O. tshawytscha (chinook salmon)	17α-Methyl-testosterone	Diet (moist)	Daily	56 84	0.2 mg/kg 1
O. keta (pink salmon)		Diet (moist)	Daily	44	0.2 mg/kg 1 10
		Diet (moist)	Daily	182	0.2 mg/kg 1
Carassius auratus (goldfish)	Testosterone propionate	ip im	Every 4 days	29 22	0.5 mg 0.2 0.5 1
	17α-Methyl-testosterone	Diet	Daily	56	1 mg/kg 10 30
	Methylandros-tenediol	im	Every 4 days	20–30	0.2 mg 0.5
	4-Chloro-testosterone acetate	im	Every 4 days	20–30	0.25 mg 0.5 0.5 0.5 1.2
		im	Every 4 days	20–30	0.2 mg 0.5 0.5 1
	Stanozolol	Diet	Daily	28 14 28	8.3–83.3 mg/ 833 833
Ictalurus punctatus (channel catfish)		Diet	Daily	56	8.3–833 mg/k
Platypoecilus maculatus		Aquarium water	Weekly	56–140	5 mg/19 fish
Lebistes reticulatus (guppy)	Testosterone propionate	ip	Twice weekly	7–120	0.03 mg
		Diet	Daily	60–105	0.5 mg/16 fis
		Diet	Daily	30	Total of 75 m
		Diet	Daily	NR	NR
		Diet	Daily	36	Total of 64.3
	17α-Methyl-testosterone	Diet	Daily	60	20–30 mg/kg
	17α-Ethynyl-testosterone	Aquarium water	Single dose	18–110	5 mg
		Aquarium water	Weekly	14–56	5 mg/16 fish
		Aquarium water	Weekly	44–50	10 mg/10 fis

Gain or loss as % of control						
Final weight	Group increment over initial weight	Statistics[a]	Initial age, weight, or length	Temperature[b] (°C)	Experimental conditions	Reference
−6.2			1.3 g	7→12→5**		
−8.1			1.4			
8.0			0.8 g	8→18.5→17		McBride and Fagerlund (1973)
17.2			0.8			
22.8			0.3 g	8.2→9.2	SW	Fagerlund and McBride (1977)
24.1			0.3			
32.9			0.3			
28.4			26.3 g	9.0→8.1	SW	
80.0			23.2			
	14.9	NS	3.4 g	Room temp.	Sexually inactive season.	Hirose and Hibiya (1968a)
	−20.1	NS	9.6–9.7 g			
	12.9	NS				
	6.2	NS				
26.3			5.8–6.7 g	18–22		Yamazaki (1976)
9.1						
−1.1						
	16.6		4.3–5.7 g	Room temp.	Sexually active season.	Hirose and Hibiya (1968a)
	1.8, 27.7					
	−1.3		4.3–6.9 g	20	Sexually active season.	
	24.7–68.3					
	51.4		10.5 g, ♂			
	35.2		10.8 g, ♀			
	7.3		11.7 g, ♀			
	37.0		3.2 g	20	Sexually inactive season.	
	55.5		3.2			
	44.8		8.9			
	25.7		3.2			
Enhanced		NS	21–64 g	21		Bulkley and Swihart (1973)
8.5		S				
Enhanced		NS				
0		NS	2–5 g	25		
−29.2 (length)			2 weeks, ♀	25		Cohen (1945)
Retarded			2 months, ♀	25.6		Eversole (1939)
Retarded			Newborn, ♀	25.6		
−5.9 (length)		S	Newborn	25		Svärdson (1943)
Retarded			Maturing, ♀	25		
	37.1 (length)	S	Sexually mature, ♀			
−18.2–19.9 (length)				21	Measured after 146 days.	Clemens et al. (1966)
Retarded			Newborn	23.3–25.6		Scott (1944)
−25 (length)			Newborn	25.6		Eversole (1941)
	Growth ceased		Sexually mature, ♀	25.6		

(Continued)

Table VIII—

Species	Steroid	Route	Frequency	Duration (days)	Dose
			Administration		
Tilapia	1-Dehydro-testosterone acetate	Diet	Daily	21	15 mg/kg
					30
					60
	17α-Methyl-testosterone	Diet	Daily	21	15 mg/kg
					30
					60
	17α-Ethynyl-testosterone	Diet	Daily	21	15 mg/kg
					30
					60
T. mossambica	17α-Methyl-testosterone	Diet	Daily	14	30 mg/kg
				21	30
				28	30
T. melanopleura	Ethylestrenol	Diet	NR	NR	NR
Mugil auratus	Testosterone	Diet	Daily	NR	1000 mg/kg
	(B) Estrogens				
S. trutta	Estradiol	Aquarium water	Daily	111	50 μg/liter
				70	300
S. gairdneri (rainbow trout)	Estrone	Diet	Daily	30–90	50–150 mg/kg
		Diet	Daily	180	50 mg/kg
				NR	
	Diethylstilbestrol	Diet (dry)	Daily	30	1.2 mg/kg
O. kisutch	Estradiol	Diet (moist)	Daily	56	1 mg/kg
					10
					50
	Diethylstilbestrol	Diet (moist)	Daily	112	1 mg/kg
					10
					50
		Diet (moist)	Daily	49	100 mg/kg
					500
		Diet	Daily	159	200 mg/kg
O. gorbuscha	Estradiol	ip	3×/week	210	1.5 μg/g BW
					15
C. auratus	Estradiol	Diet	Daily	56	1–10 mg/kg
					30
I. punctatus	Diethylstilbestrol	Diet	Daily	25	0.5 mg/kg
					4.5
					45.4
P. maculatus	Estradiol benzoate	Aquarium water	Weekly	56–140	1 mg/15 fish
L. reticulatus	Estradiol	Aquarium water	Weekly	18–110	Total of 6.5–15 mg/ 2.5 gal.

Continued

Gain or loss as % of control						
Final weight	Group increment over initial weight	Statistics[a]	Initial age, weight, or length	Temperature[b] (°C)	Experimental conditions	Reference
57.1		S	9–11 mm	21	Weighed 120 days after hormone withdrawal.	Guerrero (1975)
32.7		NS				
21.4		NS				
24.2		NS				
43.6		S				
36.7		S				
37.5		S				
7.7		NS				
21.2		NS				
82.8			9–11 mm	NR	Weighed 90 days after treatment.	Guerrero (1976)
180.0						
106.9						
0			NR	NR		Hutchison and Campbell (1964)
−12.2						Bonnet (1970)
−14.7 (length)			Alevins	8		Ashby (1957)
Retarded					High mortality.	
Inconclusive			27 days after hatching.	10–11		Scidmore (1966)
−8.2			Juveniles	8–9		Ghittino (1970)
Retarded			2–3 cm	15–16		
−10.6			13.1 g	13–15		Matty (1975); Matty and Cheema (1976 unpublished)
0.7			8.0 g	6→1.5		Fagerlund and McBride (1975 unpublished)
5.5			7.9			
−2.6			7.9			
0			0.8 g	8.5→12		
−19.4			0.7			
−44.3			0.8			
−59.0 (118 days)			0.3 g	8.5→12		
−74.3 (118 days)			0.4			
−23.9			24 g	NR		Lorz (1971)
0			Fry	Ambient	SW adapted.	Funk (1972)
Retarded					With or without salmon gonadotropin.	Funk et al. (1973)
0			5.7–7.2 g	18–22		Yamazaki (1976)
−17.8						
	−34.9	S	37.1–48.5 g	25	Sexually immature.	Bulkley (1972)
	−96.0	S				
	−102.1	S				
13.7 (length)			2 weeks, ♂	25		Cohen (1945)
0			NR	23.3–25.6		Scott (1944)

507

(Continued)

Table VIII—

| Species | Steroid | Route | Administration | | Dose |
			Fre-quency	Dura-tion (days)	
	Estrone + estriol		3 times weekly	30–150	45 rat units
	Estrone	Diet	Daily	35	Total of 300,000 International Benzoate Units.
		Diet	Daily	NR	NR
		Diet	Daily	18	Total of 100,000 IBU.
		Diet	Daily	14	Total of 50,000 IBU.
Pleuronectes platessa (plaice)	Diethylstilbestrol	Diet	Daily	70	0.6 ppm 1.2 2.4
	(C) Progestogens				
S. *trutta*	Progesterone	Aquarium water	Daily	84	50–100 μg/liter
O. *kisutch*		Diet (moist)	Daily	255	1 ppm 10 ppm
L. *reticulatus*		Injected	Weekly	56 49	0.025–0.25 mg 0.1 mg
	Pregnanediol	Injected	Weekly	56 49	0.025–0.25 mg 0.1 mg
	(D) Corticosteroids				
S. *trutta*	Deoxycorticoster-one acetate	Aquarium water	Daily	112	100–400 μg/liter
L. *reticulatus*		Injected	Weekly	56 49	0.025–0.25 mg 0.1 mg

[a] NS, not significant ($P > 0.05$); S, significant ($P \leq 0.05$).
[b] Experiments marked with the same asterisk superscripts were run simultaneously under identical conditions.
[c] NR, not recorded.

tered in high concentrations exert deleterious effects on various organs. It may be anticipated that these effects will, when the hormone reaches a certain concentration, interfere with the normal processes of these organs to the extent that they will cancel any gain in anabolic response expected by the increase in dose. This phenomenon has been observed by some investigators. Yamazaki (1976) found the effective

Continued

Gain or loss as % of control						
Final weight	Group increment over initial weight	Statistics[d]	Initial age, weight, or length	Temperature[b] (°C)	Experimental conditions	Reference
Gain			Newborn, ♂	25.6		Berkowitz (1938)
0			Newborn, ♀			
−5.6 (length)		S	Newborn, ♂, ♀	NR		Svärdson (1943)
16 (length)		S	Sexually maturing, ♂	NR		
	−75.2 (length)		Sexually mature, ♀	NR		
	+97.0 (length)	S	Sexually mature, ♂	NR		
35.7		S	8 g	15		Cowey and Sargent (1972); Cowey et al. (1973)
7.1		S				
−6.3						
Gain			Alevins	8		Ashby (1957)
0			1.4	7→12→5		McBride and Fagerlund (1976)
−10.5			1.3			Eversole (1941)
0			14 days	25.6		
0			Sexually mature.	25.6		
0			14 days	25.6		
0			Sexually mature.	25.6		
5.9 (length)			Alevins	8		Ashby (1957)
0			14 days	25.6		Eversole (1941)
0			Sexually mature.	25.6		

[d] im, intramuscular; ip, intraperitoneal.
[e] SW, seawater; salinity, 22–28‰.

concentration of 17α-methyltestosterone to be 1 ppm when fed to gold-fish. The growth rate of fish fed 10 ppm of this steroid was decreased and those on 30 ppm lost weight. McBride and Fagerlund (1973) noted that a dose of 10 ppm of the same steroid induced an increase in growth rate when fed to coho salmon for 6 weeks, but no further gain was obtained if the dose was increased to 50 ppm. At concentrations of

100 ppm or higher these fish grew at a lower rate than the controls, a response which was evident after 5 weeks of treatment. Also, mortalities occurred in the latter group (Fagerlund and McBride, 1975 unpublished observations). The negative growth response obtained by Ashby (1957) in brown trout with testosterone was probably due to the high dose (200 ppm). Evidence for this may be found in the high degree of tissue abnormalities which occurred.

The lowest dose of 17α-methyltestosterone, 0.2 ppm, tested with coho salmon was effective in some tests (Fagerlund and McBride, 1977) but not in others (Fagerlund and McBride, 1975). This may be ascribed to differences in experimental conditions, such as temperature and photoperiod.

Tilapia appear to require higher doses than salmonids. Guerrero (1975, 1976) obtained anabolic responses when 17α-methyltestosterone was fed in doses of either 30 or 60 ppm. 1-Dehydrotestosterone acetate and 17α-ethynyltestosterone were most effective at a lower dose, 15 ppm.

A number of investigators have administered the steroids by dispersing or dissolving the compounds in the aquarium water. Steroids as a rule have a low solubility in water. This, however, was not a problem in experiments with guppies and platyfish (Eversole, 1941; Cohen, 1945), since these fish readily accepted pieces of crumbled steroid tablets that were sprinkled on the water surface. Administration of the steroid in this manner is rarely applicable since most species are very selective in their choice of food. The injection technique, either im or ip, has the advantage of bringing the main portion of the dose rapidly into the circulatory system. However, by being brought more speedily into circulation the hormone may also be metabolized and eliminated from the body at a faster rate (see below).

When the economics of large-scale fish culture are considered, a serious disadvantage of the injection method of administration is the expense of either acquiring automated injection equipment or defraying the high labor costs which would be encountered when large numbers of fish are injected manually.

In this context it may be worthwhile to examine the cost of hormone supplementation of fish food. Assuming a food conversion rate of 3 : 1 (kg moist food : kg of wet fish) to produce one metric ton of fish would require three metric tons of food. If the food contains 1 ppm of 17α-methyltestosterone 3 g of hormone would be consumed at a retail cost of less than $3.00.

   d. *Effect of Environmental Conditions.* Of the three basic environmental factors which influence the anabolic effect of hormones on fish, namely temperature, salinity, and photoperiod, only the first two have received attention. The growth process in fish is highly susceptible to changes in temperature as may be expected of poikilothermic animals. Salmonids, being residents of northern regions, spend a large part of their life in cold water and the temperature range for optimum growth for most species is below 20°C. The growth rates of coho salmon fed 17α-methyltestosterone at 11.5° and 16.5°C (Fig. 9) have been compared by Fagerlund and McBride (1977). After 9 weeks of treat-

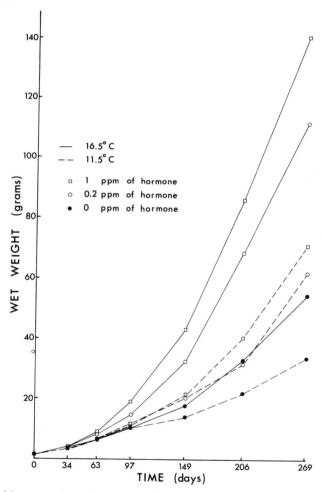

Fig. 9. Mean weights of coho salmon receiving 17α-methyltestosterone-supplemented diets while maintained in freshwater of 11.5° and 16.5°C. (From Fagerlund and McBride, 1977, *Fish. Mar. Serv. Tech. Rep. No. 716.*)

ment at 11.5°C, oral doses of 0.2 and 1 ppm induced weight gains over controls of less than 10%, but at 16.5°C the gains were 21 and 32.3%, respectively. This trend continued through the experiment. At the termination of the study, after 38 weeks of test feeding, the gain attributable to the increase in temperature was 61.8% when untreated groups were compared, 105.9% for groups receiving 1 ppm of steroid and the combined effect of the hormone at 1 ppm and a 5°C rise in temperature was a weight gain of over 300%. Unfortunately, the elevated temperature also enhanced associated side effects in the testes (Table IX).

The administration of androgenic steroids may interfere with the ability of the animal to adapt to changes in water salinity. Fagerlund and McBride (1975) found that coho salmon receiving 10 ppm of 17α-methyltestosterone orally showed an initial decrease in growth rate when transferred into seawater and finally a weight loss (Fig. 10).

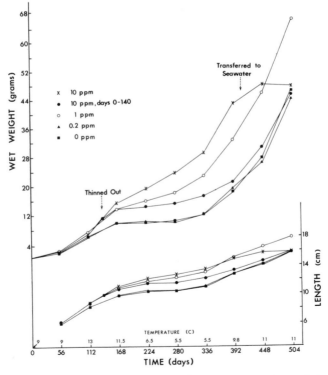

Fig. 10. Mean weights and lengths of coho salmon receiving 17α-methyltestosterone-supplemented diets. Fish were transferred to seawater at a time of year when coho smolts normally migrate to sea. (From Fagerlund and McBride, 1975, *J. Fish Biol.* 7, 305–314.)

Mortalities also occurred in this group. The growth rate of coho receiving 1 ppm of the steroid decreased slightly, following transfer into saltwater, although they still maintained a weight gain of 41% after 9 weeks in seawater (Fagerlund and McBride, 1975 unpublished observations). Saltwater adaptation during androgen administration was also studied in an experiment (Fagerlund and McBride, 1977) in which underyearling coho were acclimated to seawater 115 days after emergence from the gravel and after having received 1 ppm of 17α-methyltestosterone for 92 days (Fig. 11). These fish initially responded to the saltwater transfer with a decrease in growth rate, but 83 days later, after the density of fish in the tank was decreased, the hormone-treated fish again gained weight faster than the control

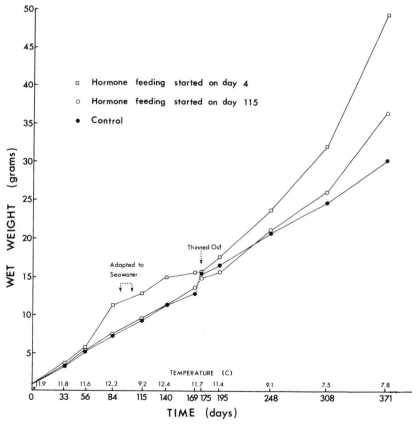

Fig. 11. Mean weights of coho salmon receiving 17α-methyltestosterone-supplemented diets. The fish were acclimated to seawater as underyearlings. (From Fagerlund and McBride, 1977, *Fish. Mar. Serv.* Tech. Rep. No. 716.)

group. The same authors (Fagerlund and McBride, 1974 unpublished observations) have also made the observation that if hormone treatment is terminated several weeks prior to seawater transfer, coho salmon smolts which have received 5 ppm of 17α-methyltestosterone do not suffer adversely from the transfer.

Pink salmon (*O. gorbuscha*), which in the wild migrate into seawater immediately after emerging from the gravel, responded well to low doses of 17α-methyltestosterone. No apparent adverse effects were noted when administration of the steroid commenced either shortly after emergence from the gravel, or at the time the fish weighed 26 g (Fagerlund and McBride, 1977). The conclusion to be drawn from the evidence available is that raised androgen levels during the time of saltwater acclimation interfere with the animals osmotic regulation, but this effect may be avoided if androgen administration is withdrawn prior to the transfer or is initiated some time after saltwater adaptation has taken place.

2. ESTROGENS

Estrogens, in particular diethylstilbestrol, have been very effective as growth promoters in cattle. It is therefore surprising that a growth response in the fish examined so far (teleosts) has been either lacking or negative. A few exceptions to this trend have been noted. A positive response was recorded by Cowey and Sargent (1972) and Cowey *et al.* (1973) in plaice (Table IX). Berkowitz (1938) working with the guppy and Cohen (1945) with the platyfish reported gains in size of males only. The males of these two latter species, however, are normally smaller than the females and the effect on size may be regarded as a feminizing effect in the same manner as the growth retardation caused by androgens in females may be regarded as a masculinizing effect.

Table IX

The Effect of Oral Diethylstilbestrol on the Growth of Small Plaice[a]

Diethylstilbestrol (mg/kg dry diet)	Initial weight		
	5 g	8 g	11 g
0.0	5.03	9.67	14.30
1.2	9.69*	15.97*	22.26*
2.4	7.39*	10.93	14.47
5.0	5.53	8.55	10.55

[a] Weight gains (in grams) of fish over a 10-week period. From Cowey and Sargent (1972, *Adv. Mar. Biol.* **10**, 383–492).

* Significantly different from control.

The contrast between the positive growth response obtained with plaice when 0.6 ppm of diethylstilbestrol was used and the generally negative response in other studies has been attributed by Cowey and Sargent (1972) to a difference in dose. However, Bulkley (1972) recorded a statistically significant weight reduction in channel catfish receiving 0.5 ppm of the same drug. Also, Fagerlund and McBride (1975 unpublished observations) did not obtain an effect with 1 ppm in coho salmon. A species-specific growth response is thus indicated.

Bulkley (1972) and Fagerlund and McBride (1975, unpublished observations) found the growth retarding effect of diethylstilbestrol in concentrations of 0.1–50 ppm to be dose dependent, while a number of authors noted that estradiol, when fed in doses ranging from 1 to 150 ppm, did not have a significant effect on growth. The decreased growth of fish receiving elevated doses of diethylstilbestrol is possibly attributable to a toxic effect, since Fagerlund and McBride observed extensive degeneration of the kidneys of coho salmon receiving a dose of 50 ppm of this hormone for 112 days. The effect of diethylstilbestrol on growth of hatchery-reared coho salmon juveniles and the subsequent return of adults was examined by Lorz (1971). Feeding of 200 ppm of the drug for 159 days resulted in a weight loss of 23.9% relative to a control group. The number of returning adults from the hormone-treated group was only 30% of the control return, but the number of returning precocious males in the hormone-treated group dropped to a quarter of the number in the control groups.

3. PROGESTOGENS AND CORTICOSTEROIDS

These steroids have received scant attention in the context of anabolic effects in fish. Neither group of compounds has been found to elicit a significant growth response with the exception of a report by Ashby (1957) of a substantial gain in length, obtained with brown trout alevins S. trutta. These fish received 50–100 μg/liter of progesterone daily in the aquarium water, a dose which is considerably higher than the doses of 1 and 10μg/g of food given by McBride and Fagerlund (1976) to coho salmon O. kisutch. Response in the latter study was absent and weakly negative, respectively. The difference in dose level between the two experiments was also evident from the powerful inhibitory effect on the reproductive system in the trout study, which contrasts with the lack of effect on gonadal tissue in the coho. It is interesting to note that the growth response to a high dose of progesterone such as the one used by Ashby was positive, while a similar dose of either estradiol or testosterone was toxic and resulted in decreased growth.

4. Growth in Length versus Growth in Weight

In the previous sections growth has been expressed as an increase in either weight = soft tissue growth or length = bone growth. The relationship of weight to length is commonly termed *condition factor* and expressed as

$$a = \frac{W}{L^b} \times 1000$$

where W is the weight in grams, L is the length in centimeters, and b is a factor that is specific for each species. For coho salmon fry this factor has been determined by Vanstone and Markert (1968) to be 3.25.

Sex hormones cause differences in the growth rates of weight and length in Pacific salmon. In coho salmon feeding testosterone, oxymetholone, 11-ketotestosterone, 4-chlorotestosterone, progesterone (McBride and Fagerlund, 1976), or 17α-methyltestosterone (Fagerlund and McBride, 1975, 1977; McBride and Fagerlund, 1976) in doses of 0.2 or 1 ppm resulted in condition factors which were inversely proportional to the dose. However, a 10 ppm dose of 17α-methyltestosterone (Fagerlund and McBride, 1975) reversed this relationship and brought about a condition factor substantially greater than those obtained with lower doses.

Fagerlund and McBride (1977) have demonstrated by statistical analysis that the decrease in condition factor occurring when the 17α-methyltestosterone dosage was increased from 0 to 1 ppm was not a direct effect of hormone treatment but rather the natural result of increasing size.

Diethylstilbestrol administered in doses ranging from 10 to 500 ppm of food caused substantially increased condition factors as a result of depressed growth in length (Fagerlund and McBride, 1975 unpublished observations).

5. Effect of Anabolic Steroids on Appetite, Food Conversion Efficiency, and Muscle Composition in Fish

Questions concerning the effects of these anabolic steroids on appetite, food conversion efficiency, and muscle composition have, to date, received only very limited attention. Fagerlund and McBride (1975) reported a decrease in the amount of flesh as well as a significant increase in flesh lipid content in juvenile *Oncorhynchus kisutch* fed a diet enriched with 10 mg/kg of 17α-methyltestosterone. Water content of the flesh was not altered. No differences from the controls were noted with lower doses of this steroid. In experiments with steelhead trout and pink salmon (Fagerlund and McBride, 1977), lower doses increased lipid and decreased moisture content significantly. On the other hand, supplementation of the diet with either testosterone or oxymetholone, at concentrations of 1 and 10 mg/kg, significantly lowered the flesh lipid content (McBride and Fagerlund, 1976). Whether these anabolic agents affect the organoleptic qualities of the flesh remains to be ascertained.

An area that remains to be explored is the increase in behavioral activity associated with the application of many of the sex steroids (Hoar *et al.*, 1952; Forselius, 1957). Whether a portion of the increased activity is translated into an increase in feeding response (i.e., appetite) warrants a close examination. Yamazaki (1976) noted that total food consumption was proportional to the final body weight in *Carassius auratus* fed diets enriched with 1–30 mg/kg of 17α-methyltestosterone for 56 days. In a hatchery trial (McBride and Fagerlund, 1976; Fagerlund and McBride, 1975 unpublished observations) a group of over 9000 coho fry, fed to satiation for 221 days a diet of moist pellets containing 1 ppm of 17α-methyltestosterone, consumed nearly the same amount of food as the control group, while gaining 44.3% in weight over the control. An improvement in food conversion efficiency from 0.292 to 0.422 kg wet weight gain/kg food was indicated.

E. Side Effects Induced by Anabolic Steroids and Related Substances

It is not our intention to review the voluminous literature concerned with the effects of sex steroids on primary and secondary sexual characteristics. For summary of this topic the reader is referred to the excellent accounts of Dodd (1955) and Hoar (1957, 1965). Here, our concern is with the possible significance of these side effects in assessing the potential for anabolic steroids in fish cultivation.

Table X summarizes the changes associated with anabolic steroids. While the investigations are relatively few in number and the data concerning the side effects are both fragmentary and often contradictory a number of important points can be raised.

First, many of the secondary sexual characteristics described are similar to those associated with normal sexual maturation. The androgens evoke a masculinizing effect in contrast to the feminizing responses to the estrogens. In those cases where a response was indicated in the gonads, the effect was generally one of suppression or degeneration. There are, however, a number of reports on induction or

Table X

Side Effects Induced by Anabolic Steroids and Related Substances

Steroid	Administration			Species	Side effects	Anabolic response[a]	Reference
	Route	Duration (days)	Dose				
(A) Androgens							
Testosterone	Alcohol solution added to aquaria water.	111	50–60 μg/liter	S. trutta	Development of germinal tissue suppressed; renotrophic changes.	–	Ashby (1957)
	Diet	244	1–10 mg/kg	O. kisutch	Darkening of skin noted in a few fish retained on highest dosage. No change noted at lower dosage. Gonads unaffected.	+	McBride and Fagerlund (1976)
Testosterone propionate	ip, im	22	0.2–1.0 mg	C. auratus	No change noted.	0	Hirose and Hibiya (1968a)
	ip	7–120	0.03 mg per injection	L. reticulatus	Injection route most effective. In females steroid induced typical black spots posterior to pectoral fins but failed to induce other typical male nuptial skin pigmentation, ova genesis suppressed. In males secondary sexual characteristics not altered, spermatogenesis inhibited with injections.	– (Females) ? (Males)	Eversole (1939)
	Diet	60–105	0.5 mg/day	L. reticulatus			
	ip	56	0.025–0.25 mg/ week	L. reticulatus	Inhibited male maturation, coloration; testes degenerated, ovary development inhibited.		Eversole (1941)

Compound	Route	N	Dosage	Species	Effects	Effect	Reference
	Diet	30 (newborn) 36 (adults)	75 mg total	L. reticulatus	In juveniles, induced secondary male sexual characteristics; accelerated development of testes, but suppressed ovary ripening. Adult females masculinized. No clear effect in adult males.	− (Newborn) − (Adult maturing females) + (Adult mature females) ? (Adult males)	Svärdson (1943)
1-Dehydrotestosterone acetate	Diet	21	15–60 mg/kg	T. aurea	No change noted.	+	Guerrero (1975)
11-Ketotestosterone	Diet	255	1–10 mg/kg	O. kisutch	Maximum dosage induced degeneration of the testes; darkening of skin color and altered head formation. Changes less pronounced at lower dosage with examples in the testes of both degeneration and spermatogenesis.	+	McBride and Fagerlund (1976)
Methenolone acetate	Diet	48	5 mg/kg	S. gairdneri (rainbow trout)	Not recorded.	0	Matty (1975)
17α-Methyltestosterone	Diet	285	1–2 mg/kg	S. gairdneri (steelhead trout)	Changes in skin color evident at higher dose; testes in state of spermatogenesis.	+	Fagerlund and McBride (1977)
	Diet	28–49	1.0 mg/kg	S. gairdneri irideus	Not recorded.	+	Yamazaki (1976)
	Diet	300–330	1.0–30.0 mg/kg	O. nerka (kokanee)	Not recorded.	+ to −	Yamazaki (1976)
	Diet	42	1.0–50.0 mg/kg	O. kisutch	Higher concentrations induced a loss in numbers of spermatogonia and a thickening of skin. Ovary unaffected.	+	McBride and Fagerlund (1973)
	Diet	504	0.2–10 mg/kg	O. kisutch	Maximum dosage resulted in increased mortalities, degeneration and possibly sterilization of testes, increase in lipid content of muscle and alterations in external features. Effects less obvious at lower dosages.	+	Fagerlund and McBride (1975)

(Continued)

Table X—Continued

Steroid	Administration			Species	Side effects	Anabolic response[a]	Reference
	Route	Duration (days)	Dose				
	Diet	35–53	100–500 mg/kg	O. kisutch	Marked degeneration noted in testes, ovary, and kidney. Pronounced changes in color of skin. Mortality high.	−	Fagerlund and McBride (1975 unpublished)
	Diet	252	0.2–1.0 mg/kg	O. kisutch	Constant water temperatures of 11.5° and 16.5°C. Maximum dose induced either spermatogenesis or degeneration in testes, a few ovotestes noted, skin thickness increased. Changes absent or marginal at lower dose.	+	Fagerlund and McBride (1977)
	Diet	221	1.0 mg/kg	O. kisutch	Ambient water temps. (2.8°–11°C). No change observed.	+	McBride and Fagerlund (1976)
	Diet	255	1.0–10.0 mg/kg	O. kisutch	Degeneration of testes; body form altered; skin color darkened. Effect less pronounced at lower dosage. Testes showed examples of both degeneration as well as spermatogenesis.	+	McBride and Fagerlund (1976)
	Diet	56–84	0.2–1.0 mg/kg	O. tshawytscha	Maximum dosage induced hypertrophy of spermatogonia cysts.	+	McBride and Fagerlund (1973)
	Diet	182	0.2–1.0 mg/kg	O. keta	Maximum dose induced spermatogenesis.	+	Fagerlund and McBride (1977)
	Diet	56	1.0–30 mg/kg	C. auratus	Not recorded.	+	Yamazaki (1976)
	Diet	60	20–30 mg/kg	L. reticulatus	Increase in percent males; behavior affected; increased mortality.	−	Clemens et al. (1966)

520

	Diet	15.0–60.0 mg/kg	T. aurea	21	All dosages led to increase in percent males; degeneration of testes at maximum dosage.	+	Guerrero (1975)
Methylandrostenediol	Diet im	30 mg/kg 0.2–0.5 mg	T. mossambica C. auratus	14–28 20–30	Increased percent males. Renotrophic changes including hemolysis. Ovary development suppressed.	+ +	Guerrero (1976) Hirose and Hibiya (1968a)
17α-Ethynyltestosterone	Diet	5–10 mg/week	L. reticulatus	14–56	Induced masculine nuptial skin coloration; accelerated development of the gonopod; accelerated spermatogenesis. Ovary development suppressed.	–	Eversole (1941)
	Added to aquaria water.	5 mg/week	L. reticulatus	18–110	Induced masculine external characteristics.	–	Scott (1944)
	Diet	5 mg/week	P. maculatus	56–140	Oogenesis suppressed; induced masculine secondary characteristics in the female (gonopodia; malelike behavior).	– (Females)	Cohen (1945)
	Diet	15–60 mg/kg	T. aurea	21	Increase in percent males at all concentrations. Testes structure normal.	+	Guerrero (1975)
Ethylestrenol	Diet Diet	2.5 mg/kg 2.5 mg/kg	S. salar S. gairdneri (rainbow trout)	Not recorded Not recorded	Suppressed smolting. Not recorded.	+ +	Simpson (1976a) Simpson (1976a)
Oxymetholone	Diet	1–10 mg/kg	O. kisutch	244	Maximum dosage induced changes in skin color and testes structure. No change with lower dosage.	+	McBride and Fagerlund (1976)
4-Chlorotestosterone acetate	im	1.0–12.5 mg per injection	S. gairdneri (rainbow trout)	30	Suppressed development of the gonads. Lower dosage increased liver glycogen level. Maximum dosage induced pycnosis of interrenal and liver cells, also minor renotrophic changes.	+	Hirose and Hibiya (1968b)

(Continued)

Table X—Continued

Steroid	Administration			Species	Side effects	Anabolic response[a]	Reference
	Route	Duration (days)	Dose				
	Diet	255	1–10 mg/kg	O. kisutch	No change noted.	−	McBride and Fagerlund (1976)
	im	20–30	0.25–1.2 mg per injection	C. auratus	Renotrophic changes; ovary development suppressed. Effects marked at higher dosage level.	+	Hirose and Hibiya (1968a)
Stanozolol	Diet	14–28	8.3–833 mg/kg	C. auratus	Maximum dose induced renal hypertrophy, hepatosomatic index unaffected.	0 to +	Bulkley and Swihart (1973)
	Diet	56	8.3–833 mg/kg	I. punctatus	Not recorded.	0	Bulkley and Swihart (1973)
Dimethazine	Diet	48	5–20 mg/kg	S. gairdneri (rainbow trout)	Renosomatic index increased, hepatosomatic index unaffected.	+	Matty and Cheema (1976 unpublished)
(B) Estrogens							
Estradiol	Alcohol solution added to aquaria water	70–111	50–300 µg/liter	S. trutta	High mortalities; development of gonad suppressed; renotrophic changes; gallbladder collapsed.	−	Ashby (1957)
	Diet	56	1–50 mg/kg	O. kisutch	Maximum dose induced hypertrophy of liver cells, variable hypertrophy of testes and reduction in numbers of spermatogonia.	−	Fagerlund and McBride (1975 unpublished)
	ip	210	1.5–15 µg/g	O. gorbuscha	Gonadosomatic index suppressed, cloaca visibly swollen, mortalities related to dose.	−	Funk et al. (1973)
	Added to aquaria water.	18–110	6.5–15 mg/week	L. reticulatus	Induced external feminization.	0	Scott (1944)
	Diet	56	1–30 µg/g	C. auratus	No change noted.	0 to −	Yamazaki (1976)

522

Estradiol benzoate	Diet	56–140	1 mg/week	*P. maculatus*	Suppressed spermatogenesis; ovotestes formed; mating behavior altered.	+ (Males) – (Females)	Cohen (1945)
Estrone	Diet	30–90	50–150 mg/kg	*S. gairdneri* (rainbow trout)	Ovotestes induced in males retained on maximum dosage.	?	Scidmore (1966)
	Diet	35 (juveniles) 18 (adults)	300,000 International Benzoate Units	*L. reticulatus*	In juveniles sexual maturity delayed. In adult males suppressed testis development and secondary sexual characteristics. In adult females inhibition of ovarian degeneration.	– (Juveniles) – (Maturing females) + (Maturing and mature males)	Svärdson (1943)
Estrone + estriol	Diet	30–120	45 Rat Units 3 times weekly	*L. reticulatus*	Treatment for full period suppressed development of male secondary sexual characteristics; inhibited testes development and induced ovotestes. Female not affected. Partial effect in males with short periods of treatment.	+ (Males) 0 (Females)	Berkowitz (1938)
Diethylstilbestrol	Diet	180	50 mg/kg	*S. gairdneri* (rainbow trout)	No change noted.	–	Ghittino (1970)
	Diet	159	200 mg/kg	*O. kisutch*	Suppressed precocious maturation of the males. No atypical histological features noted in gonads.	–	Lorz (1971)
	Diet	112	1–50 mg/kg	*O. kisutch*	Higher doses induced hypertrophy of the liver and kidney. Maximum dose resulted in marked degeneration of testes and kidney.	0 to –	Fagerlund and McBride (1975 unpublished)
	Diet	49	100–500 mg/kg	*O. kisutch*	Marked hypertrophy of the liver, extensive degeneration in the kidney and testes, ova development possibly impaired.	–	Fagerlund and McBride (1975 unpublished)

(Continued)

Table X—Continued

Steroid	Administration Route	Duration (days)	Dose	Species	Side effects	Anabolic response[a]	Reference
	Diet	25	0.5–45.4 mg/kg	*I. punctatus*	Reduced appetite; renal hypertrophy; edema; increased urine retention. Responses increased with dosage.	–	Bulkley (1972)
	Diet	70	0.6–2.4 mg/kg	*P. platessa*	Not recorded.	0 to +	Cowey et al. (1973)
(C) Progestogens							
Progesterone	Alcohol solution added to aquaria water.	84	50–100 μg/liter	*S. trutta*	Increased mortality; development of germinal tissue suppressed.	+	Ashby (1957)
	Diet	255	1–10 mg/kg	*O. kisutch*	No change noted.	0 to –	McBride and Fagerlund (1976)
Pregnanediol	Injected.	49–56	0.025–0.25 mg	*L. reticulatus*	No change noted.	0	Eversole (1941)
(D) Corticosteroids							
Deoxycorticosterone acetate	Added to aquaria water.	112	100–400 μg/liter	*S. trutta*	Gonad development suppressed, possibly an increase in percentage females.	0	Ashby (1957)
	Injected.	49	0.025–0.25 mg	*L. reticulatus*	No change noted.	0	Eversole (1941)

[a] +, positive; 0, nil; –, suppressive; ?, inconclusive.

acceleration of spermatogenesis following androgen administration. This apparent contradiction, that is, degeneration versus accelerated maturation, is consistent with the rather confused state of the literature concerning the effects of sex steroids on gonadal development (de Vlaming, 1974).

Second, the investigations of Guerrero (1975) with 1-dehydro-testosterone acetate and McBride and Fagerlund (1976) with testosterone and oxymetholone indicate that steroid-induced anabolic responses can be achieved in the absence of any obvious concomitant side effects. It should be noted, however, that in these investigations the examination of possible accessory changes was restricted to gonads and external features. One cannot discount the possibility that other structural changes did occur.

Third, a clear relationship is indicated between steroid dosage on one hand and time of induction as well as extent of responses on the other. The significance of these correlations in terms of selecting the optimum steroid dosage for growth is possibly best illustrated in the studies of Hirose and Hibiya (1968a,b) with 4-chlorotestosterone acetate, and McBride and Fagerlund (1976) with testosterone, 17α-methyltestosterone, 11-ketotestosterone, and oxymetholone. In both investigations, the respective steroid dosages which produced the greatest increases in growth responses also induced marked side effects. Increasing the intensity of feeding also increases the androgenic response at the same time as growth response is increased, as was demonstrated by Higgs et al. (1977a; see also Section VI) when food was offered to satiation during two 20-min periods daily.

A fourth point concerns the suppressive effect of many anabolic steroids on gonad development. The potential of diverting the energy normally utilized in sexual maturation to increasing the muscle component has obvious attractions. The question remains, however, whether one can, in fact, achieve this effect without inducing other deleterious side effects. Hirose and Hibiya (1968a,b) noted degenerative changes in kidney and liver structure of Carassius auratus and Salmo gairdneri exposed to the relatively high dosages of 4-chlorotestosterone acetate required to suppress gonad development. Similar steroid-induced gonad suppression was reported by Ashby (1957), Fagerlund and McBride (1975), and McBride and Fagerlund (1976). The effect of steroid withdrawal on these nongonadal side effects has received little attention. Yamazaki (personal communication) noted that the thickening of the integument in two species of Oncorhynchus (keta and gorbuscha), induced with 17α-methyltestosterone, ceased upon removal of the steroid, and shortly thereafter the skin had reverted to normal. Whether all, or just a portion, of these alterations would revert to normal once the steroid was withdrawn is not known. Fagerlund and McBride (1977) found evidence for partial cessation of spermatogenesis in steelhead trout after the hormone was withdrawn. However, deformities in the shape of the head were not reversed.

Finally, in a majority of the investigations the responses of the gonads to the anabolic steroids have been well documented. This however, was rarely the case with the secondary sexual characteristics. Indeed, the latter were often totally ignored. The fallacy here is that an assessment of the anabolic steroid can be made in the absence of detailed information concerning side effects. A pure anabolic steroid devoid of any side effects has not, to date, been documented.

The preceding items clearly identify the need for the development of a standard methodology that will allow quantitative comparisons of both the anabolic properties and side effects for the different steroids. In the rat, the finding that the involution of the muscles after castration can be reversed by the application of androgens formed the basis for the development of the anabolic–androgenic index (Eisenberg and Gordon, 1950) mentioned above.

The hypertrophy of the renal tubule epithelial cells, shown to be dose dependent with 4-chlorotestosterone acetate in *Carassius auratus* and *Salmo gairdneri* (Hirose and Hibiya, 1968a,b), offers potential as one convenient method of indexing the side effects of steroids. Another approach that may be fruitful is to index the effect of the steroids on the thickness of the skin epithelial layer. McBride and Van Overbeeke (1970) reported a highly significant increase in the thickness of the epidermis of gonadectomized adult *Oncorhynchus nerka* administered 17α-methyltestosterone, 11-ketotestosterone, estradiol, or estradiol cypionate. Similar responses were noted by Idler *et al.* (1961) with 11-ketotestosterone in maturing *Oncorhynchus nerka,* and by Yamazaki (1972) using 17α-methyltestosterone in juvenile *Oncorhynchus kisutch* and *O. gorbuscha.* Taken in conjunction with the anabolic responses (i.e., increase in body growth) it should be feasible to develop an anabolic side effect index for steroids in fish. Such an index would avoid much of the ambiguity of past studies and would be of immeasurable assistance in assessing the potential of a steroid as a growth promoter in fish cultivation.

F. Steroid Concentrations and Metabolism in Fish

The current knowledge of steroid hormones and their metabolism in fish has been admirably reviewed by Idler and Truscott (1972) in regard to corticosteroids and by Ozon (1972a,b) in regard to androgens and estrogens.

A large number of steroids representing the four groups of mammalian steroids outlined in the beginning of this section have been found in lower vertebrates and many of the metabolic pathways common to other vertebrates have also been demonstrated. Of anabolically active androgens, testosterone and 11-ketotestosterone are major androgens in a number of species (Ozon, 1972a) including salmonids.

17α-Methyltestosterone metabolism has not been studied in fish. In mammals metabolism proceeds along pathways common with those of testosterone except that dehydrogenation of the 17β-hydroxyl group does not take place.

Recently Schreck et al. (1972a) determined androgen levels in plasma of 19-month-old rainbow trout nearing maturity. Concentrations were 1.29 μg/100 ml in males and 0.89 μg/100 ml in females. The difference between the sexes was not statistically significant. When a group of 27-month-old trout were castrated the androgen level in males decreased from 2.74 to 0.68 μg/100 ml. The latter value was obtained 21 days after orchiectomy. At this time there was no decrease in females, but 42 days after ovariectomy androgen levels had decreased from 1.32 to 0.48 μg/100 ml. Schreck et al. (1972b) also noted a positive relationship between gonad development and plasma androgen concentrations in both sexes of rainbow trout. This confirms bioassay data (Idler and Tsuyuki, 1959) on testis of sockeye salmon (O. nerka) which showed androgenicity expressed as testosterone propionate equivalents rising from 41 to 133 μg/kg of testis during successive stages of spawning migration, and also data (Idler et al., 1971) on testicular and peripheral plasma androgens in Atlantic salmon (Salmo salar). Plasma androgen concentrations cited above are high compared to those in humans. Forchielli et al. (1963) found 0.56 and 0.12 μg/100 ml of plasma in males and females, respectively.

A measure of the rate of elimination of a steroid from the blood of an animal is provided by the metabolic clearance rate, which has been defined by Tait and Burstein (1964) as that volume of plasma or blood completely cleared of steroid in unit time. Fletcher et al. (1969) found metabolic clearance rates of testosterone in sexually mature male and female skate Raja radiata to be 7.93 and 10.4 ml/kg hr, respectively. Production rates, which equal rates of elimination from the body, since a steady state is assumed, were 0.38 μg/kg hr for males and 0.33 μg/kg hr for females.

Schreck (1973) injected [3H]testosterone into the epaxial muscle of rainbow trout and followed the disappearance of radioactivity in plasma and certain tissues. From the plasma concentrations during the first 24 hr after the injection the half-life of radioactivity was calculated to be 2.5 hr. [^{3}H]Testosterone (Fagerlund and McBride, 1978) and 17α-[^{3}H]methyltestosterone (Fagerlund and Dye, 1976 unpublished observations) were fed for several days to coho parr as diet supplements at the rates of 5 and 1 ppm, respectively. Half-lives of radioactivity in blood during a period of 3 days starting 16 hr after the last intake of isotope were 11.2 hr for testosterone and 13.0 hr for 17α-methyltestosterone. Corresponding values for muscle were 13.9 and 20.4 hr. The slower clearance rate of 17α-methyltestosterone, which has also been noted in humans (Quincey and Gray, 1967), may in part explain the greater growth effect of this steroid compared to testosterone.

G. Tissue Residues of Orally Administered Anabolic Steroids

To gain the acceptance of national health authorities of hormonal drug administration in fish culture, analyses must be performed to establish that concentrations of any remaining residue of the drug or any harmful metabolites in edible tissues are below those deemed tolerable. Until recently 2 ppb was commonly accepted in the United States as the tolerance for a number of hormonal drugs (Kolbye, 1975) mainly because this was the lowest enforceable level consistent with current analytical capability.

Since then more sensitive methods have pushed the limits of detection well below this concentration. Government health authorities are particularly concerned with residues of drugs known to be carcinogenic. At the present, government regulations in regard to permissible tissue levels are not well defined.

The total, combined residues of ingested hormone and its metabolites in plasma and tissues of coho salmon parr which received either 5 mg/kg of [^{3}H]testosterone or 1 mg/kg of 17α-[^{3}H]methyltestosterone in the diet are shown in Table XI. When the two steroids were fed in these concentrations, which produce good anabolic responses, residues of less than 1 ng/g (= 1 ppb) remained in any tissue 256 hr after the last feeding. The actual concentrations of unmetabolized steroids are considerably lower than the values shown in the table, since testosterone is known to undergo rapid degradation in salmon tissues, particularly the liver (Ozon, 1972a). The high concentrations of radioactivity in the gall bladder confirm the enterohepatic route as the major route of androgen elimination in this species.

The testosterone concentrations listed in Table XI may be compared with known endogenous levels of testosterone in salmonids.

Table XI

Mean Total Concentrations of Hormone and Its Metabolites in Tissues of Coho Salmon
Juveniles Fed a Diet Containing 5 ppm of [³H]Testosterone
or 1 ppm of 17α-[³H]Methyltestosterone[a]

	Testosterone		17α-Methyltestosterone	
	16 hr[b]	256 hr	16 hr	256 hr
Blood	5.9	<0.11	10.8	0.08
White muscle	0.4	<0.06	1.7	0.05
Liver	13.4	0.23	19.8	0.98
Kidney	4.4	0.11	8.7	0.23
Pyloric ceca	137.0	0.51	172.1	0.44
Gallbladder	2392.0	1.00	614.5	0.42

[a] Values, expressed as nanograms per gram (ppb) of tissue or per milliliter of blood, were calculated from tissue radioactivity. From Fagerlund and McBride (1978) and Fagerlund and Dye (1976 unpublished).

[b] Time since previous isotope intake.

Concentrations of 15.4 μg/100 ml of plasma (free testosterone + glucuronide) and 32.5 ng/g of testis (glucuronide) have been reported in sexually mature salmon (*O. nerka*) (Grajcer and Idler, 1961, 1963). Concentrations reported by Schreck *et al.* (1972a,b) in rainbow trout are discussed above.

On the basis of the 11-day concentrations listed in Table XI and half-lives determined in plasma and tissues (see above) one may anticipate that, when hormone-treated juveniles are released to the ocean and later recaptured as adults or precociously matured males, residues will be below the levels which can be detected by any known methods. If, on the other hand, the hormones are to be applied in the culture of impounded fish, which will be used for human consumption, then a hormone withdrawal period will have to be imposed, at least when synthetic 17α-methyltestosterone is used.

H. Recent Developments

The growth responses to food supplemented with two synthetic androgens have been studied in juvenile rainbow trout and Atlantic salmon parr by Simpson (1976b). Rainbow trout fed ethylestrenol at 2.5 or 12.5 mg/kg of food for 3 months gained 15.7 and 25%, respectively, more weight than controls. Fish fed 2.5 mg/kg of 17α-methyltestosterone gained 19.0% more. Some improvement of food conversion efficiency was noted in all hormone-fed groups, but the greatest increase, 17.5% over the control, was obtained with 12.5 mg/kg of ethylestrenol. Nitrogen and lipid content and dry weight of carcass as percentage of carcass wet weight were not significantly influenced by hormone treatment. However, both lipid content and weight of viscera as percentage of total body weight decreased significantly. The lowest lipid content of viscera, found in fish fed 12.5 mg/kg of ethylestrenol, was 6.9% of visceral weight, whereas control viscera contained 11.2%.

Contrary to rainbow trout, Atlantic salmon parr did not respond to 17α-methyltestosterone fed at doses of 2.5 and 12.5 mg/kg. However, groups fed ethylestrenol for 6 months at doses of 0.5, 2.5, and 12.5 mg/kg of food gained 20.8, 21.9, and 13.6%, respectively, more weight than the control group. Percentage visceral weight was significantly reduced with 12.5 mg/kg of 17α-methyltestosterone and 2.5 and 12.5 mg/kg of ethylestrenol.

Feeding of the two hormones significantly increased the gonadosomatic index (weight of gonad/body weight) of both rainbow trout and Atlantic salmon males, with the exception of the salmon group fed 2.5 mg/kg of ethylestrenol. Some advancement in development of testes occurred in all hormone-treated groups. Ethylestrenol induced dose-dependent hypertrophy of sperm ducts and methyltestosterone advanced testes to the spermatid stage.

In an extension of the Atlantic salmon study, diets supplemented with 5 mg/kg of ethylestrenol promoted growth of juveniles belonging to four individually raised families. One of these families was observed during an extended period after being divided into two groups fed either a control or a test diet. Both groups developed a bimodal weight and length distribution. Hormone treatment appeared to be most effective in promoting growth in the slower growing mode during the first year of growth. The growth of the faster growing mode was depressed by hormone treatment when the weight exceeded 10 g.

In contrast to the previous study, Saunders et al. (1977) obtained a growth response in Atlantic salmon parr by feeding lower doses of 17α-methyltestosterone. A group which received a dose of 0.2 mg/kg for 106 days while held in 15°C water weighed 23.6% more than the control group. A dose of 1 mg/kg induced a smaller but statistically significant gain in weight. The condition factor was increased significantly by the 1 mg/kg dose. Protein and lipid compositions of the muscle were not affected by the treatment, but the 1 mg/kg dose caused a small but significant decrease in moisture content. The hormone either induced spermatogenesis or reduced the number of germ cells in some males. These secondary changes were particularily evident in fish which received the higher dose.

The results obtained by Simpson (1976b) and Saunders *et al.* (1977) indicate a growth response to orally administered 17α-methyltestosterone in rainbow trout and Atlantic salmon that is weaker than that generally obtained in Pacific salmon (Table VIII). Fagerlund and McBride (1977) also observed a weak response to 17α-methyltestosterone in steelhead trout during the first 118 days of treatment, although later a good response was obtained (Table VIII). These results suggest a genus-specific difference, rainbow trout and Atlantic salmon being members of the genus *Salmo* and Pacific salmon of the genus *Oncorhynchus*.

Cheema and Matty (1976) presented the results of feeding anabolic steroids to rainbow trout fingerlings in addition to those given by Matty (1975). Dimethazine, which induced weight gains at doses of 5 and 10 mg/kg, did not change the hepatosomatic indices (weight of liver tissue/total body weight). However, the renosomatic indices of treated fish were significantly increased. The same authors found that norethandrolone fed to rainbow trout fingerlings at doses of 2.5 and 5 mg/kg for 40 days caused net weight gains of 33 and 26%, respectively, greater than that of controls. The percentage uptake into muscle protein of [14C]leucine injected intraperitoneally was significantly increased in groups which had received 5 mg/kg of either dimethazine or norethandrolone for 20 days.

The effect of anabolic steroids on appetite and food conversion efficiency of underyearling coho salmon was reported by Fagerlund *et al.* (1978). Diets containing 1 ppm of 17α-methyltestosterone or 5 ppm of testosterone were fed either to satiation or in restricted amounts. The appetite of groups fed to satiation was increased by 16.8% with 17α-methyltestosterone and by 6.2% with testosterone. Food consumption per gram of fish fed restricted rations was on an average 89.0% of that of the satiation control. Food conversion efficiency of groups fed satiation and restricted rations containing 17α-methyltestosterone increased by 10.0 and 22.1%, respectively, and of groups fed the testosterone diet by 2.6 and 8.8%, respectively. The whole body fat content of fish fed hormone-supplemented diets at either ration was significantly decreased, an observation which is in agreement with that of Simpson (1976b). The demonstrated decrease in protein requirement may be an effect of increased lipid mobilization. If this is the case, an increased amount of lipid would be available to supply energy and so spare protein for growth.

Maintaining eggs and developing larvae of *Tilapia nilotica* in water containing adrenosterone (5 mg/liter) for 3 months resulted in a significant increase in body weight (Katz *et al.*, 1976). When fingerlings were 8 months old the mean weight of the control group was 3.69

g, whereas that of the treated group was 4.08 g, an increase of 10.5%. Normal gonadal structure was not detected when hormone treatment was terminated after 3 months of treatment. When the fish were 8 months old, testicular regeneration was evident in a majority of the expected number of males, but none of the fish possessed ovarian tissue.

IV. THYROID HORMONES*

A. Introduction

The evidence to date indicates a lack of consistency in the growth response of teleosts after thyroid hormone treatment. Even within a single species opposite results have been obtained (see review in Gorbman, 1969). In the past 8 years there have been considerable advances in knowledge from studies of the fish thyroid, which help to explain these anomalies. In particular, we now know far more about the regulation of thyroid hormone secretion, the nature and circulating levels of thyroid hormones in the blood, their degradation rates, and pathways of peripheral metabolism. Factors such as route of hormone administration, nutritional state (starvation versus feeding, quantity and composition of food) and environmental temperature appear to influence thyroid hormone metabolism. A brief review of this information will be presented prior to discussion of the role of the thyroid in fish growth. Papers on the effects of antithyroid compounds, radio-thyroidectomy, and thyrotropin (TSH) on fish growth have been reviewed from 1960 to the present. Pickford and Atz (1957) and Baker-Cohen (1961) have thoroughly summarized the earlier literature. The final topics deal with the metabolic actions of thyroid hormones.

B. Biochemistry and Metabolism of Thyroid Hormones

1. THYROID HORMONE BIOSYNTHESIS AND NEGATIVE FEEDBACK CONTROL

In most teleosts the thyroid gland is not encapsulated, but consists of many saclike colloid-filled follicles lined by secretory epithelium. The follicles extend around the ventral aorta and into the branchial arches. In some species the follicles migrate to extrapharyngeal sites,

* This section was prepared by D. A. H.

particularly within the kidney (Barr, 1965; Barrington, 1975). The synthesis of thyroid hormones (see Fig. 12) within these follicles is stimulated by thyrotropin or thyroid stimulating hormone (TSH), a glycoprotein derived from the basophils of the proximal pars distalis. Thyrotropins from fish have a molecular weight of approximately 30,000 and it has been shown that they are homologous in structure with the gonadotropins (Fontaine, 1975). In the presence of TSH all aspects of thyroid function are enhanced. For example, there is increased iodide trapping by the follicle cells, oxidation of iodide to reactive iodine, and iodination of tyrosine creating first 3-monoiodotyrosine (MIT) and next, 3, 5,-diiodotyrosine (DIT). Formation of thyroxine (T_4) is thought to occur by a coupling of two molecules of DIT with subsequent loss of an alanine side chain. Triiodothyronine (T_3) results from the coupling of one molecule of MIT with one of DIT.

The release of thyroid hormones from their bond with the glycoprotein thyroglobulin, which is the main component of the colloid, is also promoted by TSH. It is generally thought that MIT and DIT do not normally leave the gland in significant quantities (Gorbman, 1969; Barrington, 1975). The secretory epithelium of the thyroid follicles in the presence of TSH increases in height and this has been used as an index of thyroid activity by many investigators. The iodothyronines released from the gland are bound to specific plasma proteins and are transported to various tissues in the body where they are metabolized (considered in detail below). Several *in vitro* studies have shown that

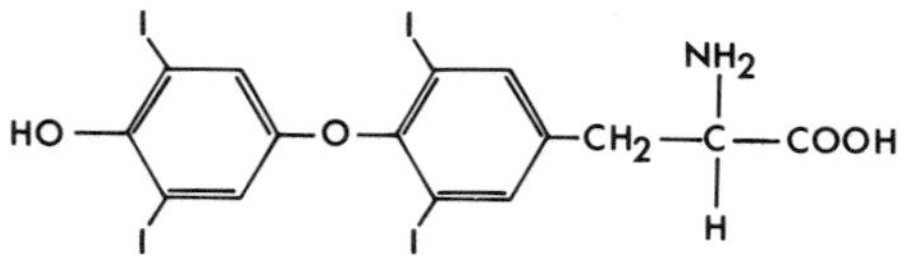

3,5,3'5' - TETRAIODOTHYRONINE (T_4)

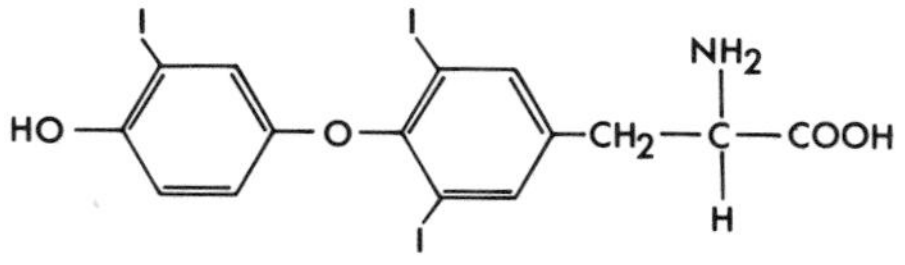

3, 5, 3' - TRIIODOTHYRONINE (T_3)

Fig. 12. Structures of vertebrate thyroid hormones.

T_4 exerts a negative feedback effect on the TSH cells in the pituitary (literature summarized in Peter, 1973). In addition to this regulatory mechanism there is evidence for hypothalamic control over TSH secretion. For many teleosts this is inhibitory in nature. This has been conclusively demonstrated in a series of excellent experiments by R. E. Peter on the goldfish (Peter, 1970, 1971, 1972, 1973; Peter and McKeown, 1975). Prior to 1973 most of the evidence for the presence of a thyrotropin inhibitory factor (TIF) was indirect. For example, teleost fishes with an ectopic pituitary transplant tended to be hyperthyroid (Peter, 1971). This was also true after electrolytic lesions were placed stereotaxically in the nucleus lateralis tuberis (NLT) pars anterior and NLT pars posterior of the goldfish hypothalamus, but not in other regions of the hypothalamus and epithalamus (Peter, 1970). However, in 1973, Peter demonstrated the presence of a factor (TIF) in goldfish hypothalamic extracts that inhibited TSH secretion as assessed from radioiodide uptake into the thyroid. The structure of TIF has not been elucidated but mammalian thyrotropin releasing hormone (TRH), somatostatin, and melanocyte inhibiting factor may have TIF activity in the goldfish (Peter and McKeown, 1975). Also, Bromage (1975) and Bromage *et al.* (1976) found evidence for inhibition of the pituitary–thyroid axis in the intact male guppy and in adult male rainbow trout after injection of TRH. Therefore, there is considerable evidence, at least for the goldfish, that thyroid hormone has a negative feedback effect at the level of the pituitary which leads to direct suppression of TSH secretion. Also, there is an effect on the hypothalamus which stimulates TIF activity. Studies on mice and rats indicate that catecholamines and sympathetic terminations in the thyroid may influence thyroid hormone secretion (Martin, 1974). Monoamines may also influence the secretion of TIF in fish (Peter, 1973).

2. BLOOD IODIDE (^{127}I) LEVELS

Levels of iodide in the blood markedly influence TSH production in fish if they are so low as to limit the formation of adequate quantities of thyroid hormone. For example, Robertson and Chaney (1953) noted an inverse relationship between cellular thyroid follicle height and blood iodide concentration. In extreme cases of iodide deficiency, goiter ensues which is characterized by thyroid hyperplasia. Marine and Lenhart (1910) and Marine (1914) were the first to observe this condition in fish (brook trout) and found that it could be alleviated by supplemental iodine in the water or diet. Gregory and Eales (1975) reported that plasma iodide of freshwater teleosts ranges from 0.5 to 2244 μg per 100 ml (μg%). Salmonid levels apparently range from 1 to 597 μg% for *Salmo* and *Oncorhynchus* genera and are sometimes greater than 2000 μg% for *Salvelinus fontinalis*. Factors contributing to variability in blood iodide levels between and within fish species include movement of fish from rich to poor iodide environments (Leloup and Fontaine, 1960); seasonal fluctuations in water iodide content (Hickman, 1962); elevated temperature, resulting in greater iodide excretion (Leloup and Fontaine, 1960; Smith and Eales, 1971); sexual maturation (Robertson and Chaney, 1953); differences in the capacity of fish to bind iodide to plasma proteins (Leloup and Fontaine, 1960; Huang and Hickman, 1968; Gregory and Eales, 1975); physical activity (Fontaine and Leloup, 1959; Higgs and Eales, 1971); and, perhaps the most important of all, dietary iodide levels (La Roche *et al.*, 1965; Gregory and Eales, 1975). Plasma iodide concentrations have a distinct effect on several radioiodide parameters of thyroid function in brook trout (Higgs and Eales, 1971). Also, it is essential to ensure, from the above discussion, that diets contain sufficient iodine (Snieszko, 1972) when investigating the influence of thyroid hormone on growth of fish.

3. NATURE, PROTEIN BINDING PROPERTIES, AND LEVELS OF CIRCULATING IODOCOMPOUNDS IN TELEOSTS

There is evidence, from several different experimental approaches, which indicates that either some or all of the thyroidal iodoamino acids can be present in the plasma of freshwater and marine teleosts. Leloup (1956) claimed the presence of labeled T_4, T_3, MIT, and DIT in the plasma of the mudskipper, *Periophthalmus koelreuteri*, after injection of radioiodide. Using thin-layer chromatography (tlc) Osborn and Simpson (1969a) found that, after ^{125}I injection, the thyroid tissue of the marine plaice, *Pleuronectes platessa*, contained labeled T_4, T_3, MIT, and DIT. However, only T_3 and T_4 were detected in the plasma. In a later study (Osborn and Simpson, 1972) using a more sensitive analytical procedure, namely, electron capture–gas liquid chromatography, they were able to establish the presence of not only T_3 and T_4 in plaice plasma, but also MIT and DIT. The level of all four compounds was noted to increase after TSH administration. Stress brought about a decline in plasma T_3 and T_4 levels. Jacoby and Hickman (1966) using the method of isotopic equilibrium ($Na^{125}I$ and ^{127}I) showed that the quantity of T_3 and iodotyrosines in the plasma of rainbow trout, *Salmo gairdneri*, equalled or exceeded the amount of T_4. This was in spite of the thyroidal T_4 concentration being approximately six times greater than the T_3 concentration.

In contrast to the above findings, Chavin and Bouwman (1965) found, after tlc of plasma from goldfish (*Carassius auratus*) not administered radioiodide, that T_4 was the only iodoamino acid present. In addition, Chan and Eales (1975) consistently demonstrated by tlc that T_4 was the main labeled iodocompound in the plasma and bile of starved TSH-treated brook trout previously injected with radioiodide. They speculated that the absence of T_3 in the plasma and bile may have been due to the high plasma iodide level. This, they thought would lead to the complete iodination of thyroidal tyrosines so that only T_4 would be synthesized.

^{125}I-Labeled L-T_4 (*T_4) tracer studies provide good evidence that the presence of T_3 in the plasma of at least some fish is the result of monodeiodination of T_4 (specific enzymatic removal of iodide from the 5' position of the T_4 molecule) during peripheral metabolism. Other deiodination products of T_4, besides T_3, have also been detected in plasma by TLC in these studies.

One of the first of these *in vivo* studies was conducted on the marine plaice by Osborn and Simpson (1969b). After intraperitoneal (ip) injection of *T_4 they recovered radioactivity in the bile, urine, and plasma in the form of T_3, 3,3',5'-triiodo-L-thyronine (reverse T_3), and 3,3'-diiodothyronine, all products of stepwise deiodination. They also noted three deamination products, namely, tetraiodothyropyruvic acid, tetraiodothyrolactic acid, and tetraiodothyroacetic acid. Higgs and Eales (1971) found that both exercised and unexercised starved brook trout had a T_3 or T_3-like substance in the plasma 25 hr postinjection (pi) of *T_4. Eales (1972a) studied the metabolism of ip-injected *T_4 in several starved freshwater teleosts acclimated at 12–13°C. In most species T_3 was detected as a plasma radioactive derivative. However, in some species at certain times pi the levels of radioactive T_3 apparently equaled or exceeded those of T_4. Thin-layer chromatography of rainbow trout plasma revealed the possible presence of other radioactive derivatives, namely, reverse T_3 and 3'-monoiodothyronine. Eales (1972a) tested for the presence of T_4 and T_3 propionate, acetate, or formate analogues in the plasma and concluded that, if formed, they were present in small concentrations and could not be considered as major products of *T_4 metabolism. Other studies on starved or fed brook trout, injected with *T_4 at various sites, have consistently established, after tlc of plasma, a distinct radioactive peak corresponding in R_f to T_3 (Eales, 1972b; Falkner and Eales, 1973;

Higgs and Eales, 1976, 1977, 1978a,b). This was also the case in starved channel catfish, *Ictalurus punctatus* (Collicutt and Eales, 1974).

The extent of binding of thyroid hormones to the plasma proteins of fish as well as the proteins responsible for this binding have been well documented for the brook trout by Falkner and Eales (1973). Coprecipitation of either $*T_4$ or ^{125}I-labeled L-$*T_3$ ($*T_3$) with plasma proteins by trichloroacetic acid (TCA) showed that over 95% of $*T_3$ or $*T_4$, with added hormone levels of at least 5 μg/ml plasma, was bound to the plasma proteins. Further *in vitro* studies using equilibrium dialysis demonstrated over 99% of $*T_4$ or $*T_3$ was bound to plasma proteins up to added hormone levels of at least 5 μg/ml. Falkner and Eales (1973) showed that $*T_3$ and $*T_4$ were mainly bound to prealbuminlike proteins *in vitro* at high hormone levels. But *in vivo* studies at more physiological levels of hormone demonstrated that β-globulinlike and slower albuminlike fractions bound substantially more hormone ($*T_3$ or $*T_4$) than the fast prealbuminlike fraction. They postulated that the prealbuminlike proteins in brook trout may have a high capacity but low affinity for T_4 and may function to buffer the plasma against excessive free T_4 or T_3 by binding them weakly. Either of the hormones could then be lost quickly to the liver with subsequent elimination from the body. On the other hand, β globulins may have greater affinity for T_4 or T_3 but lower capacity. It is not known if these findings are representative of the situation in most teleosts. There was certainly some suggestion from this study, and other work cited by Falkner and Eales that T_4 is reversibly bound to trout plasma proteins less strongly than to human plasma proteins.

Plasma or serum titers of T_4 measured either by competitive binding analysis or by radioimmunoassay, for most teleost species investigated, are usually less than 1 μg% and in many cases less than 0.5 μg% (Higgs and Eales, 1973; Collicutt and Eales, 1974; Eales, 1974; Leloup and Hardy, 1976; Brown and Eales, 1977; Hurlburt, 1977; White and Henderson, 1977). Notable exceptions to this generalization can be found in studies by Refetoff *et al.* (1970), Drongowski *et al.* (1975), and Allen (1977). Refetoff *et al.* (1970) used a modification of the Murphy and Pattee (1964) method and obtained 4.5 and 4.3 μg $T_4/100$ ml plasma for the trout, *Salmo irideus*, and perch, *Catostomus commersonii*, respectively. Drongowski *et al.* (1975) used the Tetrasorb–^{125}I T_4 diagnostic kit (Abbott laboratories) and obtained T_4 levels as high as 3.7 μg% for adult and jack coho salmon from Oregon. Allen (1977) noted that serum T_4 levels in yearling rainbow trout obtained from Michigan ranged from 1.68 to 4.68 μg%.

Circulating levels of T_3 in plasma or serum of teleosts in freshwater, as determined by radioimmunoassay, are, like T_4, usually less than 1 μg% (Leloup and Hardy, 1976; Brown and Eales, 1977; White and Henderson, 1977). In adult brook trout, T_3 titers have been shown to be consistently higher than those of T_4 over a 12-month period (White and Henderson, 1977), but in rainbow trout and eels (*Anguilla anguilla*) T_3 levels are generally lower than those of T_4 (Leloup and Hardy, 1976). Plasma levels of both T_4 and T_3 may vary on a seasonal and diurnal basis (White and Henderson, 1977) and, for teleosts in seawater, there is evidence for (Leloup and Hardy, 1976; Leloup *et al.*, 1976) and against (Dickhoff *et al.*, 1978) T_3 being the main circulating thyroid hormone.

4. PERIPHERAL METABOLISM OF T_4 AND T_3

In fish and mammals, non-protein-bound thyroid hormone is either deiodinated, with concomitant production of primarily inorganic iodide and several degradation products considered above, or eliminated from the body via the liver biliary–fecal pathway (enterohepatic system). Competition between these two pathways can exist.

a. Deiodination. At present, the sites of *in vivo* deiodination are not known. However, Law and Eales (1973) using *in vitro* methods demonstrated $*T_4$ deiodination (as assessed by the production of radioiodide using protein precipitation by TCA and chromatography) in several brook trout tissues including brain, gill, stomach, duodenum, intestine, heart, kidney, and muscle but not in liver. $*T_3$ was not an identifiable product suggesting that *in vivo* conditions are required for monodeiodination of T_4 to T_3. While data of Tata (1960) suggest T_4 deiodination by liver and muscle of plaice and brown trout this may not have been enzymatic. The extent of *in vivo* $*T_4$ deiodination in fish is considerable (Eales, 1970, 1972a, 1977a,b; Higgs and Eales, 1971; Collicutt and Eales, 1974) and in the brook trout is influenced by route of administration (Eales, 1972b; Higgs and Eales, 1976) and nutritional state (Higgs and Eales, 1977, 1978a,b).

However, negligible deiodination of $*T_3$ relative to that of $*T_4$ occurs in both starved and fed brook trout (Eales *et al.*, 1971; Higgs and Eales, 1977) (Fig. 13) and in starved rainbow trout (Eales, 1977b). In contrast, both *in vivo* and *in vitro* studies on mammals indicate more rapid deiodination of T_3 than T_4 (literature summarized in Higgs and Eales, 1977). The results from studies on amphibians are contradictory (Yamamoto *et al.*, 1966; Ashley and Frieden, 1972).

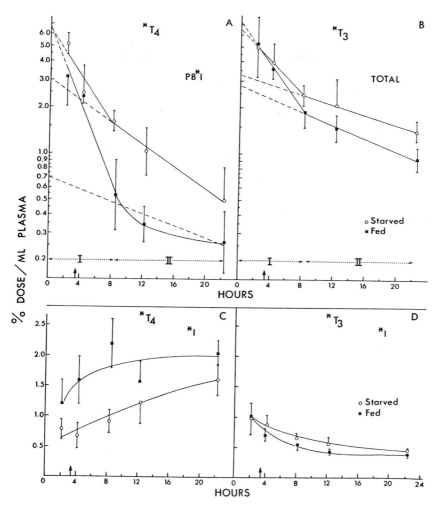

Fig. 13. Influence of starvation on the metabolism of L-thyroxine (T₄; A) and 3,5,3'-triiodo-L-thyronine (T₃; B) and on the deiodination of T₄ (C) and T₃ (D) in 1-year-old brook trout maintained between 11.4° and 12.5°C. The trout were fed daily 0.56% of their body weight [dry weight of food (g)/wet weight of fish (g)] with a high protein diet or starved for 12 days prior to injection of 1.5 μCi of radio-L-thyroxine (*T₄; phenolically labeled with ¹²⁵I; specific activity 38.2 μCi/μg) or 0.5 μCi of ¹²⁵I-labeled *T₃ (specific activity 32.6 μCi/μg) via the heart. All "fed" fish were presented food 20 hr before and 3.5 hr (↑) after injection. Curves in A and B depict loss of plasma protein-bound (PB*I) or total radioactivity (TOTAL) (% dose/ml normalized to a body weight of 100 g), respectively. (From Higgs and Eales, 1977, *Gen. Comp. Endocrinol.* **32**, 29–40.)

b. Enterohepatic System. A number of studies have conclusively established the existence of an enterohepatic system for T_4 (Leloup and Fontaine, 1960; Eales, 1969, 1970, 1972a; Osborn and Simpson, 1969b; Collicutt and Eales, 1974) and T_3 (Eales *et al.*, 1971; Higgs and Eales, 1977) (Fig. 14) in teleosts. In many starved freshwater teleosts ip injection of $*T_4$ leads to free $*T_4$ and $*T_4$-glucuronide conjugates in the bile with negligible conjugation of $*T_4$ to sulfate. A similar situation exists after $*T_3$ is injected into starved brook trout except in this case free $*T_3$ and $*T_3$-glucuronide conjugates have been identified in the bile (Sinclair and Eales, 1972). In the marine plaice, however, there is some evidence for both $*T_4$-glucuronide conjugates and $*T_4$-sulfate esters after ip injection of $*T_4$ (Osborn and Simpson, 1969b). In general, hepatic glucuronide conjugation of iodothyronines appears to be an effective mechanism for detoxifying and regulating the circulat-

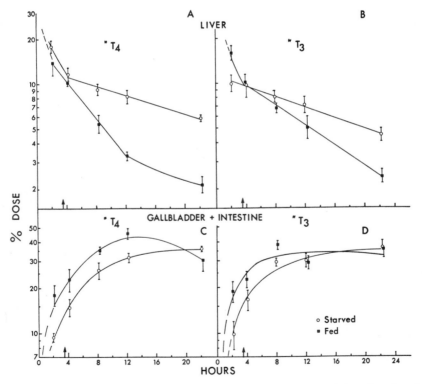

Fig. 14. Influence of starvation on the percentages of the injected dose found in the liver (A and B) and gall bladder + intestine (C and D; representative of biliary–fecal excretion) of 1-year-old brook trout at various times following injection of $*T_4$ and $*T_3$. Refer to Fig. 13 for details. (From Higgs and Eales, 1977, *Gen. Comp. Endocrinol.* **32,** 29–40.)

ing thyroid hormone level in fish in addition to the pituitary–thyroid axis. This point was stressed by Eales (1970). In mammals, iodothyronine–glucuronide conjugates can undergo intestinal hydrolysis and the liberated free iodothyronine can then either be reabsorbed into the circulation (enterohepatic cycling) or eliminated in the feces (Tata, 1964). Enterohepatic cycling does not appear to be a prominent feature in starved or fed brook trout (Eales and Sinclair, 1974) or in the starved channel catfish (Collicutt and Eales, 1974). In the latter study it was determined that less than 10% of an injected dose of $*T_4$ was excreted from the fish by pathways other than the biliary–fecal route.

5. INFLUENCE OF NUTRITIONAL STATE ON THYROID FUNCTION IN TELEOSTS

Starvation depresses thyroid function in teleosts as assessed by radioiodide uptake by the thyroid (Fontaine and Fontaine, 1956; Bonnet, 1970), thyroid histological appearance (Vilter, 1944; Ortman and Billig, 1966; McBride, 1967; Honma and Matsui, 1973), plasma hormone levels (Osborn and Simpson, 1972; Higgs and Eales, 1978a) (Table XII), plasma $*T_4$ or $*T_3$ loss (Eales and Sinclair, 1974; Higgs

Table XII
Influence of Ration on the Distribution Volume, Turnover Rate, Biological Half-Life, Metabolic Clearance Rate, and Degradation Rate of Thyroxine in 1-Year-Old Brook Trout[a]

Ration[b] (% body weight/day)	V^c (ml/100 g)	k^c (fraction/ hr)	$t_{1/2}^c$ (hr)	MCR[c] (ml/hr/ 100 g)	T_4 (ng/ml)	T_4 DR[c] (ng T_4/ day/100 g)
0	18.2	0.0980	7.07	1.78	0.47(8)	20
2.5	20.6	0.132	5.25	2.72	1.36(7)	89
5.0	10.5	0.208	3.33	2.18	3.42(6)	179

[a] Each trout was injected with 1.5 μCi of radio-L-thyroxine ($*T_4$; phenolically labeled with ^{125}I; specific activity 37.1 μCi/μg) into the circulation (Higgs and Eales, 1976). Seven to eight fish from each group were killed at 5.5, 9, 20.5, 34.5, 46, 72.5, and 96 hr after $*T_4$ injection. Disappearance of plasma protein-bound radioactivity was described by fast and slow exponents. The fast exponent, representative of $*T_4$ loss, was used to calculate kinetic parameters (Higgs and Eales, 1976). Plasma stable T_4 concentrations were determined by competitive binding analysis (Higgs and Eales, 1973). Water temperature ranged from 10.5° to 12.5°C during the experiment (from Higgs and Eales, 1978a).

[b] Dry weight of food (g)/dry weight of fish (g).

[c] V, distribution volume; k, turnover rate; $t_{1/2}$, biological half-life; MCR, metabolic clearance rate; DR, degradation rate.

and Eales, 1977, 1978a) (Fig. 13; Table XII), and peripheral metabolism of *T$_4$ and *T$_3$ (Eales and Sinclair, 1974; Higgs and Eales, 1977) (Figs. 13 and 14). A similar trend has been noted in mammals (Nathanielsz, 1970).

During starvation of brook trout both biliary–fecal *T$_4$ excretion and *T$_4$ deiodination are diminished relative to fed fish. This can account for the slower loss of *T$_4$ from the plasma of starved fish (Higgs and Eales, 1977). Feeding was found by Higgs and Eales (1977) to rapidly stimulate monodeiodination of *T$_4$ to *T$_3$ in trout previously deprived of food (Table XIII). Comparison of the metabolic clearance rates for *T$_4$ and *T$_3$ between either starved or fed brook trout indicates that injected *T$_3$ is cleared more slowly from the plasma than *T$_4$ (Higgs and Eales, 1977). These findings are contrary to the statement of Leloup and Fontaine (1960) (no data presented) that *T$_3$ is metabolized more rapidly than *T$_4$ in young Atlantic salmon, and do not agree with data for *Petromyzon planeri* (Salvatore *et al.*, 1959). The slower loss of *T$_3$ relative to *T$_4$ in the brook trout was probably a consequence of the negligible deiodination of *T$_3$ relative to that of *T$_4$ since biliary–fecal excretion of *T$_3$ and *T$_4$ was similar (Figs. 13 and 14). Higgs and Eales (1977) postulated from the above findings on the brook trout that the conversion of T$_4$ to T$_3$ may be a prerequisite for thyroid hormone action. A recent study (Eales, 1977b) conducted on 1-year-old rainbow trout supports this view because it was shown that up to 69% of T$_4$ secreted from the thyroid was converted to T$_3$.

In another study on the brook trout, Higgs and Eales (1978a) demonstrated that level of plasma T$_4$ and rate of *T$_4$ degradation were directly related to ration (Table XII). A study by Bonnet (1970) on the mullet, *Mugil auratus,* also revealed an effect of level of dietary intake on thyroid function. In addition, Higgs and Eales (1978b) found that diet composition influenced *T$_4$ metabolism in brook trout. For example, brook trout fed a high-protein and high-caloric diet had an increased rate of degradation of injected *T$_4$ relative to those fed a low-protein and low-caloric diet. Also, the high-protein fed trout were noted to have increased *T$_4$ deiodination and formation of *T$_3$ from *T$_4$, as well as elevated biliary–fecal excretion of radioactivity relative to those fed the low-protein diet. Bonnet (1970) also found an effect of food composition on mullet thyroid activity, and recently Leatherland *et al.* (1977) noted that the type of dietary fat influenced plasma T$_4$ levels in rainbow trout at 7°C but not at 19°C.

Eales and Sinclair (1974) presented two mechanisms by which food intake could stimulate thyroid function in brook trout. First, food ingestion may in some way stimulate the hypothalamo–pituitary–

Table XIII

Influence of Time of Food Presentation on Plasma Concentration of ^{125}I-Labeled T$_4$, ^{125}I-Labeled T$_3$, ^{125}I, and Plasma *T$_3$/*T$_4$ for 1-Year-Old Brook Trout at 11.4°C[a]

Group	Treatment of fish	Plasma *T$_4$ (% dose/ml plasma) X̄	SE	Plasma *T$_3$[b] (% dose/ml plasma) X̄	SE	Plasma *I[b] (% dose/ml plasma) X̄	SE	Plasma *T$_3$/*T$_4$ X̄	SE
A	Starved 7 days before injection	3.41	0.23	0.18	0.02	0.40	0.02	0.055	0.005
B	Fed daily for 6 days; last fed 20 hr before injection	3.63	0.45	0.20	0.04	0.40	0.06	0.062	0.02
C	Fed daily for 6 days; then fed 20 and 3.5 hr before injection	2.75	0.20	0.60	0.12	0.71	0.10	0.23	0.05
D	Fed daily for 6 days; then fed 20 hr before and 3.5 hr after injection	3.16	0.26	0.47	0.10	0.81	0.13	0.15	0.02

[a] *T$_4$, radio-L-thyroxine; *T$_3$, labeled 3,5,3'-triiodo-L-thyronine; *I, radioiodide. Fish were killed 4.2 hr after injection of 1.5 μCi of *T$_4$ (specific activity, 32.9 μCi/μg) into the circulation (n = 6 or 7). The daily amount of a high protein diet was restricted to 0.55% body weight [dry weight of food (g)/wet weight of fish (g)] for "fed" fish (from Higgs and Eales, 1977).

[b] One-way ANOVA showed that both plasma *T$_3$ and *I differed significantly ($P < 0.05$) between the four groups. The *T$_3$ mean for group C differed significantly ($P = 0.05$) from the *T$_3$ mean for group A [Scheffé's test (Bancroft, 1968)].

thyroid axis. This would result in increased plasma thyroid hormone levels which in turn would lead to greater thyroid hormone loss via excretion and/or degradation pathways. Second, food intake may increase biliary–fecal excretion and/or deiodination of circulating thyroid hormones so that plasma thyroid hormone levels are temporarily lowered. A negative feedback effect would then be exerted on the hypothalamo–pituitary–thyroid axis to increase thyroid hormone output to prefeeding levels. Both mechanisms may be operative.

6. Influence of Temperature on Thyroid Hormone Metabolism

Temperature influences $*T_4$ metabolism in both normal and hypophysectomized eels (Leloup, 1965) and in intact brook trout (Drury and Eales, 1968). No information is available concerning temperature effects on $*T_3$ metabolism of teleosts.

In the intact eel, Leloup (1965) observed faster disappearance of plasma protein-bound radioactivity (PB*I) after $*T_4$ injection at 24°C [biological half-life $(t_{1/2}) = 1.4$ days] relative to that at 10.5°C $(t_{1/2} = 2.5$ days). The same trend was observed in hypophysectomized fish. Peripheral metabolism of $*T_4$ (both deiodination and biliary–fecal excretion) was stimulated (no data presented) at the higher temperature in both intact and hypophysectomized fish.

Drury and Eales (1968) noted that the $t_{1/2}$ values for serum PB*I, after $*T_4$ injection of brook trout, were 18.0 days at 5°C, 3.2 days at 12°C, and 1.5 days at 15°C.

7. Rate of T_4 Secretion (Degradation) in Teleosts

Leloup and Fontaine (1960) were the first to estimate thyroid hormone secretion rates (TSR) of teleosts from an equation interrelating thyroidal radioiodide kinetic measurements with the plasma [127]I level. Their values ranged from 0.2 (0.31) to 350 (536) ng of hormonal iodine/100 g body weight/day. Values in parentheses are the T_4 equivalents of hormonal iodine obtained by multiplying hormonal iodine by 1.53. Considerable differences in TSR were noted not only between species, but also between fish of the same species at different developmental stages and exposed to dissimilar environmental conditions (different temperatures and states of physical activity). Using the T_4 substitution method (radioiodide method) Hoffert and Fromm (1959) estimated that T_4 secretion rates for 1- and 2-year-old rainbow trout at 13°C were 244 and 303 ng of L-T_4/100 g/day, respectively.

Other studies (Collicutt and Eales, 1974; Eales, 1977a,b; Higgs and Eales, 1977, 1978a,b) have measured T_4 degradation rates (TDR) of teleosts. They should be equivalent to the rate of T_4 secretion from the thyroid if the basic assumption is made that thyroid function in these fish is in a steady state. In other words, hormone loss from the blood should equal hormone entry into the blood from the thyroid. TDR or amount of T_4 undergoing degradation per unit time and body weight (ng T_4/100 g/day) is calculated as the product of the metabolic clearance rate (MCR) for T_4 (see Collicutt and Eales, 1974; Higgs and Eales, 1976, for method of determination) and concentration of T_4 (ng/ml) in the blood. With this procedure Collicutt and Eales (1974) found that channel catfish held at 12°–13°C and subjected to prolonged starvation had a TDR of 142 ng/100 g/day. Observed values of TDR for 1-year-old brook trout held between 10.5° and 12.5°C range from 20 to 179 ng/100 g/day (Higgs and Eales, 1977, 1978a,b) and are influenced by ration (Table XII).

Eales (1977a) has obtained the most accurate estimate of TDR in fish by separating T_3 and T_4 in body fluids using T_3- and T_4-specific antibodies and then analyzing T_4 kinetics on the basis of a two-compartment model. The TDR of 1-year-old rainbow trout held at 12°C and deprived of food for 3 days before *T_4 injection was noted in this study to be 51.6 ng/100 g/day.

C. Influence of Thyroid Hormone Treatment on Growth of Teleosts

Some studies have shown that the growth of *Salmo gairdneri, Salmo trutta, Salvelinus fontinalis, Oncorhynchus kisutch, Carassius auratus,* hybrids of *Xiphophorus helleri* and *X. maculatus, Lebistes reticulatus, Lepomis cyanellus,* and *Mugil auratus* can be stimulated by T_4 treatment (Table XIV). In addition, administration of mammalian thyroid powder, minced beef thyroid, or iodinated casein can enhance growth of *Salmo gairdneri, Salmo salar, Xiphophorus helleri,* and *Lebistes reticulatus (Poecilia reticulata)* (Table XIV). Four studies have demonstrated that administration of T_3 promotes growth of teleosts (Bjorklund, 1965; Bonnet, 1970; Higgs *et al.,* 1977b; Fagerlund, Higgs, and McBride, 1978 unpublished observations). Also, a recent unpublished study by T. J. Lam (1976) has shown that beef thyroglobulin mixed in the food at a dosage of 1 ppm can stimulate growth of *Siganus canaliculatus.* In contrast, several studies have not found any effects of thyroid hormone treatment on growth of *Salmo salar, Salvelinus fontinalis, Oncorhynchus keta, Platypoecilus variatus* and

Studies Assessing the Influence of Thyroid Hormone Treatment on

Species	Initial age, weight, or length	Environmental factors		Diet and ration	Hormone	Method of
		Temperature (°C)	Photoperiod			Dosage
Trout (species not given)	Fry	—	—	—	Thyroid	
Salmo irideus (rainbow trout)	Embryos	—	—	—	T_4	$1 : 1.0 \times 10^5$
Salmo irideus	—	—	—	—	T_4	—
Salmo gairdneri	E1,[e] 24 g E2, 28 g E3, 28 g E4, 21 g E5, 34 g	E1, 5–12 E2, 15–21 E3, 15–21 E4, 7–13 E5, 4–12	At least 3 hr/day light.	Beef muscle containing 0.0035 mg I/100 g. Fed 6 days/week.	Iodinated casein (7% I; 1.2% T_4) T_4 (1.2–2.4%)	—
Salmo gairdneri	Yearling	E1, 11.5–14.5 E2, 9 E3, 9 E4, 9 E5, 9	 12L : 12D 12L : 12D 12L : 12D 12L : 12D	Food in E1–E3 was minced fresh liver, spleen, and blowfly maggots. Food in E4 and E5 was lightly boiled liver embedded in gelatin. Fish in E1 fed restricted ration daily. Those in E2 fed to satiation daily. Fish. in E3–E5 fed 2 days out of every 3.	E1, thyroid E1, powder (4–5×/ week) E2, Na-L-T_4 E3, Na-L-T_4 E4, Na-L-T_4 E5, Na-L-T_4	$1 : 10$ $1 : 7.5$ $1 : 5 \times 10^5$ $1 : 5 \times 10^5$ $1 : 5 \times 10^5$ $1 : 5 \times 10^5$
Salmo gairdneri	Yearling E1, 16–21 g E2, 13–19 g	10.5		Chopped beef heart or minced liver embedded in gelatin fed daily.	Na-L-T_4	$1 : 5 \times 10^5$ (changed every third day)
Salmo gairdneri	Fry 3.6–5.1 g 69–76 mm	10	12L : 12D	Daily fed weighed meals of commercial fish pellets (22 μg I/g dry food).	L-T_4 (administered to radiothyroidectomized and untreated fish). Included irradiated and normal controls.	$1 : 1 \times 10^7$
Salmo gairdneri	Yearling 6–18 months	—	—	Fed on alternate days with trout pellets (River Pride) Fish not fed after injection of 5 μCi/g carrier free [^{35}S]sulfate.	T_4	$1 : 5 \times 10^5$

XIV

Growth of Fish with Reference to Effects on Amino Acid Metabolism

treatment			
Route	Duration (days)	Responses and remarks	Reference
d[a]	—	Body more slender.	Sklower (1927)[b]
w[c]	—	Stimulation of growth, vascular abnormalities causing death.	Baumann and Pfister (1936)[b]
ip[d]	—	Increased guanine content of skin. Stimulated incorporation of [$^{14}C_2$]glycine into newly synthesized skin guanine.	Matty and Sheltawy (1967)
d	E1, 75 E2, 42 E3, 38 E4, 75	Iodinated casein in a low iodine diet can enchance growth but effect may be dependent upon dosage, age, physiological state of fish, season, and length of treatment. Ingestion of 10–50 mg of iodinated casein/day recommended for trout of 20–65 g.	Fontaine and Baraduc (1955)
d	E5, 84	T_4 in diet stimulated growth in most experiments.	
d	59	Percent gains over initial weight were: E1, treated, 114.6 E1, control, 85.1	Barrington et al. (1961)
w	35	E2, treated, 98.3 E2, control, 40.7	
w	35	E3, treated, 68.0 E3, control, 31.2	
w	37	E4, treated, 33.9 E4, control, 29.7	
w	36	E5, treated, 44.4 E5, control, 22.4	
		Thyroid treatment also stimulated linear growth and increased silvering. T_4 treatment decreased alimentary tract lipid. Suggested that thyroid hormones either directly enhance growth or act synergistically with other hormones or exert influence on metabolic processes. Growth responses to T_4 depend upon experimental conditions.	
w	15 or 28	Growth rate (weight and length) was significantly greater for T_4-treated fish than controls. Stimulated uptake of [^{35}S]sulfate into bone and cartilage of branchial skeleton and its binding into chrondroitin sulfate. T_4 probably has a synergistic action with growth hormone.	Barrington and Rawdon (1967)
w	35	Enhanced growth (length) of control and radiothyroidectomized trout. Five weeks of T_4 treatment raised the growth rate of radiothyroidectomized trout to control levels. Concluded that thyroid hormones essential for normal growth and exert their influence early in development.	Norris (1968, 1969)
w	6–14 (prior to injection of [^{35}S]sulfate)	Stimulated significantly higher uptake of [^{35}S]sulfate into cells and matrix of branchial skeleton cartilage. Rate of excretion of [^{35}S]sulfate lower in hormone-treated fish. T_4, by enhancing sulfate uptake, could exert a growth-promoting influence upon cartilage. T_4 probably acting in conjunction with growth hormone.	Barber and Barrington (1972)

(Continued)

Table XIV—

Species	Initial age, weight, or length	Environmental factors		Diet and ration	Method of	
		Temperature (°C)	Photoperiod		Hormone	Dosage
Salmo gairdneri	—	—	—	—	Iodinated casein	
Salmo gairdneri	46–57 g	12–13	12L : 12D	Fed commercial trout pellets (Victor Fox Foods Ltd., Winnipeg, Man.) at approximately 1% wet body weight/day. Fish starved 24 hr prior to radioleucine injection.	T_4 or T_3 (monosodium) salt	$1 : 1.0 \times 10^7$ (free acid)
Salmo trutta (brown trout)	20–40 g	21	Constant illumination	Starved	Na-L-T_4	1×10^{-6} (*in vitro*)
	60–100 g	14		?	T_4	200–800 µg/ fish/week for 2 months
Salmo trutta	15–18 months (mostly sexually mature males)	10	12L : 12D	Fed meal worms and chopped ox liver *ad lib.* every 3 days.	T_4	1.25 µM 2.50 µM 3.75 µM (1 ppm) 8.75 µM 12.5 µM
Salmo trutta	Alevin 13–31 days old	—	—	—	T_3	8×10^{-7} and 4×10^{-7} dilutions of T_3
Salmo salar (Atlantic salmon)	E1, parr (5–15 cm)	E1, 8–10 E2, 10–12 E3, 9–13		E1, excess fresh ground beef liver provided every other day.	E1, Na-thyroxinate (2×/week)	200 µg
					Commercial thyroid extract (Casgrain and Charbonneau Ltd.) and beef liver.	1 : 1
					Pulverized beef thyroid and liver.	75% thyroid 25% liver
	E2, parr (5–15 cm)			E2, horse liver and spleen.	Desiccated beef thyroid mixed with diet.	1 : 1
	E3, fry (3–5 cm) Parr (5–15 cm) Smolt (15–25 cm)			E3, beef liver. Fry fed 5×/week; parr and smolts were fed 3×/week.	Desiccated beef thyroid mixed with diet.	1 : 1

treatment

Route	Duration (days)	Responses and remarks	Reference
d		Stimulated incorporation of $[1-^{35}S]$methionine into brain proteins. This was observed at 6 or 7 hr after injection of the amino acid.	Leloup *et al.* (1972)
w	1 or 6	T_4 administration for 6 days significantly increased level of radioactivity [specific activity (dpm/mg protein); L-$[1-^{14}C]$leucine injected 8 hr previously] in plasma precipitate. Also, T_4 treatment (1 or 6 days) decreased liver supernatant radioactivity, increased precipitate radioactivity, and increased incorporation (% uptake) of radioleucine into liver protein. Similar results for gill (T_4 administered 1 day). T_4 treatment (1 day) increased protein content (mg protein/g) of liver and gill. T_3-treated (1 day) fish had reduced protein concentration in liver, but higher protein concentration in gill relative to T_4 fish.	Narayansingh and Eales (1975a)
Buffer solution.		Ammonia production of gills treated with T_4 *in vitro* was not significantly different from that of untreated gills.	
		Ammonia production from isolated gills from T_4-treated trout less than that of the controls. T_4-treated trout grew larger than control group which indicates a protein anabolic effect (no data presented).	Thornburn and Matty (1963)
w (changed every 2 days)	13	Stimulated growth (length) at the 3.75 and 8.75 μM doses.	Massey and Smith (1968)
w	31	Stimulation of growth and development of lacrymal bone. Some indication that T_3 is more than 8 times as potent as T_4.	Qureshi (1976)
im^f	79	In E1, growth (weight) of thyroid treated fish less than that of controls. Pallor developed due to fewer pigment cells. Thyroid treatment resulted in protrusions on top of head, a lateral protrusion of the opercula and enophthalmos.	LaRoche and Le-Blond (1952)
d		In E2, thyroid treatment had no significant effect on growth in length of parr. Also weight gains of thyroid-treated fish less than those of control fish. Decrease in pigmentation of thyroid-treated fish noted as well as alteration in skull size. Skin of treated fish showed a marked increase in the thickness of the derma. Epidermis also thickened.	
d		In E3, only thyroid treatment of parr enhanced growth in weight. Pallor most pronounced in smolts. Derma of treated fish thicker mainly in parr and smolt.	
d	152		
	65		

(Continued)

Table XIV—

| Species | Initial age, weight, or length | Environmental factors | | Diet and ration | Method of | |
		Temperature (°C)	Photoperiod		Hormone	Dosage
Salmo salar	Parr 30–34 g	5 (winter) 10 (summer)		Minced horse liver.	Desiccated beef thyroid mixed with food.	1 : 1
Salmo salar	Parr (1 year old)	—	—	Raw minced beef liver (3×/day).	Raw minced beef thyroid mixed with food.	1 : 4
Salvelinus fontinalis (brook trout)	Hatchlings	—	—	—	T_4	$1 : 1.25 \times 10^5$
Salvelinus fontinalis	Parr (5–8 cm)	—	—	Ground beef liver 3×/week.	Dried beef thyroid mixed with food.	1 : 1
Salvelinus fontinalis L.	5–8 cm	—	—	Minced beef liver.	Desiccated beef thyroid mixed with food.	1 : 1
Salvelinus fontinalis	46.4–54.6 g	12–13	12L : 12D	Commercial trout pellets (Victor Fox Foods Ltd.) fed at approximately 1% wet body weight/day. Fish starved 24 hr prior to radioleucine injection.	T_4 (monosodium salt).	$1 : 1.0 \times 10^8$ $1 : 1.0 \times 10^7$ $1 : 2.0 \times 10^6$ (free acid)
Salvelinus fontinalis	—	12	12L : 12D	Commercial trout pellets (Victor Fox Foods Ltd.) fed. Fish were starved or fed either 0.6 or 1.2% of initial mean body weight/day.	L-T_4 (sodium pentahydrate)	$1 : 1.0 \times 10^7$ (free acid)
Oncorhnychus kisutch (coho salmon)	Fingerlings	—	—	—	Na-thyroxinate. Desiccated thyroid.	$1 : 1.125 \times 10^6$ —
Oncorhynchus kisutch	Underyearling 8 g 9 cm	10	12L : 12D	Oregon Moist Pellet (OMP-2) salmon diet fed twice daily to excess (8% wet body weight/day)	L-T_4 (sodium pentahydrate)	0.5 or 5 μg/g (2×/week) 30 μg/g (1×/3 week)

Continued

Route	Duration (days)	Responses and remarks	Reference
d	~330	No effect on growth (weight) of thyroidectomized or intact fish.	LaRoche and Le-Blond (1954)
d	69	Promoted growth (length) and silvering.	Piggins (1962)
w		Acceleration of growth rate, darkening, and spasms.	Herzfeld et al. (1931)[b]
d	56	No effect on growth. Thyroid treatment resulted in marked depigmentation and thickening of the epidermis and derma. Also head was enlarged (broader) with increased amounts of connective tissue present in the orbital space.	LaRoche and Le-Blond (1952)
d		No difference in growth (weight) between thyroid-treated and control fish. Reduction in number of pigment cells, an increase in thickness of dermis and epidermis, and an increase in size of the head explained by a hyperplasia of the connective tissue elements of the interorbital region.	LaRoche (1953)
w	1 or 3	T_4 $(1:1.0 \times 10^8)$ for 3 days decreased supernatant radioactivity in gill (specific activity; 4 hr after L-[1-^{14}C]leucine injection) and increased incorporation (% uptake) of radioleucine into plasma protein. T_4 $(1:1.0 \times 10^7)$ decreased plasma supernatant radioactivity and decreased both liver supernatant and precipitate radioactivity. T_4 $(1:2.0 \times 10^6)$ increased the radioactivity in the plasma precipitate. T_4 $(1:1.0 \times 10^7$ for 1 day) increased plasma protein concentration (mg protein/g). Also there was greater incorporation of radioactivity (% uptake) 8 hr after radioleucine injection into liver and gill protein.	Narayansingh and Eales (1975a)
w	28	Reduced loss in body weight of starved fish and increased growth of fed fish especially at higher ration. Use of non-physiological hormone dosages and improper control of ration cited as likely causes for inconsistencies in growth response to thyroid hormone treatment in other studies. T_4 treatment resulted in decreased HSI of starved fish and increased HSI in fed fish (1.2% ration). Growth promoting influence of T_4 accompanied by trend toward mobilization rather than deposition of lipid.	Narayansingh and Eales (1975b)
w	—	Slenderized shape; increased silvering.	Smith (1949)[b]
d	—		
im	84	Growth rates (weight) for T_4-treated fish (0.97–1.1%/day) were significantly higher then those of control fish (0.53–0.59%/day). A similar trend was noted for length. Increases in weight above control at 84 days ranged from 47 to 78%. Administration of higher T_4 doses caused progressive decline in condition factor of fish from control range. Cessation of hormone treatment stopped and reversed this trend. Percentages of muscle protein were decreased while lipid percentages were increased by T_4 treatment. Higher T_4 doses stimulated ovarian maturation and degranulation of pituitary somatotrops. Injection of higher dose led to pronounced growth anomalies in skull and fin structure. Fish appeared silver (all doses) and parr marks disappeared.	Higgs et al. (1976)
im			
Hormone: cholesterol implant into muscle.	84		

(Continued)

Table XIV—

| Species | Initial age, weight, or length | Environmental factors | | Diet and ration | Hormone | Method of |
		Temperature (°C)	Photoperiod			Dosage
Oncorhynchus kisutch	Yearling 10–15 g	10–10.4	Natural	Oregon Moist Pellet salmon diet fed to excess (8% wet body weight/day)	L-T$_4$ (sodium pentahydrate)	1 μg/g (1×/week)
Oncorhynchus keta	Eggs	E1, 10–13 E2, 5.5–12	—	—	Synthetic T$_4$ sodium	1 : 2.5 × 10^6 1 : 5.0 × 10^6 1 : 12.5 × 10^6 (Solutions changed weekly
Oncorhynchus keta	Alevins (treated either immediately or 1 week after hatching)	Max. (5–12) Min. (1–8.5)	—	—	T$_4$	1 : 1.0 × 10^6
Oncorhynchus keta	Larvae	—	—	—	T$_4$	1 : 1.0 × 10^6
Oncorhynchus keta	Eggs	5–11	—	—	T$_4$	1 : 1.2 × 10^7 (renewed weekly)
Carassius auratus (goldfish)	5 months– 1 year 2–10 g	20.5 and 21.2	10L : 14D	Fed an average of 1.7% body weight/day of a commercial pellet diet (Perk Foods Co., Chicago Illinois) raw minced beef liver. Fish fed 6 days out of 10.	Na-L-T$_4$ Na-L-T$_3$ Desiccated parrot fish thyroid gland.	2 μg/g bw (4×/10 days) to ♂ 2 (♂ and ♀) μg/g and 10(♀)μg/g 5 μg/g to ♂
Carassius auratus	10–15 g	20	—	Fed pablum supplemented twice weekly with shrimp meal. Also a high protein diet which consisted of a dried mixture of 40% ground beef liver, 40% salmon viscera and 20% shrimp meal. Fish starved 16–20 hr before collection of excretion products.	Synthetic *dl*-thyroxine sodium.	1 : 2.5 × 10^6 (solutions changed every third day).
Carassius auratus	10–15 g	14 or 21		"Pinch" of "bemax" fed once per day at same time.	L-T$_4$ sodium salt	1 : 2.5 × 10^6 1 : 5.0 × 10^5 1 : 1.0 × 10^6 1 × 10^{-6}

treatment

Route	Duration (days)	Responses and remarks	Reference
im	59	The growth rate (weight) of T_4-treated fish (0.93%/day) was significantly higher than that of the control fish (0.65%/day). Significant increases in length were also found. Percentage gain in weight over control at 59 days was 23.2. T_4 treatment led to a gradual increase in condition factor and greater muscle lipid content.	Higgs *et al.* (1977a)
w w w	(101 or 121 from time of fertilization)	Advanced hatching time by 4–8 days but reduced growth rate (length) of developing fish. Marked exophthalmia noted at hatching time. T_4 treatment increased pectoral fin length. Growth of body wall over yolk sac accelerated in both normal fish and fish previously treated with thiourea. T_4 reduced pigmentation and parr marks were absent. Guanine deposition was enhanced suggesting an acceleration of nucleoprotein metabolism.	Dales and Hoar (1954)
w	42 or 49	Reduced growth rate (weight), delayed rate of yolk sac absorption, broadened the head and resulted in enophthalmos. Pectoral fins were elongated. Guanine deposition stimulated, silvering.	Honma and Murakawa (1955)
w		Acceleration of metamorphosis, silvering; Increased: height of skull, elongation of opercular bones, hyperplasia of orbital connective tissue. Inhibited yolk absorption and differentiation of some viscera; retarded growth rate.	Honma (1960)
w	92	T_4 treatment of control or thiourea-treated groups advanced time of egg hatching. There was stimulation of growth of body wall over yolk sac and increased guanine deposition.	Ali (1961)
ia[g]	70 (data based on first 20 days of treatment)	T_4 and T_3 treatment increased thickness of epidermis. Paling of xanthic coloration was found. Food conversion efficiency for T_4-treated fish was 44.4%, that for T_3-treated fish ranged from 43.9 to 44.9% and that for fish receiving parrot fish thyroid extract was 32.7%. Values for control fish ranged between 30.2 to 33.0%. T_4 and T_3 treatment enhanced growth (weight and length). Administration of parrot fish thyroid extract was without effect. Suggested that effect of T_4 and T_3 on growth probably not direct but dependent upon a synergistic action, possibly with growth hormone.	Bjorklund (1958a,b, 1965)
ia	70		
ia	70		
w	7, 14, and 21. N excretion measured over 1 day.	Marked increase in nitrogen excretion (as measured by changes in the ammonia nitrogen content of the ambient water) found. Protein content of diet influenced nitrogen excretion of untreated fish.	Hoar (1958)
w		Ammonia production of goldfish immersed in T_4 solution increased at 21°C but no effect at 14°C. Immersion of fish in T_4 for 9–14 days increased free amino acid in liver and muscle, however, there was no change in muscle keto acids and a decrease in muscle creatine. Gills from fish pretreated with T_4	Thornburn and Matty (1963)

(Continued)

Table XIV—

| Species | Initial age, weight, or length | Environmental factors | | Diet and ration | Method of | |
		Temperature (°C)	Photoperiod		Hormone	Dosage
Anguilla (eel)	—	—	—	—	dl T$_4$	20 µg/day
Fundulus heteroclitus (killifish)	1.0–1.5 g	20	Natural	Fed ground quahog or brine shrimp daily.	T$_4$ or T$_3$	1 µg
Platypoecilus variatus *P. maculatus* (platyfish)	Young (sexually immature)	—	—	Ground liver and young brine shrimp (alternate days).	Desiccated mammalian thyroid powder.	"Pinch" (scattered on water daily).
Xiphophorus helleri (swordtail) and *X. maculatus* (platyfish) hybrids	Young males	—	—	—	T$_4$ mixed with food.	—
X. maculatus (BH strain)	0.5–2 months	—	—	Dried shrimp or liver-pablum (6×/week).	Desiccated thyroid tablets.	0.032–0.065 g (1×/week)
X. helleri	Juvenile	—	—	—	Beef thyroid mixed with food.	<3%
Lebistes reticulatus (guppy)	Young males	—	—	—	T$_4$	—
Lebistes reticulatus	Newborn	24–25	Light provided most of time.	Fed dried insect larvae daily.	Desiccated mammalian thyroid powder mixed with food. Synthetic T$_4$ (Roche-Organon)	1 : 1 8–15 drops/day of 2 mg/ml soln. to 5-gal. tank.
Lebistes reticulatus	Young	—	—	—	Mammalian thyroid powder.	0.2 g (2×/week)

treatment

Route	Duration (days)	Responses and remarks	Reference
		and incubated in the presence of an amino acid substrate had increased ammonia production. Addition of T_4 to muscle homogenate resulted in an 18% greater uptake of DL-[1-^{14}C]-leucine relative to that of the control. Concluded that T_4 accelerated protein metabolism from the amino acid upwards and that T_4 treatment may lead to increased passive permeability of gill cell to ammonia.	
—	7	Stimulated incorporation of [1-^{35}S]methionine into brain protein.	LeLoup *et al.* (1972)
ip (18 hr prior to injection of L-[1-^{14}C]-leucine)		Fed-untreated fish had, 6 hr after [^{14}C]leucine injection, greater incorporation of radioactivity into muscle protein than starved fish. T_3 produced a trend toward increased muscle protein synthesis but T_4 without effect. Trend toward increased protein synthesis with T_3 may be a reflection of the effects of thyroid hormones on connective tissue development as shown in other studies.	Jackim and La-Roche (1973)
Oral and w?	Several months	Marked bilateral exophthalmos, decreased growth rate (length), altered body proportions (longer in proportion to depth; fin elongation), and precocious sex maturation (indicated by early but atypical differentiation of male gonopod) noted.	Grobstein and Bellamy (1939)
d		Increased growth rate; inhibition of melanotic tumor development.	Stolk (1959)[b]
d		Thyroid treatment of radiothyroidectomized (^{131}I) fish reduced mortality. Growth (length) was greater than that of ^{131}I-treated fish and equivalent to that of untreated controls. Body shape was normal but more streamlined than that of untreated controls. Walls of ventral aorta were thickened. Stimulated maturation and development of male secondary sexual characteristics and elongation of caudal fin. No exophthalmia occurred.	Baker-Cohen (1961)
—	—	Stimulated growth, elongation of all fins, precocious development of secondary sex characteristics and exophthalmos.	Lam (1973)
—	—	Retardation of growth and earlier male maturation.	Svärdson (1943)[b]
d	60–90	No effect on growth (length) or on rate of development of male sexual characteristics. Exophthalmia not observed.	Smith and Everett (1943)
d		Thyroid treatment prevented the thiourea-induced retardation of growth and of male secondary sex differentiation.	Nigrelli *et al.* (1946)[b]

(Continued)

Table XI

| Species | Initial age, weight, or length | Environmental factors | | Diet and ration | Hormone | Meth |
		Temperature (°C)	Photoperiod			Dosage
Lebistes reticulatus	Immature (treated soon after birth)	—	—	Fed infusoria, brine shrimp and commercial food.	Desiccated mammalian thyroid powder.	1 : 1.0 ×
Lebistes	Newborn	22	—	—	T$_4$	1 : 1.0 ×
Lebistes reticulatus	Newborn	25	—	Regular flora and fauna plus finely ground dog food.	Mammalian thyroid tablets mixed with food or finely ground thyroid powder by itself.	1 : 1
Lebistes reticulatus	Young	—	—	—	Mammalian thyroid powder.	Experime group f 50 mg with fo
Poecilia		—	—	—	T$_4$	—
Lebistes reticulatus (*Poecilia reticulata*)	Young but fully mature males	26	12L : 12D	—	T$_4$	1 : 2.0 ×
Poecilia reticulata	Juvenile	—	—	—	Beef thyroid mixed with food.	<3%
Lepomis cyanellus (Rafinesque)	2–4 years E1, 13.7–13.9 g E2, 15.9–17.3 g E3, 11.4–12.2 g	23.3–25.6	8L : 16D 16L : 8D 8 L to 16 L 16 L to 8 L	Fed daily to excess with a commercially prepared dry pellet food or with frozen beef liver.	Na-L-T$_4$	100 μg/we
Mugil auratus L.	—	19–23	—	Fed 5% of body weight/ day. Diet contained 63% starch, 4% lipid, 25% protein (albumen), 0.015% vitamin premix, 6% binder.	L-T$_4$ L-T$_3$	50 mg/10 50 mg/10

a d, diet.   *b* From Baker-Cohen (1961).   *c* w, water.   *d* ip, intraperitoneal injection.

556

eatment			

Route	Duration (days)	Responses and remarks	Reference
(changed every fourth day)	90–180	Stimulated growth (length) of female and male guppies. Response of females was greater than that of males. Growth enhancement attributed to correct method of hormone administration and dosage. Fin elongation, bilateral exophthalmos, and acceleration of male gonopodium differentiation noted.	Hopper (1950, 1952)
(renewed each week)	70	Growth (length) of T_4-treated fish for first 30–40 days slower than that of controls. Then treated and controls grew at same rate. Males responded similarly to females. Advanced gonadal maturation and differentiation of male and female sexual characteristics noted. Exophthalmia not observed.	Gaiser (1952)
d	20	Thyroid hormone treatment did not abolish suppressing effects of thiourea on growth. Time of sexual differentiation of male fish delayed. Speculated that mammalian thyroid powder is either not absorbed into blood in an active form or that thyroid hormone does not effect fish growth. No growth anomalies observed.	Smith *et al.* (1953)
	60	No effect on growth (length) or on the differentiation of the testes or ovaries. Method of hormone administration may affect growth response to thyroid hormone treatment as well as experimental conditions.	Hopper (1961)
—	—	T_4 treatment of hypophysectomized *Poecilia* did not enhance growth.	Ball (1963 unpublished observations, in Ball *et al.*, 1965)
(half changed 3×/week)	28	Stimulated growth (weight) of normal male animals and abolished inhibitory effect of thiourea treatment on growth. Stimulatory effect of T_4 on growth may be due to synergism with growth hormone or to an effect on growth hormone release from acidophil cells of the proximal pars distalis.	Sage (1965, 1967)
		Promoted growth, elongation of all fins and rate of fin regeneration. Precocious development of secondary sex characteristics and exophthalmos noted.	Lam (1973)
ip	42	Enhanced growth (weight and length), increased feed intake (not significantly) and improved food conversion efficiency. Effect of thyroid on growth may be mediated through increased efficiency in food conversion.	Gross *et al.* (1963)
dry d dry d	20	Percentage increase in weight, relative to that of controls, was higher for T_4-treated fish (58.1%) than for T_3-treated fish (8.6%). T_4 treatment lowered hepatic glycogen (-35.7%), while T_3 was without effect. Both T_4 and T_3 increased liver cytochrome oxidase activity. T_4 treatment reduced liver glucose-6-phosphate dehydrogenase activity but increased malate dehydrogenase activity. The muscle free amino acid level was lowered by T_4 treatment (T_3 without effect) perhaps indicating stimulation of protein synthesis.	LeRay *et al.* (1969); Bonnet (1970); LeRay *et al.* (1970)

E, experiment (e.g., E1 = experiment 1).   [f] Intramuscular injection.   [g] Intraabdominal injection.

P. maculatus, and *Lebistes reticulatus* (Table XIV). These contradictory findings have formed the basis of considerable controversy in the literature as to whether the fish thyroid is essential for normal growth.

The causes for such discrepancies have been attributed to lack of optimal environmental or experimental conditions (Barrington *et al.,* 1961; Hopper, 1961), differences in method of hormone administration (Hopper, 1952, 1961; Smith *et al.,* 1953), dosage of hormone (Hopper, 1952; Fontaine and Baraduc, 1955; Lam, 1973; Narayansingh and Eales, 1975b), improper control of ration (Narayansingh and Eales, 1975b), and the possible administration of T_4 to fish in the presence of naturally occurring goitrogens and/or when the fish have been on low iodine intakes (LaRoche *et al.,* 1966; see also Section IV,B,2). Additional variables include season of the year and the possibility that salmonids may respond differently to thyroid hormone treatment at different stages in their life history (Fontaine and Baraduc, 1955). Some of these variables will be considered in more detail below.

The results from a study by Barrington *et al.* (1961) show that lack of optimal environmental conditions abolishes the growth response of rainbow trout to T_4 treatment. In one experiment, yearling rainbow trout at 9°C were immersed in a $1:500,000$ T_4 solution. None of the fish had less than 5 liters of water as living space. The range in mean initial weights for the groups of fish was 13.9–14.2 g and all fish were fed daily to excess. Significant enhancement of growth was obtained (Fig. 15). In another experiment, conditions were the same except diet composition was changed, fish were fed 2 days out of every 3, and mean initial group weights ranged from 27 to 27.9 g. Growth stimulation was not found (Fig. 15B). There are several possible explanations. First, reduction of food intake in the second experiment may partially account for the negative response. For example, Narayansingh and Eales (1975b) found that ration influenced the growth response of brook trout to T_4 treatment (Fig. 16). Greatest growth was noted in those fish fed the highest level of dietary intake. In addition, alterations in diet composition and ration influence T_4 metabolism in brook trout (Section IV,B,5) and it is likely that this is the case in rainbow trout. In the second experiment less conversion of T_4 to T_3 would be anticipated. If, as Higgs and Eales (1977) have postulated for the brook trout, T_3 is the predominately active form of thyroid hormone in rainbow trout, then reduced peripheral conversion of T_4 to T_3 could partly explain the lack of growth enhancement. Finally, the larger size of fish in the second experiment may have created more stress and also may have reduced the uptake of T_4 from the water. Concerning the latter point, Eales (1974) found an inverse correlation between body

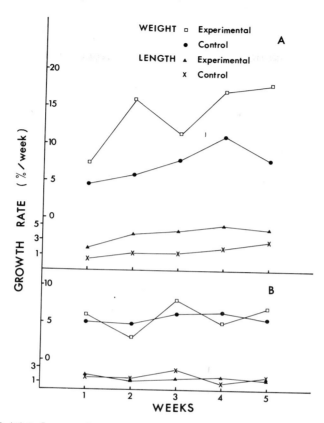

Fig. 15. (A) Influence of Na L-thyroxine (T_4) on the mean weekly specific growth rates for weight and length of yearling rainbow trout held at 9°C and maintained on a 12L : 12D photoperiod. Experimental fish were immersed in a 1 : 500,000 T_4 solution and water was changed every third day. None of the fish ($n = 12$/group) had less than 5 liters of water as living space. All fish were fed daily to excess with blowfly maggots on 2 days out of every 3 and minced liver on the third. Mean initial weights were 14.2 g (control group) and 13.9 g (experimental group). (From Barrington *et al.*, 1961.) (B) Replicate experiment of that shown in A except the trout were fed 2 days out of every 3, food consisted of lightly boiled liver embedded in gelatin, number of fish per group was 10, and mean initial weights ranged from 27.0 g (experimental) to 27.9 g (control). (From Barrington *et al.*, 1961, *Gen. Comp. Endocrinol.* **1**, 170–178.)

size of channel catfish and their plasma T_4 levels after immersion in a solution of 10 μg T_4/100 ml.

Method of thyroid hormone administration has probably had a profound influence on the type of results obtained. For example, fish receiving hormone via the oral route often fail to show a positive

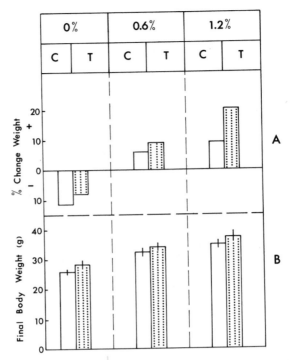

Fig. 16. Influence of ration on percentage change in body weight (A) and mean final body weight (B) for control (C) and thyroxine-treated (T) brook trout held at 12°C for 28 days. L-Thyroxine (T_4) in 0.1 N NaOH was mixed with the water to provide an ambient T_4 (free acid) level of 10 μg/100 ml. Fish were starved (0%) or fed Victor Fox dry trout pellets at 0.6 or 1.2% of initial mean body weight per day. Percentage change in body weight was calculated from the initial and final weights of each group of fish. Vertical lines represent 1 SE on either side of mean ($n = 15$). (From Narayansingh and Eales, 1975b, *Comp. Biochem. Physiol. B* **52**, 407–412.)

growth response. Thyroid hormones have been administered to fish in four different ways (Table XIV): intramuscular (im) or ip injection, implantation of T_4-cholesterol pellets into the muscle, addition of the hormone to the water, and mixing T_4 and/or T_3 and desiccated thyroid powder with the food.

It appears that the im injection and pellet implantation procedures used by Higgs *et al.* (1976, 1977a) for administering T_4 to coho salmon at 10°C lead to a sustained elevation in plasma T_4 concentration above normal circulating levels. This conclusion is based on success with the pellet procedure (Figs. 1C and 2C) and on recent findings by Hurlburt (1977) for the goldfish, and also on data of Higgs and Eales (1976) which indicate that the turnover of *T_4 in plasma of im-injected brook trout is slower than that in heart and coelom-injected trout. Higgs and Eales (1976) also found evidence of slow release of $*T_4$ from muscle to blood. The ip injection route probably does not maintain a physiological chronic elevation of plasma T_4. This method of injection leads to major surges in plasma T_4. Excess T_4 is then rapidly eliminated from the body via the biliary–fecal route (Section IV,B,4). This is the case at least in brook trout (Eales, 1974) and most likely in other species as well.

Eales (1974) has demonstrated that the immersion procedure is an effective way to increase plasma T_4 levels in brook trout and other teleosts. He noted that the response to an ambient T_4 level of 10 μg/100 ml was influenced by temperature, food intake, and species. Hence these factors should be considered before this method is employed in a growth study.

There is good indication from $*T_4$ studies on starved and fed brook trout (Eales and Sinclair, 1974) that T_4 is poorly absorbed from the gut lumen into the circulation. Eales and Sinclair (1974) state that the poor uptake of $*T_4$ from the trout intestine may be due to poor absorptive properties of the starved mucosa or binding of $*T_4$ to gut lumen contents. In fed fish, T_4 may bind to some of the dietary constituents, thereby reducing the extent of its absorption from the gut. Also, any hormone which is absorbed in such fish is rapidly cleared from the plasma. This information helps to explain why orally administered T_4, even at a dosage of 500 ppm, is ineffective in promoting growth of coho salmon (Higgs et al., 1977b) and why in many other studies relatively large amounts of thyroid hormone have been mixed with the diet to promote growth (Table XIV). In some cases, the levels of thyroid hormone in the diet have been so high as to result in pronounced growth anomalies in the skull, fins, and skin (Table XIV). Such effects have also been observed in coho salmon, but to a lesser degree, after im injection of 10 μg T_4/g/week for 84 days (Higgs et al., 1976).

While the oral route is probably not suitable for administering T_4, Higgs et al. (1977b) and Fagerlund, Higgs, and McBride (1978 unpublished observations) have found that this is not the case for T_3. Doses of T_3 as low as 4 ppm result in enhancement of growth of underyearling coho salmon held at 10°C. Therefore, it is necessary to establish the correct dosage and form of thyroid hormone for any given administration route, nutritional state, and set of environmental conditions so that growth stimulation is obtained without side effects. If thyroid hormone doses are excessive, either growth inhibition (Lam, 1973) and/or mortality will ensue.

Thyroid hormones probably do not act alone to promote growth of fish. The results of several studies suggest that T_4 can stimulate growth hormone release from pituitary somatotrops (Sage, 1967; Singh, 1971; Leatherland and Hyder, 1975; Higgs et al., 1976).

D. Influence of Thyrotropin, Antithyroid Compounds, and Radiothyroidectomy on Teleost Growth

Enomoto (1964) found a slight stimulation of growth of rainbow trout *Salmo irideus* after TSH administration (Table XV). But Pickford and Grant (1968) could not promote growth of hypophysectomized

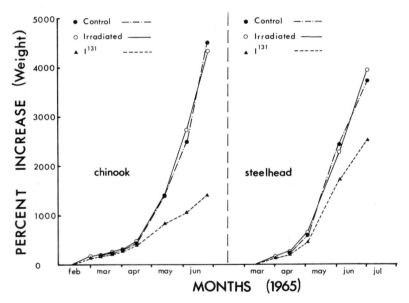

Fig. 17. Influence of radiothyroidectomy on growth (mean percentage increase) of underyearling steelhead rainbow trout and chinook salmon. Steelhead trout alevins (5-week-old, 24 mm, 0.15 g) and chinook salmon alevins (10-week-old, 31 mm, 0.3 g) were immersed in a solution of radioiodide (^{131}I; 15 μCi/ml) for 4 (steelhead) or 8 (chinook) days. A control group of untreated chinook and steelhead trout, of the same developmental stage, were exposed to a quantity of external γ radiation equivalent to that received by the radioiodide-treated fish. After thyroid destruction was complete, the growth of three groups of steelhead (760 radiothyroidectomized, 229 irradiated, 999 untreated controls) and three groups of chinook salmon (200 radiothyroidectomized, 80 irradiated, 180 untreated controls) was compared. Water temperature ranged from 8° to 22°C. Fish were fed daily to satiation on a diet of commercial fish pellets (iodine content 22 μg/g dry food) and were maintained on a natural photoperiod. (From Norris, 1966, 1969.)

male killifish with TSH. This finding is in accordance with unpublished observations of Ball (1963, cited in Ball *et al.*, 1965) who did not observe growth enhancement in the hypophysectomized guppy, *Poecilia*, after T_4 treatment. Hence, it is likely that thyroid hormones do not influence fish growth when one or more pituitary hormones are absent.

Thyroid hormone deficiency, created by the presence of glucosinolates in the diet, by administration of thiocarbamide derivatives such as thiourea and propylthiouracil, or by radiothyroidectomy, results in depression of fish growth (Higgs *et al.*, 1978b; Table XV, Fig. 17). In some studies the negative effects of either antithyroid compounds or radiothyroidectomy on growth have been abolished by T_4 treatment (Sage, 1967; Norris, 1969) (Fig. 18). This suggests that thyroid hormones are necessary for normal growth. However, these types of studies are open to some criticism. Some antithyroid compounds are antioxidants and therefore the observed depression of growth following their administration may be a toxic effect. Also, treatment of fish with large doses of radioiodide may not only destroy thyroid follicles,

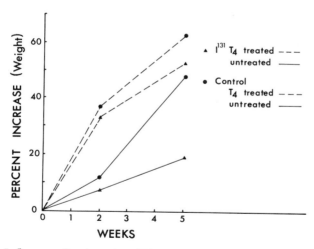

Fig. 18. Influence of L-thyroxine (T_4) on growth (mean percentage increase) of radiothyroidectomized and untreated steelhead trout. Three months after complete thyroid destruction, 10 radiothyroidectomized (69 mm, 3.67 g) and 10 untreated (73 mm, 4.74 g) steelhead trout were immersed in a solution of T_4 (1:10,000,000). Ten radiothyroidectomized (70 mm, 3.62 g) and 10 untreated (76 mm, 5.11 g) trout did not receive T_4 treatment. The four groups of fish at 10°C on a 12L:12D photoperiod were maintained in separate 80-liter aquaria. Aquaria solutions were renewed weekly and all fish were fed weighed meals of a commercial pelleted diet. Refer to Fig. 17 for additional details. (From Norris, 1966, 1969.)

Some Studies (Post-1960) That Have Assessed the Influence of Thyrotropin (TSH),

Species	Initial age, weight, or length	Environmental factors		Diet and ration	Method of	
		Temperature (°C)	Photoperiod		Hormone or antithyroid compound	Dosage
					(A) TSH	
Salmo irideus	8.2–8.6 cm 7.0–8.3 g	16		Pelleted diet (1 g/fish/day). Composition contained 40% crude protein, 2% crude fat, 5% crude cellulose, 14% crude ash, 0.1% $CaCO_3$, and 38.9% carbohydrate. Trout starved 1 day prior to injection.	TSH	0.02 IU
Fundulus heteroclitus (male, hypophysectomized)		20	8L : 16D	Fed daily with Aronson's mixture supplemented twice weekly with frozen brine shrimp.	Bovine TSH TSH-USP **(B) Antithyroid compounds**	1 mU/g 0.05 mU/g 0.1 mU/g
Salmo trutta	15–18 months (mostly sexually mature males)	10	12L : 12D	Fed mealworms and chopped ox liver every 3 days *ad lib.*	Thiourea	0.10% (13.14 mM)
Salmo gairdnerii	6 months	10		Complete test diet (Halver *et al.*, 1962).	3-Amino-1,2,4-triazole (aminotriazole).	30 mg/100 g dry diet
Oncorhynchus keta	Eggs	5–11	—	—	Thiourea	0.05% (solutions changed weekly)
Mystus vittatus (female)					Thiourea	0.03%
Lebistes reticulatus (*Poecilia reticulata*)	Young fully mature males	26	12L : 12D		Thiourea	0.1% (half of solution changed 3×/week).
Mugil auratus		19–23		Fed 5% body weight per day.	Propyl thiouracil (PTU)	1 or 5% of dry diet.
					(C) Radiothyroidectomy	
Salmo gairdnerii	Juvenile (3 g)	10		Low iodine (I) diet (0.1 μgI/g dry diet) fed to intact or ^{131}I-treated fish. I supplemented diet (6 μgI/g dry diet) fed to intact fish.	Carrier free Na ^{131}I.	100 μCi

XV

Antithyroid Compounds, and Radiothyroidectomy on Fish Growth[a]

treatment

Route	Duration (days)	Responses and remarks	Reference
im[b] (every fourth day)	27	Mean ($n = 3$ fish/group) percentage gain in weight was slightly higher (44%) than that of the controls (36%). No effect on length.	Enomoto (1964)
ip[c] 3×/week	30	Both control and TSH-treated fish decreased in weight. Significant increases in length were not found.	Pickford and Grant (1968)
w[d] (changed every 2 days).	13	Significantly decreased growth (% change) in length.	Massey and Smith (1968)
d[e]	~600	Significant retardation of growth (5 months after beginning of treatment) increased pigmentation and impairment of cephalic development.	LaRoche et al. (1966)
w	92	Similar findings to those of Dales and Hoar (1954) [i.e., thiourea reduced growth rate (length) of developing fish and retarded growth of body wall]. But, in this experiment thiourea delayed time of egg hatching after fertilization.	Ali (1961)
w		Inhibited linear growth and may have blocked growth hormone production.	Singh (1971)
w	28	Inhibited growth (weight); acidophil (STH) cells of proximal pars distalis showed signs of decreased secretion and accumulation of stainable material in the cytoplasm.	Sage (1965, 1967)
	20	1% PTU slightly reduced growth in weight (2.5% lower than controls) while 5% PTU had a greater effect (32.9% lower than controls). Liver glycogen was either reduced (1% PTU) or increased (5% PTU) relative to that of the controls. Liver cytochrome oxidase activity was increased (1% PTU) or decreased (5% PTU). Elevated free amino acid level in muscle.	Bonnet (1970); LeRay et al. (1970)
Six monthly ip injections	~750	No effect on growth during period of ^{131}I treatment. Seven months after cessation of treatment growth suppression noted (weight and length). By end of experiment growth of ^{131}I-treated fish ceased. Thyroidectomized fish had high degree of pigmentation, smaller head with reduced calcification, and greater amount of cartilage, smooth skin due to reduction in size of scales, reduced dermal connective tissue in skin, an increase in visceral fat deposition, and retarded sexual development relative to intact controls. Suggested that chronic hypothyroidism due to radiothyroidectomy impairs calcification by limiting bone matrix formation.	LaRoche et al. (1966)

(Continued)

Table XV—

| Species | Initial age, weight, or length | Environmental factors | | Diet and ration | Method of | |
		Temperature (°C)	Photoperiod		Hormone or antithyroid compound	Dosage
Salmo gairdnerii	Alevins (5-week-old, 24 mm, 0.15 g)	8–22	Natural	Fed daily to satiation on a diet of commercial fish pellets (iodine content 22 μg/g dry food).	[131]I	15 μCi/ml for 4 (*S. gairdneri*) or 8 (*O. tshawytscha*) days.
Oncorhynchus tshawytscha	Alevins (10-week-old, 31 mm, 0.3 g)					
Xiphophorus maculatus	17–56 days (BH strain) 6–8 mm			Dried shrimp or liver-pablum.	Carrier free ^{131}I, [32]P	5 mCi/200 ml for 25.5–60 hr. 1.7 mCi/200 ml for 48.5–50.75 hr.
Lepomis cyanellus	2–4 years E1, 13.7–13.9 g E2, 15.9–17.3 g E3, 11.4–12.2 g	23.3–25.6	8L : 16D 16L : 8D 8L to 16L 16L to 8L	Fed daily to excess with a commercially prepared dry pellet food or with frozen beef liver.	[131]I	50–200 μCi prior to start of experiment.

[a] Refer to Pickford (1957) and Baker-Cohen (1961) for reviews of the early literature.   [b] im, intramuscular.

but also cause radiation damage to other organs. To avoid this latter criticism Norris (1969) included irradiated controls (exposed to a quantity of external γ radiation equivalent to that received by the radioiodide-treated fish) in his study on the influence of radiothyroidectomy on growth of steelhead trout and chinook salmon. Norris (1969) concluded that growth depression of thyroidectomized fish was probably a consequence of thyroid destruction instead of a specific effect of whole-body γ irradiation. This is because the percentage increases in weight and length of both his irradiated and untreated control fish were the same throughout the experimental period (Fig. 17).

E. Effect of Thyroid Hormones on Cartilage and Bone Growth of Salmonids

Both very low and high (nonphysiological or pharmacological) circulating levels of thyroid hormone can influence skull development of salmonids. For example, La Roche *et al.* (1966) found that radio-

Continued

eatment

Route	Duration (days)	Responses and remarks	Reference
w	Several months	Significant decline in growth (weight and length) relative to that of irradiated and normal controls as early as 2 months after radioiodide treatment. Depression of growth rate probably due to thyroid destruction rather than nonspecific effect of whole-body γ irradiation. Marked retardation in head growth and flaring and reduction of the opercula noted especially in *S. gairdneri*. Also, in one experiment using *S. gairdneri*, radiothyroidectomy led to greater visceral fat deposition and to higher condition factors than noted in controls. ^{131}I treatment may have reduced appetite, efficiency of food conversion and/or intestinal uptake.	Norris (1966, 1968) Norris (1969)
w	Several months	Reduced growth in length and retarded pituitary and gonadal development. Treated fish were darker than controls. Also their livers were larger and vacuolated, probably due to fat deposition. Increased abdominal vacuolation (lipid deposition) noted. ^{32}P treatment did not inhibit growth.	Baker-Cohen (1961)
ip	42	Partial radiothyroidectomy resulted in decreased growth (weight and length) and poorer food conversion efficiency.	Gross *et al.* (1963)

p, intraperitoneal.   *d* w, water.   *e* d, diet.

thyroidectomy of rainbow trout impaired bone calcification by restricting bone matrix formation. Marked retardation of skull growth was observed. In a study on the same species, Norris (1966) also noted flaring and reduction of the opercula in addition to poorer skull growth after radiothyroidectomy. In contrast, several workers (La Roche and LeBlond, 1952; La Roche, 1953; Honma, 1960; Higgs *et al.*, 1976, 1977b) have found that administration of T_4, T_3, or beef thyroid mixed with the diet increases skull dimensions and leads to hyperplasia of connective tissue elements of the interorbital region. At the biochemical level, Barrington and Rawdon (1967) and Barber and Barrington (1972) noted that T_4 treatment of intact yearling rainbow trout stimulated uptake of [^{35}S]sulfate into the cells and matrix of branchial skeleton cartilage or bone. There is evidence from a study by Qureshi (1976) on the brown trout which indicates that T_3 is more potent than T_4 in stimulating bone development. It is probable that thyroid hormone(s) acts in conjunction with growth hormone to influence cartilage and bone development of fish (Barrington and Rawdon, 1967; Barber and Barrington, 1972).

F. Appetite, Gross Food Conversion Efficiency, and Metabolic Actions of Thyroid Hormones

Thyroid hormone treatment may enhance the growth of teleosts by increasing voluntary food intake (appetite) (Gross *et al.*, 1963; Higgs *et al.*, 1977b) and/or gross food conversion efficiency (Gross *et al.*, 1963; Bjorklund, 1965; Lam, 1976 personal communication; Higgs *et al.*, 1977b) (Table XIV). More evidence supports the latter mechanism. However, it is not clear which metabolic pathways thyroid hormones directly or indirectly influence to improve food conversion efficiency.

Thyroid hormones may have protein catabolic or anabolic actions in fish. Studies by Hoar (1958) and Thornburn and Matty (1963) showed that treatment of goldfish with T_4 increased nitrogen (ammonia) excretion. In the latter study, this effect was found at 21°C, but not at 14°C. However, Smith and Thorpe (1977) have observed that T_4 decreases nitrogen excretion in fed rainbow trout. Also, Thornburn and Matty (1963) noted that T_4 promoted *in vitro* incorporation of L-[1-^{14}C]leucine into muscle protein of goldfish. Matty and Sheltawy (1967) found that ip injection of *Salmo irideus* with T_4 enhanced *in vivo* [$^{14}C_2$]glycine incorporation into newly synthesized skin guanine. In addition, Jackim and LaRoche (1973) observed that T_3, but not T_4, stimulated *in vivo* incorporation of L-[1-^{14}C]leucine into muscle protein of the killifish. Finally, Narayansingh and Eales (1975a) reported that physiological doses of T_4 increased *in vivo* incorporation of L-[1-^{14}C]leucine into liver and gill protein of both rainbow and brook trout. Effects of T_3 on liver and gill protein concentrations were also observed (Table XIV). Thus, in some instances, more of the ingested amino acids may be channeled into protein synthetic pathways after administration of thyroid hormone.

There is evidence for an involvement of thyroid hormones in lipid metabolism of fish. Both LaRoche *et al.* (1966) and Norris (1969) found that radiothyroidectomized rainbow trout had increased visceral lipid deposition relative to intact controls. Baker-Cohen (1961) speculated that the increased liver size of radiothyroidectomized platyfish, *Xiphophorus maculatus*, was probably due to lipid deposition. There was also some indication of increased visceral lipid deposition. In contrast, Barrington *et al.* (1961) noted that rainbow trout treated with T_4 had less abdominal fat. Also Narayansingh and Eales (1975b) found decreased hepatic and visceral lipid reserves in brook trout after T_4 treatment. These findings suggest that thyroid hormones stimulate lipid mobilization in teleosts. However, some studies do not support this generalization (literature reviewed in Narayansingh and Eales,

1975b). If thyroid hormone administration normally does lead to mobilization of visceral lipid reserves in teleosts, and this is to furnish energy, then more protein would be spared for growth.

Several studies have shown that thyroid hormones influence carbohydrate metabolism in fish. Fontaine *et al.* (1953) noted that hepatic glycogen in the eel was depressed after ip T_4 injections. They also found that rainbow trout fed either iodinated casein or a diet containing added T_4 had lower hepatic glycogen relative to that of controls. LeRay *et al.* (1970) observed that T_4, but not T_3, in a diet fed to mullet lowered hepatic glycogen. In addition, cytochrome oxidase activity was increased by T_4 or T_3, while glucose-6-phosphate dehydrogenase was decreased by T_4 relative to the control level. Lowering of blood sugar by T_4 or T_3 injection of carp was found by Murat and Serfaty (1971). They also found, after T_4 injection, an increase in cardiac and muscle glycogen. Hochachka (1962) observed that both T_4 and T_3 increased the rate of gluconate oxidation by brook trout liver homogenates and slices by as much as 125%. Glucose metabolism was thought to proceed by the pentose phosphate pathway. This pathway is important in metabolism for it not only provides reduced NADP, which is required by anabolic processes outside the mitochondria, but also yields pentoses for nucleotide and nucleic acid synthesis (Harper, 1971). Some of the actions of thyroid hormones on carbohydrate metabolism reported above may reduce protein catabolism.

Higgs (1974) suggested that thyroid hormones may influence gastrointestinal function in fish. For example, in mammals it is known that thyroid hormones can stimulate gastric and intestinal mobility, enzyme synthesis, and substrate absorption rates in the small intestine, and synthesis of bile salts (Middleton, 1971). Perhaps thyroid hormones exert similar actions on the gut of teleosts.

G. Synopsis

Physiological concentrations of thyroid hormone are probably necessary for normal growth of teleosts. At present it is not known if T_4 and/or T_3 or one or more of their metabolic derivatives is the active form(s) of thyroid hormone at the cellular level in fish. Some evidence favors a role for T_3 in this regard. Administration of exogenous thyroid hormone via water, injection, or pellet procedures to fish fed to satiation with a well balanced diet most consistently results in growth enhancement. Oral administration of T_4, but not T_3, may be ineffective because of poor absorption of T_4 from the gut lumen. Both temperature and nutritional state influence the degradation rate of thyroid hormones. Therefore for any given environmental and nutritional conditions and method of hormone administration it is necessary to establish a dosage of thyroid hormone that maintains a chronic elevation of circulating levels. Too much hormone leads to growth anomalies, depression of condition factor (weight–length relationship; Table XIV), and sometimes growth inhibition and/or mortality. Thyroid hormones may enhance fish growth by improving gross food conversion efficiency. This may be due to one or more direct metabolic actions and/or to an action of thyroid hormone on synthesis and release of growth hormone from pituitary somatotrops. Growth hormone improves food conversion in yearling coho salmon (see Section II,C,2,b). Also, thyroid hormone may influence absorption of nutrients across the gut.

V. INSULIN[*]

A. Introduction

Insulin is synthesized in the β cells of the pancreas. The location of the β cells and the biological and immunological activities of the insulins in the cyclostomata, elasmobranchii, holocephali, and actinopterygii have been reviewed by Epple (1969). After reviewing the data available on changes in the islet cells during the life history of the fish, after glucose administration or starvation, administration of the β cell cytotoxin, alloxan, or treatment with exogenous insulin or glucagon Epple (1969) concluded that "the . . . observations . . . make it rather discouraging to draw conclusions on the physiological role of insulin in fishes" (p. 304). However, he was able to generalize that exogenous insulin is often hypoglycemic, the effect appearing more slowly than in mammals, that it has a variable effect on liver glycogen, that muscle glycogen is sometimes increased, and that it reduces the lipid content of the liver. Since Epple's (1969) review was published considerable progress has been made in elucidating the role of insulin in fish and several reviews have been published (Steiner et al., 1973; Epple and Lewis, 1973; Brinn, 1973; see also Chapter 2 in this volume).

B. Biochemistry

Insulin is derived from a larger single chain precursor protein, proinsulin (Steiner and Oyer, 1967) which contains 81 amino acids

[*] This section was prepared by E. M. D.

and three disulfide bonds. Proinsulin provides the correct proportions of A- and B-chains and ensures the formation of the appropriate disulfide bonds and tertiary structure; it is then largely converted by enzymatic cleavage to insulin and C-peptide prior to secretion (Steiner *et al.*, 1973). Insulin itself consists of an A-chain usually of 21 amino acids and a B-chain usually of 30 amino acid residues connected by two disulfide bonds. A third disulfide bond connects two cysteine residues in the A-chain. Steiner *et al.* (1969) reviewed the amino acid substitutions that have occurred between hagfish and human insulin and noted the conservation of the positions of the disulfide bonds and the N- and C-terminal regions of the A-chain which is related to the role of these regions in maintaining the secondary and tertiary structure necessary for biological activity.

C. Function

The homeostatic role of insulin in mammals was succinctly summarized by Tashima and Cahill (1968): After feeding, insulin (a) facilitates glucose uptake and incorporation into glycogen or conversion into lipid, (b) stimulates amino acid incorporation into protein, and (c) promotes incorporation of dietary lipid into adipose tissue. During fasting insulin regulates plasma concentrations of glucose, amino acids, and free fatty acids.

In the toadfish, *Opsanus tau*, a carnivore, insulin had a variable effect on blood glucose while it had a positive effect on the uptake of carbon from labeled glucose into muscle protein and glycogen. Insulin also had a dramatic effect on the incorporation of glycine into skeletal muscle protein and to a lesser extent muscle lipid (Tashima and Cahill, 1968). Furthermore, fish fed a diet high in protein showed an increase in serum immunoreactive insulin. Thus in this species insulin plays a more important role in protein metabolism than in glucose homeostasis. Further evidence for the role of insulin in the regulation of protein synthesis in fish was provided by Jackim and LaRoche (1973) who showed that injection of bovine insulin 1 μg/fish into *Fundulus heteroclitus* weighing 1–1.5 g resulted in a stimulation of [14C]leucine incorporation from 45.6% in controls to 73.4% 18 hr after injection. However, they caution that the rate of amino acid incorporation into protein measured in this way is probably not a valid indication of growth rate partly as a result of the high level of inexplicable individual biological variation which they encountered.

Injection of bovine or cod insulin 2 IU/kg into the pike, *Esox lucius*, lowered the plasma amino acid nitrogen level, induced a recoverable hypoglycemia, and had no effect on plasma cholesterol (Thorpe and Ince, 1974). These workers proposed that contradictory results on the effect of insulin on plasma glucose in fish may result from variability of dose, species specificity, and turnover rate of the insulin and on the initial glucose concentration. Furthermore, they suggest that insulin sensitivity may be related to the mode of life of the fish, since the pike is an intermittent fast moving predator with a high dietary protein intake.

In a similar study of the European silver eel, *Anguilla anguilla,* Ince and Thorpe (1974) observed that 2 IU/kg cod insulin lowered blood glucose, amino acid nitrogen, and cholesterol while bovine insulin at the same dosage lowered amino acid nitrogen only. Insulin was also implicated in the rapid clearance of an intraarterial load of amino acids. A recent study of the Japanese eel, *Anguilla japonica,* has shown that insulin 5 IU/kg reduces the plasma concentration of all amino acids except citrulline (Inui *et al.,* 1975).

Until recently no studies had been carried out on the role of insulin on protein metabolism of salmonids. Current studies (Matty, 1975; Matty and Cheema, 1976 unpublished observations), however, delineate the effect of insulin on *in vivo* L-[^{14}C]leucine incorporation into muscle protein in the rainbow trout (*Salmo gairdneri*). Porcine insulin (Actrapid) was injected at a dosage of 3 IU/kg simultaneously with 50 μCi/kg of [^{14}C]leucine into 10- to 15-g fish maintained at 11.5°C. The incorporation of labeled leucine into skeletal muscle protein was determined at intervals over a 48 hr period and expressed as either percentage incorporation (Jackim and LaRoche, 1973) or specific activity. The percentage incorporation increased from 39 at 3 hr to 69 at 48 hr in controls and from 45 at 3 hr to 91 at 48 hr in insulin-treated fish. Specific activity in cpm $\times$ 10^3/g protein increased from 10 at 3 hr to 21 at 12 hr and 19 at 48 hr in control fish, while it increased from 12 at 3 hr to 64 at 12 hr and 49 at 48 hr in insulin-treated fish. In both cases, the differences between control and insulin injected fish were significant at 6, 12, 24 and 48 hr (Matty and Cheema, 1976 unpublished observations).

While a significant body of evidence shows insulin to play a specific role in the incorporation of plasma amino acids into muscle protein, until recently no studies of the effect of insulin on growth in fish either on a short- or long-term basis had been published.

Recently Ludwig *et al.* (1977) injected underyearling coho salmon with bovine (Ultralente) insulin twice weekly for 70 days over a dose range of 0.32 to 10.0 IU/kg. All doses of insulin increased specific growth rates and decreased food-gain ratios relative to solvent injected controls but the differences were not significant and did not reflect a dose response relationship. A single insulin injection of 10 IU/kg into starved coho resulted in a depression in plasma glucose which returned to control levels by 24 hours indicating a response of short duration to this preparation. Future studies should explore the use of higher dosages, long acting preparations, piscine insulins and the interaction of insulin with other hormones.

VI. ADDITIVE AND SYNERGISTIC EFFECTS OF HORMONES ON TELEOST GROWTH*

Few workers have investigated the effects of hormone combinations on teleost growth. Pickford (1957) conducted the first of these types of studies on male hypophysectomized killifish held at 20°C. She showed that mammalian thyrotropin (Armour) in combination with hake pituitary growth hormone (Wilhelmi F80GH) induced a greater growth response than the effect of the growth hormone alone. In fact this combination had a synergistic effect on growth (Fig. 19) since TSH by itself did not enhance growth. This suggests that the action of thyroid hormone on growth of hypophysectomized killifish is permissive in nature. Similar findings have been obtained in mammalian studies. For example, Scow (1959) administered T_4 and growth hormone either singly or together to thyroidectomized–hypophysectomized rats and noted that in most tissues a combination of both hormones produced greater growth than did either single hormone.

Enomoto (1964) tested the effect of several hormone combinations, namely, growth hormone (GH) + thyrotropin (TSH), GH + prolactin, GH + TSH + prolactin, and TSH + prolactin, on growth of intact immature rainbow trout weighing between 6.7 and 9.3 g and held at 16°C. Unfortunately Enomoto (1964) had only three fish in each treatment group and in a number of these after 27 days of treatment one or two fish developed exophthalmos. Hence, clear trends did not emerge from this study.

In another study on male hypophysectomized killifish, Pickford et al. (1972) found that a combination of bovine growth hormone (bGH; NIH-GH-B6) and ovine luteinizing hormone (oLH; NIH-LH-SI) had a synergistic effect on growth in weight and length at 20°C. Thyroid activity, as assessed by changes in thyroid epithelial cell height, was greater in fish administered oLH than in those treated with bGH or

* This section was prepared by D. A. H.

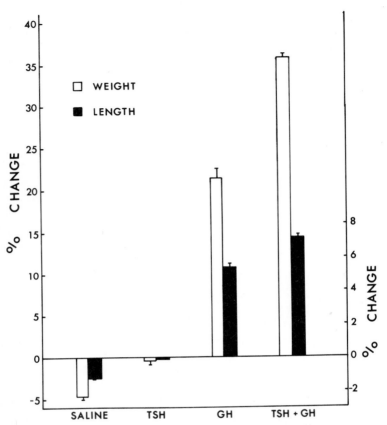

Fig. 19. Influence of purified hake pituitary growth hormone (GH), mammalian thyrotropin (TSH), and a combination of the two hormones on growth (percentage weight or
length increase) of hypophysectomized male *Fundulus heteroclitus* at 20°C, over a
period of 35 days. GH (Wilhelmi, F80GH) was injected (ip) three times per week at a
dosage of 10 μg/g body weight. TSH (Armour, Lot 317-51; 1.2 USP units/mg) was also
injected three times per week (ip) but at a dosage of 1 μg/g body weight. The same doses
were used when the combination of hormones were administered. Vertical lines indicate one standard deviation. Both GH and TSH + GH significantly enhanced growth
($P < 0.001$). (Modified from Pickford, 1957.)

saline. This suggests that oLH was functioning as a heterothyrotropic
factor. Yet oLH like TSH had no influence on killifish growth.

Recently, Clarke (1976) and Higgs *et al.* (1977a) assessed the effect
of hormone combinations on growth of intact Pacific salmon. Clarke
(1976) found that either 5 μg/g of ovine (oPrl;NIH-PS-11) prolactin, or
1 μg/g of bGH (NIH-GH-B-17) stimulated growth in weight and

length of underyearling sockeye salmon held at 17.5°C and maintained on a 12 hr photoperiod. Additional growth enhancement was not noted when bGH and oPrl were administered together (Fig. 4). It is difficult to reconcile these findings for oPrl on growth of sockeye salmon with studies which indicate that prolactin does not affect fish growth. For example, Pickford and Kosto (1957 unpublished observations) found, in many experiments involving chronic treatment of hypophysectomized killifish with oPrl, that oPrl never enhanced growth (cited in Ball *et al.*, 1965). Also, repeated injections of mammalian prolactin have never resulted in growth stimulation of hypophysectomized *Poecilia* (Ball and Slicher, unpublished observations, cited in Ball, 1965). Finally, Ball (1965) observed that surgical removal of the rostral pars distalis in *Poecilia latipinna* had no effect on growth rate. Therefore oPrl in sockeye salmon may produce its effect on growth by potentiating the effect of other endogenous hormones. Alternatively, since the structures of prolactins and growth hormones are similar (see Section II,A,6) in vertebrates it is possible that both are able to interact with the growth hormone receptor in Pacific salmon. Contamination of the prolactin preparation used by Clarke (1976) with growth hormone was probably not a factor since the growth hormone content of the prolactin administered was about one-tenth the bGH dosage.

Higgs *et al.* (1977a) treated yearling coho salmon at 10°C with bGH (NIH-GH-B-17), 17α-methyltestosterone (MT), or L-thyroxine (T$_4$) singly and in combination. Administration of each of the individual hormones and various combinations of these hormones led to pronounced differences in growth rate both in weight and length during the 59-day experimental period (Fig. 20). When all the treatment groups were ranked according to growth rate or percentage gain over corresponding control at 59 days the following sequence emerged: bGH + T$_4$ + MT > bGH + MT > bGH + T$_4$ = bGH > T$_4$ + MT > MT > T$_4$ > untreated control = injected control.

An additive effect on growth was observed for the bGH + MT combination, but not for the bGH + T$_4$ combination, in contrast to studies carried out on hypophysectomized fish. However, when MT was combined with bGH in the presence of T$_4$ a synergistic action on growth ensued (percentage gain greater than the combined effects of the individual hormones). The apparent lack of additional growth stimulation with bGH + T$_4$ above that induced by bGH was probably because the level of endogenous thyroid hormone was sufficient to enable bGH to exert its full growth-promoting effect. Addition of MT to bGH or bGH + T$_4$ most likely altered the secretion rates of other

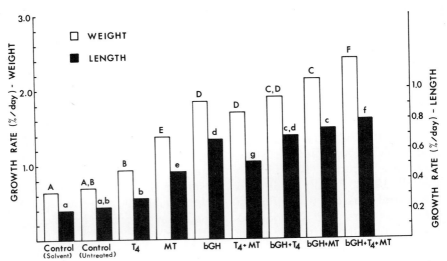

Fig. 20. Influence of combinations of bovine growth hormone [bGH; NIH-GH-B17 (0.92 IU/mg)], 17α-methyltestosterone (MT), and Na L-thyroxine pentahydrate (T₄) on the growth rate (weight and length) of yearling coho salmon (n = 60 fish/group) at 10°C over a period of 59 days. Both bGH (10 μg/g body weight) and T₄ (1 μg/g; T₄ dose refers to free acid) administered singly or in combination were injected once per week into the dorsal muscle, while MT was incorporated into the Oregon Moist Pellet diet (1 mg/kg diet). Each group of coho was presented excess ration twice daily. One control group was injected with solvent, while the other was untreated. Hormone dosage (bGH and T₄) was adjusted according to the new mean fish weight in each group at 14-day intervals. Groups with the same superscript have equal growth rates. A test of common slope showed that the slopes for ln weight or ln length of the nine groups were significantly different ($P < 0.001$). Scheffé's test with $P = 0.05$ revealed which slopes were significantly different (growth rate = slope × 100). Percentage mortality did not exceed 7% of the initial number in each group. (Modified from Higgs *et al.*, 1977a.)

endocrine glands. In mammals androgens influence the synthesis and/or release of pituitary growth hormone (see Section III).

In fish, it is known that 17α-MT may stimulate thyroid (Sage and Bromage, 1970; Van Overbeeke and McBride, 1971; Higgs *et al.*, 1977a) (Table XVI) and interrenal function (Van Overbeeke and McBride, 1971; Higgs *et al.*, 1977a) (Table VI) as well as insulin secretion from the pancreatic β cells (Higgs *et al.*, 1977a). Such is also the case following bGH treatment of coho salmon (Higgs *et al.*, 1976, 1977a) (Tables VI and XVI). Therefore both the bGH + MT and bGH + MT + T₄ combinations probably created an internal hormonal milieu most conducive for development and growth of cartilage and bone and also for muscle protein synthesis. Further enhancement of growth may be possible when these or other hormone combinations

Table XVI

Mean Thyroid Follicle Epithelial Heights of Yearling Coho Salmon Administered
Either Bovine Growth Hormone, L-Thyroxine, 17α-Methyltestosterone,
Combinations of These Hormones, or No Hormone for 59 Days

Treatment[a]	Number of fish	Thyroid follicle epithelium height[b] (μm $\pm$ 2 SE)
Control, solvent injected	5	4.97 ± 0.16*
bGH, 10 μg/g	7	5.72 ± 0.42*
T$_4$, 1 μg/g	7	4.05 ± 0.18**
MT, 1 mg/kg diet	5	6.64 ± 0.42***
bGH + T$_4$	7	5.04 ± 0.12*
bGH + MT	7	7.40 ± 0.54†
T$_4$ + MT	7	5.97 ± 0.98*,***
bGH + T$_4$ + MT	7	5.45 ± 0.49*

[a] bGH, bovine growth hormone; T$_4$, L-thyroxine; MT, 17α-methyltestosterone. Refer to Fig. 20 for additional details.

[b] Treatment groups with the same superscripts form a homogeneous subset. All other pairs of group means are significantly different (analysis of variance indicated $P < 0.001$; Student Newman–Keuls test, $P = 0.05$).

are administered in proper proportions according to environmental, nutritional, and developmental states. From mammalian studies, the hormones which have the greatest influence on cartilage and bone development are growth hormone, thyroid hormone, and androgens. Pituitary growth hormone or somatomedins mainly regulate chondrogenesis while thyroid hormones and sex steroids are more active on processes of calcification and ossification (Rappaport, 1975). A similar situation probably exists in fish (see sections on individual hormones). Some hormones such as T$_4$ may influence somatomedin activity in mammals with consequent effects on the rate of growth (Gaspard et al., 1975). There is no information available from studies on fish in this regard.

The most useful hormone combination in terms of cost and practical application to salmon enhancement and aquaculture programs may prove to be MT + thyroid hormone. Growth rates (weight) for yearling coho salmon treated with MT + T$_4$ were statistically equivalent to those of fish administered bGH (Fig. 20). Also, coho treated with either T$_4$ or MT + T$_4$ grew faster in weight than length (Fig. 5) and had a higher muscle lipid content (Table V). Therefore, both their appearance and taste may be acceptable to the consumer. However, from a practical standpoint it has been necessary to determine whether results comparable to those cited above can be obtained when MT and

T_3 are included together in the diet of salmon. This is because orally administered T_4, unlike T_3, is not effective in promoting coho salmon growth (Higgs *et al.*, 1977b). Preliminary evidence indicates that the inclusion of MT and T_3 in the diet of coho salmon held at 10°C does have an additive effect on growth, but the response depends on the respective hormones doses, fish age (size), and/or time of year (photoperiod) (Fagerlund, Higgs, and McBride, 1978 unpublished observations).

Note. The comprehensive literature survey for this chapter was completed June 1976. Certain data from recent publications have also been reviewed.

ACKNOWLEDGMENTS

The authors wish to thank Ms. Helen M. Dye for her invaluable assistance in preparing the figures and references and Mrs. Morva Young and Valerie Smalley for typing the manuscript.

REFERENCES

Adelman, I. R. (1977) Effects of bovine growth hormone on growth of carp (*Cyprinus carpio*) and the influences of temperature and photoperiod. J. Fish. Res. Board Can. 34: 509–515.

Ali, M. A. (1961). Effect of thyroxine plus thiourea on the early development of the chum salmon (*Oncorhynchus keta*). *Nature (London)* 191, 1214–1215.

Allen, D. M. (1977). Measurements of serum thyroxine and the proportions of rhodopsin and porphyropsin in rainbow trout. *Can. J. Zool.* 55, 836–842.

Aloj, S., and Edelhock, H. (1972). The molecular properties of human growth hormone. *J. Biol. Chem.* 247, 1146–1152.

Amend, D. F., and Fender, D. C. (1976). Uptake of bovine serum albumin by rainbow trout from hyperosmotic solutions—A model for vaccinating fish. *Science* 192, 793–794.

Arai, R. (1967). Androgenic effects of 11-ketotestosterone on some sexual characteristics in the teleost, *Oryzias latipes. Annot. Zool. Jpn.* 40, 1–5.

Ashby, K. R. (1957). The effect of steroid hormones on the brown trout (*Salmo trutta* L.) during the period of gonadal differentiation. *J. Embryol. Exp. Morphol.* 5, 225–249.

Ashley, H., and Frieden, E. (1972). Metabolism and distribution of triiodothyronine and thyroxine in the bullfrog tadpole. *Gen. Comp. Endocrinol.* 18, 22–31.

Baggett, B., Engel, L. L., Savard, K., and Dorfman, R. I. (1956). The conversion of testosterone-3-C^{14} to C^{14}-estradiol-17β by human ovarian tissue. *J. Biol. Chem.* 221, 931–941.

Baker, B. I., and Ingleton, P. M. (1975). Secretion of prolactin and growth hormone by teleost pituitaries *in vitro*. II. Effect of salt concentration during long-term organ culture. *J. Comp. Physiol.* 100, 269–282.

Baker-Cohen, K. F. (1961). The role of the thyroid in the development of platyfish. *Zoologica (N.Y.)* 46, 181–222.

Ball, J. N. (1965). Partial hypophysectomy in the teleost *Poecilia:* Separate identities of teleostean growth hormone and teleostean prolactinlike hormone. *Gen. Comp. Endocrinol.* 5, 654–661.

Ball, J. N. (1969). Prolactin (fish prolactin or paralactin) and growth hormone. *In* "Fish Physiology" (W. S. Hoar and D. J. Randall, eds.), Vol. 2, pp. 207–240. Academic Press, New York.

Ball, J. N., and Baker, B. I. (1969). The pituitary gland: Anatomy and histophysiology. *In* "Fish Physiology" (W. S. Hoar and D. J. Randall, eds.), Vol. 2, pp. 1–111. Academic Press, New York.

Ball, J. N., Olivereau, M., Slicher, A. M., and Kallman, K. D. (1965). Functional capacity of ectopic pituitary transplants in the teleost, *Poecilia formosa,* with a comparative discussion on the transplanted pituitary. *Philos. Trans. R. Soc. London, Ser. B* 249, 69–99.

Bancroft, T. A. (1968). "Topics in Intermediate Statistical Methods," Vol. 1, p. 129. Iowa State Univ. Press, Ames.

Barber, S., and Barrington, E. J. W. (1972). Dynamics of uptake and binding of [^{35}S]sulfate by the cartilage of rainbow trout (*Salmo gairdneri*) and the influence of thyroxine. *J. Zool.* 168, 107–117.

Barr, W. A. (1965). The endocrine physiology of fishes. *Oceanogr. Mar. Biol. Annu. Rev.* 3, 257–298.

Barrington, E. J. W. (1975). The endocrine glands of the pharynx. *In* "An Introduction to General and Comparative Endocrinology" (E. J. W. Barrington, ed.), pp. 147–184. Oxford Univ. Press (Clarendon), London and New York.

Barrington, E. J. W., and Rawdon, B. B. (1967). Influence of thyroxine upon the uptake of ^{35}S-labeled sulfate into the branchial arch skeleton of the rainbow trout (*Salmo gairdneri*). *Gen. Comp. Endocrinol.* 9, 116–128.

Barrington, E. J. W., Barron, N., and Piggins, D. J. (1961). The influence of thyroid powder and thyroxine upon the growth of rainbow trout (*Salmo gairdneri*). *Gen. Comp. Endocrinol.* 1, 170–178.

Bates, R. W., Miller, R. A., and Garrison, M. M. (1962). Evidence in the hypophysectomized pigeon of a synergism among prolactin, growth hormone, thyroxine and prednisone upon weight of the body, digestive tract, kidney and fat stores. *Endocrinology* 71, 345–360.

Baumann, G., and Pfister, C. (1936). La thyroxine provoqué chez de jeunes alevins de truite, une atrophie et une disparition des arcs aortiques, analogues a celles des embryons d'aminiotes. *C. R. Soc. Biol.* 122, 1156–1158.

Berkowitz, P. (1938). The effects of estrogenic substances in *Lebistes reticulatus* (guppy). *Anat. Rec.* 71, 161–175.

Bewley, T. A., and Li, C. H. (1975). The chemistry of human pituitary growth hormone. *Adv. Enzymol. Relat. Areas Mol. Biol.* 42, 73–166.

Bjorklund, R. G. (1958a). The biological function of the thyroid and the effect of length of day on the growth and maturation of the goldfish *Carassius auratus* (Linnaeus). Ph.D. Thesis, Univ. of Michigan, Ann Arbor.

Bjorklund, R. G. (1958b). The biological function of the thyroid and the effect of length of day on the growth and maturation of the goldfish *Carassius auratus* (Linnaeus). *Diss. Abstr.* 19, 1423.

Bjorklund, R. G. (1965). The effect of thyroid hormones on the growth of goldfish, *Carassius auratus* (L.). *Trans Ill. State Acad. Sci.* 58, 64–68.

Blake, J., and Li, C. H. (1973). Human pituitary growth hormone. 35. Synthesis and growth-promoting activity of *N*-acetyl-HGH (human growth hormone) (95–136). *Int. J. Pept. Protein Res.* 5, 123–125.

Bonnet, B. (1970). Thyroide et milieu intérieur chez *Mugil* sp. Influence de la nutrition et de l'environnement. Ph.D. Thesis, Univ. d'Aix-Marseille.

Braden, A. W. H., Thain, R. I., and Shutt, D. A. (1971). Comparison of plasma phytoestrogen levels in sheep and cattle after feeding on fresh clover. *Aust. J. Agric. Res.* **22**, 663–670.

Bray, G. A. (1974). Endocrine factors in the control of food intake. *Fed. Proc., Fed. Am. Soc. Exp. Biol.* **33**, 1140–1145.

Brazeau, P., Vale, W., Burgus, R., Ling, N., Butcher, M., Rivier, J., and Guillemin, R. (1973). Hypothalamic polypeptide that inhibits the secretion of immunoreactive pituitary growth hormone. *Science* **179**, 77–79.

Brazeau, P., Rivier, J., Vale, W., and Guillemin, R. (1974). Inhibition of growth hormone secretion in the rat by synthetic somatostatin. *Endocrinology* **94**, 184–187.

Brett, J. R., Shelbourn, J. E., and Shoop, C. T. (1969). Growth rate and body composition of fingerling sockeye salmon, *Oncorhynchus nerka*, in relation to temperature and ration size. *J. Fish. Res. Board Can.* **26**, 2363–2394.

Brinn, J. E., Jr. (1973). The pancreatic islets of bony fishes. *Am. Zool.* **13**, 653–665.

Bromage, N. R. (1975). The effects of mammalian thyrotropin-releasing hormone on the pituitary–thyroid axis of teleost fish. *Gen. Comp. Endocrinol.* **25**, 292–297.

Bromage, N. R., Whitehead, C., and Brown, T. J. (1976). Thyroxine secretion in teleosts and the effects of TSH, TRH and other peptides. *Gen. Comp. Endocrinol.* **29**, 246 (abstr.).

Brown, P. S., Giuliano, R., and Hough, G. (1974). Pituitary regulation of appetite and growth in the turtles *Pseudomys scripta elegans* and *Chelydra serpentina*. *J. Exp. Zool.* **187**, 205–216.

Brown, S., and Eales, J. G. (1977). Measurement of L-thyroxine and 3,5,3'-triiodo-L-thyronine levels in fish plasma by radioimmunoassay. *Can. J. Zool.* **55**, 293–299.

Bulkley, R. V. (1972). Diethylstilbestrol in catfish feed. *Trans. Am. Fish. Soc.* **101**, 537–539.

Bulkley, R. V., and Swihart, G. L. (1973). Effects of the anabolic steroid stanozolol on growth of channel catfish, *Ictalurus punctatus*, and goldfish *Carassius auratus*. *Trans. Am. Fish. Soc.* **102**, 444–446.

Burden, C. E. (1956). The failure of hypophysectomized *Fundulus heteroclitus* to survive in freshwater. *Biol. Bull. (Woods Hole, Mass.)* **110**, 8–28.

Cantilo, E., and Regalado, T. G. (1942). Investigaciones realizadas con el extracto anterohysofisario en el desarrollo del *Salvelinus fontinalis*. *Rev. Med. Vet. (Buenos Aires)* **24**, 323–338.

Chan, H. H., and Eales, J. G. (1975). Identification of iodoamino acids in the plasma and bile of brook trout, *Salvelinus fontinalis* (Mitchill). *Can. J. Zool.* **53**, 97–101.

Chartier-Baraduc, M. M. (1959). Influence de l'hormone somatotrope sur les teneurs en eau et en électrolytes du plasma et du muscle de la truite arc-en-ciel (*Salmo gairdnerii*). *C. R. Soc. Biol.* **153**, 1757–1761.

Chavin, W., and Bouwman, B. N. (1965). Metabolism of iodine and thyroid hormone synthesis in the goldfish, *Carassius auratus* L. *Gen. Comp. Endocrinol.* **5**, 493–503.

Cheek, D. B. (1971). Hormonal and nutritional factors influencing muscle cell growth. *J. Dent. Res.* **50**, Suppl., 1385–1391.

Cheema, I. R., and Matty, A. J. (1976). Effects of the anabolic steroids norethandrolone and dimethazine on muscle protein synthesis and growth of rainbow trout, *Salmo gairdneri*. *J. Endocrinol.* **72**, 11P–12P.

Chester Jones, I., Ball, J. N., Henderson, I. W., Sandot, T., and Baker, B. I. (1974). Endocrinology of fishes. *In* "Chemical Zoology" (M. Florkin and B. T. Scheer, eds.), Vol. 8, pp. 523–593. Academic Press, New York.

Chillemi, F., and Pecile, A. (1971). Solid-phase synthesis and bioassay by tibia test of Monotetraconta peptide 81-121 and of ditria conta peptide 122-153 of human growth hormone. *Experientia* **27**, 385–386.

Chillemi, F., Aiello, A., and Pecile, A. (1972). Synthesis of human growth hormone fragments with growth-promoting activity. *Nature (London), New Biol.* **238**, 243–245.

Clarke, W. C. (1976). Effect of prolactin and growth hormone on growth, lipid content and seawater adaptation in juvenile sockeye salmon. *Reg. Conf. Comp. Endocrinol., Oregon State Univ., Corvalis* Abstr. No. 38.

Clarke, W. C., and Nagahama, Y. (1977). Effect of premature transfer to sea water on growth and morphology of the pituitary, thyroid, pancreas, and interrenal in juvenile coho salmon *Oncorhynchus kisutch. Can. J. Zool.* **55**, 1620–1655.

Clarke, W. C., Farmer, S. W., and Hartwell, K. M. (1977). Effect of teleost pituitary growth hormone on growth of *Tilapia mossambica* and on growth and sea water adaptation of sockeye salmon (*Oncorhynchus nerka*). *Gen. Comp. Endocrinol.* **33**, 174–178.

Clemens, H. P., McDermitt, C., and Inslee, T. (1966). The effects of feeding methyl-testosterone to guppies for 60 days after birth. *Copeia* No. 2, pp. 280–284.

Cohen, D. C., Greenberg, J. A., Licht, P., Bern, H. A., and Zipser, R. D. (1972). Growth and inhibition of metamorphosis in the newt *Taricha torosa* by mammalian hypophysial and placental hormones. *Gen. Comp. Endocrinol.* **18**, 384–390.

Cohen, H. (1945). Effects of sex hormones on the development of the platyfish, *Platypoecilus maculatus. Zoologica (N.Y.)* **31**, 121–127.

Collicutt, J. M., and Eales, J. G. (1974). Excretion and enterohepatic cycling of ^{125}I-L-thyroxine in channel catfish, *Ictalurus punctatus* Rafinesque. *Gen. Comp. Endocrinol.* **23**, 390–402.

Cowey, C. B., and Sargent, J. R. (1972). Fish nutrition. *Adv. Mar. Biol.* **10**, 383–492.

Cowey, C. B., Pope, J. A., Adron, J. W., and Blair, A. (1973). Studies on the nutrition of marine flatfish. The effect of oral administration of diethylstilboestrol and cypro-heptadine on the growth of *Pleuronectes platessa. Mar. Biol.* **19**, 1–6.

Curry, D. L., and Bennett, L. I. (1973). Dynamics of insulin release by perfused rat pancreases: effect of hypophysectomy, growth hormone, adrenocorticotrophic hormone, and hydrocortisone. *Endocrinology* **93**, 602–609.

Dales, S., and Hoar, W. S. (1954). Effects of thyroxine and thiourea on the early development of chum salmon *Oncorhynchus keta. Can. J. Zool.* **32**, 244–251.

Daughaday, W. H., Herington, A. C., and Phillips, L. S. (1975). The regulation of growth by endocrines. *Annu. Rev. Physiol.* **37**, 211–244.

Demski, L. S. (1973). Feeding and aggressive behaviour evoked by hypothalamic stimulation in a cichlid fish. *Comp. Biochem. Physiol. A* **44**, 685–692.

de Vlaming, V. L. (1974). Environmental and endocrine control of teleost reproduction. *In* "Control of Sex in Fishes" (C. B. Schreck, ed.), pp. 13–83. Ext. Div., Virginia Polytechnic Inst. and State Univ., Blacksburg, Virginia.

Dickhoff, W. W., Folmar, L., and Gorbman, A. (1978). Plasma thyroxine (T_4) and triiodo-thyronine (T_3) levels and gill sodium–potassium adenosine triphosphatase (Na^+-K^+ ATPase) of smolt and parr-revertant coho salmon (*Oncorhynchus kisutch*). *West. Reg. Conf. Gen. Comp. Endocrinol., Univ. Calif., Santa Cruz* Abstr. No. 47.

DiMaggio, A., III (1960). Hormonal replacement of photoperiod as a stimulus for growth in a lizard. *Fed. Proc., Fed. Am. Soc. Exp. Biol.* **19**, Part I. (Abstr.)

Dodd, J. M. (1955). The hormones of sex and reproduction and their effects in fish and lower chordates. *Mem. Soc. Endocrinol.* **4**, 166–187.

Donaldson, E. M., and McBride, J. R. (1967). The effects of hypophysectomy in the rainbow trout *Salmo gairdnerii* (Rich.) with special reference to the pituitary-interrenal axis. *Gen. Comp. Endocrinol.* 9, 93–101.

Drongowski, R. A., Wood, J. S., and Bouck, G. R. (1975). Thyroid activity in coho salmon from Oregon and Lake Michigan. *Trans. Am. Fish. Soc.* 104, 349–352.

Drury, D. E., and Eales, J. G. (1968). The influence of temperature on histological and radiochemical measurements of thyroid activity in the eastern brook trout, *Salvelinus fontinalis* (Mitchill). *Can. J. Zool.* 46, 1–9.

Eales, J. G. (1969). *In vivo* uptake of radiothyroxine by the tissues of Atlantic salmon (*Salmo salar* L.) parr, presmolt, and smolt. *Can. J. Zool.* 47, 9–16.

Eales, J. G. (1970). Biliary excretion of radiothyroxine by the brook trout, *Salvelinus fontinalis* (Mitchill). *Gen. Comp. Endocrinol.* 14, 385–395.

Eales, J. G. (1972a). Radiothyroxine metabolism in several freshwater teleost fishes. *Can. J. Zool.* 50, 623–631.

Eales, J. G. (1972b). Influence of injection site on radiothyroxine metabolism in brook trout, *Salvelinus fontinalis* (Mitchill). *Can. J. Zool.* 50, 1337–1341.

Eales, J. G. (1974). Creation of chronic physiologic elevations of plasma thyroxine in brook trout, *Salvelinus fontinalis* (Mitchill) and other teleosts. *Gen. Comp. Endocrinol.* 22, 209–217.

Eales, J. G. (1977a). Use of thyroxine- and triiodothyronine-specific antibodies to study thyroxine kinetics in rainbow trout, *Salmo gairdneri*. *Gen. Comp. Endocrinol.* 32, 89–98.

Eales, J. G. (1977b). *In vivo* determination of thyroxine deiodination rate in rainbow trout, *Salmo gairdneri* Richardson. *Gen. Comp. Endocrinol.* 33, 541–546.

Eales, J. G., and Sinclair, D. A. R. (1974). Enterohepatic cycling of thyroxine in starved and fed brook trout, *Salvelinus fontinalis* (Mitchill). *Comp. Biochem. Physiol. A* 49, 661–672.

Eales, J. G., Welsh, L. A., and Chan, H. H. (1971). Biliary excretion of 3,5,3′-triiodo-L-thyronine-^{125}I by the brook trout, *Salvelinus fontinalis* (Mitchill). *Gen. Comp. Endocrinol.* 16, 169–175.

Eisenberg, E., and Gordon, G. S. (1950). The levator ani muscle of the rat as an index of myotrophic activity of steroidal hormones. *J. Pharmacol. Exp. Ther.* 99, 38–44.

Enemar, A., and von Mecklenburg, C. (1962). The growth response of frog tadpoles to pituitary implants from different vertebrates. *Gen. Comp. Endocrinol.* 2, 273–278.

Enomoto, Y. (1964). A preliminary experiment on the growth-promoting effect of growth hormone with thyroid-stimulating hormone and prolactin to the young rainbow trout (*Salmo irideus*). *Nippon Suisan Gakkaishi* 30, 537–541.

Epple, A. (1969). The endocrine pancreas. *In* "Fish Physiology" (W. S. Hoar and D. J. Randall, eds.), Vol. 2, pp. 275–321. Academic Press, New York.

Epple, A., and Lewis, T. L. (1973). Comparative histophysiology of the pancreatic islets. *Am. Zool.* 13, 567–590.

Eversole, W. J. (1939). The effects of androgens upon the fish (*Lebistes reticulatus*). *Endocrinology* 25, 328–330.

Eversole, W. J. (1941). The effects of pregneninolone and related steroids on sexual development in fish (*Lebistes reticulatus*). *Endocrinology* 28, 603–610.

Fagerlund, U. H. M., and McBride, J. R. (1975). Growth increments and some flesh and gonad characteristics of juvenile coho salmon receiving diets supplemented with 17α-methyltestosterone. *J. Fish Biol.* 7, 305–314.

Fagerlund, U. H. M., and McBride, J. R. (1977). Effect of 17α-methyltestosterone on growth, gonad development, external features and proximate composition of muscle of steelhead trout, coho and pink salmon. *Fish. Mar. Serv. Tech. Rep. No. 716.*

Fagerlund, U. H. M., and McBride, J. R. (1978). Distribution and disappearance of radioactivity in blood and tissues of coho salmon (*Oncorhynchus kisutch*) after oral administration of ^{3}H-testosterone. *J. Fish. Res. Board Can.* **35**, 893–900.

Fagerlund, U. H. M., Higgs, D. A., and McBride, J. R. (1978). Influence of feeding a diet containing 17α-methyltestosterone or testosterone at two ration levels on growth, appetite and food conversion efficiency of underyearling coho salmon (*Oncorhynchus kisutch*). *EIFAC Symp. Finfish Nutr. and Feed Technol., Hamburg, June 20–23, 1978* EIFAC Tech. Paper No. E/24.

Falkner, N. W., and Eales, J. G. (1973). Investigation of iodothyronine binding to plasma proteins in brook trout, *Salvelinus fontinalis*, using precipitation, dialysis and electrophoretic methods. *J. Fish. Res. Board Can.* **30**, 1131–1140.

Farmer, S. W., Papkoff, H., and Hayashida, T. (1974). Purification and properties of avian growth hormones. *Endocrinology* **95**, 1560–1565.

Farmer, S. W., Clarke, W. C., Papkoff, H., Nishioka, R. S., Bern, H. A., and Li, C. H. (1975). Studies on the purification and properties of teleost prolactin. *Life Sci.* **16**, 149–158.

Farmer, S. W., Papkoff, H., Hayashida, T., Bewley, T. A., Bern, H. A., and Li, C. H. (1976). Purification and properties of teleost growth hormone. *Gen. Comp. Endocrinol.* **30**, 91–100.

Farmer, S. W., Licht, P., and Papkoff, H. (1977). Biological activity of bullfrog growth hormone in the rat and the bullfrog (*Rana catesbeiana*). *Endocrinology* **101**, 1145–1150.

Fellows, R. E. (1973). Discussion comments. *Recent Prog. Horm. Res.* **29**, 404–408.

Fletcher, G. L., Hardy, C. D., and Idler, D. R. (1969). Testosterone production and metabolic clearance rates in sexually mature male and female skate (*Raja radiata*). *Endocrinology* **85**, 552–560.

Fontaine, M., and Baraduc, M. M. (1955). Influence de la caséine iodée et de la thyroxine sur la croissance de la jeune truite arc-en-ciel (*Salmo gairdneri*). *Bull Fr. Piscic.* **179**, 89–98.

Fontaine, M., and Fontaine, Y. A. (1956). Détermination du pouvoir thyréotrope de l'hypophyse et du milieu intérieur de téléostéens par mesure de la fixation de ^{131}I par la thyroïde de la truite arc-en-ciel (*Salmo gairdneri* Rich.). *J. Physiol. (Paris)* **48**, 881–892.

Fontaine, M., and LeLoup, J. (1959). Influence de la nage à contre-courant sur le métabolisme de l'iode et le fonctionnement thyroidien chez la truite arc-en-ciel (*Salmo gairdnerii* Rich.). *C. R. Acad. Sci.* **249**, 343–347.

Fontaine, M., Baraduc, M., and Hatey, J. (1953). Influence de la thyroxinisation sur la teneur en glycogène du foie des poissons téléostéens. *C. R. Soc. Biol.* **147**, 214–216.

Fontaine, Y. A. (1975). Hormones in fishes. *Biochem. Biophys. Perspect. Mar. Biol.* **2**, 139–212.

Forchielli, E., Sorcini, G., Nightingale, M., Brust, N., Dorfman, R. I., Perloff, W. H., and Jacobson, G. E. (1963). Testosterone in human plasma. *Anal. Biochem.* **5**, 416–421.

Forselius, S. (1957). Studies of anabantid fishes. *Zool Bijdr.* **32**, 97–597.

Frye, B. E. (1971). "Hormonal Control in Vertebrates." Macmillan, New York.

Fryer, J. N. (1977). Radioreceptor assays for teleost growth hormone and prolactin. *Am. Zool.* **17**, 857 (abstr.).

Funk, J. D. (1972). An investigation of the induction of precocious sexual maturity in juvenile pink salmon *Oncorhynchus gorbuscha.* M.S. Thesis, Univ. of British Columbia, Vancouver.

Funk, J. D., Donaldson, E. M., and Dye, H. M. (1973). Induction of precocious sexual development in female pink salmon (*Oncorhynchus gorbuscha*). *Can. J. Zool.* **51**, 493–500.

Gaiser, M.-L. (1952). Effects produits par l'administration prolongée de thiourée et de thyroxine chez *Lebistes reticulatus. C. R. Soc. Biol.* **146**, 496–498.

Gaspard, S. T., Wondergem, R., and Klitgaard, H. M. (1975). Effect of growth hormone and thyroxine on serum somatomedin and growth. *Fed. Proc., Fed. Am. Soc. Exp. Biol.* **34**, 272.

Geschwind, I. I. (1967). Growth hormone activity in the lungfish pituitary. *Gen. Comp. Endocrinol.* **8**, 82–83.

Ghittino, P. (1970). Risposta delle trote d'allevamento al dietilstilbestrolo e metiltiuracile. *Riv. Ital. Piscic. Ittiopatol.* **5**, 9–11.

Giannopoulos, G. (1973). Binding of testosterone to uterine components of the immature rat. *J. Biol. Chem.* **248**, 1004–1010.

Gloyma, R. E., and Wilson, J. D. (1969). A comparative study of the conversion of testosterone to 17β-hydroxy-5α-androstan-3-one (dihydrotestosterone) by prostate and epididymis. *J. Clin. Endocrinol. Metab.* **29**, 970–977.

Gorbman, A. (1969). Thyroid function and its control in fishes. *In* "Fish Physiology" (W. S. Hoar and D. J. Randall, eds.), Vol. 2, pp. 241–274. Academic Press, New York.

Grajcer, D., and Idler, D. R. (1961). Testosterone, conjugated and "free" in the blood of spawned Fraser River sockeye salmon (*Oncorhynchus nerka*). *Can. J. Biochem. Physiol.* **39**, 1585–1593.

Grajcer, D., and Idler, D. R. (1963). Conjugated testosterone in the blood and testes of spawned Fraser River sockeye salmon (*Oncorhynchus nerka*). *Can. J. Biochem. Physiol.* **41**, 23–30.

Grant, N. H., Sarantakis, D., and Yardley, J. P. (1974). Action of growth hormone release inhibitory hormone on prolactin release in rat pituitary cell cultures. *J. Endrocrinol.* **61**, 163–164.

Greenspan, F. S., Le, C. H., Simpson, M. E., and Evans, H. M. (1949). Bioassay of hypophyseal growth hormone: The tibia test. *Endocrinology* **45**, 455–463.

Gregory, L. A., and Eales, J. G. (1975). Factors contributing to high levels of plasma iodide in brook trout, *Salvelinus fontinalis* (Mitchill). *Can. J. Zool.* **53**, 267–277.

Grobstein, C., and Bellamy, A. W. (1939). Some effects of feeding thyroid to immature fishes (*Platypoecilus*). *Proc. Soc. Exp. Biol. Med.* **41**, 363–365.

Gross, W. L., Fromm, P. O., and Roelofs, E. W. (1963). Relationship between thyroid and growth in green sunfish, *Lepomis cyanellus* (Rafinesque). *Trans. Am. Fish. Soc.* **92**, 401–408.

Guerrero, R. D. (1975). Use of androgens for the production of all-male *Tilapia aurea* (Steindachner). *Trans. Am. Fish. Soc.* **104**, 342–348.

Guerrero, R. D. (1976). Culture of male *Tilapia mossambica* produced through artifical sex reversal. *FAO Tech. Conf. Aquacult.* FIR: AQ/Conf/76/E.15.

Hall, T. R., and Chadwick, A. (1976). Effects of growth hormone inhibiting factor (somatostatin) on the release of growth hormone and prolactin from pituitaries of the domestic fowl *in vitro. J. Endocrinol.* **68**, 163–164.

Halver, J. E., Johnson, C. L., and Ashley, L. M. (1962). Dietary carcinogens induce fish hepatoma. *Fed. Proc.* **21**, 390.

Hansson, V., Tveter, K. J., Unheim, O., and Djoseland, O. (1972). Interaction between androgen and macromolecules in male accessory sex organs of rat and man. *J. Steroid Biochem.* **3**, 427–439.

Harper, H. A. (1971). "Review of Physiological Chemistry." Lange Medical Publ., 529 pp.,

California.

Hayashida, T. (1970). Immunological studies with rat pituitary growth hormone (RGH). II. Comparative immunochemical investigation of GH from representatives of various vertebrate classes with monkey antiserum to RGH. *Gen. Comp. Endocrinol.* **15**, 432–452.

Hayashida, T. (1971). Biological and immunochemical studies with growth hormone in pituitary extracts of holostean and chondrostean fishes. *Gen. Comp. Endocrinol.* **17**, 275–280.

Hayashida, T. (1973). Biological and immunochemical studies with growth hormone in pituitary extracts of elasmobranchs. *Gen. Comp. Endocrinol.* **20**, 377–385.

Hayashida, T. (1975). Immunochemical and biological studies with antisera to pituitary growth hormones. *In* "Hormonal Proteins and Peptides" (C. H. Li, ed.), Vol. 3, pp. 42–146. Academic Press, New York.

Hayashida, T., and Lagios, M. (1969). Fish growth hormone: A biological, immunochemical and ultrastructural study of sturgeon and paddlefish pituitaries. *Gen. Comp. Endocrinol.* **13**, 403–411.

Herzfeld, E., Mayer-Umhäfer, P., and Scholz, F. (1931). Fische als Testobjekte für pharmakologische Versuche. II. Wirkung von Schildrüsen derivaten und Blutschutz. *Klin. Wochenschr.* **10**, 1908–1910.

Hickman, C. P., Jr. (1962). Influence of environment on the metabolism of iodine in fish. *Gen. Comp. Endocrinol., Suppl.* No. 1, pp. 48–62.

Higgs, D. A. (1974). Influence of nutritional state on thyroid hormone metabolism in the brook trout, *Salvelinus fontinalis* (Mitchill). Ph.D. Thesis, Univ. of Manitoba, Winnipeg.

Higgs, D. A., and Eales, J. G. (1971). Iodide and thyroxine metabolism in the brook trout, *Salvelinus fontinalis*, during sustained exercise. *Can. J. Zool.* **49**, 1255–1269.

Higgs, D. A., and Eales, J. G. (1973). Measurement of circulating thyroxine in several freshwater teleosts by competitive binding analysis. *Can. J. Zool.* **51**, 49–53.

Higgs, D. A., and Eales, J. G. (1976). Influence of injection route on radiothyroxine kinetics in brook trout, *Salvelinus fontinalis* (Mitchill). *Can. J. Zool.* **54**, 255–259.

Higgs, D. A., and Eales, J. G. (1977). Influence of food deprivation on radioiodothyronine and radioiodide kinetics in yearling brook trout, *Salvelinus fontinalis* (Mitchill), with a consideration of the extent of L-thyroxine conversion to 3,5,3′-triiodo-L-thyronine. *Gen. Comp. Endocrinol.* **32**, 29–40.

Higgs, D. A., and Eales, J. G. (1978a). Radiothyroxine kinetics in yearling brook trout, *Salvelinus fontinalis* (Mitchill), on different levels of dietary intake. *Can. J. Zool.* **56**, 80–85.

Higgs, D. A., and Eales, J. G. (1978b). Influence of diet composition on radiothyroxine kinetics in brook trout, *Salvelinus fontinalis* (Mitchill). *Can. J. Zool.* (in press).

Higgs, D. A., Donaldson, E. M., Dye, H. M., and McBride, J. R. (1975). A preliminary investigation of the effect of bovine growth hormone on growth and muscle composition of coho salmon (*Oncorhynchus kisutch*). *Gen. Comp. Endocrinol.* **27**, 240–253.

Higgs, D. A., Donaldson, E. M., Dye, H. M., and McBride, J. R. (1976). Influence of bovine growth hormone and L-thyroxine on growth, muscle composition, and histological structure of the gonads, thyroid, pancreas, and pituitary of coho salmon (*Oncorhynchus kisutch*). *J. Fish. Res. Board Can.* **33**, 1585–1603.

Higgs, D. A., Fagerlund, U. H. M., McBride, J. R., Dye, H. M., and Donaldson, E. M. (1977a). Influence of combinations of bovine growth hormone, 17α-methyltestosterone and L-thyroxine on growth of yearling coho salmon (*Oncorhynchus kisutch*). *Can. J. Zool.* **55**, 1048–1056.

Higgs, D. A., Fagerlund, U. H. M., McBride, J. R., and Eales, J. G. (1977b). Influence of orally administered L-thyroxine and 3,5,3′-triiodo-L-thyronine on growth, appetite and food conversion of underyearling coho salmon (*Oncorhynchus kisutch*). *Annu. Meeting Can. Soc. Zool., May 8–11, Univ. Victoria, B.C.* Abstr. No. 49.

Higgs, D. A., Donaldson, E. M., McBride, J. R., and Dye, H. M. (1978a). Evaluation of the potential for using a chinook salmon (*Oncorhynchus tshawytscha*) pituitary extract versus bovine growth hormone to enhance the growth of coho salmon (*Oncorhynchus kisutch*). *Can. J. Zool.* **56**, 1226–1231.

Higgs, D. A., Markert, J. R., MacQuarrie, D. W., McBride, J. R., Dosanjh, B. S., Nichols, C., and Hoskins, G. (1978b). Development of practical dry diets for coho salmon, *Oncorhynchus kisutch*, using poultry by-product meal, feather meal, soybean meal and rapeseed meal as major protein sources. *EIFAC Symp. Finfish Nutr. and Feed Technol., Hamburg, June 20–23, 1978* EIFAC Tech. Paper No. E/43.

Hirose, K., and Hibiya, T. (1968a). Physiological studies on growth-promoting effect of protein-anabolic steroids on fish. I. Effects on goldfish. *Nippon Suisan Gakkaishi* **34**, 466–472.

Hirose, K., and Hibiya, T. (1968b). Physiological studies on growth-promoting effect of protein-anabolic steroids on fish. II. Effects of 4-chlorotestosterone acetate on rainbow trout. *Nippon Suisan Gakkaishi* **34**, 473–481.

Hoar, W. S. (1951). Hormones in fish. *Univ. Toronto Stud.* Biol. Ser. No. 59 (Ontario Fish. Res. Lab. Publ. No. 71), pp. 1–51.

Hoar, W. S. (1957). The gonads and reproduction. *In* "The Physiology of Fishes" (M. E. Brown, ed.), Vol. 1, pp. 287–321. Academic Press, New York.

Hoar, W. S. (1958). Effects of synthetic thyroxine and gonadal steroids on the metabolism of goldfish. *Can. J. Zool.* **36**, 113–121.

Hoar, W. S. (1965). Comparative physiology: Hormones and reproduction in fishes. *Annu. Rev. Physiol.* **27**, 51–70.

Hoar, W. S., MacKinnon, D., and Redlich, A. (1952). Effects of some hormones on the behavior of salmon fry. *Can. J. Zool.* **30**, 273–286.

Hochachka, P. W. (1962). Thyroidal effects on pathways for carbohydrate metabolism in a teleost. *Gen. Comp. Endocrinol.* **2**, 499–505.

Hoffert, J. R., and Fromm, P. O. (1959). Estimation of thyroid secretion rate of rainbow trout using radioactive iodine. *J. Cell. Comp. Physiol.* **54**, 163–169.

Holmes, R. L., and Ball, J. N. (1974). The pituitary gland: A comparative account. *In* "Biological Structure and Function" (R. J. Harrison, R. M. H. McMinn, and J. E. Treherne, eds.), Vol. 4, pp. 1–397. Cambridge Univ. Press, London and New York.

Honma, Y. (1960). Studies on the morphology and the role of the important endocrine glands in some Japanese cyclostomes and fishes. *Publ., Dep. Biol. Univ. Niigata* pp. 1–139. (In Jpn.; Engl. summ.)

Honma, Y., and Matsui, I. (1973). Histological observation on a specimen of the Japanese eel, *Anguilla japonica*, under the long-term starvation. *Norinsho Suisan Koshusho Kenkyu Hokoku* **21**, 285–293.

Honma, Y., and Murakawa, S. (1955). Effects of thyroxine and thiourea on the development of chum salmon larvae. *Gyorvigaku Zasshi* **4**, 83–93.

Hopper, A. F. (1950). The effect of mammalian thyroid powder and thiouracil on growth rates and on the differentiation of the gonopod in *Lebistes reticulatus*. *Anat. Rec.* **108**, 554.

Hopper, A. F. (1952). Growth and maturation response of *Lebistes reticulatus* to treatment with mammalian thyroid powder and thiouracil. *J. Exp. Zool.* **119**, 205–217.

Hopper, A. F. (1961). The effect of feeding mammalian thyroid powder on growth rates of immature guppies. *Growth* **25**, 1–5.

Huang, C. T., and Hickman, C. P., Jr. (1968). Binding of inorganic iodide to the plasma proteins of teleost fishes. *J. Fish. Res. Board Can.* **25**, 1651–1666.

Hurlburt, M. E. (1977). Effects of thyroxine administration on plasma thyroxine levels in the goldfish, *Carassius auratus* L. *Can. J. Zool.* **55**, 255–258.

Hutchison, R. E., and Campbell, G. D. (1964). An attempt to increase the weight of *Tilapia melanopleura*, a fish used in fish farming, with an anabolic steroid ethylestrenol. *S. Afr. Med. J.* **38**, 640.

Idler, D. R., and Truscott, B. (1972). Corticosteroids in fish. *In* "Steroids in Nonmammalian Vertebrates" (D. R. Idler, ed.), pp. 127–217. Academic Press, New York.

Idler, D. R., and Tsuyuki, H. (1959). Biochemical studies on sockeye salmon during spawning migration. VIII. Androgen content of testes. *J. Fish. Res. Board Can.* **16**, 559–560.

Idler, D. R., Schmidt, P. J., and Ronald, A. P. (1960). Isolation and identification of 11-ketotestosterone in salmon plasma. *Can. J. Biochem. Physiol.* **38**, 1053–1057.

Idler, D. R., Bitners, I. I., and Schmidt, P. J. (1961). 11-ketotestosterone: An androgen for sockeye salmon. *Can. J. Biochem. Physiol.* **39**, 1737–1742.

Idler, D. R., Horne, D. A., and Sangalang, G. B. (1971). Identification and quantification of the major androgens in testicular and peripheral plasma of Atlantic salmon (*Salmo salar*) during sexual maturation. *Gen. Comp. Endocrinol.* **16**, 257–267.

Imura, K., Kato, Y., Ikeda, M., Morimoto, M., Yawata, M., and Fukase, M. (1968). Increased plasma levels of growth hormone during infusion of propranolol. *J. Clin. Endocrinol. Metab.* **28**, 1079–1081.

Ince, B. W., and Thorpe, A. (1974). Effects of insulin and of metabolite loading on blood metabolites in the European silver eel, *Anguilla anguilla* L. *Gen. Comp. Endocrinol.* **23**, 460–471.

Ingleton, P. M., Baker, B. I., and Ball, J. N. (1973). Secretion of prolactin and growth hormone by teleost pituitaries *in vitro*. I. Effect of sodium concentration and osmotic pressure during short-term incubations. *J. Comp. Physiol.* **87**, 317–328.

Inui, Y., Arai, S., and Yokote, M. (1975). Gluconeogenesis in the eel. VI. Effects of hepatectomy, alloxan, and mammalian insulin on the behaviour of plasma amino acids. *Nippon Suisan Gakkaishi* **41**(11), 1105–1111.

Jackim, E., and LaRoche, G. (1973). Protein synthesis in *Fundulus heteroclitus* muscle. *Comp. Biochem. Physiol. A* **44**, 851–866.

Jacoby, G. H., and Hickman, C. P., Jr. (1966). A study of the circulating iodocompounds of rainbow trout, *Salmo gairdneri*, by the method of isotopic equilibrium. *Gen. Comp. Endocrinol.* **7**, 245–254.

Jampolsky, A., and Hoar, W. S. (1954). Growth hormone from salmon pituitary glands. *J. Fish. Res. Board Can.* **11**, 57–62.

Jung, I., and Baulieu, E. E. (1972). Testosterone cytosol receptor in the rat levator ani muscle. *Nature New Biol.* **237**, 24–26.

Katz, Y., Abraham, M., and Eckstein, B. (1976). Effects of adrenosterone on gonadal and body growth in *Tilapia nilotica* (Teleostei, Cichlidae). *Gen. Comp. Endocrinol.* **29**, 414–418.

Kayes, T. (1977a). Effects of hypophysectomy, beef growth hormone replacement therapy, pituitary autotransplantation, and environmental salinity on growth in the black bullhead (*Ictalurus melas*). *Gen. Comp. Endocrinol.* **33**, 371–381.

Kayes, T. (1977b). Effects of temperature on hypophyseal (growth hormone) regulation of length, weight, and allometric growth and total lipid and water concentrations in the black bullhead (*Ictalurus melas*). *Gen. Comp. Endocrinol.* **33**, 382–393.

Kochakian, C. D., Bartlett, M. N., and Gongora, J. (1948). Effect of castration and androgens on body and organ weights, and the arginase and phosphatases of kidney and
liver of the male Syrian hamster. *Am. J. Physiol.* **153**, 210–214.

Kolbye, A. C., Jr. (1975). Zero tolerance concept as it relates to hormonal residues in
animal products. *J. Anim. Sci.* **40**, 1258–1262.

Komourdjian, M. P., Saunders, R. L., and Fenwick, J. C. (1976a). The effect of porcine
somatotropin on growth, and survival in seawater of Atlantic salmon (*Salmo salar*)
parr. *Can. J. Zool.* **54**, 531–535.

Komourdjian, M. P., Saunders, R. L., and Fenwick, J. C. (1976b). Evidence for the role of
growth hormone as a part of a "light-pituitary axis" in growth and smoltification of
Atlantic salmon (*Salmo salar*). *Can. J. Zool.* **54**, 544–551.

Komourdjian, M. P., Burton, M. P., and Idler, D. R. (1978). Growth of rainbow trout,
Salmo gairdneri, after hypophysectomy and somatotropin therapy. *Gen. Comp.
Endocrinol.* **34**, 158–162.

Krüskemper, H. L. (1968). "Anabolic Steroids." Academic Press, New York.

Lam, T. J. (1973). Experimental inductions of changes in fishes. *J. Singapore Natl. Acad.
Sci.* **3**, 188–191.

LaRoche, G. (1953). Effects de la thyroide de boeuf sur la forme de la tête de la truite
mouchetée (*Salveulinus fontinalis* L.). *Rev. Can. Biol.* **11**, 431–438.

LaRoche, G., and LeBlond, C. P. (1952). Effect of thyroid preparations and iodide on
Salmonidae. *Endocrinology* **51**, 524–545.

LaRoche, G., and LeBlond, C. P. (1954). Destruction of thyroid gland of Atlantic salmon
(*Salmo salar* L.) by means of radioiodine. *Proc. Soc. Exp. Biol. Med.* **87**, 273–276.

LaRoche, G., Johnson, C. L., and Woodall, A. N. (1965). Thyroid function in the rainbow
trout (*Salmo gairdnerii*, Rich.) I. Biochemical and histological evidence of radiothyroidectomy. *Gen. Comp. Endocrinol.* **5**, 145–159.

LaRoche, G., Woodall, A. N., Johnson, C. L., and Halver, J. E. (1966). Thyroid function
in the rainbow trout (*Salmo gairdnerii* Rich.). II. Effects of thyroidectomy on the
development of young fish. *Gen. Comp. Endocrinol.* **6**, 249–266.

Law, Y. M. C., and Eales, J. G. (1973). Deiodination of radiothyroxine by tissue homogenates of brook trout, *Salvelinus fontinalis* (Mitchill). *Comp. Biochem. Physiol. B* **44**,
1175–1183.

Leatherland, J. F. (1970). Histological investigation of pituitary homo transplants in the
marine form (*Trachurus*) of the three spined stickleback (*Gasterosteus aquleatus*
L.). *Z. Zellforsch. Mikrosk. Anat.* **104**, 337–344.

Leatherland, J. F., and Hyder, M. (1975). Effect of thyroxine on the ultrastructure of the
hypophyseal proximal pars distalis in *Tilapia zillii*. *Can. J. Zool.* **53**, 686–690.

Leatherland, J. F., McKeown, B. A., and John, T. M. (1974). Circadian rhythm of plasma
prolactin, growth hormone, glucose, and free fatty acid in juvenile kokanee salmon,
Oncorhynchus nerka. Comp. Biochem. Physiol. A **47**, 821–828.

Leatherland, J. F., Cho, C. Y., and Slinger, S. J. (1977). Effects of diet, ambient temperature, and holding conditions on plasma thyroxine levels in rainbow trout (*Salmo
gairdneri*). *J. Fish. Res. Board Can.* **34**, 677–682.

Leloup, J. (1956). Contribution à l'étude du fonctionnement thyroïdien d'un téléostéen
amphibiotique, *Periophthalmus koelreuteri. C. R. Acad. Sci., Ser. D* **242**, 1765–
1767.

Leloup, J. (1965). Métabolisme de la thyroxine chez l'anguille normale et hypophysec-
tomisée en fonction de la température. *Gen. Comp. Endocrinol.* **5**, 66.

Leloup, J., and Fontaine, M. (1960). Iodine·metabolism in lower vertebrates. *Ann. N.Y.
Acad. Sci.* **86**, 316–353.

Leloup, J., and Hardy, A. (1976). Hormones thyroidiennes circulantes chez un cyclostome et des poissons. *Gen. Comp. Endocrinol.* **29**, 258.

Leloup, J., Long, P., and Fontaine, M. (1972). Influence des hormones thyroïdiennes sur la synthèse des protéines dans le cerveau des Teleosteens. *Gen. Comp. Endocrinol.* **18**, 603.

Leloup, J., Brichon, G., and Hardy, A. (1976). Variations des hormones thyroïdiennes circulantes chez un Sélacien et des Téléostéens en fonction de l'évolution génitale et de la salinité du milieu. *J. Physiol. (Paris)* **72**, 48A.

LeRay, C., Pic, P., Bonnet, B., and Vallet, F. (1969). Aspects compares de l'activité thyroidienne sur la physiologie de *Mugil auratus* (Téléostéen Mugilidé) en croissance. *Gen. Comp. Endocrinol.* **13**, 517.

LeRay, C., Bonnet, B., Febvre, A., Vallet, F., and Pic, P. (1970). Quelques activités périphériques des hormones thyroïdiennes observeés chez *Mugil auratus* L. (Téléostéen Mugilidé). *Ann. Endocrinol.* **31**, 567–572.

Lewis, U. J., Singh, R. N., Seavey, B. K., Lasker, R., and Pickford, G. E. (1972). Growth hormone and prolactinlike proteins of the shark (*Prionace glauca*). *U.S. Fish Wildl. Serv., Fish. Bull.* **70**, 933–939.

Li, C. H. (1975). The chemistry of human pituitary growth hormone. *In* "Hormonal Proteins and Peptides" (C. H. Li, ed.), Vol. 3, pp. 1–41. Academic Press, New York.

Li, C. H., and Yamashiro, D. (1970). The synthesis of a protein possessing growth-promoting and lactogenic activities. *J. Am. Chem. Soc.* **92**, 7608–7609.

Li, C. H., Evans, H. M., and Simpson, M. E. (1945). Isolation and properties of the anterior hypophyseal growth hormone. *J. Biol. Chem.* **159**, 353–366.

Li, C. H., Dixon, J. S., and Liu, W. K. (1969). Human pituitary growth hormone. XIX. The primary structure of the hormone. *Arch. Biochem. Biophys.* **133**, 70–91.

Licht, P., and Hoyer, H. (1968). Somatotropic effects of exogenous prolactin and growth hormone in juvenile lizards. (*Lacerta s. sicula*). *Gen. Comp. Endocrinol.* **11**, 338–346.

Licht, P., Cohen, D. C., and Bern, H. A. (1972). Somatotropic effects of mammalian growth hormone and prolactin in larval newts, *Taricha torosa*. *Gen. Comp. Endocrinol.* **18**, 391, 393.

Lorz, H. W. (1971). The effects of X-irradiation, diethylstilbestrol, and size at time of release on the early sexual maturation of coho salmon (*Oncorhynchus kisutch*). Ph.D. Thesis, Oregon State Univ., Corvallis.

Ludwig, B., Higgs, D. A., Fagerlund, U. H. M., and McBride, J. R. (1977). A preliminary study of insulin participation in the growth regulation of coho salmon (*Oncorhynchus kisutch*). *Can. J. Zool.* **55**, 1756–1758.

McBride, J. R. (1967). Effects of feeding on the thyroid, kidney, and pancreas in sexually ripening adult sockeye salmon (*Oncorhynchus nerka*). *J. Fish. Res. Board Can.* **24**, 67–76.

McBride, J. R., and Fagerlund, U. H. M. (1973). The use of 17α-methyltestosterone for promoting weight increases in juvenile Pacific salmon. *J. Fish. Res. Board Can.* **30**, 1099–1104.

McBride, J. R., and Fagerlund, U. H. M. (1976). Sex steroids as growth promoters in the cultivation of juvenile coho salmon (*Oncorhynchus kisutch*). *Proc. World Maricult. Soc.* **7**, 145–161.

McBride, J. R., and Van Overbeeke, A. P. (1969). Cytological changes in the pituitary gland of the adult sockeye salmon (*Oncorhynchus nerka*) after gonadectomy. *J. Fish. Res. Board Can.* **26**, 1147–1156.

McBride, J. R., and Van Overbeeke, A. P. (1970). Effects of androgens, estrogens and cortisol on the skin, stomach, liver, pancreas, and kidney in gonadectomized adult sockeye salmon (*Oncorhynchus nerka*). *J. Fish. Res. Board Can.* **28**, 485–490.

McBride, J. R., Fagerlund, U. H. M., Smith, M., and Tomlinson, N. (1963). Resumption of feeding by and survival of adult sockeye salmon (*Oncorhynchus nerka*) following advanced gonad development. *J. Fish. Res. Board Can.* **20**, 95–100.

McCann, S. M., Fawcett, C. P., and Krulich, L. (1974). Hypothalamic hypophysial releasing and inhibiting hormones. *In* "Endocrine Physiology" (S. M. McCann, ed.), MTP International Review of Science, Physiology Series One, Vol. 5, pp. 31–65. Butterworth, London.

Machlin, L. J. (1972). Effect of porcine growth hormone on growth and carcass composition of the pig. *J. Anim. Sci.* **35**, 794–800.

McKeown, B. A., and Van Overbeeke, A. P. (1971). Immunochemical identification of pituitary hormone producing cells in the sockeye salmon (*Oncorhynchus nerka*, Walbaum). *Z. Zellforsch. Mikrosk. Anat.* **112**, 350–362.

McKeown, B. A., and Van Overbeeke, A. P. (1972). Prolactin and growth hormone concentrations in the serum and pituitary gland of adult migratory sockeye salmon (*Oncorhynchus nerka*). *J. Fish. Res. Board Can.* **29**, 303–309.

McKeown, B. A., Leatherland, J. F., and John, T. M. (1975). The effect of growth hormone and prolactin on the mobilization of free fatty acids and glucose in the kokanee salmon, *Oncorhynchus nerka*. *Comp. Biochem. Physiol. B* **50**, 425–430.

Mainwaring, W. I. P. (1972). The distribution of specific androgen receptor proteins. *J. Endocrinol.* **52**, iv–v.

Marine, D. J. (1914). Further observations and experiments on goiter (so-called thyroid carcinoma) in brook trout (*Salvelinus fontinalis*). III. Its prevention and cure. *J. Exp. Med.* **19**, 70–88.

Marine, D. J., and Lenhart, C. H. (1910). Observations and experiments on the so-called thyroid carcinoma of brook trout (*Salvelinus fontinalis*) and its relation to ordinary goiter. *J. Exp. Med.* **12**, 311–337.

Markert, J. R., Higgs, D. A., Dye, H. M., and MacQuarrie, D. W. (1977). Influence of bovine growth hormone on growth rate, appetite, and food conversion of yearling coho salmon (*Oncorhynchus kisutch*) fed two diets of different composition. *Can. J. Zool.* **55**, 74–83.

Martin, J. B. (1974). Regulation of the pituitary–thyroid axis. *In* "Endocrine Physiology" (S. M. McCann, ed.), MTP International Review of Science, Physiology Series One, Vol. 5, pp. 67–107. Butterworth, London.

Martin, J. M., and Gagliardino, J. J. (1967). Effect of growth hormone on the isolated pancreatic islets of rat *in vitro*. *Nature (London)* **213**, 630–631.

Massey, B. D., and Smith, C. L. (1968). The action of thyroxine on mitochondrial respiration and phosphorylation in the trout (*Salmo trutta* Fario L.). *Comp. Biochem. Physiol.* **25**, 241–255.

Matty, A. J. (1962). Effects of mammalian growth hormone on *Cottus scorpius* blood. *Nature (London)* **195**, 506–507.

Matty, A. J. (1975). Endocrine control of growth and protein metabolism in aquaculture. *Proc. 13th Pac. Sci. Congr.* **1**, 58.

Matty, A. J., and Sheltawy, M. J. (1967). The relation of thyroxine to skin purines in *Salmo irideus*. *Gen. Comp. Endocrinol.* **9**, 473.

Michel, G., and Baulieu, E. E. (1974). Soluble cytoplasmic androgen receptor in a striated skeletal muscle. *C. R. Acad. Sci., Ser. D* **279**, 421–424.

Middleton, W. R. J. (1971). Thyroid hormones and the gut. *Gut* **12**, 172–177.

Monfort, K. (1974). DES and beef prices. *Anim. Nutr. Health* **29**, 19.

Moudgal, N. R., and Li, C. H. (1961). Immunochemical studies of bovine and ovine pituitary growth hormone. *Arch. Biochem. Biophys.* **93**, 122–127.

Muller, E. E. (1974). Growth hormone and the regulation of metabolism. *In* "Endocrine Physiology" (S. M. McCann, ed.), MTP International Review of Science, Physiology Series One, Vol. 5, pp. 141–178. Butterworth, London.

Murat, J. C., and Serfaty, A. (1971). Variations saisonnières de l'effet de la thyroxine sur le métabolisme glucidique de la carpe. *J. Physiol. (Paris)* **63**, 80–81.

Murphy, B. E. P., and Pattee, C. J. (1964). Determination of thyroxine utilizing the property of protein-binding. *J. Clin. Endocrinol. Metab.* **24**, 187–196.

Myers, W. R., and Peterson, R. A. (1974). Responses of 6- and 10-week-old broilers to a tryptic digest of bovine growth hormone. *Poultry Sci.* **53**, 508–514.

Nagahama, Y., and Yamamoto, K. (1969). Fine structure of the glandular cells in the adenohypophysis of the kokanee *Oncorhynchus nerka*. *Hokkaido Daigaku Suisan Gakubu Kenkyu Iho* **20**, 159–168.

Nagahama, Y., and Yamamoto, K. (1970). Morphological studies on the pituitary of the chum salmon, *Oncorhynchus keta*. (1) Fine structure of the adenohypophysis. *Hokkaido Daigaku Suisan Gakubu Kenkyu Iho* **20**, 293–302.

Nagahama, Y., Clarke, W. C., and Hoar, W. S. (1977). Influence of salinity on ultrastructure of the secretory cells of the adenohypophyseal pars distalis in yearling coho salmon (*Oncorhynchus kisutch*). *Can. J. Zool.* **55**, 183–198.

Narayansingh, T., and Eales, J. G. (1975a). Effects of thyroid hormones on *in vivo* [1-^{14}C]L-leucine incorporation into plasma and tissue protein of brook trout (*Salvelinus fontinalis*) and rainbow trout (*Salmo gairdneri*). *Comp. Biochem. Physiol. B* **52**, 399–405.

Narayansingh, T., and Eales, J. G. (1975b). The influence of physiological doses of thyroxine on the lipid reserves of starved and fed brook trout, *Salvelinus fontinalis* (Mitchill). *Comp. Biochem. Physiol. B* **52**, 407–412.

Nathanielsz, P. W. (1970). The effect of diet and acute starvation on the deiodination of thyroxine and triiodothyronine in the thyroidectomized rat. *J. Physiol. (London)* **206**, 701–710.

Niall, H. D. (1971). Revised primary structure for human growth hormone. *Nature (London), New Biol.* **230**, 90–91.

Niall, H. D., Hogan, M. L., Tregear, G. W., Segre, G. V., Hwang, P., and Friesen, H. (1973). The chemistry of growth hormone and lactogenic hormones. *Recent Prog. Horm. Res.* **29**, 387–416.

Nicoll, C. S., and Licht, P. (1971). Evolutionary biology of prolactins and somatotropins. II. Electrophoretic comparison of tetrapod somatotropins. *Gen. Comp. Endocrinol.* **17**, 409–507.

Nigrelli, R. F., Goldsmith, E. D., and Charipper, H. A. (1946). Effects of mammalian thyroid powder on growth and maturation of thiourea-treated fish. *Anat. Rec.* **94**, 523.

Nixo-Nicoscio, N. Y. (1940). The influence of hormones on growth in fish. *Proc. Moscow Zool. Park* **1**, 178–184.

Norris, D. O. (1966). Radiothyroidectomy in the salmonid fishes *Salmo gairdnerii* Richardson and *Oncorhynchus tshawytscha* Walbaum. Ph.D. Thesis, Univ. of Washington, Seattle.

Norris, D. O. (1968). Effect of radiothyroidectomy and thyroxine replacement on the growth rates of salmonid fry. *J. Colo.-Wyo. Acad. Sci.* **6**, 10–11.

Norris, D. O. (1969). Depression of growth following radiothyroidectomy of larval chinook salmon and steelhead trout. *Trans. Am. Fish. Soc.* **98**, 104–106.

Olivereau, M. (1954). Hypophyse et glande thyroïde chez les poissons. Étude histophysiologique de quelques corrélations endocriniennes, en particulier chez *Salmo salar* L. *Annee Biol.* **30**, 63–80.

Olivereau, M. (1970). Cytologie de l'hypophyse auto-transplantée chez l'anguille. Comparison avec celle de *Poecilia. Colloq. Nat. CNRS* No. 927, pp. 251–260.

Olivereau, M., and Ball, J. N. (1966). Histological study of functional ectopic pituitary transplants in a teleost fish (*Poecilia formosa*). *Proc. R. Soc., Ser. B* **164**, 106–129.

Olivereau, M., and Ridgeway, G. J. (1962). Cytologie hypophysaire et antigène sérique en relation avec la maturation sexuelle chez *Oncorhynchus* species. *C. R. Acad. Sci.* **254**, 753–755.

Ortman, R., and Billig, R. D. (1966). A re-examination of the goldfish microhistometric assay method for thyrotropin. *Gen. Comp. Endocrinol.* **6**, 362–370.

Osborn, R. H., and Simpson, T. H. (1969a). The sites of iodine accumulation and the iodoamino acids in the thyroid and plasma of plaice (*Pleuronectes platessa* L.). *J. Endocrinol.* **43**, 643–650.

Osborn, R. H., and Simpson, T. H. (1969b). Thyroxine metabolism in plaice, *Pleuronectes platessa* L. *Gen. Comp. Endocrinol.* **13**, 524.

Osborn, R. H., and Simpson, T. H. (1972). Iodoamino acids of plaice plasma; the influence of TSH, stress, and starvation. *Gen. Comp. Endocrinol.* **18**, 613.

Ozon, R. (1972a). Androgens in fishes, amphibians, reptiles and birds. *In* "Steroids in Nonmammalian Vertebrates" (D. R. Idler, ed.), pp. 329–389. Academic Press, New York.

Ozon, R. (1972b). Estrogens in fishes, amphibians, reptiles and birds. *In* "Steroids in Nonmammalian Vertebrates" (D. R. Idler, ed.), pp. 390–410. Academic Press, New York.

Papkoff, H., and Hayashida, T. (1972). Pituitary growth hormone from the turtle and duck: purification and immunochemical studies. *Proc. Soc. Exp. Biol. Med.* **140**, 251–255.

Papkoff, H., and Li, C. H. (1958). The isolation and characterization of growth hormone from anterior lobes of whale pituitaries. *J. Biol. Chem.* **231**, 367–377.

Parman, A. Ü. (1975a). Effects of hypophysectomy and short-term growth hormone replacement on insulin release from and glucose metabolism in isolated rat islets of Langerhans. *J. Endocrinol.* **67**, 1–17.

Parman, A. Ü. (1975b). Changes in insulin content of, and insulin and RNA synthesis in, isolated rat islets of Langerhans after hypophysectomy and short-term growth hormone replacement. *J. Endocrinol.* **67**, 19–28.

Peter, R. E. (1970). Hypothalamic control of thyroid gland activity and gonadal activity in the goldfish, *Carassius auratus. Gen. Comp. Endocrinol.* **14**, 334–356.

Peter, R. E. (1971). Feedback effects of thyroxine on the hypothalamus and pituitary of goldfish, *Carassius auratus. J. Endocrinol.* **51**, 31–39.

Peter, R. E. (1972). Feedback effects of thyroxine in goldfish *Carassius auratus* with an autotransplanted pituitary. *Neuroendocrinology* **17**, 273–281.

Peter, R. E. (1973). Neuroendocrinology of teleosts. *Am. Zool.* **13**, 743–755.

Peter, R. E., and McKeown, B. A. (1975). Hypothalamic control of prolactin and thyrotropin secretion in teleosts, with special reference to recent studies on the goldfish. *Gen. Comp. Endocrinol.* **25**, 153–165.

Pickford, G. E. (1953a). A study of the hypophysectomized male killifish, *Fundulus heteroclitus* (Linn.). *Bull. Bingham Oceanogr. Collect.* **14**, 5–41.

Pickford, G. E. (1953b). The response of hypophysectomized male *Fundulus* to injections of purified beef growth hormone. *Bull. Bingham Oceanogr. Collect.* **14**, 46–68.

Pickford, G. E. (1954a). The response of hypophysectomized male killifish to purified fish growth hormone, as compared with the response to purified beef growth hormone. *Endocrinology* **55**, 274–287.

Pickford, G. E. (1954b). The response of hypophysectomized male killifish to prolonged treatment with small doses of thyrotropin. *Endocrinology* **55**, 589–592.

Pickford, G. E. (1957). The growth hormone. *In* "The Physiology of the Pituitary Gland of Fishes" (G. E. Pickford and J. W. Atz, eds.), Part IV, pp. 84–99. N.Y. Zool. Soc., New York.

Pickford, G. E. (1959). The nature and physiology of the pituitary hormones of fishes. *In* "Comparative Endocrinology" (A. Gorbman, ed.), pp. 404–420. Wiley, New York.

Pickford, G. E. (1963). Failure of hypophysectomized killifish, *Fundulus heteroclitus,* to respond to purified beef growth hormone in the aquarium water. *Anat. Rec.* **145**, 343.

Pickford, G. E. (1973). Introductory remarks. *Am. Zool.* **13**, 711–717.

Pickford, G. E., and Atz, J. W. (1957). "The Physiology of the Pituitary Gland of Fishes." N.Y. Zool. Soc., New York.

Pickford, G. E., and Grant, F. B. (1968). The response of hypophysectomized male killifish (*Fundulus heteroclitus*) to thyrotropin preparations and to the bovine heterothyrotropic factor. *Gen. Comp. Endocrinol.* **10**, 1–7.

Pickford, G. E., and Thompson, E. F. (1948). The effect of purified mammalian growth hormone on the killifish (*Fundulus heteroclitus* (Linn.)). *J. Exp. Zool.* **109**, 367–383.

Pickford, G. E., Wilhelmi, A. E., and Nussbaum, N. (1959). Comparative studies of the response of hypophysectomized killifish, *Fundulus heteroclitus,* to growth hormone preparations. *Anat. Rec.* **134**, 624–625 (abstr.).

Pickford, G. E., Lofts, B., Bara, G., and Atz, J. W. (1972). Testis stimulation in hypophysectomized male killifish *Fundulus heteroclitus,* treated with mammalian growth hormone and/or luteinizing hormone. *Biol. Reprod.* **7**, 370–386.

Piggins, D. J. (1962). Thyroid feeding of salmon parr. *Nature (London)* **195**, 1017–1018.

Powers, M. L., and Florini, J. R. (1975). A direct effect of testosterone on muscle cells in tissue culture. *Endocrinology* **97**, 1043–1047.

Preston, R. L. (1975). Biological responses to estrogen additives in meat producing cattle and lamb. *J. Anim. Sci.* **41**, 1414–1430.

Quincey, R. V., and Gray, C. H. (1967). The metabolism of [1,2-^{3}H]17α-methyltestosterone in human subjects. *J. Endocrinol.* **37**, 37–55.

Qureshi, F. (1976). Effect of triiodothyronine on skeletal growth of *Salmo trutta* alevin. *Experientia* **32**, 115–117.

Rappaport, R. (1975). Hormonal control of skeletal growth and maturation. *Acta Endocrinol. (Copenhagen) Suppl.* No. 199, pp. 71–72.

Refetoff, S., Robin, N. I., and Fang, V. S. (1970). Parameters of thyroid function in serum of 16 selected vertebrate species: A study of PBI, serum T_4, free T_4, and the pattern of T_4 and T_3 binding to serum proteins. *Endocrinology* **86**, 793–805.

Régnier, M. T. (1938). Action des hormones sexuelles sur l'inversion du sexe chez *Xiphophorus helleri* Heckel. *C. R. Acad. Sci.* **205**, 1451–1453.

Robertson, O. H. (1961). Prolongation of the life span of kokanee salmon (*Oncorhynchus nerka kennerlyi*) by castration before beginning of gonad development. *Proc. Natl. Acad. Sci. U.S.A.* **47**, 609–621.

Robertson, O. H., and Chaney, A. L. (1953). Thyroid hyperplasia and tissue iodine content in spawning rainbow trout: a comparative study of Lake Michigan and California sea-run trout. *Physiol. Zool.* **26**, 328–340.

Robertson, O. H., and Rinfret, A. P. (1957). Maturation of the infantile testes in rainbow trout (*Salmo gairdnerii*) produced by salmon pituitary gonadotrophins administered in cholesterol pellets. *Endocrinology* **60**, 559–562.

Sage, M. (1965). The somatotrophic cells in the pituitary of *Lebistes reticulatus* Peters. *J. Endocrinol.* **33**, v.

Sage, M. (1967). Responses of pituitary cells of *Poecilia* to changes in growth induced by thyroxine and thiourea. *Gen. Comp. Endocrinol.* **8**, 314–319.

Sage, M., and Bromage, N. R. (1970). Interactions of the TSH and thyroid cells with the gonadotropic cells and gonads in poecilid fishes. *Gen. Comp. Endocrinol.* **14**, 137–140.

Salvatore, G., Macchia, V., Vecchio, G., and Roche, J. (1959). Sur le transport de la thyroxine par les proteines du serum de l'ammocoete de *Petromyzon planeri* Bl. *C. R. Soc. Biol.* **153**, 1693–1697.

Saunders, R. L., Fagerlund, U. H. M., McBride, J. R., and Henderson, E. B. (1977). 17α-Methyltestosterone: A potential anabolic hormone in Atlantic salmon culture. *Int. Coun. Explor. Sea* C.M.1977/E:50: 8 p.

Schally, A. V., Baba, Y., Nair, R. M. G., and Bennett, C. D. (1971). The amino acid sequence of a peptide with growth hormone releasing activity isolated from porcine hypothalamus. *J. Biol. Chem.* **246**, 6647–6650.

Schmidt, P. J., and Idler, D. R. (1962). Steroid hormones in the plasma of salmon at various states of maturation. *Gen. Comp. Endocrinol.* **2**, 204–214.

Schreck, C. B. (1973). Uptake of [³H]testosterone and influence of an antiandrogen in tissues of rainbow trout (*Salmo gairdneri*). *Gen. Comp. Endocrinol.* **21**, 60–68.

Schreck, C. B., Flickinger, S. A., and Hopwood, M. L. (1972a). Plasma androgen levels in intact and castrate rainbow trout, *Salmo gairdneri. Proc. Soc. Exp. Biol. Med.* **140**, 1009–1011.

Schreck, C. B., Lackey, R. T., and Hopwood, M. L. (1972b). Evaluation of diel variation in androgen levels of rainbow trout, *Salmo gairdneri. Copeia* pp. 865–868.

Scidmore, W. J. (1966). Preliminary observations on the use of estrone for inducing sex-reversal in rainbow trout. *Minn. Dep. Game Fish, Invest. Rep.* No. 290, pp. 1–8.

Scott, J. L. (1944). The effects of steroids on the skeleton of the poecilid fish *Lebistes reticulatus. Zoologica* (*N.Y.*) **29**, 49–53.

Scow, R. O. (1959). Effect of growth hormone and thyroxine on growth and chemical composition of muscle, bone, and other tissues in thyroidectomized–hypophysectomized rats. *Am. J. Physiol.* **196**, 859–865.

Simpson, T. H. (1976a). Cited in T. Loftas, Day of the farmed Scotch salmon. *New Sci.* **70**, 20–21.

Simpson, T. H. (1976b). Endocrine aspects of salmonid culture. *Proc. R. Soc. Edinburgh, Sect. B* **75**, 241–252.

Sinclair, D. A. R., and Eales, J. G. (1972). Iodothyronine–glucuronide conjugates in the bile of brook trout, *Salvelinus fontinalis* (Mitchill) and other freshwater teleosts. *Gen. Comp. Endocrinol.* **19**, 552–559.

Singh, T. P. (1971). Growth hormone production and release following thyroxine and prolactin administration after X-irradiation and thiourea treatment in a catfish *Mystus vittatus. Int. Symp., 2nd, Growth Horm.; Excerpta Med. Found. Int. Congr. Ser.* No. 236, p. 58.

Sklower, A. (1927). Ueber den Einfluss von Schilddrüsen und thymusfütterung auf die Körperlänge und das Gewicht von Forellenbrut. *Z. Fisch. Deren Hilfswiss.* **25**, 549–552.

Smith, C. D., and Eales, J. G. (1971). Influence of acclimation temperature on rate constants for thyroid uptake of iodide by brook trout, *Salvelinus fontinalis* (Mitchill). *Can. J. Zool.* **49**, 783–786.

Smith, D. C., and Everett, G. M. (1943). The effect of thyroid hormone on growth rate, time of sexual differentiation, and oxygen consumption in the fish, *Lebistes reticulatus. J. Exp. Zool.* 94, 229–240.

Smith, D. C., Sladek, S. A., and Kellner, A. W. (1953). The effect of mammalian thyroid extract on the growth rate and sexual differentiation in the fish, *Lebistes reticulatus*, treated with thiourea. *Physiol. Zool.* 26, 117–124.

Smith, D. C. W. (1956). The role of the endocrine organs in the salinity tolerance of trout. *Mem. Soc. Endocrinol.* 5, 83–101.

Smith, M. A. K., and Thorpe, A. (1977). Endocrine effects on nitrogen excretion in the euryhaline teleost, *Salmo gairdnerii. Gen. Comp. Endocrinol.* 32, 400–406.

Smith, S. B. (1949). The effects of thyroxine and related compounds on young salmon and trout. M.S. Thesis, Univ. of British Columbia, Vancouver.

Snieszko, S. F. (1972). Nutritional fish diseases. *In* "Fish Nutrition" (J. E. Halver, ed.), pp. 403–437. Academic Press, New York.

Snyder, B. C., and Frye, B. E. (1972). Physiological responses of larval and post-metamorphic *Rana pipiens* to growth hormone and prolactin. *J. Exp. Zool.* 179, 299–314.

Solomon, J., and Greep, R. O. (1959). The growth hormone content of several vertebrate pituitaries. *Endocrinology* 65, 334–336.

Steiner, D. F., and Oyer, P. E. (1967). The biosynthesis of insulin and a probable precursor of insulin by a human islet cell adenoma. *Proc. Natl. Acad. Sci. U.S.A.* 57, 473–480.

Steiner, D. F., Clark, J. L., Nolan, C., Rubenstein, A. H., Margoliash, E., Aten, B., and Oyer, P. E. (1969). Proinsulin and the biosynthesis of insulin. *Recent Prog. Horm. Res.* 25, 207–272.

Steiner, D. F., Peterson, J. D., Tager, H., Emdin, S., Ostberg, Y., and Falkmer, S. (1973). Comparative aspects of proinsulin and insulin structure and biosynthesis. *Am. Zool.* 13, 591–604.

Stolk, A. (1959). Effect of thiouracil and thyroxine on development and growth of cutaneous melanoma in killifish hybrids. *Nature (London)* 184, 562–563.

Svärdson, G. (1943). Studien über den Zusammenhang zwischen Geschlechtsreife und Wachstum bei *Lebistes. Medd. Statens Undersök. Försöksanst. Sötvattenfisket* 21, 1–48.

Swift, D. R. (1954). Influence of mammalian growth hormone on rate of growth of fish. *Nature (London)* 173, 1096.

Swift, D. R., and Pickford, G. E. (1962a). Seasonal variations in the growth hormone content of pituitary glands of the perch, *Perca fluviatilis* L. *Gen. Comp. Endocrinol.* 2, 627.

Swift, D. R., and Pickford, G. E. (1962b). Seasonal variations in the growth hormone content of pituitary glands of the perch, *Perca fluviatilis* L. *Am. Zool.* 2, 451.

Swift, D. R., and Pickford, G. E. (1965). Seasonal variations in the hormone content of the pituitary gland of the perch, *Perca fluviatilis* L. *Gen. Comp. Endocrinol.* 5, 354–365.

Tait, J. F., and Burstein, S. (1964). *In vivo* studies of steroid dynamics in man. *In* "The Hormones" (G. Pincus, K. V. Thimann, and B. Astwood, eds.), Vol. 5, pp. 441–557. Academic Press, New York.

Talwar, G. P., Pandian, M. R., Kumar, N., Hanjan, S. N. S., Saxena, R. K., Krishnaraj, R., and Gupta, S. L. (1975). Mechanism of action of pituitary growth hormone. *Recent Prog. Horm. Res.* 31, 141–170.

Tashima, L., and Cahill, G. F. (1968). Effects of insulin in the toadfish, *Opsanus tau. Gen. Comp. Endocrinol.* 11, 262–271.

Tata, J. R. (1960). The partial purification and properties of the thyroxine dehalogenase. *Biochem. J.* **77**, 214–226.

Tata, J. R. (1964). Distribution and metabolism of thyroid hormones. *In* "The Thyroid Gland" (R. Pitt-Rivers and W. R. Trotter, eds.), Vol. 1, pp. 163–186. Butterworth, London.

Thornburn, C. C., and Matty, A. J. (1963). The effect of thyroxine on some aspects of nitrogen metabolism in the goldfish (*Carassius auratus*) and the trout (*Salmo trutta*). *Comp. Biochem. Physiol.* **8**, 1–12.

Thorpe, A., and Ince, B. W. (1974). The effects of pancreatic hormones, catecholamines, and glucose loading on blood metabolites in the Northern pike (*Esox lucius* L.). *Gen. Comp. Endocrinol.* **23**, 29–44.

Tuckmann, H. (1936). Action de l'hypophyse sur la morphogénèse et la différentiation sexuelle de *Girardinus guppii*. *C. R. Soc. Biol.* **122**, 162–164.

Umberger, E. J. (1975). Products marketed to promote growth in food-producing animals: Steroid and hormone products. *Toxicology* **3**, 3–21.

VanderWal, P., Berende, P. L. M., and Sprietsma, J. E. (1975). Effect of anabolic agents on performance of calves. *J. Anim. Sci.* **41**, 978–985.

Van Lan, V., Yamaguchi, N., and Garcia, M. J. (1974). Effect of hypophysectomy and adrenalectomy on glucagon and insulin concentration. *Endocrinology* **94**, 671–675.

Van Overbeeke, A. P., and McBride, J. R. (1967). The pituitary gland of the sockeye (*Oncorhynchus nerka*) during sexual maturation and spawning. *J. Fish. Res. Board Can.* **24**, 1791–1810.

Van Overbeeke, A. P., and McBride, J. R. (1971). Histological effects of 11-keto-testosterone, 17α-methyltestosterone, estradiol, estradiol cypionate, and cortisol on the interrenal tissue, thyroid gland, and pituitary gland of gonadectomized sockeye salmon (*Oncorhynchus nerka*). *J. Fish. Res. Board Can.* **28**, 477–484.

Vanstone, W. E., and Markert, J. R. (1968). Some morphological and biochemical changes in coho salmon, *Oncorhynchus kisutch*, during parr-smolt transformation. *J. Fish. Res. Board Can.* **25**, 2403–2418.

Van Wyk, J. J., Underwood, L. E., Hintz, R. L., Clemmons, D. R., Voina, S. J., and Weaver, R. P. (1974). The somatomedins: A family of insulin-like hormones under growth hormone control. *Recent Prog. Horm. Res.* **30**, 259–294.

Vilter, V. (1944). Comportement de la thyroide dans la metamorphose de la civelle d'Anguille. *C. R. Soc. Biol.* **138**, 615–616.

Vivien, J. H. (1941). Contribution à l'étude de la physiologie hypophysaire dans ses relations avec l'appareil génitale, la thyroïde et les corps suprarénaux chez les poissons selaciens et téléostéens. *Bull. Biol. Fr. Belg.* **75**, 257–309.

Wade, G. N. (1974). Interaction between estradiol-17β and growth hormone in control of food intake in weanling rats. *J. Comp. Physiol. Psychol.* **86**, 359–362.

Wainman, P., and Shipounoff, G. C. (1941). Effect of castration and testosterone propionate on the striated perineal musculature in the rat. *Endocrinology* **29**, 975–978.

Wallis, M. (1975). The molecular evolution of pituitary hormones. *Biol. Rev. Cambridge Philos. Soc.* **50**, 35–98.

White, B. A., and Henderson, N. E. (1977). Annual variations in the circulating levels of thyroid hormones in the brook trout, *Salvelinus fontinalis*, as measured by radio-immunoassay. *Can. J. Zool.* **55**, 475–481.

Wigham, T., Ball, J. N., and Ingleton, P. M. (1975). Secretion of prolactin and growth hormone by teleost pituitaries *in vitro*. III. Effect of dopamine on hormone release in *Poecilia latipinna*. *J. Comp. Physiol.* **104**, 87–96.

Wilhelmi, A. E. (1955). Comparative biochemistry of growth hormone from ox, sheep, pig, horse and fish pituitaries. *In* "The Hypophyseal Growth Hormone, Nature and Actions" (R. W. Smith, Jr., O. H. Gaebler, and C. N. H. Long, eds.), pp. 59–69. McGraw-Hill, New York.

Wilhelmi, A. E., and Mills, J. B. (1969). The chemistry of the growth hormone of several species. *In* "La Spécificité Zoologique des Hormones Hypophysaires et de leurs Activités," CNRS, Paris. No. 177, pp. 165–174.

Wilson, C. A. (1974). Hypothalamic amines and the release of gonadotrophins and other anterior pituitary hormones. *Adv. Drug Res.* **8**, 119–203.

Yamamoto, K., Kanski, D., and Frieden, E. (1966). The uptake and excretion of thyroxine, triiodothyronine, and iodide in bullfrog tadpoles after immersion or injection at 25° and 6°C. *Gen. Comp. Endocrinol.* **6**, 312–324.

Yamazaki, F. (1972). Effects of methyltestosterone on the skin and the gonads of salmonids. *Gen. Comp. Endocrinol., Suppl.* No. 3, pp. 741–750.

Yamazaki, F. (1976). Application for hormones in fish culture. *J. Fish. Res. Board Can.* **33**, 948–958.

Zipser, R. D., Licht, P., and Bern, H. A. (1969). Comparative effects of mammalian prolactin and growth hormone on growth in the toads *Bufo boreas* and *Bufo marinus. Gen. Comp. Endocrinol.* **13**, 382–391.

10

ENVIRONMENTAL FACTORS AND GROWTH

J. R. BRETT

I. INTRODUCTION

Although readily observed and easily measured, growth is one of the more complex activities of the organism. It represents the net outcome of a series of behavioral and physiological processes that begin with food intake (the consummation of an appetitive behavior)

FISH PHYSIOLOGY, VOL. VIII

and terminate in deposition of animal substance. The processes of digestion, absorption, assimilation, metabolic expenditure, and excretion all interplay to affect the final product. These individual functions are properly the content considered under digestion and bioenergetics (Chapters 4 and 6); however, to understand how the many environmental factors influence growth the basic rate functions should never be lost sight of, despite the general simplicity of growth measurements (Fig. 1). In so far as possible these individual functions will be introduced to provide some explanation of the response to en-

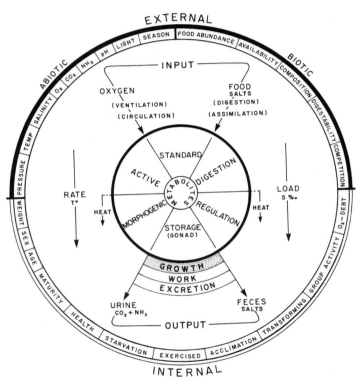

Fig. 1. Diagrammatic representation of the general metabolic system involved in growth, from input sources (food and oxygen) to output products (growth, work, excrements), showing some of the external and internal factors affecting the system (center circle). Morphogenic metabolism applies to the energy of internal work to produce body substance. Heat released from the system includes the heat increment or SDA. Temperature is shown as a major controlling factor of metabolism, and salinity as the chief energy-demanding factor in regulation.

vironmental factors, without deviating from the major purpose of examining the recorded trends in the activity of growing.

In relation to environmental effects, quite obviously growth cannot be studied without involving food consumption, even though the latter may not necessarily be measured. In this respect growing, unlike such activities as swimming or respiring, is inseparably coupled with a powerful *biotic* factor, so that any *abiotic* factor is necessarily involved in some form of interaction between the two. For example, if temperature rises, the amount of food consumed usually increases as well as the rate of digestion. Growth rate, however, may either increase or decrease depending on the nature of the *food* × *metabolism* × *temperature* relation; on analysis it can be seen that the energy demand of elevated metabolic rate could have exceeded the gain from increased food uptake resulting in reduced growth rate. Furthermore, the act of growing necessarily alters size so that another important biotic factor continually changes with time. An understanding of these inherent involvements is necessary before any clear insight of the environmental relation to the growth process can be achieved.

The general approach adopted is that elaborated by Fry (1947, 1971), who classified environmental factors not by simple type-grouping but by considering how the separate entities (temperature, oxygen, light, etc.) *acted* on any particular function or activity. Fry's extension of the classical concept of "limiting factors" (Blackman, 1905) was further enhanced by the recognition that the environment acts *through* metabolism *on* activity and not directly on the activity itself. Most of Fry's (1947, 1971) considerations were devoted to catabolic aspects of metabolism and the influence of abiotic factors. Such an approach was conceptually extended by Warren (Warren and Davis, 1967; Warren, 1971) to encompass the anabolic process of growth. Biotic factors as a group were never classified.

These basic concepts are applied and developed in the following considerations. At the outset it is necessary to define the environmental factors classified by Fry (1971) and to introduce the main terms (and abbreviations) appropriate to experimental growth studies. Fry adopted the label "factors," such as Lethal Factors, in keeping with their historical meaning as *effectors*—the influence any particular environmental entity had on the organism. Unfortunately this is now in conflict with the common use of "factors" to imply the environmental entities themselves, as used in the title of this chapter. To avoid misunderstanding, the use of "Factors" as a category of effects will be capitalized. The great advantage that such a system of classification offers comes from the small number of categories involved, and the clear distinction of action they normally imply. Any environmental factor may fall into one or more of the functional categories.

As an activity, growth must always come within the lethal limits of the environmental factor considered, so that this category (Lethal Factors) does not enter into further consideration here. Fry (1971) invokes four other categories, namely:

1. Controlling Factors, which *govern* the rates of reaction by influencing the state of molecular activation of the metabolites (e.g., temperature, pH); these operate at all levels of the environmental factor concerned
2. Limiting Factors, which *restrict* the supply or removal of metabolites, as links in the chain of metabolism [e.g., oxygen, light (as in photosynthesis)]; these become operational at a particular level of the factor, involving dependent and independent states
3. Masking Factors, which *modify* or prevent the effect of an environmental factor through some regulatory device (e.g., humidity influencing body temperature by affecting evaporative heat loss, or temperature regulation by countercurrent heat flow as in warm-bodied fish)
4. Directive Factors, which *cue* or signal the animal to select or respond to particular characteristics of the environment (e.g., temperature preference, photoperiod-induced smoltification); these may be compared to the category of "releasing mechanisms" in animal behavior (Baerends, 1971), and hormonal response may be involved

In the formulation of a growth model for salmonids, Stauffer (1973) assessed the various factors influencing growth. He concluded that any attempt at modeling must include at least the three factors, *ration, size,* and *temperature* as the most important independent variables. Elliott (1975c,d) supports this view. Although not attempting any classification, ration was considered by Stauffer to be the sole driving force, temperature the major rate-controlling force, and weight was thought to act as a scaling factor that adjusts these rates to the size of the growing individual.

Starting with the fundamental growth–ration relationship as the basis, all other factors will be examined as they affect this key "driving force" relation.

II. BASIC GROWTH RELATIONS*

A. The Growth Curve

The growth of any organism is a multiplicative process in which cell number and cell volume increase (Needham, 1964). Given an unlimited source of nutrition, growth proceeds in an almost exponential burst. The burst, however, is continuously dampened as size and age increase. Nevertheless, over short periods of time the exponential relation

$$W = ae^{gt}$$

where W = weight, t = time, and a and g are constants, closely approximates the relation (Fig. 2; see also discussion in Weatherley, 1972). The absolute rate of increase in weight (dW/dt) is greatest just at the inflection point of the sigmoid curve. However, the *relative* rate of increase (dW/dtW) is usually greatest at the youngest, smallest stage. The latter derivative is the instantaneous or *specific growth rate* represented by the slope (g) of the lines in Fig. 2B; when multiplied by 100 it is equivalent to the percentage increase in weight per unit time (usually 1 day) and is labeled "G." The latter is used throughout this chapter as "rate of growth," involving a change in weight irrespective of length (see Chapter 11).

Growth rate decreases greatly at the onset of maturity. This phenomenon can result in initially fast-growing fish, which mature at an early age, being surpassed in ultimate size by slower growing fish, demonstrating that the differences in growth rates established in young fish do not necessarily persist throughout life (Kinne, 1960).

B. Natural Environmental Relations

In nature, the timing of hatching and early growth (in temperate climatic zones) usually coincides with increasing daylength, rising temperature, and seasonal abundance of food. There is a natural compounding of these factors with a rapidly changing size of the organism. In addition, changes in prey organisms and a change in ability to seize ever larger prey introduce the variables of diet quality and quantity.

---

* The notation used in this section and elsewhere is as follows: T, temperature (°C); L, hours of light per day (photoperiod); R, ration (% weight/day); S, salinity (‰); G, % weight per day.

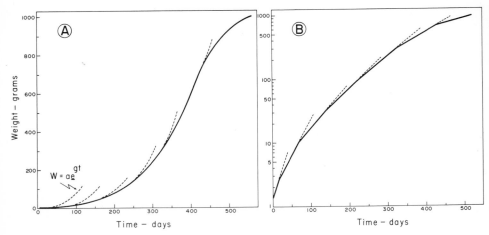

Fig. 2. (A) Generalized growth curve for a fish under constant environmental conditions, and (B) the transformed curve using a logarithmic scale for weight. Experimentally observed growth "stanzas" are indicated by straight lines in B, with their extrapolation shown as broken lines in A and B following the form of an exponential curve (Brett *et al.*, 1969—up to a weight of 100 g in B). W, weight; t, temperature, and *a* and *g* are constants. G = 100 × *g*, the specific growth rate. The broken lines simply display what the path would be if a given specific growth rate persisted. When food is not limiting, size and age progressively depress the specific growth rate, undoubtedly following a continuous curve except when developmental stages alter the pattern.

Add to this a possible factor of migrating into new environments that may involve a drastic change in salinity, it can be seen that no single, natural growth curve can reveal the underlying physiological responses to environmental factors. Nevertheless, the correlates of seasonal temperature (Fig. 3), and the ultimate damping of growth with age and maturity, show through this complexity in gross form (Larkin *et al.*, 1956; Gerking, 1966; Weatherley, 1972).

For the experimenter there is no small challenge to manipulating environmental factors in a design that is not confounded in one way or another. Kerr (1971) has elaborated on some of the inherent difficulties.

C. The Growth–Ration (GR) Curve

The overall relation of growth rate (G) to rate of food uptake (R) from minimum to maximum consumption was first depicted graphically for largemouthed bass, *Micropterus salmoides*, by Thompson (1941). Although no actual data were presented, the figure

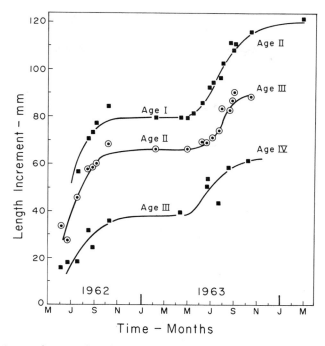

Fig. 3. Seasonal gain in length of a population of bluegill, *Lepomis macrochirus*, by age groups. The growth increment varies with season and decreases with age and size. (From Gerking, 1966.)

clearly indicated some of the simple geometry of the relation, from which certain important parameters such as optimum and maintenance ration could be derived. Earlier work of Pentelow (1939) suggests something of this relation, but only with subsequent insight is it possible to see that the feeding experiments on brown trout, *Salmo trutta,* support the general configuration.

All too little attention appears to have been paid to this fundamental relation which, together with its derived conversion efficiency equivalents, holds much of the key to understanding the action of environmental factors on growth. Figure 4 depicts the relation for sockeye fingerlings. Starting at zero ration (R_0) the GR curve rises steeply from a minimum negative value (G_{starv}) to cross the point of zero growth rate (G_0) at the maintenance ration (R_{maint}). This continued steep rise begins to flex downward such that a tangent from the origin passes through the point where the ratio of G to R is maximal, providing a measure of the optimum ration (R_{opt}). With increasing ration the GR curve flexes further, reaching a plateau of maximum growth rate

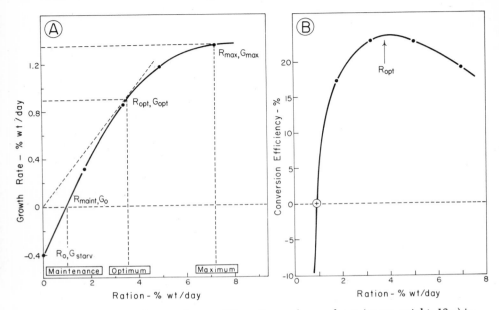

Fig. 4. (A) Specific growth rate of fingerling sockeye salmon (mean weight, 13 g) in relation to ration, at 10°C. The key parameters of the GR curve, with abbreviations, are indicated. (Data from Brett *et al.*, 1969.) (B) Conversion efficiency curve (KR) for the same data. Circled point (+) derived from A.

(G_{max}) at the point of maximum ration (R_{max}). Figure 4A is just one example of the curve. Huisman (1974) has shown that no change in shape is manifest when the specific growth rate is determined in terms of either wet weight, dry weight, or caloric content. Some difference in the value of G for a given ration occurs because the body composition changes with the level of ration, altering the protein, fat, and water ratios and hence affecting relative dry weight and energy content.

Variations in the form of the curve do occur, according to environmental conditions and species; however, the general form and indicated parameters are indisputable. Elliott (1975d) tried unsuccessfully to find a suitable linear transformation for GR curves derived from rigorously performed experiments on brown trout, and resorted to smooth curves fitted by eye. A straight line has not infrequently been used to depict the relation, especially where a considerable scatter of points exists and the growth rate for R_{max} is not well defined (e.g., Hatanaka and Takahashi, 1960; Brocksen and Cole, 1972). Stauffer (personal communication) at first accepted this linear expression because of difficulty in finding a more appropriate transformation. Subsequently the curvilinear segment was found to be fitted more closely by a sine curve, $G = a[\sin(bR + c)]$, which significantly reduced the error component for the sockeye data (Stauffer, 1973).

It is a simple and illuminating step to derive the relation for gross conversion efficiency (K), which usually traverses a tapered, dome-shaped curve (Fig. 4B). When ration is expressed in the same units as growth rate* (% weight/day), $K = G/R \times 100\%$. At R_{maint}, K must obviously be zero; similarly it is apparent that R_{opt} must define the peak of the dome where maximum efficiency occurs. It is significant to note that in the range of the dome there will be a lower ration (below R_{opt}) and higher ration (up to R_{max}) for which one and the same conversion efficiency would be obtained. This fact has confused some reporting of results. Further, there is still debate on the particular path traced by the segment of the curve beyond R_{opt}. This has been stated to follow a logarithmic decay with increasing daily food uptake, according to the equation

$$K = ae^{-bR}$$

where K is gross conversion efficiency (Paloheimo and Dickie, 1966). Greater consideration of this phenomenon is provided in Chapters 6 and 11.

Insofar as possible, each environmental factor will be examined for its effect on the GR curve and the consequence to its parameters—$G_{max}, R_{max}, R_{maint}$. This is the basic, comparative theme throughout the chapter. It should be noted that growth in nature must be responding more or less at every point along the GR curve according to feeding opportunity, with not infrequent periods of food scarcity. Hatcheries, on the other hand, tend always to be dealing with the upper end of the GR curve, above R_{opt} and G_{opt}. Experimentally it is undeniably important to examine the full GR curve over a wide range of the environmental factor(s) concerned. If that is not possible then the effect on G_{max} and its corresponding R_{max} is the most important relation to determine.

III. ABIOTIC FACTORS

A. Temperature

Since temperature acts as a Controlling Factor pacing the metabolic requirements for food and governing the rate processes involved in food processing, the GR curve responds to temperature in a varied manner as each parameter shifts. The response of fingerling sockeye salmon (mean weight = 13 g) has been studied at temper-

* Because maximum growth rate usually decreases with increasing size, this expression of relative food consumption can only be applied to fish of the same, limited weight range, or to fish that maintain the same specific growth rate over a wide range of weights.

atures from 1° to 24°C (Brett *et al.*, 1969); three cases are illustrated (Fig. 5). At a comparatively low temperature (5°C) the curve is far to the left by virtue of a low G_{starv} and small R_{maint}; it rises in a sweeping curve to a low plateau (G_{max}) in keeping with reduced daily food requirements and generally suppressed growth rate. As temperature increases to 10°C the whole curve moves to the right and is elevated with approaching optimum thermal conditions and consequent enhanced G_{max}; maintenance requirements (R_{maint}) are necessarily up. Higher temperatures, above the optimum, bring a further shift to the right as maintenance costs escalate accompanied by a lowering in G_{max} despite an increase in R_{max}; conversion efficiency has begun to decline.

The general shape of the GR curve for this species appears to shift gradually from a simple concave form at low temperatures to a sigmoid

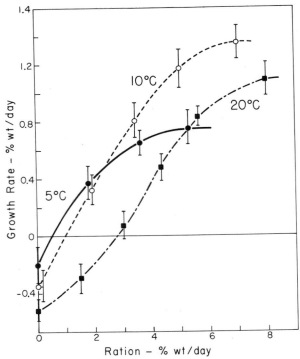

Fig. 5. The effect of temperature on the growth-ration (GR) curve for fingerling sockeye salmon. Mean weight, 13 g. Limits shown as ±2 SE. (Data from Brett *et al.*, 1969.) Maintenance requirements (R_{maint}) occur at the point of intersection of each curve with the baseline for zero growth (G_0). The highest ration at each temperature approximates R_{max}, providing a measure of the maximum growth rate (G_{max}).

shape at high temperatures; fish that are underfed at a relatively high temperature (e.g., 20°C) exhibit a great deal of searching which would compound the expenditure of energy in the lower segment of the curve causing the convex upsweep. Some of the data for brown trout at 19.5°C suggest the same relation (Elliott, 1975d). Averett (1969) also presents a set of GR curves for young coho salmon, ranging from 5° to 17°C. Although subject to considerable variability these all show the features of the sine curve applied by Stauffer (1973). At the upper end of the GR curve for carp, *Cyprinus carpio*, at 23°C a distinct inflection downward occurs, which must reflect some adverse effects on G_{max} with high ration, decreasing conversion efficiency considerably (Huisman, 1974). This is hardly apparent, if at all, for carp at 17°C.

It is these changing positions and shapes of the GR curve that dictate the temperature relations for each parameter. Most experiments on temperature effects have been involved with studying a given parameter (e.g., G_{max}, G_{starv}) without the insight of the complete GR relation. Since some parameters (e.g., G_{opt}, R_{maint}) are difficult to determine directly the advantages of interpolation from the GR curve are worth emphasizing.

1. MAXIMUM GROWTH RATE (G_{max}) × TEMPERATURE

Most effort has been devoted to studying the controlling effect of temperature on this parameter of growth (i.e., changes in G_{max} for fish fed a maximum ration). Almost all species in the young stages show a typical rapid increase in growth rate as temperature rises passing through a peak (optimum temperature) and frequently falling precipitously as high temperatures become adverse. This is well illustrated for juvenile brook trout, *Salvelinus fontinalis*, by the studies of McCormick *et al.* (1972) who also included the change in sample biomass, resulting from the onset of some mortality at high temperatures (Fig. 6). As might be expected the optimum temperature for growth of different species ranges generally upward from the ecologically cold-adapted to the warm-adapted species (Table I). One remarkable case is that for larval ciscoe, *Coregonus artedi*, a particularly temperature-tolerant member of the coregonids, which at this very early stage displays a growth rate that continues to rise almost to the lethal temperature (19.8°C) despite the onset of some mortalities above 17°C (McCormick *et al.*, 1971).

Among the highest optima in temperate freshwater are those for the channel catfish, *Ictalurus punctatus* (29°–30°C), and the pumpkinseed, *Lepomis gibbosus* (30°C) (Table I). The lethal temperature for the latter species is 34.5°C (Altman and Dittmer, 1966). Pessah and

Table I

Temperature Optima for Growth when on a High Ration[a]

Species	Opti-mum (°C)	Test interval (°C)	Salinity (‰)	Initial size (cm, g)	Comment	Reference
Salmo trutta	12.8	1.5	Fresh	10–300 g	Fed maximum ration individually	Elliott (1975c)
Oncorhynchus keta	13	3	35	Under-yearling	Size not given	Kepshire (1971)
Salvelinus fontinalis	14	3	Fresh	0.1–0.2 g	Alevin to juvenile stage; fed nauplii, liver and starter feed	McCormick *et al.* (1972)
Salvelinus alpinus	14	2	Fresh	Yearling	Diet of minced beef liver in gelatine	Swift (1964)
Pleuronectes platessa	14.2	2	Seawater	0.5–2.0 cm	Computed from observations	Jansen (1938); Ursin (1963)
Oncorhynchus nerka	15	5	Fresh	6–20 g	Moist pellet feed	Brett *et al.* (1969)
Oncorhynchus tshawytscha	15.5	2.5	Fresh	2–9 g	Moist and dry pellet feed	Banks *et al.* (1971)
Oncorhynchus gorbuscha	15.5	3	35	Under-yearling	Size not given	Kepshire (1971)
Salmo gairdneri	17.2	1.5	Fresh	0.3–3 g	Alevin to juvenile stage, brine shrimp and trout pellets	Hokanson *et al.* (1977)
Coregonus artedii	18.1	3	Fresh	0.2 g	Larval stage; nauplii	McCormick *et al.* (1971)
Morone saxatilis	24–25	2	Fresh	90–100 g	Fed on live minnows	Nelson (1975); Cox and Coutant (1975)

610

Species						Reference
Lebistes reticulatus	24–25	3	Fresh	0.7–1.8 cm	Reared from birth to adult (males); dry fish food	Gibson and Hirst (1955); Ursin (1963)
Micropterus salmoides	25	7	Fresh	8–140 g	Fed on shiners and beef liver	Niimi and Beamish (1974)
Etheostoma spectabilis	26	2	Fresh	0.6–2.1 cm	Larvae; feeding on brine shrimp	West (1966)
Micropterus salmoides	27	2	Fresh	30–240 g	Fed on live minnows	Nelson (1974); Coutant and Cox (1975)
Catostomus commersoni	27	3	Fresh	Larvae	Fed *ad libitum* on brine shrimp	McCormick *et al.* (1977)
Micropterus salmoides	27.5	2.5	Fresh	0.6–2.6 m	Fry; feeding on zooplankton	Strawn (1961)
Cichlasoma bimaculatum	28	4	Fresh	2–3 g	Fed on Tubifex	Warren and Davis (1967)
Cyprinodon macularius	28	5	15	20–30 cm	Fry; brine shrimp, white worms and fish food	Kinne (1960)
Perca flavescens	28	2	Fresh	0.4 g	Fry; brine shrimp, liver and yeast	McCormick (1976)
Ictalurus punctatus	29	2	Fresh	1.5–7.2 cm	Fry; feeding on zooplankton	West (1965)
Ictalurus punctatus	30	4	Fresh	4 g	Pelleted feed presented at 3 rates	Andrews and Stickney (1972); Andrews *et al.* (1972)
Cyprinodon macularius	30	5	35	20–30 cm	Fry; shrimp, worms and fish food	Kinne (1960)
Lepomis gibbosus	30	5	Fresh	24–34	Fed on oligochaetes	Pessah and Powles (1974)

611

[a] Temperature interval is indicated as a measure of experimental sensitivity (i.e., the smaller the interval the better the optimum will be defined). Arranged in general order of increasing temperature optimum.

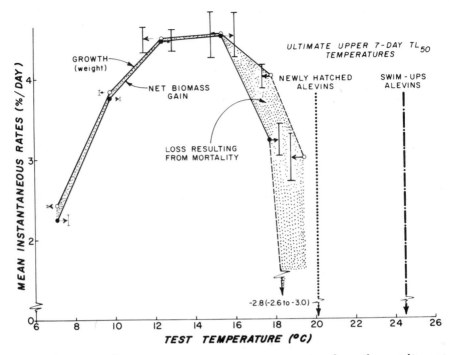

Fig. 6. Effects of temperature on mean instantaneous rates of growth, mortality, net biomass gain, and ultimate 7-day median temperature tolerance limits of brook trout, *Salvelinus fontinalis*, during their first 8 weeks after hatching. (From McCormick *et al.*, 1972.)

Powles (1974) consider that the pumpkinseed shows a form of "growth homeostasis," between 15° and 30°C, as a result of stable and similar growth rates displayed in their test lots *following* about 5 weeks of constant environmental and feeding conditions. Although the conclusion appears to be supported by the data there are a number of potentially confounding problems including progressive differences in size between lots, differential and possibly temperature-correlated hierarchial effects (for which the species is renowned), growth depensation, and the fact that temperatures in the region of the optimum may show relatively little difference in growth rate—but are likely to be significantly different in food conversion efficiency. (All of these aspects of growth are considered further in the following text.)

The records in Table I apply to different *fixed* levels of temperature held constant within experimental limits. In early tests on growth, some authors (e.g., Pentelow, 1939) relied on using the normal seasonal change in environmental temperature to conduct their "controlled" experiments. Brown (1946a) attempted to evaluate the effect of increasing and decreasing temperatures (18° to 4°C and reverse) on fingerling brown trout, *Salmo trutta*, by manipulating tank temperatures at the rate of 0.5° and 1°C per week over a 1-year period. There was no indication that either rising or falling temperature at these slow rates of change had any different effect on growth than when held at equivalent, constant, mean temperatures. The rates of temperature change were apparently well within the acclimation rate for growth.

This circumstance applies to the varying temperatures imposed under laboratory and natural stream conditions on brown trout fed to satiation (Elliott, 1975c). The resulting weight at temperatures that rose from 6.8°C in March to 12.1°C in June, and decreased from 12.9°C in August to 7.2°C in November, were estimated quite accurately by applying the developed equations of growth determined for static temperatures.

Rapid temperature change, however, results in a more complex growth response. Daily fluctuating temperatures of ±4°C, oscillating around six mean levels from 12° to 22°C, were applied to juvenile rainbow trout, *Salmo gairdneri*, by Hokanson *et al.* (1977). When the growth rates were compared with results from constant temperatures (equivalent to the mean of the fluctuating temperatures) the growth–temperature curve was shifted to the left such that the optimum mean temperature occurred 2°C below that for the constant temperatures (Fig. 7). Fluctuation resulted in growth rates which were higher than expected when in the thermal range below the "static" optimum (17°C), somewhat lower when at and above the optimum (to 21°C), and highly depressed at a mean of 22°C. The explanation of these results does not lie in the simple multiplication of the instantaneous growth rates (according to the range of temperatures experienced) and then determining the mean of these rates. It is apparent that the daily peak temperatures have a carry-over, beneficial effect when the thermal range applied occurs below the static optimum. This benefit declines where higher temperatures are involved, and comes crashing down when the peak of fluctuation enters the fringe of temperature tolerance for the species. Similar effects of daily fluctuating temperatures were obtained by Wurtsbaugh (1973) who reported negative effects on growth of trout when relatively high, oscillating temperatures were involved.

2. Optimum Growth Rate (G_{opt}) × Temperature

Few studies are complete enough to permit examining how the optimum growth rate changes with temperature. For young sockeye

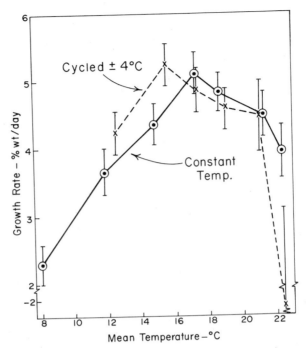

Fig. 7. Growth rate of juvenile rainbow trout in relation to constant and cycled temperatures (plotted at corresponding mean temperature). Limits shown as ±2 SE. (From Hokanson *et al.*, 1977.)

the course of this intermediate parameter of the GR curve was found to rise steadily from a position well below G_{max} at low temperatures to be almost superimposed on G_{max} at a high temperature (Fig. 8). The escalating maintenance cost with increasing temperature apparently forces G_{opt} to approach G_{max} at the upper end of the temperature tolerance scale.

3. MAXIMUM RATION (R_{max}) × TEMPERATURE

The need for meeting the increased appetite of fish with rising temperature has long been appreciated in hatchery practices. Haskell (1959) credits the introduction of the temperature-scaled feeding chart to Tunison and Deuel in 1933. In a comparison of six feeding tables for salmonids, Stauffer (1973) shows the rapid increase in recommended ration, up to a maximum of about 10% per day for 2-g salmon at 18°C (Fig. 9). Above this peak, one curve (EIFAC) inflects downward reflecting the loss of appetite that can occur at a relatively high temperature (shown to be true for brown trout above 19°C, Elliott, 1975a).

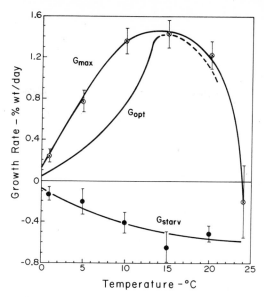

Fig. 8. Relations of various defined growth rates according to maximum ration (G_{max}, R_{max}), optimum ration (G_{opt}, R_{opt}), maintenance ration (G_0, R_{maint}), and starving (G_{starv}, R_0). Limits shown as ± 2 SE. (Data from Brett *et al.,* 1969.)

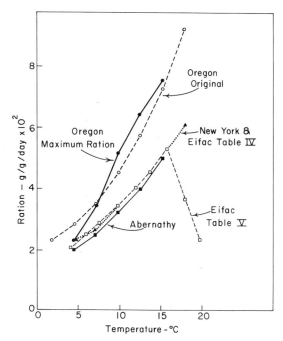

Fig. 9. The relation between recommended ration (as pelleted feed) and temperature for 1.5–2.5 g salmonids, proposed by six commercial feeding charts. Values approach R_{max}. (From Stauffer, 1973.)

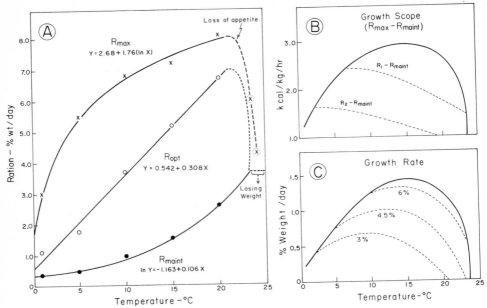

Fig. 10. (A) Relation of different levels of ration to temperature. Upper curve for R_{max}, middle line for R_{opt}, and lower curve for R_{maint}. (From Brett *et al.*, 1969.) (B) Scope for growth, derived from difference between R_{max} and R_{maint} (expressed here in kcal/kg/hr). (From Brett, 1976.) The growth scope for two hypothetical fixed rations, R_1 and R_2 (less than R_{max}), is shown by dotted lines. These may be compared with the observed growth of fingerling sockeye (C) on excess and restricted rations; much of the growth response obtained in C is predicted by the growth scope relations developed in B.

Although these charts provide the fish-culturist with a good guideline they are not necessarily right up to the maximum ration, and in some cases may be as much as one-quarter below this parameter (Stauffer, 1973). On a measured excess ration (i.e., an estimated amount over satiation) fingerling sockeye salmon increased their uptake from 3% weight/day at 1°C to 8% weight/day at 20°C, followed by a rapid decline at higher temperatures (Brett *et al.*, 1969). The rate of increase in R_{max} declined with increasing temperature (T) following a concave curve described by the equation $R = 2.68 + 1.76(\log_e T)$ (Fig. 10A).

4. MAINTENANCE RATION (R_{maint}) × TEMPERATURE

One of the earliest critical studies on the effect of temperature on R_{maint} was that of Pentelow (1939) who held brown trout, *Salmo trutta*, individually in small troughs. Although troubled by the difficulties which Brown (1946a) experienced from the constant need to adjust the ration upward or downward as the fish lost or gained weight, the deviations were not great. Other factors such as random activity, changing size, and seasonal effects were of greater consequence to the observed variability. The data were reanalyzed by Stauffer (1973) and grouped according to size and seasonal temperature (Fig. 11). Except for the lowest temperatures, where little change was noted (between 5° and 8°C), there is an exponential increase in R_{maint} with rising temperature, reaching a requirement of 4% weight/day at 15°C for the smallest size (1–2 g). A similar exponential increase was noted by Brett *et al.* (1969) for 13-g sockeye (Fig. 10A). At this weight the R_{maint} at 1°C rose from 0.3% weight/day to an estimated 3.7% weight/day at 23°C (Fig. 10). Applying the method of interpolation from GR curves, Elliott (1975d) determined the maintenance ration for brown trout. R_{maint} was found to increase exponentially from approximately 100 mg dry weight/day at 6.6°C to 350 mg dry weight/day at 19.4°C (50-g fish, Fig. 11).

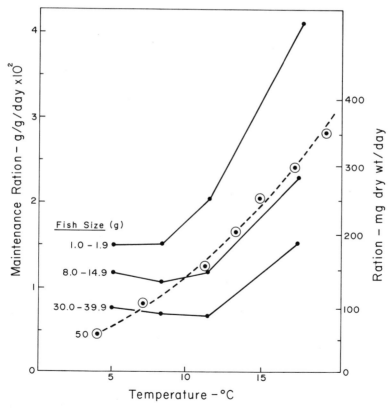

Fig. 11. Maintenance ration (R_{maint}) of three size groups of brown trout, *Salmo trutta*, in relation to temperature (solid lines). (Data of Pentelow, 1939; analyzed by Stauffer, 1973.) Values obtained by Elliott (1975d) for 50 g brown trout included for comparison (broken line), plotted as mg dry weight/day to approximately the same scale.

5. SCOPE FOR GROWTH (G_{scope}) × TEMPERATURE

The concept of *scope for activity* was first elaborated by Fry (1947) as the difference between active and standard metabolic rates; this "metabolic scope" defined the amount of energy available for activity, over the tolerable range of temperature. Fry (1947, 1971) demonstrated that the maximum sustained swimming speed of fish was closely correlated with metabolic scope and not active metabolic rate. Warren and Davis (1967) paid tribute to Fry's concept and elaborated a *scope for growth*, which they considered to be "the difference between the energy of the food an animal consumes and all other energy utilizations and losses" (p. 184). These authors recognized that the anabolic aspects of food consumption cannot be considered in the same context with the catabolic relations of active and standard metabolic rates, although one system obviously influences the other. A number of examples of scope for growth, depicted as growth rates in relation to temperature, were given by Warren and Davis (1967) in conjunction with certain associated bioenergetic relations, namely: food consumed, fecal wastes, specific dynamic action, and starving metabolic rate. However, these do not by themselves involve a calculated scope (by difference) such as Fry applied to metabolic rates. Further, it is evident that environmental factors will not act directly on growth—the pathway will be through the mechanisms of energy supply and demand influencing the scope, in a way that Warren and Davis (1967) implied but did not actually determine. Warren (1971) further elaborated on this thesis, illustrating the theoretical effects of limiting food on the energy budget and the deduced scope for growth.

Brett (1976) was able to demonstrate that the difference between R_{max} and R_{maint} gave a simple measurement of scope for growth in relation to temperature (Fig. 10B). For sockeye salmon the reason for the rapid decline in G_{max} above 8°C, in the face of increasing R_{max}, was due to the exponential increase in R_{maint} (Fig. 10A). Maximum growth is a consequence of the interplay between these two levels of ration over the tolerable range of temperature for the species. Only the food component in excess of R_{maint} is available for growth. The relation that R_{max} and R_{maint} bear to growth capacity as it is influenced by temperature is entirely analogous to the relation that standard and active metabolism bear to swimming capacity and temperature.

The effect of moderately restricted ration is to limit R_{max} at high temperatures without affecting R_{maint}. It can be deduced from such a growth-scope model (Fig. 10B) that the optimum temperature for growth would shift progressively to a lower temperature as ration was reduced—a phenomenon which has already been demonstrated for sockeye salmon (Brett *et al.*, 1969) and more recently supported by similar studies on brown trout (Elliott, 1975c).

B. Light

Studies on the influence of light on growth have not infrequently resulted in variable, complex, and confusing results. This appears to arise from the multiplicity of ways in which light can act (quality, quantity, and periodicity), its interaction with other environmental factors, particularly temperature, and the possible harmony or disharmony with internal rhythms of the fish—both daily and seasonal. Great discretion has to be exercised by the experimenter to avoid confounding the variables (light, temperature, size, season) and to avoid introducing serious uncontrolled or undocumented interfering responses involving activity levels, ion regulation, and possible induced prematuration; equal opportunity to feed must be presented in each experimental lot independent of light. Furthermore, past history and acclimation steps have an importance that has yet to be critically assessed and clarified.

Light usually acts as a Directive Factor stimulating brain–pituitary responses which radiate through the endocrine and sympathetic systems. Its natural periodicity undoubtedly induces the production of growth hormone (STH)* and anabolic steroids, and can influence locomotor activity in association with thyroid stimulation. Since injections of STH can produce great increases in growth rate (Chapter 9) light is a potentially powerful environmental factor, but this degree of influence has yet to be clearly shown experimentally despite results which are quite significant. More frequently than not, light periodicity has been manipulated in relation to the question of smolt transformation (salinity effect) or of inducing early maturation, with growth responses additionally recorded (Henderson, 1963; Pyle, 1969; Wagner, 1974). The consequent interplay of Directive and Masking Factors has tended to obscure the particular role of light on growth. In consequence of the lack of definitive experiments and the absence of testing over a sufficient range of restricted rations, only the parameters G_{max} and R_{max} of the GR curve can be given any serious consideration. From these the effect on conversion efficiency can then be considered.

Evidence for the effect of light periodicity (fixed and variable) will be examined first.

---

* Somatotrophic hormone or growth hormone (GH).

1. PHOTOPERIOD AND MAXIMUM GROWTH RATE (G_{max})

In a study of seasonal variation in the growth rate of 3-year-old, hatchery brown trout, *Salmo trutta*, Swift (1955) drew attention to two striking anomalies: (1) Growth surged forward in spring while the temperature was still cold; (2) the rate fell in summer and again in autumn when the water was still warm. This occurred despite the fact that the fish were fed daily to satiation. The high correlation of increasing daylength with growth in the spring was subsequently confirmed as the stimulating agent for endocrine activity, enhanced by rising temperature as the season progressed (Swift, 1959, 1960, 1961). The decrease in growth rate during autumn could be shown to relate to the advance of maturation accompanied by an *inflection* to falling temperature and photoperiod. But the growth slump in summer still remained open to speculation, in which the observation that change in spontaneous activity, thyroid depletion, and greater susceptibility to handling mortality might well provide some diagnostic clues.

During an investigation of the annulus formation on scales of four species of coregonids, Hogman (1968) also noted that seasonal change in growth rate was more closely related to daylength than to changes in partially controlled water temperature (Fig. 12). Rising in close

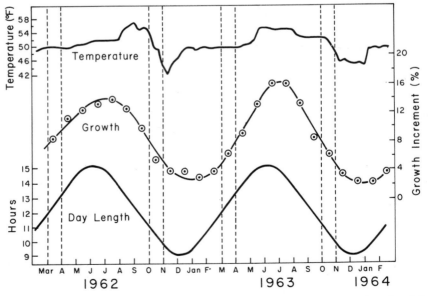

Fig. 12. Annual change in average monthly growth increments of lake whitefish, *Coregonus clupeaformis*, in relation to seasonal daylength and temperature. Fish were reared in tanks supplied with water from a natural spring. (From Hogman, 1968.)

harmony with increasing photoperiod, the peak in growth increment trailed about 1 month behind the maximum daylength; no such correlated phasing occurred in relation to some uncontrolled fluctuations in temperature (maximum change = 7°C).

These clear demonstrations of seasonal effects on growth in association with the *natural* photoperiod have only found moderate support from laboratory tests conducted under more rigorous environmental control. Indeed, early efforts by Brown (1946b) using static daylengths suggested an inverse relation of growth rate with daylength, whereas those of Anderson (1959) and Bjorklund (1958) appeared to show no relationship at all. Most studies indicate that, whatever the response, it takes at least 6–8 weeks to become manifest (e.g., Eisler, 1957).

One of the most searching enquiries was that by Gross *et al.* (1965) on the green sunfish, *Lepomis cyanellus,* using both fixed and variable photoperiods. From four 6-week experiments conducted during various seasons of the year, a comparison was made on each occasion between photoperiods of 16, 8, 8–16, and 16–8 L, all at 25° ± 1°C. A marginally significant difference in the weight gain occurred between 16 and 8 L (*P* = 0.10) with a greater significant difference obtained between 8 and 16 L, compared with 16 and 8 L (*P* < 0.05). The greatest weight gain accompanied 16 L (gain = 5.7 g) and the least with 16–8 L (gain = 4.0 g). Assuming the mean initial weight for each random sample of fish was 15 g (not given, but see Gross *et al.,* 1963) the greatest difference in achieved weight between treatments was only 8.3% in 6 weeks (difference in *G* = 0.2% weight/day). Although not analyzed by the authors it is of interest to note that for the months selected the mean weight gains for all photoperiods were greatest in early spring (increase of 6.3 g) and least in late fall (increase of 2.7 g).

By extending the latter experiment another 3 weeks, and considering the previous 6-week experimental period as acclimation time, the difference between the growth increment of the 8 L and 16–8 L was significantly increased. This led the authors to conclude that not only did increasing daylength have a stimulating effect and decreasing daylength an inhibiting effect on growth, but also that prior photoperiod history was of demonstrable importance.

In the uniform time allotted for feeding, it was further found that the sunfish consumed more food on a longer photoperiod and were also more effective at conversion, the highest average efficiency (48%) occurring on 8–16 L. The seasonal influence on conversion efficiency was even more pronounced, with the early spring period topping each of the four test seasons with an amazing average of 72% K.

Some corroborative evidence for the growth-promoting effect of long daylength was obtained by Kilambi *et al.* (1970) for channel catfish, *Ictalurus lacustris,* using a 2 × 3 factorial experiment*. Over a period of 120 days, fry (0.07 g) of this species grew at the following mean rates† (G_{max}) according to the combination of temperature and photoperiod applied:

	26°C	28°C	32°C	Mean G_{max}
10 L	2.07	1.55	2.30	1.97
14 L	2.30	2.48	2.29	2.36
Mean G_{max}	2.18	2.02	2.30	

Although the average growth rate at the lowest temperature (26°C) for both photoperiods occupied an intermediate position between the other two temperatures (28° and 32°C) the mean rate for all temperatures at 14 L was 16% higher than that at 10 L. As reported, the relation between treatments was not without odd variability in the respective order of the growth rates (of intermediate stages) during the experiment. Chance variation resulting from small samples ($n = 5$) and the fact that the optimum temperature for growth occurs at 29°–30°C (Table I) would tend to confuse the issue. Temperatures beyond the optimum or near the limit of tolerance usually suppress growth, and may also suppress photoperiod effects. The immediate prehistory of the fish was chosen as 24 L, apparently to remove any periodicity difference between lots. The consequence of this fairly extreme pretreatment is worth investigating in itself.

There was a tendency for conversion efficiency to be higher on 14 L than 10 L, but this difference diminished with time and size of the catfish.

Turning to experiments on anadromous species, Saunders and Henderson (1970) investigated the effect of natural,‡ constant (13 L), and reciprocal photoperiod on 35-g Atlantic salmon smolts, *Salmo salar.* Growth and feeding rates (G_{max} and R_{max}) were followed over a 10- to 12-month period in two successive years. At the time of downstream migration (April–May), the smolts were transferred to full seawater (30‰); this was accompanied by an increase in temperature from 10° to 15°C after which temperature was held constant. Prior history was

---

* See also studies by Huh *et al.* (1976) on yellow perch and walleye.

† Specific growth rates recalculated from length–weight equations applied to the mean starting and terminal sizes, using a least squares projection from the last five samples at each temperature.

‡ The authors have used a steady rate of increase rather than the accelerating and decelerating normal rates of increase as the season approaches the longest day.

in freshwater on natural photoperiod and seasonal temperature increase. The prime interest was environmental control of the smolting process and the possible stimulation of rapid growth in the sea, hence the experimental design. In rigorous terms it is difficult to separate seasonal, photoperiod, temperature, and size effects; however, segments of the experiment are comparable and lead to certain conclusions. At 15°C, *in seawater*, prior to June 21, the reciprocal (decreasing) photoperiod induced a small but significant increase in weight over the other photoperiod effects ($G = 1.37\%$ weight/day versus 1.29% weight/day); when applied after June 21 a lesser weight increase resulted than on natural photoperiod, but not different from a constant 13 L.

When reciprocal photoperiod was applied during 2.5 months *in freshwater* subsequent growth in saltwater was comparatively retarded ($G = 0.31\%$ weight/day). Under these circumstances no difference between natural and constant photoperiods occurred, both of which were accompanied by good growth for the size involved ($G = 0.85\%$ weight/day). Although no differences in plasma chloride concentrations could be detected in any of the fresh- and saltwater comparisons, it was apparent that the reciprocal photoperiod had affected the success of the smolting process prior to saltwater transfer, and this negative effect carried through to affect growth in salt water.

Little difference in conversion efficiency between treatments could be detected. A decrease from an average of 31 to 16% K took place as the year progressed and fish weight rose to over 700 g. Some decrease in K with increasing weight can be expected.

Knutsson and Grav (1976) examined the effects of increasing photoperiods (6–19, 8–19, and 12–19 L) on yearling Atlantic salmon held in freshwater at three temperatures (7°, 11°, and 15°C). The increasing daylengths were applied in the fall–winter season, with the inflection to 12–19 L commencing 3 months ahead of the more normal 6–19 L (at Bergen, Norway, 60°N. Lat.). The temperature effect was considerably greater than the photoperiod effect. Greatest growth occurred at 15°C on the 12–18 L, which was also the regime with most advanced, seasonal inflection.

In general, for freshwater fish, the evidence indicates that long daylength, and more especially increasing daylength applied over a number of months (particularly in the right season) is stimulating to growth. But the effects at best are not large. Decreasing daylengths have an inhibiting effect on some freshwater fish. The lack of greater induced response, compared with natural seasonal effects on normal populations (separate from temperature effects), suggests that experimental designs are somehow inadequate, or that a circannular rhythm is present and not subject to displacement.

For anadromous fish such as salmonids, if the size and season are right, decreasing photoperiod improves the smolt transformation allowing greater growth in saltwater. But this is not so if applied in a freshwater environment at a time when the seasonal norm is the reverse photoperiod (i.e., increasing daylength).

It is possible to generalize that at the time of transformation the stimulation of neuroendocrine pathways related to either sodium excretion (ACTH, prolactin), or maturation (gonadotropin), are partially and temporarily antagonistic to growth hormone production. However, as intermediate developmental stages are achieved, growth can resume to a greater degree (see Chapter 9).

2. LIGHT INTENSITY (WITHOUT PERIODICITY) AND GROWTH RATE

Although light intensity has received considerable attention in relation to both egg development (MacCrimmon and Kwain, 1969) and meristic characters (Lindsey, 1958), little research has been performed on the growth relations.

Eisler (1957) reviewed earlier literature noting that chinook salmon were reported to grow better under "dark-reared" than "light-reared" conditions, and that cod fry were attracted to artificial light. In a 12-week experiment involving exposure to four light intensities (0.02, 88, 116, and 157 f. c. fluorescent light, λ 3350–6000 Å; no photoperiod stated) Eisler reared chinook salmon fry on raw beef liver. Only the lowest intensity ("dark") showed a significant difference in growth, achieving only 68% of the weight of the "light" intensities. He concluded that high light conditions stimulated growth, apparently not conceiving that low light might have an inhibiting effect.

Kwain (1975) subjected hatched rainbow trout to three levels of light (0.2, 2, and 20 lx) at two temperatures (3° and 10°C). Significantly reduced growth rate was reported at the lowest intensity only, for both temperatures. No significant difference occurred between 2 and 20 lx. Trout reared at 0.2 lx were observed to be sluggish and not actively seeking food as at higher light intensities. If reduction in visual perception was the cause, as speculated, light intensity by itself can be dismissed as a factor impinging on growth directly.

C. Salinity

Fishes regulate their plasma ions such that the internal osmotic pressure of their body fluids is equivalent to approximately 10‰ salinity, with a range of ±2‰ depending on tolerance, regulating capacity, and environmental salinity (Holmes and Donaldson, 1969). Maintenance of internal balance in freshwater, where loss of ions and body-flooding are the major problems, is associated with the hormone *prolactin* which serves to control membrane transport and kidney function (Bern, 1975). A highly dilute urine is excreted with a concomitant small loss of salts which are replaced by the diet. Strictly freshwater fish are stenohaline (here abbreviated to Fr–St), being subject to rapid death in normal seawater.

In the marine environment where the problem of ion control is reversed, fishes survive by drinking saltwater, secreting chlorides mostly via the gills, and excreting an isotonic urine. Prolactin is suppressed; the ATPase enzymes for sodium and potassium occur abundantly in the mitochondrial-rich cells of the gills. The majority of marine fishes cannot tolerate the brackish waters of estuaries indefinitely and are therefore also stenohaline (Sa–St). Some marine species, however, are very tolerant, remaining in estuaries and occurring part way up rivers during the freshwater cycle at low tide. It is of interest to note that two species of predatory sea basses, *Dicentrarchus* sp., have been used to control excessive fry production in freshwater reservoirs (Chervinski, 1975). Such tolerant species are labeled saltwater euryhaline (Sa–Eu).

Some freshwater species, such as those occurring among the Salmonidae, commence life in freshwater, where they are spawned, and subsequently pass through a transformation (smoltification) at the time of migrating to sea—either as fry (some *Oncorhynchus* sp.) or as smolts (some *Oncorhynchus* sp. and *Salmo* sp.). These pass from freshwater stenohaline to saltwater euryhaline and are here referred to as anadromous euryhaline (An–Eu).

A few species apparently of freshwater origin* occur in a wide range of salinities, even tolerating the highly saline conditions of inland seas where evaporative loss may result in salinities of 60‰ and greater. These euryhaline fishes have been labeled Fr/Sa–Eu.

Since internal ion regulation is involved, despite the rare natural possibility of osmotic equilibrium in certain brackish waters, salinity must be classed as a Masking Factor, constantly requiring some energy expenditure associated with active transport of ions to maintain the internal melieu.

Studies on the effect of salinity on growth tend to be scattered, somewhat conflicting, and frequently lacking in any record of the internal ionic state. Although some measure of salinity tolerance may

---

* Or originally inhabiting seawater and penetrating freshwater secondarily (e.g., *Tilapia*, Chervinski, 1961).

be included, the ion regulatory capacity is almost never ascertained. These ancillary determinations are not essential to the basic observation of salinity–growth responses. But the variability in reported response has been such that uniformity of physiological state is apparently not present in many cases, and hence the explanation of some of the variability is denied. Because temperature as a Controlling Factor and photoperiod as a Directive Factor enter into the consideration, as well as prior exposure to some intermediate saline concentration (acclimation), there is need for wide-ranging systematic study of this environmental entity. As for the two previous environmental factors, the effect of G_{max} will be considered first.

1. Maximum Growth Rate (G_{max}) × Salinity

A wide variety of species have been tested for the effect of salinity on growth rate, when provided with an abundance of food (R_{max}). In particular, the anadromous fishes have received attention because of their capacity to transform from a fresh to a saltwater habitat, and their suitability for aquaculture. This is also true of at least one species of catadromous fish, the grey mullet (De Silva and Perera, 1976). Other interest has centered on the wide-ranging capacity for the estuarine and euryhaline species of freshwater origin to grow in saline environments. Since the number of searching enquiries is not great, the ecological diversity of the species studied has tended to diffuse the picture. From the compilation in Table II the greatest growth rates within species cluster either around zero salinity (freshwater), or 10 ± 2‰, or 28–35‰ (see references in Table II). These roughly distinguish between the ecologically separable freshwater stenohaline species, the anadromous species, and the euryhaline and strictly marine species (Fig. 13). On excess ration, greatest growth rate occurred at 20‰ in *Mugil cephalus*, whereas highest conversion efficiency was reported at 10‰; salinity concentration affected the level of maximum ration (De Silva and Perera, 1976).

The grouping of the freshwater species includes the presmolt stage of the anadromous fishes (e.g., *Salmo salar*). There may be some slight benefit in the isosmotic range, but above this concentration G_{max} falls off; indeed salinities in the region of 15‰ may be lethal (e.g., *Carassius auratus*) for both osmotic and ionic reasons. Such a euryhaline species as *Cyprinodon macularius* can grow comparatively well anywhere from 0 to 55‰; *Tilapia mossambica* almost matches this capacity. Chervinski (1961) reported that *Tilapia nilotica* would grow as well in 18‰ as in freshwater (avg. G_{max} = 2.4% weight/day; weight = 25 g; temperature = 24 ± 3°C), and would adapt to at least 50‰. No records

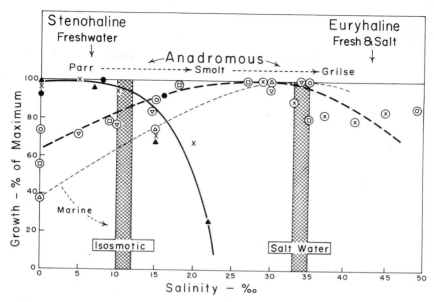

Fig. 13. Effect of salinity on growth of various fishes expressed as a percentage of the maximum rate determined for each species. Data and sources in Table II. Solid line is for stenohaline, freshwater fish, and parr stage of salmon (species: *P. reticulata, O. kisutch, S. salar*). Heavy broken line and all circled points apply to euryhaline fish distributed in freshwater and saltwater environments including the postsmolt anadromous salmonids (grilse) (species: *C. macularius, T. mossambica, T. maculatus, P. dentatus, B. icistia*). Light broken line is for an estuarine species, and indicates the likely shift in relation to a fully marine stenohaline species.

for the minimum salinity permitting growth for strictly marine species has been found; there is a suggestion in the response of *Paralichthys dentatus* at 20°C that a falling off occurs at low salinities, as depicted in Fig. 13.

At the time of smolting the salmonid fishes shift to tolerating a comparatively high salinity, accompanied by good growth capacity. Such an enhanced ability to grow appears to characterize the marine stage, quite separate from any circumstance of abundance of food. This is supported by pen-rearing studies at sea (Falk, 1968; Novotny, 1975). These general relations, however, do not take into account the interacting effect of elevated temperature which, in the case of *Cyprinodon macularius*, favors a high salinity to enhance growth whereas the reverse is true of *Trinectes maculatus* (Table II). In this species high temperature will only support good growth in freshwater. These findings serve to reveal some of the basis for the great diversity between species recorded so far.

Table

Rate of Growth (G_{max}) of Different Species of Fish

Species	Type	Initial size		Salinity acclimation		Test tempera-ture (°C)	Test duration (days)
		cm	g	‰	Days		
Poecilia reticulata	Fr–St	0.7	—	0	—	24	40
Carassius auratus	Fr–St	8.4	12.1	0	—	20	70
Oncorhynchus kisutch	Au–Eu	3.6	0.47	0	—	10	70
Oncorhynchus kisutch	An–Eu	4.2	1.01	9	2	10	70
Salmo salar	An–Eu	10.7	15	0–22	—	8–4*	126
Oncorhynchus kisutch	An–Eu	3.0	1.0	0.25	1	10	32
Oncorhynchus kisutch	An–Eu	6.0	—	0.25	1	10	32
Oncorhynchus tshawytscha	An–Eu	—	0.5	0–24	80	11	70
Oncorhynchus tshawytscha	An–Eu	—	0.8	5–25	66	12	28
Salmo salar	An–Eu	—	40	0–20	21	10	56
Oncorhynchus keta	An–Eu	—	0.35	0–28	7	15	35
Morone saxatilis	An–Eu	Fry*		0.2–4.8	5+	(18)**	7
Mugil cephalus	Ca–Eu	—	0.2	<20*	—	25	360
Cyprinodon macularius	Fr/Sa–Eu	0.8	0.02	0–55*	70	30	84
Cyprinodon macularius	Fr/Sa–Eu	0.8	0.02	0–55*	70	15	84
Tilapia mossambica	Fr/Sa–Eu	—	0.22	0–35	2	25*	56
Trinectes maculatus	Fr/Sa–Eu	3.5	0.7	0–30	4	15	7–14
Trinectes maculatus	Fr/Sa–Eu	3.5	0.7	0–30	4	35	7–14
Paralichthys dentatus	Sa–Eu	—	0.02	25–35	—*	20	4–7
Paralichthys dentatus	Sa–Eu	—	0.02	25–35	—	30	4–7
Anisotremus davidsoni	Sa–Eu	—	1.5	29–45	14	25	14
Bairdiella icistia	Sa–Eu		5.0	29–45	14	25	14

[a] Most cases are expressed as specific growth rate in weight/day, but some are in terms of length achieved (as indicated under Remarks). Selected salinity ranges are indicated, with the particular salinity shown as a superscript in parentheses. Each species is classified according to ecological type, using the abbreviations: Fr, freshwater; Sa, saltwater; An, anadromous; Ca, catadromous; St, stenohaline; Eu, euryhaline. Species that normally occur in freshwater but are found abundantly in saltwater have been labeled Fr/Sa–Eu. Because initial size, acclimation state, test

II

Exposed to Various Salinities and Fed a High Ration[a]

0	1–6	7–12	13–18	19–35	35+	Remarks	Reference
		Growth rate for salinity ranges (‰)					
1.25*	—	1.35[8]	1.25[16]	—	—	* Interpolated lengths (cm) achieved in 40 days	Gibson and Hirst (1955)
+	+[6]	—	Lethal[15]	—	—	Slight benefit in fresh at 35 days; none by 70 days	Canagaratnam (1959)
1.65	1.90[6]	2.41[12]	—	—	—	All fish fed 10% weight day; no acclimation	Canagaratnam (1959)
1.13	—	—	1.58[18]	—	—	Initial weight doubled	Canagaratnam (1959)
0.23	—	0.22[7]	0.16[15]	0.16[15]	—	* Seasonal change (not controlled), Dec. to April	Saunders and Henderson (1969)
1.60	1.65[5]	1.55[10]	1.15[15]	1.10[20]	—	Tested in Aug.–Sept. (presmolt)	Otto (1971)
0.12	0.55[5]	0.60[10]	0.10[15]	0.10[20]	—	Tested in Jan.–Feb. with larger fish (smolt)	Otto (1971)
2.7	—	—	2.6[17]	2.2[33]	—	Intermediate, increasing salinity acclimation applied	Kepshire and McNeil (1972)
—	—	—	3.0[18]	2.4[33]	—	Reduced growth rate in full seawater despite acclimation	Kepshire and McNeil (1972)
0.7	—	0.5[10]	—	0.5[20]	—	No significant difference reported (Oct.–Nov.)	Shaw et al. (1975b)
5.7	—	—	—	5.9[28]	—	Not significantly different	Shelbourn (1976)
—	9.6[6]	10.7[12]	—	9.4[20]	—	* Fry aged 5 to 63 days; ** 3 test temperatures, 12°, 18°, 24°C	Otwell and Merriner (1975)
4	—	4[10]	5[20]	0.3[30]	—	* Caught in coastal lagoons; conversion efficiency greatest at 10‰	De Silva and Perera (1976)
1.51**	—	—	1.86[15]	2.06[35]	1.78[55]	* Brood stock acclimated; ** length achieved	Kinne (1960)
1.55*	—	—	1.30[15]	1.02[35]	—	best in freshwater when at low temperature (15°C); * length achieved	Kinne (1960)
2.24	—	3.17[9]	3.96[18]	4.06[27]	3.23[35]	* Assumed room temperature (not given)	Canagaratnam (1966)
0.53	—	—	1.03[15]	1.40[30]	—	Only three fish per test used (usually)	Peters and Boyd (1972)
3.14	—	—	2.59[15]	2.09[30]	—	Large temperature effect in freshwater	Peters and Boyd (1972)
—	7.5**[5]	8.0[10]	8.7[15]	10[30]	10.5[35]	* Held through post-larval stages; ** interpolated	Peters (1971)
—	12.8*[5]	14.8[10]	15.2[15]	17.4[30]	18.2[35]	* All values interpolated from graphs	Peters (1971)
—	—	—	—	2.2[33]*	1.2[45]	* Tested at 5 salinities from 29–45‰	Brocksen and Cole (1972)
—	—	—	—	3.7[29]	3.0[45]	* Same as above	Brocksen and Cole (1972)

temperature, and duration of experiment all vary greatly between species, and can be shown to interact within species, comparison is best made as a *relative* effect of salinity on growth, as in Fig. 13. The highest value for growth in any one species is underlined. Species are arranged in ecological order, from freshwater to marine. Asterisks refer to Remarks column.

2. RESTRICTED RATIONS × SALINITY

In separate studies on the effect of salinity and of dietary sodium chloride on the growth of Atlantic salmon parr at 10°C, Shaw *et al.* (1975a,b) used daily rations from 0 to 3% weight/day for fish in the 40–60 g range. Surprisingly, no undue consequence of the greatly increased dietary loading with salt was observed at the selected environmental salinities of 0, 10, and 20‰ (cf. Basulto, 1976; Zaugg and McLain, 1969). Possibly the normal saltwater drinking rate of 5–13% weight/day (Shehadeh and Gordon, 1969) reflects such a well-developed ion secretory capacity. Despite the age and size of the Atlantic salmon, some lack of tolerance above a salinity of 22‰ was reported, with a few deaths occurring at 30‰ (in May and June). No significant effect of salinity on the GR curve could be demonstrated up to 20‰. At 30‰ the curve was shifted to the right (Fig. 14) along the abscissa in accordance with an increase in maintenance ration from 1.3% weight/day (fresh) to 3% weight/day (salt). A great depression in conversion efficiency occurred at all ration levels in saltwater. The nature of the relation suggests that the highest ration provided was in the region of R_{opt}, not R_{max}. The level of R_{max} and therefore G_{max} was not defined, yet the growth rate in freshwater was comparable with sockeye salmon G_{max} for the same weight and temperature (Brett, 1974).

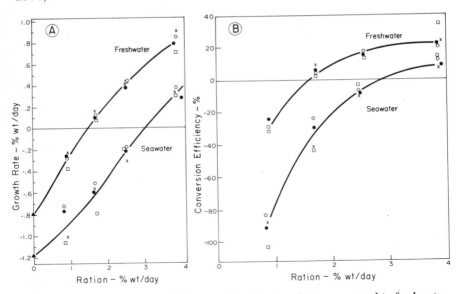

Fig. 14. (A) Growth–ration (GR) curves for Atlantic salmon parr raised in freshwater and seawater (30‰) at 10°C. (B) Derived gross conversion efficiency (K) curves. (From Shaw *et al.*, 1975a.)

In contrast to the above results with Atlantic salmon, Smith and Thorpe (1976) obtained a significant improvement in the growth rate of yearling rainbow trout, *Salmo gairdneri* (over 40 g), when acclimated to saltwater and compared with freshwater controls at 12°C. The increased growth rate only occurred in late summer following parr-smolt transformation, and at maximum ration; no advantage occurred under restricted rations. Analysis indicated that the higher growth rates in saltwater resulted from an improved ability to retain assimilated nitrogen and not from increased consumption.

The progressive increase in the salinity of the Salton Sea (37‰)* brought Brocksen and Cole (1972) to examine the effect on the growth rate of two introduced sports fish—bairdiella, *Bairdiella icistia*, and sargo, *Anisotremus davidsoni*. Each species was tested at five salinities (29–45‰) and five rations (1–10% weight/day) at 25°C. Both species grew best at salinities lower than the existing environmental level (Table II); sargo, however, required a particularly high R_{maint} at all salinities (avg. = 7.4% weight/day) which would undoubtedly affect its success. Some interaction between ration and salinity occurred for *Bairdiella*, a salinity of 37‰ being worst at high rations and best at low rations.

Peters (1971) and Peters and Boyd (1972) have used restricted rations in experiments on the combined effects of temperature, salinity, and food availability on the growth of young flatfish, *Trinectes maculatus*. By examining the response at an intermediate temperature (20°C) it is apparent that the slope of the GR curve increases with increasing salinity such that, from a common pivot point of similar maintenance rations, an increase in salinity induced a higher G_{max} when on *ad libitum* feeding. When starved, at 15°C, *Trinectes maculatus* lost weight faster in freshwater than at higher salinities; the reverse was observed at 35°C.

These illustrations serve to point up the complexity and diversity of the growth response to salinity, a circumstance that is likely to require searching physiological analysis for adequate explanation. Whatever this may be, it is of interest to note that in nature these species of flatfish occur most frequently at salinities where they grow fastest (Peters, 1971).

It can be concluded that, with the possible exception of the smolting stage in salmonids, evidence for any substantial increase in growth accompanying isosmotic conditions is lacking (isionic conditions have not been examined). Pronounced decreases in G_{max} with increasing salinity, among freshwater adapted species, appear to be the result of large increases in R_{maint}. As the limit of salinity tolerance is ap-

* Expected to be 40‰ by 1975, i.e., increasing at a rate of 1‰ per year.

proached and regulation becomes progressively inadequate, R_{max} falls to the level of R_{maint} blocking further growth.

D. Oxygen

At the time of feeding and for some hours thereafter the metabolic rate of fish is elevated (Chapter 6). Some authors deduced that for maximum rations the increased demand for oxygen could reach as high as the active metabolic rate (Paloheimo and Dickie, 1966). While this may be true in nature where the daily encounter of predator and prey can involve numerous bursts of activity, in the laboratory the maximum metabolic requirement does not appear to exceed two or three times the standard rate (about one-third active metabolism). The total daily increase in oxygen consumption is directly related to the size of the meal; like any other activity, as the load increases the energy expenditure goes up (Chapter 6). One aspect of the metabolic process, *specific dynamic action*, has not always been clearly defined and measured for fish (Warren, 1971; Beamish, 1974). Despite the sequence of complex digestion–absorption–transformation steps in food processing, environmental oxygen can be shown to act as a simple Limiting Factor, sharply curtailing growth and food conversion efficiency at critical oxygen levels, usually well below the air-saturation point. However, oxygen cannot be confined *a priori* to this single role. It is conceivable that reduced oxygen content could act as a cue (Directive Factor) for reduction of appetite—or more likely some associated change accompanying lowered oxygen, since the existence of O_2-sensors among fish has yet to be confirmed.

1. Maximum Growth Rate $(G_{max}) \times$ Oxygen

While many studies have been conducted on the effects of oxygen supply on the growth rate of fish, these are very varied in relation to duration, size, and survival of fish, nature of diet, temperature applied, and the level and precision of environmental oxygen control, such that useful systematic tabulation is difficult if the comparability is to be preserved (see Brungs, 1971; Swift, 1963, 1964; Doudoroff and Shumway, 1967, 1970; Warren, 1971; Ebeling and Alpert, 1966; Davison *et al.*, 1959). The studies involving late embryonic and early larval stages, where first feeding occurs, are complicated by the change in developmental state during the test period and the initiating of an adequate feeding response. Further, there is a bias introduced where only the size of the surviving fish can be used to determine growth.

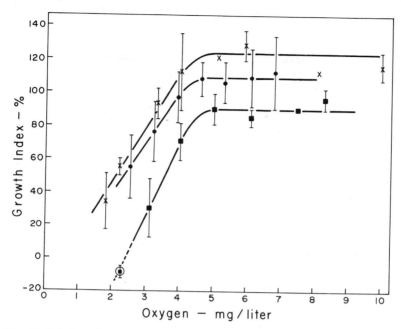

Fig. 15. Relation of oxygen concentration to growth rate expressed as the percentage of the "growth index" developed by each author. Limits = ±1 SD. Data for *Micropterus salmoides* (26°C; 2.5–4.5 g—upper curve) from Stewart *et al.* (1967), for *Cyprinus carpio* (22°C; 0.5–3.4 g—middle curve) from Chiba (1966), and for *Oncorhynchus kisutch* (20°C; 2–6 g—lower curve) from Herrmann *et al.* (1962); circled point was accompanied by significant mortalities. The vertical positioning by species has no relative significance. Lines drawn according to interpretation presented in text.

Some increased sensitivity to reduced oxygen is apparently present at the larval stage in comparison with the juvenile stage; the margin, however, is not great (Carlson and Siefert, 1974; Carlson *et al.*, 1974).

From three well-documented studies on juveniles it has been possible to determine the mean and variance of the "growth index" applied in each case (Chiba, 1966; Stewart *et al.*, 1967; Herrmann *et al.*, 1962). This was done by grouping according to concentration* intervals of 0.5 ppm O_2, and plotting the results as the mean of G_{max} against the mean value of the concentrations within each of the O_2 intervals (Fig. 15). Where only two values occurred in a given interval, the range was used. Despite the considerable deviation accompanying each mean determination (partly due to experimental variability in O_2

---

* Since fish have a great ability to extract oxygen from water, concentration is a better indicator of available amount than saturation.

level) it is clear that an oxygen concentration of close to 5 ppm is critical for growth, below which increasing suppression of G_{max} is directly proportional to decreasing O_2 concentration—a drop of 1 ppm causes a 30% reduction in growth rate. The analysis bears further comment. None of the authors would dispute the O_2-dependent segment of the curves as depicted. However, the balance of the relation might be contested, here shown as forming a plateau beyond the critical transition zone (depicting complete *independence* of O_2 concentration above 5 ppm, as in simple limiting cases). In two of the original sources (Stewart *et al.*, 1967; Herrmann *et al.*, 1962) and again in Doudoroff and Shumway (1970) the data for largemouth bass are interpreted by the authors as rising to a peak of growth close to 8 ppm (100% saturation) and falling off at concentrations far in excess of this, for example, 17 ppm at 26°C (212% saturation). It is known that excessive oxygen concentrations can even be lethal (Hubbs, 1930). However, from the variance of the small samples involved above the 5 ppm critical level, there is insufficient evidence to negate the present interpretation. Support for the latter can be gained from the fact that oxygen concentration would have to be reduced to the critical level indicated (approximately 5 ppm) in order to act as a Limiting Factor to the associated metabolic rate (see Chapter 6). It should be noted that *experimentally reduced* oxygen concentration is maintained in the complete absence of any change in the rest of the controlled environment. In nature, and in hatcheries, decreased oxygen would likely be accompanied by an increase in other environmental factors such as ammonia, urea, and nitrites which would act antagonistically to growth.

The presence of a definite upper plateau, supporting the Limiting Factor concept, is clearly depicted in the conversion efficiency relation in Fig. 16. The values obtained above 5 ppm are in keeping with the high efficiency that usually accompanies good growth when young fish are fed a nutritious diet. Tests conducted on a small number of juvenile Northern pike, *Esox lucius*, by Adelman and Smith (1970) follow the same pattern as the three species illustrated in Fig. 15. Evidence for reduction in food consumption and conversion efficiency for this species suggests a critical level between 3 and 4 ppm O_2 (at 19°C).

2. Restricted Rations × Oxygen

Since ration is the "driving force," any restriction obviously reduces the growth opportunity to a lower level; the GR curve is truncated. As was pointed out, a restricted ration is also accompanied by a

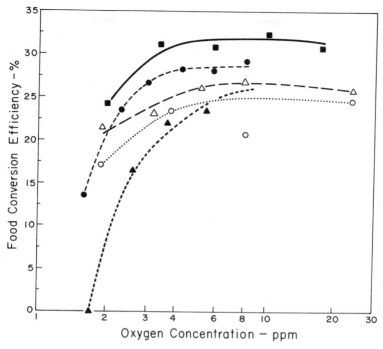

Fig. 16. Food conversion efficiency in relation to oxygen concentration. Results are for separate experiments conducted on *Micropterus salmoides* at 26°C. (From Stewart *et al.*, 1967.)

lowering of the daily metabolic rate and consequently a reduced demand for oxygen. Thus, it might be expected that the critical oxygen level would fall. Fisher (1963) demonstrated this clearly for underyearling coho salmon (Fig. 17). On a limited ration, which resulted in approximately 0.4 G_{max}, the critical oxygen level dropped to between 3 and 4 ppm. If the positions of inflection inferred by Fisher (1963) are used, intermediate levels of restricted ration would be expected to follow the paths indicated in Fig. 17 (dotted lines).

The general phenomenon witnessed under these circumstances is the interrelation of two Limiting Factors—food and oxygen—acting *in series*. Successive plateaus below G_{max} are set by the degree of restricted ration. But as oxygen is reduced, an O_2-dependent stage enters to depress the growth rate below that previously dictated by the limited ration (depicted by the slope of the line below the plateau, Fig. 17).

This phenomenon can be identified in the response of channel catfish exposed to three oxygen concentrations and two feeding re-

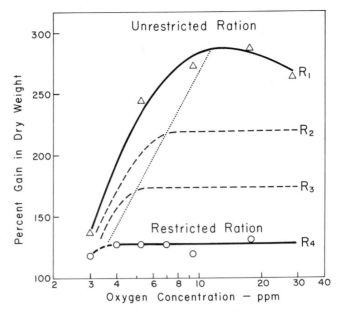

Fig. 17. Growth rate of coho salmon, *Oncorhynchus kisutch*, on unrestricted ration (R_1) and restricted ration (R_4) in relation to oxygen concentration. (Modified from Fisher, 1963; reproduced in Doudoroff and Shumway, 1967.) The dotted construction line has been added, joining the points of inflection for R_1 and R_4. The expected relations for intermediate restricted rations (R_2 and R_3) have been drawn as broken lines.

gimes (Andrews *et al.*, 1973). At 26.6°C the saturations imposed convert to the following concentrations: 100% = 7.9 ppm O_2, 60% = 4.7 ppm O_2, and 30% = 2.8 ppm O_2. On *ad libitum* feeding (R_{max}) there was a significant reduction in growth rate between each O_2 level, which would be expected for concentrations below the critical O_2 level of 5 ppm (results: G = 3.1% weight/day at 7.9 ppm, 2.7% at 4.7 ppm, and 1.8% at 2.8 ppm). When fed a fixed ration of 3% weight/day there was no significant difference between the two higher O_2 concentrations (avg. G = 1.8% weight/day); only the growth rate at 2.8 ppm O_2 was depressed (G = 1.3% weight/day). Thus, the limiting effect of reduced oxygen was shifted to a lower concentration when on restricted ration.

3. Varying Oxygen × Growth

Daily oscillations in oxygen content are not unusual in nature, frequently produced through the light cycle on photosynthesis. Stewart *et al.* (1967) subjected largemouth bass to alternately low (2–4 ppm) and higher (4–8 ppm) oxygen concentrations and showed that growth was markedly impaired. The reduction was greater than that which would have occurred if kept at the mean O_2 level, showing the detrimental consequences of the low concentrations when these were below the critical level. When the variation was entirely below the critical value (e.g., 1.8–3.7 ppm) the growth was only 32% of that expected for a concentration equivalent to the mean daily O_2 level.

Whitworth (1968) demonstrated a similar inhibiting effect on the growth of brook trout, *Salvelinus fontinalis*. When subjected to fluctuating oxygen levels (10.6 ppm reduced to either 5.3 or 3.6 ppm) the fish lost weight and were approximately 75% of the weight achieved by a control group at 10.6 ppm (for 49 days). Diurnal fluctuations from 3.0 ppm O_2 rising to either 9.5 or 18 ppm O_2, applied by Fisher (1963) to underyearling coho, resulted in an almost equally depressed growth rate, similar to that which would have occurred at fixed levels of 3.5 and 3.9 ppm, respectively.

It is apparent from these few but revealing studies on fluctuating O_2 that high concentrations above air-saturation do not confer any substantial benefit compensating for the periods of low concentration. Further, an exposure to subcritical levels of oxygen for only a portion of the day (e.g., 8–12 hr) is sufficient to depress the growth rate to that comparable with the constant low O_2 level. And this is despite feeding during the high O_2 period. There is obviously not a simple on–off effect without a carryover of serious consequences into the higher O_2 period.

E. Summary Configurations

More information is available on the effects of temperature on growth than any other abiotic factor; oxygen is next, with light and salinity not at all well documented. Within these limitations an attempt has been made to present the various patterns that the GR curves follow for each environmental factor, and the consequent path of the major parameter G_{max} (see Fig. 18). These illustrate the categories of effect (Factors) postulated by Fry (1947).

In the case of light (L) only the effects of static photoperiods are indicated, whereas dynamic states of changing photoperiod are undoubtedly more important. The consequences of the extremes (complete darkness or 24 hr light) are indicated as depressive in the former and suboptimal in the latter. Rasquin and Rosenbloom (1954) have recorded the long-term stress effects on fish held continuously in darkness.

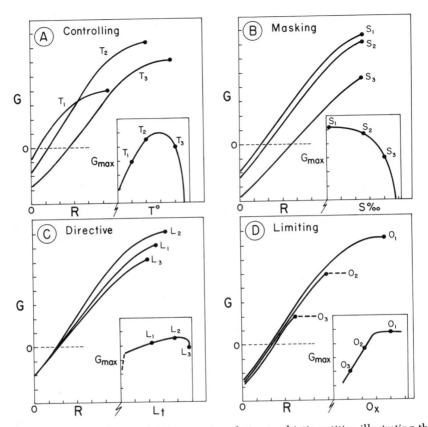

Fig. 18. Recapitulation of GR curves in relation to abiotic entities illustrating the basic forms for: (A) *temperature* (T°), a Controlling Factor; (B) *salinity* (S‰), a Masking Factor; (C) *light* (L_t = static photoperiods), a Directive Factor; and (D) *oxygen* (O_x), a Limiting Factor. Each GR curve is shown at three levels of each abiotic entity, with a point for the position of G_{max} terminating the top of the curve. The insert box shows the path of change of G_{max} as the entity increases from minimum to maximum values (e.g., as temperature increases over the tolerable range) in A. The paths of change for R_{maint} and G_{starv} can be seen by the points of intersection with the respective lines for G_0 (horizontal broken line) and R_0 (y-axis). Curves have been drawn where possible from data and figures presented; the circumstance for levels of fixed photoperiod remains speculative. G, the specific growth rate, and R, the ration, are both represented in terms of % body weight/day (or cal/kcal/day). Maximum values of G range up to 10% weight/day and of R to 30% weight/day for young fish.

By way of summary and recapitulation to this point, the GR curves can be seen to alter in four quite distinct patterns, according to the category of the abiotic factor (i.e., how the abiotic factor acts). Thus, in Fig. 18A, where temperature is acting as a Controlling Factor, the GR curve not only shifts to the right as R_{maint} increases but also rises and subsequently falls as G_{max} passes through the optimum temperature. In B, where salinity as a Masking Factor is imposing an increasing load on a freshwater species, the GR curve also shifts to the right as R_{maint} increases, but there is no optimum salinity; G_{max} falls off rapidly depressing the whole curve. In C, where light is depicted as a Directive Factor, there is a common origin of G_{starv}, little effect on R_{maint}, and an optimum daylength where the GR curve rises to a maximum in all respects. In D, where oxygen is acting as a Limiting Factor, the curves are superimposed (here shown slightly separated) and truncated at intermediate plateaus of growth according to the depression of ration induced by lowered oxygen concentrations. Any environmental factor acting within any of the four categories shown would cause the same sort of response of the GR curve.

IV. BIOTIC FACTORS

A. Ration

1. QUANTITY (R_{max})

Up to this point the use of maximum growth rate (G_{max}) obtained by providing a maximum ration (R_{max}), or something over and above that amount (*ad libitum*, or excess), has been used in a conceptual or assumed sense. Critical examination reveals a number of factors related to feeding frequency and quantity that bear on the maximum daily intake of fishes. The most important of these factors include (a) the duration of a given feeding (satiation time), (b) individual meal size (stomach capacity), (c) time between meals (feeding interval), and (d) the interaction of these. To these must be added the consequences of abiotic and biotic factors, among which temperature and fish size are of greatest importance. In practice it is not uncommon to feed very young fish almost continuously on a fine feed "mash," switching to interval feeding on progressively larger pellets as fish size increases.

It should be noted further that the conceptual approach to growth and feeding relations adopted in this chapter assumes that the prime demand for food is imposed by the maintenance requirements of the fish, with a further demand dictated by the potential growth capacity (influenced by growth hormone). These interlocking requirements set the limits of voluntary food intake, not the reverse, that is, not the case of appetite governing growth rate (compare Chapter 3).

As any experimenter will attest, when dealing with various feeding rates and groups of fish, the problem presented on restricted rations is that of providing a fair share to each ravenous member of the sample, never the total daily intake. However, as R_{max} is approached, the question of an ultimate level becomes more elusive, subject to many minor forms of disturbance, in which taste, size, shape, and movement of food particles are important as well as the elimination of any unusual sound or light stimuli. At this level of daily consumption it takes very little to put the fish off feeding to maximum capacity.

A number of species of fish have been shown to require up to an hour or more to reach satiation, for example, *Trachurus japonicus* and *Salmo gairdneri* (Ishiwata, 1968), *Oncorhynchus nerka* (Brett, 1971a), and *Salmo trutta* weighing 100 g or more (Elliott, 1975a). Other species such as the puffer, *Fugi vermiculatus,* the filefish, *Stephanolepis cirrhifer* (Ishiwata, 1968), and relatively small *Salmo trutta* (about 15 ± 10 g, Elliott, 1975a) were satiated in 15 min or somewhat less. The influence of weight and temperature on satiation time has been shown to be highly significant in the response of *Salmo trutta* (Elliott, 1975a), whereas weight did not appear to have a strong influence in *Oncorhynchus nerka* (Brett, 1971a). Differences in experimental technique may have some bearing on this apparent fundamental difference in response, namely the consequences of feeding individual fish a single food item at a fixed rate (Elliott, 1975a) and that of broadcast feeding of pellets to a school of fish (Brett, 1971a).

By contrast, maximum meal size of sockeye salmon was greatly influenced by fish weight whether on a single or multiple feeding regime, the smallest fish consuming the largest relative amount (percentage of body weight, Fig. 19). This is similarly apparent from the hatchery feeding tables mentioned previously. The relation of fish weight to maximum meal size was shown by Elliott (1975a) to be proportional to $W^{0.75}$ for brown trout. A greater exponent of weight ($W^{0.95}$), approaching weight independence, derived by Kato (1970) for rainbow trout appears to be in some contradiction to the more general findings of considerable weight-dependence suggested by hatchery feeding tables.

In addition to a weight influence, the effect of temperature on maximum meal size of brown trout was shown by Elliott (1975a) to follow four apparently distinct temperature ranges (Fig. 20). Temperatures between 13° and 18°C resulted in highest intake for a single meal. However, since rate of digestion increases with rising temperature the highest daily intake for *multiple* meals was reported at a slightly higher temperature, 18.4°C (Elliott, 1975b). Small salmonids

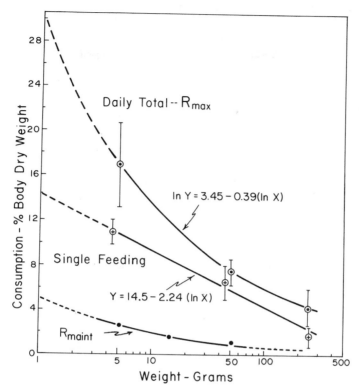

Fig. 19. Maximum food intake of sockeye salmon at 15°C on single and triple daily feedings, in relation to fish wet weight (log scale). Mean daily totals (± 1 SD) are shown with calculated extrapolations according to the equations. (Modified from Brett, 1971a.) Values for maintenance ration (R_{maint}) have been plotted in the bottom line. The difference between R_{max} and R_{maint} is the scope for growth, which can be seen to decrease with increasing weight.

(under 3 g) at a high temperature (15°C and above) can be shown to consume over 20% of their body weight per day (dry weight basis); large fish (over 1000 g) frequently require less than 1% body weight per day to meet their maximum consumption levels (cf. Fig. 19).

By studying the time sequence for return of full appetite in relation to deprivation time and rate of stomach evacuation, both Elliott (1975b) and Brett (1971a) showed that when the stomach was 75–95% empty voluntary food intake was not far from the maximum. Since this is a rate function dependent on meal size and temperature, the time sequence of feeding can be strategically manipulated to increase daily intake. As an example, restricted rations in the morning designed to

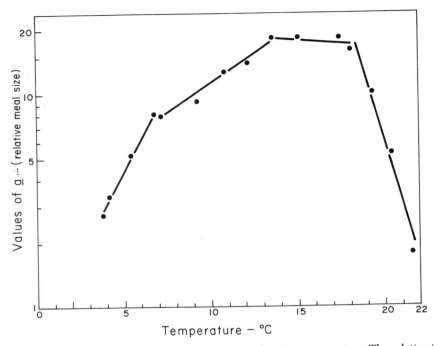

Fig. 20. Maximum meal size of brown trout in relation to temperature. The relation is shown in terms of the constant *a* from the general equation, $Q = aW^b$, relating meal size (Q) to weight (W) for single daily feedings. The relation of *a* to temperature is independent of fish weight (10–350 g). (From Elliott, 1975a.) The use of a set of straight lines was arbitrarily selected by Elliott who considered that there were four distinct temperature ranges suitable for mathematical modeling.

permit greatest intake at the *end* of an 8-hr day provide most use of the 16-hr digestive interval until next feeding opportunity. Brett (1971a) showed that on a selected 11-hr feeding interval at 15°C the sum of the two feedings was considerably greater when the first feeding was 3.4% and not the maximum intake of 4.4% (for the size of fish involved—50 g). The repeated 11-hr interval applied over a series of days was chosen to remove the possibility of endogenous or habitual rhythms of digestion interfering with the experimental design (offset from multiples of 12 hr).

Using automatic feeders that dispersed food from one to twenty-four times per day, the growth and conversion efficiency of catfish, *Ictalurus punctatus*, were discovered to be highest on satiation feeding twice per day at 28°C; the most frequent feeding gave the least benefit (Andrews and Page, 1975). Apparently, nibbling was not an effective feeding strategy for this species. This is in contrast to the behavior of young salmon where continuous feeding for 15 hr/day at 20°C produced significantly greater growth rate than feeding to satiation three times daily (Shelbourn et al., 1973).

An explanation for some of these differences has been offered by Kono and Nose (1971) who examined the effect of various feeding frequencies on six diverse species of fish. They concluded that the suitability of different time sequences was influenced by stomach size, with the smallest stomachs requiring most frequent feedings. This is exemplified by the continuous browsing behavior of surf fishes (Embiotocidae) which have a long intestine and an almost undifferentiated stomach region (De Martini, 1969).

2. QUALITY

The general form of the GR curve was presented earlier (Fig. 4), and the interrelations of restricted ration with individual abiotic factors examined. It is quite apparent that restricted ration simply cuts the curve off at the particular point of intersection for any level of ration below R_{max}. Growth would continue at a fixed rate, all things being equal. As will be seen this is not entirely the case, since increasing size would sooner or later exert a restricting effect depending on the level of reduced ration, the only exception being the maintenance ration (R_{maint}, G_0). This size involvement does not affect the self-evident conclusion that food acts as an unequivocal Limiting Factor.

The foregoing assumes that the ration is formulated as a well-balanced diet (see Chapter 1). Few nutritional experiments on fish have been concerned with comparisons over the full span of rations, or planes of nutrition, from R_{maint} to R_{max}, except some of those involving protein conversion ratios (e.g., Nose, 1963). The question of quality difference among formulated and natural diets was posed during studies on environmental effects on the growth of young sockeye salmon (Fig. 21). Greatest growth at all feeding levels was obtained with Halver's test diet, formulated with casein as the major protein source; least growth occurred on frozen marine zooplankton (mostly Calanus plumchrus, Brett, 1971b). From Fig. 21 it is apparent that as the slope of the GR curve falls, from the most effective to the least effective diet, maximum intake increases. In the present case R_{max} shifted from about 7% weight/day on the best diet to 14% per day on the worst, while the associated G_{max} fell from approximately 3% per day to 1% per day. Quite obviously this is accompanied by a great decrease in conversion efficiency.

The associated increase in R_{maint} (from 2 to 6% per day) provides some insight for the plight of the fish on a nonnutritious, low-energy

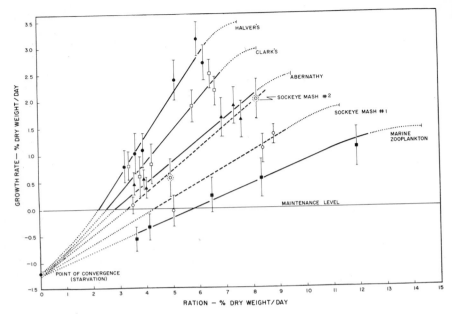

Fig. 21. Growth rate of sockeye salmon in relation to ration, for various diets. Limits: ±2 SE. All solid lines determined by least squares. Broken lines for sockeye "mash" were drawn by eye; two encircled points were for No. 2 formulation. The "point of convergence" was estimated by extrapolation. Dotted extensions of lines indicate expected shapes of curves terminating at maximum voluntary food intake. (From Brett, 1971b.)

diet. Even when such a diet was present in abundance the fish were constantly hungry, packing their stomachs to the maximum at every opportunity.

Elliott (1975a) found that naturally occurring food organisms of six varieties were consumed by brown trout to the same extent (meal size) and over the same time frame (satiation) with the exception of meal worms. The latter were much lower in water content with the result that fewer were eaten, involving a significantly shorter satiation time. All other parameters of weight relations and temperature effect on feeding remained within the 95% confidence limits developed for this species on a "standard" diet of amphipods.

B. Size

Few metabolic relations are independent of size. The anabolic process of growth is no exception. As an animal increases in size the metabolic activities are paced at a generally declining rate, frequently proportional to weight raised to the power 0.7 ($W^{0.7}$) in warm-blooded vertebrates (Kleiber, 1961). The weight relation of standard metabolism for fish is more varied with the weight exponent ranging from 0.65 to 0.85 (Beamish, 1964; Glass, 1969). The early concept that these might be explained by simple surface–volume proportionality as size increases (i.e., proportional to $W^{0.67}$) has too many exceptions to be a valid hypothesis for such a highly complex enzyme–substrate–transport–exchange system. Size has a greater restricting effect on growth rate than on metabolic rate, a difference which will be shown to account for a declining conversion efficiency with size (see Fig. 22). The separate effects of age (not maturity) independent of size have yet to be defined. However, there appears to be reason to support an independent age effect of small magnitude (Brett, 1974).

Since size has a continuous effect on growth rate throughout life, *without* an independent phase, it is difficult to conceive of size as a normal Limiting Factor, restricting the supply of metabolites at some critical point of dependence. Also, since the weight influence does not conform to the natural, physical laws of surface relations it is tempting to classify size as a Controlling Factor possibly influencing the metabolic pacing through some form of size-dependent hormonal or enzymatic control of metabolism (the "scaling effect" of Stauffer, 1973). It is not the purpose here to attempt to find biochemical support for such an hypothesis. Rather, the manner in which the GR curve responds to size will be examined and compared with the types of response already defined for abiotic factors.

1. MAXIMUM GROWTH RATE (G_{max}) × SIZE

In studies on the utilization of food by a number of marine species, Hatanaka and associates (Hatanaka and Takahashi, 1956; Hatanaka *et al.*, 1957; Hatanaka and Murakawa, 1958) showed that the growth rate of young mackerel, *Pneumatophorus japonicus*, and amberfish, *Seriola quinqueradiata*, decreased rapidly with size. Thus, at seasonal temperatures of 23 ± 2°C the growth rate of mackerel fed on anchovies (five to six times/day) fell from 9.5% weight/day at 4 g to 2.5% weight/day at 40 g. Kinne (1960) reported on the size relations affecting the increase in length of the euryhaline desert pupfish, *Cyprinodon macularius*, which displayed remarkably high growth rates soon after hatching, for example, 23% weight/day (5 mm, 20 mg, 20°C, 35‰) falling to 1.3% weight/day by the thirty-fourth week (27 mm, 220 mg).

Extensive size-effect studies have been conducted on members of the family Salmonidae, particularly influenced by the need for appropriate hatchery management practices. If the logarithm of growth rate (G_{max} as % weight/day) is plotted against the logarithm of weight (W in g) over a range of weights from 1 to 400 g, the growth rate can be seen to decrease at a fairly uniform rate (Fig. 22) (see also Kato and Sakamoto, 1969). From a compilation of these sources (Table III), Brett and Shelbourn (1975) concluded that the slope of the decreasing growth rate with size ($b = -0.41$) was characteristic of the family, with the intercept a taking on various values according to species and environmental factors (temperature, salinity). This generalization is equivalent to saying that for salmonids G_{max} is proportional to $W^{0.6}$.

Since the compilation in Table III, Elliott (1975c) has shown that brown trout are characterized by a slope of -0.33 (5.6°C) to -0.28 (19.5°C) with a mean of -0.32 for all temperatures. This is close to the value for coho salmon (-0.34) determined by Stauffer (1973), and indicates a somewhat wider range of possible slopes.

2. Growth on Restricted Rations × Size

A few studies have been conducted where it is possible to develop more-or-less complete GR curves for different weight ranges under otherwise similar conditions. Hatanaka *et al.* (1957) grouped data on young mackerel into weight ranges of 7–24, 26–48, and 50–55 g. Depicting the GR relation as a straight line (no data plotted), the slope

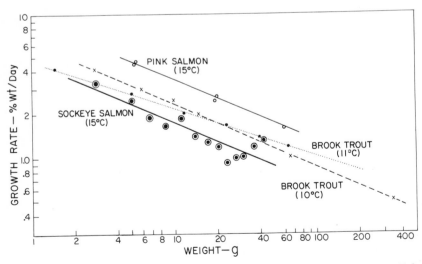

Fig. 22. The growth rate–weight relation for three salmonids. Brook trout (10°C) from Cooper (1961); brook trout (11°C) from Haskell (1959); pink salmon (15°C) from Brett (1974). (Figure from Brett and Shelbourn, 1975.)

Table III

Values of Parameters a and b for the Equation $\ln G = a + b(\ln W)$ [a]

Species	Weight range (g)	Temperature (°C)	Number (n)	Intercept (a)	Slope (b)	Source
Salvelinus fontinalis Brook trout	1.5–60	11.0	7	4.66	−0.33	Haskell (1959)
Salvelinus fontinalis Brook trout	2.5–350	10.0	8	6.49	−0.47	Cooper (1961)
Oncorhynchus nerka Sockeye salmon	1.0–30	15.0	9	5.58	−0.43	Brett *et al.* (1969); Shelbourn *et al.* (1973)
Oncorhynchus nerka Sockeye salmon	6.0–7.0	15.0	4	7.72	−0.49	Brett (1974)— wild stock
Oncorhynchus nerka Sockeye salmon	50–190	15.0	3	26.31	−0.69 [b]	Brett (1974)— cultured stock
Oncorhynchus gorbuscha Pink salmon	5.0–60	15.0	5	9.78	−0.45	Brett (1974)
Oncorhynchus nerka Sockeye salmon	0.3–75	15.5	22	5.42	−0.40	Brett (1974)— in Table IV, column 15.5°C
Oncorhynchus nerka Sockeye salmon	0.3–75	10.5	22	3.46	−0.39	Brett (1974)— in Table IV, column 10.5°C
Oncorhynchus kisutch Coho salmon	0.3–75	15.5	22	5.53	−0.34	Stauffer (1973)
Oncorhynchus kisutch Coho salmon	0.3–75	10.5	22	3.94	−0.3	Stauffer (1973)
Oncorhynchus nerka Sockeye salmon	3.0–45	15.0	13	4.47	−0.42	Brett and Shelbourn (1975)
Mean ± 2 SE					−0.43 ± 0.06	
Mean ± 2 SE [c]					−0.41 ± 0.04	

[a] G, specific growth rate (% weight/day) on maximum ration; W, weight (g). Size and source of various salmonid species indicated. From Brett and Shelbourn (1975).

[b] This stock noted as possibly atypical in Brett (1974).

[c] Mean calculated without value noted in b.

of each GR curve was shown to decrease with increasing weight. Maintenance ration (R_{maint}) fell, and the rate of loss of weight (G_{starv}) also decreased as size increased. The same general phenomenon is apparent in the studies of protein metabolism of bluegill sunfish, *Lepomis macrochirus* (Gerking, 1971). A weight-related decrease in slope between nitrogen retained versus nitrogen consumed (equivalent to GR curves) was shown to be accompanied by a decrease in the protein maintenance requirements (e.g., 0.36 mg of N/g/day for a 14-g fish decreasing to 0.26 mg of N/g/day for an 85-g fish).

By pooling the findings on sockeye salmon from a number of studies it was possible to display a set of curves for three mean sizes (Fig. 23). These confirm the general pattern of change in the GR curve as weight increases.

It can be seen that, in the interrelation of the curves, there is a similarity of patterns between size effect and temperature effect—increasing size shifts the curve down and to the left much as decreasing temperature does (below the optimum temperature). However, as will be developed under Section V, "Interaction and Optimizing," there is compelling evidence to classify size as a Limiting Factor rather than a Controlling Factor.

3. Scope for Growth (G_{scope}) × Size

Sparse as the information is, nevertheless it is apparent that the same approach to scope for growth ($R_{max} - R_{maint}$) which was developed for temperature effect can be applied to weight effect. As size increases, R_{max} falls rapidly; R_{maint} also decreases but at a slower rate than R_{max} (see Fig. 19). This results in a converging, so that G_{scope}

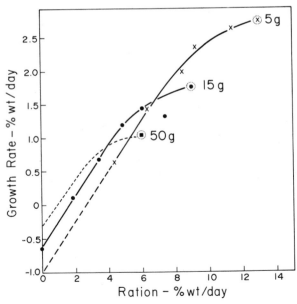

Fig. 23. The effect of size on the growth rate versus ration relation for young sockeye salmon. (Data for 5 and 15 g sizes from Brett *et al.*, 1969; Brett and Shelbourn, 1975.) Circled points representing G_{max} and R_{max} derived for each weight from Brett (1974). Only one point available for 50 g fish; expected curve drawn.

diminishes with size. An old mature fish is mainly eating for maintenance and gonad development; somatic growth is almost terminated. The diminution of growth promoting hormone that must accompany advanced age and size results in a decreased demand for food such that R_{max} approaches R_{maint}.

It is apparent that size is affecting both R_{max} and R_{maint} at the same time, and therefore operating to constrain *demand* rather than limit *supply*. This conveys a somewhat different mechanism than a simple Limiting Factor.

C. Competition

When considering the effects of abiotic factors on growth it was convenient to assume that each fish was responding as an independent entity, feeding and growing by itself without reference to the activities of any other member of the group. In truth, behavioral interaction can have great impact both masking and confounding the results of strictly physiological tests designed to seek the basic parameters of organismal growth. Perhaps the most extreme example of behavioral consequence is that of the Siamese fighting fish, *Betta splendens*, for which any close association is only resolved with the death of an opposing member. Less extreme cases of aggression have nevertheless imposed the necessity of raising each fish in a separate cell, out of visual or odor contact with its confrere (e.g., *Lepomis macrochirus*, Gerking, 1971; *Odontobutis obscurus*, Yamagishi *et al.*, 1974). The exact opposite may be true for some schooling species which are so restless and excitable away from the association of like members of similar size that feeding is disrupted and any determination of normal growth made impossible (e.g., *Clupea* sp., see Blaxter, 1970; Blaxter and Holliday, 1963). At this present stage in the elaboration of growth responses of fish it can be seen that behavioral involvements were undoubtedly the greatest obstacle in earlier research to achieving any clear understanding of environmental effects on growth. This is nowhere more apparent than in the treatise on fish physiology (Brown, 1957) where size hierarchial effects received prime attention and obviously entered into many reported experiments in a persistent, uncontrolled, if not unperceived fashion.

The interrelation of common members of a sample of fish is affected by *numbers, space, size,* and *species.* The relation that these factors bear to growth, and the extent to which they interact with each other is greatly influenced by *food availability.* In the face of the resulting competition for space and food, fishes have evolved various behavioral patterns involving defense and dominance, with their associated patrolling of territories and acts of aggression. Not only do these behavioral patterns require expenditure of energy, increasing daily maintenance requirements, but also there is evidence for suppression of growth in low-ranking members by intimidation (Wirtz, 1974) or possibly through water-borne inhibiting agents (cf. Richards, 1958; West, 1960; Yu, 1968).

When such behavioral interactions affect growth the relative difference in size of the members of a population usually increases—the large grow ever larger while the small lag further behind. This phenomenon has been called "growth depensation" and refers to the increase in variance of a size frequency distribution with time (see Chapter 11). In studies on growth variability, Yamagishi (1969) emphasized the importance of assessing the above parameter by the simple determination of the *coefficient of variation* [CV = $100 \times$ (SD/mean), in %]. In this way a measure of the degree of interaction can be assessed permitting greater insight of the factors in play; no change in CV with time signifies no significant interaction, or completely random behavioral interrelations.

Among the many experiments performed it is frequently difficult to distinguish the separate effects of numbers, space, and feeding opportunity. These factors are not infrequently correlated with each other; as numbers decrease, space and feeding opportunity increase. If the relative space configuration is maintained then the significance of numbers can be examined distinct from density, space, and configuration (e.g., two fish in a 2 liter cube versus 200 fish in a 200 liter cube). However, in the following consideration the latter distinction is not attempted.

(1) *Numbers (density)*. If it were not for behavioral relations, the number of fish that could be grown in a given volume would be directly proportional to the rate of exchange. Allen (1974) examined the growth obtained by stocking five densities (90–720 fish/m^3) of channel catfish at five different rates of flow. As the density increased, the mean weight achieved decreased. This was almost entirely due to decreasing oxygen concentrations which, as the numbers increased, fell progressively below the critical O_2 concentration level of 5 ppm.

When such obvious limitations are removed there appears to be an optimum density for some species of fish when on unrestricted rations. Brown (1946b) demonstrated this phenomenon among fingerling brown trout. At lowest densities (1 fish/50 liter) the fish did not feed as well and appeared to lack the social stimulation accompanying greater numbers. Highly crowded conditions (1 fish/3 liter) resulted in reduced food conversion efficiency and some physical interference between fish. Magnuson (1962) explored density-related growth depensation in populations of medaka, *Oryzias latipes,* after removal of the variability resulting from congenital differences. Increasing the population's size by four times, even when space was increased proportionately, reduced growth rate. This appeared to be associated with some reduction in food intake, leading Magnuson to conclude that neither a general depression in growth rate nor growth depensation occurred in this species if food was *effectively* present in excess.

More recent experiments have been conducted by Refstie and Kittelsen (1976), who followed the growth and mortality of Atlantic salmon reared from first-feeding fry to smolt at various levels of crowding. These authors noted that many studies on density effects have been performed at comparatively low concentrations, not representative of the conditions usually found in hatcheries. Using fry densities of five to 35 liters, only density effects on survival were apparent in the first 6 weeks, best survival occurring at highest density. Transferred to larger tanks at the parr stage for 30 weeks, the highest densities resulted in the lowest growth rates, without any mortalities. While it might be supposed that the main factor depressing feeding rates in this territorial species was social interaction involving dominance, no increase in the coefficient of variation occurred. It was concluded that free movement of the fish was inhibited, affecting food availability despite excess ration.

The question of how density affects growth rate when food is unrestricted appears to depend on the natural schooling relation of the species or the extent to which territorial behavior in some nonschooling species is suppressed by the sheer weight of numbers. Yamagishi (1963) observed that in a schooling race of crucian carp, *Carassius carassius,* growth depensation was significantly less than in two other nonschooling races when reared under similar conditions. This behavioral relation was further investigated in three contrasting marine species: (i) the halfbeak, *Hemirampus sajori,* a schooling fish, (ii) the red seabream, *Chrysophrys major,* a territorial fish, and (iii) the zebra sole, *Zebrias zebra,* which transforms from planktonic feeding to demersal or bottom'feeding. Among the halfbeak, growth depensation increased during earliest feeding but decreased as schooling developed. The seabream continued to increase in growth depensation, associated with aggression and cannibalism. The zebra sole displayed a sudden increase in the coefficient of variability on becoming bottom-feeders and establishing territories. These examples were considered by Yamagishi (1969) to be representative of three general types which are depicted in Fig. 24.

The shift with development to increased growth depensation of the zebra sole is reversed in the case of the ayu, *Plecoglossus altivelis,* a salmonlike fish that grazes on diatoms and algae growing on stones in rivers. As population density increases, the social structure of this species changes fron near-the-bottom territorial behavior to off-the-bottom schooling behavior (Kawanabe, 1969). Utilization of the limited food resource improves, and variability between individuals diminishes.

Density relations are of great importance in fish pond management in which carrying capacity defines the maximum weight of fish that can be sustained, and delimits the point at which no further growth is possible. Within the carrying capacity there is a high correlation between increasing density and decreasing individual growth (Hepher, 1967). In addition, it can be shown in ponds that there is a direct relation between weight gain and initial weight (for carp, Wohlfarth and Moav, 1972) although this is apparently only true during the warm, growing season. Such observations, although derived from well-designed "plot" comparisons and most meaningful in aquaculture, are the result of many interacting mechanisms which defy strict physiological interpretation.

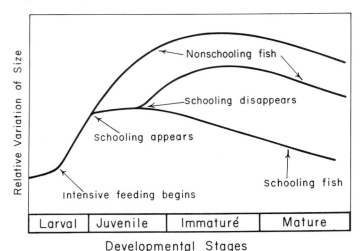

Fig. 24. General trend in relative variation of size (growth depensation) in fish characterized by different types of behavior, according to stages in development. (From Yamagishi, 1969, *Res. Popul. Ecol. (Kyoto)* **11,** 14–33.)

(2) *Competition for space.* That space, as a factor by itself, could affect growth was demonstrated in the case of female guppies, *Poecilia reticulata,* raised in aquaria of different sizes (Comfort, 1956; also cited in Brown, 1957). Greatest growth was achieved in the greatest space; if switched from small to large aquaria growth increased accordingly. Similarly, Allee *et al.* (1948) found that four green sunfish, *Lepomis cyanellus,* grouped in unobstructed aquaria containing 4 liters of water grew faster than single individuals in one-fourth that volume.

Keeping density constant (by volume), Yamagishi (1962) observed that growth rate of rainbow trout fry increased with an increase in the bottom area of the culture tanks. Fighting was intense, lasting from a fraction of a minute to 20 min. Growth was greatest when the total area exceeded the maximum size of territory that each fry could occupy.

In the presence of excess food distributed evenly, space per se was not found to be an influential factor for medaka (Magnuson, 1962). Growth depensation was no larger or smaller in populations of high density with less space per fish than among isolated control fish. By a comparison of treatments it was evident that aggressiveness in this species was a competitive mechanism for food and not for space. Even large fish had *no* competitive advantage over small fish.

(3) *Size and hierarchical effects.* Studies on the dominance interrelation between fish of different size and rank-order reveal two fundamentally different responses. Nagoshi (1967a,b) reared mixed populations of small and large guppies in 1-liter baskets using stratified rations of live zooplankton (*Cyclops vicinus*). On restricted rations a social hierarchy developed in which small fish were subordinate to large fish and comparatively suppressed in their growth rate. This difference persisted as ration increased, with only slight diminution up to an unrestricted level of feeding when size was no longer used to advantage (Fig. 25A). This may be compared with observations on the highly territorial and predacious goby, *Odontobutis obscurus,* which, when space is limited, attempt to dominate the competitor at all feeding levels (Yamagishi *et al.,* 1974). Starting with fish of almost equal size the lowest rank-order goby lost weight at the greatest rate when starving, and remained at half the G_{max} of the largest fish when prey-fish were provided in excess (Fig. 25B). The intensity of attacks by the first-ranking fish on subordinates was such that the second-ranking fish eventually surpassed the growth of the first-ranking fish without deposing this "ruler."

Kato and Sakamoto (1969) examined the effect of grading on growth rate and growth depensation among three groups of young

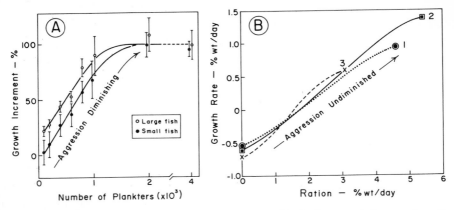

Fig. 25. (A) Relation of growth rate (% of maximum) to feeding rate in large and small guppies, *Poecilia reticulata*. (From Nagoshi, 1967a.) (B) Growth rate of gobies, *Odontobutis obscurus*, according to rank order indicated by numbers; fish No. 2 was smaller than No. 1 at start. (Data from Yamagishi *et al.*, 1974.) Aggression diminished with increased ration in A but not in B.

rainbow trout [initial sizes: large, 3.8 ± 0.13 cm (SD); medium, 3.6 ± 0.02 cm; small, 2.8 ± 0.15 cm]. Growth rate among groups declined almost uniformly with increasing size without any compensatory shift of the smaller fish, indicating true genetic differences. Depensation, however, increased in all three groups with the smallest fish exhibiting the largest change.

When food supply was limited for medaka, the larger fish were socially dominant, chasing smaller fish away from food and growing faster (Magnuson, 1962). If food was spatially localized the free-roaming hierarchial societies changed into territorial societies with the dominant defending the food areas. Magnuson concluded that aggressive behavior is a competitive mechanism that can provide the dominant animal with an advantage when food and space are limited. Aggressive behavior will disperse the competitors throughout the habitat only if food is found in all areas. Such relations would be expected to occur among fish that exhibit aggressive behavior in connection with food, in habitats containing contagiously distributed food limited in supply, and among fishes which live on the substrate or among aquatic vegetation.

In general, it can be seen that competition acts to restrict the food intake of subordinate fish (Limiting Factor—reducing R_{max}). It may be sufficiently demanding of energy expenditure to invoke a metabolic cost that significantly reduces growth (Masking Factor—elevating R_{maint}). The behavior of some species is such that competition only occurs when food is restricted (e.g., *Poecilia reticulata*); in others both Limiting and Masking Factors can reduce the scope for growth (e.g., *Odontobutis obscurus*).

D. Summary Configurations

Classifying biotic factors within the same scheme as abiotic factors (in terms of how they act) has proved possible, and reveals some new insights. None appears as a Controlling Factor, although size is not

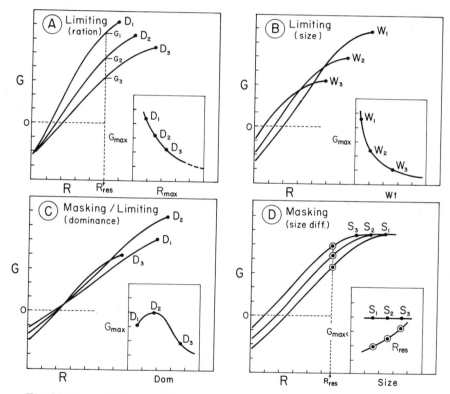

Fig. 26. Recapitulation of GR curves in relation to biotic entities, illustrating the basic forms for Limiting and Masking Factors, and the path of G_{max} according to the level of the biotic entity (boxed inserts). (A) illustrates the effects of diets of decreasing nutrient value (D_1–D_3) and the consequence of a restricted ration (R_{res}) limiting growth (G_1–G_3). (B) illustrates the effect of increasing weights (W_1–W_3) on the GR curve, and on G_{max}. (C) illustrates the effect of dominance order (D_1–D_3) requiring either increased energy expenditure (D_1), or limiting the food availability (D_3). (D) illustrates the case where size difference (S_1–S_3) between competing groups of fish only affects growth when food is limiting (R_{res}).

without some similarities (the "scaling" effect of Stauffer, 1973). Ration, an obvious Limiting Factor, has a feature of diet quality to be applied to the quantitative expression, decreasing the slope and extending the terminal point (R_{max}). Competition may evoke either one or both of Limiting and Masking effects.

These relations are assembled in Fig. 26, and bear comparison with the abiotic series (see Fig. 18). Again, it will be noted that there are only a few basic types of response of the GR curve. Ration, as a Limiting Factor, *categorically* acts like oxygen concentration by setting intermediate limits to growth, when not present in excess. Size, as will be shown, also acts to set limits on G_{max} only when ration is not limiting. Despite the complexity of behavioral relations they tend to be either Limiting by denying food to a competing member, or to be Masking by placing a metabolic burden of increased activity (attacking and/or defending) in a feeding territory.

V. INTERACTION AND OPTIMIZING

As stated in the introduction it is impossible to consider growth in relation to any one of the abiotic factors without introducing the biotic factor of *ration*. When food is present in excess (undocumented) then no account of the consequences of a given abiotic factor in relation to R_{max} and R_{maint}, and their derivative G_{scope}, is provided. Since an abiotic factor such as temperature affects maintenance metabolism it might be expected that temperature and ration would interact, affecting growth in some progressively shifting relation. Other cases of interaction have been indicated previously, including the combined effects of salinity and photoperiod, salinity and temperature, oxygen and ration, and size and ration. These couplets can be extended to multifactor combinations involving complex interactions for which the growth response could be described but which, by their very complexity, could well defy physiological explanation—a black box of stimuli producing unlimited patterns of growth response. Fortunately, the system of classifying the action of environmental factors lends insight to their combined effects.

The term "interaction" has been used so far in a general sense without any special connotation; "interrelation" could be substituted. Computational methods such as analysis of variance and covariance allow assessing the extent to which *statistical interaction* accounts for a portion of observed variability of a response. Factor combinations may also be assessed by response surface analysis. The essence of this method has been described by Alderdice (1972). *Response surface interaction* can be detected when the path of maximum response in relation to changing levels of two variables (x and y) moves along a line that is not parallel to either the x- or y-axis. The extent of this rotation is a measure of the degree of interaction. Thus, if the temperature for maximum growth shifts as ration decreases (see Fig. 27) then an *interaction* is present. This is the meaning implied henceforth. The path of maximum growth represents the optimum condition in terms of the two or more environmental factors considered; where applicable, the center of the whole configuration is the ultimate optimum.

Consideration will be given to the effects of environmental entities on growth in terms of their classified Factor relations.

A. Controlling × Limiting Factors

1. TEMPERATURE × RATION (R_{max} TO R_{maint})

Previous reference has been made to the studies on fingerling sockeye salmon in which six levels of temperature were combined with five to six levels of ration* in a multifactor experiment (Brett *et al.*, 1969). A plot of the growth rate versus temperature (Fig. 27A) reveals that as ration is reduced the optimum temperature for growth moves from 15°C for excess ration to 5°C on a restricted ration of 1.5% weight/day. In terms of surface response analysis these data are plotted as in Fig. 27B, with growth rates shown as interpolated isopleths. Each isopleth is terminated by the limits of food uptake (R_{max}) at the respective lower and upper temperature levels involved. The center appears to fall right at the peak of maximum ration (8% R at 15°C) or just outside the observed limits. A similar interaction resulting in a downward shift of the optimum temperature as ration was restricted has been obtained in preliminary experiments on growth of striped bass, *Morone saxitalis* (Cox, 1975, personal communication); Elliott (1975c) has confirmed the same phenomenon for brown trout.

Because the highest rations at any temperature do not produce the highest conversion efficiencies, the K isopleths follow a different pattern with a different center from maximum growth, as indicated in Fig. 27B. The rotation and curvature of the interaction "line" must obviously remain the same for both parameters, G_{opt} and K. The reason for the shift in optimum temperature was ascribed to the reduced demand for maintenance ration at lower temperatures, allowing a greater fraction of the available ration to be converted into growth. It was possible to conceive that the conversion efficiency of this available

* Except at 1° and 24°C where meals were not accepted over such a full range of rations.

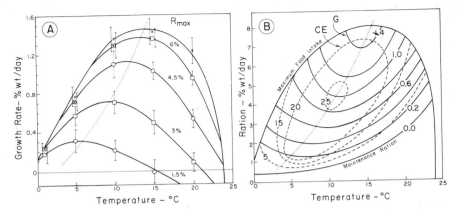

Fig. 27. Effect of restricted rations on growth rate and conversion efficiency in relation to temperature, for 13-g sockeye salmon. (From Brett *et al.*, 1969.) (A) illustrates the basic data as mean $G \pm 2$ SE for rations from maximum down to 1.5% weight/day, from which (B) the isopleths for growth rate (% weight/day) and for food conversion efficiency (CE* in %, broken lines) have been approximated. Note in A how the highest growth rate shifts to a lower temperature as ration is reduced, following the path of the dotted line (interaction), and in B that the optimum temperature for growth occurs at approximately 15°C on maximum ration, whereas the optimum temperature for the highest CE occurs close to 11°C on a ration of 4% per day (center of concentric broken lines). Dotted line in B forms the central axis of the isopleths passing through the respective centers for optimal growth rate and optimal conversion efficiency. This is comparable to the line for G_{opt} in Fig. 8.

fraction could be so reduced at lower temperatures that little growth benefit would result, or conversely that relatively high efficiency would be retained at *one* optimum temperature, suppressing any interaction. This appears to be the case for channel catfish which showed no change in optimum temperature (30°C) as ration was reduced although growth rate was greatly depressed ($3R \times 5T$ factorial experiment of Andrews and Stickney, 1972). This interpretation is supported by the generally high conversion efficiencies reported at all temperatures on a low ration of 2% weight/day, with the highest efficiency still remaining at 30°C.

2. TEMPERATURE × SIZE

The effect of these two factors acting together is taken into account in hatchery feeding charts (Haskell, 1959) which prescribe reduced ration (as % weight per day) with increasing size without altering the position of the optimum temperature. This interrelation is reflected in the growth rate table developed for sockeye (for sizes from 0.3 to 500

---

* Conversion efficiency represented by CE in (B) has been referred to as K in the text to conform with more common usage.

g), which depicts the highest growth rates remaining in the region of 15°–16°C *independent* of size (Brett, 1974). Brown trout display a similar relation, the optimum temperature for growth of all sizes tested occurring at 13°C (Elliott, 1975c). Lack of any interaction between temperature and size is, however, not the case for yellowtail, *Seriola quinqueradiata*, for which the optimum falls from 27°C for juveniles to 21°C for large adults (Oshima and Ihaba, 1969).

3. TEMPERATURE × OXYGEN

Because the metabolic rate is reduced by both low temperatures and low rations, the critical level of O_2 concentration permitting the necessary energy expenditure can be expected to be reduced by their combined influence. Furthermore, although it was pointed out that O_2 concentration was a better measure of oxygen availability than O_2 saturation, the latter is not without some influence in terms of the O_2 supply that saturation provides at different temperatures (e.g., 8 ppm O_2 = saturation at 25°C; 12 ppm O_2 = saturation at 6°C). Hence it can be predicted that reduced oxygen would affect the temperature promoting maximum growth in a combined fashion; as O_2 concentration falls the critical O_2 level permitting growth decreases with decreasing temperature, but not greatly because the necessary minimum O_2 tension (% saturation) must be maintained. This conjecture would result in an interaction configuration for limiting oxygen not unlike that for reduced rations (Fig. 27).

The only experiments performed that bear on this hypothesis are those reported by Doudoroff and Shumway (1970) on the unpublished data of Trent. Largemouth bass were acclimated to 10°, 15°, and 20°C and their growth rate determined at reduced oxygen concentrations. At the two lower temperatures the critical O_2 level was reduced from 5 ppm to about 3 ppm, supporting the above prediction (i.e., that growth is still possible at either a lowered oxygen concentration, or a lowered ration, only when temperature is reduced).

Thus the pacing effect of temperature, a Controlling Factor, raises or lowers the critical level of any Limiting Factor. The processes of growth are never free from temperature, which, as is true of any Controlling Factor, acts always in *conjunction*, never in sequence, with another environmental factor.

B. Controlling × Masking Factors

TEMPERATURE × SALINITY

As stated earlier, this combination of two environmental factors, one governing the rates of reaction (T, °C), the other exacting a toll of energy required for regulation (S, ‰), has been studied by Kinne

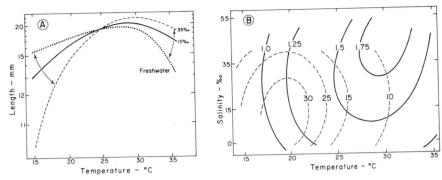

Fig. 28. Growth of the desert pupfish, *Cyprinodon macularius*, in relation to temperature at three salinities (35‰, 15‰, and freshwater). (A) from Kinne (1960). (B) illustrates isopleths at growth rates (% weight/day) with an overlay of conversion efficiencies (%, broken lines). (After Alderdice, 1972.)

(1960) on the desert pupfish, and by Peters and Boyd (1972) on the hogchoker. Only *ad libitum* feeding will be considered here.

For the pupfish, as temperature increased from a low level to a high level the accompanying salinity promoting best growth also increased from low to high, resulting in a "low-low/high-high" interaction. Alderdice (1972) analyzed these data showing that the center for maximum growth came at about 30°C at 40‰, while that for maximum conversion efficiency was close to 20°C and 15‰ (Fig. 28). Although the fish would voluntarily consume a large meal at 30°C and 40‰ (resulting in maximum growth) it was not converted with anywhere near the same efficiency as a smaller intake (R_{opt}) at lower temperature and salinity.

The hogchoker also displayed a significant temperature–salinity interaction, with the overall peak of maximum growth occurring at 25°C and 30‰, and the least recorded at 15°C and 0‰ (Peters and Boyd, 1972).

C. Limiting × Limiting Factors

RATION × SIZE

By definition, Limiting Factors cannot result in any interaction; they operate in series, like links in a chain. Optimum combinations do not exist; an ultimate maximum can occur where neither is limiting. By coupling these two entities together, ration and size, in an experiment at one temperature and one salinity it became obvious that size was acting in the role of a Limiting Factor when ration was not limit-

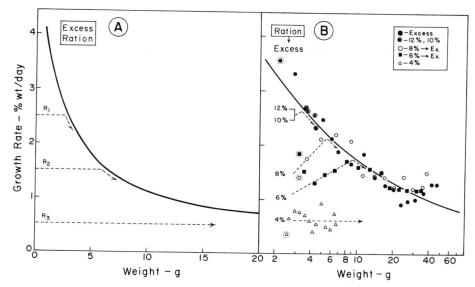

Fig. 29. The relation of growth rate to size (weight) on unrestricted and restricted rations (indicated as % weight/day). (A) depicts the model of expected responses where R_1–R_3 are limited rations. (B) displays the observed growth rates for young sockeye in freshwater at 15°C and 16 L. (From Brett and Shelbourn, 1975.) Note that weight is shown on a logarithmic scale in B, to separate points for small weights.

ing (Brett and Shelbourn, 1975). An initial high ration of 12% weight/ day fed to sockeye fry (2.4 g initial weight) soon became excessive, the growth rate falling progressively from 3.6% per day at 2.4 g to 1% per day at 37 g. This was accompanied by reduced voluntary food intake (Fig. 29). When lower levels of ration were provided *at the start* (e.g., 6 and 4% weight/day) growth rate remained constant within limits of normal variability* until reaching a limiting size at which point the growth rate followed the normal decline dictated by increasing size.

The same sort of interrelation of *ration* and *oxygen* acting as sequential Limiting Factors was depicted in Fig. 17.

D. Multifactor Effects

1. CONTROLLING × MASKING × LIMITING [$T \cdot (°C) \times S (°/oo) \times R$]

In an effort to determine how energy utilization was affected by temperature, salinity, and feeding in two estuarine fish, summer

---

* A case for possible slight increase in G can be made, which does not affect the conclusions.

flounder, *Paralichthys dentatus*, and southern flounder, *Paralichthys lethostigma*, Peters (1971) conducted a multifactorial experiment involving fifteen different treatment combinations. Five levels of ration were prescribed, from 30 to 90% of R_{max}. Since R_{max} changes with size and environmental conditions, separate, parallel determinations were made throughout the experiment to determine R_{max} at each of the temperature–salinity combinations. Fixed rations as % weight/day were therefore not used, but rather a sliding scale varying in *fixed proportion* to observed changes in R_{max}. This novel approach permitted maintaining the composite, three-variable design no matter how much or little the "parallel" control fish consumed; it does, however, add a complexity to interpretation in absolute values of ration.

Of the various configurations depicting the growth relations, one set of response surfaces has been selected (Fig. 30). These show that for *ad libitum* feeding, temperature as a Controlling Factor exerts most effect on growth. Salinity as a Masking Factor has relatively little effect but interacts in such a way that low salinities (5–15‰) at high temperatures (20°–30°C) have some suppressing effect on growth rate. The optimum combination lies outside the high temperature (30°C) × high salinity (35‰) corner of the "factor space" considered (Fig. 30C); G_{max} would be in excess of 16% weight/day. As ration is reduced, becoming limiting, growth rate decreases without altering the rotation of the temperature–salinity interaction axis. The center of optimum conditions can be seen to move such that at 60% of R_{max} it occurs at about 24°C and 21‰, with a G_{max} of close to 7% weight/day. In nature, as feeding opportunity was reduced, it would consequently be best for this species to seek intermediate temperature and salinity combinations.

Peters (1971) further notes that conversion efficiencies were gener-

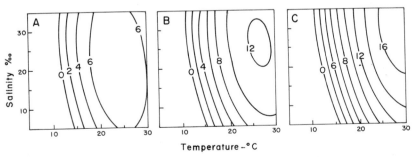

Fig. 30. Isopleths of growth rate (calories/kg/day) of *Paralychthys dentatus* in relation to salinity and temperature, for three levels of ration—(A) 60% maximum, (B) 80% maximum, (C) *ad libitum*. (From Peters, 1971.)

ally higher near 80% of R_{max} than either 60 or 100%. This would be predicted from the nature of the basic GR curve where R_{opt} is usually somewhat less than R_{max}. It would also be predicted that the limiting effect of reduced ration would be to slice the dome off the isopleth "mound" such that the lower growth rates (below 6% weight/day) would remain in a similar relation to temperature and salinity for all three configurations—which is approximately the case.

A somewhat different factorial experiment involving the same Factors but with age (= size?) as the Limiting Factor was conducted on postlarval striped bass, *Morone saxatilis*, by Otwell and Merriner (1975). Using a combination of three temperatures ($T = 12°$, $18°$, and $24°C$), three salinities ($S = 4$, 12, and 20‰) and six ages ($A = 5–63$ days), all factors and interactions were found to be significant ($P < 0.01$). Temperature, as a main factor, accounted for 83% of the variance in growth (length), with 5% attributable to age and less than 1% to salinity; first-order interaction ($T \times A$) was 3%, and second-order interaction ($T \times S \times A$) was 4% of the variability. High temperature ($24°C$) gave best growth under all circumstances; at this temperature the best combination was the younger ages (less than 28 days) and a salinity of 12‰.

2. CONTROLLING × MASKING × DIRECTING × LIMITING [T (°C) × S (‰) × L × W]

An experiment involving fixed levels of temperature × salinity × photoperiod × size has been attempted for sockeye salmon.* From an analysis of variance of the preliminary results, significant consequences of temperature on growth rate could be identified but salinity and photoperiod effects were not distinguishable from the error component, particularly at high temperatures. This was true for the two size ranges studied. It appeared that high temperature blocked the ability of photoperiod to induce smoltification, thereby eliminating potentially favorable effects of salinity.

The complexity of the system, the technical difficulties of exact control, and the need for constant attention to a large number of test tanks brought a temporary halt to this approach. Less involved combinations, using dynamic changes in photoperiod and temperature (increasing, decreasing, and static) were commenced. Preliminary results indicate that some combinations of temperature and photoperiod are synergistic; the magnitude of the photoperiod effect however is con-

---

* Unpublished records from annual reports of Pacific Biological Station, British Columbia.

siderably less than that of temperature. For sockeye of 2 g, the difference in growth rates between an increasing and a decreasing daylength was 0.44% weight/day, whereas a temperature increase from 10° to 17.5°C stimulated growth by 1.1% per day. For 4.5 g sockeye, the corresponding effects were 0.40 and 0.78% per day. A more dramatic photoperiod effect was achieved by applying photoperiod control over a period of 5 months. In this experiment, an unnatural "seasonal" cycle depressed growth rates by 1% per day. The effect of photoperiod on growth is apparently not directly related to number of hours of exposure to daylight. The direction of change in daylength is the important cue; stimulation of growth apparently occurs when the photoperiod cycle is somewhat in advance of the normal seasonal cycle.

In conclusion, these various examples serve to illustrate the forms of growth response when more than one environmental factor is involved. They undoubtedly take physiological enquiry a few steps nearer the complexity of natural environments. Also, as an exercise in the extension of the "Factors" approach to anabolic systems, involving both abiotic and biotic effects, it is apparent that justification of this system of classification is supported by virtue of the insight provided.

VI. GOVERNING MECHANISMS

It was noted at the outset that an integrated series of rate functions—feeding, assimilating, metabolizing, transforming, excreting—were the underlying mechanisms providing net energy gain and governing the rate of growth. Within the span of abiotic and biotic factors affecting these rates, temperature was identified as the only Controlling Factor, setting the pace of each function. The supply of food, oxygen, hormone, or size restrictions could act as Limiting Factors, or salinity could place some regulatory burden on the organism, but, given the supply, the power of temperature remained the greatest determinant of growth.

It is, therefore, unnecessary to seek further than the temperature relations of these metabolic functions to reveal just how they combine to regulate growth. Drawing on the various studies on sockeye salmon, the feeding, digesting, converting, and growing rates have been plotted on a common scale equating the maximum rate for each function to 100, that is, a percentage scale (Fig. 31A). The energy demands dictated by standard metabolism, feeding metabolism, and their relation to maintenance ration are depicted in Fig. 31B. From these a compos-

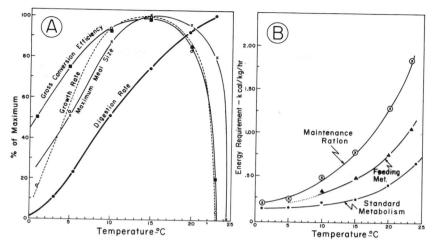

Fig. 31. The effect of temperature controlling various rate functions affecting the growth rate of young sockeye salmon. (A) illustrates the relative rates of feeding, digesting, growing and converting, plotted as a percentage of the maximum for each function. (From Brett and Higgs, 1970.) (B) illustrates the relation of upkeep to temperature, as standard metabolism, feeding metabolism, and maintenance ration, in terms of energy units (kcal/kg/hr). (From Brett, 1976.)

ite picture of the mechanisms inducing the rise and decline of maximum growth (G_{max}) with increasing temperature can be assembled. For clarity of presentation three temperature ranges will be considered: low, intermediate, and high.

a. Low Temperatures (1°–5°C)

Rate of digestion is greatly depressed, approaching zero in the region of 0°C. However, digestion shows some compensatory increase; young sockeye will grow at 1°C, consuming 1.5% weight/day. Gross conversion efficiency is only one-half the maximum, which reflects either reduced digestive efficiency or high nitrogenous excretion, or both. Rate of digestion appears to be the chief limiting factor to food consumption and hence to growth at low temperatures, with a concomitant reduction in appetite (daily meal size). Some seasonal increase in food intake may occur subsequently, indicating an adaptive capacity to take on energy and retain it for gradual digestion. All levels of metabolism (standard, routine, feeding) are a small fraction of their maximum, with the result that maintenance ration is exceedingly low (Fig. 31B).

b. Intermediate Temperatures (13°–17°C)

With rising temperature all feeding and growth functions (Fig. 31A) increase in roughly the same proportion except growth rate, which is relatively more rapid owing to an increased food conversion efficiency coupled with higher intake per meal. At 15°C an optimum is reached for three systems; only rate of digestion continues to rise. Furthermore, at 15°C a significant threshold is reached with meals being fully digested within 24 hr. Supporting metabolic rates have risen in accordance with an exponential effect of temperature, maintenance ration rising at the greatest rate.

c. High Temperatures (20°–24°C)

Rate of digestion continues to increase, abetting the ability to deal with multiple daily meals. However, appetite is reduced and conversion efficiency declines precipitously. As a result the sustained capacity to digest is not accompanied by growth, which follows conversion efficiency in a parallel decline. At 23°C high levels of standard metabolism, maintenance ration, and excretion all combine to suppress growth to the point of extinction.

VII. CONCLUDING COMMENTARY

Over the past two decades significant advances have been made in determining the effects of single and multiple environmental factors on the capacity of fish to grow. By analyzing the growth–ration (GR) response curves (where these have been recorded) an understanding of how the major parameters of growth (G_{max}, G_{opt}, G_0, G_{starv}) and of ration (R_{max}, R_{opt}, R_{maint}) respond to environmental factors has been afforded.

Conducting precise, well-controlled experiments on growth requires close attention to many aspects of biological significance; the experimenter is faced with a choice of feeding strategies and culturing conditions which can profoundly affect the outcome of the research. Food quality, quantity, and timing of presentation influence appetite, feeding reponse, and digestability; water currents, hours of light, space, density, and cover affect excitability and swimming activity thereby influencing the level of daily energy expenditure; exact measurements of food intake, fish weights, and moisture content are required before conversion efficiencies can be determined accurately. Since these normal concomitants may interact with the prescribed environmental conditions (e.g., the higher the temperature, the greater the excitability), knowledge of both the way in which environmental and behavioral factors affect growth helps to avoid pitfalls in the design of growth experiments.

Differences in behavioral response between and within tanks contributes greatly to variance, and will continue to be a challenging problem both as an object of study and as a target for improved experimental control. Considerably increased statistical sensitivity and insight can be obtained by distinctively marking each individual in a group. Assessing growth depensation is essential.

It is apparent that factorial experiments involving small numbers of combinations can provide confusing results, particularly where values of one or more of the factors occur on either side of an optimum. Interaction, rather than interrelation, will be most likely where a Controlling Factor is involved (e.g., temperature or pH). The consequence of a Masking Factor, such as salinity, is complicated by the state of neuroendocrine activity which in turn is subject to cyclic or ontogenetic change (e.g., smolting) under the influence of a Directive Factor (e.g., photoperiod, temperature cycles). The immature, stable stages of development not subject to change through differentiation (e.g., metamorphosis) are undoubtedly the best period in which to study the basic mechanisms of growth.

Since the channeling of food energy is involved, and maintenance requirements have first priority, it is fitting to reemphasize that bioenergetics and growth are inseparable—advances in each go hand in hand.

REFERENCES

Adelman, I. R., and Smith, L. L. (1970). Effect of oxygen on growth and food conversion efficiency of northern pike. *Prog. Fish Cult.* **32**, 93–96.

Alderdice, D. F. (1972). Factor combinations. Responses of marine poikilotherms to environmental factors acting in concert. *In* "Marine Ecology," Vol. 1, "Environmental Factors" (O. Kinne, ed.), Part 3, pp. 1659–1722. Wiley (Interscience), New York.

Allee, W. C., Greenberg, B., Rosenthal, G. M., and Frank, P. (1948). Some effects of social organization on growth in the green sunfish, *Lepomis cyanellus. J. Exp. Zool.* **108**, 1–19.

Allen, K. O. (1974). Effects of stocking density and water exchange rate on growth and survival of channel catfish *Ictalurus punctatus* (Rafinesque) in circular tanks. *Aquaculture* **4**, 29–40.

Altman, P. L., and Dittmer, D. S. (1966). "Environmental Biology." Fed. Am. Soc. Exp. Biol., Biol. Handbooks, Bethesda, Maryland.

Anderson, R. O. (1959). The influence of season and temperature on growth of the bluegill, *Lepomis machrochirus* (Rafinesque). Ph.D. Thesis, Univ. of Michigan, Ann Arbor.

Andrews, J. W., and Page, J. W. (1975). The effects of frequency of feeding on culture of catfish. *Trans. Am. Fish. Soc.* **104**, 317–321.

Andrews, J. W., and Stickney, R. R. (1972). Interactions of feeding rates and environmental temperature on growth, food conversion and body composition of channel catfish. *Trans. Am. Fish. Soc.* **101**, 94–99.

Andrews, J. W., Knight, L. H., and Murai, T. (1972). Temperature requirements for high density rearing of channel catfish *Ictalurus punctatus* from fingerling to market size. *Prog. Fish Cult.* **34**, 240–241.

Andrews, J. W., Murai, T., and Gibbons, G. (1973). The influence of dissolved oxygen on the growth of channel catfish. *Trans. Am. Fish. Soc.* **4**, 835–838.

Averett, R. C. (1969). Influence of temperature on energy and material utilization by juvenile coho salmon. Ph.D. Thesis, Oregon State Univ., Corvallis.

Baerends, G. P. (1971). The ethological analysis of fish behavior. *In* "Fish Physiology" (W. S. Hoar and D. J. Randall, eds.), Vol. 6, pp. 279–370. Academic Press, New York.

Banks, J. L., Fowler, L. G., and Elliott, J. W. (1971). Effects of rearing temperature on growth, body form, and hematology of fall chinook fingerlings. *Prog. Fish Cult.* **33**, 20–26.

Basulto, S. (1976). Induced saltwater tolerance in connection with inorganic salts in the feeding of Atlantic salmon (*Salmo salar* L.). *Aquaculture* **8**, 45–55.

Beamish, F. W. H. (1964). Influence of starvation on standard and routine oxygen consumption. *Trans. Am. Fish. Soc.* **93**, 103–107.

Beamish, F. W. H. (1974). Apparent specific dynamic action of largemouth bass, *Micropterus salmoides*. *J. Fish. Res. Board Can.* **31**, 1763–1769.

Bern, H. A. (1975). Prolactin and osmoregulation. *Am. Zool.* **15**, 937–948.

Bjorklund, R. G. (1958). The biological function of the thyroid and the effect of length of day on growth and maturation of goldfish, *Carassius auratus* Linn. Ph.D. Thesis, Univ. of Michigan, Ann Arbor.

Blackman, F. F. (1905). Optima and limiting factors. *Ann. Bot. (London)* **19**, 282–295.

Blaxter, J. H. S. (1970). Sensory deprivation and sensory input in rearing experiments. *Helgol. Wiss. Meeresunters.* **20**, 642–654.

Blaxter, J. H. S., and Holliday, F. G. T. (1963). The behavior and physiology of herring and other cluepids. *Adv. Mar. Biol.* **1**, 261–393.

Brett, J. R. (1971a). Satiation time, appetite, and maximum food intake of sockeye salmon, *Oncorhynchus nerka. J. Fish. Res. Board Can.* **28**, 409–415.

Brett, J. R. (1971b). Growth responses of young sockeye salmon (*Oncorhynchus nerka*) to different diets and planes of nutrition. *J. Fish. Res. Board Can.* **28**, 1635–1643.

Brett, J. R. (1974). Tank experiments on the culture of pan-size sockeye (*Oncorhynchus nerka*) and pink salmon (*O. gorbuscha*) using environmental control. *Aquaculture* **4**, 341–352.

Brett, J. R. (1976). Scope for metabolism and growth of sockeye salmon, *Oncorhynchus nerka*, and some related energetics. *J. Fish. Res. Board Can.* **33**, 307–313.

Brett, J. R., and Higgs, D. A. (1970). Effect of temperature on the rate of gastric digestion in fingerling sockeye. *J. Fish. Res. Board Can.* **27**, 1767–1779.

Brett, J. R., and Shelbourn, J. E. (1975). Growth rate of young sockeye salmon, *Oncorhynchus nerka*, in relation to fish size and ration level. *J. Fish. Res. Board Can.* **32**, 2103–2110.

Brett, J. R., Shelbourn, J. E., and Shoop, C. T. (1969). Growth rate and body composition of fingerling sockeye salmon, *Oncorhynchus nerka*, in relation to temperature and ration size. *J. Fish. Res. Board Can.* **26**, 2363–2394.

Brocksen, R. W., and Cole, R. E. (1972). Physiological responses of three species of fishes to various salinities. *J. Fish. Res. Board Can.* **29**, 399–405.

Brown, M. E. (1946a). The growth of brown trout (*Salmo trutta* Linn.). III. The effect of temperature on the growth of 2-year-old trout. *J. Exp. Biol.* **22**, 145–155.

Brown, M. E. (1946b). The growth of brown trout (*Salmo trutta* Linn.). II. Growth of 2-year-old trout at a constant temperature of 11.5°C. *J. Exp. Biol.* **22**, 130–144.

Brown, M. E. (1957). Experimental studies of growth. *In* "Physiology of Fishes" (M. E. Brown, ed.), Vol. 1, pp. 361–400. Academic Press, New York.

Brungs, W. A. (1971). Chronic effects of low dissolved oxygen concentrations on the fathead minnow (*Pimephales promelas*). *J. Fish. Res. Board Can.* **28**, 1119–1123.

Canagaratnam, P. (1959). Growth of fishes in different salinities. *J. Fish. Res. Board Can.* **16**, 121–130.

Canagaratnam, P. (1966). Growth of *Tilapia mossambica* Peters in different salinities. *Bull. Fish. Res. Stn., Ceylon* **19**, 47–50.

Carlson, A. R., and Siefert, R. E. (1974). Effects of reduced oxygen on the embryos and larvae of lake trout (*Salvelinus namaycush*) and largemouth bass (*Micropterus salmoides*). *J. Fish. Res. Board Can.* **31**, 1393–1396.

Carlson, A. R., Siefert, R. E., and Herman, L. J. (1974). Effects of lowered dissolved oxygen concentrations on channel catfish (*Ictalurus punctatus*) embryos and larvae. *Trans. Am. Fish. Soc.* **103**, 623–626.

Chervinski, J. (1961). Laboratory experiments on the growth of *Tilapia nilotica* in various saline concentrations. *Bamidgeh* **13**, 8–14.

Chervinski, J. (1975). Sea basses, *Dicentronchus labrax* (Linne) and *D. punctatus* (Bloch) (Pisces, Serranidae), a control fish in fresh water. *Aquaculture* **6**, 249–256.

Chiba, K. (1966). A study on the influence of oxygen concentration on the growth of juvenile common carp. *Bull. Freshwater Fish. Res. Lab. Tokyo* **15**, 35–47.

Comfort, A. (1956). "The Biology of Senescence." Routledge & Kegan Paul, London.

Cooper, E. L. (1961). Growth of wild and hatchery strains of brook trout. *Trans. Am. Fish. Soc.* **90**, 424–438.

Coutant, C. C., and Cox, D. K. (1975). Growth rates of subadult largemouth bass, 24–33.5°C. Environ. Sci. Div., Oak Ridge Natl. Lab., Oak Ridge, Tennessee.

Cox, D. K. (1975). Growth rate of striped bass, *Morone saxatilis*, as a function of temperature and ration. Annu. Rep. 1975. Environ. Sci. Div., Oak Ridge Natl. Lab., Oak Ridge, Tennessee. (Also personal communication.)

Cox, D. K., and Coutant, C. C. (1975). Growth–temperature response of striped bass, *Morone saxatilis*. Environ. Sci. Div., Oak Ridge Natl. Lab., Oak Ridge, Tennessee.

Davison, R. C., Breese, W. P., Warren, C. E., and Doudoroff, P. (1959). Experiments on the dissolved oxygen requirements of cold-water fishes. *Sewage Ind. Wastes* **31**, 950–966.

De Martini, E. E. (1969). A correlative study of the ecology and comparative feeding mechanism morphology of the Embiotocidae (Surf-fishes) as evidence of the family's adaptive radiation into available ecological niches. *Wasmann J. Biol.* **27**, 177–247.

De Silva, S. S., and Perera, P. A. B. (1976). Studies on the young grey mullet, *Mugil cephalus* L. I. Effects of salinity on food intake, growth and food conversion. *Aquaculture* **7**, 327–338.

Doudoroff, P., and Shumway, D. L. (1967). Dissolved oxygen criteria for the protection of fish. *Am. Fish. Soc. Spec. Publ.* No. 4, pp. 13–19.

Doudoroff, P., and Shumway, D. L. (1970). Dissolved oxygen requirements for freshwater fishes. *FAO Fish. Tech. Pap.* No. 86.

Ebeling, A. W., and Alpert, J. S. (1966). Retarded growth of the paradise fish *Macropodus opercularis* (L.) in low environmental oxygen. *Copeia* No. 3, pp. 606–610.

Eisler, T. (1957). The influence of light on the early growth of chinook salmon. *Growth* 21, 197–203.

Elliott, J. M. (1975a). Weight of food and time required to satiate brown trout, *Salmo trutta* L. *Freshwater Biol.* 5, 51–64.

Elliott, J. M. (1975b). Number of meals in a day, maximum weight of food consumed in a day and maximum rate of feeding for brown trout, *Salmo trutta* L. *Freshwater Biol.* 5, 287–303.

Elliott, J. M. (1975c). The growth rate of brown trout, *Salmo trutta* L., fed on maximum rations. *J. Anim. Ecol.* 44, 805–821.

Elliott, J. M. (1975d). The growth rate of brown trout (*Salmo trutta* L.) fed on reduced rations. *J. Anim. Ecol.* 44, 823–842.

Falk, K. (1968). Versuche zur Forellenmast in Küsten-und Binnengewässern. *Fisch.-Forsch. Wiss. Schriftenr.* 6, 93–98.

Fisher, R. J. (1963). Influence of oxygen concentration and of its diurnal fluctuations on the growth of juvenile coho salmon. M.S. Thesis, Oregon State Univ., Corvallis.

Fry, F. E. J. (1947). Effects of the environment on animal activity. *Univ. Toronto Stud. Biol. Ser.* 55, 1–62.

Fry, F. E. J. (1971). The effect of environmental factors on the physiology of fish. *In* "Fish Physiology" (W. S. Hoar and D. J. Randall, eds.), Vol. 6, pp. 1–98. Academic Press, New York.

Gerking, S. D. (1966). Annual growth cycle, growth potential, and growth compensation in the bluegill sunfish in northern Indiana lakes. *J. Fish. Res. Board Can.* 23, 1923–1956.

Gerking, S. D. (1971). Influence of rate of feeding and body weight on protein metabolism of bluegill sunfish. *Physiol. Zool.* 44, 9–19.

Gibson, M. B., and Hirst, B. (1955). The effect of salinity and temperature on the preadult growth of guppies. *Copeia* No. 3, pp. 241–243.

Glass, N. R. (1969). Discussion of calculation of power function with special reference to respiratory metabolism in fish. *J. Fish. Res. Board Can.* 26, 2643–2650.

Gross, W. L., Fromm, P. O., and Roelofs, E. W. (1963). Relationship between thyroid and growth in green sunfish, *Leopmis cyanellus* (Rafinesque). *Trans. Am. Fish. Soc.* 92, 401–408.

Gross, W. L., Roelofs, E. W., and Fromm, P. O. (1965). Influence of photoperiod on growth of green sunfish, *Lepomis cyanellus*. *J. Fish. Res. Board Can.* 22, 1379–1386.

Haskell, D. C. (1959). Trout growth in hatcheries. *N.Y. Fish Game J.* 6, 204–237.

Hatanaka, M. A., and Murakawa, G. (1958). Growth and food consumption in young amberfish, *Seriola quinqueradiata* (T. et S.). *Tohoku J. Agric. Res.* 9, 69–79.

Hatanaka, M. A., and Takahashi, M. (1956). Utilization of food by mackerel *Pneumatophorus japonicus* (Houttuyn). *Tohoku J. Agric. Res.* 7, 51–57.

Hatanaka, M. A., and Takahashi, M. (1960). Studies on the amounts of the anchovy consumed by the mackerel. *Tohoku J. Agric Res.* 11, 83–100.

Hatanaka, M. A., Sekino, K., Takahashi, M., and Ichimura, T. (1957). Growth and food consumption in young mackerel, (*Pneumatophorus japonicus*, Houttuyn). *Tohoku J. Agric. Res.* 7, 351–368.

Henderson, N. E. (1963). Influence of light and temperature on the reproductive cycle of the eastern brook trout, *Salvelinus fontinalis* (Mitchill). *J. Fish. Res. Board Can.* **20**, 859–897.

Hepher, B. (1967). Some biological aspects of warm-water fish pond management. *In* "The Biological Basis of Fresh Water Fish Production" (S. Gerking, ed.), pp. 417–428. Blackwell, Oxford.

Herrmann, R. B., Warren, C. E., and Doudoroff, P. (1962). Influence of oxygen concentration on the growth of juvenile coho salmon. *Trans. Am. Fish. Soc.* **91**, 155–167.

Hogman, W. J. (1968). Annulus formation on scales of four species of coregonids reared under artificial conditions. *J. Fish. Res. Board Can.* **25**, 2111–2112.

Hokanson, K. E. F., Kleiner, C. F., and Thorsland, T. W. (1977). Effects of constant temperature and diel fluctuation on growth, mortality, and yield of juvenile rainbow trout, *Salmo gairdneri* (Richardson). *J. Fish. Res. Board Can.* **34**, 639–648.

Holmes, W. N., and Donaldson, E. M. (1969). Excretion, ionic regulation, and metabolism. *In* "Fish Physiology" (W. S. Hoar and D. J. Randall, eds.), Vol. 1, pp. 1–89. Academic Press, New York.

Hubbs, C. L. (1930). The high toxicity of nascent oxygen. *Physiol. Zool.* **3**, 441–460.

Huh, H. T., Calbert, H. E., and Steiber, D. A. (1976). Effects of temperature and light on growth of yellow perch and walleye using formulated feed. *Trans. Am. Fish. Soc.* **105**, 254–258.

Huisman, E. A. (1974). A study on optimal rearing conditions for carp (*Cyprinus carpio* L.). Spec. Publ., Organisatie ter Verbetering van de Binnenvisserij, Utrecht.

Ishiwata, N. (1968). Ecological studies on the feeding of fishes. IV. Satiation curve. *Bull. Jpn. Soc. Sci. Fish.* **34**, 691–693.

Jansen, A. C. (1938). The growth of the plaice in the transition area. *Rapp. P.-V. Reun. Cons. Int. Explor. Mer* **108**, 104–107.

Kato, T. (1970). Studies on the variation of growth in rainbow trout, *Salmo gairdnerii*. II. Regression line of satiation amount on the body weight as an indicator of food amount. *Bull. Freshwater Fish. Res. Lab.* **20**, 101–107.

Kato, T., and Sakamoto, Y. (1969). Studies on the variation of growth in rainbow trout, *Salmo gairdnerii*. I. The effect of grading of body size on the course of growth. *Bull. Freshwater Fish. Res. Lab.* **19**, 9–16.

Kawanabe, H. (1969). The significance of social structure in production of the "ayu," *Plecoglossus altivelis*. *In* "Symposium on Salmon and Trout in Streams" (T. G. Northcote, ed.), pp. 243–251. Inst. Fish., Univ. of British Columbia, Vancouver.

Kepshire, B. M., Jr. (1971). Growth of pink, chum, and fall chinook salmon in heated seawater. *Proc. Annu. N.W. Fish Cult. Conf.*, *22nd* pp. 25–26.

Kepshire, B. M., Jr., and McNeil, W. (1972). Growth of premigratory chinook salmon in seawater. *U.S. Fish Wildl. Serv., Fish. Bull.* **70**, 119–123.

Kerr, S. R. (1971). Analysis of laboratory experiments on growth efficiency of fishes. *J. Fish. Res. Board Can.* **28**, 801–808.

Kilambi, R. V., Noble, J., and Hoffman, C. E. (1970). Influence of temperature and photoperiod on growth, food consumption, and food conversion efficiency of channel catfish. *Proc. Annu. Conf. Southeast. Assoc. Game Fish Comm., 24th* pp. 519–531. (*Aquat. Sci. Fish. Abstr.* 4, 4Q5139F, p. 231.)

Kinne, O. (1960). Growth, food intake, and food conversion in a euryplastic fish exposed to different temperatures and salinities. *Physiol. Zool.* **33**, 288–317.

Kleiber, M. (1961). "The Fire of Life. An Introduction to Animal Energetics." Wiley, New York.

Knutsson, S., and Grav, T. (1976). Seawater adaptation in Atlantic salmon (*Salmo salar*

L.) at different experimental temperatures and photoperiods. *Aquaculture* **8**, 169–·
   187.

Kono, H., and Nose, Y. (1971). Relationship between the amount of food taken and
   growth in fishes. I. Frequency of feeding for a maximum daily ration. *Bull. Jpn. Soc.
   Sci. Fish.* **37**, 169–175.

Kwain, W.-H. (1975). Embryonic development, early growth, and meristic variation in
   rainbow trout (*Salmo gairdneri*) exposed to combinations of light intensity and
   temperature. *J. Fish. Res. Board Can.* **32**, 397–402.

Larkin, P. A., Terpenning, J. G., and Parker, R. R. (1956). Size as a determinant of growth
   rate in rainbow trout *Salmo gairdneri*. *Trans. Am. Fish. Soc.* **86**, 84–96.

Lindsey, C. C. (1958). Modification of meristic characters by light duration in kokane,
   Oncorhynchus nerka. *Copeia* No. 2, pp. 134–136.

McCormick, J. H. (1976). Temperature effects on young yellow perch, *Perca flavescens*
   (Mitchell). Ecol. Res. Ser., U.S. Environ. Prot. Agency, Duluth, Minnesota.

McCormick, J. H., Jones, B. R., and Syrett, R. F. (1971). Temperature requirements for
   growth and survival of larval ciscos (*Coregonus artedii*). *J. Fish. Res. Board Can.* **28**,
   924–927.

McCormick, J. H., Hokanson, K. E. F., and Jones, B. R. (1972). Effects of temperature on
   growth and survival of young brook trout, *Salvelinus fontinalis*. *J. Fish. Res. Board
   Can.* **29**, 1107–1112.

McCormick, J. H., Jones, B. R., and Hokanson, K. E. F. (1977). White sucker (*Catostomus commersoni*) embryo development, and early growth and survival at different temperatures. *J. Fish. Res. Board Can.* **34**, 1019–1025.

MacCrimmon, H. R., and Kwain, W. -H. (1969). Influence of light on early development
   and meristic characters in the rainbow trout, *Salmo gairdneri* Richardson. *Can. J.
   Zool.* **47**, 631–63.

Magnuson, J. J. (1962). An analysis of aggressive behavior, growth, and competition for
   food and space in medaka, *Oryzias latipes* (Pisces, Cyprinodontidae). *Can. J. Zool.*
   40, 313–363.

Nagoshi, M. (1967a). Experiments on the effects of size hierarchy upon the growth of
   guppy (*Lebistes reticulatus*). *J. Fac. Fish. Prefect. Univ. Mie* **7**, 165–189.

Nagoshi, M. (1967b). On the effects of size hierarchy upon the growth of fishes. *J. Fac.
   Fish. Prefect. Univ. Mie* **7**, 191–198.

Needham, A. E. (1964). "The Growth Process in Animals." Pitman London.

Nelson, D. J. (1974). Temperature effects on growth of largemouth bass. Annu. Prog.
   Rep., pp. 28–29. Environ. Sci. Div., Oak Ridge Natl. Lab., Oak Ridge, Tennessee.

Nelson, D. J. (1975). Growth, consumption, and conversion rates as a function of temperature of subadult striped bass. Annu. Prog. Rep., pp. 45–47. Aquat. Stud., Environ.
   Sci. Div., Oak Ridge Natl. Lab., Oak Ridge, Tennessee.

Niimi, A. J., and Beamish, F. W. H. (1974). Bioenergetics and growth of largemouth bass
   (*Micropterus salmoides*) in relation to body weight and temperature. *Can. J. Zool.*
   52, 447–456.

Nose, T. (1963). Determination of nutritive value of food protein on fish. II. Effect of
   amino acid composition of high protein diets on growth and protein utilization of
   the rainbow trout. *Bull. Freshwater Fish. Res. Lab.* **13**, 41–50.

Novotny, A. J. (1975). Net-pen culture of Pacific salmon in marine waters. *Mar. Fish.
   Rev.* **37**, 36–47.

Oshima, Y., and Ihaba, D. (1969). "Fish Culture," Vol. 4, "Yellowtail—Amber Jack."
   Publ. Midori-Shobo, Tokyo.

Otto, R. G. (1971). Effects of salinity on the survival and growth of pre-smolt coho
   salmon (*Oncorhynchus kisutch*). *J. Fish. Res. Board Can.* **28**, 343–349.

Otwell, W. S., and Merriner, J. V. (1975). Survival and growth of juvenile striped bass, *Morone saxatilis,* in a factorial experiment with temperature, salinity and age. *Trans. Am. Fish. Soc.* **104,** 560–566.

Paloheimo, J. E., and Dickie, L. M. (1966). Food and growth of fishes. III. Relations among food, body size, and growth efficiency. *J. Fish. Res. Board Can.* **23,** 1209–1248.

Pentelow, F. T. K. (1939). The relation between growth and food consumption in the brown trout (*Salmo trutta*). *J. Exp. Biol.* **16,** 446–473.

Pessah, E., and Powles, P. M. (1974). Effect of constant temperature on growth rates of pumpkinseed sunfish (*Lepomis gibbosus*). *J. Fish. Res. Board Can.* **31,** 1678–1682.

Peters, D. S. (1971). Growth and energy utilization of juvenile flounder, *Paralichthys dentatus* and *Paralichthys lethostigma,* as affected by temperature, salinity, and food availability. Ph.D. Thesis, Dep. Zool., North Carolina State Univ., Raleigh.

Peters, D. S., and Boyd, M. T. (1972). The effect of temperature, salinity, and availability of food on the feeding and growth of the hogchoker, *Trinectes maculatus* (Block and Schneider). *J. Exp. Mar. Biol. Ecol.* **9,** 201–207.

Pyle, E. A. (1969). The effect of constant light or constant darkness on the growth and sexual maturity of brook trout. *Fish. Res. Bull.* No. 31, pp. 13–19.

Rasquin, P., and Rosenbloom, L. (1954). Endocrine imbalance and tissue hyperplasia in teleosts maintained in darkness. *Bull. Am. Mus. Nat. Hist.* **104,** 361–425.

Refstie, T., and Kittelsen, A. (1976). Effect of density on growth and survival of artificially reared Atlantic salmon. *Aquaculture* **8,** 319–326.

Richards, C. M. (1958). The inhibition of growth in crowded *Rana pipiens* tadpoles. *Physiol. Zool.* **31,** 138–151.

Saunders, R. L., and Henderson, E. B. (1969). Survival and growth of Atlantic salmon parr in relation to salinity. *Fish. Res. Board Can. Tech. Rep.* No. 147.

Saunders, R. L., and Henderson, E. B. (1970). Influence of photoperiod on smolt development and growth of Atlantic salmon (*Salmo salar*). *J. Fish. Res. Board Can.* **27,** 1295–1311.

Shaw, H. M., Saunders, R. L., Hall, H. C., and Henderson, E. B. (1975a). The effect of dietary sodium chloride on growth of Atlantic salmon (*Salmo salar*) parr. *J. Fish. Res. Board Can.* **32,** 1813–1819.

Shaw, H. M., Saunders, R. L., and Hall, H. C. (1975b). Environmental salinity: its failure to influence growth of Atlantic salmon (*Salmo salar*) parr. *J. Fish. Res. Board Can.* **32,** 1821–1824.

Shehadeh, Z. H., and Gordon, M. S. (1969). The role of the intestine in salinity adaptation of the rainbow trout, *Salmo gairdneri. Comp. Biochem. Physiol.* **30,** 397–418.

Shelbourn, J. E. (1976). Early growth rates of chum salmon fry (*Oncorhynchus keta*) in the laboratory in fresh and salt water. Unpublished manuscript, Pacific Biological Station, Nanaimo, British Columbia.

Shelbourn, J. E., Brett, J. R., and Shirahata, S. (1973). Effect of temperature and feeding regime on the specific growth rate of sockeye salmon fry (*Oncorhynchus nerka*), with a consideration of size effect. *J. Fish. Res. Board Can.* **30,** 1191–1194.

Smith, M. A. K., and Thorpe, A. (1976). Nitrogen metabolism and trophic input in relation to growth in freshwater and saltwater. *Biol. Bull.* (*Woods Hole, Mass.*) **150,** 139–151.

Stauffer, G. D. (1973). A growth model for salmonids reared in hatchery environments. Ph.D. Thesis, Univ. of Washington, Seattle.

Stewart, N. E., Shumway, D. L., and Doudoroff, P. (1967). Influence of oxygen concentration on the growth of juvenile largemouth bass. *J. Fish. Res. Board Can.* **24,** 475–494.

Strawn, K. (1961). Growth of largemouth bass fry at various temperatures. *Trans. Am. Fish. Soc.* **90**, 334–335.

Swift, D. R. (1955). Seasonal variations in the growth rate, thyroid gland activity, and food reserves of brown trout, (*Salmo trutta* Linn.). *J. Exp. Biol.* **32**, 751–764.

Swift, D. R. (1959). Seasonal variation in the activity of the thyroid gland of yearling brown trout (*Salmo trutta* Linn.). *J. Exp. Biol.* **36**, 120–125.

Swift, D. R. (1960). Cyclical activity of the thyroid gland of fish in relation to environment changes. *Symp. Zool. Soc. London* No. 1, pp. 17–27.

Swift, D. R. (1961). The annual growth rate cycle in brown trout (*Salmo trutta* Linn.) and its cause. *J. Exp. Biol.* **38**, 595–604.

Swift, D. R. (1963). Influence of oxygen concentration on growth of brown trout, *Salmo trutta* L. *Trans. Am. Fish. Soc.* **92**, 300–301.

Swift, D. R. (1964). The effect of temperature and oxygen on the growth rate of the Windermere char (*Salvelinus alpinus willughbii*). *Comp. Biochem. Physiol.* **12**, 179–183.

Thompson, D. H. (1941). The fish production of inland streams and lakes. *Symp. Hydrobiol., Univ. Wis., Madison* pp. 206–217.

Ursin, E. (1963). On the incorporation of temperature in the von Bertalanffy growth expression. *Medd. Dan. Fisk.- Havunders.* **4**, 1–16.

Wagner, H. H. (1974). Photoperiod and temperature regulation of smolting in steelhead trout (*Salmo gairdneri*). *Can. J. Zool.* **52**, 219–240.

Warren, C. E. (1971). "Biology and Water Pollution Control." Saunders, Philadelphia, Pennsylvania.

Warren, C. E., and Davis, G. E. (1967). Laboratory studies on the feeding bioenergetics and growth of fishes. *In* "The Biological Basis of Freshwater Fish Production" (S. D. Gerking, ed.), pp. 175–214. Blackwell, Oxford.

Weatherley, A. H. (1972). "Growth and Ecology of Fish Populations." Academic Press, New York.

West, B. W. (1965). Growth, food conversion, food consumption, and survival at various temperatures, of the channel catfish *Ictalurus punctatus* (Rafinesque). M.S. Thesis, Univ. of Arkansas, Fayetteville.

West, B. W. (1966). Growth rates at various temperatures of the orange-throat darter *Etheostoma spectabilis. Proc. Ark. Acad. Sci.* **20**, 50–53.

West, L. B. (1960). The nature of growth inhibiting material from crowded *Rana pipiens* tadpoles. *Physiol. Zool.* **33**, 232–239.

Whitworth, W. R. (1968). Effects of diurnal fluctuations of dissolved oxygen on the growth of brook trout. *J. Fish. Res. Board Can.* **25**, 579–584.

Wirtz, P. (1974). The influence of the sight of a conspecific on the growth of *Blennius pholis* (Pisces, Teleostei). *J. Comp. Physiol.* **91**, 161–165.

Wohlfarth, G. W., and Moav, R. (1972). The regression of weight gain on initial weight in carp. I. Methods and results. *Aquaculture* **1**, 7–28.

Wurtsbaugh, W. A. (1973). Effects of temperature, ration, and size on the growth of juvenile steelhead trout, *Salmo gairdneri*. M.S. Thesis, Oregon State Univ., Corvallis.

Yamagishi, H. (1962). Growth relation in some small experimental populations of rainbow trout fry, *Salmo gairdneri* Richardson, with special reference to social relations among individuals. *Jpn. J. Ecol.* (*Nippon Seitai Gakkaishi*) **12**, 43–53.

Yamagishi, H. (1963). Some observations on growth variation and feeding behavior in the fry of two races of Japanese crucian carp, *Carassius carassius* L. *Jpn. J. Ecol.* (*Nippon Seitai Gakkaishi*) **13**, 156–161.

Yamagishi, H. (1969). Postembryonal growth and its variability of the three marine fishes with special reference to the mechanism of growth variation in fishes. *Res. Popul. Ecol. (Kyoto)* **11**, 14–33.

Yamagishi, H., Maruyama, T., and Mashiko, K. (1974). Social relation in a small experimental population of *Odontobutis obscurus* (Temminck et Schlegel) as related to individual growth and food intake. *Oecologia (Berlin)* **17**, 187–202.

Yu, M.-L. (1968). A study on the growth inhibiting factors of zebra fish, *Brachydanio rerio*, and blue gourami, *Trichogaster trichopterus*. Ph.D. Thesis, Dep. Biol., New York Univ., New York.

Zaugg, W. S., and McLain, L. R. (1969). Inorganic salt effects on growth, saltwater adaptation, and gill ATPase of Pacific salmon. *In* "Fish in Research" (O. W. Newhaus and J. E. Halver, ed.), pp. 293–306. Academic Press, New York.

11

GROWTH RATES AND MODELS

W. E. RICKER

FISH PHYSIOLOGY, VOL. VIII
Copyright © 1979 by Academic Press, Inc.
All rights of reproduction in any form reserved.
ISBN 0-12-350408-2

I. MEASURING LENGTH AND WEIGHT OF FISH

A. Methods of Measuring Length

Fish lengths have been measured in many different ways (Fig. 1). The differences arise from choosing different reference points near the

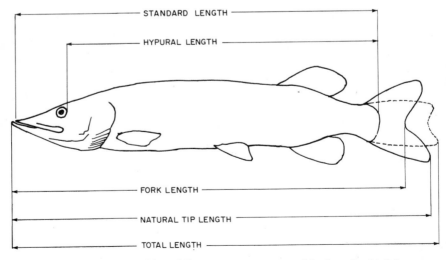

Fig. 1. Definition of five different measurements of the length of a fish.

anterior end and near the posterior end of the fish, and from using different methods of making the measurement.

Anterior reference points (A) include (1) tip of the snout; (2) tip of the snout or of the lower jaw when the mouth is closed, whichever protrudes farther; (3) anterior margin of the orbit of the eye; (4) midline of the orbit; (5) posterior margin of the orbit.

Posterior reference points (P) include (1) end of the scale covering of the body; (2) hind margin of the hypural bone of the tail (usually located by the position of a crease when the tail is bent a little); (3) tip of the shortest median ray of the tail fin; (4) tip of the longest lobe of the tail when held in a natural position; (5–7) tip of the upper (5), the lower (6), or the longer (7) lobe of the tail when squeezed into a position of maximum extension.

Methods of making the measurement (M) include (1) using calipers; (2) using a tape held along the curve of the body; (3) laying the fish on a measuring board with the front end pressed against an upright piece; (4) laying the fish on a board with a movable cross hair above it, attached to an indicator running along a scale.

In theory any combination of reference points and methods might be used, but practice is considerably more restricted. Some of the commoner combinations have special names; these are given below, together with the reference points used.

Standard length: A1, P2, and M1 or M4. Used mainly by systematists.

Median length or fork length (formerly often called total length): A2, P3, and M3 or M4. Widely used by fishery biologists for both marine and freshwater fishes.

Total length, extreme tip length: A2, P7, and M3 or M4. Widely used for freshwater fishes in the United States, and the usual "legal" length measurement there.

Postorbital–hypural length (also called simply hypural length): A5, P2, M4. Used recently by salmon biologists in British Columbia to avoid confusion caused by elongation of the snout and fraying of the tail at maturity. Anterior points A3 and A4 have been used elsewhere for the same purpose, but the hind margin of the orbit is solid bone and makes a more definite reference point.

Natural tip length: A2, P4, and M3 or M4. This measure is often used in Europe, but for every fish it is necessary to decide what tail position is natural, and different observers tend to make different choices.

In addition to the above methodological differences, length will vary with the condition of the fish, for example, whether it is alive, recently killed, after rigor mortis has set in, or at different intervals of time after preservation in formalin or alcohol.

B. Methods of Obtaining Fish Weights

In this chapter the terms mass, biomass, and weight are used interchangeably for the whole weight of a fish. There are differences in the methods and conditions under which weight is determined, just as for length. The usual procedure is to weigh the fish whole, either alive or after death. If it is alive, or preserved in liquid, some standard drip period or amount of blotting should be adopted in each experiment, but I know of no general rules about this. Also, weight can change somewhat after death with exposure to air, and also on preservation and afterward.

Stomach contents often contribute substantially to variability in weight, but usually no attempt is made to adjust for this. However, when cultured fish are weighed, greater uniformity can be achieved by always doing it at the same time of day, and the same time after the last feeding.

Another big source of variability in weight of adult fish is the seasonal sexual cycle. Particularly in females, gonad weight when the current year's eggs are nearing maturity is far greater than when the ovary is in the resting condition.

In weighing commercial catches it is sometimes necessary to use eviscerated fish, with or without the gills left in, and with or without the head left on.

C. Conversions between Length or Weight Measurements

Choice of a length must be governed by convenience and by custom, and as much uniformity as possible is desirable; nevertheless there will always be a need for conversion from one system to another. Pairs of measurements from the same fish are compared graphically or by means of a regression equation. Some important considerations here as follows.

1. Comparisons should include as wide a range of sizes as possible. Failing this, very large numbers of fish must be measured in order to obtain a reliable conversion line.
2. If X and Y represent the two measurements being compared, the regression line should pass through the point $\bar{X}, \bar{Y}$—the means of the two types of measurements—and its slope should be a functional regression, as defined in the next section. This line is symmetrical with respect to X and Y, and can be used to convert X measurements to Y, or Y to X.

3. If the intercept of the functional regression line does not differ significantly from the origin (as is usually the case), it is desirable to use the line that joins the origin to $(\bar{X}, \bar{Y})$ for conversion purposes. This means using a simple factor, which is a great computational convenience.

Conversions between different types of weights should follow the same procedure as for lengths.

D. Estimation of Weight from Length, and Length from Weight

It has been found that, within any stanza of a fish's life, the weight varies as some power of length:

$$w = al^b \tag{1}$$

$$\log w = \log a + b(\log l) \tag{2}*$$

These expressions would apply best to an individual fish that is measured and weighed in successive years of its life. This of course is rarely possible. The value of b is usually determined for a population, by plotting the logarithm of weight against the logarithm of length for a large number of fish of various sizes, the slope of the fitted line being an estimate of b. As in converting between lengths, and indeed in all biological situations where variability is for the most part inherent in the material rather than a result of errors in measuring, a "functional" line should be used: that is, the two variates must be treated symmetrically (Ricker, 1973, 1975b). Putting $\log l = X$ and $\log w = Y$, a least-squares functional line can be obtained by first dividing the values of X and Y by their respective standard deviations; when these transformed data are plotted so that one unit occupies the same distance on both axes, the functional regression line minimizes the sum of squares of the distances from the observed points to itself, and it has a slope of 1. However, it is not necessary to go to the trouble of transforming the data in this way, for it was shown by Teissier (1948) that the slope of the functional line is simply the ratio of the standard deviations of Y and X—surprising as this may seem. It is also the geometric mean of the ordinary regression of Y on X and the reciprocal of the regression of X on Y (or vice versa) and for that reason it has been called the geometric mean (GM) functional regression; other names

* The symbol ln means natural logarithm; log is used when either natural or base-10 logarithms may be employed.

are standard major axis (Jolicoeur, 1975) and reduced major axis (Teissier, 1948; Kermack and Haldane, 1950; Imbrie, 1956). The GM functional regression is also equal to the ordinary regression of Y on X divided by the coefficient of correlation between Y and X.

The two regressions are compared in Fig. 2. When variability is small, and the range of lengths and weights available is large, there is not much difference between the functional line and an ordinary regression line. But when, as in Fig. 2, neither of these conditions obtains, the difference can be considerable. In any event, the ordinary regression is always numerically smaller than the functional regression, and it is well to avoid any kind of consistent bias, however small.

The functional regression can be expressed either as the change in log weight per unit change in log length, or as the change in log length per unit change in log weight, but exactly the same line is specified in either case; thus the same line is used to convert from weight to length or from length to weight. This contrasts with the ordinary regression of

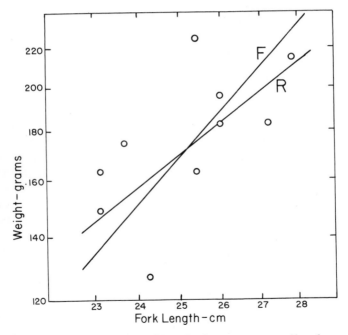

Fig. 2. Relation between weight and length of ten specimens of brook trout, plotted on logarithmic axes. The slope of the ordinary regression of weight on length (R) is 1.955, while that of the GM line (F) is 2.816. The latter is much nearer the true relationship, which is close to 3 in this species.

log weight on log length, which is a different line from the regression of log length on log weight.

When data are extensive they have sometimes been grouped into short length-classes, and the mean length and weight of each class have been used as the primary data in computing the regression line. The motive is to speed up computations, but with modern equipment this is no longer necessary, and when using means it is impossible to compute a representative standard error, nor can the functional regression be computed.

A more legitimate reason for grouping data is to distribute the observations more evenly among the range of sizes present, and so get a more representative relationship. This is best done by measuring some fixed number of fish within each of a series of short length intervals and weight intervals. The intervals used should preferably be in logarithms. However, to obtain a representative functional regression it is necessary to select half of the total sample on the basis of length and the other half on the basis of weight—otherwise there will be bias (Ricker, 1973).

When a general weight–length relationship for a population is desired, every effort should be made to obtain fish of a wide range of sizes, down to and including age 0 (unless of course the younger fish belong in a different growth stanza). When only a short range of fish sizes is available the parameters estimated can deviate importantly from the population values simply from sampling variability.

E. Numerical Representation of Growth

Growth may be described in terms of length (l) or weight (w), but for simplicity we will set down the expressions for weight only. Any measure of growth must be referred to some definite interval of time, either expressed or implied. It is when this interval is equal to one unit of whatever measure of time is being employed (days, months, years) that it becomes most appropriate to speak of a *growth rate*, although there has been no great consistency in this usage.

For an interval of time from t_1 to t_2 we may distinguish the following.

1. Absolute growth (or absolute increment), and *absolute growth rate:*

$$w_2 - w_1 \quad \text{and} \quad \frac{w_2 - w_1}{t_2 - t_1} \tag{3}$$

2. Relative growth and *relative growth rate* [often associated with the name of Minot (1891)]:

$$\frac{w_2 - w_1}{w_1} \quad \text{and} \quad \frac{w_2 - w_1}{w_1(t_2 - t_1)} \tag{4}$$

Special cases of the above rates occur when they refer to a particular instant of time rather than an interval. The absolute growth rate is then represented by dw/dt, and the relative growth rate becomes $(dw/dt)/w = dw/wdt = G$. The most appropriate name for G would be the instantaneous relative growth rate, but this is usually shortened to *instantaneous growth rate*, and is so used in this chapter. Other names are the *specific, intrinsic, exponential, logarithmic,* or *compound interest* rate. In biology this measure of growth is often associated with the name of Schmalhausen (1926). Its great advantage is that it is additive.

When growth of a fish continues at a constant instantaneous rate for a finite interval of time, its size at any time during that interval is described by an exponential curve having the formula

$$w = ae^{Gt} \tag{5}$$

where a is initial size (when $t = 0$). The absolute rate of growth at any time t during the interval is described by the slope of Eq. (5) at time t:

$$\frac{dw}{dt} = aGe^{Gt} \tag{6}$$

The constant instantaneous rate of growth is of course dw/wdt, or Eq. (6) divided by Eq. (5), which is G.

In actual work growth must be measured over an interval of time rather than at a particular instant. Let w_1 and w_2 be the weights of a fish at times t_1 and t_2; substitute each of these pairs of values in Eq. (5), take logarithms, subtract, and transpose; this gives

$$\frac{\ln w_2 - \ln w_1}{t_2 - t_1} = G \tag{7}$$

Equation (7) is the one that is usually used in practice to estimate the instantaneous rate of growth. Sometimes G is approximated by $(w_2 - w_1)/w_1$ or, somewhat more accurately, by $2(w_2 - w_1)/(w_2 + w_1)$, when $t_2 - t_1 = 1$ and the unit of time is short—1 day, for example.

For some purposes it is not even necessary that growth be exponential in order for a value of G computed from Eq. (7) to be useful. As an illustration, suppose w_2 and w_1 are measured 1 year apart; G is then the instantaneous rate of growth for that year, even though the fish's increase in weight actually followed a seasonal S-shaped cycle rather than an exponential curve. Although G cannot then be used to compute fish sizes within the year, it can be used for comparisons with other instantaneous rates on a yearly basis. For a population, for example, the instantaneous mortality rate for the year can be subtracted from the instantaneous rate of growth to give the net rate of increase or decrease in biomass for the year, regardless of how either growth or mortality are distributed throughout the year.

For a unit interval of time we can compute from Eq. (5)

$$\frac{w_2}{w_1} = e^G \qquad (t_2 - t_1 = 1) \tag{8}$$

This is sometimes called the *finite rate of growth*.

The various definitions above can also be applied to length, although instantaneous rates are only rarely used for length. As a matter of fact the instantaneous rate of increase in length and instantaneous rate of increase in weight are very similar statistics, differing only by a

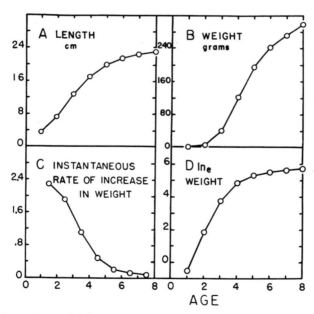

Fig. 3. Comparison of different measures of growth, for bluegill sunfish (*Lepomis macrochirus*). The instantaneous rates in panel C are the slopes of the line segments in panel D. (From Ricker, 1975a, *Bulletin of the Fisheries Research Board of Canada*, No. 191; reproduced by permission.)

constant. For a unit time interval we may combine Eq. (7) with Eq. (2), as follows.

$$
\begin{aligned}
G &= \ln w_2 - \ln w_1 \\
&= \ln a + b(\ln l_2) - \ln a - b(\ln l_1) \\
&= b(\ln l_2 - \ln l_1)
\end{aligned}
\tag{9}
$$

This provides a convenient method of estimating G from length data, provided b is known.

Figure 3 compares some of the growth indices described above.

II. ESTIMATION OF GROWTH RATES IN NATURE

A. Age from Frequency Distributions (Petersen's Method)

Samples of fish taken in nature frequently exhibit well-marked humps in their length frequency distribution, especially near the lower end of the size range. These can have different interpretations.

1. A succession of humps among the smaller fish may represent a succession of spawnings that occurred at different times during the current year.
2. The humps may represent successive broods (year-classes) of fish, spawned in successive calendar years.
3. The humps may represent year-classes spawned 2 or more years apart. This can occur when success of reproduction varies considerably from year to year.
4. The humps may represent merely random sampling variability.

If there is doubt about the nature of any series of humps, a decisive test is to take samples two or more times during a calendar year. The rate of progression of the humps with time will indicate their nature.

Frequency distributions of weight could be used in place of length, and might be superior among the older fish. However, lengths are easier to obtain, and are less subject to variation from environmental or biological causes.

To estimate growth rates, the simplest procedure, introduced by Petersen (1892), is to take the position of the mode of the successive humps (of Type 2 above) of a frequency distribution as the representative length of the year-class. If rate of growth varies from year to year, greater accuracy is achieved by sampling the same brood at the same time in successive years. After the first, second, third, or fourth years, depending on the species and the size of the samples available, year-class modes become indistinguishable from random fluctuation in the length distribution. In any event the mode is not the best measure of central tendency in length. An alternative is to fit normal distributions to the successive humps in the series. This is made easy by using probability paper, which transforms successive frequencies in a normal distribution to a straight line (Cassie, 1954). A series of humps (year-classes) gives points that describe a series of straight lines at different levels on the graph, but usually having much the same slope. For separating age groups it is usually sufficient to fit lines by eye to these points and divide the series at the middle of each transition from one straight segment to the next. By replotting cumulative percentage frequencies within each segment, the mean and standard deviation of the corresponding normal curve can be estimated, hence the curve itself if it is needed (Partlo, 1955; Tesch, 1971).

B. Age from Marks on Scales or Bones

In temperate and northern latitudes most species of fish carry a record of their age on the hard structures of their body. The most useful feature for this purpose varies with the species; scales, otoliths, fin rays, opercula and vertebrae are all used, more or less in order of decreasing frequency. In few cases is the record completely unambiguous: different biologists will often disagree as to what age should be assigned, especially for older fish. However there have been quite a number of cases where age determinations have been checked against fish of known age, or by the progression of a dominant year-class over a period of years, and the method usually has proved sufficiently reliable for routine use (Chugunova, 1959; Graham, 1929; Tesch, 1971; Van Oosten, 1929).

Not only can the current age be determined from scales or other features, but usually also the length at the end of successive annual growing seasons. This is obtained by back-calculation from measurements of the scale, using an empirically determined relation between fish length and scale radius or diameter.

C. Effects of Bias in Sampling

Samples of fish taken from the wild very rarely include all age groups in proportion to their abundance. For the most part this is because the gear used to capture them is selective by size. Even if it is not, the fish themselves tend to inhabit different parts of their environment at different ages, or within a single environment they may assemble in schools containing individuals more or less uniform in length and different from other schools. The result is that the fish toward the lower end of the length range sampled will usually include only the larger individuals of the youngest ages present. There may be the opposite bias at the upper end of the sampling range. If the mean lengths of incompletely sampled age groups at either end of the range are compared directly with the representatively sampled middle group of ages, the result is that the rate of growth is underestimated. Thus great care should be taken to exclude such ages from an analysis. To obtain an unbiased picture of growth throughout life is no easy task, and almost always requires the use of several methods of sampling.

Ideally growth should be estimated by following a single year-class of fish throughout its entire life, because growth rate may vary between broods, especially if they differ greatly in abundance. Usually, however, this is not possible, and the lengths of a succession of ages taken in a single year is used to give a general picture of growth over the time period represented.

D. Effects of Within-Age Size-Selective Mortality

Even when sampling is perfectly representative, there remains a major obstacle to an accurate appreciation of fish growth in nature. It frequently happens that the larger fish in an age group have a different mortality rate from the smaller ones: either greater or less, but usually greater. This can be detected when back-calculations of length at earlier ages are made from scales or otoliths, *using samples that are representative of the whole of each age group involved.* When a larger fraction of the larger fish die, the result is "Rosa Lee's phenomenon" (Sund, 1911; Lee, 1912), whereby the calculated average size of fish of younger ages is the smaller, the older the fish from whose scales they were calculated. Two recent reviews of this subject are by Jones (1958) and Ricker (1969).

1. *Natural* selection for size can bear more heavily on either the larger or the smaller fish. Faster-growing fish frequently tend to mature earlier and also become senile and die earlier than slower-growing fish of the same brood (Gerking, 1957). This is the principal and perhaps the only cause of natural Lee's phenomenon in unfished populations. However there are at least two possible situations that act in the opposite direction. (a) There is considerable evidence that dur-

ing the first year of life slower-growing individuals are more susceptible to predation. Such selective mortality during the first year cannot
affect calculated growths differentially, because it is only after the first
annulus is laid down that there can be any back-calculation. But if the
same situation persists into the second or later years of life it means
that, for example, the size at annulus 1 computed from fish of age 2
will tend to be *greater* than the same computed from fish of age 3. (b)
The other situation occurs when fish of both sexes are sampled and are
analyzed together, but there are in fact sex differences both in rate of
growth and in natural mortality rate. Among most flatfishes, for example, females grow faster and live longer than males. If lengths are
back-calculated from samples in which the sexes are not distinguished, the increasing representation of the faster-growing females at
older ages tends to increase the calculated mean size at younger ages,
and the result can be "reversed Lee's phenomenon."

2. Size selection by a *fishery* can also be important. The larger
members of a year-class are the first to become vulnerable to a given
type of gear, and it may be several years before the smallest members
are fully vulnerable. In sport fisheries there is often a minimum size
limit for retention of fish caught. Obviously these can be major causes
of Lee's phenomenon. It is also possible for the largest fish in a population to be less vulnerable to fishing than those of intermediate size, but
in practice this is far less important.

III. CHARACTERISTICS OF FISH GROWTH

A. Growth Stanzas

Considering the whole life of a fish, its growth can conveniently be
divided into a series of stages or stanzas,* a concept given formal
development by Vasnetsov (1953). The change from one stanza to the
next is characterized by some kind of crisis or discontinuity in
development, such as hatching or maturation, or a change of habits or
habitat. In order of decreasing severity, these may include the following.

1. A major reorganization of body structure, comparable to what
occurs in the metamorphosis of moths or wasps. Such drastic changes
are not very common among fishes, but they occur in a number of

* I use the word stanza rather than stage, to avoid confusion with developmental
stages such as are described by Ahlstrom (1943) or Pelluet (1944).

oceanic species. Eels (Anguillidae) are familiar examples, in which a flat, transparent leptocephalus is reorganized into a cylindrical pigmented elver. Somewhat less drastic is the change, by the various flatfishes, from a symmetrical pelagic fingerling to an adult form with both eyes on the same side of the head.

2. Any fairly abrupt change in body form, or in the relative lengths of appendages, or in the relative length and structure of the digestive tract. For example, Martin (1949) illustrated breaks in the slope of the plot of the logarithm of one or more linear measurements of the body or fins and the logarithm of standard length for ten species of fish, at lengths varying from 27 mm for a characin (*Brycon guatamalensis*) to 50 mm for herring (*Clupea harengus*). There was also a break at 35 mm in the slope of log weight against log length for rainbow trout (*Salmo gairdneri*), and Tesch (1971) illustrated a similar break for brown trout (*S. trutta*).

3. Major physiological changes, for example, in tolerance to temperature or salinity, accompanied by corresponding changes in endocrine and other internal organs. The diadromous fishes are the best known examples, and among them physiological changes are often accompanied by a change in form or color. For example, a young salmon, in adapting to marine life, changes the color of its sides from barred to silvery, its body becomes more elongate,and greater tolerance of salinity is acquired. Once in saltwater, growth rate increases greatly, and this is reflected in broader spaces between the circuli on the scales. Returning from the sea, salmon enter a new stanza characterized by changed color, often to bright hues, and by tolerance of freshwater, thickening of the skin, partial resorption of the scales, reduction of the digestive tract, and marked changes in external form. In Pacific salmon (*Oncorhynchus*) these changes are irreversible and the fish dies after spawning, but Atlantic salmon and others of the genus *Salmo* may recover and return for another stanza of ocean life.

4. A sudden increase or decrease in rate of growth. This is a borderline case, and whether to consider such a change the start of a new growth stanza is largely a matter of individual preference. For example, at a certain size some perch (*Perca*) shift from an insect to a fish diet and increase their growth rate rather abruptly, which can be called the start of a new stanza if you feel inclined. If growth data were available for individual fish, it is probable that the onset of first sexual maturity would be recognized as the start of a new stanza in most cases. But when averages are used, as is customary, the change in growth rate at maturity becomes blurred by the fact that different members of a brood mature at different ages.

Within any stanza of growth of animals or plants, increase in size may follow an S-shaped curve. This was originally suggested for plants by Sachs (1874) and is often called a Sachs cycle. The lower part of the S may (or may not) approximate to an exponential curve, while the upper asymptotic part may reflect preparations for the next stanza—unless the fish is already in its final stanza. However the complete Sachs cycle need not always be present, and Hayes (1949) points out that the time at which the inflexion point occurs differs greatly, depending on whether it is length or weight that is under consideration (compare A and B in Fig. 3). This difference is of course a direct consequence of the relationship between length and weight described by Eq. (1).

B. The Seasonal Growth Cycle

Another characteristic of the growth of many fishes in nature is a marked seasonal variability. This is universal outside of tropical regions, and is by no means rare within them, where it is usually related to seasonal rainfall. One thinks of such extremes as the Alaska blackfish (*Dallia*) that hibernate like frogs at the bottom of tundra ponds, or the African lungfishes (*Protopterus*) that retreat into a cocoon far down in the mud during the dry season.

Under less severe conditions there have been a number of investigations showing that growth tends to follow the cycle of the seasons, usually faster in summer and slower in winter. An example at random is Alexander and Shetter's (1961) study of brook and rainbow trout (*Salvelinus fontinalis* and *Salmo gairdneri*) in a Michigan lake. Growth was very slow from late December to early April, but did not stop entirely. Bluegill sunfish (*Lepomis macrochirus*) ceased to grow during cool weather in several Indiana Lakes (Gerking, 1966), while in ponds carp (*Cyprinus carpio*) sometimes lose weight in winter. Similarly, a species that must endure summer temperatures considerably greater than its preferred temperature may feed little and stop growing at that season; indeed, growth may slow down even if the fish continue to feed at their maximum rate for the prevailing temperature (see Sections V and VI).

However, the seasonal cycle of growth is not always, perhaps never, wholly under temperature control. The whitefish (*Coregonus clupeaformis*) studied by Hogman (1968) provide an example (see Chapter 10, Fig. 12). Their growth was closely correlated (with a time lag of about 1 month) with the seasonal cycle of daylength, and was only weakly correlated with the temperature of the spring water in which they were held. It seems possible also that an internal seasonal rhythm may play a role in regulating growth rate, though I know of no experiments or observations that bear on this, apart from the seasonal sexual cycle of mature fish.

C. Shape and Variability of Frequency Distributions of Length and Weight

In temperate and northern latitudes the distribution of lengths within a single brood (year-class) of fish is usually unimodal, apart from random variation, and frequently it is reasonably close to a normal or Gaussian distribution. Exceptions occur when spawning takes place at intervals over a considerable period of time, as in anchovies (*Engraulis*); or when a few individuals in a brood start to grow exceptionally rapidly by reason of preempting favorable territory, or because they turn cannibal on their siblings, or both; the largemouth bass (*Micropterus salmoides*) is an example.

Very often, however, not only is the length frequency distribution approximately normal at the end of the first year, but it remains so throughout life. This might appear to conflict with the widespread occurrence of greater mortality rates among the larger members of a brood. However Jones (1958) showed that size-selective mortality does not necessarily change the shape of the length distribution of a year-class. When the gradient of instantaneous mortality rate within a year-class is linear with respect to length, an originally normal length frequency distribution will remain normal no matter how severe the mortality gradient (Fig. 4). In nature, of course, the length–mortality relation need not be exactly linear, but it requires a marked deviation from linearity to produce appreciable skewness in the derived length distribution (Ricker, 1969). Jones (1958) also showed that with any linear length–mortality relation the original variability of the frequency distribution is conserved (Fig. 4).

If the distribution of lengths of an age group of fish conforms to a normal curve, then its weight distribution will not be exactly normal, and vice versa (Fig. 5). This follows directly from the weight–length relationship of Eq. (1). The relation between variability in length and in weight is somewhat less obvious. For example, if all the fish in a year-class increase in length by the same absolute amount, their variability in length will remain the same, but their variability in weight will increase greatly (Fig. 5).

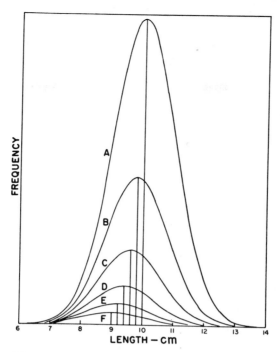

Fig. 4. Normal length distribution curves having a standard deviation of 10 mm. Each curve is obtained by subjecting the fish in the next larger curve to an instantaneous mortality rate that increases by 0.02 per millimeter of length, averaging 0.75. (From Ricker, 1969, *Journal of the Fisheries Research Board of Canada;* reproduced by permission.)

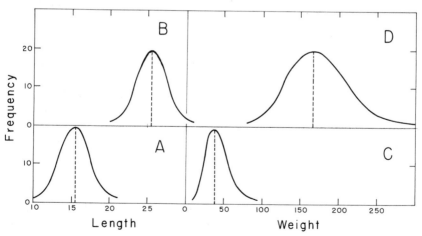

Fig. 5. Frequency distributions of length and weight in a model year-class of fish in successive years of life. From A to B the fish of all sizes have increased in length by 10 units, thus maintaining their original normal frequency distribution. The corresponding distributions of weight are shown in C and D.

In practice, the variability in length of a year-class usually increases somewhat during the first few years of life; then it stabilizes,
and it may decrease in old age. The phase of increasing variability is
sometimes called *growth depensation*, because the longer fish are
becoming even longer relative to their smaller congeners. The later
stage is *growth compensation*, when the smaller fish start to catch up
to the larger ones. Variability in weight, however, continues to increase for some time after variability in length has started to decrease.
Thus growth compensation in terms of weight starts much later than
the same phenomenon in terms of length, and sometimes does not
occur at all.

D. Isometric and Allometric Growth

Even within a single growth stanza, different parts of the body of a
fish may grow at different rates. This can frequently be described by
an expression similar to Eq. (1) or (2):

$$l_2 = al_1^b \tag{10}$$

$$\log l_2 = \log a + b(\log l_1) \tag{11}$$

where l_1 and l_2, lengths of any two body parts (l_1 is often the standard
length of the body); a, a constant equal to the value of l_2 when $l_1 = 1$; b,
the exponent that indicates the direction and speed of any change in
body form. When $b = 1$, growth is *isometric:* The body parts being
compared are growing proportionally. When $b > 1$ the length l_2 is
increasing faster than l_1, and vice versa; in either case growth is *allometric.*

Zar (1968) and others have pointed out that fitting Eq. (11) by
least-squares is not the same thing as obtaining a least-squares fit to
Eq. (10), and nowadays a least-squares fit to Eq. (10) can be obtained
iteratively by computer. Nevertheless the latter procedure is not appropriate here, for two reasons. In the first place, there is a general
tendency for natural variability to increase as size increases, so that the
logarithmic transformation tends to stabilize variance and so make a
least-squares fit more appropriate. Second, the variability in the data
will be almost entirely natural, with very little contribution from errors in the measurements (assuming these are made with ordinary
care), and the body parts measured cannot be categorized as "dependent" and "independent." Hence the line must be fitted symmetrically with respect to both variates, and this requires a functional
regression, specifically, the GM functional regression described in Sec-

tion I,D, which is always numerically greater than the ordinary regression.

For example, suppose l_2 is eye diameter and l_1 is head length. If a functional regression indicates that $b = 1$, eye and head are growing proportionally. But if an ordinary regression of $\log l_2$ on $\log l_1$ is fitted to the same data, b will be less than 1 and the eye will appear to be decreasing in size relative to the head. Because ordinary regressions have been used for this relationship up to recently, there must be many erroneous interpretations of this type in the literature. The magnitude of the correlation coefficient between l_2 and l_1 indicates the difference between the two regressions. The difference is small if the variability in the $l_2 : l_1$ ratio between different fish of the same size is small, and also if the range of fish sizes being compared is large. But increasing the number of fish measured, within the same size range, will not tend to make the two regressions more alike.

When comparing weight and length the situation is similar. A functional slope or exponent $b = 3$ of the line relating $\log w$ and $\log l$ indicates isometric growth, in which weight increases as the cube of length. When $b > 3$ the fish is increasing in weight (presumably also in volume) at a greater rate than required to maintain constant body proportions, and vice versa. While many fishes grow approximately isometrically during their final growth stanza, values of b up to about 3.5 have been observed in some species. Sometimes, also, values less than 3 are observed, usually in populations where the larger individuals lack a suitable food supply.

Lumer (1937) pointed out that Eq. (10) implies that the instantaneous rates of increase in the lengths of two body parts must either remain constant or change at the same instantaneous rate. The initial growth rates of the two parts, however, must then differ if $b \neq 1$.

Although growth conforming to Eq. (10) is very common, it is not universal. Some internal organs reach their maximum size, or even regress, before adult body size is achieved, not to mention the seasonal development of the gonads. Actually there is some ambiguity as to whether the term "allometric" should apply only to cases that conform to Eq. (10) or whether it can be used for any kind of nonisometric growth.

E. Effects of Size-Related Mortality on Estimates of Growth Rate

In Section II,C mention was made of the effect of sampling bias in producing erroneous estimates of growth rate, and how this can be avoided if it is possible to take samples using gears of different selec-

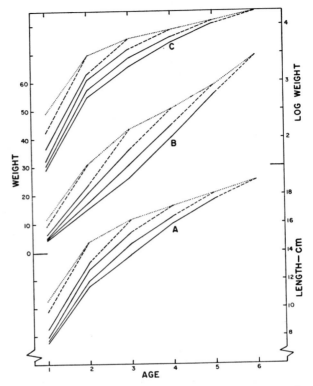

Fig. 6. Growth of a population of ciscoes (*Coregonus artedii*), somewhat idealized, in terms of length (A), weight (B), and natural logarithm of weight (C). Solid and broken lines represent back-calculated sizes from the age at which they terminate. The dotted line joins the observed sizes at successive ages. (From Ricker, 1969, *Journal of the Fisheries Research Board of Canada*; reproduced by permission.)

tivities. However, the effect of within-age selective mortality on the population, as distinct from the sample, is less easily disposed of. Whether it is a fishery or natural causes that make the faster-growing fish die sooner than slower-growing ones, the result is a serious complication of the growth picture. Figure 6 shows the kind of analysis that is possible when back-calculated lengths of fish of a series of ages are available. For the fish taken at each age there is a separate growth line, which typically lies below the line for the next younger age. (If at the younger ages the smaller fish of a brood experience the greater mortality, the picture can become quite complicated.) The growth rates that are most nearly representative of the population that exists during each year's interval are indicated by the final segment of each line, shown broken in Fig. 6.

On the other hand, if back-calculated lengths are not available, and what is known is only the average size of the fish present in successive years, the computed growth rate is less than the actual at all ages, as shown by the fine dotted lines in Fig. 6. This latter has been called the *population growth rate,* in contrast to the true average growth rate of the fish themselves (Ricker, 1975a, p. 217), and the difference between the two can be quite large.

We are left with a very inconvenient conclusion, which has not yet been widely appreciated. Usually it is impossible to represent true growth rates and the true size of the fish at successive ages in natural populations by a continuous line or continuous series of lines on a graph. If fish sizes are plotted, the slopes of the lines joining the points will underestimate growth rate, as do the fine dotted lines in Fig. 6. But if a continuous series of lines were to be plotted with the correct slope for the prevailing growth rate each year, the fish sizes indicated would be increasingly greater than actual average size as age increases.

IV. GROWTH MODELS RELATED TO AGE

A. Growth during Early Life

The early development of fish eggs and young has been studied at constant temperature under laboratory conditions by several authors, usually in terms of weight. For short periods of time, growth can be described by the exponential curve of Eq. (5). Exponential curves can also be used to describe any growth sequence, by dividing the latter into short segments. These may represent either natural periods of exponential growth at different rates, or merely arbitrary time divisions. Hayes and Armstrong (1943) show data on growth of embryos of salmon (*Salmo salar*), plotted on a semilogarithmic scale (Fig. 7). With a little imagination four linear segments can be distinguished, and if these could be related to embryological events they might serve as the dividing points for as many growth stanzas. On the other hand, a single smooth curve would describe the data about as well.

An alternative expression to describe early growth is

$$w = a(t - t_0)^b \tag{12}$$

$$\log w = \log a + b[\log(t - t_0)] \tag{13}$$

where w, weight at time t; t_0, an initial time chosen so as to provide best agreement to the formula; a and b, constants. This expression,

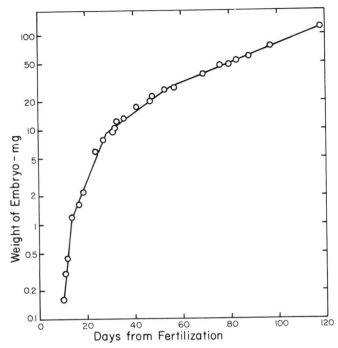

Fig. 7. Growth of embryos of Atlantic salmon, plotted with a logarithmic ordinate. The four straight lines indicate possible successive stanzas of simple exponential growth. (From Hayes and Armstrong, 1943. Redrawn by permission of the National Research Council of Canada from the *Canadian Journal of Research*, Volume 21, Section D, pp. 19–33.)

without the t_0, was proposed by Friedenthal for the growth of mammalian embryos, and it was later applied to other organisms. The modified form shown above was proposed by MacDowell *et al.* (1927), and it was applied to salmon development by Hayes and Armstrong (1943). Figure 8 shows the same data as Fig. 7, plotted in this manner and fitted with the double logarithmic straight line.

Gray (1928), however, observed that growth of embryos of trout (*Salmo trutta*) declined greatly during the last 20 days before hatching, suggesting a complete S-shaped Sachs cycle before the next growth stanza began. To fit this with a single line would require one of the curves described in the sections below—probably a Richards curve.

The straight line of Eq. (13) was found by Allen (1951, p. 121) to describe the growth of brown trout in the wild during their first 450 days after hatching. Considering the changes in temperature and other

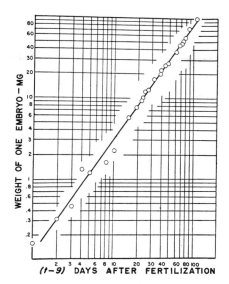

Fig. 8. Double logarithmic plot of the data of Fig. 7, with the time axis originating 9 days after fertilization, when the embryonic axis is established. (From Hayes, 1949, Fig. 7; reproduced by permission.)

conditions over so long a period, this is quite remarkable, and perhaps fortuitous, characteristic only of the Horokiwi Stream and others of its type.

B. Growth Modeled by Successive Exponential Segments

The most convenient way to model a fish's growth is to compute the instantaneous rate of growth for successive time intervals. In this form growth rate can be compared directly with mortality rate using simple subtraction. The shorter the intervals at which observations of size are available, the more accurate will be the resulting representation. During the phase of increasing rate of growth (in length or weight), the exponential curves will agree better with observation because they are concave upward; when growth rate starts to decrease, the graph of size against age is convex upward and so has the opposite curvature to that of the corresponding exponential segments. However, by using short time intervals the difference between the observed and computed lines can easily be made so small as to be of no consequence. If size is known only at yearly intervals, segments less than a year long are pointless because of the seasonal variation in growth rate.

It is not necessary to actually compute the exponential curves when using successive instantaneous rates of growth [G in Eqs. (5)–(7)]. For example, the increase in biomass of a fish during a unit time interval, or the production of a population, is equal to the instantaneous rate of growth times the mean biomass present (Ricker, 1946). In this way Ricker and Foerster (1948) used a series of instantaneous growth rates in making a computation of sockeye production in a small lake, and Ricker (1945, 1958, 1975a) used the same plan to compare growth and mortality rate and to compute the catch taken from a unit weight of recruits to a fishery.

In dealing with natural populations, rather than individual fish, a computation using successive exponential segments has the important advantage that the true growth rate, as defined in Section III,E, can be combined directly with actual population biomass in calculating the production of the population. This is not possible with the growth curves of Sections IV,D–IV,I, which are always based on the size of the surviving fish at each age, and accordingly yield only minimum estimates of growth rate and hence of production.

C. General Characteristics of Curves Applied to the Final Stanza of Life

Growth curves for fish in the wild are usually fitted to data on size at yearly intervals. Either length or weight can be fitted, but length is usually easier because the inflection point for length has usually (not always) been passed by age 1, so it is only the part of the curve having decreasing curvature that needs to be described by a formula. By contrast, the absolute rate of increase in weight often continues to increase for several years before decreasing; hence in order to fit the whole of a weight curve it is necessary to find an S-shaped curve having the correct curvature on both sides of the inflection point.

Whether length or weight data are to be fitted with a curve will depend on the quality of the data available, but if the weight : length relationship of Eq. (1) is known, the one can always be computed from the other. The curves most frequently fitted to size data are almost all *bipartite* in their general differential form (Fletcher, 1973, 1975). Rate of increase in size is proportional to the difference between a positive constant times size already achieved, ay, and some function of that size, $f(y)$. When $f(y) < ay$, this differential form is

$$\frac{dy}{dt} = ay - f(y) \tag{14}$$

When $f(y) > ay$, it becomes

$$\frac{dy}{dt} = f(y) - ay \tag{15}$$

All the curves described by Eqs. (14) and (15) have an upper asymptote, and most are either asymptotic to the time axis or are at some point tangent to it. One or more of the asymptotic curves described below may be fitted to at least the upper portion of a series of annual observations of either length or weight. The symbols l, L_∞ are used for length, end w, W_∞ for weight. Formulas are given in terms of one or the other, as seems most appropriate; but as a matter of observation *all these curves have given reasonable fits to both length and weight data*, but not necessarily on the same species, and not necessarily for the complete range of ages.

The only criteria for choosing a growth curve that have proved valid are goodness of fit and convenience. Historically, however, most of the curves in use have been proposed along with some mathematico–physiological theory as to how growth might be regulated, and much ingenuity has been expended in trying to relate them to growth processes (Pütter, 1920; Brody, 1927; von Bertalanffy, 1934; Parker and Larkin, 1959; Taylor, 1962; Laird *et al.*, 1965; Ursin, 1967; Zweifel and Lasker, 1976). Although none of these theories has been demonstrated to have any biological basis, it is of interest to mention them when introducing each of the curves in turn.

All of the curves in use may be written with a variety of different parameters, and in different ways. Rather than aiming at a uniform manner of presentation, the forms shown are those most commonly encountered in the literature on growth, particularly of fishes and fish populations. Relationships among the different parameters are indicated, and their derivation from the basic differential form.

D. Logistic Growth Curve

Figure 9 is an example of a logistic curve, which also represents the "autocatalytic law" of physiology and chemistry. It is one of several bipartite expressions proposed by Verhulst (1838) as possible descriptions of the succession of age frequencies in human populations. Its differential is of the type of Eq. (14), with $f(y) = by^2$. Using weight symbols, and putting $W_\infty = a/b$ and $g = a$, Eq. (14) becomes

$$\frac{dw}{dt} = gw - \frac{g}{W_\infty} w^2 = \frac{gw(W_\infty - w)}{W_\infty} \tag{16}$$

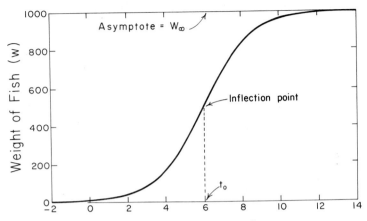

Fig. 9. Example of a logistic growth curve.

The first form shown above shows that the absolute rate of increase in weight, plotted against weight, will be an inverted quadratic parabola. The second form of Eq. (16) illustrates the rationale of this curve as applied to growth; divided by w, it becomes the instantaneous rate of increase in weight, and this is evidently proportional to the difference between the asymptotic weight W_∞ and the actual weight. Two integral forms of Eq. (16) are

$$w = \frac{W_\infty}{1 + e^{-g(t-t_0)}} \tag{17}$$

$$w = \frac{W_\infty}{1 + ce^{-gt}} \tag{18}$$

where w, weight at any time t; W_∞, asymptotic weight; g, instantaneous rate of growth when $w \to 0$; t_0, the time at which the absolute rate of increase in weight begins to decrease, that is, the inflection point of the integral curve, or the maximum of the differential Eq. (16); c, e^{gt_0}.

A logistic curve has the t-axis as its lower asymptote, when $t \to -\infty$. The point of inflection is at $t = t_0$, $w = W_\infty/2$, exactly half-way between the two asymptotes. The two halves of the curve are symmetrical, or rather, antisymmetrical. The constant t_0 adjusts the time scale so that time is in effect measured from the inflection point. The instantaneous rate of growth at the inflection point is $dw/wdt = g/2$.

The logistic curve of Fig. 9 has the parameters $W_\infty = 1000, g = 0.8$, $t_0 = 6$, and $c = 121.5$.

Ricklefs (1967) showed that the logistic curve described well the growth in weight of three species of birds. In ichthyology it has been used mainly to describe the increase in weight of *populations,* rather than individuals (Graham, 1935; Schaefer, 1954). In that context the differential form of the curve is called the Graham surplus production curve, or the Graham–Schaefer curve (Ricker, 1975a, p. 310).

E. Gompertz Growth Curve

Apart from seasonal variations, a fish's rate of increase in biomass typically decreases throughout life, or at any rate throughout its last growth stanza. If the instantaneous rate of decrease in the instantaneous rate of increase is constant, it leads to the type of curve shown in Fig. 10, which was originally proposed by Gompertz (1825) to describe a portion of the distribution of ages in human populations. A number of investigators have interpreted this curve as reflecting the activity of two different types of regulatory factors during growth. For example, Laird *et al.* (1965) hypothesize that "the interaction between the two opposing, genetically programmed, processes of exponential growth and of exponential decay of the specific growth rate, results in the familiar properties of growth of warm-blooded animals . . . (1) growth toward a final limiting size, (2) a limiting size that is characteristic of the species or breed . . . , (3) a specific growth rate that decreases constantly as the organism ages, and (4) a decreasing rate of this decrease" (p. 244). In the notation of Eqs. (21)–(23) below, the original instantaneous rate of increase, when $w = w_0$, is represented by

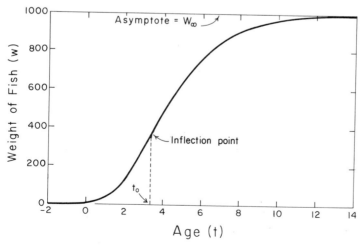

Fig. 10. Example of a Gompertz growth curve.

the product kg, while g is the instantaneous rate of decrease of that rate.

Weymouth and Thompson (1930) applied the Gompertz curve to the growth of a bivalve, Silliman (1967) used it for fishes, and Zweifel and Lasker (1976) argue that it should be the preferred model for fish growth—including larval growth. On the theoretical side, however, it is very difficult to picture how development could reach a satisfactory conclusion (a weight within the normal range for the species), if it were wholly controlled by two rates fixed in the zygote, in the absence of some kind of feedback mechanism to relate growth to size already achieved. And as a matter of observation, we know that growth rate is influenced by diet and exercise in both warm-blooded and cold-blooded animals. Also, the guinea pig seems to be the only known case where growth in weight from early embryo to adult can be described by a simple Gompertz curve (Laird et al., 1965, Fig. 1). For other mammals and birds Laird et al. found that there is a break at birth or hatching; in addition, there is a non-Gompertzian, in fact more or less linear, phase of growth from adolescence to adult size, for which they invented a special, but unconvincing, explanation. Among primates, and man especially, the growth pattern is even more complicated (Laird, 1967). Similarly, Zweifel and Lasker (1976, Fig. 2) found that not one but two Gompertz cycles were needed to describe just the larval growth of anchovies (*Engraulis*). Obviously something considerably more sophisticated than a fixed initial pair of rates must be in control.

To date, then, no satisfactory evidence has been presented that the Gompertz curve is closer to developmental realities than are the other curves in use.* However this does not detract from its usefulness.

The differential form of the Gompertz curve is of the type shown in

---

* Zweifel and Lasker's (1976) enthusiasm for the Gompertz curve has even led them to question the appropriateness of the allometric formula for relating weight to length. They point out that if both length and weight of a given group of fish conform to Gompertz curves, their weight–length relationship cannot be exactly of the type described by Eq. (1), except in two special cases. On the basis that "all experimental evidence indicates that both length and weight can be described by a Gompertz-type curve," they propose a more complex weight–length relationship in their expression (4). No references are given to this experimental evidence; in fact their paper does not cite even one case where Gompertz curves have been fitted to both length and weight of the same group of fish. We may reasonably doubt that existing agreements of lengths and weights to the Gompertz curve are either numerous enough or exact enough to make us adopt a new weight–length relationship on that basis. Zweifel and Lasker's Fig. 1 shows both their new relationship and the ordinary regression of log weight on log length, for anchovies of about 0.02 to 4 mg. The new relationship is slightly the better fit, but a functional regression line would be better than either (Section I,D).

Eq. (14), with $f(y) = by(\ln y)$; in weight symbols it is

$$\frac{dw}{dt} = aw - bw(\ln w) \tag{19}$$

Putting $g = b$ and $W_\infty = e^{a/b}$, this takes the form

$$\frac{dw}{dt} = gw(\ln W_\infty - \ln w) \tag{20}$$

Equation (20) has some similarity to the logistic of Eq. (16): Here the instantaneous growth rate is proportional to the difference between the *logarithms* of the asymptotic size and the actual size.

Three commonly used integral forms of Eq. (20) are

$$w = w_0 e^{k(1-e^{-gt})} \tag{21}$$

$$w = W_\infty e^{-ke^{-gt}} \tag{22}$$

$$w = W_\infty e^{-e^{-g(t-t_0)}} \tag{23}$$

where w, biomass at any time t; w_0, biomass at time $t = 0$ (*not* $t = t_0$); W_∞, asymptotic biomass; g, the instantaneous rate of growth when $t = t_0$; k, a dimensionless parameter, such that kg is the instantaneous growth rate when $t = 0$ and $w = w_0$; t_0, the time at which the (absolute) growth rate starts to decrease, that is, the inflection point of the curve. Evidently Eq. (22) can be derived from Eq. (21) by using the transformation $W_\infty = w_0 e^k$. Similarly Eq. (23) can be derived from Eq. (22) using $k = e^{gt_0}$.

The point of inflection of the Gompertz curve is at $t = t_0$ and $w = W_\infty/e$. Hence it is situated $1/e = 0.3679$ of the distance from the t-axis to the asymptote. At large negative values of t, w asymptotically approaches the t-axis. The instantaneous growth rate at inflection is $dw/wdt = g$.

The Gompertz curve of Fig. 10 has the parameters $W_\infty = 1000$, $k = 5.436$, $g = 0.5$, $w_0 = 4.357$, $t_0 = 3.386$.

In population dynamics studies the Gompertz curve has been used successfully to describe fish growth by Silliman (1967), who says that its form makes it particularly convenient for use with an analog computer.

F. Pütter Growth Curve No. 1

Pütter (1920) was apparently the first to introduce this curve in biology. Brody (1927, 1945) applied it to the growth in weight of domestic animals beyond the inflection point and, applied to length, it has had an energetic proponent in von Bertalanffy (1934, 1938, 1957). In fishery literature it is usually called the Bertalanffy or Brody–Bertalanffy curve, but it seems more appropriate to give it the name of the very original and perspicacious scientist who first proposed it.

Von Bertalanffy distinguished three metabolic types in the animal kingdom: (1) where metabolism is proportional to surface area or to the $2/3$ power of weight; (2) where it is proportional to weight; (3) where it is intermediate between these situations. On the basis of oxygen consumption experiments with guppies (*Lebistes*), and a few other species examined by Jost, he assigned fishes to the first category above; although by 1957 data for many other species were available which indicate that most fishes belong to the intermediate type (Winberg, 1956). He then endeavored to establish "a definite and strict connection between metabolic types and growth types, in consequence of a general theory of growth which establishes rational quantitative laws of growth and indicates the physiological mechanisms upon which growth is based" (von Bertalanffy, 1957, p. 223). For this general theory he adopted Pütter's suggestion that rate of increase in biomass (w) is proportional to the difference between a rate of anabolism that is proportional to the $2/3$ power of biomass and a rate of catabolism that is proportional to biomass. This implies a relationship of the form of Eq. (15) with $f(y) = by^{2/3}$; or in weight symbols

$$\frac{dw}{dt} = bw^{2/3} - aw \tag{24}$$

Dividing through by $w^{2/3}$, and assuming that weight (w) is proportional to the cube of length (l), we obtain the corresponding expression in terms of length

$$\frac{dl}{dt} = b - al \tag{25}$$

where a and b are positive constants, different from those in Eq. (24). Eq. (25) might be regarded as a bipartite expression of the type of Eq. (15), in which $f(y) = by^0$.

How exactly von Bertalanffy related the concepts of anabolism and catabolism to his metabolic types is never made clear. By "catabolism" he did not mean total metabolism as measured by oxygen consumption; rather he tried to limit it to the breakdown of body tissue. His attempt to measure this catabolism in terms of nitrogen excreted by unfed animals is not convincing: breakdown of fat tissue is not accounted for, and there is no demonstration that values obtained for unfed fish apply to more normal conditions.

"Anabolism," in von Bertalanffy's scheme, would presumably be the sum of catabolism as defined above plus the increase in body size over a given period of time—both expressed in the same units, such as calories. The idea that anabolism, however defined, is proportional to the ⅔ power of weight may have derived from an assumption that the surface area of the gut would limit absorption of nutrients and hence additions to biomass, but there is no basis for such an idea. For example, Szarski *et al.* (1956) have shown that the absorptive area of the gut of the bream (*Abramis*) increases more rapidly than the body's surface area, in fact about as fast as biomass, and other similar findings are cited by Parker and Larkin (1959). Apart from that, area of the gut wall could be a factor limiting food absorption and hence "anabolism" only if food were always available in excess, which is far from being the case in nature.

The fanciful speculations above have been summarized here because they have been rather widely quoted as a solid theoretical basis for the Pütter No. 1 growth curve. In fact, however, neither theory nor data are available to indicate that any one of the asymptotic curves should be preferred to any other, except on purely empirical grounds.

1. PÜTTER'S EQUATION

Putting $K = a$ and $L_\infty = b/a$, Eq. (25) becomes

$$\frac{dl}{dt} = K(L_\infty - l) \tag{26}$$

showing that the (absolute) rate of increase in length is proportional to the difference between the asymptotic length L_∞ and the actual length. This can be integrated to a form used by Pütter (1920) and (in weight symbols) by Brody (1927)

$$l = L_\infty - ce^{-Kt} \tag{27}$$

where l, length of the fish at any time t; L_∞, asymptotic length; K, a parameter that governs the rate at which increase in length decreases, the *Pütter growth coefficient* (also called the Brody coefficient), which has the dimensions of 1/time; c, a parameter equal to the difference between L_∞ and the value of l when $t = 0$. An expression of identical form in chemistry is known as the equation of monomolecular reactions.

An example of Eq. (27) is Curve No. 1 in Fig. 11. By using the transformation

$$t_0 = \frac{\ln(c/L_\infty)}{K} \tag{28}$$

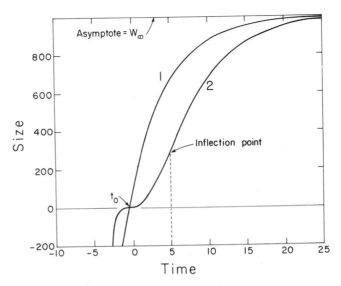

Fig. 11. Examples of Pütter growth curves Nos. 1 and 2 (see Sections IV,F and IV,G).

Eq. (27) can be changed into the form used by von Bertalanffy (1934), in which two parameters are in the exponent

$$l = L_\infty(1 - e^{-K(t-t_0)})$$ (29)

The parameter t_0 is the time at which the fish would have had zero length if it had always grown according to Eq. (29).

Fabens (1965) has an excellent exposition of the meaning of the terms in Eq. (27), and Allen (1950) applied it to some New Zealand brown trout populations, notably those of the Horokiwi Stream (Allen, 1951, p. 123). Subsequently the curve received extensive usage in the Eq. (29) form after it was adopted in Beverton and Holt's (1957) monograph of fish population dynamics.

The parameters of Curve No. 1 of Fig. 11 are $L_\infty = 1000, K = 0.2$, $c = 897.0$, and $t_0 = -0.5435$.

2. FORD'S EQUATION

Independently of other workers, Ford (1933) developed a curve on the basis of empirical observations on the growth of herring

$$l_{t+1} = L_\infty(1 - k) + kl_t$$ (30)

The parameter k, called *Ford's growth coefficient*, can be estimated from the slope of a GM functional line fitted to a graph of one year's length against that of the previous year. Equation (30) can be developed from Eq. (29) by inserting lengths for two successive years, l_t and l_{t+1}, in Eq. (29), subtracting the former from the latter, and putting $k = e^{-K}$. Equation (30) can be used to estimate L_∞, but it does not of itself provide the complete curve (29) because t_0 is lacking. Another useful relationship is

$$l_{t+2} - l_{t+1} = k(l_{t+1} - l_t) \tag{31}$$

3. WALFORD LINES

Walford (1946) was the first to plot a graph of l_{t+1} against l_t, obtaining the straight line indicated by Eq. (30). This line has slope k, and has become known as a *Walford line*. The graph is particularly useful for eliminating from a series of age–length data the points that do not fit the assumptions required for application of Eqs. (29) and (30). Figure 12 shows several Walford lines.

4. CHAPMAN'S MODIFICATION

Chapman (1961) suggested using an expression obtained by subtracting l_t from both sides of Eq. (30); after rearrangement this gives

$$l_{t+1} - l_t = L_\infty(1 - k) - l_t(1 - k) \tag{32}$$

Here a regression of $l_{t+1} - l_t$ on l_t has a slope of $-(1 - k)$; its ordinate intercept is $L_\infty(1 - k)$, and its abscissal intercept is L_∞. Again both L_∞ and k can be estimated by fitting a functional straight line.

5. WEIGHT DATA

While the Pütter No. 1 curve is commonly fitted to length data, Brody (1927, 1945) and a few other authors have fitted it directly to weight data. It can be shown that the value of K so estimated is often practically identical with that obtained from the lengths of the same group of animals (Ricker, 1958, p. 200), though the value of t_0 is quite different. However, this procedure precludes using weights at ages that lie below the inflection point of the weight–age curve.

G. Pütter Growth Curve No. 2

The cube of Eq. (27) or (29) has sometimes been called the Bertalanffy growth curve, for example, by Ricklefs (1967) and Fletcher (1975), and indeed it comes more directly from the basic Eq. (24).

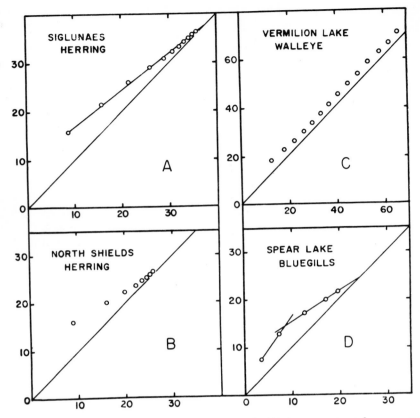

Fig. 12. Examples of Walford lines, l_{t+1} against l_t. (A) An approximately asymptotic line for *Clupea harengus*; (B) a line that becomes parallel to the diagonal at older ages; (C) a line parallel to the diagonal, for *Stizostedion vitreum*; (D) a line that describes the three right-hand points, but is not applicable at the younger ages, for *Lepomis macrochirus*. Both axes are in cm; the first point on each graph represents length at age 2 plotted against length at age 1. (From Ricker, 1975a, *Bulletin of the Fisheries Research Board of Canada*, No. 191; reproduced by permission.)

However, it has a very different shape from the No. 1 curve in the region between the time axis and the asymptote, so it is here considered separately. Since biomass is commonly proportional to a power of length that is close to 3, this curve is better written with weight symbols. Putting $W_\infty = (b/a)^3$ and $K = a/3$, Eq. (24 becomes

$$\frac{dw}{dt} = 3Kw[(W_\infty/w)^{1/3} - 1] \qquad (33)$$

Two integral forms of Eq. (33) are

$$w = (W_\infty^{1/3} - ce^{-Kt})^3 \tag{34}$$

$$w = W_\infty(1 - e^{-K(t-t_0)})^3 \tag{35}$$

where w, mass at any time t; W_∞, asymptotic mass; K, the Pütter coefficient governing the rate at which increase in mass decreases; t_0, the point where the curve becomes tangent to the time axis; c, $W_\infty^{1/3}e^{Kt_0}$. Curve No. 2 in Fig. 11 is of this type. The symbols K and t_0 estimate the same quantities as in a corresponding length curve fitted by Eq. (27) or (29), provided w is proportional to l^3. This curve has an inflection point where

$$t = \frac{-\ln(W_\infty^{1/3}/3c)}{K} = t_0 + \frac{\ln 3}{K} \tag{36}$$

$$w = \frac{8W_\infty}{27} = 0.2963W_\infty \tag{37}$$

The instantaneous rate of growth at the inflection point is $dw/wdt = 3K/2$.

Expressions (34) and (35) become tangent to the t-axis at $t = t_0$, then fall away steeply as t becomes smaller (Fig. 11), but the portion below the time axis is of course not involved in growth modeling.

The parameters of curve No. 2 in Fig. 11 are $W_\infty = 1000$, $K = 0.2$, $c = 8.970$, and $t_0 = -0.5435$.

H. Johnson's Growth Curve

Johnson (1935) and Schumacher (1939) independently proposed this curve, and Krüger (1962, 1964, 1965, 1973) applied it to fish growth using the name "reciprocal function." However he refrained from suggesting any theoretical basis for its applicability, pointing out only that it provides some good fits to data. The general differential form of the Johnson curve is

$$\frac{dy}{dt} = y[a - b(\ln y)]^2 \tag{38}$$

Putting $l = y$, $L_\infty = e^{a/b}$ and $g = b^2$, Eq. (38) becomes

$$\frac{dl}{dt} = gl(\ln L_\infty - \ln l)^2 \tag{39}$$

Equation (39) resembles the Gompertz Eq. (20), but here the instantaneous rate of growth is proportional to the *square* of the difference between the logarithms of asymptotic size and actual size. An integral form of Eq. (39) is

$$l = L_\infty e^{-1/g(t-t_0)} \tag{40}$$

$$\ln l = \ln L_\infty - \frac{1}{g(t - t_0)} \tag{41}$$

where l, length at any time t; L_∞, asymptotic length; g, a parameter with the dimensions of 1/time ($1/g$ = Krüger's a); t_0, the point at which the curve meets the time axis; t_0 usually lies to the left of the vertical axis (i.e., is negative).

The inflection point of a Johnson curve is at

$$t = t_0 + \frac{1}{2g}; \qquad l = L_\infty e^{-2} \tag{42}$$

Thus the inflection point is situated $e^{-2} = 0.1353$ of the distance from the time axis to the asymptote. The instantaneous growth rate at the inflection point is $dl/ldt = 4g$. Figure 13 shows a Johnson curve in which $L_\infty = 1000$, $g = 0.4343$, and $t_0 = -5$.

As the Johnson curve approaches the time axis it becomes tangent to it when $t = t_0$. Then it immediately rises vertically

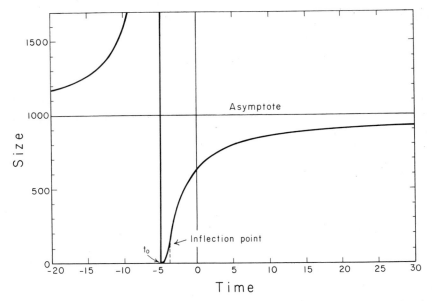

Fig. 13. Example of a Johnson growth curve.

to an indefinitely large positive value, from which it descends, as t continues to decrease, until it approaches the asymptote L_∞ from above when $t \to -|\infty$ (Fig. 13). Of course t values less than t_0 are not involved in describing fish growth.

I. The Richards Function

·By adding an additional parameter, flexibility can be obtained in respect to the position of the inflection point of a bipartite curve, in combination with different types of curvature. Such a four-parameter function was proposed by Richards (1959) as a general growth curve, and it has been discussed by various authors since; what follows is adapted from Fletcher (1975). The general differential form of the Richards function is

$$\frac{dw}{dt} = c_1 w + c_2 w^n \tag{43}$$

where n is positive and c_1 and c_2 can be either positive or negative. Given that a and b are positive coefficients, when $c_1 = +a$, $c_2 = -b$ and $n > 1$, Eq. (43) is of the type of Eq. (14), and the integral form is

$$w^{1-n} = \frac{b}{a} + K e^{a(1-n)t} \qquad (n > 1) \tag{44}$$

With $c_1 = -a$, $c_2 = +b$, and $n < 1$, Eq. (43) is of the Eq. (15) type and the integral becomes

$$w^{1-n} = \frac{b}{a} - K e^{a(n-1)t} \qquad (0 < n < 1) \tag{45}$$

For the range that is of interest here $(0 < w < W_\infty)$, Eqs. (44) and (45) have been written so that the integration constant K is positive. In both these expressions the upper asymptote is

$$W_\infty = (b/a)^{1/(1-n)} \tag{46}$$

The inflection point is at

$$w = W_\infty n^{1/(1-n)} \tag{47}$$

The absolute growth rate (slope) at inflection is

$$\frac{dw}{dt} = \pm \frac{a(1-n)}{n} \left(\frac{bn}{a}\right)^{1/(1-n)} = m \tag{48}$$

where the negative sign applies to Eq. (44) and the positive to Eq. (45). The instantaneous growth rate at inflection is equal to Eq. (47)

divided into Eq. (48), which is $dw/wdt = \pm a(1 - n)/n$, the sign being such that the rate is positive.

The Richards function has usually been written with separate equations for $n > 1$ and $n < 1$, as above. However, Fletcher (1975) has derived a single expression which takes care of the whole range of possible values of n:

$$w^{1-n} = W_\infty^{1-n} + K[\exp(-tmn^{n/(n-1)}/W_\infty)] \qquad (49)$$

where W_∞ is the asymptote and m is the slope of the curve at the point of inflection, as defined in Eq. (48).

In theory the inflection point of a Richards curve can be located at any position between the time axis and the asymptote. However Fletcher (1975) points out that inflections close to the upper asymptote require very large values of n and are not very practical.

Chapman (1961) first suggested that Eq. (43) might be used to describe the growth of fishes. By incorporating Eq. (1) he developed its integral form in terms of length. Pella and Tomlinson (1969) used the Richards function to describe the growth in weight of fish *popula-tions*. They applied it to Silliman and Gutsell's aquarium populations of guppies, and to the stock of yellowfin tuna (*Thunnus albacares*) of the eastern Pacific Ocean.

J. Asymptotic Growth: Is It Real?

The logistic, Pütter, Johnson, Gompertz, and Richards formulas all imply that the increase in size of a fish is asymptotic; that is, size will tend toward some fixed limit no matter how long the fish lives. In practice it is an average asymptotic size that is estimated; for individual fish the asymptote may be greater or less than average. Thus asymptotic size, however estimated, should not be called the maximum size of fish in the population concerned; usually a few old individuals will be found that are considerably larger than the asymptotic size computed for the population, particularly in terms of weight. In any event, the magnitude of any computed asymptote depends partly on what growth function is used to estimate it. For example, Krüger (1969, p. 212) fitted both a Pütter No. 1 and a Johnson curve to Ketchen and Forrester's data for female "English" sole (*Eopsetta jordani*). Agreement with observation was slightly better for the Johnson curve, but both fits were satisfactory. However the Pütter asymptote was at 600 mm, whereas the Johnson asymptote was 739 mm—a considerable difference.

Knight (1968) has contended that for fishes asymptotic growth is a mathematical fiction rather than a real phenomenon. Krüger (1969), on the other hand, maintains that it is real, even while pointing out that estimates of the position of the asymptote are subject to wide variation depending on the curve fitted. Knight is right in this sense, that no matter to what age a fish's size is observed to have an asymptotic trend, we can never be sure that it would continue in that fashion if the fish were to survive longer.

A factor that may contribute to a spurious appearance of asymptotic growth is the within-age size-selective mortality discussed in Section III,E. Nearly all the data fitted by asymptotic formulas have been "population" growth curves, the points fitted being the mean size of the survivors at successive ages. Parker and Larkin (1959)* have compared back-calculated lengths of chinook salmon (*Oncorhynchus tshawytscha*) with observed lengths of maturing specimens. The right-hand panel of Fig. 14 shows growth as computed from scales of maturing fish of different ages. These have the pattern, invariable in salmon and very common in other fishes, of slower growth by the fish that survive the longest. The corresponding Walford lines are in the left panel, and only the broken "population" line, formed by joining the observed size of maturing fish in successive years, shows any tendency to approach the diagonal. If that were the only line available, growth would (erroneously) be considered to be asymptotic. A special study would be needed to ascertain how generally this effect contributes to "asymptotism" in other species.

There are also examples of Walford lines that tend to curve parallel to the diagonal at the oldest ages represented (Fig. 12), in spite of being based on surviving fish. Thus it is by no means established that asymptotic growth is characteristic of fishes generally.

Our conclusion must be that the existence of asymptotic growth to an indefinitely great age can be neither proved nor disproved; nor need all kinds of fish be the same in this respect. Also, the question is not one of any great importance. Whether asymptotes really exist or not, asymptotic formulas are a convenient way of modeling many observed growth series, and we may expect them to be used into the indefinite future.

---

* Parker and Larkin's "concept of growth" has not been considered here because it relates growth to size already achieved rather than to age, and because it has apparently not been used in any practical manner. However their paper contains much useful information and stimulating discussion.

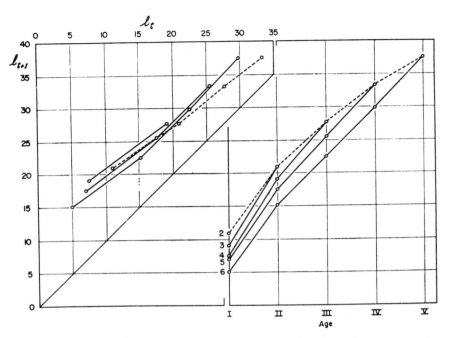

Fig. 14. Right panel: Computed fork lengths of chinook salmon that mature at the successive ages shown by arabic numerals at the left; the final season of growth has been omitted because it is not a full year. Left panel: The solid lines are Walford lines for the same salmon using back-calculated lengths from fish of ages 4–6. The broken line is a Walford line based on the lengths of the fish at their last scale annulus prior to maturity. (From Parker and Larkin, 1959, *Journal of the Fisheries Research Board of Canada;* reproduced by permission.)

K. Fitting Growth Curves

The logistic, Gompertz, Johnson, and the two Pütter curves all have three parameters to be estimated from the data. Many methods of estimation have been proposed, but they all involve successive approximation either by hand or by computer. Here we will describe briefly a simple noncomputer method available for each curve.

For the Pütter No. 1 curve, a simple and adequate method of fitting was described by Beverton (1954) (see also Ricker, 1958, 1975a). A trial estimate of the asymptote (L_∞) is obtained from a Walford line $(l_{t+1}$ against l_t), fitted freehand or by a functional straight line, taking care to eliminate any points at the lower or upper end that do not conform to the line. To estimate t_0, Eq. (29) may be written as

$$\ln(L_\infty - l) = \ln L_\infty + Kt_0 - Kt \qquad (50)$$

Thus the observed values of l are subtracted from the trial L_∞ and the natural logarithms of these differences are plotted against age (t). This line will, in general, have some curvature, but this is sensitive to the trial value chosen for L_∞. Additional trials will reveal the best (straightest) line, which can be selected sufficiently well by eye. Recalcitrant values at either the lower or the upper end should be rejected, the former because they indicate that the relationship is not appropriate for the younger ages, the latter perhaps for a similar reason, but also because they will usually be based on few fish. From the line finally chosen, the slope is a new estimate of K, and t_0 can be calculated from the ordinate intercept, which is equal to $\ln L_\infty + Kt_0$. If this line is fitted by least squares, it is of course the ordinary, not the functional, regression that should be used. Another method is described by Rafail (1973), and for computer use a program is given by Allen (1967) and Fabens (1965), while Marquardt's algorithm is recommended by Conway et al. (1970).

For the logistic, Gompertz, and Pütter No. 2 curves, Ricklefs (1967) published tables, reproduced in Table I, of "conversion factors" whereby the appropriateness of a given curve for the data at hand may be tested using a trial estimate of the asymptotic value, W_∞ (or L_∞). Observed figures are expressed as a percentage of the trial asymptote, and the corresponding conversion factors are plotted against age. If the line is curved, and if no larger or smaller trial asymptote decreases the curvature, the formula chosen is not suitable for the data. If a substantial part of the curve is straight, the formula is suitable, and additional trial values of the asymptote are used until the maximum number of points fit the line. The slope (S) of this "converted growth curve," fitted by eye or by an ordinary regression line, is equal to $1/W_\infty$ times the differential (absolute growth rate) of the corresponding curve at its inflection point. The value of t at the inflection point is the point at which the converted growth curve meets the t-axis; call this X. The curve parameters are then computed as follows.

Logistic [Eqs. (16)–(18)]: $g = 4S$; $t_0 = X$; $c = e^{gX}$

Gompertz [Eqs. (20)–(23)]: $g = Se$; $t_0 = X$; $k = e^{gX}$

Pütter No. 2 [Eqs. (33)–(35)]: $K = 9S/4$ *; $t_0 = X - (\ln 3)/K$;
$$c = W_\infty^{1/3}/3e^{-KX}$$

Johnson [Eq. (40)]: $g = Se^2/4$; $t_0 = X - 1/(2g)$

It frequently happens that more than one of the three-parameter curves above will fit a given body of data adequately, and only rarely

* Ricklefs (1967) inadvertently wrote this as $4S/9$ (his Eq. 7).

Table I

Ordinates of Straight Lines That Correspond to Ordinates (w/W_∞ or l/L_∞)
of Four Growth Curves[a]

	0.00	0.01	0.02	0.03	0.04	0.05	0.06	0.07	0.08	0.09
Logistic curve										
0.0		−1.149	−0.973	−0.869	−0.795	−0.736	−0.688	−0.617	−0.611	−0.578
0.1	−0.549	−0.523	−0.498	−0.475	−0.454	−0.434	−0.415	−0.396	−0.379	−0.363
0.2	−0.347	−0.331	−0.316	−0.302	−0.288	−0.275	−0.261	−0.249	−0.236	−0.224
0.3	−0.212	−0.200	−0.188	−0.177	−0.166	−0.155	−0.144	−0.133	−0.122	−0.112
0.4	−0.101	−0.091	−0.081	−0.070	−0.060	−0.050	−0.040	−0.030	−0.020	−0.010
0.5	0.000	0.010	0.020	0.030	0.040	0.050	0.060	0.070	0.080	0.090
0.6	0.101	0.112	0.122	0.133	0.144	0.155	0.166	0.177	0.188	0.200
0.7	0.212	0.224	0.236	0.249	0.261	0.275	0.288	0.302	0.316	0.331
0.8	0.347	0.363	0.379	0.396	0.415	0.434	0.454	0.475	0.498	0.523
0.9	0.549	0.578	0.611	0.647	0.688	0.736	0.795	0.869	0.973	1.149
Gompertz curve										
0.0		−0.562	−0.502	−0.462	−0.430	−0.403	−0.381	−0.360	−0.341	−0.323
0.1	−0.307	−0.291	−0.276	−0.262	−0.249	−0.236	−0.223	−0.210	−0.198	−0.187
0.2	−0.175	−0.164	−0.153	−0.142	−0.131	−0.120	−0.110	−0.099	−0.089	−0.079
0.3	−0.068	−0.058	−0.048	−0.038	−0.028	−0.018	−0.008	0.002	0.012	0.022
0.4	0.032	0.042	0.052	0.062	0.073	0.083	0.093	0.103	0.114	0.124
0.5	0.135	0.146	0.156	0.167	0.178	0.189	0.201	0.212	0.223	0.235
0.6	0.247	0.259	0.272	0.284	0.297	0.310	0.323	0.337	0.351	0.365
0.7	0.379	0.394	0.410	0.425	0.442	0.458	0.476	0.494	0.512	0.532
0.8	0.552	0.573	0.595	0.618	0.643	0.668	0.696	0.725	0.757	0.791
0.9	0.828	0.869	0.914	0.965	1.024	1.093	1.177	1.284	1.435	1.692
Pütter No. 2 curve										
0.0		−0.380	−0.348	−0.323	−0.302	−0.284	−0.268	−0.252	−0.238	−0.224
0.1	−0.211	−0.198	−0.186	−0.174	−0.163	−0.151	−0.140	−0.129	−0.119	−0.108
0.2	−0.098	−0.087	−0.077	−0.067	−0.057	−0.046	−0.036	−0.026	−0.016	−0.006
0.3	0.004	0.014	0.024	0.034	0.044	0.054	0.064	0.074	0.084	0.095
0.4	0.105	0.115	0.126	0.136	0.147	0.158	0.169	0.180	0.191	0.202
0.5	0.213	0.225	0.236	0.248	0.260	0.272	0.285	0.297	0.310	0.323
0.6	0.336	0.349	0.363	0.377	0.391	0.406	0.421	0.436	0.452	0.468
0.7	0.484	0.501	0.519	0.537	0.556	0.575	0.595	0.616	0.637	0.660
0.8	0.683	0.708	0.733	0.761	0.789	0.820	0.852	0.886	0.924	0.964
0.9	1.008	1.056	1.110	1.171	1.241	1.342	1.425	1.554	1.736	2.045
Johnson curve										
0.0		−0.153	−0.132	−0.116	−0.102	−0.090	−0.078	−0.067	−0.056	−0.046
0.1	−0.036	−0.025	−0.015	−0.005	0.005	0.015	0.025	0.035	0.045	0.055
0.2	0.066	0.076	0.087	0.098	0.109	0.120	0.131	0.143	0.155	0.167
0.3	0.179	0.192	0.204	0.218	0.231	0.245	0.259	0.274	0.289	0.304
0.4	0.320	0.336	0.353	0.371	0.389	0.407	0.426	0.446	0.467	0.488
0.5	0.510	0.533	0.557	0.582	0.608	0.635	0.663	0.692	0.723	0.755
0.6	0.789	0.825	0.862	0.901	0.942	0.986	1.032	1.081	1.133	1.188
0.7	1.247	1.310	1.377	1.449	1.527	1.611	1.702	1.801	1.908	2.026
0.8	2.155	2.298	2.457	2.635	2.834	3.060	3.319	3.617	3.964	4.375
0.9	4.867	5.469	6.222	7.189	8.478	10.283	12.990	17.502	26.525	53.592

[a] The ordinates of the observed curves are indicated by the sum of corresponding entries in the top row and the left-hand column. Figures for the logistic, Gompertz, and Pütter No. 2 curves are from page 982 of Ricklefs (1967), and are used by permission of Dr. R. E. Ricklefs and *Ecology*. They are Copyright 1967 by The Ecological Society of America. Figures for the Johnson curve have been computed using a formula derived by Dr. Jon Schnute: $-4e^{-2}[0.5 + (\ln l/L_\infty)^{-1}]$.

will none of them prove suitable. For such cases the four-parameter Richards function should almost always provide a good fit. However, to fit it requires two-stage iteration that is best done by computer, although Richards (1959) describes a manual trial and error method. Computer programs are given by Nelder (1961) and Causton (1969).

L. Curves Fitted to the Seasonal Cycle of Growth

A typical seasonal cycle of fish growth in temperate or chilly regions resembles an S-shaped Sachs cycle (Section III,A). Thus the S-shaped curves, particularly the logistic, Gompertz, or Pütter No. 2, are the most appropriate candidates for describing the seasonal course of growth. There are apparently not many examples of such fittings, but Miura *et al.* (1976) used logistic curves to describe seasonal growth in length of four age groups of redspot salmon (*Oncorhynchus rhodurus*).

Lockwood (1974) fitted Pütter No. 1 curves to the seasonal increase in length of trout (*Salmo trutta*) and plaice (*Pleuronectes platessa*). Although the fit was reasonable from the inflection point onward, the early-season phase of accelerating growth could of course not be described by this curve.

Pitcher and MacDonald (1973) suggest two methods of modifying the Pütter No. 1 curve so that it will describe seasonal increase in length over several years. The first uses a cosine function to switch growth off in winter and turn it on again in spring. The second incorporates a sine function that produces a smoother seasonal pattern; however, it includes some shrinkage in winter, which could be appropriate for weight data but less so for length. In either case two parameters are added to the basic Pütter expression, making five in all—which is not excessive considering that several years of growth are described. Either curve gave a reasonable fit to data on growth of a population of minnows (*Phoxinus phoxinus*). However the practical value of these expressions remains obscure.

V. GROWTH IN RELATION TO TEMPERATURE

A. Maximum Size and Age

It has long been known that many fishes tend to live longer and grow larger in the cooler parts of their range, the increase in age being usually more striking than the increase in size. This is true in spite of

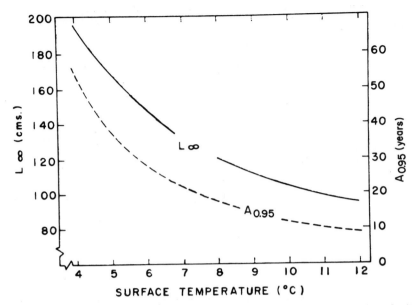

Fig. 15. Relation between ocean temperature and the Pütter asymptotic lengths (L_∞), and an index of maximum age $(A_{0.95})$, for a series of cod populations. (From Taylor, 1958; reproduced by permission.)

the fact that for most fishes both maximum age and maximum size must be defined somewhat arbitrarily. On a basis of straightforward sampling, the larger the sample, the greater will be the maximum age or size observed; however, if samples of several thousand large individuals have been examined, the oldest and largest ones obtained can reasonably be considered of the maximum age and size for purpose of discussion. In a similar manner we might use 100 years and 230 cm as limits that are not often exceeded by man.

Another arbitrary but useful upper limit of size is the asymptotic length or weight indicated by one of the growth curves, and the only one that has been used comparatively is the Pütter No. 1 (Taylor, 1958, 1959; Beverton and Holt, 1959). For an upper age limit Taylor uses the age at which, according to the curve, 95% of the asymptotic growth will have been realized (Fig. 15).

The slower growth of fish at more northern latitudes is not unexpected. Like other physiological processes, growth is affected by body temperature, which in most fishes is close to the ambient temperature. Relating growth rate to temperature is complicated by a strong negative relation between growth rate and fish size. The first attempts to circumvent this made use of the inverse relationship; that is, the time required to attain a given size or stage of development was related to temperature.

B. Hyperbolic Relationships

The simplest temperature relationship, and the one most used to date, is a rule that is said to date from Réaumur in 1735 (Hayes, 1949). Development from fertilized egg to hatching requires a fixed number of degree-days (day-degrees, Tagesgrade) and, within limits, is independent of the temperatures that actually prevail. The limits for most trout and salmon are about 3°–14°C (Fig. 16). If temperature does not vary, this implies direct proportionality between absolute rate of growth and Celsius temperature, and the equation is the hyperbola

$$t = \frac{K}{T} \tag{51}$$

where t, time in days needed to complete a given stage of growth; T, temperature in degrees Celsius; K, number of degree-days required.

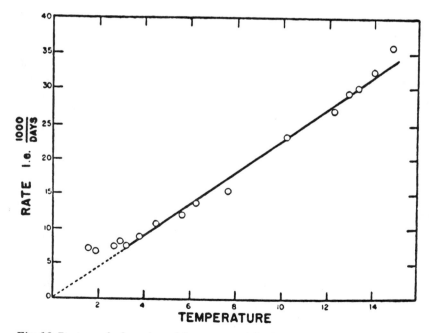

Fig. 16. Reciprocal of number of days required for development of brook trout eggs (fertilization to hatching) plotted against water temperature. (From Hayes, 1949; reproduced by permission.)

If temperature varies, Eq. (51) becomes

$$\sum_{i=1}^{t} T_i = K \tag{52}$$

where T_i is the temperature on day i.

A modification of Eqs. (51) and (52) has been proposed in which the temperature is measured from a point other than 0°C. If T_0 is the new base temperature, the expressions become

$$t = \frac{K}{T - T_0} \tag{53}$$

$$\sum_{i=1}^{t} (T - T_0)_i = K \tag{54}$$

This makes the hyperbolic relationship applicable, approximately at least, to a much larger number of species. For example, Lasker's (1964) observations on incubation times of Pacific sardines (*Sardina caerulea*) between 12° and 20°C follow the relationship fairly well, with $T_0 = 8°C$ and $K = 474$ degree-hours; the extremes of 11° and 21°C are more aberrant.

It is also possible to relate size directly to cumulative degree-days, and when length is used the relationship can be close to linear, even in the wild. For postembryonic growth of mussels (*Mytilus edulis*) near Copenhagen Boëtius (1962) obtained a perfectly straight line over a period of 2 years, using 0°C as the base temperature. Ursin (1963) shows several examples which have some curvature, and he was able to get a straight line for North Sea plaice (*Pleuronectes platessa*) over 4 years' time by making some major adjustments to the temperatures used in calculating the degree-days.

C. Janisch's Catenary Curve

When the hyperbolic degree-day formula came to be applied to experimental data involving a wide range of temperatures, it failed badly; it was found that there was an optimum temperature for growth, beyond which growth slowed down—even when the fish were given all the food they would eat.

Janisch (1927) found that a catenary curve would fit such observations; it has the form

$$t = t_0[\cosh k(T - T_0)] \tag{55}$$

where t, time in days required for growth to a specified size or stage; t_0, time required for that growth when $T = T_0$; T, observed temperature during time t; T_0, the temperature at which growth is fastest; k, a parameter that may be called *Janisch's temperature coefficient*. A catenary is the curve described by a uniform chain or rope suspended between two points at the same elevation. There is no obvious reason why it should describe the rate of growth of fishes or anything else, but this is not an obstacle to employing it if it proves empirically useful.

Curve B in Fig. 17 is a catenary for which most rapid development (50 days) is at 10°C and the temperature coefficient is $k = 0.22$. The limbs of the curve on either side of the minimum are shown of equal length, but in practice the right limb is much shorter, being curtailed by death of the fish at higher temperatures. Curve A in Fig. 17 is the hyperbola for the degree-day rule, when development takes 50 days at 10°C. The two curves have a generally similar shape from about 3° to 9°C. They can be made to provide very similar values within that range by using $k = 0.263$ for Janisch's coefficient in the catenary, while retaining the 10°C optimum point.

Ursin (1963) points out that the hyperbola of Eq. (53), approximating part of the left half of the catenary, can be computed from the

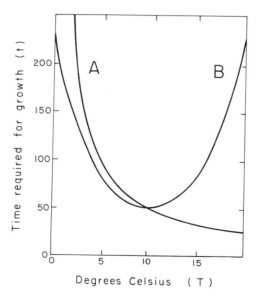

Fig. 17. Example of a catenary curve of duration of growth against temperature, with most rapid growth at 10°C (B); and the hyperbola that passes through the minimum point of the catenary (A).

tangent of the left inflection point of the bell-shaped reciprocal of the catenary; it has the form

$$(T - T_0 + 2.296/k)t = 2t_0/k \qquad (56)$$

where T_0, t_0, and k are all as in Eq. (55). Such an hyperbola passes through the inflection point but of course does not pass exactly through the optimum point.

Ursin (1963) also attempted to combine the catenary temperature relationship with the Pütter No. 1 growth curve. He accepted von Bertalanffy's theoretical background for the Pütter curve and postulated that "anabolic and catabolic processes separately conform to catenary curves" (p. 2). For the resulting abstruse speculations the reader must be referred to Ursin's paper. Like von Bertalanffy, he failed to estimate anabolism and catabolism quantitatively, let alone test their agreement with the catenary relationship, yet he felt able to reach such conclusions as, for example, that the cod is "a fish in which anabolism is more affected by temperature than catabolism" (p. 15).

D. Parabolas

A more familiar curve that provides just as good a fit to most or all of the growth–temperature data available is the quadratic parabola. Here it takes the form

$$t = t_0 + k(T - T_0)^2 \qquad (57)$$

The symbols t, t_0, T, and T_0 all have the same meaning as in Eq. (55), but the coefficient k of course has a different value.

Bělehrádek (1930) suggested generalizing the above as the so-called "power parabola":

$$t = t_0 + a(T - T_0)^n \qquad (58)$$

This provides a growth–temperature relationship flexible enough to take care of almost any set of observations, but at the expense of using four parameters.

E. Elliott's Growth Curves

Undeterred by the obvious complexity of the relation between growth, temperature and size, Elliott (1975a) has constructed a set of growth curves on the basis of his experiments with brown trout (*Salmo trutta*) of 12 to about 250 g, fed an excess ration at different temperatures for up to 42 days. During this short period of time growth was adequately represented by an exponential curve. For trout of a given size he found that the relation between instantaneous growth rate (G) and Celsius temperature (T) could be approximated by four straight lines, each of the form

$$G = a + bT \tag{59}$$

In the range 3.8°–12.8°C b was positive; in the range 12.8°–13.6°C it was close to zero, although this short interval is ignored; for 13.6°–19.5°C b was negative; and for 19.5°–21.7°C it was negative with a steeper slope.

At any given temperature Elliott found that a plot of the logarithm of instantaneous growth rate (G) against logarithm of mean biomass $(\overline{w})$ produced a straight line (Fig. 18); when antilogged, the resulting equation is

$$G = p\overline{w}^{-q} \tag{60}$$

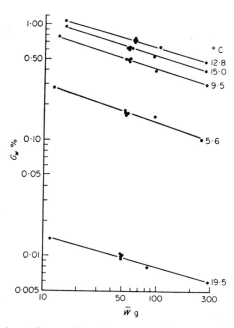

Fig. 18. Double logarithmic relation between relative growth rate, as percent per day (ordinate), and the mean weight of brown trout fed unlimited rations (abscissa), at different temperatures. (From Elliott, 1975a; reproduced by permission of Blackwell Scientific Publications from the *Journal of Animal Ecology*.)

[Elliott used an arithmetic mean of initial and final weight, which is adequate for the short time spans he used. For a longer period of exponential growth, say n time units in length, the true mean is $w_0(e^{Gn} - 1)/Gn$, where w_0 is the initial weight.]

Equations (59) and (60) were then combined by computing a and b in Eq. (59) for a 1 g trout, and combining it with Eq. (60) when $\bar{w} = 1$ g; this gave three expressions of the form

$$G = (a + bT)\bar{w}^{-q} \tag{61}$$

Finally, growth was put into the general form

$$\frac{dw}{dt} = (a + bT)w^{1-q} \tag{62}$$

$$w_t = [q(a + bT)t + w_0^q]^{1/q} \tag{63}$$

where w_0 is initial weight and a, b, and q are the same parameters as in Eq. (61), for the different temperature ranges. Comparison of observed final sizes with those calculated from Eq. (63) shows very close agreement.

Equation (63) is an elegant summary of Elliott's experimental data but, unreasonably no doubt, we experience a feeling of disappointment. It applies to only a short period of time, it sheds no light on the nature of growth processes, nor is it an expression that can be used routinely to represent fish growth. Elliott is careful to restrict its applicability to brown trout fed excess rations under his own experimental conditions. Even if this form of expression were to prove generally applicable, the experimentation needed to determine the parameters makes it impractical to employ it in any general fashion.

F. Zweifel and Lasker's Temperature Functions

Zweifel and Lasker (1976) combined the Gompertz curve with temperature and developed two functions to describe growth in relation to temperature. Using length symbols (l instead of w) in Eq. (21) above, the parameter g was itself conceived as a variable that changes with Celsius temperature according to a Gompertz curve of the form of Eq. (21)—see Zweifel and Lasker's expressions (5) and (5a). Finally, they permitted the origin of the temperature scale to vary, and so obtained a six-parameter expression [their Eq. (5b)].

Considered empirically, these functions provided good fits to data for growth of yolk sac larvae of Pacific sardines (*Sardina caerulea*) and the eggs of anchovies (*Engraulis mordax*). However, the large number of parameters makes these expressions cumbersome and presents an onerous problem of curve fitting, while at the same time it is not obvious that they are either theoretically enlightening or practically useful.

G. Brett's Tabular Presentation

Without trying to fit any mathematical curve to his data, Brett (1974, Table 4) tabulated the expected instantaneous growth rates of sockeye salmon (*Oncorhynchus nerka*) fed to satiation three times a day in freshwater. The table is for sockeye from 0.3 to 500 g, and temperatures from 3° to 18°C. In the absence of a meaningful or even a convenient mathematical relationship—whose discovery becomes more and more unlikely as the years go by—this type of presentation emerges as the most useful and informative way to summarize growth–temperature relationships. Figure 19 illustrates the nature of the trends.

By analogy with Section V,C, the reciprocals of the instantaneous growth rates (1/G) can be plotted. The curves obtained are concave upward, but neither a parabola nor a catenary will describe them at all closely. A reasonably good fit can be made of an ordinary parabola to

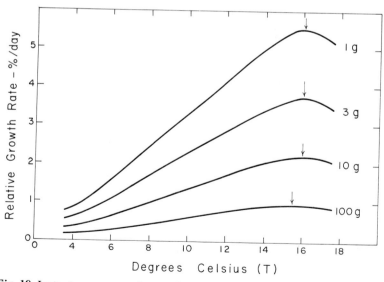

Fig. 19. Instantaneous growth rate plotted against temperature, for sockeye salmon of different sizes fed to satiation three times a day. (Data are from the smoothed values in Table 4 of Brett, 1974.)

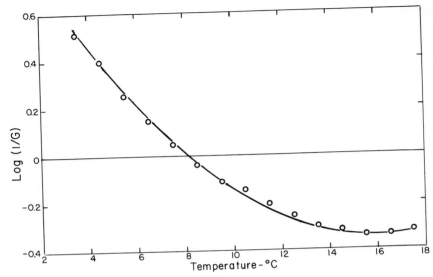

Fig. 20. Relation between the logarithm of the reciprocal of the instantaneous rate of growth of 10-g sockeye salmon and temperature, fitted with a parabola. (Data from Table 4 of Brett, 1974.)

the *logarithm* of the reciprocal of the instaneous growth rate (Fig. 20), which is a relationship of the form

$$\log(1/G) = a + b(T - T_0)^2 \qquad (64)$$

T_0 is the temperature for maximum rate of growth, a is the corresponding value of $\log(1/G)$, and b is a constant to be found by trial. Using base-10 logarithms, the curve in Fig. 20 has $a = -0.344$, $b = 0.0056$, $T_0 = 16.0$. But why go to this trouble? Nothing is gained over using Brett's Table 4 directly.

VI. GROWTH IN RELATION TO RATIONS

A. Empirical Representations

The relation between the food consumed by fish and the resulting growth is a matter of great practical importance, so it has been the subject of numerous experiments, among which Ivlev's (1955) work is outstanding. Two recent comprehensive studies are by Brett *et al.* (1969) on sockeye salmon and by Elliott (1975b) on brown trout. Both authors present their information in a series of tables and graphs illustrating various aspects of the phenomenon, particularly its relation to temperature. One of these is shown in Fig. 21. The technique used by Brett *et al.* differed from that of Elliott and most other authors in that they imitated natural conditions to the extent of making their fish swim continuously at a moderate rate. The ordinary procedure has been simply to let the fish perform only voluntary movements in aquaria or ponds.

The most interesting, albeit predictable, result that emerges from all this work is that an increase in temperature will increase growth rate only if it is accompanied by an increase in food consumption more than sufficient to meet the energy requirements of the automatic increase in basal metabolism and of a probable increase in activity. If temperature increases but food consumption does not, growth rate will decrease and may even fall to zero or beyond. For example, in Fig. 21 the ration that permits the maximum rate of conversion of food to growth at 5°C does little more than maintain body size at 20°C. More generally, when food is scarce fastest growth will occur at a lower temperature than what is optimum when food is available in excess.

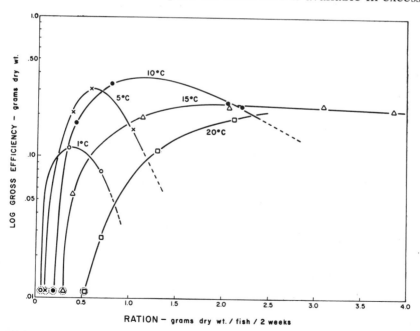

Fig. 21. "K-lines" at five temperatures for sockeye salmon forced to swim continuously at a moderate rate. Circled points are interpolated. (From Brett *et al.*, 1969, *Journal of the Fisheries Research Board of Canada;* reproduced by permission.)

The sections to follow give brief accounts of a number of attempts to describe the relation between growth rate and ration by mathematical expressions, and even to expand this into general accounts of trophic relationships in nature.

B. Stauffer's Sine Curve

Stauffer's (1973) approach to this problem was to find the best relationship between ration and instantaneous growth rate at a given temperature, in terms of two or more of four critical points in the growth–ration relationship. These points are as follows, described in terms of ration (R), increase in mass (Δw), and instantaneous growth rate (G).

No food:	$R = 0$, Δw and G are negative
Maintenance ration:	R small, Δw and $G = 0$
Optimum ration:	$\Delta w / R$ (Ivlev's K_1) is a maximum
Maximum ration:	R and G have their maximum value

Stauffer examined three possible mathematical relationships between G and R: a straight line, the Michaelis–Menton curve, and a sine curve. He tested each of them against some data of Brett *et al.* (1969) and found best agreement with a sine curve of the form

$$G = G_m \times \sin\left(\frac{\pi}{2} \times \frac{R - R_n}{R_m - R_n}\right) \tag{65}$$

where G, instantaneous growth rate at ration R; G_m, maximum instantaneous growth rate; R_m, maximum ration; R_n, maintenance ration. The fit was good at 5° and 10°C, but much less so at 15° and 20°C (Fig. 22).

Although Stauffer (1973, p. 29) states that "this algebraic expression does not explain the growth phenomenon," he also makes the claim that "it does contribute to the understanding of growth," but fails to indicate how.

In practice, there can be considerable difficulty in evaluating G_m and R_m, for the amount of food that fish will consume when not particularly hungry can vary a good deal with frequency of feeding and with slight changes in environmental conditions, outside stimuli, and so on (Brett, 1974).

C. Elliott's Combined Formula

Elliott (1975b) attempted the more ambitious task of combining growth, ration, and temperature into a single formula. He derived a

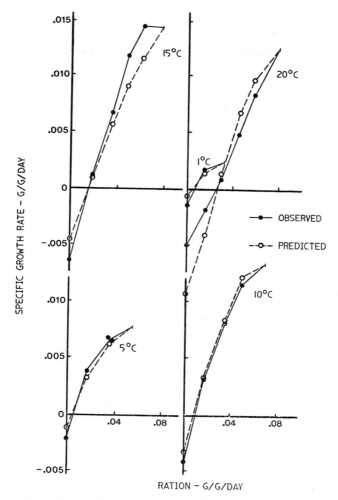

Fig. 22. Observed growth rates of sockeye salmon in relation to ration at different temperatures (solid lines), and growth rates computed from Stauffer's sine formula (broken lines). Both rations and growth are in grams per gram of mass of the fish per day. (From Stauffer, 1973; reproduced by permission.)

five-parameter expression that will describe the situation for his experimental brown trout, provided temperature is broken into three separate ranges; thus, fifteen parameters are required in all. The analysis is an impressive *tour de force* but, in contrast to Elliott's excellent empirical data, it is difficult to picture how his formula can be anything more than a mathematical curiosity of no real utility.

D. Paloheimo and Dickie's Analysis

Paloheimo and Dickie (1965, 1966a,b) made a general review and analysis of data on fish growth, apparently in an effort to penetrate more deeply into the physiological processes involved. Their approach is based on two empirically determined relationships. The first of these is a logarithmic relation between metabolism (T) and biomass (w) or, better, body caloric content, at any given temperature:

$$\log T = a + b(\log w) \tag{66}$$

This relationship has been observed in many experiments with many species of fish and other animals, and it has been generally familiar to ichthyologists since the appearance of Winberg's (1956) comprehensive review. The slope b has usually been estimated from the ordinary regression of $\log T$ on $\log w$, but Ricker (1973) showed that it is more appropriate to use a functional regression, which gives metabolism and biomass the same mathematical position in the relationship.

For fish held in respirometers performing voluntary movements only, Winberg's summary indicates considerable variation between experiments and possibly between species, though it is difficult to be sure that different kinds or sizes of fish are being held under equivalent experimental conditions. However, a general average slope for all species of $b = 0.80$ at 20°C was obtained, and this value was adopted by Paloheimo and Dickie. More recent experiments summarized by Glass (1969) show that this exponent tends to decrease with increase in temperature over the range 10°–25°C, with very few exceptions*; at 10°C it may average more than 1. In any event a functional regression line should be used to estimate the exponent from Eq. (66), and applying this method to the various experiments summarized by Winberg increased the general average exponent from 0.80 to 0.85 at 20°C (Ricker, 1973). There is also evidence that b becomes considerably

---

* Glass computes exponents in two ways: by the usual logarithmic formula, and also by a nonlinear iterative method that requires successive approximations by computer. He points out that the iterative method is correct if the error term in the function $T = cw^b$ is additive, but that the logarithmic method is correct if the errors are multiplicative. However, in this as in most biological situations the errors are in fact multiplicative, approximately at least, as shown by the fact that variability becomes larger absolutely among larger fish. Thus it would be wrong to adopt the cumbersome iterative procedure. In the examples Glass used, the estimates by iteration were sometimes larger, sometimes smaller, than the logarithmic estimate: the average for all species and temperatures was 0.880 for the former and 0.938 for the latter; for the 20°C experiments only, the corresponding figures were 0.788 and 0.833.

larger—approaching 1—when the fish are in active motion (Brett and Glass, 1973).

The other basic relationship used by Paloheimo and Dickie is between the growth efficiency coefficient (K) and food ration (r). Ivlev (1945) defined a coefficient K_1, equal to increase in mass divided by food consumed, and a coefficient K_2 for which the divisor is food absorbed by the alimentary canal*; but since these two are approximately proportional, Paloheimo and Dickie refer to both by the symbol K. A graph of K against r has an ascending left limb starting from a positive value on the r-axis, a maximum, and a descending right limb (Fig. 22). Paloheimo and Dickie discovered that in most published experiments the right limb could be described by a straight line in a semi-logarithmic presentation:

$$\log K = a - br \qquad (67)$$

They refer to this as the "K-line," and it plays a major role in their subsequent theoretical development.

Two aspects of the K-line deserve discussion. One is Paloheimo and Dickie's (1966a) conclusion that for any given absolute ration the value of K is independent of the biomass of the fish. However, data bearing on this generalization are scanty and far from conclusive, consisting essentially of one of Kinne's experiments with a small warmwater cyprinodont. In any event there must be definite limits to its applicability, for a ration that would merely maintain weight $(K = 0)$ for a fish of 1 kg would permit rapid growth $(K > 0.2)$ in one of 10 g. Also, Paloheimo and Dickie state that "body weight is the more efficient predictor of growth efficiency," which seems to contradict their generalization.

Kerr (1971a) has emphasized that in nature the nutritional–physiological situation for fish that have to obtain food by active foraging is likely to be different from that of aquarium fish, and from the ecological point of view it would seem to make little difference whether K is governed by size of ration or fish size or both, for in nature the two are highly correlated: big fish tend to eat more than small fish. Something that must be considered, of course, is that when a fish matures, a considerable part of its food becomes channeled into the ovaries and testes, so eggs and milt should be included in the new

---

* Ivlev also proposed a K_3 coefficient, but this was shown by Winberg (1956) to be an artificial quantity based on erroneous interpretation of experimental data (see Ricker, 1966, p. 1750, footnote 10).

body substance when computing K. If growth is estimated from yearly weighings, or from scale annuli, it will not include the sexual products, and a K so computed may decrease abruptly at maturity.

A puzzling aspect of Eq. (67) is the fact that in some cases the K-line extends linearly back to very small rations. In fact, in none of the experiments cited by Paloheimo and Dickie is there much trace of the left-hand or ascending portion of the K-line, yet it is obvious that the line must have its origin at the maintenance ration—where none of the food ingested is used for growth and $K = 0$. In some experiments the smallest ration used was apparently well above maintenance, but this is not true of those presented in Figs. 6, 7, and 9 and one series in Fig. 10 of Paloheimo and Dickie (1966a). The K-lines that can be computed from Fig. 1 of Elliott (1975b) are similar, although for two of the curves there is a point on the ascending limb. All these experiments were based on the weight rather than the caloric content of the fish, so there may be a plausible explanation: the nutritional conditions experienced by the fish prior to the experiment may have exerted an influence on into the experimental period. For some time prior to an experiment the fish to be used are commonly fed to satiation or nearly so, and so will have the high fat and protein content, and correspondingly low water content, that is characteristic of such a regime. On reduced rations percentage of water increases substantially (Brett et al., 1969; Elliott, 1975b). It seems possible that growth processes such as cell division that are involved in an increase in size have a certain momentum, and do not stop immediately when a fish is suddenly put on low rations. Thus the fish may for a while continue to increase in length and weight by increasing the water content of its body. In the case of LeBrasseur's young chum salmon (*Oncorhynchus keta*) graphed by Kerr (1971a, Fig. 3), another factor may have entered: The smallest fish may still have contained a quantity of egg yolk, which too would make for a spuriously large K. None of the experiments from which K-lines have been estimated appear to have been started with fish that were previously held on a maintenance ration. It would be instructive to perform such an experiment and find whether, under such circumstances, the K values computed from weight would at first increase only quite gradually with size of ration. If so, then long straight K-lines would be the exception rather than the rule, and, insofar as they are an essential part of Paloheimo and Dickie's theoretical framework, the latter may be of limited applicability to conditions in the wild.

Notice also that the K-lines of Fig. 21 are not of the sort that Paloheimo and Dickie consider typical. Rather, they increase gradually to their maximum, and in fact at 20°C the maximum was not reached at the largest ration the fish would accept. The reason for the absence of any major straight section in these curves must lie in some aspect of the experimental conditions used by Brett *et al.* (1969). Their sockeye were made to swim continuously at a moderate rate, simulating the active life of sockeye in the wild.

It is outside the scope of this chapter to review Paloheimo and Dickie's analysis in detail. However, in spite of the ingenuity and persistence that are exhibited in the development of various aspects of their work, one finishes reading their papers with a feeling of anticlimax. After their bold start from observational data, we had hoped that they would be able to shed greater light on growth processes and, just possibly, produce a new growth equation involving parameters convenient for routine use.

E. Kerr's Analysis

Starting where Paloheimo and Dickie leave off, Kerr (1971a,b,c) has extended their work, taking into consideration the various components of metabolism and the size, density and distribution of food organisms. Much ingenuity is expended in this analysis, and various aspects of it can be subjected to additional experimental testing in future. As with the Paloheimo–Dickie scheme, no outline of the development can be given here. Figures 2 and 3 of Kerr's (1971c) paper compare a computed growth curve with an observed one for the Lake Opeongo lake trout (*Salvelinus namaycush*), both before and after the trout obtained an additional and larger prey species, the cisco (*Coregonus artedii*). However, agreement is obtained by postulating prey densities (different for each of four size categories of trout) that will generate the two growth curves observed, and there is no single mathematical expression that describes either curve.

F. Ursin's Analysis

Another ambitious theoretical analysis of fish growth processes has been made by Ursin (1967). He works from and toward Pütter's generalized growth equation

$$\frac{dw}{dt} = Hw^m - kw^n \tag{68}$$

where w is mass at time t, and H, m, k, and n are coefficients. Whereas Pütter (1920), von Bertalanffy (1934), Beverton and Holt (1957), and others assume that $m = \frac{2}{3}$ and $n = 1$ in Eq. (68), Ursin estimates these parameters from published experimental data, using a variety of techniques and some rather uncertain assumptions. These all give values of n that are less than 1, with a mean about 0.83; while estimates of m are more variable, with 0.6 as a possible representative figure. However there is no attempt to demonstrate that m and n are constants that represent physiological realities reflecting basic protoplasmic processes. Rather, they are artificial, though not necessarily useless, parameters that emerge automatically once the form of Eq. (68) and various ancillary hypotheses have been assumed.

In any event Ursin's analysis, like those of Elliott, Kerr, and Paloheimo and Dickie, has not produced any relationship for everyday use based on physiologically meaningful concepts. It now seems safe to conclude that no such simple relationship exists.

VII. SUMMARY

Starting with the fertilized egg, the growth of a fish can be divided into a series of stanzas, marked off by developmental or ecological crises such as hatching, transformation from larva to fingerling, migration from salt to fresh water or vice versa, or maturation. Within each stanza growth usually follows an S-shaped "Sachs cycle," although the relative duration of the two branches of the S can vary greatly, and they are very different depending on whether length or weight is under consideration. A distinction must be made between individual growth and the increase in mean length or weight of a year-class, because of the widespread occurrence of mortality that is selective by size. Either the larger or the smaller members of a brood may suffer the greater mortality. Mortality from predation may be more severe among the smaller members of a brood, especially during the first year of life. As the fish grow older and larger, both internal mechanisms and the effects of fishing usually make for greater mortality among faster-growing individuals (Lee's phenomenon), with the result that estimates of growth rates from a comparison of the surviving fish at successive ages are too small, sometimes much too small. However, in many species true growth rates for individual fish can be calculated at yearly intervals from marks on scales or bones, and can be used, for example, to compute the production of the species in its habitat.

The numerical description of growth that is most convenient for combination with other parameters of the stock or of the environment is the instantaneous or specific growth rate; if growth is changing rapidly, it must be measured at short intervals of time. Over somewhat longer periods a double logarithmic relationship may describe growth satisfactorily, particularly during incubation and larval life, but it is difficult to combine this with other parameters. A variety of descriptive asymptotic curves are also available, including the logistic, Gompertz, Pütter Nos. 1 and 2, Johnson, and Richards. These have been used mainly to model growth at yearly intervals, but two have been applied to the seasonal course of growth. Although there have been several suggestions, none of these curves has been demonstrated to have any physiological basis that might qualify it as *the* growth curve for fishes generally, and frequently two or more of them will describe satisfactorily the same series of observations. When such curves are applied to the mean size of the fish in a population at successive ages, rather than to an individual fish, they all have the limitation that they describe only the size achieved by the *survivors* at each age. Thus they cannot be used to measure the true mean growth rate of the fish in the population, in the very common situation where faster-growing individuals tend to die earlier than slower-growing ones.

For fish of a given size, consuming a fixed ration greater than maintenance, the instantaneous growth rate at first increases with temperature, reaches a maximum, then decreases to zero. Continued temperature increase with the same ration causes loss of weight and eventually death. This complex relationship can be simplified by using its reciprocal—the length of time required to achieve a given size—and in that form it has, in some cases, been fitted reasonably well by a parabola or a catenary. Even a simple hyperbola may describe a considerable portion of the range of observations, and this is the basis of the degree-day rule used to predict hatching times of fish eggs.

Several rather complex curves have been used to describe the relation between growth and quantity of food consumed, but none shows real promise of having wide usefulness. To date, the most convenient form for presenting data on growth in relation to temperature or rations is an empirical table of smoothed values, or the corresponding graph.

ACKNOWLEDGMENTS

I am greatly indebted to Dr. J. R. Brett of the Pacific Biological Station, Nanaimo, British Columbia, for encouragement to write this chapter, for copies of and references to pertinent literature, and for penetrating comments. Drs. R. I. Fletcher of the University of Washington and J. Schnute of the Pacific Biological Station have given invaluable assistance with the mathematical aspects of the various growth curves. Dr. G. D. Stauffer of the U.S. National Marine Fishery Service, La Jolla, made available his thesis and additional references.

REFERENCES

Ahlstrom, E. H. (1943). Studies on the Pacific pilchard or sardine (*Sardinops caerulea*). 4. Influence of temperature on the rate of development of pilchard eggs in nature. *U.S. Fish Wildl. Serv., Spec. Sci. Rep.—Fish*. No. 23, 1–26.

Alexander, G. R., and Shetter, D. S. (1961). Seasonal mortality and growth of hatchery-reared brook and rainbow trout in East Fish Lake, Montmorency County, Michigan, 1958–59. *Pap. Mich. Acad. Sci., Arts Lett*. **46**, 317–328.

Allen, K. R. (1950). The computation of production in fish populations. *N.Z. Sci. Rev*. **8**, 89.

Allen, K. R. (1951). The Horokiwi stream: A study of a trout population. *N.Z. Fish. Res. Div. Fish. Res. Bull*. No. 10.

Allen, K. R. (1967). Computer programs available at St. Andrews Biological Station. *Fish. Res. Board Can. Tech. Rep*. No. 20.

Belehrádek, J. (1930). Temperature coefficients in biology. *Biol. Rev. Biol. Proc. Cambridge Philos. Soc*. **5**, 30–58.

Beverton, R. J. H. (1954). Notes on the use of theoretical models in the study of the dynamics of exploited fish populations. *U.S. Fish. Lab., Beaufort, N.C*. Misc. Contrib. No. 2.

Beverton, R. J. H., and Holt, S. J. (1957). On the dynamics of exploited fish populations. *U.K. Min. Agric. Fish., Fish. Invest*. **19**, 1–533.

Beverton, R. J. H., and Holt, S. J. (1959). A review of the lifespans and mortality rates of fish in nature, and their relation to growth and other physiological characteristics. *In* "The Lifespan of Animals." *Ciba Found. Colloq. Ageing* **5**, 142–177.

Boëtius, I. (1962). Temperature and growth in a population of *Mytilus edulis* (L.) from the northern harbour of Copenhagen (the Sound). *Meddelelsen Danmarks Fiskeri-og Havunders*. **3**, 339–346.

Brett, J. R. (1974). Tank experiments on the culture of pan-size sockeye (*Oncorhynchus nerka*) and pink salmon (*O. gorbuscha*) using environmental control. *Aquaculture* **4**, 341–352.

Brett, J. R., and Glass, N. R. (1973). Metabolic rates and critical swimming speed of sockeye salmon (*Oncorhynchus nerka*) in relation to size and temperature. *J. Fish. Res. Board Can*. **30**, 379–387.

Brett, J. R., Shelbourn, J. E., and Shoop, C. T. (1969). Growth rate and body composition of fingerling sockeye salmon, *Oncorhynchus nerka*, in relation to temperature and ration size. *J. Fish. Res. Board Can*. **26**, 2363–2394.

Brody, S. (1927). Growth rates. *Mo. Agric. Exp. Stn., Bull*. No. 97.

Brody, S. (1945). "Bioenergetics and Growth." Reinhold, New York.

Cassie, R. M. (1954). Some uses of probability paper in the analysis of size frequency distributions. *Aust. J. Mar. Freshwater Res*. **5**, 513–522.

Causton, D. R. (1969). A computer program for fitting the Richards function. *Biometrics* **25**, 401–409.

Chapman, D. G. (1961). Statistical problems in dynamics of exploited fish populations. *Proc. Berkeley Symp. Math. Stat. Probab., 4th* pp. 153–168.

Chugunova, N. I. (1959). "Handbook for the Study of Age and Growth of Fishes." Akad. Nauk Press, Moscow. (English transl., "Age and Growth Studies in Fish." Off. Tech. Serv., Washington, D.C., 1963.)

Conway, G. R., Glass, N. R., and Wilcox, J. C. (1970). Fitting nonlinear models to biological data by Marquardt's algorithm. *Ecology* **51**, 503–507.

Elliott, J. M. (1975a). The growth rate of brown trout, *Salmo trutta* L., fed on maximum rations. *J. Anim. Ecol.* **44**, 805–821.

Elliott, J. M. (1975b). The growth rate of brown trout, *Salmo trutta* L., fed on reduced rations. *J. Anim. Ecol.* **44**, 823–842.

Fabens, A. J. (1965). Properties and fitting of the von Bertalanffy growth curve. *Growth* **29**, 265–289.

Fletcher, R. I. (1973). A synthesis of deterministic growth laws. Univ. Rhode Island Sch. Oceanogr., Kingston, Rhode Island.

Fletcher, R. I. (1975). A general solution for the complete Richards function. *Math. Biosci.* **27**, 349–360.

Ford, E. (1933). An account of the herring investigations conducted at Plymouth during the years from 1924 to 1933. *J. Mar. Biol. Assoc. U.K.* **19**, 305–384.

Gerking, S. D. (1957). Evidence of aging in natural populations of fishes. *Gerontologia* **1**, 287–305.

Gerking, S. D. (1966). Annual growth cycle, growth potential, and growth compensation in the bluegill sunfish in northern Indiana lakes. *J. Fish. Res. Board Can.* **23**, 1923–1956.

Glass, N. R. (1969). Discussion of calculation of power function with special reference to respiratory metabolism in fish. *J. Fish. Res. Board Can.* **26**, 2643–2650.

Gompertz, B. (1825). On the nature of the function expressive of the law of human mortality, and on a new mode of determining the value of life contingencies. *Philos. Trans. R. Soc. London* **115**, 515–585.

Graham, M. (1929). Studies of age determination in fish. Part II. A survey of the literature. *U.K. Min. Agric. Fish., Fish. Invest.* **11**(2).

Graham, M. (1935). Modern theory of exploiting a fishery, and application to North Sea trawling. *J. Cons., Cons. Perm. Int. Explor. Mer* **13**, 76–90.

Gray, J. (1928). The growth of fish. II. The growth rate of the embryos of *Salmo fario*. *Br. J. Exp. Biol.* **6**, 110–124.

Hayes, F. R. (1949). The growth, general chemistry, and temperature relations of salmonid eggs. *Q. Rev. Biol.* **24**, 281–308.

Hayes, F. R., and Armstrong, F. H. (1943). Growth of the salmon embryo. *Can. J. Res., Sect. D* **21**, 19–33.

Hogman, W. J. (1968). Annulus formation on the scales of four species of coregonids reared under artificial conditions. *J. Fish. Res. Board Can.* **25**, 2111–2122.

Imbrie, J. (1956). Biometrical methods in the study of invertebrate fossils. *Bull. Am. Mus. Nat. Hist.* **108**, 211–252.

Ivlev, V. S. (1945). The biological productivity of waters. *Usp. Sovrem. Biol.* **19**, 98–120. [English transl., *J. Fish. Res. Board Can.* **23**, 1727–1759 (1966).]

Ivlev, V. S. (1955). "Experimental Ecology of the Feeding of Fishes." Pishchepromizdat, Moscow. [English transl., Yale University Press, New Haven (1961).]

Janisch, E. (1927). Das Exponentialgesetz als Grundlage einer vergleichenden Biologie. *Abh. Theorie Org. Entwicklung* **2**, 1–371.

Johnson, N. O. (1935). A trend line for growth series. *J. Am. Stat. Assoc.* **30**, 717.

Jolicoeur, P. (1975). Linear regressions in fishery research: Some comments. *J. Fish. Res. Board Can.* **32**, 1491–1494.

Jones, R. (1958). Lee's phenomenon of "apparent change in growth rate," with particular reference to cod and plaice. *Int. Comm. Northwest Atl. Fish. Spec. Publ.* **1**, 229–242.

Kermack, K. A., and Haldane, J. B. S. (1950). Organic correlation and allometry. *Biometrika* **37**, 30–41.

Kerr, S. R. (1971a). Analysis of laboratory experiments on growth efficiency of fishes. *J. Fish. Res. Board Can.* **28**, 801–808.

Kerr, S. R. (1971b). Prediction of fish growth efficiency in nature. *J. Fish. Res. Board Can.* **28**, 809–814.

Kerr, S. R. (1971c). A simulation model of lake trout growth. *J. Fish. Res. Board Can.* **28**, 815–819.

Knight, W. (1968). Asymptotic growth: an example of nonsense disguised as mathematics. *J. Fish. Res. Board Can.* **25**, 1303–1307.

Krüger, F. (1962). Über die mathematische Darstellung des tierischen Wachstums. *Naturwissenschaften* **49**, 454.

Krüger, F. (1964). Neuere mathematische Formulierungen der biologischen Temperaturfunktion und des Wachstums. *Helgol. Wiss. Meeresunters.* **9**, 108–124.

Krüger, F. (1965). Zur Mathematik des tierischen Wachstums. I. Grundlagen einer neuen Wachstumsfunktion. *Helgol. Wiss. Meeresunters.* **12**, 78–136.

Krüger, F. (1969). Das asymptotische Wachstum der Fische—ein Nonsens? *Helgol. Wiss. Meeresunters.* **19**, 205–215.

Krüger, F. (1973). Zur Mathematik des tierischen Wachstums. II. Vergleich einiger Wachstumsfunktionen. *Helgol. Wiss. Meeresunters.* **25**, 509–550.

Laird, A. K. (1967). Evolution of the human growth curve. *Growth* **31**, 345–355.

Laird, A. K., Tyler, S. A., and Barton, A. D. (1965). Dynamics of normal growth. *Growth* **29**, 233–248.

Lasker, R. (1964). An experimental study of the effect of temperature on the incubation time, development and growth of Pacific sardine embryos and larvae. *Copeia* pp. 399–405.

Lee, R. M. (1912). An investigation into the methods of growth determination in fishes. *Publ. Circonstance, Cons. Perm. Explor. Mer* **63**.

Lockwood, S. J. (1974). The use of the von Bertalanffy growth equation to describe the seasonal growth of fish. *J. Cons., Cons. Perm. Int. Explor. Mer* **35**, 175–179.

Lumer, H. (1937). The consequences of sigmoid growth for relative growth functions. *Growth* **1**, 140–154.

MacDowell, E. C., Allen, E., and MacDowell, G. G. (1927). The prenatal growth of the mouse. *J. Gen. Physiol.* **11**, 57–70.

Martin, W. R. (1949). The mechanics of environmental control of body form in fishes. *Univ. Toronto Stud., Biol. Ser.* No. 58. (Also called *Publ. Ont. Fish. Res. Lab.* No. 70.)

Minot, C. S. (1891). Senescence and rejuvenation. First paper. On the weight of guinea pigs. *J. Physiol. (London)* **5**, 457–464.

Miura, T., Suzuki, N., Nagoshi, M., and Yamamura, K. (1976). The rate of production and food consumption of the biwamasu, *Oncorhynchus rhodurus*, population in Lake Biwa. *Res. Popul. Ecol. (Tokyo)* **17**, 135–154.

Nelder, J. A. (1961). The fitting of a generalization of the logistic curve. *Biometrics* **17**, 89–110.

Paloheimo, J., and Dickie, L. M. (1965). Food and growth of fishes. I. A growth curve derived from experimental data. *J. Fish. Res. Board Can.* **22**, 521–542.

Paloheimo, J., and Dickie, L. M. (1966a). Food and growth of fishes. II. Effects of food and temperature on the relation between metabolism and body size. *J. Fish. Res. Board Can.* **23**, 869–908.

Paloheimo, J., and Dickie, L. M. (1966b). Food and growth of fishes. III. Relations among food, body size and growth efficiency. *J. Fish. Res. Board Can.* **23**, 1209–1248.

Parker, R. R., and Larkin, P. A. (1959) A concept of growth in fishes. *J. Fish. Res. Board Can.* **16**, 721–745.

Partlo, J. M. (1955). Distribution, age and growth of eastern Pacific albacore (*Thunnus alalunga* Gmelin). *J. Fish. Res. Board Can.* **12**, 35–60.

Pella, J. J., and Tomlinson, P. K. (1969). A generalized stock production model. *Inter-Am. Trop. Tuna Comm., Bull.* **13**, 420–496.

Pelluet, D. (1944). Criteria for the recognition of developmental stages in the salmon (*Salmo salar*). *J. Morphol.* **74**, 395–407.

Petersen, C. G. J. (1892). Fiskensbiologiske forhold i Holboek Fjord, 1890–91. *Beret. Dan. Biol. Stn. 1890–1891* **1**, 121–183.

Pitcher, T. J., and MacDonald, P. D. M. (1973). Two models for seasonal growth in fishes. *J. Appl. Ecol.* **10**, 599–606.

Pütter, A. (1920). Wachstumsähnlichkeiten. *Pfluegers Arch. Gesamte Physiol. Menschen Tiere* **180**, 298–340.

Rafail, S. Z. (1973). A simple and precise method for fitting a von Bertalanffy growth curve. *Mar. Biol.* **19**, 354–358.

Richards, F. J. (1959). A flexible growth function for empirical use. *J. Exp. Bot.* **10**, 290–300.

Ricker, W. E. (1945). A method of estimating minimum size limits for obtaining maximum yield. *Copeia* No. 2, pp. 84–94.

Ricker, W. E. (1946). Production and utilization of fish populations. *Ecol. Monogr.* **16**, 373–391.

Ricker, W. E. (1958). Handbook of computations for biological statistics of fish populations. *Bull., Fish. Res. Board Can.* No. 119.

Ricker, W. E. (1966). Annotations to Ivlev's paper "The biological productivity of waters." *J. Fish. Res. Board Can.* **23**, 1727–1759.

Ricker, W. E. (1969). Effect of size-selective mortality and sampling bias on estimates of growth, mortality, production and yield. *J. Fish. Res. Board Can.* **26**, 479–541.

Ricker, W. E. (1973). Linear regressions in fishery research. *J. Fish Res. Board Can.* **30**, 409–434.

Ricker, W. E. (1975a). Computation and interpretation of biological statistics of fish populations. *Bull., Fish. Res. Board Can.* No. 191.

Ricker, W. E. (1975b). A note concerning Professor Jolicoeur's comments. *J. Fish. Res. Board Can.* **32**, 1494–1498.

Ricker, W. E., and Foerster, R. E. (1948). Computation of fish production. *Bull. Bingham Oceanogr. Collect.* **11**, 173–211.

Ricklefs, R. E. (1967). A graphical method of fitting equations to growth curves. *Ecology* **48**, 978–983.

Sachs, J. (1874). Über den Einfluss der Lufttemperatur und des Tageslichtes auf die stündlichen und täglichen Änderung des Langenwachstums der Internoden. *Arb. bot. Inst. Wurzburg* **1**, 99–192.

Schaefer, M. B. (1954). Some aspects of the dynamics of populations important to the management of the commercial marine fishes. *Inter-Am. Trop. Tuna Comm., Bull.* 1(2), 27–56.

Schmalhausen, I. (1926). Studien über Wachstum und Differenzierung. III. Die embryonale Wachstumskurve des Hünchens. *Wilhelm Roux' Arch. Entwicklungsmech. Org.* 109, 322–387.

Schumacher, F. X. (1939). A new growth curve and its application to time-yield studies. *J. For.* 37, 819–820.

Silliman, R. P. (1967). Analog computer models of fish populations. *U.S. Fish Wildl. Serv., Fish. Bull.* 66, 31–46.

Stauffer, G. D. (1973). A growth model for salmonids reared in hatchery environments. Ph.D. Thesis, Univ. of Washington, Seattle.

Sund, O. (1911). Undersökelser over brislingen i norske farvand vaesentlig paa grundlag av "Michael Sar's" togt 1908. *Aarsberet. Nor. Fisk.* (1910) 3, 357–410.

Szarski, H., Delewka, E., Olechnowiczawa, S., Predygier, Z., and Slankowa, L. (1956). Uklad trawienny leszcza (*Abramis brama* L.). [The digestive system of the bream.] *Stud. Soc. Sci. Torun., Sect. E* (*Zool.*) 3, 113–146.

Taylor, C. C. (1958). Cod growth and temperature. *J. Cons., Cons. Perm. Int. Explor. Mer* 23, 366–370.

Taylor, C. C. (1959). Temperature and growth—the Pacific razor clam. *J. Cons., Cons. Perm. Int. Explor. Mer* 25, 93–101.

Taylor, C. C. (1962). Growth equations with metabolic parameters. *J. Cons., Cons. Perm. Int. Explor. Mer* 27, 270–286.

Teissier, G. (1948). La relation d'allométrie: sa signification statistique et biologique. *Biometrics* 4, 14–18.

Tesch, F. W. (1971). Age and growth. *In* "Methods for Assessment of Fish Production in Fresh Waters" (W. E. Ricker, ed.), pp. 98–130, Int. Biol. Program, Handbook No. 3, 2nd ed. Blackwell Scientific, Oxford and Edinburgh.

Ursin, E. (1963). On the incorporation of temperature in the von Bertalanffy growth equation. *Medd. fra Danmarks Fiskeri- og Havundersøgelser* 4, 1–16.

Ursin, E. (1967). A mathematical model of some aspects of fish growth, respiration and mortality. *J. Fish. Res. Board Can.* 24, 2355–2453.

Van Oosten, J. (1929). Life history of the lake herring (*Leucicthys artedi* LeSueur) of Lake Huron, as revealed by its scales with a critique of the scale method. *U.S. Bur. Fish. Bull.* 44, 265–448.

Vasnetsov, V. V. (1953). Developmental stages of bony fishes. *In* "Ocherki po Obshchim Voprosam Ikhtiologii," pp. 207–217. Akademiya Nauk Press, Moscow (in Russian).

Verhulst, P. F. (1838). Notice sur la loi que la population suit dans son acroissement. *Corres. Math. Phys.* 10, 113–121.

von Bertalanffy, L. (1934). Untersuchungen über die Gesetzlichkeit des Wachstums. *Wilhelm Roux' Arch. Entwicklungsmech. Org.* 131, 613.

von Bertalanffy, L. (1938). A quantitative theory of organic growth. *Hum. Biol.* 10, 181–213.

von Bertalanffy, L. (1957). Quantitative laws in metabolism and growth. *Q. Rev. Biol.* 32, 217–231.

Walford, L. A. (1946). A new graphic method of describing the growth of animals. *Biol. Bull.* 90, 141–147.

Weymouth, F. W., and Thompson, S. H. (1930). The age and growth of the Pacific cockle (*Cardium corbis* Martyn). *Bull. U. S. Bur. Fish.* 46, 633–641.

Winberg, G. G. (1956). "Rate of Metabolism and Food Requirements of Fishes."

Nauchnye Trudy Belorusskogo Gos. Univ. Minsk, 253 pp. (English version, *Fish. Res. Board Can. Transl. No.* 194, 1960.)

Zar, J. H. (1968). Calculation and miscalculation of the allometric equation as a model in biological data. *Bioscience* **18**, 1118–1120.

Zweifel, J. R., and Lasker, R. (1976). Prehatch and posthatch growth of fishes—A general model. *U.S. Fish Wildl. Serv., Fish. Bull.* **74**, 609–621.

AUTHOR INDEX

Numbers in italics refer to the pages on which the complete references are listed. The asterisks that precede similar names indicate that the spellings are different translations of the same author.

A

Abe, S., 355, 357, *393*
Abraham, M., 531, *587*
Abrahamsson, T., 221, 240, *254*
Abramoff, P., 361, 362, *393, 394, 396, 402*
Abramova, N. B., 262, *276*
Ackman, R. G., 48, *58*
Adam, H., 173, *241*
Adelman, I. R., 473, *578*, 634, *667*
Adron, J. W., 6, 8, 12, 18, 26, 28, 29, 42, 46, 47, 50, *58*, 60, 66, 130, *154*, 213, *241*, 509, 514, 524, *581*
Aftergood, L., 45, *58*
Ahlskog, J. E., 132, *153*
Ahlstrom, E. H., 689, *738*
Ahsgren, L., 355, *400*
Aida, T., 367, 369, *393*
Aiello, A., 464, *581*
Akhalkatis, R. G., 266, *275*
Albert, D. J., 132, *153, 158*
Albertini-Berhaut, J., 188, *241*
Alderdice, D. F., 656, 660, *667*
Alex, M., 182, *252*
Alexander, G. R., 691, *738*
Alexander, R. McN., 163, *242*, 310, *344*
Alfin-Slater, R., 45, *58*
Al-Hussaini, A. H., 162, 170, 180, *242*
Ali, M. A., 78, *113*, 553, 565, *578*
Allard, D. R., 240, *248*
Allee, W. C., 653, *667*
Allen, D. M., 537, *578*
Allen, E., 698, *740*
Allen, K. O., 650, *667*
Allen, K. R., 104, *113*, 698, 708, 717, *738*
Allen, O. W., 56, *62*
Allendorf, F. W., 356, *398*, 408, 409, 410, 411, 413, 414, 421, 425, 426, 431, 437, 438, 439, 440, *449, 452, 454*

Alliot, E., 181, 182, 183, *242*
Aloj, S., 464, *578*
Alpert, J. S., 632, *670*
Altman, P. L., 291, 292, *344*, 609, *667*
Altukhov, Y. P., 408, 430, *449*
Amend, D. F., 412, *450*, 465, *578*
Ames, W. E., 439, *454*
Anders, A., 370, 371, 372, *393*
Anders, F., 370, 371, 372, *393*
Anderson, R. O., 621, *668*
Andersson, B., 130, *153*
Andrews, J. W., 8, 9, 12, 13, 41, 42, 46, 49, 51, 54, 55, 57, *58, 59, 64, 65, 66, 68*, 611, 636, 642, 658, *668*
Angelescu, V., 170, *242*
Annison, E. F., 305, *346*
Antelman, S. M., 143, *157*
Aoyama, T., 198, *242*
Applegate, R. L., 108, *118*
Arai, R., 501, *578*
Arai, S., 2, 3, 27, 54, *59*, 63, 572, *587*
Arcangeli, A., 168, *242*
Arees, E. A., 132, *153*
Ariens, E. J., 216, *242*
Ariens Kappers, C. U., 139, *153*
Armstrong, F. H., 697, 698, *739*
Armstrong, R., 8, *69*
Armstrong, R. H., 195, *242*
Aronson, L. R., 134, *153*
Arthur, D. K., 204, *242*
Asano, N., 356, *398*
Ashby, K. R., 501, 503, 507, 509, 510, 515, 518, 522, 524, 525, *578*
Ashley, H., 538, *578*
Aspinwall, N., 423, *450*
Atema, J., 80, 102, *113*
Aten, B., 571, *595*
Atherton, L. M., 355, 383, 390, *396*, 418, *454*

745

Chao, L. N., 203, *245*
Chapman, C. J., 84, *115*
Chapman, D. G., 709, 714, *739*
Chapman, G. A., 284, 318, *349*
Chapman, G. B., 173, *251*
Chappell, J. B., 18, *64*
Charipper, H. A., 555, *591*
Charles, M. A., 141, 142, 143, *155*
Charlton, C. B., 148, *154*
Chartier-Baraduc, M. M., 465, 467, 478, 484, *580*
Chavin, W., 536, *580*
Cheek, D. B., 490, *580*
Cheema, I. R., 490, 491, 503, 507, 522, 531, 572, *580*
Chen, M. S., 5, 6, 7, 8, *66*
Chen, S. H., 411, *450*
Chen, T. R., 355, 356, 359, 365, 366, 372, 373, 378, 381, 385, 393, *395, 396*
Cherfas, B. I., 408, *450*
Cherfas, N. B., 364, 391, *396*
Chervinski, J., 625, 626, *669*
Chesley, L. C., 175, 184, *245*
Chesney, E. J., Jr., 329, *346*
Chester Jones, I., 191, 257, 485, *581*
Chestukhin, A. V., 272, *274*
Chiang, W., 318, 340, *346*
Chiarelli, A. B., 354, 356, 364, 367, 372, 373, 374, *396*
Chiba, K., 204, *245*, 633, *669*
Chillemi, F., 464, *581*
Chiou, J. Y., 22, 58, *60, 66*
Cho, C. Y., 11, 57, *60, 62*, 339, *346*, 542, *588*
Choe, T. S., 271, *277*
Choudhury, R. B. R., 44, *66*
Christian, L., 381, 388, *401*
Christiansen, F. B., 419, *450, 451*
Chuecas, L., 48, *60*
Chugunova, N. I., 687, *739*
Chujyo, N., 231, *250*
Cimino, M. C., 362, 363, 364, 376, *396*
Clark, E., 358, 360, *396*
Clark, J. L., 571, *595*
Clarke, W. C., 457, 460, 462, 471, 479, 481, 484, 485, 489, 574, 575, *581, 583, 591*
Claro, R., 198, 211, *257*
Clayton, J. W., 410, 411, *451*
Clem, W. L., 365, *396*
Clemens, H. P., 500, 501, 505, 520, *581*
Clemens, W. A., 284, 309, *348*

Clemmons, D. R., 492, *596*
Clinebell, J. R., 8, *69*
Coates, J. A., 54, *61*
Coates, P. M., 411, *451*
Coble, D. W., 101, *117*
Cocchi, D., 141, *156*
Coche, A. G., 104, *114*
Cohen, D. C., 462, *581, 589*
Cohen, H., 500, 505, 507, 510, 514, 521, 523, *581*
Cole, R. E., 606, 629, 631, *669*
Collicutt, J. M., 537, 538, 540, 541, 545, *581*
Colot, S. E., 263, *277*
Comfort, A., 653, *669*
Confer, J. L., 81, 82, *114*
Connolly, J. J., 356, *400*
Conte, F. P., 52, *60*
Conway, G. R., 717, *739*
Cooper, E. L., 646, 647, *669*
Cooper, W. E., 128, *155*
Corbet, P. S., 73, *114*
Corbit, J. D., 123, *154*
Corner, E. D. S., 174, *245*
Corrodi, H., 237, *245*
Costa, R. R., 107, 108, *114*
Cousineau, G. H., 263, *275*
Coutant, C. C., 610, 611, *669*
Cowey, C. B., 1, 4, 5, 6, 8, 12, 13, 18, 19, 20, 21, 22, 26, 28, 29, 42, 43, 46, 47, 48, 49, 50, 55, *58, 60, 66, 68*, 130, *154*, 213, *241*, 318, 319, *346*, 509, 514, 515, 524, *581*
Cox, D. K., 610, 611, 657, *669*
Crabtree, B., 28, *65*
Crawford, M. A., 46, *67*
Crawford, R. B., 263, 268, *274, 277*
Creac'h, P. V., 162, 179, 181, *245*
Critchlow, V., 132, *157*
Crosby, E. C., 139, *153*
Cross, C. G., 101, *114*
Crossman, E. J., 72, 73, 76, 98, 102, 103, 118, 389, *394*
Croston, C. B., 181, *245*
Crow, J. F., 360, *396*, 434, 435, *451*
Csengeri, I., 40, *61*
Cucchi, C., 356, 367, *396*
Cuellar, O., 388, *396*
Cullen, J. M., 90, *117*
Cummins, K. W., 107, 108, *114*
Curry, D. L., 490, *581*

D

Daan, N., 192, 196, 200, 210, *245*
Dabrowska, H., 57, *60*
Dahlstedt, E., 176, *245*
Dales, S., 553, 565, *581*
D'Ancona, U., 358, *396*
Dandrifosse, G., 180, 183, *245, 253*
Dangel, J. R., 385, *396*
Daniel, E. E., 216, *245*
Darby, W. B., Jr., 356, *400*
Darlington, J. P. E. C., 101, 108, *117*
Darnell, R. M., 73, 74, *114*, 192, 195, *246*,
    361, 362, *393, 394, 396, 400, 402*
Daughaday, W. H., 465, 492, *581*
David, H., 174, *246*
Davidson, E. H., 263, 267, 268, *274, 275*
Davies, P. M. C., 280, 316, 322, *346*
Davies, W., 240, *248*
Davis, G. E., 298, 305, 315, 316, 329, *346*,
    352, 601, 611, 618, *674*
Davison, R. C., 632, *669*
Davison, W., 283, 286, *347*
Davisson, M. T., 355, 383, 390, *396*
Davson, H., 190, *246*
Dawson, I., 177, *246*
Dean, W. F., 10, *60*
De Ciechomski, J. D., 97, *114*
De Gasquet, P., 28, *63*
De Gier, J., 41, *68*
Degkwitz, E., 52, *61*
De Groot, S. J., 78, 80, *114*, 165, 170, 171,
    193, 196, 203, *246*
De Haën, C., 182, *246*
De Kruyff, B., 41, *61*
de la Higuera, M., 28, 29, *60*
Delanoy, R. L., 80, 82, 85, *114*
Delewka, E., 707, *742*
de Ligny, W., 408, *451*
DeLoach, H. L., 57, *65*
Delong, D. C., 3, *61*
De Martini, E. E., 643, *669*
Demel, R. A., 41, *68*
Demski, L. S., 133, 134, 136, 137, *154*,
    488, *581*
Denton, E. J., 174, *245*
Denton, T. E., 354, 355, 356, 367, 372, 373,
    374, 377, 378, 385, 386, 388, *395, 396,*
    398
Dépêche, J., 262, *274*
De Ruiter, L., 85, *114*

De Silva, S. S., 626, 629, *669*
De Torrengo, M. A. P., 39, *65*
De Torrengo, M. P., 39, *61*
Deuchar, E. M., 271, *274*
deVlaming, V. L., 147, 148, *154*, 525, *581*
Dexter, R. P., 355, *405*
Deyoe, C. W., 55, *61*, 195, 214, *255, 257*
DeYoung, A., 142, *159*
Dickhoff, W. W., 538, *581*
Dickie, L. M., 72, *118*, 123, *157*, 290, 327,
    329, 336, *350*, 607, 632, *673*, 732, 733,
    734, 735, 736, *741*
DiMaggio, A., III., 462, *581*
Ditlevsen, E., 367, 369, 371, *405*
Dittmer, D. S., 291, 292, *344*, 609, *667*
Dixon, J. S., 464, *589*
Djoseland, O., 498, *584*
Dobbs, R. E., 177, *259*
Dobzhansky, T., 381, 384, *396*
Dockray, G. J., 176, *243, 246*
Dodd, J. M., 517, *582*
Dodes, L. M., 262, *274*
Dodson, S. I., 108, *113*
Dollar, A. M., 383, 385, *403*
Donaldson, E. M., 28, *62*, 130, *155*, 457,
    460, 462, 465, 567, 469, 474, 475, 477,
    478, 479, 480, 481, 483, 484, 485, 586,
    488, 489, 490, 507, 522, 525, 551, 553,
    560, 561, 562, 567, 568, 574, 575, 576,
    582, 584, 585, 586, 625, *671*
Donaldson, L. R., 434, *451*
Dontsova, G. V., 266, *275*
Doo, W. K., 238, *260*
Dorfman, R. I., 496, 527, *578, 583*
Dosanjh, B. S., 563, *586*
Doudoroff, P., 123, *158*, 333, *351*, 632, 633,
    634, 635, 636, 659, *669, 670, 671, 673*
Drewry, G., 355, 361, *396*
Dreyer, N. B., 220, *246*
Driedzic, W. R., 28, *61*
Drongowski, R. A., 537, *582*
Drury, D. E., 544, *582*
Duncan, A., 4, *61*
Dunn, I. G., 101, 108, *117*
Dunn, R. S., 128, *159*
Dunn, V., 191, *252*
Dupree, H. K., 9, *61*
Durbin, A. G., 322, *346*
Dye, H. M., 130, *155*, 460, 462, 465, 467,
    469, 474, 475, 477, 478, 479, 480, 481,
    482, 483, 484, 485, 486, 487, 488, 489,

SYSTEMATIC INDEX

Note: Names listed are those used by the authors of the various chapters. No attempt has been made to provide the current nomenclature where taxonomic changes have occurred.

A

Abramis, 707
  A. brama, 204
Acerina, 175
Acipenser ruthenus, 390
Agnatha, 353, 373, 462
Agriopus, 172
Aholehole, *see Kuhlia sandvicensis*
Alewife, *see Alosa pseudoharengus*
Alopias vulpes, 178
Alosa, 164
  A. pseudoharengus, 105–106
  A. sapidissima, 309
Ameiurus nebulosus, 295
Amia, 172
A. calva, 194
Ammotretis, 222, 226
Amphioxus, 173, 176
Amphiprion, 174
Amyda japonica, 238
Anarrhichas lupus, 178, 191
Ancanthophthalmus khulii, 388
Anchiovella, 79
Anchovy, 31, *see also Engraulis*
Angel fish, *see Holocanthus bermudensis*
Anguilla, 168, 172, 175–176, 191–192, 226, 457, 459, 554, 569
  A. anguilla, 38, 174, 177, 224, 356, 367, 538, 572
  A. japonica, 2, 3, 195, 356, 572
  A. rostrata, 422
  A. vulgaris, 139
Anguillidae, 690
Anisotremus davidsoni, 628, 631
Apeltes quadracus, 366
Aphyosemion, 384
Aplocheilus, 384
Ariomma, 166
Arocheilus alutaceus, 75

Asellus, 93
Astatoreochromis, 96
Atherina pantica, 196
Ayu, *see Plecoglossus altivelis*

B

Balistidae, 94, 100
Balistopus
  B. fuscus, 100
  B. undulatus, 76, 100
Barbus
  B. barbus, 388
  B. liberiensis, 195, 201
Bass
  largemouth, *see Micropterus salmoides*
  sea, *see Dicentrarchus*
  smallmouth, *see Micropterus*
  striped, 32, *see also Morone saxatilis*
Bathylagidae, 365, 378, 388
Bathylagus, 373
  B. milleri, 365
  B. ochotensis, 365
  B. stilbius, 365
  B. wesethi, 365
Batrachus, 172
Bavidiella icistia, 627–628, 631
Belone belone, 204
Betta splendens, 649
Blackfish, Alaskan, *see Dallia*
Blennius, 208, 223, 226
  B. pholis, 169, 203, 206, 210, 225, 238, 314
Blenny, 76, 164, 170
Bluegill, *see Lepomis macrochirus*
Boreogadus raida, 292
Bothidae, 170
Botia macracantha, 388
Brachyistius frenatus, 198, 202
Brama, 164

D

E

F

G

SUBJECT INDEX

A

Absorption, 190–192, 214–215
Acetylcholine, 223
Adenosine triphosphate, 231, 269
Adrenaline, 220–221
Adrenergic innervation, gut, 221–223
α-Adrenoceptor, 226, 231
β-Adrenoceptor, 226, 231
Aerobic metabolism, 281
Age estimation, 686–689
Alleles, null, 417
Amines, storage, 175
Amino acids
  catabolism, 4, 17–21
  deficiencies, 4, 30
  energy source of, 26
  essential, 2–4, 6, 8, 20–21, 56
  and food intake, 85, 142, 146
  incorporation of, 15–16, 20–21, 177, 490,
    568, 571
  requirements, 4
  sulfur containing, 3–4, 8
  supplements in diet, 8–11
  and weight regulation, 142, 146, 487
Ammonia, 286
Androgens, 495–496, 500–506, 518–522
Aphagia, 131
Apomixis, 362
Appetite, 213, 486–488, 516–517, 568–569
Ascorbate, 51–52, 57
Assimilation efficiency, 192
Automixis, 362

B

Barbel, 163–164
Behavior
  aggression, 150, 649–655
  brain regions in feeding, 131–140
  hierarchy effect, 149–151
  hormone effects on, 517
  and social facilitation, 151
Bile, 184–185, 188

Bisexuality, 364–373
Body size, 645
Brain
  control of feeding, 133–140
  hypothalamus, 131–140
  telencephalon, 134, 140
Brockerhoff hypothesis, 32
Breeding, 390, 432–435, 443–447

C

Calcium, 54–56, 58, 144, 191
Caloric values, 285–286, 289
Calorimetry, 282–284
Carbohydrates
  absorption of, 191
  digestion, 189
  energy source, 22
  gluconeogenesis, 26–30
  metabolism, 283, 569
  storage, 287
  utilization of, 22–26, 268–270
Carbon dioxide, excretion of, 53
Carnivores, 335–340
Catecholamines, 240, 458, 498
Chemoreception, 79–80
Chitinase, 178, 180, 183, 190
Cholecystokinin, 176
Cholesterol, 41
Cholinergic innervation, gut, 222–223
Chromatophores, 485
Chromosomes
  aneuploidy, 390
  best sources, 354–356
  changes, 384–387
  morphology, 378–381
  number, 374, 382–383, 385–386
  polymorphism, 381–384
  polyploidy, 387–390
  sex, 366–367
Chylomicrons, 34
Clone, 362
Circulation, blood, 178
Coefficient of variation, 650